T0330168

Critical Systems Thinking and the Management of Complexity

Critical Systems Thinking and the Management of Complexity

Michael C. Jackson

Registered Offices
John Wiley & Sons, Inc., 111 River Street, Hoboken, NJ 07030, USA
John Wiley & Sons Ltd, The Atrium, Southern Gate, Chichester, West Sussex, PO19 8SQ, UK

Editorial Office
9600 Garsington Road, Oxford, OX4 2DQ, UK

For details of our global editorial offices, customer services, and more information about Wiley products visit us at www.wiley.com.

Wiley also publishes its books in a variety of electronic formats and by print-on-demand. Some content that appears in standard print versions of this book may not be available in other formats.

Library of Congress Cataloging-in-Publication Data applied for
ISBN: 9781119118374

Cover design by Wiley
Cover image: © laflor/Getty Images-group of people, © John Lund/Getty Images-medal gears, © DNY59/Getty Images-man looking at wall, © Brasil2/Getty Images-landfill, © DurkTalsma/Getty Images-iceberg

Set in 10/12pt WarnockPro by SPi Global, Chennai, India

Printed and bound by CPI Group (UK) Ltd, Croydon CR0 4YY

C9781119118374_190724

To Pauline, Christopher and Richard:

This book is what I think. It is not all I know.
That, I hope, I have conveyed to you in other ways.

Dust as we are, the immortal spirit grows
Like harmony in music; there is a dark
Inscrutable workmanship that reconciles
Discordant elements, makes them cling together
In one society.

Wordsworth (The Prelude, 1850)

Contents

Preface

There is a considerable debate about how to describe the modern world. Alternatives include the following: a global village, postindustrial society, consumer society, media society, network society, risk society, late capitalism, high modernity, postmodernity, liquid modernity, and the information age. To some, the new names just signal the rapid acceleration of changes in society that started to emerge between the sixteenth and eighteenth centuries. To others, we have crossed a threshold and entered a completely new era. What no one doubts is that things have become much more complex. We are entangled in complexity.

An IBM survey of more than 1500 Chief Executive Officers worldwide states:

> The world's private and public sector leaders believe that a rapid escalation of 'complexity' is the biggest challenge confronting them. They expect it to continue – indeed, to accelerate – in the coming years.
>
> *(2010)*

An OECD report begins:

> Complexity is a core feature of most policy issues today; their components are interrelated in multiple, hard-to-define ways. Yet governments are ill-equipped to deal with complex problems.
>
> *(2017)*

At the global level, economic, social, technological, and ecological systems have become interconnected in unprecedented ways, and the consequences are immense. We face a growing set of apparently intractable problems, including the nuclear threat; continual warfare; terrorism; climate change; difficulties in securing energy, food, and water supplies; pollution; environmental degradation; species extinction; automation; inequality; poverty; and exclusion. Attempts to provide solutions to these ills only seem to make matters worse. Unpredictable "black swan" events (Taleb 2007), like the fall of the Soviet Union, 9/11, and the financial crisis, have become frequent and have widespread impact. On top of this, there are fewer shared values that help tame complexity by guaranteeing consensus. At the more local level, leaders and managers, whether operating in the private, public, or voluntary sectors, are plagued by interconnectivity and volatility and are uncertain about how to act. They have to ensure that objectives are met and that processes are efficient. They also have to struggle with complex new technologies and constantly innovate to keep ahead of the competition and/or do more with less. They have to deal with increased risk. Talented employees have to be attracted, retained, and inspired, and the enterprise's stock of knowledge captured and distributed so that it can learn faster than its rivals. This requires transformational leadership and the putting in place of flexible, networked structures. Changes in the law and in social expectations require managers to respond positively to different stakeholder demands and to monitor the impact of their organization's activities. They have to manage diversity and act with integrity.

Various authors have sought to summarize what they see as the key features of the complex world in which we live. Boulton et al. (2015, p. 36) provide some valuable generalizations, seeing it as:

- *Systemic and synergistic*: interconnected and resulting from many causes that interact together in complex ways
- *Multiscalar*: with interactions across many levels
- Having variety, diversity, variation, and fluctuations that can give rise to both resilience and adaptability
- *Path-dependent*: contingent on the local context, and on the sequence of what happens
- *Changing episodically*: sometimes demonstrating resilience, at other times "tipping" into new regimes
- *Possessing more than one future*: the future is unknowable
- Capable of self-organizing and self-regulating and, in some circumstances, giving rise to novel, emergent features

Warfield (2002) sets out 20 "laws of complexity," emphasizing that 70% of these result from the nature of human beings. For him, it is our cognitive limitations, dysfunctional group and organizational behavior, differences of perception ("spread-think"), and the conflict we engage in that have to be overcome if we are to get to grips with complexity.

Whether complexity arises from systems or from people, decision-makers are finding that the problems they face rarely present themselves individually as, for example, production, marketing, human resource, or finance problems. They come intertwined as sets of problems that are better described as "messes" (Ackoff 1999a). Once they are examined, they expand to involve more and more issues and stakeholders. Rittel and Webber (1981) call them "wicked problems" and argue that they possess these characteristics:

- Difficult to formulate
- It is never clear when a solution has been reached
- They don't have true or false solutions, only good or bad according to the perspective taken
- A solution will have long drawn out consequences that need to be taken into account in evaluating it
- An attempted solution will change a wicked problem so it is difficult to learn from trial and error
- There will always be untried solutions that might have been better
- All wicked problems are essentially unique; there are no classes of wicked problems to which similar solutions can be applied

- They have multiple, interdependent causes
- There are lots of explanations for any wicked problem depending on point of view
- Solutions have consequences for which the decision-makers have responsibility

Summarizing, they describe the difficulties "wicked problems" cause decision-makers as follows:

> The planner who works with open systems is caught up in the ambiguity of their causal webs. Moreover, his would-be solutions are confounded by a still further set of dilemmas posed by the growing pluralism of the contemporary publics, whose valuations of his proposals are judged against an array of different and contradicting scales.
>
> *(Rittel and Webber 1981, p. 99)*

What help can decision-makers expect when tackling the "messes" and "wicked problems" that proliferate in this age of complexity? They are usually brought up on classical management theory that emphasizes the need to forecast, plan, organize, lead, and control. This approach relies on there being a predictable future environment in which it is possible to set goals that remain relevant into the foreseeable future; on enough stability to ensure that tasks arranged in a fixed hierarchy continue to deliver efficiency and effectiveness; on a passive and unified workforce; and on a capacity to take control action on the basis of clear measures of success. These assumptions do not hold in the modern world, and classical management theory provides the wrong prescriptions. This is widely recognized and has led to numerous alternative solutions being offered to business managers and other leaders, for example, lean, six sigma, business analytics, value chain analysis, total quality management, learning organizations, process reengineering, knowledge management, balanced scorecard, outsourcing, and enterprise architecture. Occasionally, they hit the mark or at least shake things up. It is sometimes better to do anything rather than nothing. Usually, however, they fail to bring the promised benefits and can even make things worse. They are simple, "quick-fix" solutions that flounder in the face of interconnectedness, volatility, and uncertainty. They pander to the notion that there is one best solution in all circumstances and seek to reduce complex problems to the particular issues they can deal with. They concentrate on parts of the problem situation rather than on the whole, missing the crucial interactions between the parts. They fail to recognize that optimizing the performance of one part may have consequences elsewhere that are damaging for the whole. They often fail to consider an organization's interactions with its rapidly changing environment. Finally, they don't acknowledge the importance of multiple viewpoints and internal politics. Fundamentally, and in the terms used in this book, they are not systemic enough. In the absence of more thoroughly researched ways forward, however, managers are left to persevere with their favorite panacea in the face of ever diminishing returns or to turn to whatever new fad has hit the market.

This book proposes systems thinking as the only appropriate response to complexity. In systems thinking, the study of wholes, and their emergent properties, is put on an equal footing with the study of parts. The approach also insists that a wide variety of stakeholder perspectives is considered when engaging with problem situations. It has a long history, but it is only recently that it has become possible to recommend systems thinking to leaders and managers as the cornerstone of their practice. This is because the philosophy and theory have now been translated into useful and usable guidelines for action. It possesses a range of methodologies that can be used to confront different aspects of complexity according to the circumstances. In its most advanced form, the systems approach encourages the employment of a variety of methodologies in combination to manage "messes" and "wicked problems." Critical systems practice informs this way of working and demonstrates how decision-makers can achieve successful outcomes by becoming "multimethodological."

The genesis of the book goes back to the early 1980s when Paul Keys and I, at the University of Hull, established a research program to inquire into the theoretical coherence and practical value of different systems approaches to management. One outcome was a much cited paper (Jackson and Keys 1984), which outlined a "system of systems methodologies." The research continued in the late 1980s and I wrote *Systems Methodology for the Management Sciences* (1991a), which provided an overview and evaluation of various strands of systems thinking and sought to provide a theoretical justification for critical systems thinking and the meta-methodology of "Total Systems Intervention" (TSI). In the same year, Bob Flood and I published a popularizing text, called *Creative Problem Solving: Total Systems Intervention*, which was the first practical guide to using different systems approaches in combination. *Creative Problem Solving* did well. However, in some important respects, it was flawed. Having completed another major theoretical tome in 2000 – *Systems Approaches to Management* – I became confident that I had done enough additional research to generate new thinking about the difficult issues surrounding the combined use of systems methodologies to ensure successful interventions. Again, I wanted to make the results of the work available in a more popular format. The outcome was *Systems Thinking: Creative Holism for Managers* (2003), which provided a richer array of background material, a more thorough analysis of the various systems methodologies and their strengths and weaknesses, and new material advocating a creative way of using systems approaches in combination. Fortunately, the book found a ready audience and was widely read and used by managers, researchers, and students. It has been translated into Chinese, Japanese, Russian, and Spanish. I promised, in its preface, that it would be my last book.

Times change and I decided to completely update and rewrite *Systems Thinking: Creative Holism for Managers*. My reasons are threefold. First, a lot of excellent research has been undertaken in the field since 2003 and I wanted to acknowledge and take account of that in developing my own conclusions. Much of the research relates to specific areas of systems thinking, and I will make reference to these contributions in the relevant chapters. Suffice it to say, at this point, that the research communities around complexity theory, system dynamics, organizational cybernetics, soft systems methodology, and critical systems thinking have been particularly active. Of the texts covering the wider field, I need to mention a few. From the Open University, that long-time bastion of systems thinking, have come Reynolds and Holwell, eds, *Systems Approaches to Managing Change: A Practical Guide* (2010), and Ramage and Shipp *Systems Thinkers* (2009). They are both very good. The former has an introduction to the various systems approaches and covers five methodologies in chapters written by their originators and/or advocates. The latter provides brief summaries of the work of 30 leading systems thinkers and an extract from the work of each. We are all grateful to

Gerald Midgley (2002) for his four volumes of collected papers on "Systems Thinking." Comprehensive and well-edited, I have benefited from their existence throughout the writing of this book. Stowell and Welch (2012) cover the ground but with something of a bias toward soft systems thinking. Of the more specialized texts, Capra and Luisi's *The Systems View of Life: A Unifying Vision* (2014) provides an excellent overview of systems thinking in the physical and life sciences. It was a constant companion for most of my time writing the book although, I hope, I was eventually able to add to its conclusions by paying more attention to the social sciences. As will become obvious, my thinking, since 2003, has been influenced by a more careful reading of Luhmann (e.g. 2013). The volume I enjoyed reading most, in preparing the book, was Pickering's *The Cybernetic Brain: Sketches of Another Future* (2010), covering "British cybernetics" in the 1960s. I guess that is because I am a child of that decade when, in MacDonald's words: "The Beatles *felt* their way through life, acting or expressing first, thinking, if at all, only later" (2008, p. 22). It was not only the Beatles.

Second, as my thinking developed, I came up with new ways of explaining the material and a different understanding of what is useful to decision-makers and what is not. This altered my perception of the best way to structure the book and what to include. There is more upfront on basic philosophy as I have come to recognize the significance, for example, of Kant in orientating the systems worldview. I appreciate the value of complexity theory as a description of the world, and regard it as complementing and enriching the earlier systems view. On the other hand, complexity theory has failed to come up with anything resembling a practical methodology to address the issues it identifies. I do not, therefore, include a chapter on complexity theory in the "systems practice" section. In terms of the individual methodologies that are included, I have found space for chapters on "The Vanguard Method" and "Socio-Technical Systems Thinking." The Vanguard Method earns its place because of the popularity it has attained, especially in local government. The Socio-Technical approach played an important role in the early days of applied systems thinking and could have been included in the previous book. It shouldn't have fallen out of favor. I have dropped the chapter on "Postmodern Systems Thinking." In the crude terms of the previous book, I now see it as a retreat from the problems posed by complex-coercive situations rather than as an attempt to do something about them. I continue to employ ideas from postmodernism when it seems helpful. There are 10 individual systems approaches covered. They are, I think, the ones that are the most philosophically sound and thoroughly researched, and which have a good track record of application. Of course, there is a lot of subjectivity in this choice. I made a determined effort to "inhabit" and believe in each of the 10 methodologies during the weeks I was writing about it. I tried to become a Vanguard Method person, a system dynamics advocate, a soft systems thinker, and so on, for that period. It is up to the reader to decide whether I succeeded. Finally, there is more on "Critical Systems Thinking." There is a separate chapter on critical systems theory and its use in other management subdisciplines; a chapter on the variety of multimethodological approaches; and a chapter on my own latest thinking on "Critical Systems Practice."

The third reason for doing a new book is personal. In 2011, I was diagnosed with neuroendocrine cancer. This is incurable, once it has spread, but it usually gives you some time. Steve Jobs died of the disease the same year I was diagnosed. As a fellow sufferer, Alan Rodger, quipped: "Of all the things for me to have in common with the multi-billionaire, world-renowned genius, it had to be his illness." I was lucky that they could operate and I had most of my insides removed. Until recently, I did not think I would survive long and writing a book seemed low on the list of priorities (give me Hull Kingston Rovers, Hull City, and Yorkshire cricket for entertainment any day!). However by 2017, and despite another operation for a recurrence, it seemed I might still have a few years left. I just started writing. I hope you enjoy *Critical Systems Thinking and the Management of Complexity*.

Three apologies before I pass on to some acknowledgments. First, John Pourdehnad counseled me against using the phrase "the management of complexity" in the title. In his view, we need to "navigate" through complexity; we can't manage it. This is a good point and one with which I largely agree. However, I decided to keep the title as it is. There are some aspects of complexity that we can "manage"; the book is primarily for managers, broadly defined; and managing can carry the meaning of "handling," "coping," and "getting by," as well as controlling. Second apology: in a book covering this much ground I was driven, necessarily, to make use of a lot of secondary sources. I can claim to have read most of the original material at some time in my career, and only hope that has helped me to choose my secondary sources well. Third, the way the material is arranged in the book emphasizes some of the connections between authors and ideas and puts others into the shade. I have thought this through carefully and done the best I can to highlight the most significant linkages. I apologize for not doing better. There is a lot of work still to do.

I am grateful to the following for their permission to reproduce previously published material: Random House for Figure 7.1; Vanguard Press for Figure 10.1; SNCSC for Figure 10.2 and Table 18.1; Productivity Press for Figure 11.4; Plenum Press for Figure 16.6; and Elsevier for Figure 16.7.

I have been lucky to make and retain friends from school, from the universities I attended and the places where I have worked. They will know who they are because they will receive a signed copy of this book – whether they like it or not! I am grateful to them. Thanks to those who helped me in my systems career, especially Peter Checkland and the late Russ Ackoff, and to others with whom I have worked closely in developing systems ideas, particularly Paul Keys, Bob Flood, Ramses Fuenmayor, Amanda Gregory, Angela Espinosa, and Gerald Midgley. My thinking has also benefitted significantly from exchanges with various "sparring partners" for whose work I have the greatest respect – John Mingers, Werner Ulrich, Richard Ormerod, and Ralph Stacey. I have been influenced by the work of Jonathan Rosenhead and Colin Eden from the "Soft-OR" community. I am grateful to the many staff, acknowledged in Chapter 19, who worked with me in the Centre for Systems Studies. I was lucky to tutor some excellent masters' students. The contributions of Said Medjedoub, Joseph Ho, Mary Ashton, Ellis Chung, Steve Green, and Raj Chowdhury are referred to in the book. The work of many of my PhD students is acknowledged in the text and all of them contributed to the thinking: Mo Salah, E.A. Youssef, D.P. Dash, Giles Hindle, Nasser Jabari, Martin Hall, Alejandro Ochoa-Arias, Bridget Mears-Young, Amanda Gregory, Luisa Garcia, Andres Mejia, Alvaro Carrizosa, Maria Ortegon, Beatriz Acevedo, Clemencia Morales, Roberto Palacios, Catherine Gaskell, Gokhan Torlak, and Luis Sambo. Thanks to those who helped me to establish and make a success of Hull University Business School between 1999 and 2011. It was a huge endeavor, a fantastic learning experience, and great fun. Bill Walsh was an excellent chair of its Advisory Board for many years. Thanks to

Dr. Andrew Chen who has generously established an annual lecture in my name at Hull. Special acknowledgment is due to my surgeon at St James Hospital, Leeds, Professor Peter Lodge, whose knowledge and skills have ensured I am still here. The members of the "Old Gits' Club" have never been completely convinced by systems thinking and keep my feet on the ground. Nevertheless, one of them, David Tucker, helped me to improve the manuscript. As ever, I owe so much to my immediate family. The book, as with all the best things in my life, would not have been possible without the unwavering support of my wife Pauline. It is dedicated to her and my two sons Christopher and Richard. I can also announce that, with the birth of Freddie to Christopher and Tess, the next generation of Jacksons has started to arrive. Kelly passed away, so Molly is now the dog enjoying the walks on Beverley Westwood. I have enjoyed the process of thinking through and writing the book in Beverley; the North Yorkshire Moors; and Blanca, Spain. But it is definitely my last book.

Beverley, 2 October 2018   *Michael C. Jackson*

Introduction

The book is divided into four parts.

Part I considers the development and impact of systems ideas in four broad disciplinary areas: Philosophy (Chapter 1), the physical sciences (Chapter 2), the life sciences (Chapter 3), and the social sciences (Chapter 4). This theoretical background is necessary because it provides an introduction to the language of systems thinking and to the key concepts it employs. In the case of the social sciences, for example, a number of the systems thinkers studied in Parts II–IV have either developed their systems approaches with the help of social theory or, at least, related their work to social theory. This is significant because it can provide a basis for critique. The strengths and weaknesses of the different systems methodologies are related to the particular social theories they endorse. The intention in Part I is to make the absorption of the philosophical material as painless as possible for the reader and only to introduce those aspects of theory essential for understanding the practical systems approaches that are covered later.

Part II considers the development of systems thinking as a separate transdiscipline. Transdisciplines are unconstrained by normal academic boundaries and can recognize "messes" and "wicked problems" and not just, for example, individual marketing, production, human resource, and finance problems. Chapters 5 and 6 outline the emergence and significance of general systems theory and cybernetics, the two intellectual pillars on which systems thinking rose to prominence in the mid-twentieth century. Chapter 7 covers complexity theory, another transdiscipline that has come to the fore more recently. Complexity theory offers a complementary approach to systems thinking, adding to its theoretical armory and providing some new concepts that are appropriate for describing contemporary organizations and society.

Part III of the book turns to systems practice and the way systems ideas can be put to use in dealing with the problems posed by complexity. It begins by providing, in Chapter 8, an overview of applied systems thinking in the form of an updated "system of systems methodologies" (SOSM). Following this orientation, Part III is divided into sections, emphasizing that different types of systems approach have different visions of where the main sources of complexity arise. This broad division offers a starting point for discussion. There are six sections:

- Systems approaches for technical complexity (Type A)
- Systems approaches for process complexity (Type B)
- Systems approaches for structural complexity (Type C)
- Systems approaches for organizational complexity (Type D)
- Systems approaches for people complexity (Type E)
- Systems approaches for coercive complexity (Type F)

Using these headings for guidance, we consider (Chapters 9–18) 10 of the most significant attempts that have been made to construct a systems approach capable of improving the practice of management. The 10 methodologies outlined make use of the systems theory and concepts presented in Parts I and II. The manner in which they use systems ideas and the range of concepts employed are however different – in particular, in terms of what they regard as the most important aspects of the manager's task. There will be howls of anger that the different systems approaches are being "pigeon-holed." But we have to start somewhere. I will be absolutely clear about my starting point. The individual chapters will detail how the different approaches diverge from the broad distinctions initially employed and how some have evolved in an attempt to tackle other aspects of complexity. Each of the 10 approaches is presented in terms of its history, philosophy, and theory, methodology and methods, and examples of application are provided. The theoretical considerations set out earlier in the book are used to provide a critique of each approach.

One conclusion from Part III is that the different systems approaches emphasize and seek to address different aspects of complexity. Another is that they are heavily influenced by different philosophies and social theories and their particular strengths and weaknesses stem in part from the theoretical assumptions they take as their starting point. It follows that we have the best chance of managing complexity overall if we can understand and capitalize on their different strengths and compensate for their different weaknesses by using them in combination. This way of looking at things is called critical systems thinking and is the focus of Part IV of the book. Critical systems thinkers argue that the different systems methodologies and methods must be employed together, creatively and in a theoretically informed way, to improve leadership and managerial and organizational performance. Part IV has three chapters. Chapter 19 looks at the theory that underpins critical systems thinking and its relevance for the management sciences generally. Chapter 20 considers some different ways that have been developed for using systems approaches in combination. My own latest version of "Critical Systems Practice" is set out in Chapter 21.

The book ends with a short conclusion.

In this introduction, I have sought to make clear the structure of the book and the logic underlying that structure. This is summarized in Table 1.

Table 1 The structure of the book.

Introduction		
Part I: Systems Thinking in the Disciplines		Chapter 1: Philosophy
		Chapter 2: The Physical Sciences and the Scientific Method
		Chapter 3: The Life Sciences
		Chapter 4: The Social Sciences
Part II: The Systems Sciences		Chapter 5: General Systems Theory
		Chapter 6: Cybernetics
		Chapter 7: Complexity Theory
Part III: Systems Practice		Chapter 8: A System of Systems Methodologies
	Type A: Systems Approaches for Technical Complexity	Chapter 9: Operational Research, Systems Analysis, Systems Engineering (Hard Systems Thinking)
	Type B: Systems Approaches for Process Complexity	Chapter 10: The Vanguard Method
	Type C: Systems Approaches for Structural Complexity	Chapter 11: System Dynamics
	Type D: Systems Approaches for Organizational Complexity	Chapter 12: Socio-Technical Systems Thinking
		Chapter 13: Organizational Cybernetics and the Viable System Model
	Type E: Systems Approaches for People Complexity	Chapter 14: Strategic Assumption Surfacing and Testing
		Chapter 15: Interactive Planning
		Chapter 16: Soft Systems Methodology
	Type F: Systems Approaches for Coercive Complexity	Chapter 17: Team Syntegrity
		Chapter 18: Critical Systems Heuristics
Part IV: Critical Systems Thinking		Chapter 19: Critical Systems Theory
		Chapter 20: Critical Systems Thinking and Multimethodology
		Chapter 21: Critical Systems Practice
Conclusion		

Part I

Systems Thinking in the Disciplines

Mark this well, you proud men of action: You are nothing but the unwitting agents of the men of thought who often, in quiet self-effacement, mark out most exactly all your doings in advance

(Heine 1834)

Part I traces the emergence of systems thinking in philosophy, the physical sciences, the life sciences, and the social sciences. The reason for concentrating on these broad fields of knowledge is that it demonstrates the necessity of systems thinking for making intellectual progress in a wider context than that of individual disciplines. A downside is that individual disciplines impacted by systems thinking, such as geography and political science, are ignored if not central to that purpose. Chapter 1 is a review of the long engagement that has taken place between philosophy and systems thinking. Chapter 2 looks at the physical sciences, the refinement of the "scientific method," and at how that method (based on "reductionism") enabled spectacular progress to be made in science and technology in the seventeenth, eighteenth, and nineteenth centuries. It notes, however, that newer discoveries in general relativity, quantum mechanics, and chaos theory are leading to a rethink of the traditional scientific method and requiring the physical sciences to embrace systems ideas. In contrast to the physical sciences, the life sciences, specifically biology and ecology, seemed to require a commitment to systemic thinking from their early days. As a result, they have provided a rich resource of systems concepts and played a major part in establishing systems thinking as a "trans-discipline." This is the topic of Chapter 3. In Chapter 4, the focus is on social theory, a field that makes significant use of systems ideas developed elsewhere but has also come up with its own original contributions to the systems approach. The treatment of theoretical matters in Part I is designed to illuminate and guide the practical employment of the systems methodologies that are detailed in Part III.

Critical Systems Thinking and the Management of Complexity, First Edition. Michael C. Jackson.
© 2019 John Wiley & Sons Ltd. Published 2019 by John Wiley & Sons Ltd.

1

Philosophy

In the case of all things which have several parts and in which the totality is not, as it were, a mere heap, but the whole is something beside the parts …

(Aristotle n.d., 350 BCE, VIII: line 1)

1.1 Introduction

Fritjof Capra (1975) has, for some time, been pointing to similarities between the holistic understanding of the world supplied by Eastern philosophy and the findings of modern science. Churchman regarded the *I Ching,* with its emphasis on dynamic changes of relationship between interconnected elements, as presenting the oldest systems approach (Hammond 2003, p. 13). Boulton et al. (2015) claim Daoism, with its sense of interconnection and co-creation, as a precursor of complexity theory. This book will restrict itself to the Western intellectual tradition. It is upon Western sources that systems practitioners have, probably to their detriment, almost exclusively drawn. As with so much in this tradition, we owe the first attempts to use systems ideas to the ancient Greeks. von Bertalanffy (1971) and Prigogine (1997) cite the pre-Socratic philosopher Heraclitus as an influence. More specifically, Aristotle (n.d.) 350 BCE was the first to imply that "the whole is more than the sum of its parts." Indeed, he reasoned, the parts only obtain their meaning in terms of the purpose of the whole. The parts of the body make sense because of the way they function to support the organism. Individuals can only find meaning in helping the state to achieve its purpose. The other great master in the Greek philosophical tradition, Plato, also found value in employing systems ideas across different domains. There is a Greek word *kybernetes* meaning the art of steersmanship. The word referred principally to the control of a vessel, but Plato (1999, pp. 230–231) used it to draw comparisons with steering the ship of state. Both uses imply regulation, which is why the name cybernetics was given to the new science of "communication and control" in the 1940s.

Critical Systems Thinking and the Management of Complexity, First Edition. Michael C. Jackson.
© 2019 John Wiley & Sons Ltd. Published 2019 by John Wiley & Sons Ltd.

1.2 Kant

Moving forward two millennia, to the latter part of the eighteenth century, we reach Immanuel Kant. Kant is often seen as the greatest philosopher of the modern era and provided the Enlightenment with its motto: *Sapere aude!* (Dare to know!). Knowledge should be based solely on reason rather than superstition and tradition. Kant's work is significant for systems thinking for three reasons. First, he thought that science could obtain true knowledge, as it had with Newtonian physics, and he wanted to show why this was the case. He also wanted to understand the limitations of science. The second reason lies in his interest in "organicism" as a complementary approach to mechanistic thinking, especially in the study of nature. Third are his arguments about the capacity of humans to generate principles of moral conduct because, uniquely, they possess "the autonomy of the will."

In his *Critique of Pure Reason*, Kant sought to expose the shortcomings of both "rationalism" and "empiricism" as approaches to gaining knowledge. Rationalists, such as Descartes, believe that it is possible to employ cogent thinking alone to arrive at knowledge about the nature of things. In Kant's view using rational thought on its own leads to contradictions, for example, to proofs that God exists and doesn't exist. Reason has to be grounded in experience if it is to yield true knowledge. Empiricists (e.g. Locke, Berkeley, and Hume) believe that all knowledge has to be derived directly from experience through the senses. Kant thought that this was too subjective and opened the door to skepticism because our senses can easily deceive us. We need something more certain to rely on. Kant used the famous phrase: "Thoughts without content are empty, intuitions [perceptions] without concepts are blind" (quoted in Kemp 1968, p. 16).

If we are to overcome the weaknesses of rationalism and empiricism, we require, Kant says, a revolution in philosophy akin to that of Copernicus in cosmology (Kemp 1968). Instead of seeing knowledge as dependent upon our minds representing what actually exists in reality, we should see it as based upon what we perceive conforming to the nature of the mind. It is because all human minds structure the experiences they receive in a particular way that shared perceptions and knowledge are possible. This notion of mind as the creator of reality becomes clearer if we consider the latest brain research. According to Armson:

> My senses receive 400 thousand million bits of data every second. My brain only deals with 2000 bits per second so I only notice a very small fraction – a half a millionth of one percent – of what I see, hear and smell. More extraordinary still is the observation that the 100 bits per second that trigger my visual perception are not enough to form any image of what is going on around me. My brain fills in the deficiency. It is hard to defend any claim to an objective view under such circumstances.
>
> *(Armson 2011, loc. 975)*

The world does not present itself to us as already organized. The mind must play an active role for humans to experience it as they do.

In pursuing this argument, Kant requires a distinction between "phenomena," things as they appear to our senses, and "noumena," things as they actually are in themselves. Knowledge is possible because there is an inevitable correspondence between our minds and things as they appear to us. This arises because our minds structure the sense impressions we receive in order that we can perceive them in the first place. Far from the mind being a *tabula rasa* (blank slate) upon which reality writes its script, it actually provides the framework that makes experiences possible. According to Kant, the human mind possesses "sensibility," which delivers experiences, and "categories" which organize those experiences and provide understanding. There are two elements of sensibility, space and time, which supply the mind with perceptions. There are 12 ordering categories, which Kant derives from Aristotle's logic, with four broad classes of quantity, quality, relation, and modality, each divided into three subclasses. Examples of the categories are "substance" and "cause." The idea of substances with attributes and the idea of universal causation are not given to us in experience but are provided by the mind and impose order on our perceptions. Since these structural features of the mind are innate in human beings, the world appears to all people in essentially the same form. William Golding's (1955) novel *The Inheritors* is a brilliant attempt to capture what the world might have looked like to Neanderthal people in contrast to our world. In the case of the Neanderthal mind, the sensibilities and categories are not quite fully established.

In short, we can only have the experiences we have because of our minds, and so there is a necessary correspondence between the structure of the mind and the way the world appears. Logic, mathematics, and sciences such as physics, Kant argues, also depend on the concepts of space and time and the 12 categories, and it is this that makes it possible for them to be successful and to add to our stock of knowledge. They are able to produce knowledge that is universally true. This is the case even though we can never have access to the external world that provides the things we sense, i.e. the noumena or things in themselves. We will never know about the world of noumena. We are human beings who observe the world through our senses so we can only ever know things as they appear. Scientific knowledge is only possible because it restricts itself to elucidating what the mind makes available through the senses.

We now have a reason, although admittedly a topsy-turvy one, for accepting the knowledge produced by science. But what are, for Kant, the limits of scientific knowledge? Clearly, it carries no weight in fields such as psychology or in answering metaphysical questions about the existence of God, the immortality of the soul, and free will. This is because the subject matters of psychology and metaphysics lie beyond what we can observe with our senses and so what science can explore on the basis of space and time and the categories. As Kant argued, thoughts without content are empty. We can prove anything and so are led into contradiction. That does not mean that reflection on these matters is pointless, just that in the case, for example, of seeking principles to guide human conduct, we have no choice but to venture beyond the knowledge that science can provide.

It is now possible to consider the second reason for Kant's importance for systems thinking. At the time that he was beginning his philosophical work, the mechanistic view, insisting that all life forms had remained the same since their creation, was being questioned from an "organicist" perspective. Kant was much influenced by this thinking and agreed that it was impossible to provide a mechanical account of organic processes such as change, growth, and development. The vitality and diversity of nature seemed to require a different kind of explanation that accepted the emergence of new and more complex organisms. As he wrote: "Are we in a position to say: *Give me matter and I will show you how a caterpillar can be created?"* (quoted in Mensch 2013, loc. 234). Kant was now in a dilemma because his arguments for what constituted scientific knowledge, later set out in the "Critique," only permitted mechanical explanations. There seemed to be a requirement for organicist thinking in the "life sciences" but using it meant it was impossible to attain the same certainty as in mathematics and physics.

Kant returned to this issue in earnest in his *Critique of Judgement*. In terms of scientific reasoning, he argued, it is indeed impossible to support organicism because this would take us "beyond the mechanism of blind efficient causes" (quoted in Kemp 1968, p. 114). On the other hand, using a simple example, biologists are not going to get very far in studying the human heart if they restrict themselves to the question of "how did this come about?" and ignore the question of "what is this for?" Teleological explanation employing a form of causality directed to ends, in this case the parts serving the purposes of the whole, is essential. Kant's solution is to argue that, while it cannot be fully justified, it seems to be of considerable heuristic value to assume purposiveness of this kind in nature. Organicism is essential to pursuing studies in the life sciences even if it does take us "beyond the world of sense" (Kant, quoted in Kemp 1968, p. 114). This will always be the case because of the unavoidable limitations on our thinking. Nature is just too complex to be encompassed by the human mind. Kant is here anticipating a conclusion of Checkland's (see Chapter 16) that systemicity is best seen as an epistemological device to inquire into the world rather than assumed to be a characteristic of the world. As Mensch has argued, not all of Kant's followers were quite as theoretically scrupulous:

> Convinced of nature's vitality, naturalists and philosophers would make use of Kant's work as they saw fit. The most significant transformation of Kant's work concerned the use of transcendental principles themselves, since these tools for *thinking* about nature would be subsequently ascribed to nature itself.
>
> *(Mensch 2013, loc. 420)*

As an aside, Mensch (2013) has argued that the organicist perspective had a major impact on Kant even in the *Critique of Pure Reason*. Alongside its job of constructing experiences, Kant thought, the mind must also integrate them. Ultimately this depends on the self, a single consciousness to which thoughts, reflections, and intuitions are related (Kemp 1968). In creating such a unity, the mind must operate according to some sort of organic logic. According to Mensch,

> … like an organism, cognition functioned [for Kant] as a set of parts whose thoroughgoing connection realized unity even as the grounds of that unity preceded it. This was a different logic at work … it was a reflexive logic according to which the unity of apperception was both cause and effect of itself, or, as Kant would put it in another context, both author of and subject to its own laws.
>
> *(Mensch 2013, loc. 374)*

This is an interesting foretaste of the notion of the mind, or "psychic system," as a self-producing system that will be discovered in Luhmann's systems theory in Section 4.7.

On the subject of human behavior, and here we come to his third great contribution to systems thinking, Kant faced a problem even more severe than he had encountered with nature. According to science, the self must be subject to the laws of causality just as are all other phenomena in the realm of appearances. In order to uphold the notion of morality, however, we need to believe in the existence of free will. In the *Critique of Practical Reason*, Kant sets out his solution. As phenomena, humans are subject to causal determinism. However as "things in themselves" (noumena), beyond the reach of scientific knowledge, it is completely legitimate to regard them as possessing freedom:

> We have in the world beings of but one kind whose causality is teleological, or directed to ends, and which at the same time are beings of such character that the law according to which they have to determine ends for themselves is represented by themselves as unconditioned and not dependent on anything in nature, but as necessary in itself. The being of this kind is man, but man regarded as noumenon.
> *(Kant, quoted in Kemp 1968, pp. 120–121)*

Having demonstrated that it is possible to believe in free will, Kant argues that it is essential to do so. Although it may not be theoretically justifiable, from a practical point of view, it is necessary to believe in freedom of choice just because morality depends on it. On the basis of such "practical reason," it then becomes possible to establish proper rules of human conduct. In order to be sure that they are acting morally rather than according to their individual desires, humans must be able to universalize their actions. Kant's famous "categorical imperative" follows: "Act only on that maxim which you can at the same time will to be a universal law." (This is sometimes formulated as always treating people as ends in themselves, never as means to an end.) So, for example, borrowing money without intending to pay it back fails the test because it treats the lender as a means and undermines trust and the possibility of future borrowing for other humans. It also turns out that belief in God and an immortal soul is rational from the perspective of practical reason. Only God can guarantee a fair correspondence between the virtue we display and the rewards we receive, and only immortality can provide for its delivery. Freedom, God, and immortality may not be laws of nature, but they are powerful laws of morality.

Before leaving this section, it is worth reflecting on the considerable impact Kant's philosophy continues to have on Western culture. There are, for example, numerous speculations on what it might be like to escape the world of appearances and see "things in themselves" and various proposals as to how this might be achieved. Not all acknowledge Kant, but it is Kant's influence that is at work. Huxley (1959) experimented with the drug mescaline to try to break through the "eliminative" function of the brain and sense organs. Castaneda sought the guidance of a Yaqui Indian, don Juan, to get beyond the *tonal*, "the organizer of the world," and to witness the *nagual* (the Yaqui equivalent of noumena) and is warned, along the way, that:

> Ordinarily, if an average man comes face to face with the *nagual* the shock would be so great that he would die.
> *(Castaneda 1974, p. 174)*

In the science fiction film *The Arrival* (Denis Villeneuve, director, 2016), it takes mastering an alien language to allow humans to escape the tyranny of space and of time as a linear phenomenon. Adam Roberts' novel *The Thing Itself* speculates that artificial intelligence (AI) might achieve what humans can't:

> And we've discovered that, once you abandon the notion of trying to *copy* human consciousness, AI is really quite easy to achieve .... You've done this? ... Sure ... A rational, sentient, intelligent consciousness, unfettered by the constraints of space and time? One that can see into the Ding an Sich [thing in itself]? .... Essentially, yes. Pretty much.
>
> *(Roberts 2015, p. 92)*

Returning to the argument of the book, Kant's philosophy provided a kind of inverted justification for what mathematicians and physicists were doing in their own fields. They were gaining knowledge by learning how the human mind structures reality. It also gave a warning to scientists who sought to extend the mechanical model into the domains of the biological, human, and social domains. These warnings were rarely heeded as many sought to increase the scope of the scientific method even as far as psychology and sociology. For the moment, we shall continue to explore how later philosophers engaged with Kant's conclusions on the limitations of the human mind and how they impact on what we can know with certainty.

1.3 Hegel

Hegel, writing at the beginning of the nineteenth century, criticized Kant for his a-historical account of mind. For Hegel, the mind gives rise to reality but, at the same time, is itself historically conditioned. The mind is the driving force of history but has tended to externalize itself in an alienated way in which customs and institutions seem to stand above and control human action. However, during the Enlightenment, thinking has progressed to the stage where it is able to understand its true destiny. "The history of the world," Hegel wrote, "is none other than the progress of the consciousness of freedom" (quoted in Honderich 1995, p. 339). It was now possible for humans, with their common capacity for reason, to take control of history and build a truly free community to which they can all assent because it is a rational expression of their will. Thought frees itself from history and becomes capable of determining its future course. Society ceases to be alien and hostile to people because it is a reflection of their rational intentions.

In Hegel's "absolute idealism," the dualisms of mind and nature and subject and object are overcome because there is only mind and mind determines reality. The process by which mind is able to overcome its historical limitations and gain a holistic understanding of itself is called "dialectical." Comprehension of the whole, "the absolute," is gained through a systemic unfolding of partial truths in the form of a thesis, an antithesis, and a synthesis, which embraces the positive aspects of the thesis and antithesis and goes beyond them. Each movement through this cycle, with the synthesis becoming the new thesis, gradually enriches our grasp of the whole system. An example given in Honderich (1995, p. 342) is of "the customary morality of ancient Greece [as] the thesis, the

Reformation morality of individual conscience its antithesis, and the rational community [as] the synthesis of the two."

1.4   Pragmatism

One way of reading the work of the pragmatist philosophers, Pierce, James, and Dewey
(writing in the United States in the late nineteenth and the early twentieth century), is
as a response to Kant's concern about how to proceed in the noumenal world, beyond
the sway of science as he strictly defined it. This realm is vast, embracing such matters
as the behavior of organisms, the free will of humans, the purpose of social organizations, morality, esthetics, as well as all aspects of theology. The pragmatists found a clue
in Kant's very definition of pragmatic belief. He had written that "contingent belief,
which yet forms the ground for the effective employment of means to certain actions, I
entitle *pragmatic belief*" (quoted in Honderich 1995, p. 710). Their response was to seek
a justification for belief and action in terms of its practical effectiveness. They differed
among themselves, however, about the scope of this justification and about who should
decide whether the standard set had been met. Pierce, the most restrictive of the three,
felt that it could be used by scientists to extend their knowledge by taking predictive
success as the main criterion for deciding between competing theories. James felt that
the justification could be extended to the rightness of actions as judged by individuals.
Dewey's interest was in the resolution of problems, which meant knowledge was confirmed only when it was recognized by the community as being successful in transforming practice so as to overcome problems. Ormerod precisely summarizes the situation:

> Peirce's pragmatism is scientifically élitist, James's is psychologically personalis-
> tic, Dewey's is democratically populist.
>
> *(Ormerod 2006, p. 893)*

Let us follow the thinking of James (particularly as described in Passmore 1970), who
was the main popularizer of pragmatism with his 1907 book *Pragmatism: A New Name
for Some Old Ways of Thinking*. James rejected, as did all the pragmatists, what Dewey
called the "the spectator theory of knowledge," which presented knowledge as a passive
reflection of some external reality. To him the world had a "concatenated unity," experienced by human beings not as divided into parts but as a "stream of consciousness."
Individuals had to impose a structure upon the wholeness of everything – upon the
"blooming, buzzing confusion," as he described a baby's first experience of the world. The
question for James, therefore, became what was the best way of using concepts to create
order. Since reality is not static but in process, the pragmatic answer is to employ ideas
that are effective in the long run in helping realize our goals and objectives and so bring
benefits. True beliefs are those that prove useful over time as judged by the individuals
concerned. James is excited about the free will his account lends to human beings. Because
reality is in the making, they have the capacity to change and improve the world in which
they live. "The greatest discovery of my generation," everyone quotes him as saying, "is
that a human being can alter his life by altering his attitudes of mind." This freedom is
constrained, however, because all ideas and actions are subject to an empirical reality
check. If their consequences do not prove fruitful, then they will have to be abandoned.

1.5   Husserl and Phenomenology

Husserl is another philosopher who has had a major influence on systems thinkers. He wrote his major works on "phenomenology" in the early years of the twentieth century. The term phenomenology indicates that his interest was in phenomena, "things as they appear to our senses" (as Kant would say), rather than in speculating about any independent reality. Indeed, to develop a science of pure consciousness, which is the aim of phenomenology, it is necessary to "bracket" our "natural attitude" that things like trees and tables exist and cause our sensations. Once this is achieved, it becomes possible to go directly "to the things themselves" and begin a rigorous investigation of the common features of all acts of consciousness and of how the mind constitutes and experiences the world.

For Husserl, all conscious mental activity, whether linked to sensory perception, the imagination, or our emotions, is thinking about something. Philosophy is about uncovering how the mind addresses and gives meaning to the world through "intentionality." In his later work, this thinking took him closer to Hegel's philosophy (see Honderich 1995) as he became interested in the historicity of consciousness. He began to see experiences as conditioned by the traditions and social context of the time. They are part of a "life-world," which we share with others and largely take for granted. Science itself emerges from this life-world and is dependent on it for the research it does, the evidence it collects, the experiments it conducts, and the way it interprets its results. It has, however, been losing touch with the life-world since its "mathematization," following Galileo, and has nothing to say now about the really important issues of concern to humankind:

> In our vital need … science has nothing to say to us. It excludes in principle precisely the question which man, given over in our unhappy times to the most portentous upheavals, finds the most burning: questions about the meaning or meaninglessness of this whole human existence.
>
> *(Husserl 1970, p. 6)*

Husserl's attention to how individuals actually experience the everyday world was attractive to many philosophers and established a phenomenological tradition (see Bakewell 2017). His even more famous or infamous protégé, Heidegger, shaped phenomenology into an investigation of "being" and especially the mode of "being-in-the-world" in a particular social context. In his later work, Heidegger became concerned with a change in "intentionality" – in the way the mind relates to things as its objects. The proper purpose of human consciousness is to "reveal" the nature of being and the tools of phenomenology can be employed to this end. In contemporary society, however, this possibility is endangered by the advance of an "enframing" mentality. This is clearly expressed in modern technology, which presents both nature and human beings as a "standing reserve," ready to be used for some instrumental purpose. Commenting on the relevance of this thinking to the Internet, Bakewell comments:

> Later Heideggerians, notably Hubert Dreyfus, have written about the internet as the technological innovation that most clearly reveals what technology is. Its infinite connectivity promises to make the entire world store-able and available, but, in doing so, it also removes privacy and depth from things. Everything, above all ourselves, becomes a resource, precisely as Heidegger warned.
>
> *(Bakewell 2017, p. 324)*

A particular strand of phenomenology, pioneered by Jean-Paul Sartre and Simone de Beauvoir, became known as "existentialism" and gained significant popular appeal. Sartre read the notion of "intentionality" as suggesting that the mind has immense freedom to interpret the world as it wishes. Individuals are influenced by biology and social conditioning, but they have no predefined nature. "Existence precedes essence," and people are free to decide how to live and act. This radical freedom is frightening to many, and they reject the responsibilities it brings by taking on a ready-made role in the life-world. Sartre explored, in novels and plays as well as philosophical writings, how it is possible for individuals to escape their apparent destiny and live free of "bad faith." Later, reacting to the Second World War and the events surrounding it, Sartre's primary concern became how we should use our freedom. Thus began a life-long endeavor to fuse elements of phenomenology and Marxism into a practical program of action. Deciding that "truth" could only be established by looking at the world through "the eyes of the least favored" or to "those treated the most unjustly" (Bakewell 2017, p. 271), Sartre was led to support a variety of radical causes and groups and to reject the Nobel Prize in literature. Meanwhile his partner in life and in developing existentialism, Simone de Beauvoir, was using "applied existentialism" to explore the history of patriarchy and how it plays out as individual women lead their lives from birth to old age. Her great book *The Second Sex* argues that a female is not born but becomes "a woman" as she takes on the dominant male perspective and sees herself through the "male gaze." Women need to stop seeing themselves as "objects" and assert their subjectivity. They can then confront the world as it really is for them, break out of gendered roles, and change their lives. Bakewell considers *The Second Sex* as the most influential work to emerge from the existentialist movement and as deserving of a place alongside those of Darwin, Marx, and Freud as "one of the great cultural re-evaluations of modern times" (2017, p. 216). It is difficult to argue with this conclusion. The two central notions of "applied existentialism" – of always siding with the oppressed and of liberating "slaves" from the perspective of their "masters" – made existentialism popular with anti-colonialist writers and campaigners such as Albert Memmi and Frantz Fanon. Both were championed by Sartre and de Beauvoir. Black American writers, such as Richard Wright and James Baldwin (who was also gay), turned up in Paris to absorb the doctrine of existentialism and to experience a freedom denied to them in their homeland. Another influential French existentialist, Maurice Merleau-Ponty, contributed to systems thinking in a different way. He introduced the idea of the human body, with its hands, feet, etc., as a primordial and permanent condition of experience. He also argued, following the *Gestalt* psychologists, that we make sense of the world through unified and meaningful "wholes" rather than clearly delineated individual perceptions. The wholes we construct are ever changing according to the intentionality guiding our observations.

1.6 Radical Constructivism

Radical constructivism may be little more than a footnote in the history of mainstream philosophy, but it has contributed significantly to the development of second-order cybernetics (see Section 6.4). It is associated primarily with the work of von Glasersfeld, who was writing in the late twentieth century. von Glasersfeld took his inspiration from the genetic epistemology of Piaget. Piaget's theory stated that cognitive development in children occurs as mental processes reorganize themselves as a result of the interaction between biological maturation and environmental experience. "Intelligence organizes the world by organizing itself," he wrote (quoted in von Glasersfeld 1984, p. 5). At various stages of their development, children engage with the world and understand it differently. For von Glasersfeld (1984), this was confirmation of Kant's argument that our minds do not reflect some external reality but construct that reality from what is provided by experience. In contradistinction to Kant, however, von Glasersfeld goes on to emphasize the considerable freedom this provides to human beings because of the extremely rich raw material that the experiential world provides. The world out there presumably imposes some boundaries on what is possible, but these are very broad. We are not governed in our thinking by immutable Kantian "categories" but only by the opportunities presented by the history of the construction process to date. In this respect, von Glasersfeld argues,

> ... the theory of evolution can serve as a powerful analogy: the relation between viable biological structures and their environments is, indeed, the same as the relation between viable cognitive structures and the experiential world of the thinking subject. Both *fit* – the first because natural accident has shaped them that way, the second because human intention has formed them to attain the ends they happen to attain; ends that are the explanation, prediction, or control of specific experiences.
>
> *(von Glasersfeld 1984, p. 4)*

Just as the natural environment allows for many types of organism, so the experiential is forgiving of many ways of understanding and being in the world. von Glasersfeld criticizes pragmatist philosophers because they foster the temptation to seek access to an "objective" world on the basis of "effectiveness" but "effectiveness" he argues "is a judgement made within a domain of experience which itself was brought forth by an observer's activity of distinguishing" (1990, p. 3). In radical constructivism, viability replaces truth as the key concept because, however much we push against the world "out-there," all we can ever learn is whether the cognitive apparatus we have developed provides one *fit* among all those that might be possible. Returning to Piaget, a very young child possesses a viable cognitive structure even though it is incapable of logical or abstract thought. The child learns through physical actions and monitoring their results to construct a relatively stable world in which its needs are met. It does not gain objective knowledge about reality. The degree of cognitive freedom implied by radical constructivism leads von Glasersfeld to stress, again and again, the personal responsibility we all have for our words and deeds.

1.7 Conclusion

The philosophical ideas set out above are those that are of most relevance for examining the different ways of using systems ideas in management that will be considered in Part III of this book. For the moment, we pass onto other disciplinary areas and detail their impact upon the development of systems thinking. In a sense, though, the next three chapters are a continuation of the debate with Kant's philosophy. His insistence on

Newton's mechanical model as the exemplar of knowledge in the physical sciences was not seriously challenged until the genesis of general relativity and quantum mechanics. His organicist approach to the life sciences continued to be influential until Darwin's theory of evolution provided an alternative to teleological explanations in that field. Social sciences, such as psychology and sociology, can be regarded as correctives to the notion of the "autonomy of the will" that underpins his reflections on proper human conduct.

2

The Physical Sciences and the Scientific Method

If we possessed a thorough knowledge of all the parts of the seed of any species of animal (e.g. man), we could from that alone, by reasons entirely mathematical and certain, deduce the whole figure and conformation of each of its members
(Descartes, *Oeuvres* 1897, iv, 494, originally from the 1630s)

2.1 Introduction

Newton's *Principia* (1687) set out his laws of motion, his theory of universal gravitation, and a new cosmology. It was the apotheosis of the Scientific Revolution, which had begun with Copernicus' challenge to Ptolemy's earth-centered model of the universe. That revolution produced a huge growth of knowledge in fields such as mathematics, physics, astronomy, chemistry, and biology. To many scientists, it seemed that everything would soon be known. Further, the technologies that stemmed from the advances in science transformed the world in which we live. There were revolutions in agriculture and industry leading to massive increases in productivity and the rapid growth of urbanization. These, together with better prevention and treatment of disease, led to population growth and longer life expectancy. Roads were improved, canals and railways were built, and shipping lines developed, increasing trade and making the world a smaller place. Despite frequent wars and the continuance of poverty, it seemed that progress toward a better society was being made and that this would continue.

The achievements of science in opening up the physical world to our understanding are said to stem from the method it employs to gain knowledge. Sir Hermann Bondi, the distinguished mathematician and astronomer, declared: "There is no more to science than its method …" (Lewens 2015, loc.140). It is this scientific method that we must explore to reveal why it is so successful in enabling mastery of many aspects of the physical world and what limitations there are to its proper use in this domain and in other fields.

Critical Systems Thinking and the Management of Complexity, First Edition. Michael C. Jackson.
© 2019 John Wiley & Sons Ltd. Published 2019 by John Wiley & Sons Ltd.

2.2 The Scientific Method and the Scientific Revolution

An excellent account of the evolution of the scientific method, in terms of its relevance to systems thinking, is provided by Checkland (1981). It begins with the pre-Socratic philosophers of ancient Greece who, eschewing explanations relying on magic or the Gods, employed "rational thought" to develop and defend their conclusions. Further contributions were made by the great Muslim thinkers who rediscovered the significance of Aristotle's work and who made advances in mathematics and optics. The mediaeval alchemists added to the mix with their zeal for experimentation.

The decisive moment came in the early seventeenth century through the efforts of Francis Bacon and Galileo. Bacon thought deeply about how to do science. In his view, scientists should give precedence to studying nature directly and not through the works of Aristotle. He advocated close observation of the facts, the development of hypotheses from those facts and directed experiments to test the hypotheses. The results of the experiments should be recorded and reported so that progress was cumulative. In this way, science could help improve man's lot on earth (Chalmers 1982; Checkland 1981). Galileo's practice as a scientist gave substance to Bacon's words. He was wedded to the facts obtained from observation and defended what he saw. The moons of Jupiter existed, he insisted, and were not aberrations of his telescope because if they were aberrations he would see moons around other planets as well (Chalmers 1982). He established the experimental method as the norm in science; most famously measuring the time it took spheres of different weights to reach the ground when dropped from the top of the Leaning Tower of Pisa. His experiments were designed to test particular hypotheses. The details and results were written up so that they could be repeated and validated by other scientists in different places. Where he went beyond Bacon's thinking was in his commitment to mathematics as the language of science. If at all possible, theories and experimental demonstrations were expressed in mathematical terms.

In the early years of the seventeenth century, therefore, a best way of doing science, a well-defined "scientific method," was established. This can be summarized as having five steps. First a part of reality, of interest to the scientist, is separated from the rest (in the scientist's mind) and observed. Second, on the basis of numerous observations, a hypothesis is constructed, in mathematical terms if possible, setting out how some of the variables that make up that part of reality behave. This move from a finite number of observations to a possible universal law is known as induction. Third, deductions are made and predictions formulated about how the variables will behave in the future. Fourth, carefully devised experiments are conducted to test the predictions and the results of these measured. The experiments must be clearly described so they can be repeated by other scientists. Finally, the results are analyzed and conclusions reported setting out whether the experiments confirm or disprove the hypothesis. On this basis, the progress of science is guaranteed.

In Descartes' opinion, the success of this scientific method was due to its "reductionism." A part of the real world is isolated from the rest for study and then broken down into separate objects or variables for further analysis. The logic of mathematics is used to build back up to an understanding of the whole. He used his considerable philosophical weight in support of this approach. Writing in 1637, he argued that, if he wanted to understand the world and the problems it posed, it was essential to proceed by the method of reductionism

> ... to divide each of the difficulties that I was examining into as many parts as might be possible and necessary in order best to solve it [and] beginning with the simplest objects and the easiest to know ... to climb gradually ... as far as the knowledge of the most complex.
>
> *(Descartes 1968, pp. 40–41)*

The validity of this mechanistic perspective, in which the whole is no more than the sum of its parts, seemed to be confirmed, later in the century, when Newton used the scientific method to realize his supreme achievement of uniting terrestrial and celestial mechanics. For example, his hypothesis about gravity as a universal force was refined into equations, which enabled accurate predictions to be made about the movement of objects on earth and planets in the sky. The universe was like clockwork set in motion and sustained by God and followed entirely predictable rules that could be understood by humans. As Buchdahl summarizes it:

> [Newton's] synthesis of empirical data and abstract mathematical relations which are here united to lead to accurately verifiable observations, impressed [his] contemporaries by seemingly bestowing the certainty of mathematics upon man's knowledge of physical phenomena, and gave them a new sense of power over nature.
>
> *(Quoted in Checkland 1981, p. 44)*

The success of the Scientific Revolution led to the almost complete dominance of Cartesian mechanism during the eighteenth and nineteenth centuries. Scientists refined Newtonian mechanics to provide precise explanations of the behavior of solids, liquids, and gases, and phenomena such as sound and the tides. In 1814, the mathematician Laplace was able to assert that Newton's laws could in principle be used to predict everything for all time as long as the current position and velocity of all the particles in the universe were known (Mitchell 2009, p. 19).

The most significant extension of science in the nineteenth century followed in the same tradition. It was achieved by applying Newtonian mechanics to thermal phenomena and "treating liquids and gases as complicated mechanical systems" (Capra and Luisi 2014, p. 32). This gave rise to "thermodynamics" and the discovery of its two fundamental laws: the law of the conservation of energy and the law of the dissipation of energy. The second of these is particularly important for our purposes since it argues that all isolated systems, such as the universe, inevitably dissipate mechanical energy and move from order to disorder. In the jargon, entropy, as a measure of disorder, gradually increases as useful energy is lost in the form of friction or heat. In the words of Capra and Luisi: "The entire world-machine is running down and will eventually grind to a halt" (2014, p. 33). This, of course, causes a problem for those who perceive the living and social worlds as apparently increasing in order and complexity.

The development of thermodynamics was made possible by the invention of statistical mechanics, essentially a combination of statistics and probability theory with Newtonian mechanics (Capra and Luisi 2014, p. 104). Newton's equations of motion were notoriously difficult to solve when applied to more than a few bodies exhibiting regular behavior. In fact, at the time, even three bodies precluded precise solutions. So how could scientists proceed when confronted by, for example, gases with millions of molecules? A solution was found by the physicists Ludwig Boltzmann and James Clerk Maxwell. They accepted that it was impossible to predict the behavior of all the molecules individually but reasoned that, as each molecule acted independently, statistical methods could be applied to predict their average behavior. Because of the myriads of molecules involved, this corresponded almost exactly to their actual overall behavior. Thus, they were able to derive the laws of motion of gases by using the average behavior of molecules.

By the end of the nineteenth century, therefore, as Capra and Luisi say,

> ... scientists had developed two different mathematical tools to model natural phenomena – exact, deterministic equations of motion for simple systems; and the equations of thermodynamics, based on statistical analysis of average quantities, for complex systems.
>
> *(2014, p. 104)*

They could be forgiven for believing that they were involved in an enterprise that was steadily discovering the truth about the world. Further, and despite Kant's cautionary warnings, they thought that the scientific method that was bringing such success in physics could be extended with equally positive results to other fields. Dalton's work on the physical behavior of gases led, in the nineteenth century, to the formulation of an atomic theory of chemistry with the promise of explaining all chemical phenomena using physics (Capra and Luisi 2014, p. 30). It surely would not be long before biology, and perhaps even social systems, would succumb to scientific explanations and be seen as nothing but complicated expressions of the laws of physics. The physicist Albert Michelson proclaimed in 1894 that

> ... it seems probable that most of the grand underlying principles have been firmly established and that further advances are to be sought chiefly in the rigorous application of these principles to all phenomena which come under our notice.
>
> *(Quoted in Mitchell 2009, pp. ix–x)*

Systems ideas were pushed to the margins, championed only by a few in the life sciences and those artists, writers, poets, and philosophers working under the banner of "romanticism."

The revolutions in thinking in the physical sciences in the twentieth century, however, have shattered the self-assurance of scientists and led to doubts being expressed about the nature and scope of application of the scientific method as previously described. It is to these matters that we now turn.

2.3   The Physical Sciences in the Modern Era

Gleick (1987) has argued that twentieth-century science will be remembered for three things: relativity, quantum mechanics, and chaos theory. What all three announce and share in common is a revolutionary transformation of understanding. Scientists have had to abandon the mechanistic and deterministic assumptions underlying the Newtonian worldview and embrace a perspective that is more systemic in character.

Einstein revealed his general theory of relativity in 1915 following long reflection on the contradictions that existed between Newton's law of gravity and his own special theory of relativity. Newton saw the universe as a great empty box within which gravity operates to determine the motion of objects. He conceived of gravity as a force of attraction possessed by all objects with its power dependent on how far apart they are and their mass. As Rovelli (2015, p. 5) suggests, this is a mechanical world where objects eternally travel on long precise trajectories, determined by gravity, in geometrically immutable space. In Einstein's vision, however, space-time and matter/energy are inextricably linked. Large objects, like the sun, bend space-time around themselves. The warped space-time causes the path of objects to be curved because this is the shortest route it offers. Planets revolve around the sun because space is curved. The curvature of space and gravity are the same thing. At the heart of Einstein's theory is a loop in which, in the words of the physicist John Wheeler: "Space-time tells matter how to move; matter tells space-time how to curve" (quoted in The Economist 2015). This magnificent theory, captured in elegant mathematics, was soon confirmed by experiments demonstrating that the sun does cause light to deviate and that time does pass more quickly higher up than closer to the earth. Later, it led scientists to discover black holes and to determine that the universe was born with the "big bang" and was continuing to expand. In short, says Rovelli,

> … the theory describes a colourful and amazing world where universes explode, space collapses into bottomless holes, time sags and slows near a planet, and the unbounded extensions of interstellar space ripple and sway like the surface of the sea.
>
> *(2015, p. 9)*

At once, Einstein had surpassed Newton's theory, hitherto regarded as the pinnacle of scientific achievement, and called into question Kant's philosophy, dependent as it was on Newton's mechanics and Euclid's geometry. Perhaps science could see beyond the sensibilities and categories as envisaged by Kant. Einstein had produced a theory of the universe as a more organic entity, more interconnected and dynamic, than anything previously envisaged.

The next revolutionary development in twentieth-century physics, quantum mechanics, was a step too far even for Einstein. In general relativity, everything is still certain. There is an objective reality "out there" that behaves in a deterministic fashion independently of how it is observed. In quantum theory, nothing is stable and particles can be both here and there simultaneously depending on their interrelationships. This theory, of matter and energy at the atomic and subatomic levels, originated in the work of Max Planck and Einstein himself and found its fully developed form in the results published by Niels Bohr and Werner Heisenberg in 1925. Their equations describe a world completely at odds with Newton's deterministic perspective. According to the widely accepted "Copenhagen" interpretation of the results, incorporating Heisenberg's famous "uncertainty principle," particles do not exist in any definite "place." Much to Einstein's frustration, quantum mechanics allowed God "to play dice." Electrons come in and out of existence as they jump from one interaction to another and it is impossible to be certain where they will reappear. It is only possible to calculate the probability of them surfacing in one place or another. Indeed, they only seem to come into existence when interacting with other systems, for example, measuring instruments used by observers. Furthermore, the interactions do not have to be "local." "Spookily," in Einstein's view, particles are "entangled" with one another such that if one changes, there can be an immediate impact upon another even across vast distances:

> No one knows how it works. The entangled particles are chained together by a connection that we don't understand. They may be one particle that manifests in our world in two separate places. They may even be, by some hidden, contorted geometry of space, right next to each other.
>
> *(Brooks 2017, p. 164)*

It is impossible to predict the position of each particle but only to evaluate the "quantum state" of the system as a whole. Rovelli is convinced that "reality is only interaction" (2015, p. 18). Despite Einstein's doubts, it is now fully accepted by scientists that quantum mechanics provides an accurate account of the behavior of tiny particles and forces. And the theory is used in an increasing number of real-world applications, for example, to build ultra-precise clocks, unbreakable codes, better microscopes, and super-powerful computers. If quantum mechanics deals in probabilities that is not because it lacks knowledge, it is because that is the way the world is.

In the 1960s and 1970s, Gleick's third hallmark of twentieth century science began to take shape in the form of chaos and complexity theory. Chaos and complexity theory lays claim to being the science of the global nature of systems and is considered fully in Chapter 7. For the present, it is enough to highlight how the theory aims to extend the reach of classical science. Classical science sought to discover orderly, regular patterns of behavior, based on cause–effect relationships, from which fixed laws could be derived. This emphasis, according to Gleick, led it to ignore "the irregular side of nature," those things that were erratic and discontinuous. These phenomena remained puzzling to science. They appeared as "monstrosities." However, around 50 years ago, a small number of scientists came face to face with the "erratic side" of nature and began to think along different lines. Two discoveries earned chaos theory an important place in science. First, it was found that complex and unpredictable behavior could emerge in systems constructed on the basis of entirely deterministic equations. There was no need to introduce any probabilistic element. Second, it was a common finding of the early pioneers that there is considerable order underlying chaos. Chaos did not therefore, as in everyday language, imply anarchy. In the zone between stability and instability, general patterns of behavior emerge even if the specifics are unpredictable.

In the modern era, it is clear that science has abandoned mechanism and embraced concepts such as relationships, indeterminacy, and emergence. It is no longer at loggerheads with systems thinking. Indeed, the physical sciences have undergone their own systems revolution and are now able to make a contribution to systems thinking in other disciplines.

2.4   The Scientific Method in the Modern Era

The revolution in the way that the physical world is understood necessitated further reflection on the nature of the scientific method. The most famous twentieth-century philosopher of science is Karl Popper. To continue the quote from Sir Herman Bondi: "There is no more to science than its method, and there is no more to its method than Popper has said" (quoted in Lewens 2015, loc. 140). To Popper, it was obvious that Einstein had not come up with his theory of general relativity by carrying out repeated observations. Rather it had emerged almost fully formed, as a brand new way of seeing the world, from a series of thought experiments. In Popper's account, this could not be otherwise because theories always precede observations. Scientists are guided by the theory they currently hold in their minds (remember Kant and Hegel) to make certain observations, to see things in certain ways, and to interpret their observations in terms of the theory (Chalmers 1982; Lewens 2015). Science starts for Popper not with repeated observations but with bold conjectures about the nature of some aspect of reality, often provoked by problems encountered by earlier theories. It all begins with an imaginative leap.

Once that has occurred, the scientific method kicks in. Einstein's theory of general relativity was able to provoke such a revolution in science because it was so clearly formulated, in the language of mathematics, and so productive of new hypotheses that it laid itself open to experimental testing and therefore to the possibility of falsification. There are two points to make here. First, for Popper, bona fide science must be falsifiable. Astrology is not science because its statements are so vague that they do not lend themselves to testing. Marxism is not a science because it is so flexible that no contrary evidence seems able to refute it. If the working class does not rise in revolution, it is apparently because they suffer from false consciousness. Second, while the classical version of the scientific method depends upon induction as its criterion of truth, the more supportive evidence the better, Popper replaces this with falsifiability as a criterion of plausibility. Induction can never lead to certainty because there might always be black swans around the corner. Neither can falsifiability, but if one theory is more falsifiable than its rivals, and succeeds in making more accurate predictions, it is likely to be nearer to the truth. For Popper, science as a process proceeds by conjecture and refutation to arrive at the best available theories. Progress is made.

Kuhn (1970) took Popper's notion of the theory dependence of observations to its logical conclusion (a conclusion that Popper avoided) to develop his highly influential account of the structure of scientific revolutions. Kuhn argues that scientists, their thinking governed by some current theory, find it hard to see contradictory evidence and, even if they do so, are likely to try to reconcile it with their existing preferences rather than use it to challenge the theory. Newton's laws failed to predict accurately the movements of Uranus. Scientists did not abandon his laws but sought an explanation in the existence of another planet. This is not necessarily a bad thing. It led to the discovery of Neptune and that was then seen as a triumph for the theory.

Looking at the actual history of science, Kuhn argued that it proceeds through cycles of "normal science" and "revolutionary science." A science becomes established when a scientific community embraces a single "paradigm" as the basis for their work. A paradigm is a set of general theoretical assumptions, laws, and techniques, to which they give their adherence. A period of normal science ensues as scientists accept the paradigm uncritically and explore the possibilities offered by it. Kuhn calls this "puzzle-solving" and sees it as essential since it allows the working out of the detail of a theory. Eventually, however, anomalies arise, which go to the heart of the existing paradigm and are impossible to reconcile with it. A period of uncertainty follows. However, it requires the appearance of a rival paradigm before scientists begin to abandon their existing theories. Science then enters its revolutionary stage. Eventually, if the scientific community as a whole shifts its thinking, the old paradigm is abandoned and the new one elevated to dominant status. There follows another period of normal science until a new batch of significant anomalies arises. This seems a compelling account of, for example, the dramatic shifts from Aristotle's to Newton's and then to Einstein's view of the world and the long periods of calm in between.

Where Kuhn becomes controversial is in his insistence that there are no logical reasons for choosing one paradigm over another. Paradigms offer different, incompatible ways of viewing the world and, because the world does not exist independently of the way we view it, all claims about reality are dependent upon the paradigm employed. To illustrate how the world can change according to the paradigm you adopt, Kuhn describes his encounter with Aristotle's *Physics* developed around 350 BCE. At first he found it incomprehensible but:

> Suddenly the fragments in my head sorted themselves out in a new way, and fell into place together. My jaw dropped, for all at once Aristotle seemed a very good physicist indeed, but of a sort I'd never dreamed possible.
>
> *(Quoted in Lewens 2015, p. 88)*

Paradigms are "incommensurable" and, ultimately, the "religious conversion" involved in embracing a new paradigm happens for psychological or sociological rather than objective reasons. An individual scientist may see the new paradigm as offering better career opportunities or a scientific community as bringing it closer to political and economic power. Scientists, forced to provide a rationale, may cite simplicity or greater predictive power but these reasons in themselves simply reflect the values of their particular scientific community. Kuhn, however, denies being a "relativist." Science does not progress toward the truth but as paradigm replaces paradigm, through a process akin to natural selection, some improvements in problem-solving capability do seem to accrue. Kuhn describes his writings "as a sort of post-Darwinian Kantianism" (Lewens 2015, p. 95). We see the world through the particular paradigm that inhabits our minds but paradigms, like species, do evolve.

In spite of this small concession, most natural scientists remain aghast at Kuhn's conclusions. Despite Kant's great work, they tend to be "realists" in the sense that they believe the real world exists independently both of themselves and their theories about it. They think they are gaining knowledge about how that real world works. After all, didn't Newtonian physics put a man on the moon and hasn't science advanced even more since Newton? In Hilary Putnam's words:

> The positive argument for realism is that it is the only philosophy that doesn't make the success of science a miracle.
>
> *(Quoted in Lewens 2015, p. 118)*

Scientists who adhere to this position have Roy Bhaskar and his theory of "critical realism" on their side. In *A Realist Theory of Science*, Bhaskar accepts (this account follows Mingers 2006) that knowledge production is the work of humans and, in "the transitive domain," is cognitively and culturally conditioned. In doing so, he avoids the "naïve realism" of the traditional scientific method and accepts that our knowledge is "relative" in this domain. At the same time, he wants to demolish the more extreme relativism stimulated by Kuhn. It is an "epistemic fallacy" he argues, encouraged by philosophy since Kant, to concentrate solely on the ways humans understand and arrange the world through their thoughts and customs. This can only yield an anthropomorphic view. Instead we should ask the question "What must reality be like for scientific knowledge to be possible?" If we do so, by prioritizing ontology over epistemology, we can establish that there is indeed a real world of objects and structures, independent of our observations, which causes what happens and what doesn't happen.

Bhaskar calls the real world, which exists independently of humans and their theories, the "intransitive domain of knowledge." In this domain, entities and structures, often unobservable, can generate events, some of which we observe. Occasionally, this will include the kind of regularities that the traditional scientific method seeks out. Bhaskar postulates a stratified reality made up of:

- The "real," the whole of reality with, at its base, objects and structures with enduring properties that can cause things to happen
- The "actual," events (and nonevents) generated by the causal mechanisms in the real, including the empirical and
- The "empirical," those events that happen to be experienced by humans

Scientists must proceed by "abduction" to postulate what hypothetical causal mechanisms might give rise to observed phenomena. The experimental method then allows them to identify the correct generative mechanism at work and to eliminate other alternative explanations along the way. A scientist, as a human being, might bungle a particular experiment but cannot change the laws of nature and so cannot cause the results. The justification for critical realism's account of a stratified reality, and faith in scientific experiments, comes primarily from the success of the technologies that have been derived from this thinking and method. The success suggests that science can provide rational grounds for choosing between competing theories and does enable correct causal laws to be identified. Bhaskar is challenged by critics who believe that his acceptance of relativity in the transitive domain fatally undermines any certainty he can have about the existence of a real world of objects, structures, and generative mechanisms. Critical realism is relevant to the social sciences, and the argument is pursued in that context in Chapter 4.

It was the hubristic claim of scientists, toward the end of the nineteenth century, that they had developed a method capable of uncovering "the truth." The scientific method, rigorously applied, would enable them to find out everything about the nature of reality. Philosophers of science in the modern era have been led to question this. Science, at the end of the twentieth century, was left with two magnificent but contradictory accounts of the world, general relativity and quantum mechanics, and this has led to numerous theories seeking to reconcile them. Du Sautoy (2016) has identified seven "edges," including chaos and uncertainty, that science can never cross and will always impose limits on what we can know with certainty. Perhaps science was just lucky in what it chose to study in the early days of the scientific revolution. As Rapoport has it:

> Fortunately for the success of the mechanistic method, the solar system ... constituted a special tractable case of several bodies in motion.
>
> *(Quoted in Weinberg 2011, loc. 289)*

Today, hardly anyone claims that science is close to understanding everything, even in its favored domains of physics, cosmology, and chemistry.

2.5 Extending the Scientific Method to Other Disciplines

Physicists have been tempted to claim that the method that brought them success in their field should be extended to other disciplinary areas. This is captured in a remark reputed to have been made by Lord Rutherford: "All science is either physics or stamp collecting." In fact, as we go up the levels of system complexity, through biological, ecological, human, and social systems, the problems encountered by mechanism just seem to grow. Aristotle's insight that the whole is more than the sum of its parts becomes more pertinent. This is considered in detail in the next two chapters on the life and social sciences. A brief synopsis here will not, however, be out of place.

Checkland (1981) characterizes the scientific method, drawing upon both the traditional and Popperian versions, as possessing three key elements: reductionism, repeatability, and refutation. In the case of reductionism, it is not clear how researchers in the life and social sciences can separate out a part of reality, be sure they have identified the key variables and, from studying those variables, work up to an understanding of the whole. With ecological systems, involving multiple organisms and their environments, the interconnectivities can be such that the relevant, significant variables are hard to find and certainly hard to separate out. The same goes for studying, for example, crime and low educational attainment in the social arena. These areas of research struggle to identify obvious boundaries and discrete elements. Furthermore, even if the whole could be decomposed and the parts subject to further analysis, there is a danger of missing what is most essential about such complex systems. In an organism, for example, the relationships between the parts seem to be at least as important as the nature of the parts themselves. Indeed, new properties that are not present in the parts emerge from the way the parts are organized, for example, life itself. Repeatability and refutation are also difficult to enact in the life and social sciences. With ecological and social systems, it is usually impossible (and often unethical) to carry out experiments of the kind recommended by the scientific method. The real world cannot be dragged into the laboratory for study. Nor can experiments be easily repeated because the systems of interest tend to change rapidly, not least in response to the experiments. It is impossible to replicate the exact same conditions for the experiments. The theory will need to change to keep up and that makes refutation a difficult business. In social systems, the situation is exacerbated because humans are self-conscious and have free will (at least as noumena) so it is necessary to take into account different beliefs and purposes, the danger of self-fulfilling prophecies, and the capacity of humans to refute any prediction made about them. It seems that the domain of application of the scientific method is much smaller than many had thought and hoped.

There is a further set of issues. The scientific method and its associated technologies have yielded some undoubted benefits, but they seem to demand "mastery" over the areas to which they are applied. As Heidegger (1978) has argued, in "The Question Concerning Technology," they seek to "enframe" both nature and human beings, reducing them to a "standing reserve" on call for technological purposes. This leads to unforeseen consequences, which the scientific method struggles to deal with because they too pose "wicked problems" involving many interconnectivities and stakeholders. These unintended consequences are the "second-order effects" (Weinberg 2011) of the scientific, industrial, and information revolutions, and arguably they have generated the great majority of the intractable issues that decision-makers face in the complexity age, for example, climate change, environmental degradation, pollution, inequality, exclusion, poverty, energy, food and water shortages, the danger of global recession, a possible global epidemic, nuclear proliferation, terrorism.

2.6 Conclusion

So when reductionism and the scientific method cannot cope, indeed make things worse, what alternative is there? The answer provided by this book is to look to systems thinking as a complementary approach. The reasons are clear. It is holistic, takes seriously the idea that the whole can be more than the sum of its parts, and considers the consequences that flow from this. It accepts that there will be multiple perspectives on any problem situation. Indeed, it believes that alternative viewpoints are to be encouraged. There is much to be gained from systems thinking both for our understanding of the systems we hope to manage and for us as human participants in those systems. It was the romantic poet Wordsworth (1814) who put it best:

> For was it meant
> That we should pore, and dwindle as we pore,
> For ever dimly pore on things minute,
> On solitary objects, still beheld
> In disconnection dead and spiritless,
> And still dividing, and dividing still
> Break down all grandeur …

3

The Life Sciences

When we try to pick out anything by itself, we find it hitched to everything else in the universe

(John Muir, nineteenth century campaigner for national parks in the US, quoted in Wulf 2015, p. 321)

3.1 Introduction

The doctrine, promoted by Descartes, that sees the workings and behavior of all organisms (except that driven by the human mind) as explicable by mechanical principles had some success in the life sciences. He was, for example, able to point to Harvey's achievement in describing how the heart pumped blood to the body and brain. By the second half of the eighteenth century, however, the mechanical model was beginning to break down both as a way of classifying and of understanding life forms. The sheer variety and vitality of life led to the development of an alternative perspective on nature in the writings of romantic poets, philosophers, and natural historians. This perspective, known as "organicism," regards life as a special phenomenon that cannot be understood simply by using the laws of physics. Kant was a supporter of the new thinking, but he struggled with the "teleological" explanations that the life sciences seemed to demand i.e. causes understood as fulfilling some end purposes. He could accept that it was useful to see organisms as having parts that functioned to ensure their survival. But, for him, there was an unfortunate corollary – the life sciences would never be scientific because they strayed beyond what could be understood through the "mechanism of blind efficient causes." His conclusion was that it was sensible to orientate our studies in this domain on the basis of organicism but that we could not then claim to know things with certainty. We can trace how these ideas were worked through in later centuries in what became biology and ecology.

3.2 Biology

The problems posed to the traditional scientific method by biological phenomena are severe. Organisms possess highly interrelated parts and seem to defy some of the laws of physics, for example, demonstrating the characteristic emergent behaviors

Critical Systems Thinking and the Management of Complexity, First Edition. Michael C. Jackson.
© 2019 John Wiley & Sons Ltd. Published 2019 by John Wiley & Sons Ltd.

associated with life. That said, the history of biological science can be seen as a series of pendulum swings between the dominance of reductionist and holistic explanations, and the outcome of that battle is still not determined.

Goethe, although better known now for his literary output, was in his day a serious contributor to debates about the biological structure of plants and animals. Much influenced by Kant and organicism, he took the view that all organisms were driven by "vital forces" that provide them with a general form that is further shaped by their environment. Summarizing, Wulf writes:

> Goethe then went onto explain his belief that – contrary to Descartes's theory that animals were machines – a living organism consisted of parts that only function as a unified whole. To put it simply, a machine could be dismantled and then assembled again, while the parts of a living organism worked only in relation to each other. In a mechanical system the parts shaped the whole while in an organic system the whole shaped the parts.
>
> *(2015, p. 32)*

Thinking that put the organism first continued to flourish in the nineteenth century, and the physiologist Denis Noble (2008) credits Claude Bernard as being the first systems biologist. Bernard insisted on the importance of both the relationships an organism has with its environment and the *milieu intérieur* of the organism itself; its own internal environment of relationships between organs and tissues. It is the ability of the organism to maintain a stable internal environment, in the face of external perturbations, that allows it to have an autonomous existence.

About the same time as Bernard was writing, however, the pendulum began to swing back in the direction of mechanistic explanations. The stimulus, as Capra and Luisi (2014) see it, was the perfection of the microscope. This allowed Virchow to establish cell theory in its modern form and Pasteur to pioneer biochemistry. Biologists came to believe that cells were the basic building blocks to which living organisms could be reduced. It seemed only a matter of time before biological phenomena yielded to explanations in terms of the chemical and physical laws that governed behavior at the cellular level.

Things did not quite turn out this way, however. As Capra and Luisi write:

> Cell biology made enormous progress in understanding the structures and functions of many of the cell's subunits, but it still has revealed very little about the coordinating activities that integrate those operations into the functioning of the cell as a whole.
>
> *(2014, p. 37)*

It remained impossible to conceive, from a mechanistic perspective, how apparently identical cells developed and differentiated themselves to undertake specialized functions in the organism (as muscle cells, blood cells, bone cells, etc.). The door was ajar for further attempts to develop biology as a holistic science. Initially, these took two directions and there were disputes between "vitalist" (Goethe's work offered a foretaste) and "organismic" biologists. Driesch, experimenting with sea urchin eggs at the end of the nineteenth century, found that when he divided an embryo, at the two-cell stage of its development, the remaining cells still yielded complete if smaller organisms. In later research, he found that if he grafted the future tail of a newt embryo onto where its leg would normally grow, the putative tail grew into a leg. Although impossible from the point of view of causal-mechanical logic, the parts seemed to "know" about the whole they were involved in producing. Driesch concluded that life involved some "vital force" beyond the physical. Summarizing the difference between this position and that of later organismic biologists, Capra writes:

> Vitalists assert that some nonphysical entity, force, or field, must be added to the laws of physics and chemistry to understand life. Organismic biologists maintain that the additional ingredient is the understanding of 'organization', or 'organizing relations'.
>
> *(1996, p. 25)*

The organismic standpoint, not surprisingly, proved more fruitful for making progress in biological science and for the advancement of systems thinking.

Checkland (1981) discusses Smut's *Holism and Evolution* (1926), Broad's *The Mind and its Place in Nature* (1923), and Woodger's *Biological Principles* (1929), as important milestones in the establishment of organismic biology. These works were particularly significant because they detailed, respectively, the concepts of "organized complexity," "emergence," and "hierarchy," which were essential in the early development of systems thinking. Smuts, who was actually responsible for coining the term "holism," said this about the nature of organized complexity:

> Every organism, every plant or animal, is a whole with a certain internal organization and a measure of self-direction.... A whole is a synthesis or unity of parts, so close that it affects the activities and interactions of those parts ... their independent functions and activities are grouped, correlated and unified in a structural whole.
>
> *(Quoted in Checkland 1981, pp. 78–79)*

Broad used the term "emergent properties" for those characteristics that arose at certain levels of organized complexity but did not exist at lower levels. For example, organisms exhibited properties, those of life itself, which were not reducible to their parts. Woodger insisted that a hierarchy of stable levels of organized complexity exists in nature, each giving birth to emergent properties and each requiring different types of explanation, because

> ... a physiologist who wishes to study the physiology of the nervous system must have a level of organization above the cell level *to begin with*. He must have at least the elements necessary to constitute a reflex arc, and in actual practice he uses the concepts appropriate to that level which are not concepts of physics and chemistry.
>
> *(Quoted in Checkland 1981, p. 79)*

Thus, emergent properties at the level of the organism give rise to the need for biology as a separate discipline. They cannot be understood just in terms of physics and chemistry. Rapoport puts it like this:

> Biology would be crippled if it did not depend on concepts outside the scope of physical concepts: organism, life, birth, death, sex, viability, adaptation, behavior, cell, organ, evolution, species, genus, class, phenotype, genotype, mutation, selection, clone, embryo, etc., etc..... Biological processes are simply too complex to yield to the analytic method.
>
> *(1968, pp. xvi–xvii)*

The best known of the organismic biologists today is Ludwig von Bertalanffy. This is because, as we shall see in Chapter 5, he extended the systems ideas he originally developed in biology and made them relevant to other fields through his "general system theory." In the process, he became one of the founding fathers of systems thinking as a trans-discipline. Here we are concerned with his writings on biology. The failure of physics to explain biological systems, von Bertalanffy argued, was because it only dealt with isolated systems, "closed" to their environments. Closed systems obey the second law of thermodynamics, increasing in entropy and reaching an equilibrium state where no energy can be obtained from them (see Section 2.2). The universe is presented as a machine that is gradually running down to randomness or disorder. However, von Bertalanffy asserted, many systems are "open systems" importing matter and energy from, and exporting them to, their environments:

> However, we find systems which by their very nature and definition are not closed systems. Every living organism is essentially an open system. It maintains itself in a continuous inflow and outflow, a building up and breaking down of components, never being, so long as it is alive, in a state of chemical and thermodynamic equilibrium but maintained in a so called steady state which is distinct from the latter .... Obviously the conventional formulations of physics are, in principle, inapplicable to the living organism *qua* open system and steady state ...
>
> *(1971, p. 38)*

Open systems can temporarily defeat the second law of thermodynamics. Living off their environments, importing complex molecules high in free energy, they can evolve toward states of increased order and complexity. Organisms, for example, can maintain themselves in a dynamic state far from "true" equilibrium, constantly changing while retaining their basic form. Many have argued (e.g. Emery 1969; Lilienfeld 1978) that von Bertalanffy's 1950 article *The Theory of Open Systems in Physics and Biology* (1969), which first rigorously distinguished closed and open systems, established systems thinking as a modern intellectual movement.

von Bertalanffy's importance to biology rests on his insistence that organisms must be grasped as a whole and that their behavior cannot simply be reduced to the laws of physics and chemistry. According to Drack (2009), he developed and emphasized three essential organismic principles. First, organisms are open systems capable of maintaining themselves in a dynamic state far from true equilibrium. Second, in this state, they can exhibit progressive internal organization of parts (differentiation, specialization, hierarchy). Third, they can protect their own integrity in the face of environmental disturbances by reaching the same final state in different ways from different initial conditions. This final characteristic was called "equifinality."

Working about the same time as the organismic biologists, and expanding Bernard's concept of the *milieu intérieur,* was the American physiologist Walter B. Cannon. His major work, *The Wisdom of the Body* (1932, revised 1939), was concerned with the ability of organisms, and particularly our own bodies, to persist over many decades while consisting of extraordinarily unstable material and being open to the environment. For Cannon, living systems are marvelous in that they

> ... may be confronted by dangerous conditions in the outer world and by equally dangerous possibilities within the body, and yet they continue to live and carry on their functions with relatively little disturbance.
>
> *(1939, pp. 22–23)*

He refers to the processes that maintain stability as *homeostatic*; examples being the self-regulating mechanisms controlling glucose concentrations, body temperature, and the acid–base balance. He also speculates that it might be useful to examine industrial, domestic, and social organizations in the light of the organization of the body.

von Bertalanffy's and Cannon's work was warmly welcomed in other fields, such as engineering and management, but the immediate impact on biology was negligible. This was because, as Capra (1996, p. 77) argues, the 1950s saw yet another turn toward mechanism in the discipline. Research in molecular biology led to the discovery of the structure of DNA and, eventually, to the unraveling of the genetic code. Just as, in the nineteenth century, it had seemed that biological phenomena could be fully explained by laws pertaining to the cellular level, now it appeared that they could be understood in terms of the molecular structure of the gene. The gene became the new elementary unit of "reductionist" biology and:

> The exclusive focus on genes ... largely eclipsed the organism from the biologists' view. Living organisms tended to be viewed simply as collections of genes, subject to random mutations and selective forces in the environment over which they have no control.
>
> *(Capra and Luisi 2014, p. 42)*

Richard Dawkins (1976) book *The Selfish Gene,* arguing for genetic determinism, was the radical intellectual expression of this movement and the Human Genome Project its crowning glory.

Molecular biology has indeed taught us a great deal about what the body is made of at the molecular level but the promise that it would yield a full understanding of organisms, their health and behavior, has not yet been realized. According to Capra:

> While biologists know the precise structure of a few genes, they know little of the ways in which genes communicate and cooperate in the development of an organism. In other words, they know the alphabet of the genetic code but have almost no idea of its syntax. It is now apparent that most of the DNA – perhaps as much as ninety-five percent – may be used for integrative activities about

which biologists are likely to remain ignorant as long as they adhere to mechanistic models.

(1996, p. 77)

It appears that complex biological behavior actually emerges from the interactions arising in networks of multiple genes connected through numerous feedback loops. To take an example, the causes of cancer and heart disease cannot be attributed to just one, or even a few malfunctioning genes. As Strohman, writing in 1997, explains: "In the case of coronary heart disease, there are more than 100 genes identified as having some interactive contribution" (quoted in Capra and Luisi 2014, p. 324). The late twentieth and early twenty-first centuries have, therefore, heard new calls for respect to be given to the whole organism. Goodwin (1994), in his early career an austere mathematical biologist, argues for a more holistic approach in which organisms are seen as irreducible wholes giving rise to structures that cannot be understood simply in terms of genes. The distinguished physiologist Denis Noble has argued that because there are feedbacks between different levels of organization, there is downward as well as upward causality governing the way genes operate, and has championed the cause of "systems` biology":

> It's difficult to define precisely .... But if you look at molecular biology as breaking Humpty Dumpty into as many pieces as possible, then systems biology is about trying to put him back together again. And that's actually a great deal more difficult. It's about recognizing that every physical component is part of a system and that everything interacts.
>
> *(2008)*

Welcoming a report on systems biology from the Royal Academy of Engineering and the Academy of Medical Sciences, he argues that:

> Combinatorial explosion means that a fully bottom-up understanding of life will probably always elude us. This is where systems biology and the merger of engineering and biology come in. The study of the interactions between the components of a system – rather than the components themselves – can be pursued at all levels of biological organization, from gene-protein networks up to the whole organism. A basic principle of engineering is central: investigate the principles of organization at each chosen level using the tools appropriate to that level.
>
> *(Noble 2007)*

One biological theory that rejects genetic and environmental determinism, and places organisms at the center of the stage as active players, is the theory of *autopoiesis* developed by Maturana and Varela. The term *autopoiesis* derives from the ancient Greek for "self-making." Maturana and Varela argue that, in order to answer the question "what is life?," biologists must concentrate on the circular processes through which organisms ensure their own continued self-maintenance. In their view, living beings are self-producing systems constituted by a network of biochemical production processes in which the components involved interact to produce the network, which in turn produces them. At the same time, these processes create a boundary that defines the system in relationship to its environment and is essential to the maintenance of the mutual interactions that produce the system:

> ... the autopoietic organization is defined as a unity by a network of components which (1) participate recursively in the same network of productions of components which produced these components, and (2) realize the network of productions as a unity in the space in which the components exist.
>
> *(Varela et al. 1974, p. 188)*

It may seem that autopoiesis is a theory of closed systems. This is not the case. According to Maturana and Varela (1992), autopoietic systems have both "organization" and "structure." Organization denotes those relations that must exist among the components of a system for it to be of a particular class. This must remain invariant if the living system is to maintain its identity. Its structure, however, defined as the physical form the components and relations actually take in a particular unity, can change without the system ceasing to exist. In other words, two unities of the same class must have the same organization but may have different structures. The fact that the structure can change allows the system to build relationships with its environment. Although it is organizationally closed, it is structurally open to exchanges of energy and matter. Maturana concludes that "a dynamic composite unity is a composite unity in continuous structural change with conservation of organization" (1987, p. 335). Even in terms of change of structure, however, autopoietic systems are structure-determined rather than externally determined systems. This means that the nature of any change is determined internally by the current structure of the unity, itself the result of a history of previous structural changes, and not by an "independent external agent":

> ... an external agent that interacts with a composite unity only triggers in it a structural change ... nothing external to them can specify what happens to them .... It follows from this that composite unities are structure determined systems in the sense that everything is determined by their structure.
>
> *(Maturana 1987, pp. 335–336)*

Living systems are autonomous in the sense that it is their own organization and structure that ultimately determine their behavior. The environment can only disturb them. It is the systems themselves that "decide" what structural changes, if any, take place. This leads to a very different account of system–environment relations to that prevalent in open systems theory, where the environment is seen as dominant. Maturana and Varela's account is known as "structural coupling." Even though they are organizationally closed and structure-determined, autopoietic systems must establish appropriate relationships with their environments to ensure access to those elements that permit the processes of production of components to take place. The environment, as we know, cannot direct or specify changes in the unity. Nevertheless, it does "trigger" events that bring about structure-determined changes. On this basis, the interactions between the organism and its medium can achieve some stability over time. If this occurs, the unity and its environment become "structurally coupled," mutually influencing one another, and the unity can preserve its identity. Two organisms can also become structurally coupled, in which case

... the result ... is a consensual domain, that is, a domain of behavior in which the structurally determined changes of state of the coupled organisms correspond to each other in interlocked sequences.

(Maturana 1975, p. 326)

There is another aspect to the theory of autopoiesis, which is equally challenging to traditional ways of thinking. For Maturana and Varela, living systems are identical to cognitive systems and so cognition itself can be explained using the same theory. Because autopoietic systems are structurally determined, and have an invariant organization, they decide not just how they will react to environmental perturbations but also what perturbations they notice and respond to. They are involved in creating their environments and, as they become structurally coupled with them, they come to know them. This is well put by Capra:

The structural changes in the system constitute acts of cognition. By specifying which perturbations from the environment trigger its changes, the system 'brings forth a world', as Maturana and Varela put it. Cognition, then, is not a representation of an independently existing world, but rather a continual *bringing forth of a world* through the process of living. The interactions of a living system with its environment are cognitive interactions, and the process of living itself is a process of cognition. In the words of Maturana and Varela, 'to live is to know'.

(1996, p. 260)

Maturana, with Lettvin, McCulloch, and Pitts, had conducted early work on "what the frog's eye tells the frog's brain," and this convinced him that living systems have no direct access to an independent external world. They can only see and respond to what is made available to them by the self-organizing and self-referential nervous systems they have developed through their biological evolution in interaction with their environments. Frogs do not see flies, rather they recognize patterns of moving shadows, which enable them to catch flies. The theory of autopoiesis, developed with Varela, provides a rigorous working through of this early thinking. He was able, with confidence, to extend his conclusions to all living systems. Because of the process of circular organization, living systems, even without nervous systems, are still cognitive systems. Bacteria, for example, bring forth a world in which they distinguish heat, light, and chemical differences. As living systems co-evolve further, in interaction with their environments, nervous systems and brains develop and more sophisticated forms of knowing, learning, and remembering become possible. Eventually, as Capra remarks:

At a certain level of complexity, a living organism couples structurally not only to its environment but also to itself, and thus brings forth not only an external but also an inner world. In human beings, the bringing forth of such an inner world is intimately linked to language, thought, and consciousness.

(1996, pp. 262–263)

Even at the human level, however, there is no access to an objective external world. Knowledge remains embodied in particular human beings. We have to be aware that the world changes according to who observes it and that "anything said is said by an observer" (Maturana and Varela 1980). This thinking resonates with Kant's work and with the thinking of von Glasersfeld, as outlined in Section 1.6, and provides support from biology for their conclusions.

When they argue that *all* living systems are cognitive systems, Maturana and Varela are using the word cognition without the connotation that mental processes are necessarily present. In one way, this is insightful because it facilitates an explanation of the origins and development of more sophisticated forms of cognition, involving consciousness, at higher evolutionary levels. On the other hand, it does lead some inattentive acolytes into believing that entities such as lower organisms and forests are capable of thinking.

Autopoiesis has been important in the development of systems thinking as we shall see, in Section 6.4, when reviewing second-order cybernetics. It has also had a significant impact in disciplines such as sociology, law, family therapy, information systems, and cognitive science (Mingers 1995). Maturana and Varela themselves are skeptical of extensions of their theory of autopoiesis beyond the biochemical domain. They prefer the term "operational closure" when referring to systems, in other fields, which exhibit autonomy and self-maintenance but do not meet the exact requirements for the circular production of components. Social systems, for example, seem capable of maintaining their identity over time but, if you take individual human beings to be their components, it is not clear how these are "produced," in the strict sense, by the social systems of which they are parts. Each case will be taken on its merits, with particular attention given to Luhmann's novel theory of social systems as autopoietic systems of communication in Section 4.7.

3.3 Ecology

Ecology studies the relationships between organisms and between organisms and their environments. To think this way is to challenge the mechanistic view, dominant in the seventeenth century, that organisms are independent species that have remained the same since they were created by God. According to that perspective, it is the role of the life sciences simply to classify them. Ecology sought from its beginnings to grasp interconnectivity, emergence, and organism–environment interdependency. It was born taking a systems approach.

The most important figure in the development of ecology, in the "invention of nature" as Wulf's (2015) biography has it, was Alexander von Humboldt. Humboldt was a friend of Goethe and was immersed in the philosophy of Kant. From Goethe, he took the idea of organisms driven by internal, vital forces, and impacted by their environments, and applied this to the whole of the natural world. He owed to Kant the notion of the mind as a shaper of what we perceive and derived from this his insights into the interrelationships that exist between the Self and nature and between imagination and science. In this organicist context, he was able, according to Wulf's excellent account to

> … [revolutionize] the way we see the natural world. He found connections everywhere. Nothing, not even the tiniest organism, was looked at on its own. 'In this great chain of causes and effects', Humboldt said 'no single fact can be considered

in isolation." With this insight, he invented the web of life, the concept of nature as we know it today.

(2015, p. 5)

Plants and animals were seen as closely interrelated and dependent upon each other in an intricate web of life. Through his extensive travels, he came to appreciate the existence of climate zones across continents, with the species of plants and animals in different zones determined by climate and altitude:

> Humboldt saw "unity in variety." Instead of placing plants in their taxonomic categories, he saw vegetation through the lens of climate and location: a radically new idea that still shapes our understanding of ecosystems today.
>
> *(Wulf 2015, p. 89)*

Remarkably for his time, von Humboldt also recognized the destructive impact that humans could have on the environment if they did not see that they too are a part of the web of life:

> When he listed the three ways in which the human species was affecting the climate, he named deforestation, ruthless irrigation and, perhaps most prophetically, the "great masses of steam and gas" produced in the industrial centres.
>
> *(Wulf 2015, p. 213)*

Finally, it is worth noting that many of his ideas about how to conduct research into complex real-world phenomena anticipate those of general systems theory a century later (see Chapter 5). Studying nature as a whole requires a multidisciplinary approach, he believed, and he encouraged just this among fellow scientists. Once a set of ideas was found useful in one discipline, like the life sciences, he was convinced that they could profitably be employed elsewhere. He believed that his systems perspective had something to say about language, the universe, and "global patterns" in which colonialism, slavery, and economics are all linked. His *magnum opus, Cosmos: A Sketch of the Physical Description of the Universe* (first volume published in 1845), strove to present a unified view of the universe as an ordered system just at a time when other scientists were retreating into their separate disciplines.

von Humboldt's exhilarating work set ecology along a systemic path. The next important step was the combination of von Humboldt's conception of nature as an interconnected whole with Darwin's theory of natural selection; the theory that finally provided a nonteleological explanation of the evolution of new life forms. This achievement was due to the zoologist Haeckel and announced in his *General Morphology of Organisms* (1866). In that book, Haeckel named his field of study "ecology" (the first use of the term) from the Greek word *oikos*, meaning household. Its task was to study the co-operating and conflicting relationships between the constituents of nature's household, as you might a family occupying a single dwelling. In Wulf's words:

> Haeckel took Humboldt's idea of nature as a unified whole made up of complex interrelationships and gave it a name. Ecology, Haeckel said, was the 'science of the relationships of an organism with its environment'.
>
> *(2015, p. 307)*

Like von Humboldt, Haeckel was inspired by the beauty and vitality of natural forces and by the interdependence of the human mind and the cosmos. His *Art Forms in Nature* (published between 1899 and 1904) heavily influenced the *Art Nouveau* style, encouraging the use of natural forms and motifs in urban settings. He became an adherent of "monism" arguing, in particular, that the organic and inorganic worlds could not be separated.

The most significant development in twentieth-century ecology has been the formulation of the concept of an "ecosystem." The word was first used by the botanist Tansley, in 1935, to refer to communities of living organisms and their physical environments (air, water, soil, etc.) interacting as a system. This idea was popularized and further developed by the brothers E. P. Odum and H. T. Odum (see Ramage and Shipp 2009). The Odums were determined to find a means of understanding the networks of interaction that gave rise to the complex behavior of ecosystems. E. P. Odum's *Fundamentals of Ecology* (1953) was a popular textbook, which introduced flow diagrams as a means of charting interrelationships in ecosystems. H. T. Odum's *Systems Ecology: An Introduction* (1983) was able to make use of what by then were established concepts in general systems theory and cybernetics (see Chapters 5 and 6). von Bertalanffy's theory of open systems provided him with the insight that ecosystems could best be viewed in terms of the transfer and transformation of energy in the system. Cybernetics suggested that these energy flows could give rise to control mechanisms through the interplay of feedback loops. In taking this perspective, he was aware that he was using a "macroscope" rather than a microscope:

> Bit by bit the machinery of the macroscope is evolving in various sciences .... Whereas men used to search among the parts to find mechanistic explanations, the macroscopic view is the reverse. Men, already having a clear view of the parts in their fantastically complex detail, must somehow get away, rise above, step back, group parts, simplify concepts, interpose frosted glass, and thus somehow see the big patterns.
>
> *(H. T. Odum, quoted in Ramage and Shipp 2009, p. 89)*

The Odums continued to keep up-to-date with the latest developments in systems and complexity theory and E. P. Odum, citing Prigogine on "dissipative structures"(see Section 7.3) in 1992, saw an ecosystem as "a thermodynamically open, far from equilibrium, system" (quoted in Ramage and Shipp 2009, p. 90).

When von Humboldt described the Earth as "a natural whole animated and moved by inward forces" (quoted in Wulf 2015, p. 7), he was anticipating, by more than a century, another significant development in the life sciences: the *Gaia* hypothesis that the Earth itself is a living system. James Lovelock's theory, named in honor of the Greek goddess of the earth, was conceived when he was helping NASA with the design of instruments to detect life on Mars. He reasoned that the impact of any life should be traceable in the atmosphere; just as the gases produced by plants (especially oxygen) and other organisms are obvious in the Earth's atmosphere. Extending this insight, Lovelock began to see all the organisms and inorganic elements of earth and its atmosphere as closely integrated, through feedback loops, in a self-regulating system. A moment of inspiration led him to entertain a remarkable possibility:

> The Earth's atmosphere was an extraordinary and unstable mixture of gases, yet I knew that it was constant in composition over quite long periods of time. Could

it be that life on Earth not only made the atmosphere, but also regulated it – keeping it at a constant composition, and at a level favorable for organisms?

(Lovelock, quoted in Capra 1996, p. 102)

While the hypothesis was attractive to romantics who, by now, had seen pictures taken from space of the beautiful blue and white globe that is Earth, it was too much for the natural scientists. When they took notice at all, they condemned it as teleological. How could natural processes be shaped by the purpose of preserving life?: "Are there committee meetings of species to negotiate next year's temperature?" (quoted in Capra 1996, p. 107). There is evidence for the hypothesis, however. The heat of the sun has increased by 25% since the beginnings of life on Earth, but the temperature on the surface has remained reasonably stable and suitable for life. With the help of the microbiologist Lynn Margulis, Lovelock began to turn his hypothesis into a theory by identifying the exact nature of the complex feedback loops, involving both organisms and inorganic matter, which enable the Earth to regulate temperature, the oxygen in the atmosphere, the salinity of the oceans – all those variables essential for maintaining the conditions for life. The planetary system, he demonstrated, was an evolving, self-regulating system in which life, at the very least, has an important role to play in creating the conditions for its own existence. Life is in no way as passive as Darwin painted it. To the charge of teleology Lovelock and a colleague, Andrew Watson, responded by producing, a computer simulation of a simple Gaia system, called *Daisyworld*, which in Capra and Luisi's opinion makes it

... absolutely clear that temperature regulation is an emergent property of the system that arises automatically, without any purposeful action, as a consequence of feedback loops between the planet's organisms and their environment.

(2014, p. 165.)

The Gaia theory has given rise to the subdiscipline of "Earth System Science" and aspects of it are sufficiently rigorous to permit experimental testing. Nevertheless, controversy remains. It appears mystical to regard the Earth itself as a living organism and even more so if, as some argue, it is viewed as capable of consciously taking decisions to make conditions appropriate for life forms. So, how does this thinking stand up when confronted by the theory of autopoiesis, the standard for "life" provided by Maturana and Varela? The components of the Earth system, as we have seen, are tightly coupled to a distinct boundary, the atmosphere, and Margulis is happy that "the planetary patina – including ourselves – is autopoietic" (quoted in Capra and Luisi 2014, p. 351.). But the production processes whereby the planet replaces organisms and its inorganic components (of oceans, soil, and air) do differ from those that take place in cellular networks, if only in the extremely long timescales involved. It is safest to regard the planetary system not as literally "living" but as a complex network of components, some living some not, which is capable of self-regulation because of the interacting feedback loops in which those components are involved.

The idea that other, more restricted ecosystems are "alive" has, of course, a long history. Paul Kingsnorth describes the Lani of West Papua as seeing the rainforest as a great being to whom they sang and which could sing back. For him:

That the world is a machine is one story; that the world is alive and aware is another. The latter story has probably been taken for granted by the majority of human societies throughout history. The former has only really taken root in ours: post-Enlightenment, industrial western culture. The results of it – climate change, mass extinction, factory farming – should be enough to make us wonder if this story is badly constructed, badly told, or just plain wrong.

(2016)

Kingsnorth has sought, in two powerful novels (2015, 2017), to reorientate our thinking so that the landscape is regarded as a character, an actor in human affairs. This perspective on nature has recently received a boost from scientific work on plant and tree communication. Brooks (2016) summarizes the argument put forward by Wohlleben, in his 2015 book *The Hidden Life of Trees*, to the effect that trees are essentially social beings communicating through a variety of olfactory, visual, and chemical impulses. In particular, they use the *mycelia*, huge underground fungal networks, dubbed the "wood-wide web," linking their roots to generate a kind of "collective fungal consciousness." This is employed by plants and trees to warn of predators, to show care for each other, to protect their young, and to support their sick or dying brethren. Certain species, however, are not quite as co-operative as others. There has been some formal recognition of this thinking in, for example, the decision in New Zealand to grant "all the rights, duties, and liabilities of a legal person" to the Te Urewera National Park and the Whanganui River and its tributaries. The theory of autopoiesis may give us pause when considering whether ecosystems are actually "alive," principally because their boundaries do not seem sufficiently fixed. Nevertheless, it remains clear that one of humanity's most pressing problems is how long ecosystems, and indeed the planet itself, can remain self-regulating in a manner suitable for life under the unique pressures placed on them by the human race.

The Kogi people, living in seclusion deep in Colombia's Sierra Nevada de Santa Marta mountains, have twice now, in 1990 and 2013, made deliberate contact with the outside world. The reason was to warn their "Younger Brothers," the people of the industrialized world, that their exploitation, devastation, and plundering of "The Great Mother," on whom we all depend, will result in our destruction. Perhaps, the scientists are at last catching up. It has been proposed that we have entered a new geological epoch, "The Anthropocene," which is distinguished by the increasingly dominant impact that human activity is having upon the Earth's geology, ecosystems, and atmosphere. In order for a new geological epoch to be defined, there must be a signal, occurring globally, that will be identifiable in deposits in the future geological record. There are many candidates including radioactive elements, carbon emissions, plastic pollution, aluminum and concrete particles, and nitrogen and phosphate residues in soils stemming from artificial fertilizers. Chris Rapley, a climate scientist, says:

The Anthropocene marks a new period in which our collective activities dominate the planetary machinery. Since the planet is our life support system – we are essentially the crew of a largish spaceship – interference with its functioning at this level and on this scale is highly significant. If you or I were crew on a smaller spacecraft, it would be unthinkable to interfere with the systems that provide us with air, water, fodder, and climate control. The shift into the Anthropocene tells us that we are playing with fire, a potentially reckless mode of behavior which we are likely to come to regret unless we get a grip on the situation.

(2016)

There was, according to Rob Cowen, "a belief among certain Native Americans that the cry of the owl was its mournful remembrance of a golden age when men and nature lived in harmony" (2015, p. 71). We will need to listen carefully. The human impact on biodiversity, climate, drainage, deforestation, and the increase in pollution will test Gaia to the ultimate over the coming decades.

3.4 Conclusion

In this chapter, we have drawn heavily on the excellent books by Capra (1996) and Capra and Luisi (2014). These authors believe that, if we are to tackle the massive problems confronting humankind, we need a new vision of reality that has life at its center. This paradigm shift in science, at its deepest level, involves "a perceptual shift from physics to the life sciences" (Capra and Luisi 2014, p. 15). Benefitting from the new understanding that has been obtained of ecosystems as autopoietic networks and dissipative structures, we can achieve and act on the basis of what Capra calls a "deep ecological awareness." Drawing on a distinction first made by Arne Naess, he defines "shallow ecology" as human-centered, placing humans as above or outside nature. Deep ecology, by contrast, sees the world

> ... as a network of phenomena that are fundamentally interconnected and interdependent. Deep ecology recognizes the intrinsic value of all living beings and views humans as just one particular strand in the web of life.
>
> *(Capra 1996, p. 7)*

On the basis of five principles that he sees as forming the pattern and structure of ecological systems, Capra (1996, pp. 290–295) is led to propose new social arrangements, which will promote "maximum sustainability." The first of these principles is "interdependence." We have to understand that each element of an ecosystem is interrelated with others in an extremely complex network of relationships. Human communities also need to nourish the multiple, mutual relationships between members. The second principle relates to the "cyclical nature" of ecological processes. Waste from one species is food for another and so there is no waste in the whole. This is seen as offering an extremely valuable lesson to human communities and should lead us to question the current way businesses interact and the economy functions. The third principle stems from the sun being the primary source of "energy flow" to ecosystems. This indicates that solar energy is the only form of energy that can maintain our human communities without pollution and that to ignore this would be disastrous. A fourth principle builds on the "co-operation and partnership" that are fundamental to ecosystems because each element within the web of life contributes to the sustenance of the whole. In human communities, co-operation and partnership have become subordinated to values of competition, expansion, and domination. Taking a lesson from ecology will help us value co-operation and partnership more highly and help conservation of the entire community. Finally, we should recognize that ecosystems are "flexible" and encompass "diversity," promoting their resilience in the face of environmental changes. Managers of social systems must pay attention to keeping a range of variables within their critical limits, not just try to optimize one, and should encourage the free exchange of ideas and a variety of opinions. In short, it is Capra's argument that

> ... the survival of humanity will depend on our ecological literacy, on our ability to understand these principles of ecology and live accordingly.
>
> *(1996, p. 295)*

A number of these principles have been given a practical orientation for the business world by Gunter Pauli, with whom Capra has sometimes co-written, in his book *Blue Economy: 10 years – 100 innovations – 100 million jobs* (2010), originally a report to the "Club of Rome" international think-tank. In short, Pauli seeks to provide guidelines for stimulating economic activity that is consistent with a sustainable society. These guidelines are derived from studying nature's ecosystems. He espouses a range of strategies: augmenting those global assets that meet basic human needs; creating jobs by identifying entrepreneurial opportunities that tackle environmental issues; building chains of businesses that make use of each other's "waste"; involving all citizens; increasing resilience by ensuring multiple cash flows; and thinking and acting locally.

Capra's work is insightful but, in my view, needs supplementing with a richer appreciation of the peculiarities of social systems if it is to become possible to implement his solutions for the benefit of all. This matter is addressed by Capra in the brilliant film *Mindwalk* (Bernt A. Capra, director 1990) for which he wrote the script. In the film, set at the Mont Saint-Michel, three characters look at mankind's future through the lens of Capra's ecological systems theory. Capra's ideas are powerfully presented by the character Sonia played by Liv Ullman. They are challenged throughout by a poet (played by John Heard) and a disillusioned US presidential candidate (Sam Waterston). Capra has related to me how he put all the criticisms he could find of his work into the mouths of these two characters. What a brave thing to do! The politician claims to understand the issues that Sonia raises but wants to know how her new way of thinking about them can be translated into action. In his view, they are just ideas unless they can somehow be transferred into mainstream politics. Sonia needs to get her hands dirty in the real world of wheeling and dealing and political compromise. He asks her to join his staff to help get the ideas across, but she is reluctant. The poet argues that, in the end, even Sonia's new vision of a world of creative, interconnected, self-organizing systems is distorting. It cannot come close to capturing the richness of life. Life is infinitely more than all the abstruse theories about it: "Where are the people in your system Sonia," he asks, "the ones with longings, the ones you love?" This is a hard question for Sonia, who has a difficult relationship with her own daughter. In a sense, both critics are posing the same problem. Ecological thinking may provide a grand metaphor for understanding human and social systems, but it cannot capture their essence. It cannot explain what lived experience feels like from the inside and, in that case, it cannot understand how social change is brought about. Sonia's ideas are wonderful but not practical. It is clear that Capra has thought deeply about these matters, but I do not think that he has come close to addressing them. Capra and Luisi (2014) devote a chapter to social thought, but it seems to be more for the purpose of bolstering their argument against mechanism than searching for whatever new insights the social sciences can provide. They argue for a paradigm shift from physics to the life sciences, but this is not enough. In order to avoid the reductionism inherent in using models from the life sciences to understand social systems, we also need to take account of the special characteristics of social systems and the new emergent properties that they exhibit. I seek to provide this understanding in Chapter 4.

4

The Social Sciences

We build scientific theories to organize and manipulate the world, to reduce phenomena into manageable units. Science is based on reproducibility and manufactured objectivity. As strong as that makes its ability to generate claims about matter and energy, it also makes scientific knowledge inapplicable to the existential, visceral nature of human life, which is unique and subjective and unpredictable ...
(Kalanithi 2016, pp. 169–170)

4.1 Introduction

The most reductionist of the social sciences is orthodox economics, which has, for much of its life, sought scientific respectability by trying to emulate physics. The elementary unit of neoclassical economics is the "econ" (Thaler 2015), a travesty of a human being portrayed as a mathematical calculator, possessing perfect information and always seeking to optimize utility regardless of how others might think and behave. The actions of such "econs" are as determined and predictable as matter and can be incorporated into abstract mathematical models of the economy, as a mechanical system, in which a few variables can be identified and used to describe and predict behavior. The failure of orthodox economics to predict the financial crash of 2008 led Andrew Haldane, chief economist of the Bank of England, to admit: "It's a fair cop to say that the profession is to some degree in crisis." Attempts are, in fact, being made in "behavioral economics" to pay greater attention to humans as they actually are. Go too far in this direction, however, and economics ceases to be economics as it enters the territory of competitor social sciences such as psychology and sociology. Better for economists to remain economists but be much more self-reflective about the limitations of the approach they use.

By contrast, psychology, at least when pursued as a social science, has been readier to embrace holism. Sigmund Freud, the father of psychoanalysis, saw the human personality as an energy system, comprised of the *id, ego,* and *super-ego,* which had to be brought into a harmonious balance to achieve psychological well-being. More significant, for the later development of systems thinking, was the work of the *Gestalt* psychologists. Koffka, Wertheimer, and Koehler, writing in the early twentieth century, reacted against the atomistic psychology of the time, which linked particular sensations to particular

Critical Systems Thinking and the Management of Complexity, First Edition. Michael C. Jackson.
© 2019 John Wiley & Sons Ltd. Published 2019 by John Wiley & Sons Ltd.

physical stimuli, and sought to understand the holistic processes whereby the mind is able to bring order to the chaos of the reality with which it is confronted. There are links to Kant and Husserl and similarities with pragmatist philosophy. The German word *Gestalt*, meaning shape or form, refers to the patterns employed by the mind to organize what is perceived. In Koffka's words, "the whole is something else than the sum of its parts" (quoted in Ramage and Shipp 2009, p. 260) and Gestalt theorists deem the patterns more important than the individual elements. When we apprehend a set of dots, it is some pattern to the arrangement of dots that we see before the individual marks. Eventually, the innate laws used by the mind to generate wholes were categorized into a number of gestalt laws of perceptual organization. Koehler insisted that, on this basis, Gestalt theory could be generalized beyond psychology to philosophy, the arts, and the sciences.

Nothing more will be said about economics or psychology. In this chapter, the focus of attention is on social theory and, when helpful, on organizational analysis. There are two reasons for this. Firstly, social and organizational theorists are divided into warring factions in terms of their understanding of the nature of social reality. In Kuhn's terms (see Section 2.4), they inhabit incompatible "paradigms." Moreover, whereas in the physical sciences, it is reasonable to regard Newton's physics as replacing Aristotle's and, in turn, being replaced by Einstein's, paradigms in sociology continue to co-exist as rival interpretations. Different systems approaches embrace these competing paradigms and we need to understand the impact this has on them. Secondly, some social theorists confronting human subjectivity, a level of complexity beyond even the life sciences, have responded by developing radically different methods of enquiry, which they regard as more appropriate than the scientific method for the social domain. Certain systems thinkers hoping to apply their insights to management have done the same. It is, therefore, social and organizational theory that can provide most insight into the different systems methodologies considered in Part III.

In exploring different sociological paradigms, we begin with Dawe's (1970) distinction between the "sociology of social systems" (called below "functionalism") and the "sociology of social action" (called below "interpretive social theory"). Concepts such as conflict, critique, and power are then introduced through consideration of the "sociology of radical change." There are further sections on postmodernism and poststructuralism, "integrationist social theory," Luhmann's highly original "social systems theory," and action research. The broad categories are indicative only and variations are explained as they arise.

4.2 Functionalism

According to Burrell and Morgan (1979), the functionalist paradigm in social theory regards social reality as having a hard, objective existence and of consisting of systems and structures that determine the behavior of individuals. For this reason, social reality is amenable to the usual methods of the natural sciences. It is observed for evidence of regularities and scientific tests are employed to establish causal relationships. The main purpose of functionalism is to understand how social order is maintained.

The French philosopher Auguste Comte was the first to argue for "sociology" as a new science, calling it initially "social physics." He was writing in the early nineteenth century. His search for patterns in the way society behaved anticipated later developments in the functionalist paradigm and, in setting out his doctrine of "positivism," he provided it with an epistemology and methodology. Positivism privileges scientific knowledge and seeks out general laws in the natural and social world, emphasizing empirical observation and quantification. Following on from Comte's work, the functionalist paradigm developed in two directions, one favoring a mechanical-equilibrium model of how order is maintained and the other taking an organismic perspective on the issue.

Vilfredo Pareto (1848–1923) was the originator of the first variant. In Aron's words, he used "a simplified model comparable to the simplified model of rational mechanics" (1967, p. 174) to explain society as a system in equilibrium. A number of key variables in a state of mutual dependence determine the movement of society. At the surface, significant change may appear to take place as different elite groups succeed one another in power. These changes are, however, merely the result of temporary fluctuations in the relationships between the key variables. Equilibrium will reassert itself sooner or later and so social stability is maintained. Of particular significance in diffusing Pareto's ideas in the United States was the biochemist L.J. Henderson, who worked alongside the physiologist Cannon (see Section 3.2) at Harvard. From his Harvard base, Henderson created a "Pareto Circle," which involved and heavily influenced thinkers such as Parsons, Roethlisberger, Barnard, and Miller, to mention only those considered elsewhere in this book. According to Henderson (1970, originally 1938–1942), the components of social systems, together with their properties and relations, exist in a "state of flux." The connections and constraints resulting from the interactions between them ensure, however, that equilibrium always reasserts itself and stability is maintained in the long run. Even if they are disturbed from outside they are, rather like a boxer's punch-ball, able to return to their original states.

The central figures in developing the organismic perspective in sociology were Spencer (1820–1903) and Durkheim (1858–1917). Drawing on an analogy with biological systems, they reasoned that society was a complex system made up of interconnected parts functioning in ways that contribute to the maintenance of the whole. Such systems are capable of adapting in response to environmental and other changes. Spencer (1969), writing at about the same time as Darwin, was interested in how societies as a whole evolve. In order to be successful in adapting to their environments, and therefore to survive in the long term, they need specific characteristics. The "survival of the fittest" ensures that only those societies that develop such characteristics prosper. Too much government regulation, in his view, hinders a society in the battle of the "survival of the fittest." Spencer also exploited, at every opportunity, the comparison between parts of a society and organs in the body. In Durkheim's sociology, social order is the most important functional prerequisite of society and has to be supported by appropriate forms of the division of labor, by other structural elements, and by shared societal values (Durkheim 1933). His focus was on "social facts" operating at the societal level, which constrained individuals and could be linked causally to other social facts using the positivist method (Durkheim 1938). Thus suicide rates are seen, by Durkheim, not as a phenomenon related to individual psychology but as generated by forms of social control and cohesion. From sociology, the organismic analogy passed into anthropology and was given coherent theoretical expression by Malinowski and Radcliffe-Brown as "structural-functionalism" (Craib

1992). The recurrent activities in a society, its structures, function to meet the survival needs of that society.

In the 1940s and 1950s, one version of sociological systems theory came to dominate American sociology. This was Talcott Parsons' "equilibrium-function model," as Buckley (1967) names it. This comprehensive model is an attempt to combine the mechanical-equilibrium model, structural-functionalism and, drawn from Weber (see next section), the idea that social systems are made up of the actions of individuals. In practice, in Parsons' theory, individual action is so circumscribed by the structures people inhabit that it is the two analogies we have been considering to date that hold center stage. As Craib puts it:

> Parsons sees a social system of action as having needs which must be met if it is to survive, and a number of parts which function to meet those needs. All living systems are seen as tending towards equilibrium, a stable and balanced relationship between the different parts, and maintaining themselves separately from other systems.
>
> *(1992, p. 39)*

The most famous aspect of Parsons' equilibrium-function model is his elaboration, with Smelser (1956), of the four functional imperatives that must be adequately fulfilled for a system by its subsystems if it is to continue to exist. The first letters of these four imperatives, adaptation, goal attainment, integration, and latency (or pattern maintenance), make up the well-known AGIL mnemonic. Due to the recursive character of systems, this AGIL scheme can be employed to analyze and link the various levels of system – from the societal to the organizational to the individual personality system. The meaning of the terms that make up AGIL is as follows:

A = Adaptation: the system has to establish relationships between itself and its external environment

G = Goal attainment: goals have to be defined and resources mobilized and managed in pursuit of those goals

I = Integration: the system has to have a means of coordinating its efforts

L = Latency (or pattern maintenance): the first three prerequisites for organizational survival have to be solved with the minimum of strain and tension by ensuring that organizational "actors" are motivated to act in the appropriate manner

The elegance of Parsons' thinking can best be grasped if we turn to his study of organizations. The defining characteristic of formal organizations for Parsons (1956) is their primacy of orientation to the attainment of a specific goal. The goals of organizations can, following the functionalist logic, be directly related to the needs of the wider society and organizations classified on that basis. So there are:

- Economic organizations, like business firms, oriented to the adaptive function
- Political organizations, like government departments, oriented to the goal-attainment function
- Integrative organizations, like those of the legal profession, oriented to the integrative function

- Latency organizations, like churches and schools, oriented to the pattern maintenance function

In organizations, as in society, order is maintained by a value system that inculcates shared norms among participants. To ensure harmony, organizational values have to be congruent with the central value system of society internalized by individuals during the socialization process, e.g. education. Equilibrium should then be assured since organizations can legitimate themselves in their members' eyes in terms of the function they perform for society. The main source of strain occurs if the central value system of society begins to change. In such circumstances, there is "moving equilibrium" and organizations must adapt in the direction of a new type of stability.

Finally, on Parsons' thinking, he sees the management task in organizations as differing depending upon at which of three levels it operates. At the "technical system level" it is concerned directly with the transformation process; at the "managerial level" with integrating technical-level activities and mediating between these; and at the "institutional level" it integrates the organization with the wider community it is supposed to serve (Parsons 1960).

In the 1960s, "the times they were a changing" and Parsons' work became deeply unpopular. In that radical decade, it was seen as placing too much emphasis on social order and, therefore, as unable to explain conflict and social change. The "Students for a Democratic Society" organization in the United States recognized their enemy, and Parsons' influence, in their "Port Huron Statement" of 1962:

> The vast majority of our people regard the temporary equilibriums of our society and world as eternally-functional parts.
>
> *(Quoted in Bell 2013, loc. 3892)*

Parsons' reward for lecturing to an audience at the London School of Economics was to be surrounded by naked young women carrying placards bearing women's liberation slogans (Hamilton 1983).

Turning briefly to functionalist organization theory, we find the same machine and organism metaphors prevailing. The "classical management theory" that dominated the early part of the twentieth century was reductionist to its core. Frederick Taylor's "scientific management," which required the breaking down and standardization of all work tasks, was the exemplar. The mechanical-equilibrium model came to the fore in the 1930s in two publications from members of Henderson's "Pareto Circle"; Barnard's *The Functions of the Executive*, in 1938, and Roethlisberger and Dickson's *Management and the* Worker, in 1939. Barnard viewed organizations as "co-operative systems" and used the model to advise executives on how they should sustain them in equilibrium by the careful manipulation of inducements to stakeholders. Roethlisberger and Dickson employed the ideas to explain the findings of the famous Hawthorne experiments, so crucial at the beginning of the "human relations" movement. If a worker is knocked out of equilibrium, it will trigger complaints about the work situation and/or reduced work effectiveness. This can come about through the influence of a variety of interconnected factors, ranging from those pertaining to the individual, to those concerned with work conditions, and to social conditions outside the factory. Gradually, as the century progressed, the organismic analogy began to displace the mechanical-equilibrium model.

Selznick (1948) agreed with Barnard that organizations were co-operative systems but insisted that they were also "adaptive structures" and had to change their goals and themselves in response to environmental circumstances. To Selznick, it appeared that many of the adjustments made by organizations, in response to both internal and external conditions, took place independently of the consciousness of the individuals involved. Organizations are primarily oriented to their own survival and their behavior is best understood in terms of how they meet their survival needs. This requires a structural-functionalist analysis. In the 1950s, von Bertalanffy's rigorous working out of the idea that organisms are open systems became thoroughly absorbed into organization theory (see Section 5.6) and, for a couple of decades, the organismic analogy had the field virtually to itself.

Another theory that hopes to understand how social order arises and render social reality amenable to objective, scientific study goes by the name of "structuralism." It differs from the version of functionalism considered so far in its rejection of positivism. Rather than search for and chart empirical correlations, structuralism seeks to reveal patterns or structural relationships operating below the surface level and to use the knowledge obtained to explain the social activities we observe. These deep structures are hidden from human social actors who, nevertheless, have to make use of them and so are determined in their behavior by them. Structuralism originated in the linguistics of Ferdinand de Saussure (1857–1913). Saussure (Culler 1976) insisted that the signs we use in language to denote things in the world are essentially arbitrary. They do, after all, differ from language to language. The meaning of a sign is instead determined by its relationship to other signs. Language is, therefore, a "system of differences" that imposes order on the world. The task of linguistics, for Saussure, is to study language as a system and uncover the structural features of which users are unaware but which shape their sense of reality. Claude Levi-Strauss gained an immense academic reputation, in the 1960s and 1970s, by developing this approach in his own field of anthropology. He was also the first theorist to use the term "structuralism." Levi-Strauss studies the apparently arbitrary cultural activities and myths of "primitive" societies and seeks to demonstrate that there are fundamental structures that provide a pattern to the chaos observable at the surface. Acknowledging a debt to the *Gestalt* psychologists, Levi-Strauss (1968) relates these fundamental structures back to the structures of the human mind, although not the individual human mind. In Ricoeur's words, this is Kantianism without a transcendental subject (Pettit 1977, p. 78). The structures appear to operate on the basis of binary oppositions and give rise, in "primitive" societies, to cultural categories such as hot and cold, male and female, culture and nature, raw and cooked, and marriageable and tabooed women. Unconscious of what they are really doing, "primitive" people struggle with these categories and try to find meaning by reconciling the oppositions through a dialectical process of thesis, antithesis, and synthesis. The fact that it is the human mind that gives rise to the structures underpinning mythological systems means that "civilized" societies must fundamentally be patterned in the same manner as "primitive societies," although drawing, obviously, on a radically different environment. Roland Barthes, in his book *Mythologies* (1973), takes this point and uses a structuralist approach in attempting to reveal the governing structures underlying the modern "myths" of Western capitalist society. When transferred to literary criticism, structuralism leads to the infamous notion of the "death of the author," an idea made famous by Barthes, because it implies that the meaning of a literary text is best understood on the basis of its structure (intertextual connections, narrative structure, recurrent patterns) rather than the intentions of the author. Structuralism came under fire from such as Sartre (from the phenomenological tradition) who put more of a premium on human agency and, as will become apparent in Section 4.5, from "poststructuralist" thinkers.

4.3 Interpretive Social Theory

Interpretive social theory turns the functionalist approach on its head and starts from social action rather than the social system. According to Burrell and Morgan (1979), it views social reality as having a subjective existence as the product of the action and interaction of individual human actors. Social structures arise as a result of recurrent forms of interaction. The natural scientific method is abandoned and knowledge is sought by attempting to get as close as possible to the social actors in order to understand the perspectives they bring to bear in creating social reality.

Dilthey (1833–1911) was the first theorist to offer sociology this alternative direction. Heavily influenced by Kant, he rejected the notion that human actions are governed by cause and effect relations and can therefore be explained using the natural scientific method. Instead, he regarded human behavior as unpredictable and unique to individuals and wanted to ground the human sciences on "hermeneutics," the theory of interpretation, which he felt he could use to grasp the motivations that drive social action. The method of *verstehen*, or empathetic understanding, is recommended as a means of gaining insight into human intentions and how they give rise, through a process of "objectification," to cultural artifacts. This requires continuously going round "the hermeneutic circle" and gaining increased understanding of the relationship between the parts and wholes that constitute social reality (Checkland 1981). Dilthey also introduced the concept of *Weltanschauung* into social theory. *Weltanshauungen* are world-images constructed on the basis of our views of the world, our evaluation of life, and our ideals. Common types tend to recur and are therefore significantly implicated in objectification.

If Dilthey established that there was an interpretive option, Max Weber (1864–1920) forced sociology, through his wide-ranging and influential *oeuvre*, to take it seriously. Rather than using "system" and "structure" as starting points, he argued, sociology should be based upon the study of social action:

> Sociology … is a science which attempts the interpretive understanding of social action in order thereby to arrive at a causal explanation of its cause and effects …. Action is social in so far as, by virtue of the subjective meaning attached to it by the acting individual (or individuals), it takes account of the behaviour of others and is thereby oriented in its course.
>
> *(Weber 1964, p. 88)*

The possibility of interpretively understanding the subjective meaning behind social action gives students of society, Weber thought, an advantage over those working in the natural sciences because the latter can have only external knowledge of their subject matter. To model sociology on the natural sciences to the exclusion of this "inner-understanding" or *verstehen* can only impoverish it. That said "meaningfully adequate" interpretations of social action, as the quote above implies, need to be complemented by considerations of "causal adequacy." Weber (1949) thought that the gap between the two forms of inquiry could be bridged by using "ideal-types." Ideal-types are not "ideal" in the utopian sense nor are they descriptions of reality. They are theoretical constructs that offer a "one-sided," accentuated view of a portion of reality that is both adequate at the level of meaning, in that it embodies possible forms of action, and can also be incorporated into cause–effect explanations. Weber's work is full of examples of ideal-types, for example, bureaucracy, Calvinism, feudalism, capitalism. Again, they are not descriptions of reality but precisely and unambiguously defined constructs that can be compared to reality in order to establish its similarities and divergencies.

The "phenomenological sociology" of Alfred Schutz (1899–1959) can be seen as a synthesis of the work of Weber and of Husserl, the philosophical originator of phenomenology (see Section 1.5). Schutz felt that Weber had not stated clearly enough the essential characteristics of *verstehen*, or of subjective meaning or, indeed, of action, and sought to use the phenomenological method to bring clarity to these concepts by closely interrogating people's immediate experiences of daily life and the common-sense knowledge with which they operate. Craib summarizes:

> He attempted to show how we build our knowledge of the world from a basic stream of incoherent and meaningless experience. We do this through a process of 'typification', which involves building up classes of experience through similarity .... Action and social action thus become things that happen in consciousness: we are concerned with acts of consciousness rather than action in the world, and the social world is something which we create together.
>
> *(1992, p. 99)*

This notion of the "social construction of reality" was popularized by Berger and Luckmann (1971), who see social reality as something that has to be constantly produced and reproduced by individuals in interaction. It might appear to exist "out-there" but it is actually an "ongoing human production." From this perspective, Durkheim's suicide rates are not "social facts" but are the result of "negotiated meanings" constructed, for example, when loved ones and officials are complicit in concealing the true cause of death in societies where a social stigma is attached to suicide (Douglas 2015). Another significant off-shoot from Schutz's thinking is "ethnomethodology." Associated particularly with the work of Garfinkel (1984), "ethnomethodology" focuses on the routine methods people employ to bring about social order. Of particular interest is the so-called "breaching experiment" where taken for granted routines are deliberately breached by the ethnomethodologist in order to reveal the work involved in this process, e.g. by acting as a stranger in your own household or cheating at board games. It will be clear that, for phenomenological sociologists and ethnomethodologists, the study of society must start from individual consciousness and understanding and from the typifications people share in constructing social reality. Indeed all that social scientists can bring are "second-order" descriptions of these primal typifications.

Our final "interpretive" tradition in sociology stems from the "symbolic interactionism" of George Herbert Mead (1863–1931). Mead is often classed, with Pierce, James, and Dewey, as one of the four key figures in American pragmatism and so the philosophical lineage of symbolic interactionism is clear. For Mead, society comes from micro-level interactions between people making judgments about what has been useful in the past and what might be in the future. Human beings have language at their disposal and this interaction can take very sophisticated forms as each agent uses vocal symbols to call forth a response in the other while, at the same time, anticipating that response. This "conversation of gestures" also allows minds to develop and self-consciousness to arise when individuals take the perspective of the other and become capable of seeing themselves as objects. Individuals are socialized when they can conceive of the attitude that their social group, the "generalised other," takes toward them and regulate their conduct accordingly (Stacey 2003, p. 322). Of course, there will be various "generalised others" available and individuals can take on many "roles" in relation to different "generalised others." Thus, mind, self, and society all emerge as a result of social processes of communication, from the bottom-up. Mead was keen that the theory of symbolic interactionism should be combined with active field work to study and address real social problems.

The occasion when the "interpretive" position in sociology challenged the hegemony of functionalism in United Kingdom organization theory can be dated precisely to 1970 and the publication of David Silverman's *The Theory of Organisations*. In this book, Silverman launched a damning critique of systems theory in its functionalist form and proposed, as an alternative for studying organizations, an "action frame of reference" derived from Weber and Schutz. This alternative was presented as an "ideal-type" constituted by seven propositions, which are summarized in Table 4.1.

As Dawe (1970) argues, there is no necessary postulate of consensus or co-operation in social action theory. In practice, however, theorists of the interpretive persuasion do seem to share with functionalists an overriding interest in how social order is constructed and maintained. This is enough for Burrell and Morgan (1979) to class the

Table 4.1 Silverman's ideal type of action theory.

1. The social sciences and the natural sciences deal with entirely different types of subject matter
2. Sociology is concerned with understanding action rather than observing behavior
   Action arises from meanings that define social reality
3. Shared orientations become institutionalized and can be experienced by later generations as social facts
4. While society defines man, man also defines society. Particular constellations of meaning have to be continually reaffirmed in everyday actions
5. Through their interactions men can modify, change, and transform social meanings
6. Explanations of human actions must take account of the meanings of those involved in the social construction of reality
7. Positivistic explanations asserting that action is determined by external constraining forces are inadmissible

Source: Adapted from Silverman (1970).

two together as constituting the "sociology of regulation" in opposition to the "sociology of radical change," which is now considered.

4.4 The Sociology of Radical Change

The sociology of radical change portrays society as divided by structural inequalities that give rise to conflict between different groups, which in turn leads to change. Any cohesion is achieved only because of the power some groups have over others. Marxism and critical theory are the most important sources for the sociology of radical change. The work of the later Marx is "objectivist" in nature seeing the contradictions that exist in society as almost inevitably provoking radical change. The work of the early Marx and the critical theorists is more "subjectivist," emphasizing the need for those suffering from the inequalities to become aware of the reality of their situation and to reshape the social system more in their own and the general interest. The job of a theorist embracing the sociology of radical change is to offer a "critique," which reveals the nature of society to the disadvantaged and provides them with the means to take action.

Marx, as is well known, turned Hegel's dialectic on its head, and saw it as operating not in the realm of thought but in history itself, specifically in the history of class struggle. As he and Engels have it, at the beginning of *The Communist Manifesto*, first published in 1848:

> The history of all hitherto existing society is the history of class struggles. Freeman and slave, patrician and plebeian, lord and serf, guild-master and journeyman, in a word, oppressor and oppressed, stood in constant opposition to one another, carried on an uninterrupted, now hidden, now open fight .... Our epoch, the epoch of the bourgeoisie, possesses, however, this distinctive feature: it has simplified the class antagonisms. Society as a whole is more and more splitting up into two great hostile camps, into two great classes directly facing each other: Bourgeoisie and Proletariat.
>
> *(1967, pp. 79–80)*

Marx's early writings revolve around a critique of alienated labor in capitalist society (Marx 1975a, originally 1844). Work, he argues, should be as natural to people as rest and play. However, because there is private ownership of the means of production, the owners, the bourgeoisie, are alienated by worship of money, while workers are alienated because they only involve themselves in the labor process because they are forced to in order to sustain their physical existence. They do not control the production process and the objects they make become a power over them in someone else's hands. The workers will eventually come to resent the "actuality of an inhuman existence" that they face, take over the means of production, and create a communist society. Everyone will be able to realize their essential being, creative and many-sided, through co-operative labor. In his later work, Marx's vision became less humanistic and more objectivist and determinist in nature. The three volumes of *Capital* (1961, originally 1867) seek to provide a "scientific" explanation of how the economic base of society determines its characteristics, the nature of conflict, and how change will come about.

The political and ideological "superstructure" of a society is conditioned by the economic base and simply reflects the interests and ideas of the ruling class. The capitalist economic system inevitably leads to conflict between capitalists and workers because the former can only create wealth by extracting "surplus value" from the efforts of the latter. The whole system is seen as subject to increasing crises of overproduction and falls in profit. Wages are driven down and unemployment escalates. This exacerbates class conflict. Eventually the oppressed workers see through the ideologies of the ruling class, become conscious of their true interests, overthrow capitalism in a revolution, and bring a communist system into being. Althusser and Balibar (1970), much influenced by structuralism, read Marx's final writings as granting more autonomy to the superstructural "instances" such as politics, ideology, and theory. In the social totality, each instance develops unevenly, has its own contradictions and, although the economy always has the last word, can become temporarily dominant in the social formation. Revolutionary change will only occur when the contradictions in the different instances coincide. Granting some freedom to the superstructural instances is, however, at the expense of any freedom that might be attributed to human agency. Following the structuralist logic, history becomes a "process without a subject," the result of the relations between the "relatively autonomous" instances and the contradictions internal to each.

Critical Theory, associated with the Frankfurt Institute for Social Research, can be seen as a reaction to the fact that the overthrow of capitalism forecast by Marx did not occur, at least in the developed Western economies. The ultimate aim of the "Frankfurt School," which began its work in the 1920s, was still as Horkheimer put it "man's emancipation from slavery" (1976, p. 224), but a step backward seemed to be necessary in the form of a research program that accounted for the failure of Marx's prediction. The concern of Horkheimer, Adorno, and Marcuse became to explain exactly how capitalism manages to survive by means such as promoting the dominance of instrumental reason, drawing on powerful forms of socialization, colonizing culture and the mass media, and encouraging "false needs" among passive consumers. In summary, Craib argues:

> The Frankfurt theorists are concerned with the way the system dominates: with the ways it forces, manipulates, blinds or fools people into ensuring its reproduction and continuation.
>
> *(1992, pp. 210–211)*

The most influential modern thinker associated with the Frankfurt School is the German political philosopher, Jurgen Habermas. He is the critical theorist covered here in most depth because his work has had the closest engagement with systems thinking. However, it is broad in scope, and his conclusions have changed over time, so I have had to be selective in choosing which aspects of his thought to highlight based on what is most relevant to later discussions.

Habermas' writings are redolent of Kant and the Enlightenment in that he wants to transform and improve society by spreading the human potential for reason. Indeed, he sees the Enlightenment as an "unfinished project," in need of some correction but worthy of support and continuance. A number of important themes emerged in his inaugural lecture at the University of Frankfurt in 1965 (Habermas 1970). According to Habermas, human beings possess two fundamental, species-driven cognitive interests that direct their attempts to acquire knowledge: a *technical* interest in achieving goals through prediction and control and a *practical* interest in ensuring mutual understanding. There also exists a third, although derivative, cognitive interest in the analysis of power, called the *emancipatory* interest. Power can interfere with the proper functioning of the other two interests by, for example, preventing open and free discussion. The aim of the emancipatory interest is to free humans from the constraints imposed by power relations so that they can successfully pursue their interest in prediction and control, where this is relevant, and learn to take charge of their own destiny through the full development of the practical interest. Corresponding to the three cognitive interests are three different types of knowledge. The "empirical analytical" sciences support the technical interest in prediction and control. They have the advantage of referring to something in the world and their conclusions are deemed true if the claims they make about the world prove to be accurate. The test is that they can recommend actions that work. Truth, following Pierce's pragmatism, is seen as arrived at through debate and a consensus among the scientific community. The "hermeneutic" sciences correspond to the practical interest and seek to access meaning and improve mutual understanding. They have no objective world to refer to but can give rise to claims of "rightness" if they promote unfettered discourse and all those involved come to agree on some universal moral standard. Supporting the emancipatory interest are the "critical" sciences, which seek to ensure the proper operation of the other two interests and point out the dangers when they are inappropriately used. Habermas (1974), in his early work, compared the use of the critical sciences in the social domain to a psychoanalytic encounter between analyst and patient. Social scientists bring knowledge from the outside – in this case Marxism – which provides causal explanations to oppressed groups in society about why they are in the position they find themselves. If successful, they liberate the subjects from forces they could not understand and increase the area over which they have rational mastery. Success is measured by the extent to which the subjects recognize themselves in the explanations offered and become equal partners in the dialogue. In his later work, Habermas abandons the notion that social theory can provide an external standpoint from which to assess the nature of society. The critical standard becomes the degree to which debate between different viewpoints is fairly organized.

In applying his analysis, Habermas (1975) argues that, in advanced capitalist societies, the technical interest has come to dominate at the expense of the practical interest and the knowledge produced by the empirical-analytical sciences is seen as the prototype of all knowledge. Practical problems about what ought to be done are defined as technical problems to be tackled by scientific experts outside the realm of public discussion. The state sees its function as steering society and overcoming the periodic crises of capitalism. Politics is defined as the task of ensuring that the social system runs smoothly. Recently, commenting on the European Union, Habermas (2011, 2016) emphasizes his support for the broad European ideal as a possible extension of democratic politics to tackle issues beyond the reach of any nation state. But he thoroughly condemns the élite, technocratic, and market-driven manner in which it operates and the way it reduces European citizens to mere spectators; something in which politicians and the mass media seemingly acquiesce. Habermas regards Luhmann's systems theory (see Section 4.7) as the prime ideological reflection of the predominance of the technical interest:

> This theory represents the advanced form of technocratic consciousness, which today permits practical questions to be defined from the outset as technical ones, and thereby withholds them from public and unconstrained discussion.
>
> *(Quoted in Frisby 1976, p. xxxii)*

To Habermas, this is anathema. The knowledge produced by the empirical-analytical sciences is essential to the development of the forces of production and the "steering" capacities of society but must be restricted to the appropriate subsystems. Extending rationality in the domain of social interaction, in the institutional realm of society, depends upon other criteria drawn from the hermeneutic and critical sciences. Questions of what we should do, or might do, are logically independent of questions about the development of productive forces or about system integration and cannot be reduced to them. Where possible, therefore, Habermas wants to reduce the area of social life where people are subject to technical reason and increase the realm of the hermeneutic, where rational intentions are formulated and realized. To facilitate this ambition, he makes his most important theoretical contribution: his theory of "communicative competence."

Whereas Marx had based his early critique of capitalist society on alienated labor, Habermas takes a linguistic-communicative turn and grounds his critique on the type of alienation that occurs in the sociocultural life world, on what he calls "distorted communication." Communicative competence now becomes the focus of attention as the means by which individuals can reach mutual agreement and free themselves from constraints. A critique based on communicative competence is possible because, Habermas argues, a commitment to mutual understanding is prefigured in all communicative interaction and it is a normal expectation that participants are willing to enter into discourse to defend disputed positions. It simply remains for him to describe the conditions for an "ideal speech situation" from which true agreements can emerge. Summarizing greatly, and guided by McCarthy (1973), the structure of communication is free from constraint when all participants have equal chances to initiate, perform, and perpetuate speech acts and there is equality of opportunity to give reasons for or against statements, explanations, interpretations, and justifications. In particular, there should be no barriers to communication erected and enforced by power relations. Where these exacting conditions are met, an ideal speech situation pertains, the force of the better argument prevails, and the consensus emerging will be rationally motivated and genuine. Such circumstances may be rare but this does not detract from the usefulness of Habermas' conceptualization of the ideal because it can be used to unmask "systematically distorted communication" where unequal chances to participate in dialogue, deriving from an unequal distribution of power, determine that a false consensus emerges. For Habermas, therefore, progress toward a rational society is measured by the extent to which communicative competence is achieved. This in turn depends on the establishment of certain social conditions related to freedom and justice. In particular, the "public sphere" must be re-energized as an arena where democratic discussions can take place and genuine agreements can be reached.

Two remarks will be made in concluding these reflections on the "sociology of radical change." The first concerns the weakening of the binary divide between bourgeoisie and proletariat upon which Marx based his theory of class conflict in capitalist society. There have been changes in ownership patterns, with the state and pension funds now playing a significant role, and a growth in the numbers of individual shareholders. Those working for a living are a more diffuse group because of burgeoning public employment, the rise of powerful professional groups, the expansion of the middle class, the development of the service sector, the decline of heavy industry, and divisions in the working class itself. Class conflict remains and class, it can be argued, still plays a predominant role in determining life chances, but things are much more complicated than they were. Secondly, more attention is now given to other divisions and inequalities in society, which can lead to conflict and change. In 1922, Weber, while acknowledging the significance of class struggle, identified "status honor" as another source of division and conflict in society. Status honor for Weber (1948), while often coinciding with class position, arose from noneconomic qualities such as prestige, race, caste, professional groups, and religion. Different status groups arise from "the house of honor" and pursue their own lifestyles and interests, often conflicting with those of other groups. Today, gender, race, sexuality, religion, disability, and age are all recognized as sources of inequality. Sociological theories have developed that urge society to change in order to liberate oppressed groups from the social structures that have historically prevented them gaining equal access to resources, status, and power, and from participating fully in society. Varieties of feminist theory address gender oppression. Queer theory seeks to question the notion of stable sexual identities. In general, although there is a long way to go, capitalism seems able to reform itself in ways that deal with these inequalities and incorporate hitherto disadvantaged groups. This may be a sign of its progressive and liberating character or may be, as Marxists would argue, because it has more significant class inequality to protect. It remains a matter of dispute. At the level of the "world system," Wallerstein (2004) argues, inequalities continue to get worse as the dominant capitalist countries take advantage of their control of most of the world's capital and technology, and their cultural hegemony, to control trade, determine economic agreements, and set prices, thereby exploiting "peripheral" countries for raw materials and labor.

4.5 Postmodernism and Poststructuralism

Overwhelmingly, the social theorists considered so far can be seen as working in the tradition of the Enlightenment. They are committed to sweeping away the myths and prejudices that bound previous generations and using reason to understand and improve the world. They share a belief in rationality, truth, and progress. This position has been labeled by postmodernists as "modernist". Postmodernists attack the whole Enlightenment rationale and want to puncture the certainties of modernism. In doing so, they embrace a position that, if not "subjectivist," is certainly "relativist" in nature.

Lyotard, in his book *The Postmodern Condition* (1984), recognizes two major manifestations of modernism in social theory. These can be called, following Cooper and

Burrell (1988), "systemic modernism" and "critical modernism." Systemic modernism, as its name suggests, is identified with the systems approach as a means of both understanding society and programming it for more effective performance. Parsons (see Section 4.2) and Luhmann (see Section 4.7) are regarded as representative theorists. In this approach, the system stands as the subject of history and progress, following its own logic to increase "performativity," in terms of input–output measures, and handle environmental uncertainty. Systemic modernism relies on the scientific method to discover what is logical and orderly about the world and to assist with prediction and control. Humanity is dragged in the wake of the system as questions about efficiency and stability replace those about truth, falsity, and justice. Critical modernism is identified with theories that seek to explain history in terms of the accumulation of learning and the progressive liberation of humanity from constraints so that it can assume mastery and take on responsibility for its own destiny. Hegel and Marx are in this category and Habermas (see Section 4.4) is seen as the archetypal contemporary representative of the approach, proposing a unified theory of knowledge linked to different human interests and aiming his whole project at human emancipation directed by universal consensus arrived at in the "ideal speech situation." Lyotard opposes all the "grand narratives" proposed by modernist thinkers. Science is, in his view, only one kind of "language game" with limited relevance to social affairs. The new physics demonstrates, as in quantum theory, that the quest for precise knowledge about systems is misguided. And the attempt to limit individual initiative to serving current system imperatives destroys exactly the novelty a system needs to adjust to its environment. Nor is it easy, in Lyotard's view, to sustain the modernist notion that language is transparent and oriented to achieving consensus. There are many language games, obeying different rules, characterized by struggle and dissension, and this is necessary to promote innovation, change, and renewal. We have, therefore, to be tolerant of differences and of multiple interpretations of the world, and we must learn to live with the incommensurable since there is no meta-theory that can reconcile or decide between different positions. Postmodernism, indeed, thrives on instability, disruption, disorder, contingency, paradox, and indeterminacy.

We now turn to the contributions of Derrida and Foucault, two of the most famous postmodern theorists (see the chapters by Hoy and Philp in Skinner 1985). Both are also commonly referred to as "poststructuralists" as their work emerged out of but then transformed the structuralist theory considered in Section 4.2. Structuralism tends to regard the underlying structures that govern surface activity as "fixed" and as having an "objective" status. Poststructuralism suggests that structures are more unstable and fluid. They condition the way we think but can give rise to multiple meanings.

Derrida accepts Saussure's conclusion that linguistic meaning derives from the structure of language itself so that, rather than simply mirroring objects, language creates objects. He goes much further, however, in embracing a relativistic position. Once the relationship between signs and what is signified in the world is broken, it appears to Derrida that it must be possible to create an infinite number of relational systems of signs from which different meanings can be derived. To take the distinctions made in any particular discourse as representative of reality is an illegitimate privileging of that discourse, which involves hiding other possible distinctions. Derrida's "deconstructive" method seeks to reveal the deceptiveness of language and the work that has to go on in any "text" to hide contradictions (which might reveal alternative readings) so that a certain unity and order can be privileged and "rationality" maintained. The shift to the study of the structure of language and away from the intentions of the speaker, as the route to discovering the meaning of "texts," puts Derrida at the forefront of the post-modernist assault on humanism. In his view, it is discourse that speaks the person and not the person who uses language. In the contemporary world, where there are many possible discourses, the idea of an integrated, self-determining individual becomes untenable. From this follows a rejection of the notion of historical progress, especially with humans at the center of it.

In his early work, Foucault conducts an "archeological" investigation of discursive formations in different human sciences, such as medicine, psychiatry, and criminology (see Philp in Skinner 1985). In his view, every field of knowledge is constituted by sets of classificatory rules, which determine whether statements are adjudged true or false in that field. The discursive formations and classificatory rules that govern a discipline will alter over time, but there is no reason to believe that the current arrangements give rise to more "objective" statements than earlier ones in the sense that they mirror reality more closely. The idea of the accumulation of knowledge is rejected by Foucault. So is the notion of a constant human subject who can autonomously engage in pro-moting emancipation. Individuals have their subjectivities formed by the discourses that pertain at the time of their birth and socialization. These not only structure the world but shape individuals in terms of their identity and ways of seeing. To help make this point, Foucault uses a passage from Borges:

> This passage quotes a "certain Chinese encyclopaedia" in which it is written that "animals are divided into: a. belonging to the Emperor, b. embalmed, c. tame, d. sucking pigs, e. sirens, f. fabulous, g. stray dogs, h. included in the present clas-sification, i. frenzied, j. innumerable, k. drawn with a very fine camelhair brush, l. et cetera, m. having just broken the water pitcher, n. that from a long way off look like flies".
>
> *(Foucault 1973, p. xv)*

What this reveals to us, Foucault comments, is both the stark impossibility of think-ing *that* and the limitations of our own system of thought.

For Foucault, discourses are not simply "free-floating" as they appear to be in Derrida. In his later writings, he emphasizes the need to study the power relations with which they are inextricably connected and gives the name "genealogy" to the accounts he offers of the power struggles involved as particular forms of discourse compete for dominance. For example, *I Pierre Riviere* (Foucault 1982) sets out the power dynamics underlying the competition between psychiatry and criminal jus-tice to explain a brutal murder in France. Medicine was at the point of getting its own custodial institutions and it was essential, to a group of leading Paris psychiatrists, that their discourse triumphed over that of the legal establishment and that Riviere was affirmed as mad. If particular discourses come to the fore because of power relations, they also embody knowledge and, Foucault argues, knowledge offers power over oth-ers. In the modern era, the human sciences have created human "subjects" in such a way as to make them available for stricter discipline and control by society. Discourses, therefore, play a role in establishing patterns of domination, benefiting the meanings favored by some while marginalizing the voices of others. This explanation of the power/knowledge relationship, owing much to Nietzsche, is Foucault's most valuable contribution to social theory. Discourses depend upon power relationships. On the other hand, they carry power in the way they make distinctions and so open or close possibilities for social action. A claim to power can, therefore, be seen as present in any claim to knowledge. Power, understood in this way, is omnipresent in social relations. Foucault's genealogy is aimed at unmasking the pretensions of all "totalising discourses." It dismisses their claims to provide objective knowledge. In particular, it offers criticisms directed at the power/knowledge systems of the modern age and support for "subjugated knowledge." In this way, a space is opened up which makes resistance possible, albeit on a local basis in response to specific issues. By paying attention to difference at the local level, to points of continuing dissension, it become feasible to give a voice back to those silenced or marginalized by the dominant discourses.

It is worth concluding this section by referring to a series of lectures by Habermas (1987) responding to the postmodern attack on his position and elaborating a critique of various postmodernist thinkers. In each case he shows that the theorist he is critiquing has something valid to say but exaggerates it out of all proportion. Derrida concentrates on the problems that exist in using language to achieve mutual understanding. Habermas is prepared to acknowledge they exist but details all the positive aspects of language for learning and dealing with problems in the world. Foucault focuses on certain dysfunctions associated with rationalization processes in society but ignores, Habermas argues, the achievements of those same forces. In short, Habermas recognizes that postmodernists have something to say but believes that we should renew and revitalize the Enlightenment vision rather than abandon it. More reason is needed rather than less.

4.6   Integrationist Social Theory

Integrationist social theorists seek to reconcile some of the divides in sociology. They concentrate on the contrasting views that have arisen about the relative importance of "social facts" and human agency and also, sometimes, try to resolve the debate around social order and radical change. In many ways their work is an extended elaboration of Marx's famous statement, of 1852, that:

> Men make their own history, but not of their own free will; not under circumstances they themselves have chosen but under given and inherited circumstances with which they are directly confronted.
>
> *(Marx 1973, p. 146)*

An attempt will be made to capture the essence of the approach here.

Walter Buckley (1976) takes his inspiration from general systems theory (see Chapter 5), which he sees as an essentially "process-conscious" approach. In mechanical and organic systems, he argues, the ties linking components tend to be rigid, concrete, direct, simple, and stable. In complex adaptive systems, like organizations and societies, however:

> Transmission of energy along unchanging and physically continuous links gives way in importance to transmission of information via internally varying, discontinuous components with many more degrees of freedom.
>
> *(Buckley 1976, pp. 184–185)*

In these circumstances, "structure" becomes a theoretical construct used to refer to the relative stability of "underlying, ongoing micro-processes." To understand society, we need a "process" approach that concentrates on the actions and interactions of the components through which structures arise, persist, and change. Buckley calls this process of structure elaboration "morphogenesis." It is essential because complex adaptive systems can only survive by adapting their structures in response to internal and external changes. As part of the process, individuals and groups become linked in different types of "communication nets," which can form structures characterized by any of "cooperation," "competition," or "conflict." According to Buckley, his theory can balance and integrate structural and process analysis and use the same variables to explain both stability and change.

Giddens' theory of "structuration" is frequently cited as seeking to unite the structural and social action perspectives (e.g. Haralambos and Holborn 1995). Structuration theory explores the "duality of structure" whereby structures are both created by social action and, at the same time, make social action possible. Structure does not exist "outside" people, it is inseparable from people's practices, and so agents have the capacity, during interaction, to "make a difference," to either reproduce or change it. However, in doing so, they are constrained because, whether they are aware of it or not, they have to draw upon the practices and resources made available to them by the structure of the existing social system. To his critics (see Inglis and Thorpe 2012, pp. 230–231), by treating agency and structure as inseparable, Giddens fails to give enough independence to structure, which, they argue, pre-exists and weighs down heavily on individuals. As a result, he is viewed as significantly overestimating human agency. By contrast, another candidate integrationist, Pierre Bourdieu, is accused by critics of leaning toward the "structure" side of the divide (see Inglis and Thorpe 2012, pp. 211–225). Bourdieu's key concept is "habitus," the typical ways (usually unconscious) of thinking, feeling, acting, and experiencing shared by a social group. A group's habitus is a reflection of its life-conditions and the socialization its members undergo. It describes both the way society impacts and fashions human actions and, within limits, provides space for human agency. In whatever fields of social activity people are involved, they compete with each other drawing upon their resources of economic (monetary), social (networks), and cultural (mainly educational) "capital." Bourdieu's research suggests that there remains a high degree of class-based domination in contemporary society and so it is social class that primarily determines the amount of access an individual has to the relevant capital necessary to be successful in each field. To critics, Bourdieu is simply replaying a neo-Marxist, objectivist record in which exploitative class relationships are reproduced with the help of the socialization process. Bourdieu argues that his theory leaves room for social agency and we would recognize such agency through empirical research if it occurred.

For example, if it was easy for agents to move from one class to another the evidence would be there in high social mobility. In short, he says,

> … the degree to which the world is really determined is not a question of opinion: as a sociologist, it's not for me to be 'for determinism' or 'for freedom', but to discover necessity, if it exists, in the places where it is …
>
> *(Bourdieu, quoted in Inglis and Thorpe 2012, p. 225)*

Bourdieu hopes that his sociology can be used to unmask the power of élite groups and that this will lead to change in society.

"Critical realism," discussed in the context of the physical sciences (in Section 2.4), is equally relevant, Bhaskar claims, to the social domain as long as appropriate adjustments are made. It is reintroduced at this point because it offers a form of integrationist social theory. According to Bhaskar (see Mingers 2006, p. 25), we must accept that social structures only exist as a result of social action and interaction, are localized in particular cultures for restricted periods of time, and are "open" and therefore cannot be made subject to scientific experiments. Nevertheless, social structures are real and possess their own "emergent properties" separate from the interaction that produces them. Individuals are born into an already existing society and provided with social activities that have to be learned and a variety of social roles to assume as, for example, family members, employees, or representatives of trade unions. There are reasons therefore, just as in the natural sciences, to look for the underlying causal mechanisms that give rise to recurrent social activity even while recognizing that it is this activity (conscious or otherwise) that itself ensures the continuance or otherwise of social structures as generative mechanisms. Bhaskar's theory, therefore, recognizes a "dualism" of two distinct and equally real entities – social structures and human agency. The word "critical" in critical realism refers to the idea that social science is always evaluative and offers an explanatory critique of the current order.

Margaret Archer, who worked for a time with Bourdieu, embraces a form of critical realism heavily influenced by Buckley's theory of morphogenesis, which also seeks to deal with the structure/agency divide. Indeed she sees it as being at the core of her project. She is of significant interest here because of the conceptual clarification she brings to the issue. Archer (2017) labels those social theorists, such as functionalists and deterministic Marxists, who grant no autonomy to agents as "downwards conflationists." Those who grant no causal powers to structure, such as interpretative sociologists, are "upward conflationists." Giddens, who others find guilty of upward conflation, is a "central conflationist" who sees agency and structure as co-constitutive and amalgamates them in his concept of the "duality of structure." However, for Archer, this strategy provides

> … no analytical grip on *which* is likely to prevail under what conditions or circumstances…. In other words, the 'central notion' of the 'structuration' approach fails to specify when there will be 'more voluntarism' or 'more determinism'.
>
> *(Archer 2017, p. 106)*

Archer's solution is to accept Bhaskar's notion that structure and agency are two distinct although interdependent features of social reality, with the former being constituted from the latter but having emergent properties and causal powers. She also adds a timeline. Thus, for her, there is a "morphogenetic sequence" in which pre-existing structural conditions govern social interaction, which in turn leads to the reproduction or elaboration of the structure. This sequence can be broken up analytically and the role of structure and agency in maintaining and changing society studied separately. Only then can the influence of both, and their interplay, be given proper consideration. For example, in a particular situation we can investigate how and the degree to which a particular structural context constrains the subsequent behavior of agents. However intertwined structure and agency may be in practice, we must adhere to "analytic dualism" when studying the structure/agency relationship or our analysis will inevitably collapse into either functionalism or interpretivism.

4.7   Luhmann's Social Systems Theory

In Luhmann's (2013) view, contemporary sociology is in crisis because it remains in thrall to the "old European thought" of Durkheim, Weber, and Marx. He has no doubt that his own rigorously constructed theory of society, heavily reliant on modern systems theory and cybernetics, can provide a better alternative. This alternative is of the most far-reaching kind and challenges dominant assumptions such as that society consists of human beings and their interrelations, that it is integrated by consensus, that it consists of regional and territorially limited units, e.g. countries, and that it can be observed from the outside. His own investigations, instead, reveal the need for a transition to "… a radically antihumanist, a radically antiregionalist, and a radically constructivist concept of society" (Luhmann 2006a, p. 238).

According to Luhmann (the following account also draws upon Borch 2011; Moeller 2006, 2012), between the sixteenth and eighteenth centuries, society began to change from the stratified form of differentiation, the focus of Marxist theory, toward functional differentiation. This signaled the arrival of modernity. In contemporary society, a number of significant function systems can be identified: economy, politics, law, education, science, art, religion, sport and, a rising star, mass media. They all serve functions for society, e.g. politics promotes collectively binding decisions, but they do not coalesce into a unified whole. For Luhmann, modern society is less than the sum of its parts. Society is also "decentered" because there is no hierarchy of function systems. Further, it is also a "world society," since most of the function systems transcend geography and operate on a global basis. Society is not moving beyond functional differentiation and so the idea of postmodernity is a myth. Nevertheless, the advance of functional differentiation does require a radical change in theoretical attitude:

> If we see stratification we will tend to see … injustice, exploitation and suppression …. If, on the other hand, we see functional differentiation, our description will point to the autonomy of the function systems …. Then, we will see a society without top and without centre; a society that evolves but cannot control itself.
>
> *(Luhmann, quoted in Moeller 2012, loc. 85)*

Luhmann studied under Parsons (see Section 4.2), and respects his sociology, but feels he has to break with Parsons' theory of social systems in two ways to better understand modern function systems.

First, he argues, the idea that human actors are the components of social systems, that "action is system," needs revisiting (Luhmann 2013, p. 7). Inspired by Maturana and Varela's theory of autopoiesis, Luhmann wants to find a correlate in the social domain for the biochemical circularity that produces the operationally closed systems of life. He concludes that the only possible candidate is "communication." Unlike action, communication is clearly a social operation because it "involves or implies … a simultaneous presence of at least two consciousness systems for its emergence" (Luhmann 2013, p. 211). In Luhmann's theory of social systems, therefore, communication replaces action as the basic operation that gives rise to operational closure:

> The idea of system elements must be changed from substances (individuals) to self-referential operations that can be produced only within the system and with the help of a network of the same operations (autopoiesis). For social systems in general and the system of society in particular the operation of (self-referential) communication seems to be the most appropriate candidate.
>
> *(Luhmann 1989, pp. 6–7)*

Communicative events make up "networks" in that they are constantly referring to previous communications and necessarily lead on to others. Mingers provides an example:

> In the law, a legal communication might be the judgement of a court. It contains a particular selection of information … is presented in a particular way … and … interpreted in particular ways. The judgement as a whole leads to further communications, both directly through its consequences and indirectly as part of case law.
>
> *(1995, p. 143)*

Social systems are, therefore, operationally closed systems maintained by an ongoing flow of communications.

Second, Luhmann finds it necessary to abandon the whole/parts distinction embodied in Parsons' work. To develop something better, he again draws upon Maturana and Varela's theory of autopoiesis. In the process of self-production, operationally closed systems define themselves in relationship to their environments. If they did not do so, they would be overwhelmed by complexity. Operational closure ensures that they are constituted to take notice of only a part of the environment and this makes knowledge possible for them; ironically because they are protected from having direct access to "reality." Social systems are, therefore, "cognitive systems," which, as we saw in Section 3.2, create their own reality. Different social systems will do this in different ways rather than act as parts of a whole. In Luhmann's conception, society must be considered not as a unity but in terms of "difference," especially the distinction between system and environment:

> The theory must change its direction from the unity of the social whole as a smaller unity within a larger one … to the difference of the systems of society and environment…. More exactly, the theme of sociological investigation is not the

system of society, but instead the unity of the difference of the system of society and its environment.

(Luhmann 1989, p. 6)

Function systems, as social systems, construct reality in different ways and each creates its own specific communicative operations exploring the broad "conditions of possibility" made available by language.

This is redolent of the thinking of von Glasersfeld, as well as Maturana and Varela, and Luhmann self-identifies as a "radical constructivist." To refine his "difference theoretical" approach and escape the biological terminology of autopoiesis, he turns to the mathematician George Spencer-Brown, another member of the "second-order cybernetics" tradition broadly defined (see Section 6.4). Spencer-Brown declares that the theme of his book *Laws of Form*, published in 1969, "is that a universe comes into being when a space is severed or taken apart" (quoted in Borch 2011, p. 51). The act of severance leads to a distinction between the thing that is distinguished and its context; between system and environment in Luhmann's theory. Function systems are networks of communication that distinguish certain communications to which they will give intense attention while all others, being in their environments, are treated with indifference. They do this, echoes of Levi-Strauss's structuralism here, using binary codes that determine their area of interest. Thus the function of science is to generate new knowledge and it operates on the basis of the distinction true/false. It is indifferent to whether something is legal or illegal. That is the concern of law. Economics is interested in the distinction profitable/not profitable, politics in government/opposition, art in beautiful/ugly, education in good grades/bad grades, sport in winning/losing, and the mass media in information/noninformation. Once established, function systems develop programs that enable them to apply their codes correctly as with, for example, the theories and methods of science, and institutions come into being to facilitate their work. Clashes between function systems occur as, for example, when the US legal system objected, in 2018, to President Trump treating who should be admitted into the country as a political issue.

For Luhmann, therefore, highly differentiated function systems, interpreting the world according to their own logics, have largely replaced other differences, such as class, religion, race, sex, and region, as the defining feature of modern society. The Enlightenment project can be condemned to history because no "overarching reason" exists on which to base a critique of the existing order and because the power of human agency is significantly downgraded. People cannot pretend to be in a position to steer society and mold it according to their intentions. Certain positives emerge from Luhmann's new vision. The separation of function systems acts as a bulwark against totalitarianism and as a counter to totalizing discourses, which can wreck havoc in the name of universal liberation. There are also important negatives. The decentering of society makes it virtually impossible to mount a co-ordinated response to "grand societal challenges" such as climate change because the function systems see the issues differently, operate on different timescales, and can only provide partial solutions (Luhmann 1989). In his later writings, Luhmann also warns that the globalization of functional differentiation is bringing a new meta-code, of inclusion/exclusion, to the fore. Global action systems aim at all-inclusion but there are increasing numbers of people who fail to meet their requirements and individuals excluded from one function system are likely to fail the test of others:

> No education, no work, no income, no regular marriages, children with no birth certificate, no passport, no participation in politics, no access to legal advice, to the police, or to the courts – the list can be extended and it concerns, depending on the circumstances, all marginalizations up to total exclusion.
>
> *(Luhmann 2006b, p. 270)*

Further, groups clinging on to old identities perhaps religious, and excluded from communication by the new societal configuration, can turn to violence to assert themselves. Critics of Luhmann will, of course, continue to argue that some systems in the social whole, usually the economy, dominate others or, at least, agree with Habermas that their rationality inappropriately pervades the rest. Bourdieu offers a compelling case that stratification rather than functional differentiation continues to dominate society, with social class determining the access individuals and groups have to the rewards of the different function systems. It is indeed strange that Luhmann points to multiple exclusions at the bottom of function systems but not multiple inclusions at the top. In Owen Jones' view:

> As well as a shared mentality, the Establishment is cemented by financial links and a 'revolving door' culture: that is, powerful individuals gliding between the political, corporate and media worlds – or who manage to inhabit these various worlds at the same time.
>
> *(2014, p. 6)*

Although, in Luhmann's theory, systems are highly differentiated from one another – for one social system all the others are in its environment – relations do develop between them. To explain how this can occur, Luhmann again draws upon the work of Maturana and Varela, this time making use of the concept of "structural coupling." Social systems are operationally closed, and therefore develop according to their own structural logics, but they can be perturbed or "irritated" by other systems in their environments in ways that bring about structure determined changes. Over time, frequent irritations between two social systems can lead to them continuously resonating with one another and becoming "structurally coupled" in the sense that their relationship achieves some stability and they come to rely on each other. The association between the function systems of politics and economics, for example, is signaled by taxes and central banks. Both function systems retain their overall autonomy but integration leads to a reduction in the freedom each has individually.

Luhmann's social theory has provoked significant controversy. If communication constitutes the distinctive operation of social systems, as he argues, then human beings, as both living and psychic systems, are not components of social systems but in their environment (Luhmann 2013, p. 188). This conclusion has brought him grief from critics who regret that this makes his theory antihumanist. For Luhmann, however, it is an inevitable result of the shift from the old "action is system" type of sociology to the more rigorous and fruitful system/environment version. Human beings are operationally closed "psychic systems," constituted by the thoughts and feelings that go round in consciousness. Social systems can provoke thoughts and feelings in psychic systems but cannot know or determine them. This is something to welcome, Luhmann says. Surely we do not want social systems that have direct access to our minds? Social systems are also operationally closed, sustained by ongoing networks of communication. Psychic systems can irritate them but whether there is any impact, and if so what the impact is, depends entirely on the readiness of the communication system to pay attention and respond at the time. That said, Luhmann is clear that the autopoiesis of social systems does depend on the existence of living and psychic systems. For social systems to survive, a consciousness has to engage in the process of communication, paying attention to it and showing a willingness to continue it. He uses the term "interpenetration" to describe the extremely close structural coupling that exists between psychic and social systems. Mind and society share the medium of meaning and co-evolve using language as a coupling mechanism. Psychic systems provide social systems with communications they can make use of to sustain and develop themselves. Social systems provide psychic systems with things to think about and, in modern, functionally differentiated society, the opportunity to adopt multiple identities. The fact that social and psychic systems are operationally closed to each other does, however, give Luhmann pause when considering, for example, efforts by the state to reform criminals and change the attitude to work of the continuously unemployed. Social systems can "irritate" psychic systems but certainly not control the way welfare efforts are understood and received. He worries that the cost to the welfare state of trying to ensure the inclusion of all citizens impinges on the boundary with the economy and impacts its effective functioning. This argument has been read as supportive of a neo-liberal agenda.

Luhmann's work is equally far-reaching when applied to social theory as to society itself. Here he takes inspiration from the founding father of second-order cybernetics, Heinz von Foerster (see Section 6.4), and again from Maturana and Varela. von Foerster complemented the original emphasis of cybernetics on observed systems of communication and control with the insight that it also needed to study "observing systems." Maturana and Varela's study of living systems led them, as we saw in Section 3.2, to the conclusion that "anything said is said by an observer." In Luhmann's view, operationally closed psychic and social systems, including social theorists and theories, are themselves "observers" and should be studied as such. Philosophers and theorists, such as Parsons, Marx, Husserl, and Habermas, seem to think that they are "first-order" observers with a grasp of social reality, which they can understand better than ordinary people and other theorists. In fact, they are simply "observers" who create the reality they theorize about using their favored distinctions, often the out-moded subject/system divide. Social theory, in Luhmann's view, must give up on ontological certainty and become the study of how first-order observers observe:

> Whether or not philosophers accept this, we will henceforth always be dealing with a world description that filters the presentation of facts, including purposes, action potentials, and so forth, by indicating a reference to an observer .... I have good reasons ... for the assumption that the observation of the observers – that is to say, the shift from a consciousness of reality to a description of descriptions, or the perception of what others say or do not say – has become the advanced

mode of perceiving the world in modern society …. It is no longer necessary to know how the world is, as long as we know how it is observed and as long as we find orientation at the level of second-order observation.

(Luhmann 2013, pp. 99–100)

"Second-order" observation represents a shift from ontology to epistemology and means that we give up on making claims about the nature of social reality, concentrating instead on how different social theorists and theories construct societal issues and problems from the distinctions they employ. Using second-order observation we are able to understand how the first-order theory we are studying observes, and what it sees and possibly does not see. Luhmann accepts, of course, that every second-order observation will itself carry the imprint of the distinctions it decides to make and that this is equally true of his own systems theory based on the system/environment divide. If however we accept, as we should, that social theories can have an impact on society, then the "modesty" of second-order observation has more to recommend it than the "know-it-all" first-order theories that claim to possess the truth about how things are and need to be.

4.8 Action Research

The primary purpose of most of the social research we have been studying, functionalist, interpretive, radical, integrationist, or whatever, is to produce knowledge about the social world rather than to help anyone who might make use of the research. Occasionally there is reference to a broad group of potential users, such as the working class in Marxism, and sometimes the hope is expressed that the theory will lead to appropriate change in society, as with Bhaskar and Bourdieu, or be used to address social problems, as with Mead. Generally, however, it is simply assumed that knowledge about how society operates must be a good thing in itself and it is left to others to make what use of it they can for tackling the problems they face. It is refreshing therefore to find another type of research in the social sciences called "action research," which, as Rapoport defines it, aims to contribute

> … *both* to the practical concerns of people in an immediate problematic situation and to the goals of social science by joint collaboration within a mutually agreed ethical framework.
>
> *(1970, p. 499)*

This kind of research has had a significant impact upon applied systems thinking as Part III of the book will show.

The term "action research" was coined around 1944 by the social psychologist Kurt Lewin. Lewin firmly believed that research in the social sciences should serve social practice: "Research that produces nothing but books will not suffice" (Lewin 1967, p. 443). This means that any research programme set up in an organization

> … must be guided closely by the needs of that organization, and must help define those needs more specifically.
>
> *(Lewin 1967, p. 444)*

The research must have the full co-operation of those engaged in the social situation of interest and they will also be involved in monitoring and assessing the results obtained. In Lewin's view, this does not necessarily lead to any loss in scientific rigor and so the outcomes can, at the same time, enhance knowledge in the social sciences.

Lewin was much influenced by Gestalt psychology and believed that social situations should be assessed as totalities of mutually interdependent facts rather than by breaking them down into their constituent parts. Indeed, because of the close interrelationships between the large numbers of elements involved in social systems, it is only possible to come to an understanding of a system of interest by trying to change it. To ensure scientific rigor, this demands a close analysis of the initial social situation, clearly documented action to bring about a desired change and continuous monitoring of the effects, and careful analysis of the end results of the action. Action research offers a significant break from the usual presumption in natural science and most social science that the experimenter, to ensure objectivity, must take every precaution to ensure that they do not themselves impact on the outcome of the experiment. The action researcher accepts that any research on social systems has the capacity to change them, as recognized to some degree in social science since the Hawthorne experiments, reported by Roethlisberger and Dickson (1939), and makes a virtue of this by trying to ensure that the change that is brought about benefits those involved. There is a further consequence, which is that values and power, including the researcher's own, are inevitably central in action research. In Lewin's words, the action researcher

> ... has to learn to understand how scientific and moral aspects are frequently interlocked in problems, and how the scientific aspects may still be approached. He has to see realistically the problems of power, which are interwoven with many of the questions he is to study, without his becoming a servant to vested interests.
>
> *(1967, p. 444)*

4.9 Conclusion

This has been a long chapter but patience will be rewarded later in the book when the relevance of social theory for applying systems thinking to address real-world problem situations becomes apparent. Briefly, the social sciences are strong on theory but are weak on practice. Social scientists rarely seem to draw out the implications of their work in terms of specific guidance for what can and should be done to improve organizations and society. The systems practitioners studied later are, by contrast, dedicated to practice but often neglect theory. It is obvious that any attempt to change social systems must rest upon assumptions, conscious, or otherwise, about the nature of social reality. If systems practitioners fail to reflect upon the theoretical assumptions they make, they deprive themselves of learning, from practice, how useful or otherwise their implicit theories are and, as a result, miss out on the opportunity offered to rethink their practice. Parts III and IV hope to demonstrate how powerful a combination social theory and systems practice can be.

One important conclusion that can immediately be drawn is that, to deal with the massive problems confronting humankind, we need more than the perceptual shift from physics to the life sciences that Capra and Luisi (2014) advocate. There is much to learn from biology and ecology, including much that the social sciences do not touch upon. Nevertheless, as this chapter demonstrates, the social sciences point to "emergent properties," which give rise to new issues that only come to the fore at the societal level of complexity. These desperately require our attention. They include:

- The need to maintain order and manage systems in which components are linked by the transmission of information rather than fixed energy links, making process at least as important as structure
- The importance of meaning and the way it influences human intentions, purposes, actions, and interactions
- The need to establish a rational consensus, or at least an accommodation, between individuals and groups with different perspectives so that decisions can be made and action can be taken
- "Social facts," or social structures, or "function systems," which may initially emerge from human action and interaction but then escape our control
- Structural inequalities, which can give rise to conflict, including issues of class, gender, race, globalization
- The exercise of power by some social groups over others and the role of power/knowledge
- Poverty and the issue of exclusion
- The way that social theories can themselves play a role in shaping the social world they describe

A further perceptual shift is needed, which embraces the lessons of the social sciences. It is the argument of Parts III and IV that systems thinking is making that shift and, in doing so, is developing in a manner that can significantly improve our chances of success in managing the enormous challenges that we face.

Before pursuing that argument, however, consideration must be given to the attempts made by systems thinking to establish itself as a science in its own right; as a "trans-discipline" with "organised complexity" as its subject matter. In doing so, it has sought to contribute to established disciplines in the physical, life, and social sciences and to unite the scientific endeavor for the betterment of mankind.

Part II

The Systems Sciences

Now I want to talk about the other significant historical event which has happened in my lifetime, approximately in 1946–7. This was the growing together of a number of ideas which had developed in different places during the Second World War. We may call the aggregate of these ideas cybernetics, or communication theory, or information theory, or systems theory ... . All these separate developments in different intellectual centres dealt with communicational problems, especially with the problem of what sort of a thing is an organized system

(Bateson 1973, p. 450)

A paper by Weaver (2003, originally 1948) helps us to clarify the subject matter of these "systems sciences." Weaver argues that the traditional scientific method has been successful in fields characterized by quantitative and logical problems where its mathematical tools can gain purchase. In the seventeenth, eighteenth, and nineteenth centuries it was able to tackle problems of *organized simplicity* involving a very small number of objects related in predictable ways (simple, deterministic). Weinberg (2011) calls this a region of machines or mechanisms. The problems it poses yield to the classical mathematical tools of calculus and differential equations. Newtonian mechanics provides the exemplar. In the late nineteenth century, with the advent of statistical mechanics (see Section 2.2), science was able to broaden its scope to problems of *unorganized complexity* consisting of huge numbers of components exhibiting a high degree of unpredictability (complex, random). This is a region of aggregates, of gasses and populations (Weinberg 2011). It can be tamed by statistics and probability theory and the equations of thermodynamics are the exemplar. The two sets of mathematical tools are, therefore, complementary. Unfortunately, as Klir (2001) notes, they address only the extremes of the scales of complexity and randomness and the great majority of real-world problems are located somewhere in between. This is illustrated in the figure below.

Critical Systems Thinking and the Management of Complexity, First Edition. Michael C. Jackson.
© 2019 John Wiley & Sons Ltd. Published 2019 by John Wiley & Sons Ltd.

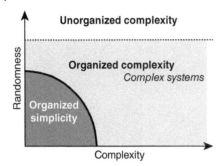

Three classes of systems that require distinct mathematical tools.

Weaver comments:

> One is tempted to oversimplify, and say that scientific methodology went from one extreme to the other – from two variables to an astronomical number – and left untouched a great middle region. The importance of this middle region, moreover, does not depend primarily on the fact that the number of variables involved is moderate .... The really important characteristic ... which science has as yet little explored or conquered, lies in the fact that these problems ... show the essential feature of *organization*. In fact, one can refer to this group of problems as those of *organized complexity*.
>
> *(2003, p. 380)*

Organized complexity, the great yawning gap in the middle, throws up problems that are too complex for analysis and too organized for statistics. They are problems that require us to deal simultaneously with a sizeable number of factors interrelated into an "organic whole." This is the region of "systems" (Weinberg 2011) in which the traditional methods of science are simply not suitable. Weaver provides, as examples, environmental problems, the study of aging, diverse problems associated with modern technology and medicine, how currency can be wisely and effectively stabilized, how the behavior of organized groups of people can be explained, what sacrifices of present self-interest are necessary to bring about a "stable, decent and peaceful world" (Weaver 2003, p. 381).

Let us consider the 2014 Ebola outbreak in West Africa. Epidemiology has traditionally relied upon both deterministic and statistical models to make its predictions about disease spread; either seeking to categorize a population into a very small number of "compartments," and charting the linear interactions between these using differential equations, or trying to determine the average behavior of individuals and using probabilities to make calculations. Unfortunately, disease transmission systems exhibit neither organized simplicity nor unorganized complexity, and so the models have been found wanting. It was clear to Pruyt et al. (2015) that the Ebola outbreak was characterized by organized complexity and that modeling needed to be based upon a systems approach. In particular, the normal factors that might be taken into account in transmission models should be complemented with a host of psychological and sociocultural effects that play an equally significant role in "organizing" the system. To name just a few: fear-induced contact rate reduction, fear-induced increases in levels of hygiene, indigenous protocols for epidemics and burials, fear of dying in quarantine, learning and the accompanying attitude change, fleeing the region, uprisings. It is possible to add, with an awareness of Luhmann's sociology (see Section 4.7), the role that the various "observations" made by governments, the World Health Organisation (both locally and internationally), aid agencies, the media and drug companies play in creating the organized complexity.

Problems of this type, Weaver insists, which predominate in the life, behavioral, social, and environmental sciences, require

> ... science to make a third great advance, an advance that must be even greater than the ... conquest of problems of simplicity or the ... victory over problems of disorganised complexity. Science must, over the next 50 years, learn to deal with ... problems of organized complexity.
>
> *(2003, p. 341)*

It is this challenge that the systems sciences have embraced in the form of general systems theory, cybernetics, and complexity theory.

5

General Systems Theory

Tektology deals with organizational experiences not of this or that specialized field, but of all these fields together. In other words, tektology embraces the subject matter of all the other sciences and of all the human experience giving rise to these sciences, but only from the aspect of method, that is, it is interested only in the modes of organization of this subject matter

(Bogdanov 1922, quoted in Gorelik 1975a, p. 348)

5.1 Introduction

Bogdanov's three volume *Tektology*, published in Russia between 1912 and 1917, anticipated many of the themes that later became associated with general systems theory (GST). In particular, he emphasized that the subject matter of tektology, as the "universal science of organization," was "organization" in general and that its ambition was to be "trans-disciplinary," i.e. relevant to all branches of knowledge. Although it is clear to Gorelik (1975b) that the "conceptual part" of GST is all present and correct in *Tektology*, Bogdanov's work had little influence on its further development. This was in part because it was suppressed by the Soviet authorities and in part because, despite the publication of a German edition of *Tektology* in 1928, it was largely ignored in the West until 1975, when Gorelik (1984) translated a substantial portion of the text into English. It is, therefore, to Ludwig von Bertalanffy that the credit goes for being the founding father of GST. In Section 3.2, von Bertalanffy's influential contribution to organismic biology was reviewed. This work culminated in his seminal article on open systems published in 1950. In this chapter, we look at his vitally important efforts in the field of general system theory, at the work of his collaborators and the foundation of the Society for General Systems Research (SGSR), at two alternative futures mapped out for the approach by Miller and Boulding, and at the influence of GST particularly in organization and management theory.

5.2 von Bertalanffy and General System Theory

von Bertalanffy first articulated the need for a general system theory at the University of Chicago in 1937 but did not formally announce it as a research project, or publish his ideas, until shortly after the Second World War. A collection of his essays, first

Critical Systems Thinking and the Management of Complexity, First Edition. Michael C. Jackson.
© 2019 John Wiley & Sons Ltd. Published 2019 by John Wiley & Sons Ltd.

published in 1968 (von Bertalanffy 1971), is a key resource for discovering who influenced him and understanding the nature and scope of his thought. The collection is called *General System Theory* although, even by that time, the plural general system<u>s</u> theory had become the more normal descriptor for the approach.

In this collection, von Bertalanffy acknowledges that the systems concept has a long history and draws upon the work of many systems thinkers whose contributions to their own disciplines we have already encountered. Thus, he references Heraclitus (philosophy) for his dynamic conception of reality; Hegel (philosophy) and Marx (sociology) for the dialectic; Goethe, Bernard, and Cannon (biology) for the dynamic structure of living systems, the "organismic conception" and homeostasis respectively; and the "preliminary" general systems work of the Gestalt psychologists. He also compliments Weaver for recognizing that the basic problem posed to modern science is "organized complexity." What has come to the fore as "the system problem," von Bertalanffy (1971, pp. 16–17) argues, is a recognition of the limitations of the analytical methods of science, first enunciated by Galileo and Descartes, to systems in which there are strong interactions between parts whose relations are nonlinear, i.e. when, as it is often described, "the whole is more than the sum of its parts." Entities of an essentially new sort, systems, have entered the scientific discourse and cannot be explained in terms of the traditional paradigm. We are in the throes of a Kuhnian scientific revolution (see Section 2.4) and require a new formulation of science based on an organismic outlook of "the world as a great organization" (von Bertalanffy 1971, p. xix).

It is apparent, he argues, that the scientific study of "wholes" has become essential across a wide range of disciplines. This is the domain of GST:

> General system theory, therefore, is a general science of 'wholeness' .... In elaborate form it would be a logico-mathematical discipline, in itself purely formal but applicable to the various empirical sciences.
>
> *(von Bertalanffy 1971, p. 36)*

von Bertalanffy, therefore, conceives of GST as a new scientific worldview concerned with the laws that apply to systems behavior in general. Such a science is possible, and will be particularly fruitful, because of the large number of parallelisms or isomorphies that appear across systems whatever their component parts:

> Thus, there exist models, principles and laws that apply to generalized systems or their subclasses, irrespective of their particular kind, the nature of their component elements, and the relations or 'forces' between them. It seems legitimate to ask for a theory, not of systems of a more or less special kind, but of universal principles applying to systems in general. In this way we postulate a new discipline called *General System Theory*. Its subject matter is the formulation and derivation of those principles which are valid for 'systems' in general.
>
> *(von Bertalanffy 1971, p. 31)*

The study of general systems is to focus on such principles as

> ... growth, regulation, hierarchical order, equifinality, progressive differentiation, progressive mechanization, progressive centralization, closed and open

systems, competition, evolution toward higher organization, teleology, and goal-directedness.

(Hammond 2003, p. 119)

Although, as we can see, von Bertalanffy derives many of his insights from his biological work, he believes that they can be transferred to other disciplines because the principles are not specific to biology. They are general system principles that apply to complex systems of all types, whether they are of a physical, biological, or social nature. The principle of progressive differentiation, for example, is ubiquitous in biology, psychology, and sociology. According to von Bertalanffy, GST is not just possible, it also fulfills a real and urgent need. The sciences have become increasingly specialized and scientists in different disciplines find it difficult to communicate with one another. GST can provide a much broader and better framework for the unification of disciplines than the reductionism that comes from following in the footsteps of physics. He is now able to explain the major aims of GST (von Bertalanffy 1971, p. 37):

- There is a general tendency toward integration in the various sciences, natural and social
- Such integration seems to be centered in a general theory of systems
- Such theory may be an important means of aiming at exact theory in the nonphysical fields of science
- Developing unifying principles running "vertically" through the universe of the individual sciences, this theory brings us nearer to the goal of the unity of science
- This can lead to a much-needed integration in scientific education

A main thrust of von Bertalanffy's thinking is to reject the reductionism involved in explaining biology purely in terms of physics. He is equally keen to protect the autonomy of the human and social sciences and condemns simplistic attempts to apply concepts from biology to psychological and social phenomena. These higher levels of complexity give rise to their own emergent properties. A systems-theoretical reorientation of psychology leads away from "the robot model of human behavior" toward a new image of the human being as an "active personality system," inner-rather than outer-directed, and creating its own universe:

> Emphasis [is] on the creative side of human beings, on the importance of individual differences, on aspects that are non-utilitarian and beyond the biological values of subsistence and survival – this and more is implied in the model of the active organism.

(von Bertalanffy 1971, p. 204)

Once we reach the social level, a world of symbols, values, social entities, and cultures emerges. Humans are, through language, both symbol-dominated and symbol-creating beings; they are conditioned by structures and cultures but essential to their preservation and transformation. General system theory, the open-system concept in particular, incorporates maintenance and change equally, as Buckley appreciated in developing his social theory (see Section 4.6).

These considerations lead von Bertalanffy to what he calls a "perspective" philosophy. He agrees with Kant that our perceptions do not give us direct access to "truth" or "reality." However, he dissents from Kant's thesis that all rational beings share the same common intuitions (space and time) and categories through which they construct the world. Instead, he sees the way we view reality as dependent on a "multiplicity of factors" of a biological (how we have evolved), psychological, cultural, and linguistic nature (von Bertalanffy 1971, p. xxi). The "organizational and functional plan" of an organism determines what is a "stimulus" for it and what it will ignore (von Bertalanffy 1971, p. 241). Written in 1955, this predates the theory of autopoiesis by 20 years. People with different personalities, speaking different languages, and socialized into different cultures will harbor alternative, sometimes conflicting, perspectives on the nature of the world. No perspective can claim absolute value and this is equally true of science:

> Against reductionism and theories declaring that reality is 'nothing but' (a heap of physical particles, genes, reflexes, drives, or whatever the case may be), we see science as one of the 'perspectives' man with his biological, cultural and linguistic endowment and bondage, has created to deal with the universe he is 'thrown in', or rather to which he is adapted owing to evolution and history.
>
> *(von Bertalanffy 1971, p. xxi)*

GST, von Bertalanffy believes, emphasizing autonomy, creativity, and dynamism, offers greater opportunities for addressing humanistic concerns and social ills than does a science that identifies itself with physics. This said, he is also keen to indicate limits to the relativism apparent in his perspectivist position (von Bertalanffy 1971, pp. 252–261). During their evolutionary journey, he argues, human forms of experience have had to adapt to the real world to ensure our survival and so we should expect at least a sufficient correspondence between "appearance" and "reality." Further, science is "de-anthropomorphizing" itself. Human observers are being replaced by recording instruments and human intuition by systems of mathematical relations. These enable science to expand the observable and eliminate the restrictions imposed on the mind by the Kantian forms of intuition and categories. Science has become a thinking machine in its own right and the evidence that it can produce theories that at least approximate to reality is there in the convergence of research results. GST over time may, by looking through its systemic lens, also yield useful knowledge about structural uniformities existing at different levels of the "real." Here, he finds parallels with the "French structuralism" of, for example, Levi-Strauss (von Bertalanffy 1971, p. xvii). Ultimately, however, for von Bertalanffy, all knowledge can only mirror certain aspects of reality.

Georgiou (1999) locates von Bertalanffy's perspectivism within the main phenomenological tradition. He believes that critical systems thinkers have been wrong to ignore it and turn to Habermas for inspiration. von Bertalanffy has much to say about issues of critical awareness, social awareness, and human emancipation (see Chapter 19). Further, his contribution opens the way for a Sartrean input to critical systems thinking. Sartre articulated many of the same concerns and battled to elucidate ethics from an ontological position (see Section 1.5). Pouvreau (2014) similarly recognizes similarities between von Bertalanffy's perspectivism and the "pluralism" inherent in critical systems thinking. Both recommend employing different conceptual schemes, which express distinct viewpoints on the order that exists in "reality," as a useful approach for helping people orientate themselves and learn about what is possible in the world.

5.3   von Bertalanffy's Collaborators and the Society for General Systems Research

von Bertalanffy's main collaborators in establishing and developing GST were the physiologist Ralph Gerard, the mathematician Anatol Rapoport, the economist Kenneth Boulding, and the psychologist James Grier Miller. In 1954–1955, von Bertalanffy found himself on a fellowship at the Center for Advanced Study in the Behavioral Sciences (CASBS), in Stanford University, where Boulding, Gerard, and Rapoport were also based. It was the first year of operation of CASBS, which had been set up, with funding from the Ford Foundation, to promote interdisciplinary research into the scientific study of human behavior – an area of investigation deemed of extreme importance at the time of the cold war. As Hammond (2003) details, in her excellent book *The Science of Synthesis*, the four immediately recognized the similarity of their intellectual concerns and, in the fall of 1954, proposed the formation of a "Society for General System Theory" at the annual convention of the American Association for the Advancement of Science. Its principle aims were stated as:

- To investigate the isomorphy of concepts, laws, and models in various fields, and to help in useful transfers from one field to another
- To encourage the development of adequate theoretical models in fields that lack them
- To eliminate the duplication of theoretical efforts in different fields
- To promote the unity of science through improving the communication among specialists

The Society was eventually established in 1956 before undergoing a name change, in 1957, to the "Society for General Systems Research" (SGSR). Boulding was chosen as its first president (1957–1958) and the SGSR began to hold annual conferences and to publish a *General Systems Yearbook*. In 1988, the name of the Society was altered to the "International Society for the Systems Sciences" (ISSS) and it continues its activities to this day. The fifth collaborator, Miller, actually invented the term "behavioral science," in 1949, to reflect the goal of integrating study of the biological, psychological, and social dimensions of human behavior. He participated in the planning for CASBS and in its work and, though not directly involved with the formation of SGSR, was closely associated with it thereafter. In 1956, Miller founded the journal *Behavioral Science*, which, under its new name *Systems Research and Behavioral Science*, remains the premier journal in the field.

In setting out what happened to GST in later years, it is useful to follow a distinction made by Hammond (2003) between those pioneers who put an emphasis on discovering "isomorphisms," similarities at different levels of organization, and those who stress the importance of "emergence," the new properties that arise at higher levels of complexity. In the first camp are Gerard and Miller, whose somewhat "organismic" models can be seen as reductionist when applied to social systems. In the second, Hammond has von Bertalanffy, Rapoport, and Boulding, who are as interested in the new characteristics that emerge at different system levels as they are in the similarities. In placing von Bertalanffy in the second camp, she highlights his belief that

> … the most important characteristic of open, living systems was their capacity for creativity and self-transcendence.
>
> *(Hammond 2003, p. 133)*

In my view, von Bertalanffy's work can be read as giving succor to both camps, though I accept that Hammond provides a useful corrective by paying proper attention to the often neglected side of his contribution. The distinction she makes is crucial for Hammond because it illuminates the social implications of GST. Hammond's study began, as she states in the prologue, as a response to Lilienfeld's (1978) critique of systems theory. Lilienfeld argued that systems theory is a conservative ideology designed to serve an emerging technocratic élite. Hammond can see the force of this argument when applied to Gerard and Miller, who offer "totalising" visions, which lend themselves to the mission of social control. The argument is misplaced, she argues, in the case of von Bertalanffy, Rapoport, and especially, Boulding. Their work, she believes, offers a fundamentally different direction for systems thinking, one that is inclusive, participatory, and humanistic. It is no surprise to Hammond that Gerard and Miller were conservatives while von Bertalanffy, Rapoport, and Boulding were all more politically radical, opposing, for example, the Vietnam War. I will continue the story by looking at the different directions Miller and Boulding took GST.

5.4 Miller and the Search for Isomorphisms at Different System Levels

In the conclusion to *Living Systems*, the massive tome in which he pursues his vision, Miller writes:

> All nature is a continuum. The endless complexity of life is organized into patterns which repeat themselves at each level of system.
>
> *(1978, p. 1025)*

The book itself is an ambitious attempt to justify this conclusion and to unify scientific knowledge across the levels of biological and social systems.

Miller's theory of "living systems" proceeds empirically to describe "concrete" systems, which he is certain exist because we can actually observe their structures and processes. This can be contrasted with Parsons' reliance on abstract concepts, such as social role, to build his model of social systems. Living systems exist for Miller (and his wife Jessie) at eight hierarchical levels (Miller 1978; Miller and Miller 1990):

- Cell
- Organ
- Organism
- Group
- Organization
- Community
- Society
- Supranational Systems

At every level, living systems are structured in exactly the same way, possessing 20 "critical subsystems," which carry out specific processes that are essential for life (Miller 1978; Miller and Miller 1990). In brief, these subsystems allow the system to process matter–energy (ingestor, convertor, extruder, etc.) and transmit information (decider, channel and net, decoder, etc.). Miller regards living systems theory (LST) as a scientific theory from which testable hypotheses can be derived. *Living Systems* contains a list of 173 testable cross-level hypotheses and others have been added since. It also implies an approach to solving problems. Miller and Miller describe an appropriate methodology:

> It involves observing and measuring important relationships between inputs and outputs of the total system and identifying the structures that perform each of the [20] sub-system processes .... The flows of relevant matter, energy, and information through the system and the adjustment processes of subsystems and the total system are also examined. The status and function of the system are analyzed and compared with what is average or normal for that type of system. If the system is experiencing a disturbance in some steady state, an effort is made to discover the source of the strain and correct it.
>
> *(1995, pp. 25–26)*

There must surely be issues around determining what is "average or normal" for a particular type of system. Nevertheless, examples are given of apparently successful living systems applications in hospitals, a psychiatric ward, public schools, a public transportation system, the US Army, and IBM.

An even more ambitious route "toward a science of systems" has been charted by the biologist Len Troncale (2006). Whereas Miller sets out a framework on the basis of which isomorphies can be discovered, the 8 hierarchical levels and 20 critical subsystems, Troncale builds his GST upward from the actual processes he observes are essential in all successful systems. He has found over 100 such processes. These processes are then linked in a network, "the system of system's processes" (SSP), on the basis of postulated interactions between them:

> So the resulting network is proposed as an expression of all known interactions that are required for a system to survive sufficiently long for humans to recognize it.
>
> *(2006, p. 306)*

The focus on interactions as well as processes, Troncale suggests, provides an additional level of isomorphic comparison and greater descriptive detail to his theory. Further, while Miller only considers isomorphies pertaining to biological and social systems, and is content to explore them if they cross just a couple of levels, Troncale insists that systems processes and their interactions must be identifiable at all known or knowable hierarchical levels (including the physical) to qualify as isomorphies in his scheme.

In reflecting upon the obstacles blocking the progress and acceptance of GST, Troncale (2003) calls for more rigor: more precisely defined concepts, demonstrable in all disciplines, use of the full set of isomorphies, specific linkages between isomorphies, etc. What he does not contemplate is that this version of GST may simply be barking up the wrong tree. Hammond (2003, p. 184) summarizes the criticisms from the other camp. Both Boulding and Rapoport regard Miller's LST as too "organismic," with the same structure of processes replicated at each level. It cannot cope with the complexity at higher system levels where meaning and values become defining features. A concentration on the system-maintaining functions of individual biological organisms prevents sufficient attention being given to the different perspectives and interests that exist in social organizations and fails to deal adequately with conflict and change. It is simply confusing to regard socially organized aggregates as living entities. Society, Boulding argues, is best seen as an "ecosystem of interacting biological and human artifacts" (quoted in Hammond 2003, p. 184).

Despite these criticisms, another group of scholars (David Rousseau, Jennifer Wilby, Julie Billingham, and Stefan Blachfellner) came together in 2013 with a view to revitalizing the quest for a "general systemology." Their precise aim was

> ... to rekindle the enthusiasm for establishing a scientific general systems theory and operationalizing it as a transdisciplinary methodology for effective and efficient exploration, design, and theory building, and as a framework for visualizing the submerged land connecting our islands of knowledge.
>
> *(Rousseau et al. 2018, pp. ix–x)*

At the 2015 meeting of the ISSS, they launched a manifesto calling for a "general systems transdisciplinarity," which bridges the gap between physical, biological, and social systems and contributes to the greater good (Rousseau et al. 2016). This "general systemology" (Rousseau 2017a) will be built on the discovery of scientific systems principles, which, in the manner of "critical realism" (see Section 2.4), are seen as approximately capturing the systemic nature of objective reality and the way it generates the systems we actually observe in the world. A coherent set of such explanatory (not just descriptive) principles would form the core of a foundational GST. Rousseau accepts that little progress has been made to date on discovering principles of this type but believes that a careful analysis of how the more mature specialized sciences have evolved can provide a template GST can employ to build its own trans-disciplinary knowledge base. Once scientific systems principles are discovered, they can provide the basis for systems engineering to become a science-based technology capable of realizing optimal systemic interventions. Rousseau (2017b) has reported on three candidate systems principles: "conservation of properties," "universal interdependence," and "complexity dominance." The outcomes of the first five years of the groups' efforts are now available in a book (Rousseau et al. 2018). It remains unclear whether the three candidate principles, or any others that might be discovered, will prove relevant for managers confronted by organizational and societal issues. Given the track record of previous efforts at this type of GST, it seems, unfortunately, to be highly unlikely.

5.5 Boulding, Emergence and the Centrality of "The Image"

Of all the pioneers of GST, it is Boulding who pays most attention to the specific, unique characteristics of systems at different levels of complexity and to specifying an "appropriate epistemology" for each level. The argument is set out in his seminal 1956 paper, *General Systems Theory – The Skeleton of Science* (1968). According to Boulding, there are two complementary ways of developing GST. Either you attempt to construct a theory of general phenomena found in many different disciplines (the road taken by Miller, for example) or, as he preferred, you seek

> ... to arrange the empirical fields in a hierarchy of complexity of organization of their basic 'individual' or unit of behavior, and ... try to develop a level of abstraction appropriate to each.
>
> *(1968, p. 5)*

In pursuit of his preferred approach, Boulding produced a nine-level hierarchy of levels of real-world complexity, stretching from structures and frameworks, at the simplest level, to transcendental systems at the most complex. Modifications to the hierarchy have been proposed by von Bertalanffy (1971, pp. 26–27) and, in the light of the theory of autopoiesis, by Mingers (1997a), but it continues to be useful as the intuitive guide that Boulding intended. The hierarchy is summarized in Table 5.1.

Reviewing his hierarchy, Boulding notes that the characteristics of lower level systems can be found in higher level systems, for example, aspects of levels 1–6 in level 7 (people). Each level, however, presents emergent properties that cannot be understood simply in terms of the theoretical concepts employed successfully at lower levels – hence the need for new disciplines like biology, psychology, anthropology, and sociology at more complex system levels. This reminds us of the danger of reductionism; of employing a level of theoretical analysis below the level of complexity of the empirical phenomenon of interest. For example, it is wrong to treat an organization as though it is a machine or an organism, although it will share some of the characteristics of machines and organisms. Concluding his review, Boulding uses the hierarchy to point to gaps in

Table 5.1 A summary of Boulding's (1968) hierarchy of complexity.

1.	At level 1 are structures and frameworks, which exhibit static behavior and are studied by verbal or pictorial description in any discipline; an example being crystal structures
2.	At level 2 are clockworks, which exhibit predetermined motion and are studied by classical natural science; an example being the solar system
3.	At level 3 are control mechanisms, which exhibit closed-loop control and are studied by cybernetics; an example being a thermostat
4.	At level 4 are open systems, which exhibit structural self-maintenance and are studied by theories of metabolism; an example being a biological cell
5.	At level 5 are lower organisms, which have functional parts, exhibit blue-printed growth and reproduction, and are studied by botany; an example being a plant
6.	At level 6 are animals, which have a brain to guide behavior, are capable of learning, and are studied by zoology; an example being an elephant
7.	At level 7 are people who possess self-consciousness, know that they know, employ symbolic language, and are studied by biology and psychology; an example being any human being
8.	At level 8 are sociocultural systems, which are typified by the existence of roles, communications and the transmission of values, and are studied by history, sociology, anthropology, and behavioral science; an example being a nation
9.	At level 9 are transcendental systems, the home of "inescapable unknowables," which no scientific discipline can capture; an example being the idea of God

our knowledge, especially our lack of adequate systems models much above level 4. A key issue in understanding and predicting system behavior at higher levels of complexity is the intervention of "the image" into the chain of causality. As we ascend system levels, brains develop, which organize information intake into a knowledge structure or image. Behavior results from the structure and setting of the image rather than directly from some stimulus. Human images are highly complex and, furthermore, have a self-reflective quality; people not only know but know that they know. Despite these difficulties, Boulding notes a surprising twist. An "inside track" exists to a kind of knowledge relevant at the people and sociocultural levels:

> Nevertheless as we move towards the human and societal level a curious thing happens: the fact that we have, as it were, an inside track, and that we ourselves are the systems which we are studying, enables us to utilize systems which we do not really understand.
>
> *(1968, p. 9)*

This is very important. In contrast to other pioneers, such as Gerard and Miller, Boulding is arguing that GST cannot just be about extending the reach of the traditional scientific approach to higher levels of complexity. At these levels, a very different kind of knowledge is necessary and must be sought using an approach more in tune with phenomenology and pragmatism (see Chapter 1) and with interpretive social science (see Section 4.3). This insight had a decisive influence on Vickers, who knew Boulding well, and through him on the development of "soft systems methodology" (see Chapter 16). It warrants more consideration here.

In the final two weeks of a year at CASBS (1954–1955), Boulding dictated the draft of his most influential book *The Image* (Hammond 2003, p. 217). In *The Image* (1961), he rejects any correspondence theory of truth and charts, predating Maturana and Varela's similar account of living systems as cognitive systems (see Chapter 3), the growing sophistication of the images systems have of their environments and, eventually, of themselves as well. Even at the control system level, a thermostat has an image consisting of facts about the temperature of the outside world and a value for an ideal temperature. Simple organisms have knowledge in the form of structured images of their environments, which can change with experience. Animals possess brains fed by specialized sense receptors, such as eyes and ears, show awareness, exhibit emotion, and are capable of sophisticated learning. At the people level, the human brain organizes the information it receives into extremely rich and complex images, which are massively enhanced by language. These images consist of facts, or what is believed to be true about the world, which are inextricably linked to values applying standards of good and bad. They have conscious, unconscious, and subconscious elements. Whatever is made available biologically is built upon as a result of experiences including, particularly, socialization in the family and at school. Our images constitute our subjective knowledge and govern our behavior:

> I have suggested that one of the basic theorems of the theory of the image is that it is the image which in fact determines what might be called the current

behavior of any organism .... The image acts as a field. The behavior consists in gravitating toward the most highly valued part of the field.

(Boulding 1961, p. 115)

Images are generally resistant to change and ignore messages that do not conform to their internal settings. Sometimes, however, they do react and can alter in an incremental or even revolutionary manner. Humans can talk about and share their images and, in the symbolic universe they create, reflect upon what is and what might be. This "brings the actor into the act" (Boulding 1961, p. 175) capable, in part at least, of molding the future. Common images or "universes of discourse" come into being and make society, which itself, in turn, shapes individual images. By metaphor and analogy, we can speak of organizations and society as a whole as having images or, better, an inventory of images some of which will conflict. With this argument for the importance and generality of "the image" in place, Boulding declares that we need a science of images, *Eiconics.* This can support and help to develop disciplines like psychology (the Gestalt school) and social psychology (Mead), which have taken an *eiconic* direction, and give a wake-up call to those, such as economics, which lack any theory of the image and will not progress until they get one. The entire book, of course, is dedicated to arguing that GST can itself only develop if it has "the image" at the center of its concerns.

5.6 The Influence of General Systems Theory

Boulding notes that the greatest impact of GST was in business schools (Hammond 2003, p. 249) where its insights were held to be directly relevant to the management of organizations. By the 1960s, it had become thoroughly absorbed into organization theory, with the rich armory of concepts surrounding the open-system notion complementing those of the structural-functionalism championed by Selznick and Parsons (see Section 4.2). Katz and Kahn's 1966 *The Social Psychology of Organizations* (1978, 2nd edn) is the classic expression of this new development. Katz and Kahn begin by pointing out the advantages of their theory. Traditional approaches take a closed-system perspective on organizations. It is much better to start by looking at organizations as open systems. Organizations should be seen as entities in close interrelationship with their environments, taking in inputs and transforming them into outputs. These outputs, in the form of products, can provide the money for new inputs, so that the cycle can begin again. The main purpose is to maintain a steady state and to survive. Reviewing and building on von Bertalanffy's findings, Katz and Kahn have it that 10 characteristics define all open systems (including, of course, organizations):

- The importation of energy from the external environment
- The throughput and transformation of the input in the system
- The output, which is exported to the environment
- *Systems as cycles of events*: the output furnishes new sources of energy for the input so the cycle can start again
- *Negative entropy*: open systems "live" off their environments, acquiring more energy than they spend

- *Information input, negative feedback, and a coding process*: systems selectively gather information about their environments and also about their own activities (so they can take corrective action)
- *The steady state and dynamic homeostasis*: despite continuous inflow and export of energy, the character of the system remains the same
- *Differentiation*: open systems move in the direction of differentiation and structure elaboration (e.g. greater specialization of functions)
- Integration and co-ordination to ensure unified functioning
- Equifinality

Other aspects of the Katz and Kahn model closely follow Parsons' thinking. For example, five subsystems are recognized that must be present to meet the organization's functional needs – production, supportive, maintenance, adaptive, and managerial.

Following on from Katz and Kahn's work, countless books were published looking at management from an open system's perspective and, in the guise of "contingency theory" (see Kast and Rosenzweig 1981), this became the dominant paradigm in organization theory in the 1970s and beyond. An organization is viewed by contingency theory as a center of mutual influence and interaction between the subsystems essential for its survival (goal, human, technical, and managerial), the variables of size and structure, and its environment. Contingency theory postulates that the effective performance of an organization is contingent upon getting an organizational structure appropriate to size, the demands of the subsystems, and the environment with which it and they interact. The methodology employed in research is "scientific." Empirical evidence is collected across large numbers of organizations in the hope of finding predictive correlations between the key contingency variables, e.g. an organization employing a particular technology will always perform well if it adopts a certain set of structural arrangements. Some interesting conclusions have been drawn, for example, Burns and Stalker's (1961) correlation of "mechanistic" structures with success in stable environments and looser, "organic" structures with success in turbulent environments. Overall, however, the results have been disappointing, the approach has been severely criticized, usually for lacking an "action" dimension, and it is now little used (Jackson 2000, pp. 109–117). Sociotechnical systems thinking, which also draws upon open-system theory, but marries it with action research, has had a longer lasting and more significant impact on systems thinking applied to management, as will be seen in Chapter 12.

5.7 Conclusion

Comparing his GST to cybernetics, the topic of Chapter 6, von Bertalanffy (1971, p. 21) sees many parallels but regards his approach as broader and more relevant to the biological and social levels of complexity. Cybernetics, he argues, offers an extension rather than a replacement for the mechanistic view. In doing so, it presents a "special case" of GST. In open systems, the primary source of order comes from the dynamic interplay of processes. The control mechanisms that cybernetics studies, based on fixed structural arrangements of the feedback type, are actually only secondary regulations superimposed on the primary processes (von Bertalanffy 1971, pp. 15, 43, 48). von Bertalanffy does, however, accept that Ashby's deductive approach to cybernetics, starting from the set of all conceivable systems, offers a legitimate alternative to his own empirical approach (1971, p. 101). An excellent account of the relationship between von Bertalanffy's thinking and all the different branches of cybernetics is provided by Drack and Pouvreau (2015). von Bertalanffy was writing too early to take account of the excitement that later surrounded complexity theory (see Chapter 7). He does, however, comment favorably on Prigogine's early work on nonequilibrium thermodynamics, which seeks to introduce more rigor and refinement to the theory of open systems.

6

Cybernetics

Here I need only mention the fact that cybernetics is likely to reveal a great number of interesting and suggestive parallelisms between machine and brain and society. And it can provide the common language by which discoveries in one branch can readily be made use of in the others

(Ashby 1956, p. 4)

6.1   Introduction

Norbert Wiener defined cybernetics as the study of "control and communication in the animal and the machine" and used this definition as the subtitle of his book *Cybernetics* (1948). It provides the clues as to its origins in biology and in engineering. As soon as Wiener begins to consider regulation in biological systems, he turns to Cannon's research on *homeostasis*. The same is true of Ashby in *An Introduction to Cybernetics* (1956), that other foundational text of the trans-discipline. Cannon's work in biology was touched upon in Section 3.2. We will now look at the contribution of engineering to the development of cybernetics by considering two key concepts developed in that discipline as it sought to enhance its ability to design and operate complex machines.

The first of these concepts is "negative feedback control." Otto Mayr (1975) has charted the historical origins of feedback devices applied in water clocks, thermostats, and windmills. Ktesibios of Alexandria, in the third century BCE, used a float valve, similar to the mechanism employed in modern flush toilets, to enable his water clock to keep the time. The first automatic temperature regulators were invented in England, by the Dutch engineer Cornelis Drebbel in the 1620s, and applied to a chemical laboratory furnace and a chicken incubator. There were also several feedback mechanisms devised, in the eighteenth century, by engineers seeking to control the performance of windmills. Their work culminated in the invention of centrifugal pendulums, which could regulate the speed of windmills by altering the sail area. It was a centrifugal "governor" of this type that James Watt was able to adapt for steam engines, in 1788, and which became the most famous feedback control system of all. The purpose of Watt's governor was to keep the engine working at constant speed whatever external changes occurred in load or steam pressure. The centrifugal fly-ball governor, driven by engine speed, was connected to a steam inlet valve, which increased the steam supply with

Critical Systems Thinking and the Management of Complexity, First Edition. Michael C. Jackson.
© 2019 John Wiley & Sons Ltd. Published 2019 by John Wiley & Sons Ltd.

falling speed and decreased it with rising speed. If, for example, the speed dropped, because the load on the engine was increasing, the governor would function to open the valve further so that just enough steam was supplied to maintain a constant speed.

In 1868, James Clerk Maxwell wrote his paper "On Governors," which began:

> A Governor is a part of a machine by means of which the velocity of the machine is kept near uniform, notwithstanding variations in the driving-power or the resistance.

He proceeded to give the first theoretical account of how steam engine governors operate, using differential equations to detail the workings of the fly-ball control system. This might have marked the beginning of mathematical control theory but, for reasons explored by Bennett (2002), the paper was largely ignored until Norbert Wiener drew attention to it in *Cybernetics*. Since then interest in control theory has grown exponentially and control engineering has broken away from its origins in mechanical engineering to establish itself as another subdiscipline of engineering.

The second key concept is the "black box technique," which comes from electrical engineering. The term "black box" was first used in the mid-1940s in electronic circuit theory and marked a move away from network analysis toward network synthesis as a means of understanding the behavior of electronic circuits. Network analysis requires the engineer to predict behavior by examining all the components of the network. In network synthesis, by contrast, effort is concentrated on getting the desired response from the system by manipulating its inputs. The engineer regards the network as a "black box" that you do not have to enter in order to get the information required for control purposes. For example, an engineer can determine what a sealed black box does by applying voltages, shocks, or other disturbances to its input terminals and observing what occurs at the output terminals. To understand the behavior of the black box, it is enough to manipulate its inputs and record the corresponding outputs. The early cyberneticians saw themselves as surrounded by extremely complex systems, like brains and firms, which they could not analyze in precise detail but whose behavior they wished to control. They readily adopted the available terminology. Systems in which the elements and processes could be easily observed and defined were seen as "transparent" and regarded as "white" or "transparent" boxes. Cybernetics was concerned with black boxes, systems that could not be examined and described in any precise fashion. The black box technique provided a means of gaining enough understanding of how these systems behaved, in order to control them, even if it was impossible to grasp the exact nature of the causal links leading to that behavior. Today, the black box technique is most frequently heard of in connection with the higher level testing of software applications, avoiding having to look into the internal workings; and in marketing, enabling consumer behavior to be managed without looking into the processes at work in the human mind.

In seeking to elucidate the development and nature of cybernetics itself, it is useful to distinguish three strands of the trans-discipline with very different philosophical underpinnings. For two strands, I employ the usual labels of "first-order cybernetics" and "second-order cybernetics," which are too well embedded to do anything about. But, much influenced by Pickering's excellent book *The Cybernetic Brain* (2010), I add a third category: "British cybernetics." This does justice to a group of British cyberneticians whose work, primarily concerned with "the adaptive brain," actually offers the greatest insight to managers – as will be seen in Chapter 13 when we study Beer's "viable system model."

6.2 First-Order Cybernetics

In 1943, Rosenblueth, Wiener, and Bigelow published the paper that is often regarded as the founding document of cybernetics (Rosenblueth et al. 1968). Its findings derived from two research projects, with Wiener involved in both. The first with the physiologist Rosenblueth, a student of Cannon, concerned the role of feedback in human and animal physiology. The second, with the mathematician and electrical engineer Bigelow, was a wartime project to improve the accuracy of anti-aircraft weaponry by making use of feedback revealing the difference between the predicted pattern of motion of an airplane and its actual motion. On the basis of their work, the three authors had

> ... become aware of the essential unity of the set of problems centering about communication, control, and statistical mechanics, whether in the machine or in living tissue.
>
> *(Wiener 1948, p. 19)*

The 1943 paper discusses how systems are able to engage in purposeful, goal-seeking behavior in the face of changes in the environment. The authors argue that: "All purposeful behavior may be considered to require negative feedback" (Rosenblueth et al. 1968, p. 222). In other words, if a goal is to be attained, it is necessary for a system to make use of signals about deviations from the goal in order to direct its behavior. This is true both for "servomechanisms," machines with in-built purposeful behavior, and a living organism such as a cat, which seeks to catch a mouse by constantly monitoring its own position in relation to the mouse and adjusting its behavior accordingly. The authors argue that their analysis re-establishes "teleology" as a scientific concept. It had become discredited because it seemed to imply a cause subsequent in time to a given effect. If instead it is given the limited meaning of purposeful behavior controlled by feedback, then it no longer offends against scientific principles. Also mentioned in the article is "positive feedback," which adds to input signals rather than correcting them. The paper has been subject to much criticism over the years. Suffice it to say that it aids clarity to make a distinction between the "purposive" behavior of a system that has a goal built into it by a designer or ascribed to it by an observer (e.g. an organism seeking survival) and self-conscious "purposeful" behavior in the Kantian sense. It is best, with Kant, Ackoff (1999a, p. 52), and Checkland (1981, pp. 316–317), to reserve the term "purposeful" for actions willed by human beings.

World War II had provided the impetus for scientists to work together on applied, interdisciplinary problems and the Rosenblueth et al. paper, introducing concepts apparently applicable across the physical, life, and social sciences, appeared at the most opportune of times. Various discussions took place among interested parties until, in 1946, the neurophysiologist Warren McCulloch made arrangements with the Josiah Macy foundation for an organized series of meetings devoted to the problems of feedback. These meetings, chaired by McCulloch, went at first under the title of the

"Conference for Circular Causal and Feedback Mechanisms in Biological and Social Systems." Ten conferences were held, between 1946 and 1953, bringing together in intense and frequently bad-tempered discussions luminaries such as McCulloch, Wiener (the dominant voice), Ashby, Bigelow, Shannon, von Foerster, the psychologist Kurt Lewin, and the anthropologists Gregory Bateson and Margaret Mead. But they still did not have a name for their area of study and Wiener was becoming concerned that this was preventing progress toward a unified terminology and literature around the issues that interested them. In 1947, therefore, he introduced the term "Cybernetics" to refer to the entire area of work surrounding control and communication theory. As was mentioned, in Section 1.1, this designation was derived from the ancient Greek word *kybernetes*, meaning the "art of steersmanship," which had been employed by Plato to refer both to the piloting of a vessel and to steering the "ship of state." From *kybernetes* came the Latin *gubernator* and hence the English "governor," a word used for political leaders and also applied to Watt's automatic regulating device. The term *kybernetes* had been used in the nineteenth century by Ampere in the context of political science, but this was unknown to Wiener. In any case, in 1947, it seemed wholly appropriate and met with almost immediate approval. In 1949, at the suggestion of Heinz von Foerster, the title of the Macy events was changed to the "Conference on Cybernetics."

In 1948, Wiener's *Cybernetics* appeared, establishing the famous definition of the field as the science of control and communication in the animal and the machine; a definition that was already too narrow as Wiener and others were extending its insights to human concerns. Cybernetics was a true trans-disciplinary science, Wiener argued, because it dealt with general laws that governed control processes whatever the nature of the system under governance. In understanding control, whether in the mechanical, biological, or human realm, the idea of negative feedback is crucial. A negative feedback control system is characterized by a closed-loop structure. It operates by the continuous feedback of information about the output of the system. This output is compared with some predetermined goal, and if the system is not achieving its goal, then the margin of error (the negative feedback) becomes the basis for adjustments to the system designed to bring it closer to realizing the goal. A simple closed-loop feedback system is represented in Figure 6.1.

It seems that, for the negative feedback system to maintain control, four elements are required:

- A desired goal, which is conveyed to the comparator from outside the system
- A sensor (a means of sensing the current state of the system)
- A comparator, which compares the current state and the desired outcome
- An activator (a decision-making element that responds to any discrepancies discovered by the comparator in such a way as to bring the system back toward its goal)

This kind of control system is extremely effective since any movement away from the goal automatically sets in motion changes aimed at bringing the system back on course. A domestic thermostat, for example, measures the temperature of a room, compares it to a previously prescribed temperature, and if there is a difference, either turns on or off the boiler until (within its capacities) the desired temperature is reached. Even if it gets hotter or colder outside, the temperature of the room is maintained. Homeostasis in

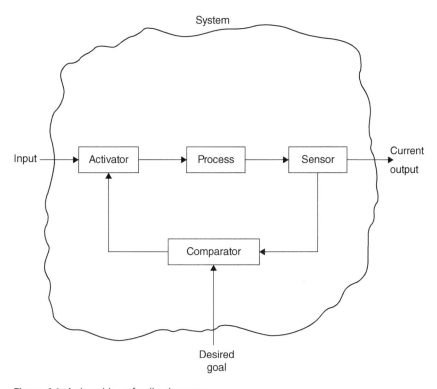

Figure 6.1 A closed-loop feedback system.

the body is achieved by negative feedback, so that the body is able to maintain stability in spite of extensive shifts in outside circumstances. An example is the homeostatic process by which humans maintain their body heat at around 37 °C. Picking up a pen from a desk, constantly registering the discrepancy between the position of the hand and the pen, involves negative feedback. The second term in Wiener's definition of cybernetics, "communication," is significant because control depends on the communication of information. If control is to be achieved, then communications must flow within the system and between the system and its environment. Thus the theory of control can be seen as part of the theory of messages. In developing this aspect of their work, cyberneticians were able to draw on the 1949 volume *The Mathematical Theory of Communication*, by the communications engineers Shannon and Weaver. Finally, Wiener warns that a negative feedback control system must operate rapidly and continuously. If there are delays, then oscillations or "hunting" can occur. We have all experienced jumping around in a shower as we over-correct the flows of hot and cold water because our original adjustments have, from our perspective, taken too long to bring about the desired effect.

It should be noted that Figure 6.1 represents only a simple, first-order negative feedback system (Schoderbeck et al. 1985). More sophisticated, second-order systems are capable of considering and choosing between various methods designed to bring the system back toward its goal. Still more sophisticated, third-order systems can change the goal state itself in response to the feedback received. In this case, the goal is determined inside the system rather than externally as in Figure 6.1. Finally, successful management usually relies on "feedforward" as well as feedback information. Attempts are made to predict disturbances before they actually affect the organization so that action can be taken in advance of a deviation actually being registered.

When the word feedback was first used by radio engineers, in 1920, to describe "undesirable parasitic connections in a wireless amplifier that resulted in local oscillations," it was positive feedback to which they were referring (Bennett 2002, p. 33). As we saw, Rosenblueth et al. noted the existence of positive feedback but, thereafter, it tended to be neglected by the early cyberneticians. The positive feedback process is one where the output is fed back to the input but, rather than reducing any divergence from the goal, it produces a further movement in the direction in which the output is already moving. In a seminal paper, written in 1963, Maruyama (1968) berated cyberneticians for focusing on deviation–counteracting relationships and all but ignoring mutual causal processes which are deviation amplifying:

> Such systems are ubiquitous: accumulation of capital in industry, evolution of living organisms, the rise of cultures of various types, interpersonal processes which produce mental illness, international conflicts, and the processes that are loosely termed as 'vicious circles' and 'compound interests'; in short, all processes of mutual causal relationships that amplify an insignificant or accidental initial kick, build up deviation and diverge from the initial condition.
>
> *(Maruyama 1968, p. 304)*

Maruyama argues that both deviation–counteracting processes, known as "morphostasis," and deviation–amplifying processes, called "morphogenesis," are well known. However, in cybernetics, study of the first type has been predominant. He therefore recommends calling this "the first cybernetics" and giving the name "the second cybernetics" (not to be confused with "second-order cybernetics") to the newer study of deviation–amplifying relationships. He goes on to examine the importance of the relationships between positive and negative feedback loops, suggesting

> ... a society or an organism contains many deviation-amplifying loops as well as deviation-counteracting loops, and an understanding of a society or an organism cannot be attained without studying both types of loops and the relationships between them.
>
> *(Maruyama 1968, p. 312)*

It is enough to record at this point that the deviation–amplifying form of feedback played just as important a role in later systems thinking as it did in the careers of rock musicians such as Jimi Hendrix. Indeed, questions about the behavior of systems containing many interacting loops were already being asked and answered in the related field of "system dynamics," as we shall see in Chapter 11.

Wiener believed that his cybernetics constituted a significant break from the Newtonian world view and, certainly, his cybernetic perspective differs markedly from Newton's clockwork representation of the solar system. The feedback loops he studies operate in a circular fashion such that each element in a system has a causal effect on the next and the final element feeds back to impact the first. This transmission of information allows organization to arise through a process of self-regulation:

> In control and communication we are always fighting nature's tendency to degrade the organized and to destroy the meaningful; the tendency for entropy to increase.
>
> *(Wiener, quoted in Hammond 2003, p. 64)*

The type of cybernetics described so far is often referred to as "first-order cybernetics." First-order cybernetics (see Umpleby 2016) is the "cybernetics of observed systems"; an engineering approach that seeks to provide objective knowledge about feedback loops and concentrates on promoting our ability to exercise control. This characterization of Wiener's work is somewhat unfair as he devoted a book to human beings, the limits imposed on them by their nervous systems, their capacity to learn, and how cybernetics can be used to benefit society (Wiener 1950). Nevertheless, it is true that Wiener's writings are best seen as extending rather than replacing the traditional model of science. They join von Bertalanffy's theory of "open systems" (and Darwin's theory of natural selection) in helping to provide a nonmystical explanation of purposive behavior in complex systems, this time both mechanical and living. They are partial, at best, as an explanation of human, purposeful behavior. By the mid-1970s, those of a technical persuasion, interested in pursuing cybernetics as a form of engineering, began to disperse into more narrowly defined fields such as control engineering, computer science, information theory, management science, robotics, and artificial intelligence. Cybernetics itself was left to more radical thinkers interested in exploring the "cybernetics of observing systems," i.e. "second-order cybernetics."

6.3 British Cybernetics

British cyberneticians, particularly Ashby, Bateson, Pask, and Beer, helped to establish both first- and second-order cybernetics and were heavily influenced by these traditions. Nevertheless, British cybernetics had its own identity nurtured in the Ratio Club, founded in 1949 as a small informal dining circle of interested scientists (Alan Turing was a member), and at the Namur conferences on cybernetics, which began in 1956. It also developed a distinctive philosophy, reflected in its enthusiastic embrace of what Pickering (2010) calls the "performative idiom." This came about because the founders of cybernetics in Britain were primarily interested in the human brain and saw it as working in a particular way. The brain, for them, is not primarily a cognitive organ that seeks true representations of an objective reality. Rather it performs an active function, linking our sensory inputs to our motor organs. It is also intrinsically subjective, caring only about things that relate to its own purposes. It is an "embodied organ" that has evolved to help human beings adapt and learn about situations that they have not encountered before. It enables us to get on with a world that we cannot know. Knowledge, therefore, is not something fixed. It is always in the making, emerging from interactions between systems and forever leaving new things to be discovered. This appreciation of how the brain functions provides insight into how it should be studied. It is impossible to gain exhaustive knowledge of this most complex of systems in any traditional scientific sense. Instead, you can only observe it in interaction with other systems and see how it performs. For Pickering cybernetics, built upon this type of model, offers a kind of "ontological theater," an exciting exploration of practice in its own right. Rigid, controlled experiments are abandoned in favor of pursuing a "dance of agency" in a manner that is entirely "nonmodern." The injunction is to stage performative events, see where they take you and enjoy the ride. Pickering recognizes elements of "pragmatism" (see Section 1.4) in this emphasis on situated rather than "true" knowledge but feels that in pragmatism practice is used to decide between knowledge claims rather than being pursued as a primary endeavor. A more helpful comparison can be made by drawing upon some work of Heidegger. Heidegger recognizes a dominant, modern ontology in which science provides the knowledge through which technology "enframes" the world and puts it at our disposal. By contrast, Pickering (2009) argues, the nonmodern paradigm of British cybernetics offers an alternative ontology of "unknowability" and "becoming" and encourages us to engage in "revealing" what the world has to offer. This is considerably less dangerous than the project of "enframing," which convinces us that we can command and control social and ecological systems. As Pickering continually emphasizes, the philosophy and personality of British cybernetics happened to chime remarkably well with the spirit of the "swinging sixties" when it was, perhaps, at its most influential. We are now in a position (drawing heavily on Pickering 2010) to see how the ideas developed and came to fruition in the work of Walter, Ashby, Beer, Pask, Bateson, and Laing.

Grey Walter was an American-born neurophysiologist who was brought to England early in life and carried out his main research at the Burden Neurological Institute in Bristol from 1939 to 1970. A leading figure in electroencephalogram (EEG) research, his contributions to cybernetics began in 1948 with the construction of his first electromechanical robot "tortoise." The hope was that the tortoise would mirror the adaptive capacity exhibited by the brain and so conclusions could be drawn about how the brain was structured. Increasingly sophisticated tortoises were built that were capable of scanning, exploring, and reacting to their environments. Walter had, therefore, a simple model of the adaptive brain, arrived at not by working up from the properties of individual neurons but by building a machine and seeing what it did. A further finding was that even the minimal interconnections between the parts of his tortoises led to complex, emergent behavior. Tortoises equipped with a light on their nose, and responsive to other light sources, would begin "twittering" and "jigging" if they observed themselves in a mirror and engaged in a kind of "mating dance" when they came across others of their kind. None of this could be predicted in advance. You could only learn about it by watching the performance. Like modern, automatic vacuum cleaners and grass cutters, tortoises returned to their source of electricity when their power ran low.

Walter's other great contribution to cybernetics, and to the 1960s, was his research into "flicker" – the effects on the brain produced by flickering lights. Pursuing his EEG research, in 1945, Walter discovered that staring with closed eyes at an electronic stroboscope, with the light flickering near the alpha frequency of the brain, induced colorful, dream-like hallucinations. This was evidence that the brain did not simply "represent" the outside world. It also provided Walter with another means of exploring the behavior of the brain without digging down to its fundamental units. He proceeded to experiment and found that the stroboscope could evoke responses in the brains of normal people that were identical to those found in epileptic patients. A more complicated version of the stroboscope fed back the emergent brainwaves to the strobe altering its behavior and so on round the loop. This "feedback flicker" was the first manifestation of the "dance of agency" that Pickering sees as central to the mature form of British cybernetics. The strobe investigated the brain's possibilities and the brain explored the strobe's potential as they interacted in a decentred combination. Walter's popular 1953 book *The Living Brain* described his flicker experiences and related them to the altered brain states achieved by Eastern mystics. The book was eagerly read by the writer William Burroughs, guru to the beat generation, who persuaded an acquaintance to make a stroboscope suitable for private use. This became known as the "dream machine" (Ter Meulen et al. 2009). Encouraged by Burroughs, the beat poet allen Ginsberg first took LSD when hooked up to a feedback controlled flicker device. Timothy Leary discussed flicker as another route to LSD-type experiences. Strobe lighting became a pervasive backdrop to the psychedelic sixties.

Walter's even more influential contemporary in the first phase of British cybernetics was W Ross Ashby. Ashby was a psychiatrist who spent his career working in UK mental institutions (including one unfortunate spell as Walter's boss at the Burden) until, in 1961, he joined von Foerster at the University of Illinois. His cybernetics grew out of his professional interest in understanding what went wrong in the pathological brain. His insight was that pathological brains somehow lose the capacity to adapt to their environments. To try to discover how this could happen, he decided to build a machine that would model the behavior of a normal brain with the usual ability to learn and therefore adapt. Ashby built his first "homeostat," as his machine was called, in 1948, demonstrated it at the 9th Macy Conference in 1952, and published a book about it, *Design for a Brain*, in the same year. The homeostat was an electromechanical device constructed from four surplus RAF bomb control units linked in dynamic feedback interrelations. It was, in Wiener's words (quoted in Pickering 2010, p. 93), "one of the great philosophical contributions of the present day." The machine proved to be "ultrastable" in that it was capable of finding and maintaining a state of dynamic equilibrium with its environment. It responded to fluctuations in its environment by conducting an autonomous trial and error search of possible configurations before eventually settling down in a field comfortable for its own critical variables and appropriate to the environment. This ability to "self-organize," and discover new states of equilibrium, was far beyond the capacity of Walter's tortoises, which simply adapted to their spatial environments. In Pickering's words, the homeostat confronted with other homeostats in its environment

> … stages for us a vision of the world in which fluid and dynamic entities evolve together in a decentered fashion, exploring each other's properties in a performative back-and-forth dance of agency.
>
> *(2010, p. 106)*

Ashby soon discovered, as had Walter, that his simple machines could give rise spontaneously to highly complex behavior. It was impossible to predict what combinations of homeostats would end up doing. There was no point trying to gain knowledge of the elementary parts and then using scientific design principles to get some predefined output. You just had to put the parts together and see what happened. This is the essence of what Pickering describes as the ontology of "unknowability" and "becoming."

In a second, highly influential book, *An Introduction to Cybernetics* (1956), Ashby's ambitions escalate as he seeks to set out a program for the future development of the whole trans-discipline of cybernetics. Science, he argues, limits itself to the domain of simple, reducible systems where its methods can cope. Cybernetics comes into its own when confronted with complex systems where the parts are closely interlinked and a change in one part immediately causes alterations in others, perhaps a great many. Cybernetics, indeed, studies complexity in its own right and

> ... offers the hope of providing effective methods for the study, and control, of systems that are intrinsically extremely complex.
>
> *(Ashby 1956, pp. 5–6)*

To be successful, it needs to become a theory of all possible "machines," whether electronic, mechanical, neural, or economic, i.e. not restricted to mechanical systems with Newtonian dynamics. A "framework" is required that can order, relate, and understand all the individual machines that emerge after regulation and control has exercised constraint on what is theoretically possible. This is another, rather abstract way of pursuing general systems theory, as von Bertalanffy (1971, p. 101) conceded, although preferring himself to start from actually existing systems. It is then necessary to find out all the possible behaviors individual systems can produce and, drawing upon our knowledge of the outstanding regulative powers of the brain, enumerate principles for designing and controlling all forms of complex machine. Ashby deals with three principles at some length – the black box technique, negative feedback, and the "law of requisite variety." We have considered negative feedback previously and so will concentrate on the other two here.

As we saw, the concept of the "black box" arose in electrical engineering to describe a system whose internal mechanisms cannot be examined to reveal what is producing its behavior. Ashby takes the view that black boxes are a pervasive feature of all disciplines, for example, when a clinical psychologist examines a brain-damaged patient. In fact, he argues, "the theory of Black boxes is practically co-extensive with that of everyday life" (Ashby 1956, p. 110). Most of us are content to treat our cars as black boxes, happy to get something close to the output we want by manipulating a couple of pedals and the steering wheel. A mechanic digs deeper but remains content to treat the parts of the engine as black boxes and does not concern herself with the interatomic forces holding the particles of metal together. We necessarily use black boxes as part of our performative engagement with the world. With exceedingly complex systems, like brains and economies, our knowledge is clearly incomplete as is demonstrated by the fact that, from our point of view, they exhibit emergent properties. We have to treat them as black boxes and the cybernetician simply asks that we acknowledge we are only studying selected aspects of them and play close attention to the way we get our information about them. According to Ashby, the way *not* to proceed in approaching an exceedingly complex system is by analysis. If we take a complex system apart, we find that we cannot reassemble it in a way that produces the same pattern of behavior. Instead of analysis, therefore, the black box technique of input manipulation and output classification should be employed. Initially, if nothing is known about the box, random variations of input will be as good as any. As regularities become established, a more directed program of research can be conducted. The aim is to record, over a long period of time, the sequence of input and output states. By this procedure, an experimenter can infer the constraints under which the system operates and build a useful, if partial, model of it. There are, however, problems with the black box technique because a particular experiment will often change a system to such an extent that it cannot be returned to its original state for further experiments.

Ashby defines "variety" as the number of distinguishable states a system can exhibit. It is, therefore, a measure of complexity. It is also a subjective concept, depending on the observer. A particular football team's variety will be much greater to the manager of an opposing team than to someone assessing it for a draw on the football pools. The problem for would be regulators, as Ashby's "law of requisite variety" has it, is that only "variety can destroy variety." To keep a system stable, the variety of the control mechanism must be equal to or greater than the variety of the system to be controlled. So, if a machine has 20 ways of breaking down, the regulator needs to be able to match this with 20 different responses. Sometimes exhibiting requisite variety is easy enough. If I am engaged in a game of noughts and crosses (tic-tac-toe) and am reasonably skilled, I can always exhibit enough variety to prevent my opponent from winning. But what if we are faced with systems exhibiting apparently massive variety? The answer is that we must either reduce the variety of the system we are confronting (variety reduction) or increase our own variety (variety amplification). This process of balancing varieties is known as "variety engineering" and will be discussed in relation to Beer's "viable system model" in Chapter 13. An example might be useful now, however. If I am manager of a relatively low-variety football team, such as Hull City, that is facing a high-variety team, like Barcelona (our aspirations have risen recently), and I want to win, I have to engage in variety engineering. I must amplify the variety of the Tigers, perhaps by employing improved tactics or by entering the transfer market. Alternatively, I can reduce the variety of Barcelona by allocating someone to take their best player out of the game or by getting information about their pattern of play, thereby making it more predictable. If, as Conant and Ashby (1970) argue, "every good regulator of a system must be a model of that system," then success will depend on having a useful "black box" model of the way Barcelona plays.

Pickering (2010) regards Walter and Ashby as occupying a half-way house in the development from first-order cybernetics to the fully mature form of British cybernetics found in the work of, for example, Bateson and Pask. Sometimes they talk as if they are using their models to build a scientific picture of the brain, to reveal what is inside the black box, rather than just employing them in performative engagement to find out about some things the brain can do. Certainly, they had enough confidence in their models to endorse what we would now regard as fairly barbaric psychiatric practices. Electro-convulsive therapy (ECT) and lobotomy were encouraged, employing analogies that suggested it was akin to applying a drastic shock to a machine to make it responsive again or to severing a few electrical connections. The less said the better, especially given the primitive nature of "tortoises" and "homeostats" as models. These could, after all, only "learn" on the basis of a severely limited variety pool.

While Walter and Ashby were occupied building machine models of the brain, Stafford Beer was busy searching for some equivalent to the adaptive brain to use in his "management cybernetics," first outlined in 1959 in *Cybernetics and Management* (Beer 1967). He could employ the usual principles of cybernetics – feedback, black box technique, variety engineering – to sketch out how a "cybernetic factory" might be organized but doubted that the mechanical machines so far used as analogues for the brain could ever become adaptive enough to serve a firm operating in a rapidly changing environment. They were always likely to be more deterministic and less probabilistic than a biological model would suggest is necessary:

> It is this kind of thinking which has led me to make an almost unbounded survey of naturally occurring systems in search of materials for the construction of cybernetic machines .... Any colony of living creatures, for example, which it is possible to affect by some input analogous to an industrial input, which then behaves as a Black Box for resolving the environmental disturbance created by this input and which finally exhibits some other mode of behavior which is capable of interpretation as an output, would be potentially a cybernetic machine. It would have the advantages of being itself a living system, capable of regeneration, growth and adaptation.
>
> *(Beer 1967, p. 162)*

Beer goes on to discuss colonies of *Daphnia*, tiny animals found in ponds, and a kind of "fungoid" machine he was developing with Pask as possible regulators for his cybernetic factory. Pickering (2009) describes another experiment in which Beer sought to test whether an entire pond ecosystem could act as adaptive brain for a factory. Ponds live in a state of dynamic equilibrium with their environments. A factory might be connected to a pond so that if the factory became unstable it impacted on the pond's equilibrium, the pond would naturally reconfigure to a new equilibrium and could signal to the factory how it too should respond to changing circumstances. There was no need for human intervention. A problem arose, however, with the connections. It proved impossible to design signals from the factory that interested ponds. We will deal later (Chapters 13 and 17) with other, more successful attempts by Beer to provide organizations with the ability to remain viable in turbulent environments. For the moment it is worth standing back, with Pickering (2009), and admiring the sheer *chutzpah* of such experiments. Instead of pursuing the long scientific detour of trying to build a representational model of how a factory behaves and using that to manage it, Beer simply borrows an existing adaptive system from the natural world as a ready-made regulator. Pickering links this confidence that nature can help to Beer's *hylozoism*; the belief that living and nonliving things obey a single set of laws and are essentially the same except, perhaps, in their degree of organization. This is no doubt encouraged by the basic premise of cybernetics that the same rules govern all "machines" but it goes much further in Beer. To him, people and their science are grossly inadequate for grasping the mysteries of the universe. Perhaps mind is everywhere. As Pickering says:

> Here Beer goes beyond his awe at the sheer excess of matter over representation to emphasise once again a spiritually-charged wonder at matter's *performativity* and, especially its computational performance.
>
> *(2009, p. 488)*

It is worth looking to the natural world to see if what we want is already there. This idea seems much less bizarre today than when Beer was carrying out his original experiments. Pickering (2009) cites a 2005 project involving a cockroach-controlled robot. Today "bio-mimicry," the study of nature's designs, organisms, and ecosystems to solve human problems, and "bio-engineering" are flourishing. Examples include using a kingfisher's beak as a model for the redesign of the nose of the Shinkansen 500-series train; marine life-forms as a template for futuristic submarine designs; scales on a butterfly's wings as a model for solar panels; running shoe soles that replicate the traction functions of a mountain goat's hoof; deer's antlers as a basis for extremely tough, resilient materials; surgical staples inspired by porcupine quills; a surgical glue based on the sticky secretions of marine worms; synthetic jellyfish-like tentacles to capture circulating tumor cells. For Jeffrey Karp, a bioengineer at MIT, the appeal of bio-inspiration is the idea that

> … every living creature that exists today is here because it tackled an incredible number of challenges. And those that haven't have quickly become extinct …. In essence, we are surrounded by solutions. Evolution is truly the best problem-solver.
>
> *(Quoted in Parker 2016)*

Pask's projects on learning systems, the "musicolour machine" and The Fun Palace, illustrate perfectly what Pickering (2010) calls the "dance of agency," as two complex systems interact in dynamic equilibrium, exploring each other's possibilities and, in the process, come up with something new and unexpected. He developed keyboard training devices that could respond to the learner's rate of progress and continually return to areas of weakness. A modern analogue is the "duolingo" application on my iphone, which I am gradually getting the hang of while it rapidly learns that I am never going to be any good at Spanish. Musicolour was a machine that used musical sounds to control a light show that itself changed on the basis of what had gone before and, therefore, encouraged the musicians to further innovation. Syd Barrett, the founder of Pink Floyd, explained the impact of a similar arrangement:

> It's quite a revelation to have people operating something like a light show while you're playing as a direct stimulus to what you're playing. It's rather like audience reaction except it's on a higher level.
>
> *(Quoted in Chapman 2010, p. 198)*

In 2017, the band New Order toured with a "responsive" stage set designed by the "relational artist" Liam Gillick.

The Fun Palace, initiated by the architect Cedric Price and the radical theater producer Joan Littlewood, was one of the great unrealized projects of the 1960s. Individuals including Buckminster Fuller, Yehudi Menuhin, and Tony Benn volunteered to help. Pask was asked to collaborate and organized a Fun Palace Cybernetics Subcommittee, which provided intellectual clout (Mathews 2005). The idea was to construct a space for the arts and sciences that would encourage the personal development and unleash the creativity of people and, in turn, could respond to the changing demands made upon it. This clearly could not be a conventional building. It would have to allow permanent dismantling and reassembly in response to the shifting cultural and social demands that it promoted in its users. It needed to be a socially interactive machine. Though never built, it provided the model for the 1976 Centre Pompidou in Paris. In 2014, inspired by the centenary of Littlewood's birth, the Fun Palace idea was revived with events planned every October. In October 2016, 292 "pop-up" Fun Palaces were set up in the United Kingdom and internationally, led by local people. The importance of Pask's cybernetic approach to architecture and urban design is widely recognized today and continues to inspire (see Werner 2017).

Bateson and Laing (a fuller account of their cybernetics can be found in Pickering 2010) are associated by many with the "anti-psychiatry" movement. Madness is seen as just another form of adaptation, if an unfortunate one, made by the self in seeking to find some sort of equilibrium with its surroundings. The hierarchical relationship between psychiatrist and patient, accepted by Walter and Ashby, is replaced by a more even-handed arrangement. Bateson encourages unstructured, open-ended sessions in which psychiatrist and sufferer make reciprocal adaptations. R.D. Laing, an archetypal 1960s figure, established the infamous Kingsley Hall community where psychiatrists and those labeled mad lived together with artists, dancers, and others and interacted on an equal basis. Residents were to undergo a transformative experience, which could include "normal" individuals going "mad," in the search for a joint equilibrium that might reveal something new about human potentiality.

6.4 Second-Order Cybernetics

Second-order cybernetics, or "the cybernetics of cybernetics," emerged as a recognizable focus of intellectual activity between the late 1960s and mid-1970s. Its institutional home was the Biological Computer Laboratory (BCL), which operated at the University of Illinois between 1958 and 1975. The founder and director was Heinz von Foerster, a physicist, philosopher, and cybernetician, who had been heavily involved in the Macy conferences as, from the sixth event, secretary and editor of the proceedings. The BCL built the first parallel computer, was home to the first major conference on self-organizing systems in 1959, and hosted, on the staff or in residence, figures such as Ashby, Beer, Pask, Maturana, and Varela.

"Cybernetics of cybernetics" was a phrase first used in 1968 for the title of a paper by Margaret Mead. The paper was based on her keynote address at the founding of the American Society for Cybernetics in 1967 and the title was provided by von Foerster. The idea was that cybernetics should engage in some self-reflection and use its own methods on itself. This was an obvious next step for cybernetics, which had long observed the circularity inherent in feedback systems. It was natural to extend the study of circularity to the relationship between the observing system and the observed system (Glanville 2003a). von Foerster first used the phrase "second-order cybernetics" in 1974 and, in his edited collection of that year *Cybernetics of Cybernetics*, carried forward the cybernetic critique of cybernetics and provided an explanation in more general terms of what was going on:

> I submit that the cybernetics of observed systems we may consider to be first order cybernetics; while second order cybernetics is the cybernetics of observing systems.
>
> *(Quoted in Clemson 1984, p. 246)*

In von Foerster's view, the traditional scientific method, in its search for "objectivity," had created a cognitive blind spot:

> 'The properties of the observer shall not enter the description of his observations'. But I ask, how would it be possible to make a description in the first place if not the observer were to have properties that allows for a description to be made? .... We have to ask a new question: 'What are the properties of an observer?'
>
> *(Quoted in Ramage and Shipp 2009, p. 185)*

It is no good pretending that you can exclude the observer from the domain of science. There has to be a recognition that the observer matters. This is, of course, a generalization of what Heisenberg (see Section 2.3) had discovered in the physical sciences and Roethlisberger and Dickson (see Section 4.8) had observed in researching organizational behavior. von Foerster liked to justify his conclusion with an appeal to neurophysiology. As Clemson puts it:

> One of von Foerster's favourite points is that the human nervous system deals in undifferentiated electrical impulses .... The nervous system has no way of distinguishing the various sensations except in terms of pattern of input arrival. All of our perceived world is built out of these undifferentiated electrical impulses. We literally construct our own reality out of an undifferentiated flow of identical electrical impulses.
>
> *(1984, p. 10)*

It follows that "black boxes" are created by observers who come to possess a shared understanding of reality primarily through social learning processes. Cybernetics must pay attention to the circular processes in which observers interact with what they observe.

von Foerster's insight helped to establish a common language for various scholars who had been thinking along the same lines. His review of Spencer-Brown's 1969 book *Laws of Form* brought the British mathematician, who had already been influenced by R.D. Laing, firmly into the second-order cybernetic fold. It remains in dispute whether Spencer-Brown was a genius or a charlatan. He was certainly an eccentric, suing Beer for wrongful dismissal when he was sacked for not performing in a job Beer had provided for him, and taking the Longleat Estate to a residential property tribunal when it tried to raise the rent on a cottage the Marquess of Bath had supplied to him and which he had brought into a state of disrepair. What is not in doubt is that the implications of his idea that the most basic act of epistemology is the drawing of a distinction that makes a difference had a tremendous influence on von Foerster, Maturana, Varela, and Luhmann (see Section 4.7). It was also clear to von Foerster that there were parallels between second-order cybernetics and the "radical constructivism" that von Glasersfeld had been developing for some time based upon Piaget's work (see Section 1.6). This link provided welcome support from the classical philosophical tradition to the shift taken by second-order cybernetics. In return, the radical constructivist position received validation from second-order cybernetics, particularly from the work of Maturana and Varela. It is their contribution that we will now consider.

Maturana had worked with the physiologist Warren McCulloch, organizer and chair of the Macy conferences, and it is McCulloch's notion of "experimental epistemology," how the brain actually knows the world, that helps explain the association between second-order cybernetics and the work of Maturana and Varela. Maturana and Varela offered a theory of cognition that was consonant with von Foerster's work on observers but had the added benefit of being derived from experimental studies of the brain and nervous system. When we left them, in Section 3.2, they had reached the conclusion that cognition does not provide an objective picture of some external "real world" but continually brings forth a world as organisms undertake their normal processes of living: "living systems are cognitive systems, and to live is to know" (Maturana 1987, p. 357). Spending time at the BCL, Maturana and Varela benefitted from their exposure to the thinking of von Foerster, Spencer-Brown (see especially, Varela, *Principles of Biological Autonomy*, 1979), and von Glasersfeld, and were able to further refine their theory of "autopoiesis and cognition" (Maturana and Varela 1980).

For Maturana and Varela, all living systems, with or without a nervous system, are cognitive systems because in the process of maintaining their organization, they bring forth a world which they can perceive and in which they can behave or act. Consciousness is a special type of cognition that comes later, as the emergent property of increasingly complex cognitive processes that arise after a long history of evolution. As we saw in Section 3.3, there are disputes about whether organisms such as trees possess consciousness and whether it is a feature of ecosystems and *Gaia* or even, if you are impressed by Beer's *hylozoism*, of matter generally. For Maturana and Varela, consciousness requires advanced brains and nervous systems capable of complex, dynamic neural activity providing for a wider range of perceptual experiences, emotional responses, basic mental images, and enhanced reasoning. Consciousness enables a living system to engage in more sophisticated interactions with its experienced world but does not provide any more access to "objective reality." It is a quality we commonly attribute to mammals, birds, and some other vertebrates. We do not, however, grant these animals "self-awareness" except, perhaps, in the immediate "here and now" (see Capra and Luisi 2014, p. 260). Self-awareness is a more sophisticated form of consciousness that we primarily associate with humans, although sometimes allowing that other great apes, dolphins, whales, elephants (and my dog) may have it to a degree. Peter Godfrey-Smith makes an excellent case for cephalopods – octopuses, squids, and nautiluses – which, he argues

> ... are an island of mental complexity in the sea of invertebrate animals, an independent experiment in the evolution of large brains and complex behavior.
>
> *(Quoted in Hoare 2017, p. 7)*

Self-awareness is a more elaborate form of consciousness possessed by a reflecting subject and based on a clear sense of the self with a past and a future:

> Reflective consciousness involves a level of cognitive abstraction that includes the ability to hold mental images, allowing us to formulate values, beliefs, goals, and strategies.
>
> *(Capra and Luisi 2014, p. 260)*

There are obvious parallels with Boulding's account of the increasing sophistication of "the image" as we rise up the levels of complexity in his hierarchy (see Section 5.5).

For Maturana (1987), the emergence of self-awareness cannot be separated from the evolution of language. It is a social as well as biological phenomenon. The brain and nervous system have to develop the necessary "diversity and plasticity" to enable language to happen but it does not take place in the brain. It depends on living systems interacting recurrently to constitute a consensual domain of interaction or "linguistic domain" (Maturana 1987, p. 359). Participants in a consensual domain of interaction co-ordinate both their actions and the distinctions they make in relation to the environment. Language arises when living systems become capable of using symbols – words, gestures, and signs – as effective tools for doing this. As Maturana proclaims:

> Accordingly, I claim that the phenomenon of language takes place in the co-ontogeny of living systems when two or more organisms operate, through their recurrent ontogenic consensual interactions, in an ongoing process of recursive consensual co-ordinations of consensual co-ordinations of actions or distinctions.
>
> *(1987, p. 360)*

The establishment of language as a recursive consensual domain also makes it possible to distinguish between the makers of distinctions and the actual distinctions they make. Individuals become aware of themselves as observers and watch what they do. Try this:

> When this happens self-consciousness arises as a domain of distinctions in which the observers participate in the consensual distinctions of their participations in language through languaging. It follows from this that the individual exists only in language, and that the self exists only in language, and that self-consciousness as a phenomenon of self-distinction takes place only in language.
>
> *(Maturana 1987, p. 364)*

Human beings, therefore, exist in language. It arises from our co-existence and orientates our behavior. As the consensual co-ordination of actions and distinctions, it facilitates co-operative activity and brings forth the world of objects that we share, whether houses, people, or elementary particles. We exist as self-conscious beings in language. Language is the distinctive cognitive domain of humans, providing us with possibilities for a wide range of interactions, but it remains a closed operational domain. It emerges from and enables the praxis of living but cannot give us access to some external world beyond language: "Everything said is said by an observer to another observer that could be him or herself" (Maturana 1987, p. 332). Expanding this argument, Maturana claims that science is just another form of "languaging," which can help explain human experience but not provide objective knowledge. The statements it makes depend for their acceptance on a consensual domain of standard observers not on any correspondence with objective reality.

It is reasonable to ask whether all this amounts to much. Isn't it something we have heard before but dressed up in even more convoluted language? von Glasersfeld (1990) accepts that many of the ideas are already present in philosophy post-Kant (and we might add in Kuhn's work, see Section 2.4, and in interpretive social theory, see Section 4.3) but suggests that they have previously always been marginalized. He believes that the philosophical climate may now be right to hear them again, presented in a different form. Umpleby argues that Maturana and Varela's arguments carry particular weight because they rest on biological foundations. They demonstrate that "observations independent of the characteristics of the observer are not physically possible" (2016, p. 456). Here we will try to gauge the significance of the ideas, and of second-order cybernetics as a whole, by considering their implications for practice.

Given their shared epistemological conclusions, it is not surprising that the pronouncements for action delivered by von Glasersfeld, Maturana, Varela, and von Foerster are similar. In Section 1.6, von Glasersfeld was found arguing, on the basis of his "radical constructivism," that we must take full personal responsibility for our words and deeds. We cannot hide behind some notion of "objective knowledge" to guide our actions. From Maturana's (1987, p. 332) perspective, "reality" is brought forth by the distinctions made by observers and so there will be as many "truths" as observers can bring forth in their distinctions. All are equally valid and those who make claims to objectivity are simply making demands for "obedience to knowledge" (a surprisingly Foucaulvian turn of phrase, see Section 4.5). Disagreements among different observers in the "multiverse" must be solved not by making an appeal to an independent reality but by consensus. Our continued coexistence demands it. Habermas would be pleased (see Section, 4.4). Furthermore,

> ... everything that we do becomes part of the world that we live in as we bring it forth as social entities in language.
>
> *(Maturana 1987, p. 378)*

Not all worlds are equally pleasant to live in. For that reason, nothing we do as human beings is trivial. Human responsibility is total. There is much of the same in von Foerster. In his view, we create the world through the observations we make and so cannot escape responsibility for these observations. We have to make a choice and live by it. We can even choose to accept that there is a real world external to us, wonder at it and ourselves, and search for meaning. In Glanville's words:

> This is, I believe, the most significant thing von Foerster did, and what, in the end he cared most about. It is also what divides his deeply humane, human, and humanist (second-order) cybernetics from the mechanical control systems that so often still parade as cybernetics.
>
> *(2003b, p. 88)*

There are, of course, arguments that can be made against the "subjectivist" position of second-order cybernetics. One can be raised from within the cybernetics tradition itself. The concept of the "embodied mind," endorsed by Bateson and Varela, indicates that our biology imposes limitations on what we are able to think. It suggests that the very "structures of our bodies and brains determine the concepts we can form and the reasoning we can engage in" (Capra and Luisi 2014, p. 273). To others, taking a more "objectivist" stance, second-order cybernetics is not only wrong but annoying:

> Frankly to us this seems rubbish. Frogs and humans see the world in different ways but the world is there to be seen and shapes what we see.
>
> *(Byrne and Callaghan 2014, loc. 769)*

Another, more sophisticated, objection is put by Poerksen, who objects to the implication in von Foerster's theory that observers are free to decide what distinctions they want to make:

> My objection is that the world never is – in Spencer-Brown's terminology – an *unmarked space*, but that we are all pressured in many ways, and even condemned, to reproduce the distinctions and views of our own groups, of parents, friends, and institutions .... Here is my counter-thesis: In the act of observing, we reproduce either old orders or systems of distinctions, or we develop new ones from or against them. Therefore the freedom and arbitrariness of constructions is massively reduced.
>
> *(2003, pp. 22–23)*

von Foerster accepts that communality leads to a reduction in arbitrariness but insists, *contra* Luhmann (see Section 4.7), that individuals make society and still have freedom to choose. He takes this position, he says, because it is the only way he can make sense of personal responsibility. In his view, people can reinvent themselves and employ new distinctions. They can opt out of the social networks in which they find themselves and seek to construct new realities through communality: "The kinds of dances one chooses along this way may be infinitely variable" (von Foerster, quoted in Poerksen 2003, p. 23). There are always opportunities for action and we should try and keep these visible. That is why von Foerster says that his ethical imperative is: "Act always so as to increase the number of choices." Second-order cybernetics often seems to take us back to the beginnings of the debate in social theory.

Stuart Umpleby (2016) argues that, in his later years, von Foerster wanted to do more than simply insist we pay greater attention to observers. He intended a fundamental revolution in scientific methodology. Second-order cybernetics, he believed, signaled the need for a paradigm shift. Traditional science had sought to exclude observer effects from the research process. Second-order cybernetics yields an understanding of biology and cognition that demonstrates why this is impossible. The scientific enterprise needs to bring observers in rather than seeking to ignore them. Luhmann, as we witnessed in Section 4.7 (see also Beyes 2005), took von Foerster's message to heart and applied a second-order critique to systems theory, including his own version of it. Also exposed by the shift from first- to second-order cybernetics, according to von Foerster, were

> ... two fundamentally different epistemological, even ethical positions, where one considers oneself, on the one hand, as an independent observer who watches

the world go by; or on the other hand, as a particular actor in the circularity of human relations.

(von Foerster Quoted in Umpleby 2016, p. 459)

Henceforth, researchers should see themselves as "standing inside" rather than "standing outside" science. Umpleby suggests that science needs to take heed of action research if it wants to learn how to do "science from within" and commends the work of the Institute of Cultural Affairs (ICA) for its use of participatory methods in social research. An explicit example of the use of second-order cybernetics in the field is Restrepo et al.'s (2016) co-construction with dairy farmers in Kenya of a new way of looking at and controlling milk production. Müller and Riegler (2016) are in tune with Umpleby in their determination to take on von Foerster's challenge and extend the insights of second-order cybernetics to "second-order science." They propose a general methodology of science based on the "new epistemic mode" of doing science "from within" and set out a convincing agenda for increasing the influence of second-order cybernetics. A website, www.secondorderscience.org, defines first-order science as "the science of exploring the world" and second-order science as "the science of reflecting on these explorations." It goes on, ambitiously, to suggest a second-order science "process" involving six steps: question uncritically examined presuppositions; add observers; be reflexive; respect pluralism; do meta-analysis; avoid hubris. This is all very well, but one has to think that second-order cybernetics would save itself a lot of time learning about how to implement "science from within" if it recognized that this is exactly what "soft systems thinking" (see Chapters 14–16) has been doing since the 1970s. And it could rapidly extend the scope of the paradigm shift it advocates if it acknowledged the second-order critique of different systems approaches that "critical systems thinking" has been consciously and systematically engaging in since the early 1980s (see Part IV). It has a lot of ground to make up.

6.5 Conclusion

There is a case to be made that the three strands of cybernetics are not as different as they have been painted. Certainly, a number of key figures were influential in all of them. McCulloch started the Macy conferences and chaired them, was Beer's cybernetics guru, and orientated second-order cybernetics with his concept of "experimental epistemology." Ashby was authoritative across the board. Perhaps all that British and second-order cybernetics achieved was to complete the project of first-order cybernetics by opening up for exploration another circular relationship, this time between observer and observed. Pickering (2010) is well aware that the work of many of the individual cyberneticians cannot easily be pigeon-holed into any one strand. Nevertheless, he is convinced that the different theoretical positions he identifies as underpinning the three strands have a huge impact on the way that cybernetics is practiced and on its ability to yield new and useful applications. In his words "ontology matters." He is scathing, for example, on where the theoretical assumptions of second-order cybernetics lead:

> So I think that second-order cybernetics has talked itself into a corner in its intensified emphasis on epistemology .... To see cybernetics as being primarily

about epistemology is to invite endless agonising about the observer's personal responsibility for his or her knowledge claims. Fine. But the other side of this is the disappearance of the performative materiality of the field .... Tortoises, homeostats, biological computers, Musicolour machines, adaptive architecture – all of these are just history as far as second-order cybernetics is concerned. We used to do things like that in our youth; now we do serious epistemology.

(2010, p. 26)

I side with Pickering on this, not least because of the evidence available from my own field of management. First-order cybernetics has added valuable concepts such as feedback to the lexicon but the "management cybernetics" with which it is associated has done little to challenge the paradigm of classical management theory. Umpleby (2016) and Müller and Riegler (2016), while identifying a bright future for second-order cybernetics, confess that it has yet to realize its potential and has, to date, had no sustainable consequences for other disciplines. Beer's "organizational cybernetics," however, nourished in the "performative idiom" of British cybernetics, has had a tremendous impact. As we shall see in Part III, it offers a complete alternative to traditional management theory and provides, in the form of the "viable system model" and "team syntegrity," two of the most powerful approaches available for managing complexity.

7

Complexity Theory

But it is nevertheless incumbent upon us to push the scientific paradigm as far as possible to determine where the boundaries are and not be deterred by the specter of overwhelming complexity and diversity

(West 2017, loc. 4658)

7.1 Introduction

It has been convincingly argued by Gleick (1987) that "chaos theory," the forerunner of "complexity theory," is one of the three landmark achievements of twentieth-century science (see Section 2.3). There were, however, two earlier scientific discoveries that set the scene for the emergence of chaos theory. The first was due to Poincaré (1854–1912) and has frequently led to him being declared the founding father of chaos and complexity theory. The mathematician Poincaré was seeking to solve the so-called "many-body" problem in celestial mechanics. Newton's laws were fine for predicting the behavior of two bodies exerting gravitational forces on one another, but it was extremely difficult to extend them even to three bodies subject to mutual attraction. Poincaré did not completely succeed in providing a solution but he thoroughly grasped the nature of the problem he was up against:

> But even if it were the case that the natural laws had no longer any secret for us, we could still only know the initial situation approximately. If that enabled us to predict the succeeding situation with the same approximation, that is all we require, and we should say that the phenomenon has been predicted, that it is governed by laws. But it is not always so; it may happen that small differences in the initial conditions produce very great ones in the final phenomenon. A small error in the former will produce an enormous error in the latter. Prediction becomes impossible …
>
> *(Poincaré, quoted in Mitchell 2009, p. 21)*

It doesn't matter if the laws of motion are known perfectly, small differences in the initial positions, masses, and velocities of objects can lead to significant differences in the way the system behaves in the future. He had discovered "sensitive dependence on initial conditions," one of the signatures of chaos theory. Unfortunately, the

Critical Systems Thinking and the Management of Complexity, First Edition. Michael C. Jackson.
© 2019 John Wiley & Sons Ltd. Published 2019 by John Wiley & Sons Ltd.

sophisticated mathematics and the computational tools that would have enabled Poincaré to develop his findings were not available at that time.

The second was Darwin's theory of evolution set out, in 1859, in *On the Origin of Species*. The theory stated that random variations in the traits possessed by individual organisms, passed on by inheritance, can lead to a particular species flourishing because the new traits are well adapted to the environment. Evolution proceeds not on the basis of determinism but as a result of the "natural selection" of the outcomes of chance events. Furthermore, living systems are able to become more complex and intricate, defying the second-law of thermodynamics. There is all the appearance of design but no need for a grand designer. Darwin's ideas lit a slow fuse that eventually had an immense impact on the way science is perceived. The pragmatist philosopher Charles Pierce (see Section 1.4), writing in 1891, captures it well:

> Now the only possible way of accounting for the laws of nature and the uniformity in general is to suppose them results of evolution. This supposes them not to be absolute, not to be obeyed precisely. It makes an element of indeterminacy, spontaneity, or absolute chance in nature.
>
> *(Quoted in Boulton et al. 2015, p. 63)*

In describing contemporary "complexity theory," in this chapter, I follow Stacey et al. (2000) in identifying and describing three distinct strands: chaos theory, the theory of dissipative structures, and the study of "complex adaptive systems." We then turn to its impact. The claim of complexity theory to be a universal science of systems is supported by its relevance to disciplines as diverse as meteorology, chemistry, geology, evolutionary biology, ecology, economics, and management. The purpose of this book determines that emphasis is placed on its lessons for management. Most models of organization can be seen as emphasizing order and regularity. In contrast, complexity theory focuses on disorder, irregularity, and randomness – aspects of organizational life of equal importance to managers but largely ignored in the literature. It takes instability and change seriously, sometimes pointing to an underlying order in the disorder or, at the least, providing advice on how to cope with unpredictability. Offering help when it seemed that none was available, complexity theory has proved attractive to management thinkers. They have, however, interpreted it in many different ways and a "second-order" analysis is necessary to determine exactly what is being proposed and what benefits might ensue. Finally, we consider the vexed question of the relationship between complexity theory and systems thinking. Some complexity theorists acknowledge their debt to earlier systems thinkers: Prigogine to von Bertalanffy; Kaufmann to Ashby and McCulloch. More often, complexity theory likes to present itself as a "radical break" with what has gone before and, as a result, fails to acknowledge the achievements of the systems and cybernetics traditions that, particularly in terms of practical advice to managers (as will be witnessed in Parts III and IV), are far ahead of their upstart rival.

7.2 Chaos Theory

The pioneer in the development of chaos theory was the meteorologist Edward Lorenz (see Gleick 1987). Lorenz was working on the problem of long-range weather forecasting. In the winter of 1961, using a basic digital computer, he was running a weather simulation based on just a few relatively simple, nonlinear equations. Intent on studying one particular weather sequence at greater length, and in a hurry, Lorenz took a short-cut and started the run again mid-way through. Using information from the earlier print-out, Lorenz re-entered the conditions at that point using the three decimal places shown. The computer's memory, however, stored six decimal places. Given that the difference was only one part in a thousand, he assumed that the new run would exactly duplicate the old. To his amazement, however, the new weather pattern rapidly diverged from that shown in the previous run and within a few months all resemblance had been lost. Lorenz had discovered that tiny changes in the initial state of a complex system can significantly and unpredictably alter its long-term behavior. Following his 1972 paper "Does the Flap of a Butterfly's wings in Brazil Set off a Tornado in Texas?" (and encouraged by the shape of the "Lorenz attractor" – see Figure 7.1) this sensitive dependence on initial conditions became known as the "butterfly effect." Dr. Ian Malcolm, the chaos theorist in the Jurassic Park franchise, explains it as minor events having very major unexpected consequences. An unforeseen storm sets everything in motion on Isla Nublar.

It is the ubiquity of nonlinear relationships that makes prediction impossible in complex natural and social systems. In linear relationships, a change in the value of a causal element brings about a proportionate change in an effected element (small changes produce small effects, etc.). Furthermore, causes are seen as acting independently. The results of different causes can simply be added to deduce what will happen in the whole, which is, therefore, equal to the sum of its parts. This is the logic underpinning Newtonian science. When there are nonlinear relationships, however, which is usual in reality, changes in output are disproportionate to changes in input and causes are inter-linked in complex feedback structures (Strogatz 2004). Positive (reinforcing) and negative (self-regulating) feedback loops abound. The behavior of the whole system becomes impossible to predict on the basis of its parts and emergent properties are displayed.

Lorenz had discovered that the way weather systems develop is extremely sensitive to changes in initial conditions. It was also clear to him that there was a good deal of order underlying the apparent chaos. Weather systems with slightly differing initial conditions might not repeat the same behavior over time but they do demonstrate similar patterns. Even if the weather is notoriously fickle in England, it does not bring severe drought, or monsoon, or frequent hurricanes. Unpredictability could encompass a surprising degree of order. Lorenz now started experimenting with sets of equations describing the behavior of somewhat simpler systems such as a waterwheel and a convection system. In each case, there was sensitive dependence on initial conditions but also a recognizable pattern. By the 1960s, advances in mathematics and computing made it possible to "solve" nonlinear equations by trial and error and to plot all possible states of a complex system graphically. When the behavior of the waterwheel was mapped, a remarkable result was observed. The output always remained within the limits of a double spiral curve. This was novel because previously only two types of behavior had been investigated mathematically – stable equilibrium and regular periodic oscillation. Lorenz's simple sets of equations behaved in an "a-periodic" fashion, never settling down to a steady state and never repeating themselves exactly. Nevertheless, they were clearly "attracted" to a particular pattern of behavior. For this reason, Lorenz called the image he had produced the "Lorenz attractor" (see Figure 7.1).

Since Lorenz's experiments, a-periodic behavior has been thoroughly investigated and it has been postulated that all manner of natural and social systems are governed by

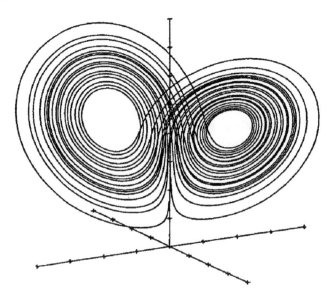

Figure 7.1 The Lorenz attractor. Because the system never exactly repeats itself, the trajectory never intersects itself. Instead, it loops around and around for ever. Motion on the attractor is abstract, but it conveys the flavor of the motion of the real system. For example, the crossover from one wing of the attractor to the other corresponds to a reversal in the direction of spin of the waterwheel or convecting fluid. *Source:* From Gleick (1987). Reproduced with permission of Random House.

what are now called "strange attractors." A strange attractor seems to keep the trajectory followed by an otherwise unpredictable system within the bounds of a particular pattern without requiring it ever to exactly repeat itself. Strange attractors also produce "self-similar behavior," giving rise to the same pattern at whatever scale their effects are examined.

Lorenz published his early results in a meteorological journal in 1963. The next decade or so was marked by individual scientists unearthing similar findings. The behavior of complex systems in a range of disciplines, although apparently describable in terms of a few simple equations, was found to be characterized by the emergence of unpredictability governed by a considerable degree of order. Robert May is worthy of some attention because he extended the application of chaos theory beyond physical systems. May was a biologist interested in predicting the behavior of biological populations over time. The equations describing such systems were simple if populations were allowed to rise indefinitely. However, once predators and diminishing food supplies were included, behavior became impossible to predict. The study of biological populations required mathematical description using nonlinear differential equations (May 1974). Experimenting with different inputs to his equations, May discovered that once growth rate passed a certain level, there was a bifurcation in his results indicating that the population could be in two different states. Raising growth rate further led the outcome to jump between four different values. And if the rate was increased still further, the predictions bifurcated again, yielding eight possible states. Bifurcations became quicker and quicker until chaos suddenly appeared and it became impossible to predict the outcome. Close inspection of a graph of the results did, however, show patterns of order emerging in the chaos. The graph also demonstrated the property of "self-similarity," containing inside an exact copy of itself.

Another, independent investigation into self-similarity and the emergence of order was undertaken by the mathematician Benoit Mandelbrot and his work has had an immense influence on the development of chaos and complexity theory. Mandelbrot built on the findings of Lewis Fry Richardson, who had discovered the "coastline paradox." This stated that the measured length of the coastline increases with the resolution level of the measuring instrument you are using. If you use a 100-mile ruler and insist that both ends touch the coast, most of the meanderings and wiggles will be missed and a relatively short length recorded. Increase the resolution level of your ruler by making it shorter, a foot perhaps, and the coastline will appear much longer. Mandelbrot recognized that such "crinkliness" and roughness existed throughout nature and that Euclidian geometry, which describes ideal shapes, such as circles, squares, and cubes, cannot cope with it (Lesmoir-Gordon et al. 2009). A cloud, for example, is indescribable in traditional geometric terms. It eventually dawned on Mandelbrot that the irregular phenomena he was studying all had something in common. They were "fractals"; a term he coined, in 1975, to describe a system in which the parts, at any scale, look approximately the same as the whole. A cauliflower is one of his favorite examples. Break it into smaller pieces and each piece resembles a smaller version of the whole. You can continue this process for a long time before revealing any new pattern or detail. Coastlines, clouds, galaxies, mountain ranges, trees, ferns, cauliflowers, broccoli, snowflakes, brains, river deltas, lightning, circulatory systems, etc., all exhibit self-similarity. This insight led Mandelbrot to propose "fractal geometry," a mathematical language capable of describing the complex, irregular shapes found in nature. In this geometry, although there is no definite answer to the question about the length of a coastline, it is possible to assign a number to its "jaggedness." Capra and Luisi explain:

> We can understand this idea intuitively by realizing that a jagged line on a plane fills up more space than a smooth line, which has dimension 1, but less than the plane, which has dimension 2. The more jagged the line, the closer its fractal dimension will be to 2. Similarly, a crumpled-up piece of paper fills up more space than a plane but less than a sphere. Thus, the more tightly the paper is crumpled, the closer its fractal dimension will be to 3.

> *(2014, p. 118)*

The fractal dimension of the British coastline turns out to be approximately 1.58, whereas the figure for the Norwegian coast is approximately 1.7. Nature seizes the opportunity provided by fractal dimensions to, for example, cram 100 000 km of arteries, veins, and capillaries, if laid end-to-end, into the human body and increase the effective surface area of the lungs almost to the size of a tennis court (West 2017). Using Mandelbrot's insights, scientists have been able to model the complex shapes found in nature by constantly repeating the geometric operation necessary to create a particular fractal at smaller and smaller scales. It was found that you could magnify the border of a cloud 10 million times and still be looking at the same general pattern. A recognition that "strange attractors" possessed all the characteristics of fractals sealed the marriage between the two bodies of work.

T.Y. Li and J.A. Yorke provided the name "chaos" for the new findings in a paper they published in 1975 and, by the late 1970s, the chaos theory movement began to take shape. Of particular significance was the establishment, in 1977, of the Dynamical Systems Collective at Santa Cruz College in the United States, headed by the physicist Robert Stetson Shaw. The group carried out work on "strange attractors" in a variety of contexts. A member of the Collective, the physicist and mathematician Norman Packard, is credited, together with Chris Langton, who was working independently on information systems at the University of Arizona, with recognizing the importance of the "edge of chaos" (see Gleick 1987). The edge of chaos is a narrow transition zone between order and chaos that is extremely conducive to the emergence of novel patterns of behavior. A system driven to the edge of chaos is likely to exhibit spontaneous processes of self-organization. Stacey (1996) uses the "edge of chaos" concept to suggest how creativity and learning can be promoted in organizations. He notes that all complex adaptive systems can operate in one of three zones: a stable zone, an unstable zone, and at the edge of chaos. In the stable zone they ossify; in the unstable zone they disintegrate; but at the edge of chaos, spontaneous processes of self-organization occur and innovative patterns of behavior can emerge. This, therefore, seems to be the best place for organizations to be. At the edge of chaos, in a state of "bounded instability," they display their full potential for inventiveness and innovation.

According to Sardar and Abrams, chaos theory

> ... presents a universe that is at once deterministic and obeys the fundamental physical laws, but is capable of disorder, complexity, and unpredictability. It shows that predictability is a rare phenomenon operating only within the constraints that science has filtered out from the rich diversity of our complex world.
> *(2008, p. 6)*

In the real world, even simple nonlinear systems can display uncertain, complex behavior that makes anything beyond short-term prediction impossible. However, the "chaos" that ensues is not chaos in the everyday meaning of the word. Chaos theory has revealed that, between order and complete disorder, a "hidden order" can appear. In this middle-ground, behavior never repeats itself exactly but it is drawn to a "strange attractor," which seems to set limits to what is possible. In these circumstances, patterns can be recognized and it becomes possible to predict the overall "shape" of what can happen. All snowflakes have six sides while still all being different. Newer developments in complexity theory mean that, nowadays, the term chaos theory tends to be restricted to the study of the nonlinear dynamics observed in natural systems. In meteorology, physics, chemistry, and biology, the behavior of systems can often be modeled on the basis of a small number of variables with fixed interactions. In social and ecological systems, with huge numbers of elements (people in a city, species in a forest), with different and evolving characteristics, and impacted by numerous internal and external changes, it is much more difficult to discern the influence of strange attractors (Boulton et al. 2015). Peter Senge with his "system archetypes" has led the search in social systems (see Chapter 11). Nevertheless, as we saw with the "edge of chaos," chaos theory can still provide some insightful metaphors. Certainly, small, apparently innocuous decisions often seem to have significant consequences for how our lives turn out. As Bob Dylan (2001) regrets, reprising an old prisoners refrain:

> Only one thing I did wrong
> Stayed in Mississippi a day too long.

7.3   Dissipative Structures

The concept of "dissipative structures" was originated by Ilya Prigogine, a Belgian physical chemist who, in 1977, won the Nobel Prize in Chemistry for his work on nonequilibrium thermodynamics. In elaborating his idea, Prigogine was inspired by Darwin's evolutionary theory to cross swords with both Newton's mechanics and the second law of thermodynamics (see Section 2.2). In Newtonian, indeed Einsteinian, mechanics, the universe is regarded as both deterministic and reversible. Exactly reversing the motion of every particle at some instant would see the world returning to states it had occupied at earlier times (Smith 1984). This is very much contrary to our everyday experience. I have frequently knocked a beer glass off a table in a pub but, unfortunately, have never seen it rise back onto the table having reassembled itself and recovered its contents. The second law of thermodynamics, by contrast, accepts that irreversible change occurs but only in the direction of maximum disorder. This leaves no room for biological and social evolution, which is clearly accompanied by increased levels of organization.

Prigogine wanted to extend science to incorporate both irreversible processes and increased complexity. In doing so, he drew upon the work of von Bertalanffy, who had provided part of the answer by arguing that the second law applied only to closed systems (see Section 3.2). For Prigogine (Prigogine and Stengers 1984), generalizing von Bertalanffy's insight, traditional science had focused for too long on systems in a state of "thermodynamic equilibrium" and so could not explain how systems evolve and change. In his view, systems are continuously subject to internal and external "fluctuations" and, as a result of positive feedback loops, these fluctuations can be powerful enough to drive systems far from equilibrium. In this state, a system might be expected simply to disintegrate. Prigogine showed that, under certain conditions, systems are able to pass through randomness and achieve a new level of order as "dissipative structures" – so called because they require energy from the outside to prevent them from dissipating. An example is provided in the press release for his Nobel Prize:

> The most well-known dissipative structure is perhaps the so-called Benárd instability. This is formed when a layer of liquid is heated from below. At a given temperature heat conduction starts to occur predominantly through convection, and it can be observed that regularly spaced, hexagonal convection cells are formed in the layer of liquid. This structure is wholly dependent on the supply of heat and disappears when this ceases.
>
> *(Nobel Prize Committee 1977)*

Rather than focusing on the inevitability of decay as systems run down to a state of maximum entropy, as classical thermodynamics does, the theory of dissipative structures highlights the capability of open systems to evolve toward greater complexity through spontaneous self-organization and to maintain themselves in unstable conditions far from equilibrium. They do this through the continual exchange of matter and energy with their environments. Dissipative structures, it follows, arise from the interaction of a given system with its surroundings. They are islands of order in a sea of disorder, increasing their own complexity while increasing entropy overall (Prigogine and Stengers 1984).

Recognizing that nonlinear equations could be employed to explain the behavior of the chemical systems he was researching, Prigogine was able to draw on ideas familiar from chaos theory (sensitivity to initial conditions, bifurcation points) to develop his thinking. He could also use mathematics to argue for the importance of dissipative structures in other domains – in "biochemistry, biological aggregation, animal societies, sociology" (Prigogine 1976). All open systems are seen to be subject to instabilities produced by internal and external fluctuations. Sometimes negative feedback loops can maintain them in a stable but fluctuating state. At other times, positive feedback loops amplify fluctuations and drive the system to a critical bifurcation point where a new emergent structure of increased complexity comes into being. Systems can pass through disorder before spontaneously self-organizing into new states dramatically different from what has gone before. Prigogine and Stengers describe the emergence of "chemical clocks," chemical reactions that behave in a coherent, rhythmical fashion:

> … in a chemical clock all molecules change their chemical identity *simultaneously*, at regular time intervals …. A new type of order has appeared …. We can speak of a new coherence, of a mechanism of 'communication' among molecules. But this type of communication can arise only in far-from-equilibrium conditions.
>
> *(1984, p. 13)*

What emerges depends on the existing form of the system but also on the effects of whatever novel developments and variations occur at particular times and places on its journey. Systems are determined in their behavior but not predictable. The future is under construction and chance has a role to play. So the processes we observe in nature are indeed irreversible.

Prigogine (1976) states that his theory of dissipative structures brings the physical, life, and social sciences together within the framework of an extended thermodynamics:

> All these problems have a common element: time. Maybe the orientation of my work came from the conflict which arose from my humanist vocation as an adolescent and from the scientific orientation I chose for my university training. Almost by instinct, I turned myself later toward problems of increasing complexity, perhaps in the belief that I could find there a junction in physical science on one hand, and in biology and human science on the other.
>
> *(Quoted in Ramage and Shipp 2009, p. 233)*

Biological, ecological, and social systems are open, are replete with disorder and demonstrate irreversible change, and Prigogine believes his "new science" can provide an understanding of their behavior. It is clear to him that living organisms are dissipative structures, capable of maintaining and developing their complexity as long as they continue to receive inputs of energy. Evolution produces unpredictable novelty. At crucial bifurcation points, the micro details of interaction between organisms and their environments determine the path taken. Boulton et al. (2015) use Prigogine's concepts to describe how mature forests can become resilient enough to survive large fluctuations, internal or external, and self-organize into new, unpredictable forms dependent on the particularities of the situation. They also seek to explain, as we shall see in Section 7.5, how the theory of dissipative structures is relevant to social systems and management. Prigogine is exhilarated by his own vision of the universe – one that is changing and developing and in which human beings have choices. This is surely a better prospect, he believes, than the alternatives envisaged by Newton and by the second law of thermodynamics. All very well, thinks Smith (1984), as long as it is good science as well.

7.4 Complex Adaptive Systems

The year 1984 saw the establishment of the Santa Fe Institute, in New Mexico. It has since become the most famous center for research into complex systems. The Santa Fe Institute (SFI) was the brainchild of "a diverse interdisciplinary group of twenty-four prominent scientists and mathematicians" who wanted to pursue theoretical research outside traditional disciplinary boundaries (Mitchell 2009). By 1984, the term "chaos theory" was already giving way to the grander conception of "complexity theory" and the SFI explicitly stated its desire to promote the "emerging syntheses in science" through a new interdisciplinary research area called complexity theory. Holland (2014) provides a distinction, seeing chaos theory as restricted to the study of "complex physical systems" containing elements with fixed properties, while the broader research area of "complexity theory" also embraces "complex adaptive systems," which consist of agents that are capable of learning and adapting as they interact with other agents. In complex adaptive systems, the interactions of the agents generate emergent and self-organizing behavior. Mitchell proposes a definition of a complex system of this type as

> ... a system in which large networks of components with no central control and simple rules of operation give rise to complex collective behavior, sophisticated information processing, and adaptation via learning or evolution.
>
> *(2009, p. 13)*

Today, the SFI remains a successful, independent, nonprofit organization engaging in theoretical research, providing educational programs, and running a business network. It declares itself as "the world headquarters for complexity science" and sees its mission as "searching for order in the complexity of evolving worlds":

> Our researchers endeavor to understand and unify the underlying, shared patterns in complex physical, biological, social, cultural, technological, and even

possible astrobiological worlds. Our global research network of scholars spans borders, departments, and disciplines, unifying curious minds steeped in rigorous logical, mathematical, and computational reasoning. As we reveal the unseen mechanisms and processes that shape these evolving worlds, we seek to use this understanding to promote the well-being of humankind and of life on earth.

(Santa Fe Institute 2017)

The researchers consist of a small number of resident faculty, with support from postdoctoral fellows, and a large number of visiting scholars. During its history, the SFI has made important contributions to the study of complex adaptive systems of a physical, computational, biological, ecological, and social nature. Its current research themes are: complex intelligence, natural, artificial, and collective; complex time; invention and innovation; limits; mental models of complexity; and law, history, and regulation. The Institute has also given birth to a number of spin-off companies including the *Bios Group* (associated with Kauffman – helping companies manage projects and supply chains), *Swarm Development Group* (associated with Langton – agent-based modeling tools), and *Prediction Company* (associated with Packard – forecasting techniques for financial markets). We concentrate on the three contributions of the SFI that are most relevant to management: "fitness landscapes," agent-based modeling, and "scale theory."

Stuart Kauffman, a medical doctor and theoretical biologist, was a key member of the SFI from 1986 to 1997. Influenced by the cyberneticians McCulloch and Ashby, he is interested in building computer models to illustrate how order might arise spontaneously in biological systems. The models consist of networks of interconnected nodes, which can either be "on" or "off." He found that by varying the number of nodes, the connections, and the rules of interaction, he could simulate self-organization in the network and the genesis of different kinds of emergent behavior (Ramage and Shipp 2009). The most successful example is the "fitness landscape." This started as an application of the model to the interaction and evolution of diverse organisms, where an organism reaching a high peak in the "landscape" represents evolutionary success while being stuck in a valley poses a threat to survival. However, Kauffman (1995) is clear that the same principles apply to other "agents" (protein molecules, species, companies, technologies) in other contexts (ecosystems, economies), and so a general description is in order. The idea is that the environment of any "agent" will contain numerous other agents all engaged in co-evolution. Kauffman envisages the overall environment as a heaving landscape. It shifts around, throwing up peaks of different heights separated by valleys, as a result of the nonlinear interactions between agents possessing different attributes. An agent has to try to navigate this landscape, which is unpredictable from its point of view. If an agent finds itself on a high peak, then it can gaze contentedly at competitors. If it is in a valley, it is in a poor situation. Luck might drive it up a local peak, representing a rise in fitness, but it may still only be in the foothills compared with the high peaks it can now see in the distance. Furthermore, seeking to reach those distant peaks is treacherous both because it will have to cross a fitness valley and because the landscape is ever-changing as a result of the adaptive moves of its co-evolutionary partners. Even if the agent attains a high peak, it will need to be watchful because, as everyone else clambers onto it, their "weight" will force the peak down. Kauffmann argues that networks that operate near the "edge of chaos," between order and disorder, provide the best environment for solving problems and for further evolution, and regards this as a good justification for liberal democracy.

Kauffman believes that his work points in a spiritual direction that can leave us feeling "at home in the universe." In his view, the evolution of life and of complex organisms appears to owe as least as much to self-organization as to the chance of Darwinian natural selection. The development of complex life-forms is to be expected. An argument that began with von Bertalanffy, and was elaborated by Maturana and Varela, and Prigogine, finally finds its full philosophical expression in Kauffman:

> Order, vast and generative, not fought for against the entropic tides but freely available, undergirds all subsequent biological evolution. The order of organisms is natural, not merely the unexpected triumph of natural selection …. If true, we are at home in the universe in ways not imagined since Darwin stood natural theology on its head with his blind watchmaker.
>
> *(Quoted in Ramage and Shipp 2009, p. 244)*

We cannot expect a biologist to tell us whether self-conscious human beings, psychic and social as well as biological entities, should rest content with this.

Much of the work of the SFI has consisted of using computational techniques to conduct "agent-based modeling" of the behavior of complex systems. Agent-based modeling works from the "bottom-up" and seeks to explain the behavior of the whole in terms of the rules of interaction of the "agents" that constitute the system. This differs from chaos theory and the theory of dissipative structures, which use macro-equations to construct models of the whole system (Burnes 2005). The hope is that, although complex systems usually contain large numbers of agents, it might be possible to discover simple rules that govern their local interactions and produce the emergent structures or patterns of behavior that are observed. Indeed, it does seem possible to model the amazing flocking behavior of birds as a network of moving agents, called Boids in the simulation, if each Boid obeys the same three simple rules (see Stacey 2003, p. 239):

- Maintain a minimum distance from other objects in the environment including other Boids
- Match velocities with other Boids in the neighborhood
- Move toward the perceived center of mass of the Boids in the neighborhood

There is no overall plan at the system level. Flocking emerges as an attractor simply as a result of the Boids interacting according to the local rules.

Another related source of inspiration for agent-based modelers, at the SFI, has been the self-organizing behavior of social insects such as ants, bees, termites, and wasps. In colonies or swarms, these insects display a kind of "collective intelligence," even though there is no overall control and no individual agent understands what is going on. Ant colonies, for example, prosper in the face of competition for resources, react to predators, divide up the labor required to support the colony, and forage for food in an optimal manner. The numerous interactions that occur between agents give rise to these emergent behaviors as a result of simple rules based on chemical signals, e.g. pheromone trails left by ants stimulate nest-mates to choose the same path. Thus, a combination of natural self-organizing processes, and evolution operating at the level of the colony, has produced "super-organisms" that enjoy a rich relationship with their complex environments. Peter Miller, then a senior editor at National Geographic, popularized these findings in his book *The Smart Swarm: How to Work Efficiently, Communicate Effectively, and Make Better Decisions Using the Secrets of Flocks, Schools, and Colonies* (2011). His claim is that studying how social insects function in colonies or swarms can help humans manage their own complex problems. One example shows how algorithms derived from the behavior of ant colonies enabled *Air Liquide* (North America) to optimize various business processes such as the routing of trucks to factories, production levels at different plants, and which customers to serve first.

The agent-based models considered so far assume that all agents are the same, and remain so, and that the rules of interaction between them stay fixed. Such models are an improvement on equilibrium models, which require reality to move to a stable state. But, possessing the internal capacity to explore just one "attractor," they can still only produce a limited range of behaviors. Nothing truly novel occurs. They are not suitable, therefore, for describing most of the scenarios in the life and social sciences where we witness systems changing radically and unpredictably over time and exhibiting surprising emergent properties. The solution, of course, is to loosen the constraints imposed upon both agents and rules of interaction until a model can be constructed that more accurately resembles the real world. Miller and Page (2007) methodically set out the various steps that can be taken. Heterogeneous agents, or classes of agents, possessing different information resources, can be used in the model. Agents can be provided with the capacity to learn and so adapt their behavior based on experience. They can be provided with motivations and even be required to dislike each other. More elaborate rules of interaction between agents and with the artificial environment can be introduced. The ways of interacting can be allowed to evolve. In these circumstances, the interactions become highly nonlinear and the system as a whole changes and evolves as it moves unpredictably from one attractor to another. SFI researchers have used more sophisticated models based on more intelligent agents and more complex agent dynamics to simulate the behavior of, among other things, traffic patterns, the internet, ecosystems, forest fires, epidemics, technological developments, pandemics, the stock market, and the whole economy. These models are often an improvement on what was available before. It is surely better to represent an economy as a complex adaptive system that gives rise to emergent properties as a result of the interactions between multiple, diverse agents than to picture it as a system in equilibrium. Nevertheless, the ability of the models to describe and explain, let alone help us manage, actual, complex biological, ecological, and social systems remains in doubt. The problem is well captured in the science fiction novel *The Cyberiad*:

> .... Trurl decided to silence [Klapaucius] once and for all by building a machine that could write poetry. First Trurl collected eight hundred and twenty tons of books on cybernetics and twelve thousand tons of the finest poetry, then sat down to read it all. Whenever he felt he just couldn't take another chart or equation, he would switch over to verse, and vice versa. After a while it became clear to him that the construction of the machine itself was child's play in comparison with the writing of the program. The program found in the head of an average poet, after all, was written by the poet's civilization, and that civilization was in turn programmed by the civilization that preceded it, and so on to the very Dawn of Time, when those bits of information that concerned the poet-to-be were still

swirling about in the primordial chaos of the cosmic deep. Hence in order to program a poetry machine, one would first have to repeat the entire Universe from the beginning – or at least a good piece of it.

(Lem 1975, p. 43)

And what would be the point of doing so? While being supportive of the construction of quantitative agent-based models that make their assumptions clear, Miller and Page recognize that models that seek to be comprehensive are inevitably "big and messy," and

... substituting a real world that is tough to understand for an equally confounding artificial one may not be all that helpful.

(2007, p. 199)

Our final drink at the Santa Fe saloon comes in the form of Geoffrey West's book on "scaling" – how things change with size. West, a theoretical physicist, served as SFI president from 2005 to 2009. In the book, he reasserts the Institute's ambition to develop a grand unified theory of complex adaptive systems and wonders:

Could there conceivably be a few simple rules that all organisms obey, indeed all complex systems, from plants and animals to cities and companies?

(2017, loc. 147)

The hope is that, by studying the regularities apparent in how things scale with size, he will be able to find some generic laws that help in understanding the "coarse-grained" behavior of all such complex adaptive systems. His approach is therefore, like that of chaos theory and of Prigogine, to search for laws that operate directly at the system level rather than, as agent-based modelers propose, work from the bottom up.

West acknowledges two theoretical precursors to his thinking. The first, Galileo, recognized the importance of scaling. For example, there are limits to which an animal, tree, or building can be "scaled up" before it collapses under its own weight. This, as West explains, is because weight "scales" as a cube of linear dimensions while strength increases only as a square. The second, D'Arcy Wentworth Thompson, provided the inspiration for thinking that there could be a proper science of organisms. A professor of natural history, he published *On Growth and Form* in 1917. This book, famous in its time, is attracting renewed interest and its centenary was celebrated in editorials, abridged versions, and a facsimile edition of the original (Rose 2017). Thompson argued that the remarkable shapes and forms taken by living organisms can be explained in terms of simple mathematical laws. The spirals of a sunflower seed head, the whorls of a pine cone, and the spirals of a snail shell are all arranged in a Fibonacci series. Other natural phenomena are shown to follow what we would now call fractal logic. He reasoned that there must be some physical forces at play that bring forth the patterns that we observe. The followers of Darwin saw natural selection as the only driving force. Thompson insists that there are constraints imposed by physical laws on the outcomes to which evolution can give rise. Only certain structures and forms can possibly evolve.

West begins his search for the scientific laws underpinning the growth and shape of biological systems, cities, and companies by providing some examples of scale regularities. It turns out that any animal that is twice the size of another one requires only 75% more food and energy to sustain itself. A city double the size of another in population requires only 85% of the infrastructure (length of roads, electrical cables, water pipes, and number of gas stations). These examples of "sublinear scaling" reflect "economies of scale." The city is also likely to produce 15% more patents and wealth, have 15% more restaurants and, unfortunately, 15% more crime and flu than if the scaling had been linear. These are examples of super-linear scaling. Such "power laws" seem to operate across taxonomic groups of animals and, in the case of cities, across national boundaries and historical periods. There are, however, outliers. For example, crime tends to be lower in Japan so, although the 1.15 power law holds in both Japan and the United States, you cannot predict crime in a Japanese city on the basis of US data. In order to develop a proper science of organisms, cities, and companies, it is now necessary to discover the physical laws that generate scaling regularities. According to West, all complex adaptive systems are networks that have to integrate their components and service them efficiently at all scales, and it is the logic imposed by network dynamics that constrains how such systems grow and the form they take:

> It is the generic, physical, geometric, and mathematical properties of these network systems that underlie the origin of these scaling laws …
>
> *(West 2017, loc. 574)*

   The network properties that are at work here are "space filling" (the "tentacles" of the network have to spread to all parts); "invariance of terminal units" (providing "ready-made" basic building blocks, e.g. the capillaries of the circulatory systems of all organisms are approximately the same size, as are electrical outlets in buildings in a city); and "optimization" (e.g. use less energy for system maintenance so that more is available for sex, reproduction, child rearing, etc.). As was noted, in Section 7.2, fractal dimensions can promote the effectiveness of terminal units in space-filling networks such as the circulatory and respiratory systems. It is from such mathematical laws, seized upon by natural selection, that the power laws are derived. West suggests that one use for the power laws might be to work out which cities are performing well and which are not, for example, in curbing crime, according to the scale regularities. While this might be interesting, such variations must also undermine any belief we have in the predictive power of the "scientific laws" to which network dynamics give rise. Other Santa Fe researchers, such as Bettencourt, use the knowledge they believe they possess of cities as complex adaptive systems to offer planning advice and suggest designs "that are generative of the whole but not prescriptive of the parts" (2013, p. 10). In his book, West has even bigger fish to fry. In biological systems, network dynamics produce sublinear scaling, slowing down the pace of life. Social networks, however, seem to produce super-linear power laws, probably because as group size increases the number of possible social interactions increases more rapidly than the numbers in the group. This leads to an exponential growth in economic activity (and social ills) that is unsustainable in the context of limited resources. Although social networks also promote creativity, it seems unlikely that the rate of innovation can keep up and prevent eventual collapse.

   What you make of this will depend in part on how accurate and general West's scaling regularities turn out to be. Human beings already seem to have disproved his "scaling regularity" between the number of heartbeats in a lifetime and the weight of an animal by doubling their lifespan over the past hundred years. It will also depend on how interesting you regard the results as being. Even if the heartbeat/weight regularity can be made to fit some notional, "normal" human lifespan, we are likely to be more concerned with the factors that cause variations from the norm. It seems to me that West's "coarse-grained" analysis misses out most of the fine-grained differences that make us human. Contemplate the following:

> We may look different, dress differently, speak different languages, and have different belief systems, but to a large extent our biological and social organization and dynamics are remarkably similar.
>
> *(West 2017, loc. 4848)*

7.5 Complexity Theory and Management

The world that managers inhabit can often be random, unpredictable, high risk, and messy. Complexity theory seems to offer a better description of this reality than other management theories. It also promises to find order in the midst of all the chaos. This combination has proved irresistible to management writers, many of whom regard the insights of complexity theory as being of great importance for management theory and practice. Rosenhead, however, has his doubts:

> It hardly needs saying that there is no formally validated evidence demonstrating that the complexity theory-based prescriptions for management style, structure and process do produce the results claimed for them.
>
> *(1998, p. 10)*

In the absence of reliable evidence, advocates of complexity theory tend to rely on the "authority of science" to make the case for them. This is hardly compelling. To produce a convincing argument for the generality of their theories, Rosenhead suggests, they would first have to demonstrate that they apply to the natural systems that they have been widely used to investigate. He concedes that there is some solid evidence here – the weather, ecological cycles, chemical clocks, etc. – but not enough to enable us to conclude that such results apply to all natural systems facing similar conditions. John Maynard Smith, a distinguished evolutionary biologist, argues that the behavior exhibited by Prigogine's dissipative structures in chemistry describes "the *kind* of process developmental biologists should be looking for," but comments:

> It is nice to know that the processes we are thinking about do not contradict the laws of thermodynamics .... What I cannot tell is whether the new thermodynamics is going to be of any more detailed use in analyzing development. At present I confess I do not see how.
>
> *(1984, pp. 4–5)*

In any case, not all nonlinear dynamical systems do exhibit chaotic behavior. Depending on the equations and the relationships between and strength of the feedback loops, some do settle down to a state of stable equilibrium. Furthermore, many of the results cited are the outcomes of computer simulations rather than empirical observations. Such demonstrations are suggestive but cannot be proof that actual observed behavior is caused by the laws built into the computer program. Byrne and Callaghan (2014) comment on the miniscule number of papers in the *Journal of Artificial Societies and Social Simulation* and in *Computational and Mathematical Organization Theory* that employ evidence-based modeling based on real data. If complexity theory is unproven in the natural sciences, where it originated, why should we believe that it has a lot to offer managers who operate within social systems?

Perhaps, Rosenhead suggests, complexity theory can provide "analogies" that are relevant in the social domain. This, following Brodbeck, would require:

> ... (a) that the natural scientific domain of complexity theory is better understood than that of management; (b) that there are concepts in the first domain which have been clearly put in one-to-one correspondence with similarly precise equivalents in the second; and (c) that connections (especially causal ones) between groups of concepts in the first domain are implicitly preserved between their equivalents in the second.
>
> *(Quoted in Rosenhead 1998, p. 14)*

In reviewing the evidence, Rosenhead feels that complexity theory is not yet mature enough to provide a reliable source for analogies and whatever equivalences might be claimed. For example, the simple existence of nonlinear feedback between elements is too general and undemanding to carry weight. Johnson and Burton share Rosenhead's view that the argument from analogy does not work for organizational systems:

> Nothing about real social systems fits within [chaos theory] limitations ... all of the systems that organizational researchers study are complex and open to numerous outside influences.
>
> *(1994, pp. 323–324)*

Social systems, it seems, pose particular problems for complexity theory. As Byrne and Callaghan argue:

> ... any general complexity social science has ... to allow for structures with causal powers and it has to address human agency as capable of transcending narrow rules for behavior.
>
> *(2014, loc. 1372)*

Taking human agency first, the fact that humans possess self-consciousness and free will, and act on the basis of sophisticated, internal "images" of the world, embracing both facts and values (see Section 5.5), means that the behavior of social systems cannot be explained in the same way as that of natural systems. Humans are self-reflective, they formulate intentions and purposes, and learn and adapt. They are capable of reacting against and usually disproving any law that is said to apply to them. Byrne and Callaghan also insist that we look closely at how complexity theorists treat the agency/structure relationship, which absorbs so much attention in social science. In most cases, they see structure as micro-emergent but as possessing no independent reality and causal powers of its own. By contrast, many sociologists tend to see humans as born into social structures, which constrain life opportunities and socialize individuals in ways that make it more likely that their agency will reproduce rather than change existing arrangements (see Chapter 4). Causality operates from the whole to the parts as well as vice versa.

Byrne and Callaghan (2014, loc. 963–979) use a distinction originally drawn by Edgar Morin to explain the problems posed for complexity theory in the social realm. To be successful, it has to shift from looking at "restricted complexity" to considering "general complexity." Restricted complexity remains in thrall to the epistemology of traditional science and sees complexity as emerging from the interactions of simple components or agents. It assumes that complexity can be explained on the basis of the development of nonlinear equations or a few rules governing the interactions of the agents. Weather systems may be difficult to predict but meteorologists know the basic equations that underlie them. The flocking behavior of starlings may appear unfathomable but only three rules need be known to construct accurate computer simulations. The primary concern is with deterministic chaos. This is not the world encountered by managers. Managers confront "general complexity" in which human actors are themselves complex entities and there are mutual interactions between the micro and macro levels of the system, with both possessing causal power. Their world is one of "nested but interpenetrating systems with causal powers running in all directions" (Byrne and Callaghan 2014, loc. 1122). Miller and Page (2007) have set out the steps modelers need to take to approach an understanding of general complexity, and sophisticated complexity thinkers such as Kauffman (with diverse agents producing fitness landscapes that certainly impact their options) have taken many of them on board, but still complexity theory does not come close. As suggested previously, in this chapter, it is an impossible task. Let us allow a Zen Buddhist, Haemin Sunim (2017, p. 120), a final attempt to suggest why:

> The whole universe is contained
>
> in an apple
>
> wedge in a lunch box.
>
> Apple tree, sunlight, cloud, rain,
>
> earth, air,
>
> farmer's sweat are all in it.
>
> Delivery truck, gas, market,
>
> money,
>
> cashier's smile are all in it.
>
> Refrigerator, knife, cutting board,
>
> mother's love are all in it.
>
> Everything in the whole universe
>
> depends
>
> on one another.

Now, think about what exists in

you.

The whole universe is in us.

We must conclude that complexity theory falls short in its attempts to demonstrate that there are either scientific laws or analogies that can help managers understand and improve the performance of the complex adaptive systems with which they have to deal. There remains the possibility that it can provide illuminating metaphors. Under this reading, it loses any prescriptive force and must compete, on the basis of vividness, resonance, and traction, with other possible metaphors. If the insights it offers are as novel as claimed, however, it should have no difficulty on this score. Begun puts the case, arguing that

> ... chaos and complexity theory invite us to explore the 95% of the organizational world that we have avoided because it is too dark, murky, and intimidating. Or, our theories and methods simply have not allowed us to see it. Integration of chaos and complexity theory into organization science will fertilize the soil of the discipline's weed patch of theories ... allowing some flowers and fruits to grow.
>
> *(1994, p. 334)*

Rosenhead (1998) agrees that metaphors derived from complexity theory can be helpful when they challenge the classical view that consensus in organizations is a good thing. Shared vision can lead to groupthink. Disputes can help promote the creativity and learning necessary for organizational viability and effectiveness. There is, in fact, a huge literature seeking to apply metaphors from complexity theory to management. I will present a few examples. Stacey (2003, pp. 261–290) examines a larger sample.

Wheatley notes that complex systems are extremely sensitive to small changes and argues that this makes it pointless managers attempting to plan, organize, and control their enterprises in detail. However, they need not worry that things will fall apart. Rather, they should trust, indeed delight, in chaos:

> It is time to take the world off our shoulders, to lay it gently down and look to it for an easier way. Lessons are everywhere .... Nature is abundantly littered with examples and lessons of order. Despite the experience of fluctuations and changes that disrupt our plans, the world is inherently orderly. And fluctuation and change are part of the very process by which order is created.
>
> *(Wheatley 1992, pp. 17–18)*

This, though, does not mean that managers can leave everything to chance. They have to propitiate favorable conditions for creativity and learning.

Kuhn (2009, p. 12) argues that looking through a "complexity lens" is useful for understanding and managing organizations because it:

- removes simplistic hopes of an ordered and controlled existence in which we can fashion organizations to our own image

- offers a means of discerning and identifying underlying patterns of order, providing a richer appreciation of situations
- introduces us to *potentiality*, by showing how complex behavior results from interactive local relationships and suggesting how we can influence what emerges

Morgan (1997), aligning complexity theory with his "flux and transformation" metaphor, suggests a three-stage methodology for its use. Stage 1 consists of understanding the attractor pattern determining the current behavior of the organization and the reasons why it is dominant. If the pattern is not desirable from the organization's point of view, then change must be brought about in order to ensure the system shifts to another pattern. Making the change is Stage 2. With sufficient awareness of points of maximum leverage we can, by making relatively small changes, break the cycle and bring about more fruitful ways of functioning. Stage 3 requires the new attractor pattern to be stabilized while, at the same time, ensuring that it does not lock the organization, in the long term, into routine forms of action.

Stacey (1996) echoes Wheatley, Kuhn, and Morgan in arguing that, although we should trust chaos, it sometimes needs a helping hand. He thinks that managers need to steer organizations to the "edge of chaos" because that is where complex adaptive systems fully display their potential for creativity and innovation (see Section 7.2). The edge of chaos demands that there be a creative tension between an organization's "legitimate" and "shadow" systems. The legitimate system must provide clear guidelines, authorize appropriate structures and procedures, and contain anxiety among the personnel. It is vital in promoting efficiency but, if it becomes too dominant, it can constrain all opposition, prevent questioning of objectives, kill learning, and stop change. The shadow system must give rise to a diversity of perspectives. It is the source of innovation, contention, and political struggle as different groups engage in dialogue and learning, and entertain alternatives to the status quo. If it becomes too dominant, however, anarchy can result. For Stacey, the new thinking offered by complexity theory is about

> … sustaining contradictory positions and behavior in an organization … positively using instability and crisis to generate new perspectives, provoking continual questioning and organizational learning through which unknowable futures can be created and discovered.
>
> *(1992, p. 17)*

Organizations that operate "far from equilibrium" can embark on exciting and unpredictable new journeys.

A nicely measured interpretation of the value of Prigogine's theory of dissipative structures to the social sciences, and to strategic and project management, is offered by Boulton et al. In their view, Prigogine's models, embracing internal and external "fluctuations" and microscopic diversity, enter the "realm of evolutionary complex models" that are relevant in the social as well as the natural sciences. The details are different, because of human self-consciousness and free will, but the concepts do transfer across. They therefore encourage us to embrace a complexity "mind-set," and argue that concepts such as "far from equilibrium," "path-dependence," "self-organization," and "emergence" can be very illuminating about the nature of organizations:

> A complexity view sees the organization as ongoing, evolving, organic, and reflexive – continually interacting with its environment. The organization is in a continual state of emergence, and its future is path-dependent. That is to say the scope for change in the organization is constrained by past actions and decisions and yet still able to change. It is not a static entity that can be viewed as a stable 'structure' of roles, reporting lines, and accountabilities. It certainly cannot be viewed as a machine or an 'instrument' of those who think they can control it. We cannot 'pull' a lever of change and assume we can predict the ultimate impact of this intervention.
>
> *(Boulton et al. 2015, pp. 169–170)*

Managing organizations, seen in this light, requires flexible leadership. It is still necessary to establish and pursue clear goals. Self-organization cannot be relied upon to take us where we want to go: "there is nothing self-evidently 'good' about self-organization" (Boulton et al. 2015, p. 236). However, we must constantly be on the look-out for unanticipated changes and plan to be more flexible in responding to the unexpected. This translates into six recommendations for strategic management (pp. 164–167):

- Build a portfolio – your cash cow will die sooner than you expect
- Foresight the future, imagining a number of possibilities and scanning for indicators of major change
- Weave clear intentions globally but plan locally to take account of local conditions
- Cast around for growing shoots
- Experiment, seize opportunities, and adapt to changing circumstances
- Find the middle ground between persistence and agility

Boulton et al. (2015, p. 236) identify the philosophy underpinning their version of complexity management as "pragmatism" (see Section 1.4). Knowledge about the behavior of dissipative structures can only ever be local and contextual and so we must content ourselves with trying to find out "what works" using an action research approach (see Section 4.8) based on trial and error and reflection.

It is clear, even from our small sample, that the attempt to apply complexity ideas to management produces a variety of results. In trying to make sense of this, we need to carry out a "second-order" analysis that applies complexity theory to itself. The aim is to uncover the reasons why observers, using complexity ideas to inform their thoughts on management, characterize complexity in the ways that they do. In part, no doubt, this will be due to chance – to the complexity ideas they have access to at the time of committing their thoughts to paper. It will also depend on purpose. Tsoukas and Hatch (2001) argue that behind every narrative describing a system as complex must be a "narrator" constructing a story for a purpose. They suggest a form of "second-order complexity" that concerns itself with narrators and the subjective accounts of complexity that they produce. This enables them to identify a dominant, logico-scientific type of complexity theory used in organizational analysis and to argue that alternative accounts of the nature of complexity, and the way it is relevant to organizations, are available. They set the scene but don't take us very far in analyzing the alternatives. A more productive "second-order" analysis is possible if we follow Stacey and his collaborators (Stacey et al. 2000; Stacey 2003; Stacey and Mowles 2016) – henceforth referred to just as Stacey in the text – in exploring the theoretical assumptions about human beings and social reality, implicit or explicit, that different observers bring to the task of applying complexity theory to management. This reveals that there is indeed an order underlying the apparent chaos.

Stacey argues that much of the work of management complexity writers fails to capitalize on the radical potential of complexity theory because it is stuck in the "dominant discourse on management" as expressed in scientific management and systems thinking. He includes his own early work, mentioned above, in this critique. In these writings, managers (and their advisors) are pictured as standing above their organizations making rational choices about goals and how to achieve those goals. They are able to employ command and control because organizations are seen as operating on the basis of scientific laws (scientific management) or as behaving in an ordered way as a result of the interactions between their parts (systems thinking). Thus a management complexity writer, under the sway of systems thinking, might argue that organizations should be steered to the "edge of chaos" because at that point the parts will naturally interact in such a way as to make the whole more effective. Stacey notes a contradiction in this thinking because it postulates that managers function according to a "rationalist teleology," freely choosing purposes, while organizations obey causal laws (scientific management) or display "formative teleology," evolving according to a pattern set by an existing but hidden order (systems thinking). Stacey regards both aspects of this thinking, which he sees as originating in a confused understanding of Kant's philosophy, as wrong when applied to social systems. Managers (and their advisors) are participants in organizations, not apart from them, and organizations emerge as a result of the relationships between all their purposeful participants (including managers), which can give rise to radical and unpredictable change as well as continuity. Sophisticated thinkers, such as Kauffman and Prigogine, Stacey admits, have made significant progress in freeing complexity theory from the traditional management paradigm, but still do not go quite far enough. They continue to talk as though they are modeling "systems" and do not grasp the full complexity that comes about when "systems" develop a dynamic of their own as a result of multiple human purposes and the relationships to which these give rise.

To realize the full potential of complexity theory for management, Stacey reasons, we must rethink what its implications are when we free it completely from systems thinking and the traditional scientific paradigm. We can do this if we shift perspective and take a "process view" rather than a "systems view" when making use of complexity ideas. Those who succeed bring a much more radical version of complexity theory to the fore:

> These voices emphasize the radically unpredictable aspects of self-organizing processes and their creative potential. These are the voices of decentered agency, which talk about agents and the social world in which they live as mutually created and sustained. This way of thinking weaves together relationship psychologies and the work of complexity theorists who focus on the emergent and radically unpredictable aspects of complex systems.
>
> *(Stacey et al. 2000, p. x)*

Taking such a process view means giving up on the idea that organizations actually *are* complex adaptive systems. In exchange, we liberate the complexity sciences to provide a rich source of metaphors more appropriate for the social domain because they foreground the "transformative teleology" that organizations frequently exhibit.

Three key figures for Stacey, in developing his process perspective, are Hegel, Mead, and Elias. Hegel's philosophy (see Section 1.3) provides a broad historical account of the social evolution of thought and individual autonomy through the dialectical process. Mead (see Section 4.3) was influenced by Hegel but turned his attention to investigating social processes at the micro-level. His "symbolic interactionism" sought to show that both society and self-awareness develop as a result of human interaction and reflection conducted in the medium of language. Norbert Elias offers a sociology of process constructed on the basis of individual interactions, which give rise to unintended outcomes because people must pursue their intentions in relationship with each other in a group or power configuration. Stacey's own extension of this mode of thinking is called the study of "complex responsive processes of relating" (CRP). According to this perspective, the individual and the social are the same phenomenon, forming and being formed by each other in processes of action and interaction. Organizing consists of the continuous interactions between individual humans, all capable of making choices, as they relate to each other through communication. Through this ongoing process of communicating and relating, they transform both their environments and their own identities. Organizational strategy, the way in which an organization's identity emerges, can be explained in the same way:

> Strategy ceases to be understood as the realisation of someone's intended or desired future state for the whole organisation. It ceases to be understood as the intentional design and leveraging of whole organisational learning and knowledge-creating systems. Instead, strategy is understood as evolving patterns of simultaneously collective and individual identities. Evolving identities are understood to emerge in the local communicative interacting, power-relating and ideology-based choices of the people who constitute an organisation.
>
> *(Stacey and Mowles 2016, p. 301)*

We can pick out some important points. Traditional management theory pictures managers as standing outside their organizations and determining their destiny. Stacey (2003, pp. 359–360, 383) regards them as participants in self-organizing processes leading to an unknowable future. For Stacey, the organization itself is seen as being in a process of perpetual construction based on the micro-interactions between humans (including managers and their advisors) as they relate to each other through symbols. These symbols, usually in the form of language, consist of gestures and responses interwoven with emotions. They form themes that produce patterns of conversation and power relations, enabling some things to be said, done, and thought while preventing other things from coming to the fore. Such patterns act rather like strange attractors in complex adaptive systems. Themes often serve as ideologies in support of the status quo but they can also justify undermining it. In either case, they tend to work by dividing experience into sameness and difference; who is included and who excluded. Themes guide interactions and self-organization in ways that can furnish continuity but also, sometimes, bring about radical transformations in organizational and individual identity, acting as a source of novelty and creativity. Organizations, therefore, are arenas of "transformative teleology" where strategies, goals, and values emerge from self-organizing complex responsive processes. If the quality of relationships is good in an organization, there will be a sufficient diversity in organizing themes and this will permit free-flowing conversations to take place. If trust is built, change can be embraced without provoking too much anxiety among the participants.

In recent work, Stacey and Mowles place greater emphasis on the nature of the "narrative themes" that emerge from the patterning of communicative interaction and the way in which they become influential in organizations. These "first-order abstractions" seek to provide an account of and give meaning to what people are actually doing and seeking to achieve. They reflect people's natural desire to talk about their collective experience in an abstract way:

> We do not simply experience one thing after another in a random disconnected way but, rather, we experience some kind of continuity, some kind of ongoing unity, in what we and others are doing in an organisation and the wider society it is involved in. This experience of unity requires acts of human imagination in which we generalise population-wide patterns, imaginatively ascribe a unity to them and then idealise this 'whole'. This imaginative activity is of the greatest importance because, in articulating social objects and cult values, it is repeatedly taken up by many, many people in the particularity of their local interactions.
>
> *(Stacey and Mowles 2016, p. 419)*

Leaders and managers take advantage of the power these "abstractions" can exert to reinforce their own position. Strategies, goals, visions, and missions represent ideologies that perform this role. Managers may also make use of more formal "second-order abstractions." These "models" or "maps" provide them with tools to use to steer their organizations from a distance. In all cases, however, abstractions can only be influential if they resonate with people and are taken up in the myriad of local interactions that occur in the organization and, even then, they will be interpreted according to the diverse intentions of the individuals involved.

In considering the implications of this theory for how managers (and their advisors) might act in organizations, Stacey regards his most important contribution as helping managers to "think differently." Rather than see themselves as external controllers of their organizations, they should see themselves as participants joining in existing conversations and attempting to learn something about the prevailing and emerging themes that are being expressed in the conversations:

> There is no intention to design anything, improve it, or make it right or more creative. Instead the intention is the same as that of other participants, namely to understand what they are all doing together, what they are talking about and why.
>
> *(Stacey 2003, p. 401)*

The focus of attention shifts away from trying to achieve a specific purpose to the actual quality of participation in the ongoing conversations. It is important that active "forums" or "gatherings" emerge spontaneously around people's interests, enthusiasms, or frustrations and that the discussions that occur have an "everyday quality," messy and meandering, but also engaging and exploratory so that it is clear that the future is still under construction. Any attempt to structure these meetings is thought to detract from their potential to produce new thinking. Effective managers (and advisors) should be aware that they are managing in relationship with others and be self-reflective about the way they themselves are acting and how others are reacting to them. They can then join in the conversational processes that are taking place as fully and responsively as possible. Of course, they will want what they say and how they behave to contribute to the conversations being free-flowing. They will seek to shift the conversation away from routine, repetitive themes and to encourage emerging themes expressing diversity and novelty. It will assist them in doing this if they have an awareness of power relations, covert politics, and unconscious group processes. Managers need to be comfortable knowing that there is a lot that they cannot know. In these circumstances, the best course of action can be keeping options open for as long as possible or simply acting as best one can, on an ethical basis, recognizing that the outcome is uncertain. Discussions about change always provoke anxiety, so managers must build sufficient trust among participants to allow difficult conversations to take place. Managers working from the CRP perspective know that thinking and talking are action. Nothing else is required and so they do not have to concern themselves with a further act of implementation. The spirit of all this is nicely captured by Shaw in discussing her proposal for a consultancy with an American-owned multinational with which she had already had a minor involvement:

> I called Donald to get his reaction to my contract proposal. 'Fine', he said. 'Go ahead. I don't quite see what you are going to do, but we're all happy for you to start'. 'We will start by phoning people and talking, trying to enter the networks and conversational life of the organization'. He chuckled. 'I recognize that's what you're already doing'.
>
> *(Quoted in Stacey 2003, p. 447)*

Developing our own second-order analysis, making use of the divisions in social theory described in Chapter 4, it is clear that Stacey's severest criticism of the way complexity theory is applied to management is reserved for those writers who understand it through the lens of the functionalist paradigm. Such theorists, as we have seen, assume that complex adaptive systems have an objective existence independent of observers and study how order arises and can be maintained using traditional scientific methods. Broadly, Stacey believes he can liberate complexity theory to realize its full potential in the domain of social systems if he frees it from any taint of functionalism and harnesses it to the interpretive paradigm. Interpretive social theorists such as Mead, one of Stacey's guiding lights, start from the position that social reality is the product of the action and interaction of individual social actors. Society emerges from recurrent forms of social action. As a result, understanding social systems requires the researcher to appreciate the perspectives and intentions of the social actors responsible for their construction. Stacey and Mowles describe their theory this way:

> It supports the call for an activity-based perspective ... which focuses attention on the micro, local interactions of people in an organisation. Furthermore, it offers a different understanding of the link between micro and macro – the macro emerges in the micro. In fact, the distinction between the macro and the micro as different ontological levels dissolves and with it further taken-for–granted

> distinctions between intention and emergence, unpredictability and order, individual and organisation also dissolve.
>
> *(2016, p. 298)*

There are influences from integrationist social theory (see Section 4.6) in Stacey's later work, for example, in the growing concern with the influence of "first-order abstractions" and the interest shown in Bourdieu's concept of *habitus*, but overwhelmingly his analysis and prescriptions for action rest comfortably alongside those of the interpretive social theorists outlined in Chapter 4. As Zhu (2007) argues, Stacey ultimately adopts an "all-process-no-structure" position in which social structure is dissolved into "social practice" and loses any causal power of its own.

This begs the question of how many more sociological paradigms complexity theory can embrace in trying to extend its range of applications to management. The answer is, quite revealingly, all of them. Walby (2007) wants to align complexity theory to the "sociology of radical change." In her view, it provides a conceptual toolkit that can help improve the way we understand power and inequality. In particular, it enables us to theorize better the intersection of "multiple complex inequalities" such as class, gender, and ethnicity. Instead of seeing these inequalities as "nested" in a pre-specified hierarchy, as is the case in much sociology, they can be conceptualized as separate but overlapping and interrelated co-evolving sets of social relations. Cilliers (1998) believes that looking at complexity from a postmodern or poststructuralist perspective helps illuminate complexity theory's value to the social sciences. Postmodernism suggests that modern society is diverse, volatile, and contingent and that this is accompanied by a widespread skepticism concerning rationality, truth, and progress. It becomes immediately apparent that chaos theory, which interests itself in "deterministic chaos" and is therefore of a "modernist" persuasion, is of little use. On the other hand, a complexity theory that emphasizes agency, unpredictability, emergence, and structure elaboration, and which recognizes that knowledge can only ever be local, contextual, and specific to place and time, has much to recommend it. Combining insights from this strand of complexity theory, postmodernism, and poststructuralism, encourages the use of an action research approach to engage with specific complex systems. Byrne and Callaghan (2014) reject functionalist and postmodern versions of complexity theory and advocate a synthesis of complexity theory with Bhaskar's critical realism. Only "complex realism" is deemed to be up to the task of understanding social systems characterized by the "whole-part" mutual causal relations captured in Morin's notion of generalized complexity. The work of Bourdieu and Archer, classified under "integrationist social theory" in Section 4.6, is regarded as coming closest to what is required. They describe a social world in which there is room for human agency but it is heavily circumscribed by the causal power of the existing social structure and culture. Individual actors are therefore more likely to reproduce current power relationships and inequalities through their actions and interactions than to change them. Complexity thinking, Byrne and Callaghan argue, highlights the need for a participatory practice along the lines recommended by Paulo Freire in which social actors, in dialogue with researchers, come to understand the mechanisms producing the disadvantages to which they are subject and recognize what is required to change them.

It is obvious, from this analysis, that complexity theory is unable to transform the social sciences by providing an account of social reality that is superior to existing social theories. In particular, it is unable to overcome the problem posed by the existence of competing and apparently irreconcilable paradigms of analysis in social theory. As Zhu remarks with regard to Stacey's management complexity theory:

> Is it not the case that, in the end, adopting or rejecting Stacey's 'responsive process perspective' amounts more to choosing between available social theories than learning from complexity theories that are still evolving?
>
> *(2007, p. 458)*

This has important implications in terms of its usefulness for managers who are left asking what version of management complexity theory can be believed. As Ortegón-Monroy (2003) has argued, on the basis of a critical systems analysis, complexity theory remains too theoretically incoherent when applied to social systems to act as a guide to practice. Using a case study, she reveals some of the paradoxes this can present. For example, it can leave managers trying to reconcile "functionalist" versions of complexity theory, which inevitably end up recommending the use of command and control to remove existing structures of command and control, with "interpretive" versions, which forbid any such strategy. The best complexity theory seems able to achieve in the social sciences is to add some fruitful new imagery to each of the existing paradigms. And even here we need to be wary of conceding too much novelty to the approach. Carrizosa and Ortegon (1998) argue that the features of organizational life that complexity theory claims to highlight can be perfectly expressed using already existing organizational metaphors. Certainly, previous work on informal groups, group working, open systems, emergence, self-organization, and "turbulent field" environments (see Chapter 12) seems to cover much of the territory that complexity theory wants to claim as its own.

7.6 Complexity Theory and Systems Thinking

Stacey et al. (2000, pp. 9, 11, 158) argue that their theory of "complex responsive processes of relating" (CRP) offers a radical alternative to systems thinking. It "challenges" the "limited ways of understanding" organizational life offered by systems thinking and enables complexity theory to find an original voice. It represents a "decisive move away from systems thinking." Others, however, maintain that Stacey and his collaborators are only able to sustain their argument by misrepresenting systems thinking. Luoma et al. (2011) are adamant that Stacey misses the mark in his criticisms of systems thinking. Zhu (2007, p. 459) is frustrated by Stacey's "selective reading of systems thinking" to serve his own purpose. The evidence is examined by setting out Stacey's main criticisms of systems thinking and seeing whether they hold against three variants of the systems approach: "functionalist," "soft," and "critical." The reader will need to revisit this brief examination of Stacey's argument having absorbed the fuller account of applied systems thinking provided in Parts III and IV of this book.

Stacey regards systems thinking as being the main element in the "dominant discourse" on management. Because it requires the drawing of boundaries, always leaving something outside the system, it encourages an approach to management based on two conflicting "teleologies." On the one hand, managers are seen in terms of a "rationalist teleology," which assumes that they can direct their organizations from the outside using rational means to oversee and improve the interactions of the parts. On the other, organizations are presented as systems governed by a "formative teleology," which sees them as unfolding their potential in a predictable manner. Human purpose is therefore excluded from the organization itself. This discourse leads to managers having an illusion that they are in control but being constantly disappointed when their predictions are falsified by experience and their plans come to nothing. It also blinds them to the main sources of creativity and novelty existing in their organizations – the freedom and agency of people in the organization and the self-organizing capacity that arises from their interactions and relationships. It becomes impossible to grasp how diversity, conflict, and change occur:

> The dominant discourse takes its own ideology for granted, and through its mainstream theories largely sidelines matters to do with interdependence, unpredictability, power, ordinary day-to-day activities such as conversation and politics, and the richer fabric of social life.
>
> *(Stacey and Mowles 2016, p. 204)*

A more realistic picture of what happens in organizations, Stacey argues, can only be achieved if we abandon the notion that human action takes place in a system. This requires a radical change in mind-set to a process theory based on a unifying "transformative teleology," i.e. to CRP. Here the focus is on the micro-interactions and relationships between people and how these give rise to the emergent and creative potential of organizations as they reach out to a future that is unknowable because it is under perpetual construction. Managers should see themselves as participants in these self-organizing processes, seeking influence through participation rather than design. Their visions, strategies, and change programs are merely "gestures," often powerful it's true, which can impact but certainly not determine the responses and interactions of others.

Stacey's critique would carry some weight if the only form of systems thinking available were "functionalist." For our purposes here, I include in this category "hard systems thinking," first-order cybernetics, traditional system dynamics, and some applied systems approaches, deriving from general system theory, such as contingency theory. Hard systems thinking tends to posit experts sitting outside the system under investigation and advising managers what to do on the basis of a mathematical model that claims to represent the logical relationships governing the behavior of the system. First-order cybernetics situates goal-setting as outside the system (see Figure 6.1) and seeks to put in place negative feedback arrangements to ensure achievement of that goal. Traditional system dynamics wants to advise managers what to do using models that capture the interaction of the parts in multiple positive and negative feedback loops and indicate key leverage points for changing things that are revealed by the models. Contingency theory claims to produce scientific knowledge that can tell managers what structural arrangements to put in place to ensure effective performance given the demands of the key organizational subsystems and the environment.

Even in the case of this functionalist systems thinking, it is reasonable to ask in what circumstances it can be useful. Despite the weaknesses he perceives it having, Stacey does grant that it has made a "significant contribution" to management theory and practice (Stacey et al. 2000, p. 80). It is deemed good for ensuring continuity and designing repetitive processes. It is praised for help in designing regulatory processes and pointing out that interactions between the parts of systems can lead to unintended and unexpected consequences. John Seddon's "Vanguard Method" (see Chapter 10), which we might expect to incur Stacey's wrath, is regarded as an effective way to design flows of activity (Stacey and Mowles 2016, p. 194). It is acceptable, apparently, because the system concept is here used as a "tool" by people in communicative interaction as a means of orientating their work. "System" here refers to a set of activities that people need to undertake to achieve particular objectives rather than to the organization as a whole with people as its parts. Once this is granted, however, the same argument applies to most functionalist systems approaches. We do not have to "move away" and abandon them. They should be included in the managerial toolkit with a note advising when it is right and proper to use them. There will be occasions when the employment of particular types of expertise can be justified and can be shown to be useful. Operational Research models can teach us how to optimize queuing, inventory, allocation, routing, etc., systems. First-order cybernetics provides the expertise for the design of effective control systems. Traditional system dynamics can help suggest, if only in the short term, some potential unexpected consequences that might arise from intervening in a system. In all of these cases, as well as with Vanguard Method projects, expertise can provide something significantly more than would be likely arrived at through "complex processes of relating." As Stacey and Mowles sometimes concede:

> Predictable and repetitive areas are indeed very prominent and important aspects of organizational life.
>
> *(2016, p. 97)*

Functionalist systems approaches must be respected as providing an effective way of managing these important aspects. Our concern is that while this is occasionally acknowledged, it is more usually downplayed in Stacey's pronouncements.

In moving on to discuss Stacey's critique of the "soft" and "critical" variants of systems thinking, we need to note that his appreciation of these newer strands has become more nuanced with time. He now calls them "second-order systems thinking" and accepts that they have moved away from the dominant management discourse "to some extent" because of the attention they give to participation, relationships, culture, politics, power, and ideology. However,

> ... the fundamental assumptions to do with systems and the primacy of the individual remain intact in the critique mounted by second-order systems thinking.
>
> *(Stacey and Mowles 2016, p. 44)*

In Stacey's view, the second-order perspective abandons the "realist" position espoused by traditional systems thinking, which represents organizations as actual systems "out-there" waiting to be engineered. Instead, it adopts the "idealist" contention that "system" is a mental construct that can be usefully applied to organizations. It benefits from this shift because the question of where the boundary of the system is to be drawn becomes a subjective matter. The boundary becomes flexible. More people can be involved in deciding what the system of concern is to be and the boundary can be enlarged to incorporate things such as culture and politics. However, Stacey believes, second-order systems thinking still treats organizations "as if" they are systems and retains the fundamental divide between autonomous agents outside the system determining what is going to be done and a system deemed to be incapable of generating anything truly novel and innovative itself. The system boundary may be extended, but someone now has to decide who should participate in drawing the boundary and what should be included in the system of concern. Let us see whether these points do indeed tell against soft and critical systems thinking.

Stacey gets soft systems thinking wrong because he feels it treats organizations "as if" they are systems that can be managed externally. This is not the case. Soft systems thinkers offer the notion of "system" to all stakeholders of an organization as a construct that they can employ to enhance their creativity and improve their chances of achieving what they want through their organizations. Churchman sees "system" as a concept that can be used to increase human purposefulness by helping facilitate a never-ending process of learning conducted through the dialectical method. Checkland dismisses the notion that "organizations" *are* systems. He is horrified that his work is associated with systems ideas that he explicitly rejects:

> *Complexity and Management,* an otherwise interesting book, thus offers a compendium of the systems ideas which had to be explicitly rejected in developing SSM [Soft Systems Methodology]: the world as a set of systems, and intervention in the world as a process of designing or redesigning systems.
>
> *(Checkland and Poulter 2006, p. 151)*

He reiterates that his soft systems approach views social reality as constructed through a never-ending process of interaction between actors with multiple values and interests. He takes this to be the position that Stacey and his collaborators endorse and suggests that SSM, properly understood and used, has much to offer their version of the complexity sciences. Ackoff talks of "social systems" but insists that these consist of purposeful individuals who can use the system concept of "idealized design" to think through and pursue the future they most desire. His maxim "plan or be planned for" is not one that would be adopted by someone wishing to design a system for other people. Examples could also be provided from the "second-order cybernetic" and "soft system dynamics" traditions that would show Stacey's critique to be misguided in relation to them.

According to Luoma et al. and Zhu, Stacey is trying to make the case for the uniqueness of his own CRP approach by "freezing" the concept of system so that it can only be used to take an externalist position:

> In systems thinking thus conceived, the manager's focus is on the organization as an object to be manipulated and controlled .... This way of perceiving systems thinking, while polemically powerful, clearly misrepresents the field as a whole.
>
> *(Luoma et al. 2011, p. 5)*

> Stacey allows only one, selected, 'standard' meaning of the concept [system], regardless the diverse understanding, applications and effects of the concept in many, many different contexts and historical moments that have already happened.
>
> *(Zhu 2007, p. 460)*

It is ironic that Stacey, so keen to liberate complexity theory from functionalism, is determined to keep systems thinking confined within that particular paradigm. Once we reject his attempt to freeze the meaning of systems thinking and accept that "system" is a conceptual construct that can be put to different uses, it is clear, as Luoma et al. argue, that there are few theoretical differences between Stacey's "complex responsive processes" approach and soft systems thinking. This is not surprising because they both draw on related philosophical and sociological traditions: Stacey on Hegel, pragmatism and process sociology; soft systems thinkers on Hegel, pragmatism, phenomenology, and interpretive sociology. So Stacey's work and that of soft systems thinkers is compatible. Indeed:

> If one takes the term systems thinking to include the entire family of systems-based problem-solving methodologies, it provides a host of approaches to *organizing* as opposed to theories *of* organizations. Many of these methodologies might well prove useful in a number of local processes and 'joint actions' and as part of productive 'complex responsive processes'.
>
> *(Luoma et al. 2011, p. 7)*

One would think that this would be clear to Stacey now he has come to recognize that people naturally think, talk, and orientate their actions and interactions in terms of "first-order abstractions" (Stacey and Mowles 2016). They converse and conduct themselves as if "wholes," such as families, groups, organizations, states, actually exist. Managers, too, are recognized as operating with "second-order abstractions," like models and maps, many of them systemic in nature. In such circumstances, it seems perverse not to recognize the importance of and make use of systems approaches, which can help illuminate, facilitate, and improve "complex processes of relating," which are already frequently conducted using the language of "wholes" (Luoma et al. 2011). Rosenhead (1998) makes a similar point when discussing Stacey's explicit rejection of tools from the "soft operational research" tradition, such as "scenario planning," "robustness analysis," "group decision support systems," and "problem structuring methods," which he feels would have value even in a chaotic universe.

Stacey's suspicions about soft systems thinking are exacerbated because he believes it encourages practitioners to overspecify what should happen in an intervention:

> Systems practitioners think of themselves as facilitators who structure, shape and guide workshops and other intervention events using the methodologies of systems thinkers.
>
> *(Stacey and Mowles 2016, p. 448)*

The CRP practitioner, by contrast, believes that creativity and novelty have more chance of emerging if meetings are underspecified:

> Here the practitioner joins a group of people as a participant in their conversations, seeking to understand something of the organising themes that are emerging in these conversations .... The practitioner has no intention of creating the right conditions for better conversations or identifying the right people to be

involved in them. There is no intention to design anything, improve it, or make it right or more creative.

(Stacey and Mowles 2016, p. 446)

To assess these two positions, we need to consider the nature of the "structure" soft systems thinkers seek to provide to events. Essentially, as we shall see in Chapters 14–16, the specifications made are meant to allow for the greatest possible participation, to facilitate open discussion about future possibilities, to aid creativity, and to clarify arguments. They also require the systems practitioner to be explicit about what she is bringing to the party at any point. Drawing inputs from its "emancipatory" variant, systems thinking can also offer "team syntegrity" (see Chapter 17) to ensure that the participation is democratically organized, and "critical systems heuristics" (see Chapter 18) to ensure diversity and space for the concerns of potentially disadvantaged stakeholders. One would think that Stacey would welcome this kind of structuring as it is designed to enhance the very "complex responsive processes of relating" that his theory advocates. Instead, he seems so desperate to avoid anything that might smack of "orchestration" that he is happy to leave everything to chance and to remain vague about the role of the CRP practitioner who is, after all, doing something just by being present. Given the choice, I prefer an approach that provides more structure. We have all been in meetings that have turned out to be pointless because they are badly chaired. Occasionally, we have attended meetings that are extremely productive because they have been well managed by a chair providing everyone with the opportunity to speak, encouraging new avenues of thought and aware of personality, gender, power, and other factors, which might influence the manner in which people contribute. Systems thinking provides a variety of "specifications," which can assist with creative thinking and the open and free sharing of opinions about what is to be done to deal with current issues and to ensure involvement in planning for the future.

I think that underlying these different approaches to "intervention" is a deeper disagreement about the purpose of human activity in organizations. Stacey and Mowles declare:

> To summarize, organisations exist to enable joint action and people can only act jointly through their relationships with each other.
>
> *(2016, p. 453)*

However, Stacey always thinks of interaction and relationships as being "for the sake of emerging identity and difference realized in the living present" (Stacey et al. 2000, p. 8). This is important in organizations but so is the pursuit of goals. Any definition of an organization always alludes to it being a group of people undertaking joint activity in order to achieve a purpose. People establish their identities and differences through their involvement in a variety of primary social groups, such as families, friends, clubs, and religious associations. They continue to develop them in secondary social groups; that is, in organizations like universities and companies. But, unlike primary groups, secondary groups don't see their main aim as being about creating individual identities and developing relationships for their own sake (Cooley 1909). Rather they exist to promote joint activity that is goal-oriented. In most cases, individuals joining secondary groups accept this and are willing to sacrifice the expression of some aspects of their individual personalities to ensure the organized pursuit of goals. Systems thinkers are unashamedly on the side of those who want to get things done.

We are close to being able to establish the respective strengths of CRP and soft systems thinking. Stacey provides an extremely sophisticated account of a very important and much neglected feature of organizational life. It is a reading that is more revealing of that aspect than anything offered by soft systems thinkers. Nevertheless, having given his description, he is at a loss as to how it can help people improve their organizations. Soft systems thinkers accept, implicitly or explicitly, that organizations are in some respects pretty much as Stacey describes them. But, they have also been successful in developing methodologies that both respect the important features that Stacey highlights and, at the same time, can assist organizations to take purposeful action. That is not to say that close attention to Stacey's CRP couldn't help them to improve those methodologies further.

Stacey's criticisms of "critical systems thinking" also miss their target. To understand why, we need to look separately at critical systems theory and critical systems practice (CSP). Critical systems theory is immune from his strictures because it does not seek to argue that one approach is better than another in providing an understanding of organizations. Rather, it offers a second-order critique with the different aim of paying respect to all the various strands that make up systems thinking. Sometimes these place the observer outside the system, sometimes inside. Critical systems theory operates on the basis of "critical" and "social" awareness (Jackson 1991a, pp. 184–187). Critical awareness considers the theoretical assumptions of the different approaches, seeking to uncover what aspects of complexity they are able to address and which they hide. Social awareness demands attention to the social context in which the methodologies are used and the consequences of use in different contexts. Stacey can argue with the accuracy of the analyses performed, as he occasionally does, but it misses the point to claim that critical systems theory provides a poorer account of the nature of organizations than CRP. In carrying out its second-order analysis, critical systems theory makes use of tools such as the "system of systems methodologies" (SOSM, see Section 8.3) and accounts of different social theories, such as provided in Chapter 4. As Stacey notes, the SOSM cannot provide comment on his CRP which is a process approach that eschews systems concepts. This is true because the SOSM was devised only for the purpose of critiquing systems methodologies. The account of different sociological paradigms can however, as we have seen, provide a theoretical positioning of CRP. It is an "interpretive" approach that, broadly, privileges social action and interaction over social structures and the social system. Because it has succeeded in populating a much wider range of paradigms, systems thinking has a good deal more to offer, both theoretically and practically, than CRP, which promotes only one.

CSP addresses what to do with the outputs of critical systems theory in order to bring about "improvement" broadly defined. It concludes that is best to use the variety of systems approaches in combination, being careful to remain true to their different theoretical assumptions. It is legitimate for Stacey to criticize this and he does. In his view, it is better to conduct ourselves according to what CRP tells us because it alone is theoretically correct. He found that in early versions of CSP, such as "total systems intervention" (Flood and Jackson 1991a), systems practitioners seem to occupy a position outside a problem situation to which they seek to bring relief by choosing and applying an appropriate combination of systems methodologies. This is an impossible position to achieve, and was certainly a weakness, but it has long been abandoned. In more recent statements of CSP (Jackson 2000, 2003), it is made clear that users cannot stand outside the paradigms, but must learn to think in terms of each of them in order to take full advantage of the various strengths of the related systems methodologies and methods they can employ in a problem situation. The choice of appropriate systems approaches is determined by users taking account of their particular situation and what they hope to achieve. It can only be seen as successful or not in terms of judgments made about what actually occurs during and following use. Users have to be quick on their feet and good at shifting their perspectives, and the systems methodologies and methods they are using, depending upon what happens. To Zhu, this "pragmatic" way of proceeding provides CSP with a distinctive advantage over CRP. CRP pursues theorizing for its own sake because it wants to justify itself on philosophical grounds. Users of CSP are informed about the theoretical perspectives underpinning different systems methodologies and methods but can then get on and use them in practice, discovering whether they work for them in their particular context:

> Very few of us, I suspect, really care whether our actions are compatible with Kant or Hegel. We do care, however, about concrete, detailed, practical proposals. When alternatives emerge, we as a community debate them, not in terms of Philosophy or Teleology, but in terms of the various situated, practical attractions, advantages and disadvantages.
>
> *(Zhu 2007, p. 454)*

Stacey argues that CSP is impossible to put into effect because of the difficulty individuals face in changing whatever theoretical "lens" they are familiar with. This is a reasonable point but one that I will leave until Chapter 21.

Thinking beyond Stacey's interpretive reading of complexity theory, we have already noted other authors providing functionalist, radical change, postmodern, and integrationist perspectives on how they think complexity theory should be used in the social sciences. It is apparent that complexity theory is unable to offer an overarching theory that can overcome the paradigm divides. Like systems thinking, it is exploring the different sociological paradigms, seeing what it can learn from them and what it can add theoretically and practically. The question then becomes whether systems thinking or complexity theory is doing a better job of this from the point of view of managers. It is clear that systems thinkers have been at it longer. Complexity theory is following a path already well trodden by systems theorists (Jackson 2003, p. 125), replicating paradigm shifts that have already taken place in systems thinking (Midgley and Richardson 2007). Systems thinkers have, as a result, achieved a head-start on complexity theorists in at least three respects. First, since the advent of its critical variant, systems thinking has been relating systems approaches to social theory in a self-conscious, second-order, systemic endeavor, and much progress has been made. Secondly, driven by a practical emphasis on doing things in order to bring about improvement, systems thinking has developed a range of methodologies that are well enough articulated to lend themselves to testing by researchers and application by managers and other practitioners. Thirdly, CSP has sought to develop means for using systems methodologies and methods in combination, taking advantage of the different strengths they possess. So, while we should note complexity theory's capacity, with its rich array of insightful concepts, to add to what has already been achieved, we must also recognize that there is a great deal to be done before it can offer as much to decision-makers as systems thinking.

7.7 Conclusion

One of the most exciting conclusions we can draw from complexity theory, according to Kaufmann, is that we get "order for free." But the question surely is whether that order is desirable or not from the point of view of human beings. It is worth thinking on.

I would agree with Johnson that the interaction of skilled jazz musicians can, apparently spontaneously, produce music of a sublime quality. In his view, this is because modern jazz is the closest musical equivalent to complexity:

> Modern jazz involves a spontaneous interaction of a collection of objects (i.e. musicians). It exhibits surprising emergent phenomena in that it is improvised, and hence what emerges in a given solo is a product of the actual feedback which that soloist receives at that moment in time. It is also an open system in that its best performances arise in an environment with audience feedback .... Above all, it has no 'invisible hand' such as an orchestral conductor or an existing piece of melody that all the players are simply repeating ... and if you ever happen to see a transcription of Michael Brecker's tenor saxophone solos ... you will see fractals dancing before your eyes.
>
> *(2009, pp. 63–64)*

It is, perhaps, even more exciting to find, in Fischlin et al., the notion of musical improvisation, with its complexity implications, linked to the establishment of civil, human, and environmental rights and acting as a model for "relational co-creation." In their view, it epitomizes a democratic, humane, emancipatory practice capable of helping to create new, unexpected, and productive relations between people:

> We believe that cultivating the capacity for action, a capacity inculcated by improvisation, and channeling it toward meaningful, embodied rights outcomes is one of the ways we can make things better .... Improvisation promotes personal confidence and makes people accustomed to taking action, to activating their agency publicly and in relation to others.
>
> *(2013, p. xiv)*

However, we must surely ask questions about whether such improvisation can be relied upon to produce outcomes that all will regard as beneficial. We have been warned by Stacey et al. (2000) and Boulton et al. (2015) that self-organization, left to its own devices, can be a force for bad as well as good. Curtis (2011) recounts an interview, on BBC *Newsnight*, in which a spokesman for a direct action group was asked whether she condemned the violence in London for which the group had been responsible. She replied that:

> We are a network of people who self-organise. We don't have a position on things. It's about empowering the individual to go out there and be creative.

Curtis suggests that the belief that self-regulation, or self-organization, is a good thing can be dated back to Smuts work on "organized complexity" and "holism," which we mentioned briefly in Section 3.2. Smuts, of course, ruled South Africa for the British Empire with an iron fist. He was accused, at the time, of the "abuse of vegetational concepts" (the organismic analogy) to justify the oppression of black people. And H G Wells argued that he was seeking to use an apparently scientific theory about order in nature to justify order in the British Empire. Curtis goes on to describe various attempts at self-organization that have gone wrong and led to bullying, coercion, sexual oppression, and power struggles – from hippy communes to formal experiments such as *Biosphere 2* in the early 1990s. And, I agree with Rosenhead (1998) that we should have severe reservations when complexity thinking is transferred to financial markets and economies to suggest that there is no alternative to the price mechanism for ordering our affairs. In 1936, Friedrich Hayek wrote:

> How can the combination of fragments of knowledge existing in different minds bring about results which, if they were to be brought about deliberately, would require a knowledge on the part of the directing mind which no single person can possess?
>
> *(Quoted in Metcalfe 2017, p. 30)*

Hayek, of course, became the guru of Thatcher and Reagan, and is regarded as the father of modern neoliberalism. Apparently, the market is obedient to natural laws and simply dictates that there must be deregulation, a shrinking state charged with ensuring there is no interference, and gross inequalities between rich and poor. The 2007–2008 financial melt-down was an "automatic mechanism of adjustment," which had to occur and which produced consequences that we will just have to live with. This is pushing the boundaries of complexity theory as "science" too far; robbing human beings of the possibility of using public deliberation, enacted democratically through politics, to decide on what are desirable social ends.

Part III

Systems Practice

The philosophers have only interpreted *the world, in various ways; the point is to change it*

(Marx 1975b, originally 1845)

Gibbons et al. identify two types of research. In Mode 1 research, the problems tackled are defined by the particular scientific interests of an academic community and the primary aim is to gain new knowledge for its own sake. It is a

> ... complex of ideas, methods, values and norms that has grown up to control the diffusion of the Newtonian model of science to more and more fields of enquiry and ensure its compliance with what is considered sound scientific practice.
>
> *(1994, p. 167)*

By contrast, Mode 2 type research is carried out in the context of application and is produced to satisfy the demands of particular users. Tranfield and Starkey (1998) argue that management research should adopt a Mode 2 orientation, positioning itself in the social sciences as equivalent to engineering in the physical sciences and medicine in the biological sciences. In fact, and to its great credit, systems thinking applied to management has already occupied this space. From the first formulation of approaches such as operational research (OR), systems analysis, and systems engineering, in the 1940s and 1950s, the emphasis has been on "clients" and their requirements. This has remained fundamentally the case, although "problem owners," as distinct from clients, and the "affected but not involved" now also fall within the remit of systems thinking.

The difference between these two types of research can be illustrated by considering the approaches of two pairs of researchers undertaking work in prisons. Operating in the Mode 1 manner, Genders and Player describe the intention behind their research in HM Prison Grendon as being "to capture certain fundamental truths" about the institution. The problem for them was that the prison simply wouldn't stay stable enough for them to make accurate measurements. It fell into such a state of "crisis" that the research

Critical Systems Thinking and the Management of Complexity, First Edition. Michael C. Jackson.
© 2019 John Wiley & Sons Ltd. Published 2019 by John Wiley & Sons Ltd.

had to be abandoned lest it contribute to the problems and made "proper" scientific results impossible:

> The prevailing sense of crisis distracted staff from their usual business and, in some respects, distorted the processes which the research was designed to study. The air of despondency led to the expression of jaundiced views during the interviews, which marked a notable break with the earlier responses, and largely reflected a transient and ephemeral reaction to the issues of the moment. In consequence, interviews had to be halted until the institution regained a sense of equilibrium.
>
> *(Quoted in Gaskell 1997, p. 175)*

Catherine Gaskell and I, by contrast, were clear from the outset that studying "the management of change in Hull Prison" required us to work with the Governor and get involved in the change process:

> We had come to characterize [the world in which prison research takes place] as rapidly changing; as complex; as involving unique sets of local circumstances; as hard to experiment on while retaining 'objectivity'; and as involving difficult relationships between researchers and practitioners.
>
> *(Gaskell et al. 1996)*

We declared, in the *Prison Service Journal*, that ours would be a Mode 2 intervention, designed to:

- Assist in the creative construction of the world of prisons and the prison service
- Avoid reductionism; treat problems systemically
- Accept research has only local plausibility and can bring only local improvements
- Accept all research is observer-dependent and changes the system concerned
- Merge roles; practitioners becoming part of the research team and researchers sharing the responsibility for outcomes

Remarkably, the Home Office didn't take away our grant.

In developing Mode 2 research, systems practitioners have embraced trans-disciplinarity and aspects of action research (see Section 4.8). Because applied systems research revolves around real-world problems, it is unable to rely on any single discipline to provide it with the theory to guide its practice. The real world is not divided up into academic disciplines. Systems practice is therefore, of necessity, trans-disciplinary, drawing as seems appropriate on systems concepts derived from philosophy, the physical, life, and social sciences, and upon the frameworks of systems ideas developed in general systems theory, cybernetics, and complexity theory. To compensate for the absence of a well-articulated theory relevant to all problem situations, it has concentrated on developing "systems methodologies," which certainly carry theoretical assumptions but are more concerned with providing principles for the use of methods and techniques to ensure that they bring about desirable improvements in specific circumstances. These methodologies are exactly what are required in Mode 2 knowledge production. They act as a kind of transferable problem-solving capability and, being more generally applicable than theories, can be transported from one unique problem situation to another.

Many systems practitioners embrace action research and accept that research on social systems will inevitably lead to change occurring. Checkland and Holwell refer to conversations with Geoffrey Vickers on this matter:

> ... he used to point out that while Copernicus and Ptolemy offer very different hypotheses about the basic structure of our solar system, we know that, irrespective of whether the sun or the earth is at the centre of the system, the actual structure is entirely unaffected by our having theories about it. Whereas when Marx propounds a theory of history this changes history!
>
> *(1998, p. 19)*

Systems practitioners tend to see themselves as fully involved, with other participants, in a social process that will change the problem situation they are engaging in. Indeed, they make a virtue of this fact rather than trying to hide it under a veil of "objectivity." They must, therefore, take some responsibility for the practical outcomes that arise and seek to ensure that these represent "improvements." Of course, they should also reflect on what happens in their interventions to improve their methodologies for the next time they are used. As this part of the book will show, some systems practitioners have been more conscientious than others in this respect. Checkland (see Chapter 16) built an exemplary action research program and was successful in continually enhancing his soft systems methodology. Systems thinkers should also seek, as required by the principles of action research, to learn from their interventions about the usefulness, or otherwise, of any theoretical assumptions underpinning their methodologies. There is a good example from Soft-OR showing what can be achieved. Eden and Ackermann (2018) detail an extended action research program in which they successfully developed a theory alongside a Soft-OR methodology. Systems practitioners still have much to do in this respect. The maturation of critical systems thinking, which makes it a priority to interrogate the explicit and implicit philosophies and theories underlying systems methodologies, should speed up progress. Parts III and IV of the book are designed to assist future systems practitioners to reap the full benefits of action research.

Part III of the book describes and critiques 10 applied systems approaches demonstrating, hopefully, how and why they are significant from the point of view of decision-makers. The 10 approaches are divided into 6 types in terms of whether they are most appropriate for dealing with technical, process, structural, organizational, people, or coercive complexity – a broad categorization, which is modified as necessary as the individual methodologies are considered. Hard systems approaches (Chapter 9) see the world in terms of technical complexity; the Vanguard Method (Chapter 10) seeks to tame process complexity; system dynamics (Chapter 11) addresses structural complexity; sociotechnical systems thinking (Chapter 12) and organizational cybernetics (Chapter 13) concentrate on organizational complexity; soft systems approaches (Chapters 14–16) emphasize people complexity; and team syntegrity (Chapter 17) and critical systems heuristics (Chapter 18) try to combat aspects of coercive complexity. These 10 systems approaches are, in my view, the most philosophically sound and thoroughly researched of those on offer and all have a good track record of application. Each approach is reviewed in approximately the same way. First, a prologue is offered, which sets the scene by suggesting what the approach can achieve. Second, a description of the approach is provided in terms of historical background, any explicit philosophy/theory underlying the approach, the methodology used to translate the philosophy/theory into practical application, and the methods – different models, tools, and techniques – usually associated with the approach. Methodology refers to the logical principles that govern the use of methods in order that the philosophy/theory embraced by the approach is properly respected and appropriately put into practice. It is not detachable from the philosophy/theory of a particular systems approach. Methods are concerned with achieving more specific procedural outcomes, are detachable, and can be used in the service of other systems approaches with varying degrees of success or failure. Third, examples are offered of the approach in action, concentrating particularly on how the methodology and associated methods contribute to the outcomes. Fourth, a critique of the approach is provided, setting out its particular strengths and weaknesses. This is conducted on the basis of the philosophies and the social theories described earlier in the book, as well as the revised "system of systems methodologies" developed in Chapter 8. The critique enables us to understand how adhering to a particular theory facilitates the ability of a methodology to achieve certain things for managers while constraining it in other respects. A fifth section in each chapter is reserved for any further comments I judge to be useful and/or interesting. Finally, attention is devoted to what exactly each systems approach offers to managers in terms of improving their ability to handle complexity.

Before embarking on our examination of the 10 different systems approaches, we devote Chapter 8 to constructing a revised "system of systems methodologies." This provides a preliminary overview of the development of systems practice and the role that can be played by the individual methodologies.

8

A System of Systems Methodologies

If one introduces an observer, a speaker, or simply one to whom one attributes a statement, one relativizes ontology .... One always faces the questions of who says a particular thing, and who does something, and from which system perspective the world is seen in a particular way (and no other)

(Luhmann 2013, p. 99)

8.1 Introduction

Systems thinkers claim that "the whole is more than the sum of its parts" and that "everything is interconnected." To some (Phillips 1976; Bryer 1980), this leaves systems thinking open to the criticism that it is hopelessly idealistic and unworkable. It suggests that you must always start with the whole system because you need to know everything in order to know anything. Further, unless you do know the whole system, you cannot justify taking action because you can never anticipate the results. As Ulrich (1981a) argues, however, such critics are "blaming the messenger for the bad news." The mature response is to regard understanding and improvement of the "whole system" as a theoretical ideal that forces us to reflect critically on the provisional nature of our existing knowledge and the inevitable lack of comprehensiveness of our actual systems designs. We can work with limited conceptions of the whole as long as we are transparent about what we are doing and make it easy to reflect on the limitations of our knowledge and interventions. What we have to do is find ways of proceeding given that it is always going to be impossible to achieve comprehensive understanding, however theoretically necessary and practically desirable it might be.

If this is the case, it becomes essential to adopt a "critical systems thinking" or "second-order" perspective on systems practice, highlighting the ways different systems thinkers observe the social world and seek to change it. The details of this are explained in the next section. The chapter continues by constructing a revised "system of systems methodologies" (SOSM) that can, with the help of the philosophical and theoretical foundations established earlier, make transparent the strengths and limitations of the 10 systems methodologies reviewed in Chapters 9–18. We then employ the SOSM to orientate the reader by using it to consider the development of applied systems thinking. Finally, the reason for dividing the 10 methodologies into 6 categories is explained.

Critical Systems Thinking and the Management of Complexity, First Edition. Michael C. Jackson.
© 2019 John Wiley & Sons Ltd. Published 2019 by John Wiley & Sons Ltd.

8.2 Critical or "Second-Order" Systems Thinking

A self-conscious critical approach to systems thinking developed in the 1980s when critical systems thinkers established "critical awareness" and "social awareness" as two of the central tenets of their enterprise.

A commitment to critical awareness is evident in the writings of Mingers (1980, 1984) and Jackson (1982a) when they ask questions about the social theory on which soft systems thinking rests and how this impacts upon its ability to bring about change. A typical argument is that the particular theoretical assumptions on which soft systems thinking is based prevent it from seeing and getting to grips with the unequal power resources that exist in social systems and constrain the participation that soft methodologies value so highly. To conduct critique of this kind, it was necessary to take an overview of different possible ways of understanding and intervening in organizations. This was based initially on social theory. For example, Burrell and Morgan's (1979) book on sociological paradigms and organizational analysis, and Morgan's (1986) examination of "images" of organization, enabled critique of the assumptions different systems approaches make about social reality and social science. Following the articulation, by Jackson and Keys (1984), of the first version of the SOSM, this was added to the armoury of critical systems thinkers to interrogate how "first-order" systems practitioners must be observing the social world in trying to change it in certain ways. A series of critiques of different systems approaches was undertaken, culminating in a review of five strands of systems thinking – "organizations as systems," "hard," "cybernetic," "soft", and "emancipatory" – from the point of view of relevant social theory and the SOSM (Jackson 1991a). Once in possession of critical awareness, systems thinking could not return to its "dogmatic slumbers," as the growing body of work in critical systems thinking attests (Jackson 2000, 2003; Flood 1999; Midgley 2000; Mingers 2006).

As well as allowing the theoretical underpinnings of available systems approaches to be unearthed, social theory drew attention to the importance of the social context in which methodologies are used. This type of understanding has been labeled "social awareness" (Jackson 1991a). There are inevitably societal and organizational pressures that lead to certain systems theories and methodologies becoming popular for observing social systems and guiding interventions at particular times. Rosenhead and Thunhurst (1982) offer a "materialist" account of the development of operational research in terms of wider social processes and the history of capitalism. A link between the neoliberal agenda and the popularity of complexity theory was noted in Chapter 7. Li and Zhu (2014) trace the fate of soft operational research in China in the light of the particular historical and institutional circumstances of that country. Social awareness also asks systems practitioners to contemplate the consequences of use of the approaches they employ. As Li and Zhu argue: "OR is not merely a neutral tool for solving technical problems, but a world-building discourse that shapes society" (2014, p. 427). The choice of a hard systems approach to conduct a project implies that the pursuit of one goal is being privileged. Is this a goal general to all stakeholders or simply the most powerful? Similarly, the use of soft systems methodologies, which are dependent upon free and open debate to justify their results, may have unintended social consequences if the conditions for such debate are absent. The reflections of postmodern thinkers on the power/knowledge nexus add to our awareness of how important it is to consider the social context in which systems methodologies are employed.

In Section 4.7, we saw Luhmann advocating a similar enterprise to that undertaken by critical systems thinkers. Drawing upon the work of von Foerster, Maturana and Varela, and Spencer-Brown, he makes a case for basing sociology on "second-order" observation. Instead of pretending to observe the social world directly, he maintains, we should observe how other observers view social reality. What we need is a "description of descriptions." At the beginning of his *Introduction to Systems Theory* (2013, pp. 3–24), Luhmann subjects Parsons' sociological theory to a critical interrogation, which reveals:

- All that can be achieved by means of this theory pattern
- The consequences to which the design of a particular theory leads
- The contrast between this design and other theory patterns, such as, say, a dialectical theory

The purpose of such second-order observation is to uncover the distinctions with which a particular observer works in order to understand how these distinctions allow the observer to see some things and not others:

> Second-order observation is observation of an observer with a view to that which he cannot see … we become interested in the distinctions with which the observed observer works, and in how he divides up the world, and in what he considers important (or not) in which situations.
>
> *(Luhmann 2013, p. 112)*

Luhmann (2013, p. 113) suggests that explicitly describing situations in which particular characters cannot see certain things began with Cervantes. The reader of *Don Quixote* is aware of the extent to which the chivalric code constrains the thinking and actions of the eponymous hero of the story.

As a "radical constructivist," Luhmann does not believe that there is any meta-position, high above reality and looking down, that a second-order observer can adopt to reveal the blind-spots of other social theorists. There can be no ontological certainty and so any second-order observation is also simultaneously a first-order observation subject to its own biases resulting from the distinctions that it employs:

> For this reason, there are only shifts between that which one sees and that which one does not see, but there is no comprehensive enlightening or scientific elucidation of the world as the totality of things or forms or essences that could be worked through piece by piece, not even if the task is seen as infinite.
>
> *(Luhmann 2013, p. 105)*

Further, for Luhmann, sociologists who build social theories inevitably become internal observers of society. They intend to communicate and so participate in the construction of society's image of itself and its concerns and problems. This is one reason why it is impossible to arrive at an "objective" account of social reality:

> How is it possible to arrive at the objective knowledge of reality if one knows that stating one's prognosis changes reality? How is this possible if one knows, for example, that a Keynesian economic policy, once it has been formulated and

implemented, creates the anticipation of such policy and thereby the expectation of inflation?

(Luhmann 2013, p. 119)

The observations of theorists, therefore, have consequences of which they should be aware. Second-order observation can help us understand what happens when "organizations, families, or therapies" are planned on the basis of various first-order theoretical distinctions.

Luhmann's arguments for second-order observation mirror those of critical systems thinking, although not all critical systems thinkers are ready yet to accept ontological relativism. He is firmly with critical systems thinkers in arguing for the need to shift between different theoretical dispositions in order to open up new vistas and possibilities. This is difficult, he argues, because:

> Fully constructed theories are complicated formations. In a sense, they are works of art. It is difficult to get involved with them and still know how one will relativize all that again or how one will be able to detach oneself when the time comes.
>
> *(Luhmann 2013, p. 254)*

For it to be possible at all, we need to make our theoretical decisions and their consequences thoroughly transparent.

Edwards (2014) uses the label "metatheorizing" for the kind of critical or second-order observation conducted by critical systems thinkers and Luhmann. Meta-theory does not undertake empirical observations of the "real world" but is built and tested on "middle-range" theories, seeking to reveal their ontological, epistemological, methodological, and ethical dimensions. It does not seek to replace middle-range theories but to critically evaluate the assumptions and conceptualizations on which they are constructed. Critical systems theory is, he says, an exemplar of such research because the meta-theoretical devices it employs

> ... i) often emerge out of a process of theory building that integrates, synthesises or constructively analyses other theories and methods, ii) are themselves used to systematically review and critique other theories and iii) include new conceptual lenses that explore new ways of understanding.
>
> *(Edwards 2014, p. 725)*

Such meta-theoretical research is important because:

> It provides opportunities for finding connections and differences between other theories, for assessing current theoretical developments as well as potential avenues for future research and for reviewing and critically analysing the basic assumptions and conceptual lenses employed by groups of researchers.
>
> *(Edwards 2014, p. 740)*

It is also intensely practical and ethical because it engages critically with "big ideas," the way they are developed, enacted, and can give rise to problems such as economic inequality, biodiversity collapse, and global warming. Meta-theoretical research may seem abstract but shunning it because of this

... effectively leaves the field of big ideas, and their enactment at the global level, open to the world of multinational corporations, global media companies and the economic forces of globalization.

(Edwards 2014, p. 732)

Edwards' preferred type of meta-theory does not assume a "view from nowhere" when it is interrogating middle-range theories. It accepts that its own analyses will always be partial, reflecting the meta-theorists own experiences and biases. Instead, it seeks to offer a more encompassing perspective from "somewhere else" while, at the same time, being as rigorous and transparent as possible about its own underlying assumptions and partialities.

The next section sets out to compare some meta-theoretical frameworks relevant to systems practice and to develop one as our preferred vehicle for reviewing and critiquing systems methodologies in the following chapters.

8.3 Toward a System of Systems Methodologies

8.3.1 Preliminary Considerations

Three factors are of particular significance in choosing a suitable vehicle for conducting our second-order critique of systems methodologies. The first is that the framework is absolutely transparent about the assumptions underlying its own construction. The second is that the framework is both comprehensive and refined enough to encompass the 10 systems methodologies that we are deeming the best on offer. Only then can we complete the investigation of similarities and differences between all existing approaches that Edwards (2014) sees as essential to good meta-theoretical research in a field. Finally, the vehicle we choose must be able to act as a direct link to the accounts of relevant philosophy and theory provided in Part I and especially to the review of social theory since our concern is with the management of social systems. This is certainly not straightforward. We are, after all, not just concerned with interrogating other social theories, as in Luhmann's case, but with examining a range of systems methodologies designed to conduct Mode 2 type interventions in the real world. Before we can use our account of social theory in the critique, we need to uncover the often implicit assumptions made by systems approaches about the social reality that they are seeking to change. The preferred framework must, therefore, be capable of bridging the gap between social theory and the action orientation of systems methodologies.

It will be enlightening to start by examining examples of existing meta-theoretical frameworks in the systems sciences. One each is chosen from cybernetics, general systems theory (GST), and complexity theory.

8.3.2 Beer's Classification of Systems

Beer is explicit in stating that his 1959 classification of systems is "arbitrary" in respect of "real-world" systems and that the distinctions it draws must be seen as "hazy" (see Beer 1967, p. 12–19). Its purpose is to act as a starting point for thinking about the "province" of cybernetics. To that end, the classification groups systems according to the types of control, and therefore the scientific methods, to which they are amenable. As can be seen in Table 8.1, a grid is constructed on the basis of two axes. The horizontal axis is concerned with complexity and divides systems into *simple* (but dynamic), *complex* (highly elaborate and interconnected but describable), and *exceedingly complex* (cannot be described in a precise and detailed fashion). The vertical axis is interested in whether systems can be defined as deterministic or probabilistic. *Deterministic* systems are predictable because the parts interact in a predetermined fashion. *Probabilistic* systems cannot be predicted precisely, at least given currently available knowledge.

Examples are then provided for the six types of system that the grid describes (see Table 8.1). A window catch is a *simple deterministic* system that displays predictable dynamic behavior. An electronic computer is a *complex deterministic* system; complex but entirely predictable. Tossing a coin is a *simple probabilistic* system. A stock-control system is *complex probabilistic*. It can give rise to involved behavior, but this is describable in terms of mathematical statistics. The *exceedingly complex deterministic* box is empty because, according to Beer, all deterministic systems can eventually be described completely. In the *exceedingly complex probabilistic* category are systems such as the economy, the brain, and the company, which we can never expect to describe fully or to yield to prediction.

Beer argues that both simple and complex deterministic systems can be handled using the traditional scientific method filtered through applied subjects such as production engineering. The extension of the scientific method achieved through applied statistics also brings simple probabilistic systems into its domain. His own interest is in complex probabilistic systems, which he regards as falling in the realm of operational research, and exceedingly complex probabilistic systems, the province of cybernetics.

Beer is clear about the purpose of his classification. He sees it as providing a device for thinking about the nature of cybernetics. He is transparent about how his grid is constructed. The classification serves his purpose well. In seeking to distinguish cybernetics from traditional science and operational research, and establish its unique area of application, he extends his reach beyond just systems thinking. In that respect, the classification mirrors that of Weaver (see Part II) in incorporating "organized simplicity," which can be managed using classical mathematical tools, and "disorganized complexity," with which statistics and probability can cope, as well as "organized complexity," the sphere in which the systems sciences become essential. It cannot, however,

Table 8.1 Beer's arbitrary classification of systems.

Systems	Simple	Complex	Exceedingly complex
Deterministic	• Window catch • Billiards • Machine-shop lay-out	• Electrical digital computer • Planetary system • Automation	Empty category
Probabilistic	• Penny tossing • Jellyfish movements • Statistical quality control	• Stockholding • Conditioned reflexes • Industrial profitability	• The economy • The brain • The company

Source: Adapted from Beer (1967).

provide a basis for a critical examination of contemporary systems methodologies. We are left with just the one distinction between operational research, applicable to complex probabilistic systems, and cybernetics, suitable for exceedingly complex probabilistic systems, to serve us in interrogating systems methodologies. Beer is proud of the radical interdisciplinarity of his cybernetic approach. He sees it as "deliberately impertinent toward the narrow-minded specialist" (1967, p. 18). This is all very well, but it leads him to ignore the distinctive emergent properties, arising in human and social systems, which other systems thinkers see as crucial because they give rise to new forms of complexity that they believe are more important than those that are shared with mechanical and biological systems. For the same reason, it is obvious that Beer's classification cannot provide us with an easy link to the social theory that we want to include in our second-order critique.

8.3.3 The Original "System of Systems Methodologies"

The "system of systems methodologies" (SOSM) originated in a 1984 paper by Jackson and Keys called "Towards a System of Systems Methodologies." The starting point was an "ideal-type" grid of problem contexts. The concept of an "ideal type," derived from Weber (1949), establishes that the grid presents some abstract, logical classes of problem context. It does not seek to describe actual problem contexts, which, as was stated explicitly, will look different to different observers and even to one observer taking a look for different purposes. As its name suggests, the SOSM was designed to classify existing systems methodologies and to probe the relationships between them.

In its early manifestation, the grid underpinning the SOSM consisted of a four-celled matrix defined by dimensions of systems complexity (mechanical or systemic) and decision-makers' perspectives (unitary or pluralist). Over time, the grid was developed until it took on the stable form presented in Figure 8.1 (Jackson 1991a, 2000). The two axes, of *systems* and *participants* (replacing decision-makers) on which the grid is built, represent the two major sources of complexity that face systems practitioners. The vertical axis details increasing complexity on a continuum from *simple* to *complex*. Simple systems can be characterized as having a few components or subsystems that are involved in a small number of highly structured interactions. They tend not to change much over time, being relatively unaffected by the independent actions of their parts or by environmental influences. Complex systems can be characterized as having a large number of components or subsystems that are involved in many loosely structured interactions the outcome of which is not predetermined. Such systems adapt and evolve over time as they are affected by their own purposeful parts and by the turbulent environments in

		Participants		
		Unitary	Pluralist	Coercive
Systems	**Simple**	Simple–unitary	Simple–pluralist	Simple–coercive
	Complex	Complex–unitary	Complex–pluralist	Complex–coercive

Figure 8.1 Jackson's extended version of Jackson and Key's "ideal-type" grid of problem contexts.

which they exist. The horizontal axis presents a continuum of increasing complexity deriving from an increasing divergence of values and/or interests between those concerned with, or affected by, a problem situation. The terms *unitary, pluralist,* and *coercive* are employed. Participants defined as being in a unitary relationship have similar values, beliefs, and interests. They share common purposes and are all involved, in one way or another, in decision-making about how to realize their agreed objectives. Those defined as being in a pluralist relationship differ in that, although their basic interests are compatible, they do not share the same values and beliefs. Space needs to be made available within which debate, disagreement, even conflict, can take place. If this is done, and all feel they have been involved in decision-making, then accommodations and compromises can be found. Participants will come to agree, at least temporarily, on productive ways forward and will act accordingly. Those participants defined as being in coercive relationships have few interests in common and, if free to express them, would hold irreconcilable values and beliefs. Compromise is not possible and so no agreed objectives can direct action. Decisions are taken on the basis of who has most power and various forms of coercion employed to ensure that the powerful prevail. Combining the "systems" and "participants" dimensions, divided as suggested above, yields six "ideal-type" problem contexts: *simple–unitary, simple–pluralist, simple–coercive, complex–unitary, complex–pluralist,* and *complex–coercive.*

The next step in building the SOSM was to relate existing systems methodologies to the problem contexts defined by the grid. One attempt to achieve this can be found in Figure 8.2. Hard systems thinking was equated to simple–unitary assumptions about problem contexts because it takes as a given that it can establish agreed objectives for the system of concern and can make a quantitative model of that system. System dynamics, organizational cybernetics, and complexity theory approaches, it was argued, were based upon complex–unitary assumptions. Soft systems approaches were identified with simple–pluralist and complex–pluralist problem situations. Emancipatory systems approaches were related to simple–coercive contexts. A rationale was provided for postmodern systems thinking as attempting to come to terms with the massive complexity of complex–coercive contexts. In general, there seemed to be a good match between the problem contexts defined by the grid and the assumptions made by existing systems methodologies. It appeared that the developers of different systems approaches were governed, consciously or otherwise, by their preferred ideal-type

		Participants		
		Unitary	**Pluralist**	**Coercive**
Systems	**Simple**	Hard systems thinking	Soft systems approaches	Emancipatory systems thinking
	Complex	System dynamics Organizational cybernetics Complexity theory		Postmodern systems thinking

Figure 8.2 Systems approaches related to problem contexts in the system of systems methodologies. *Source:* Adapted from Jackson (2003).

perspectives on the nature of problem contexts. The SOSM proved useful for classifying systems methodologies, indicating their various strengths and weaknesses, and pointing the way to pluralism in systems thinking if a way could be found to use the methodologies in combination to complement one another.

The original version of the SOSM was clear about its purpose and transparent about how it was constructed. Its purpose was to classify existing systems approaches, relate them to each other, and do so using a language appropriate to the action orientation of the methodologies. The axes making up the grid were derived from Ackoff's (1974a,b) thinking about the growing complexity of problem situations, as we move from the "machine" to the "systems" age (systems dimension) and from the industrial relations literature (participants dimension). The grid was constructed with the intention of embracing existing systems methodologies and not surprisingly, therefore, covers the ground in that respect. As the systems landscape moved, the grid was suitably adjusted. When the SOSM was born, in 1984, there seemed to be no systems methodologies based on the assumption that problem contexts were coercive. From their reading of the sociological literature, the developers of the SOSM knew there was a gap but were unable to point to any remedy (Jackson and Keys 1984, p. 483). In fact, the major text on the first significant "emancipatory" systems approach, Ulrich's (1983) "critical systems heuristics," had just been published. Later renditions of the grid explicitly included "coercive" problem contexts to reflect this. The grid concentrates on the area of "organized complexity," in Weaver's nomenclature, and so can employ distinctions that are relevant purely to systems thinking. The particular distinctions used successfully allow important similarities and differences between methodologies to be exposed. That said, in retrospect, it is easy to see that the simple/complex distinction of the vertical axis is too broad to get to the heart of the differences between hard systems thinking, system dynamics, and organizational cybernetics.

The construction of the SOSM recognized the significance of emergent properties at higher levels of complexity by providing human and social complexity with their own, "participants" dimension. All the systems sciences have, over the course of their evolution, felt obliged to pay some attention to "people complexity," for example, with Boulding's work in GST, von Foerster's second-order cybernetics, and Stacey's version of complexity theory. Those attempting to use systems ideas in practice have, of necessity, had to give it very high priority. The SOSM responds fully to these issues of emergence with its horizontal dimension. The separation implied by the two axes is nevertheless contrived in the sense that, in the real world, technical, organizational, human, and social aspects of complexity are all entangled. It does, however, allow full consideration to be given to the soft and emancipatory systems methodologies that prioritize people and power complexity. It also ensures that an easy link can be made to social theory – something that has to be desirable when we are considering systems methodologies seeking to improve practice in the social domain. For example, Habermas' sociology (see Section 4.4) was used to enhance the arguments presented in the name of the SOSM (see Jackson 1991a). The "systems" dimension can be seen as corresponding to what Habermas describes as the "technical" interest that the human species has in the prediction and control of natural and social systems. The participants dimension corresponds to the "practical" and "emancipatory" interests that humans have in securing and expanding the possibilities of mutual understanding among all those involved in social life. Thus, it could be argued, systems methodologies have a complementary role to play in all aspects of societal improvement. Lukes' (1974) reflections on the "three dimensions" of power were employed to enhance understanding of coercive contexts. The first face of power, decision-making power, is expressed through political action and is easy to recognize. The second face, non-decision-making power, is more subtle and occurs when the powerful set the agenda to prevent certain issues ever being discussed. "Ideological" power, the third dimension, occurs when there is such an influence over the thoughts and desires of the oppressed that they do not recognize their own true interests – like the people caught in The Matrix. If the first dimension is associated with simple–coercive contexts, and the second and third with complex–coercive, it becomes clear that very different responses are required in the two situations (see Jackson 2000).

8.3.4 Snowden's Cynefin Framework

The *Cynefin* framework was developed by Snowden and Kurtz from the consultancy and action research they conducted at IBM's Institute of Knowledge Management (Kurtz and Snowden 2003). It gained popularity following a 2007 article, by Snowden and Boone, published in the Harvard Business Review. The framework is now widely employed in projects by Snowden's consultancy company *Cognitive Edge.*

Cynefin is a Welsh word meaning "habitat" or "the place of our multiple affiliations." The framework is described as a "sense-making" device, which helps people arrive at a shared understanding of the complexities they face and how to respond to them. The building blocks of the framework are four main variants of behavior that, according to Kurtz and Snowden (2003), systems display depending on the relationships between their structural elements. These exist on a continuum running from simple to chaotic but are usually positioned on a grid, as in Figure 8.3. The two axes are *degree of central direction* and *strength of connections among constituent components.* Strong central direction and weak connections among components produce *simple* "domains." Strong

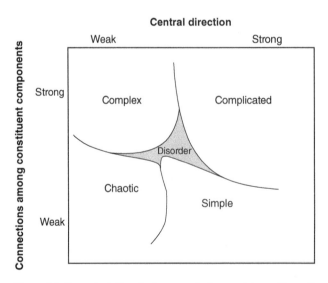

Figure 8.3 Snowden's Cynefin framework. *Source:* Adapted from Kurtz and Snowden (2003).

central direction and strong connections among components produce *complicated* domains. *Complex* domains come about when central direction is weak and the connections between components strong. *Chaotic* domains arise when both central direction and the connections among components are weak. The different characteristics of the four domains can now be described. Simple domains exhibit linear cause and effect relationships, which are easily identifiable and lead to predictable outcomes. In complicated domains, cause and effect relationships can be identified but are often separated in time and space and linked in chains that are difficult to fully understand. In complex domains, there are so many agents and relationships that it is impossible to predict the outcome of their interactions. The innumerable cause and effect concatenations do, however, produce emergent patterns of behavior that can be discerned in retrospect. Chaotic domains display no visible relationships between cause and effect. Turbulence reigns and no patterns show themselves.

On the basis of this analysis, Snowden and Boone (2007) suggest that leaders should adopt different approaches to decision-making depending on the complexity of the circumstances they face. In simple "contexts," we are dealing with "known knowns." There is unlikely to be disagreement about what needs doing. Leaders can *sense, categorize,* and *respond* using best practice or techniques such as process reengineering. Complicated contexts are the realm of "known unknowns." The cause and effect relationships are not obvious and different ways forward might be proposed. Leaders need to sense, *analyze,* and respond, making use of experts who can help them understand the behavior of the system of interest using approaches such as systems thinking. In complex contexts, we are confronted by "unknown unknowns." There is constant flux and nothing is predictable. A more experimental approach is required, opening up discussion and encouraging diversity until favorable "attractors" begin to emerge. Leaders should *probe* first, then sense, and respond. Kurtz and Snowden provide the example of managing playtime at a kindergarten:

> Experienced teachers allow a degree of freedom at the start of the session, then intervene to stabilize desirable patterns and destabilize undesirable ones; and, when they are very clever, they seed the space so that the patterns they want are more likely to emerge.
>
> *(2003, p. 466)*

Chaotic contexts are the domain of "unknowables." There is no point in waiting in the hope that manageable patterns of behavior will arise. Leaders must *act,* then sense, and respond. It may require a strong top–down approach to transform the context away from the chaotic domain and back to the complex where patterns begin to emerge. Snowden and Boone suggest that successful leaders know how to identify the context they are in and can change their decision-making styles to fit that context. They should also be proactive and use their knowledge of "Cynefin dynamics" to guide their management practice. If they understand how the relationships between structural elements affect system behavior, they can bring about changes of context to take advantage of opportunities or avoid threats, e.g. visiting the chaotic context to "break up a strong but unhealthy stability" (Kurtz and Snowden 2003, p. 478).

There is a fifth context in the middle of the Cynefin framework labeled *disorder.* This is the domain of "uncertainty about uncertainty." It comes into play when individuals and groups seek to promote their own interpretations of the context they are in and come into conflict with others holding alternative interpretations. It reminds us that the framework does not seek to capture "objective" features of the real world but to act as a "sense-making" device to achieve clarity about how the world is being framed at a particular place and time. Individual leaders can employ it to reflect upon their own perceptions of the context they are in. Groups of executives can use it to help negotiate between different perspectives held about the nature of a context. Some might see it as "complicated," others as "complex." The Cynefin framework provides a shared vocabulary for debating these issues. All viewpoints are to be respected and, indeed, may be more or less appropriate for different aspects of the context.

The inspiration for Snowden's framework comes from his years of experience using narrative theory and complexity theory applied to knowledge management, decision-making, and strategy. During this time, he was led to question whether the common-place assumptions of "order," "rational choice," and "intentional capability" held in all contexts. His terminology derives from complexity theory, although using the label "chaotic" for contexts where there is no order of any kind panders to the everyday rather than the technical use of that term in chaos theory. There is also a lack of clarity about how exactly the structural arrangements, defined by the axes, give rise to the types of system behavior described in the four domains.

Kurtz and Snowden suggest that "systems thinking" is appropriate for complicated contexts where cause and effect are separated over time and space but order can be discovered with the help of experts. This is an excellent description of how "system dynamics" frames the world but completely inaccurate for most of the other systems methodologies. Their failure to appreciate the very different framings employed, for example, by hard, soft, and emancipatory systems approaches, means that they ignore methodologies that could contribute to management in the three other contexts and fail to recognize the need to extend the framework to encompass aspects of complexity fore-grounded by those approaches, e.g. people and power complexity. Kurtz and Snowden actually provide a good account of the "important contextual differences between human organizations and those of ant colonies" (2003, p. 464) – humans have more than one identity, have free will, and are aware of and act on global as well as local patterns. But they only seem to embrace these fundamental, emergent characteristics in describing the fifth context. And, even here, human disagreements get filtered through whether contexts are seen as simple, complicated, complex, or chaotic. What about differences of values, beliefs, purposes, and interests? A related consequence is that applied systems/complexity theorists hoping to benefit from the findings of social theory will find themselves disappointed by the Cynefin framework. It reflects only one, or at best two, of the many sociological perspectives we saw on offer in Chapter 4. Managers who know and use the framework will appreciate the need to view contexts through different framings and adopt different modes of intervention, but they are sent out into the world with a restricted appreciation of social reality and a very limited toolkit compared to connoisseurs of the SOSM.

8.3.5 A Revised "System of Systems Methodologies"

The reader will not be surprised to learn that it is the SOSM that we will rely on as our vehicle for the critical examination of the 10 systems methodologies. It is transparent about how it was constructed and the two axes employed continue to receive support from other systems thinkers looking to classify systems approaches according to their underlying assumptions and what they are most useful for. Bawden recognizes two major transitions in the history of systems thinking, one toward holism and the other toward pluralism. Reynolds and Holwell (2010) relate these to the two dimensions of the grid. Williams and Hummelbrunner (2010) insist that there are three ways of understanding what is involved in thinking systemically: understanding relationships, taking account of multiple perspectives, and showing awareness of how power can impact the drawing of boundaries to the advantage of some groups and disadvantage of others. This is another, fuller restatement of what is implied by the vertical and horizontal axes of the SOSM. Dent and Umpleby (1998) consider various traditions in systems theory and cybernetics using six concepts, which correspond to the SOSM's "systems" dimension (holism, relationships, environment, indeterminism, causality, and self-organization) and two captured by the "participants" dimension (observation and reflexivity). Support for having separate axes for what are actually interdependent aspects of complexity ("systems" and "participants") comes from sociology in the form of Archer's work on "analytic dualism" (see Section 4.6). Archer argues that it is necessary to keep "structure" and "agency" distinct for analytical purposes, however intertwined they are in reality.

It is, however, appropriate to make three modifications to the original grid of problem contexts to ensure it is fit for purpose. First, the distinction between "complicated" and "complex" contexts, made by the Cynefin framework and now widely accepted in the systems and complexity literature, does bring added resolution to the vertical dimension. In the "unitary" column, it helps point to the distinctive features of system dynamics as opposed to hard systems thinking and organizational cybernetics. The refinement it offers is also of use in looking at "complicated" versions of "pluralist" and "coercive" contexts. It will be adopted. Second, although the "participants" dimension was always defined to include the "affected but not involved," changing the label of this axis to "stakeholders" makes this clearer. That concept is also more widely used and better understood. Mitroff and Linstone provide a good definition:

> Stakeholders are any individual, group, organization, institution that can *affect* as well as *be affected* by an individual's, group's, organization's, or institution's *policy* or *policies*.
>
> *(1993, p. 141)*

Finally, if the "systems" dimension is turned upside down, compared to earlier versions of the grid, we have the more esthetically pleasing panorama of complexity increasing diagonally from bottom left to top right. The revised "ideal-type" grid stands, therefore, as in Table 8.2.

This framework is comprehensive and refined enough to have something useful to say about the 10 systems methodologies considered in the following chapters. It covers all the ground of Beer's classification (as far as that is relevant to "organized complexity") and the Cynefin framework, and much more besides. The "ideal-type" grid is useful in facilitating links with all types of social theory and enables advantage to be taken of these connections to deepen the critique of the systems methodologies. It is even able to relate to the varieties of complexity theory discussed in Section 7.5 – "interpretive,"

Table 8.2 A revised "ideal-type" grid of problem contexts.

		Stakeholders		
		Unitary	Pluralist	Coercive
Systems	Complex	Complex–unitary	Complex–pluralist	Complex–coercive
	Complicated	Complicated–unitary	Complicated–pluralist	Complicated–coercive
	Simple	Simple–unitary	Simple–pluralist	Simple–coercive

"radical change," "postmodernist" and "critical realist" – about which Snowden's complexity-driven but restrictive framework must remain silent. Of course, the ultimate justification of the revised SOSM rests on how successful it proves to be in bringing to fruition our critical, second-order interrogation of the various systems methodologies. It is happy to be judged on:

- The clarity it brings to how different systems thinkers see the world when seeking to intervene in it using their methodologies
- Its ability to critique the explicit and implicit theoretical assumptions made by systems thinkers
- Its ability to reveal similarities and differences between the systems methodologies examined
- Its contribution to suggesting what are the main strengths and weaknesses of the best known and most useful systems methodologies
- Its contribution to demonstrating how different systems methodologies address different aspects of complexity
- Its respect for pluralism and the pointers it provides for a pluralist approach to systems intervention, making use of all the systems methodologies as appropriate
- Its ability to provoke reflection on the circumstances in which systems methodologies are used and the consequences of methodology choice
- The respect it pays to the different systems methodologies in recognizing their validity in appropriate circumstances
- The modesty it evokes among systems thinkers as it reveals that all systems methodologies have their blind-spots

There will always be those who are horrified by grids whether made up of four, six, or nine cells. They believe that matrices disfigure the things they seek to classify in order to promote a simplified view of how things are. This, they argue, is both dishonest and dangerous. I believe such criticisms are misplaced. As Beer stated in presenting his classification, grids of the type we have been considering are a starting point for thinking not an end point. We have to start somewhere and, with the SOSM, I am seeking to be absolutely clear about the sources of my reflections on systems methodologies. The chapters on the individual methodologies will detail when there are problems with the broad distinctions initially made and how particular systems approaches have changed in an attempt to tackle aspects of complexity different from those for which they were originally designed. I can also point to Edwards' (2014) argument that frameworks act as an important protection of diversity by continually drawing attention to all the perspectives available. They prevent currently less popular approaches from being ignored or subsumed by the dominant mode of thinking and acting.

Reynolds and Holwell (2010, p. 12) suggest two causes of difficulties for those who want to use the SOSM. The first stems from the grid, in their view, implying that any problem situation in the real world can be easily identified with one of the cells of the matrix, e.g. as complex–pluralist. But the SOSM does not imply this. The SOSM framework is designed to classify systems methodologies on the basis of their assumptions, not to categorize actual problem-contexts. The SOSM has been constructed to embrace a wide range of perspectives that others employ when viewing problem-contexts. In particular, it provides an aerial map of how the main social scientific and systems traditions of thought have pictured social reality. It makes no assumptions of its own. Its purpose is to reveal and examine the assumptions made about the nature of problem-contexts by systems thinkers as they have explored the territory defined by the map. The SOSM does its job if it helps explicate the nature of the different systems approaches and points to their different strengths and weaknesses. In response to an earlier attempt to do this (Jackson 2003), the originators of the methodologies almost universally told me they thought I had done a good job on other peoples' work. Less good, they opined, on their own. I'll take that. Reynolds and Holwell's second presumed cause of difficulties is that the SOSM "pigeon-holes" individual methodologies and thereby suggests that they cannot evolve and that they can only be used in one way and so are only suitable for specific types of situation. These criticisms, deriving originally from Wendy Gregory, are also adopted by Midgley (2000, p. 238). The critics argue that systems methodologies can evolve and develop and argue that "fixing" them prevents us seeing this. Furthermore, they note that different systems practitioners use methodologies in different ways and for different purposes so they are capable of being used in a wider range of contexts than implied by the grid. These points of criticism miss the mark when aimed at the SOSM. The SOSM starts by trying to discover and reveal the assumptions that the original or traditional versions of the methodologies make but is open to these changing over time or being varied by different users. In these cases, far from pigeon-holing methodologies, the SOSM makes it possible to see exactly how methodologies have changed by charting their migration across the territory mapped by the grid (as we shall see with "Soft-OR" and "qualitative system dynamics") and to pin down the different assumptions that different practitioners employ when using methodologies in different ways (as we shall do with organizational cybernetics). I might also add that methodologies are designed by their originators to make practical their considered beliefs, implicit or explicit, about the nature of social systems and how they should be managed. These should not be jettisoned lightly. Methodologies can lose their original, often unique, strengths if someone tries to shift their assumptions too much. We may not then be able to get much out of using them. The SOSM can help explain why.

The SOSM can now be used with confidence to orientate some initial thoughts about the development and nature of the various strands of applied systems thinking. As mentioned in the preface, I will be reviewing what I consider to be the 10 most significant systems approaches from the point of view of management practice. For the reasons given I have, since the construction of Figure 8.2 in 2003, dropped "complexity theory" and "postmodern systems thinking" from consideration as systems methodologies and added "The Vanguard Method" and "Socio-Technical Systems Thinking." I will hold back from firmly and pictorially positioning systems methodologies on the grid of problem contexts until they are discussed in detail in the following chapters.

8.4   The Development of Applied Systems Thinking

The "ideal-type" grid of problem contexts that constitutes the SOSM is useful in helping us to understand the development of applied systems thinking since its origins. It enables us to grasp the variety of responses made by systems practitioners in their attempts to tackle what they recognize as increasingly complex problem situations – complex in terms of both "systems" and "stakeholders." Using it, we can discern a pattern in the history of applied systems thinking.

When systems practitioners bring together various systems ideas and techniques in an organized way, and employ them to try to improve a problem situation, they are said to be using a "systems methodology." The attempt to devise such methodologies began around the time of the Second World War. It was during World War 2, and its immediate aftermath, that the methodologies of Operational Research, Systems Analysis, and Systems Engineering were born. These three methodologies taken together were later labeled "hard systems thinking" because of what they all take for granted (Checkland 1981). In terms of the SOSM, it is clear that they assume that problem contexts are simple–unitary in character and recommend intervening accordingly. This is hardly surprising given the circumstances in which they were developed. If you are trying to win a war or are engaged in postwar reconstruction, it is completely reasonable to believe in a shared, readily identifiable goal. Later in the 1960s and 1970s, when hard systems approaches were taken into universities to be further "refined" by academics, an original bias toward quantification became an obsession with mathematically modeling the system of concern. To think that this is possible and useful requires you to assume that the system you are dealing with is relatively simple. Hard systems thinking, therefore, lies toward the bottom left corner of the grid of problem contexts because it makes simple–unitary assumptions. It takes it as given that stakeholders agree on objectives and that systems are simple enough to be modeled mathematically. These assumptions have, as we shall see, served it well, enabling practitioners of the approach to tackle a whole variety of technical issues.

Unfortunately, because of the assumptions embedded within them, difficulties arise when attempts are made to extend the range of application of hard systems approaches. If the system of concern reveals itself during an intervention to be more complicated or complex, then traditional mathematical models will fail to capture the multiple interactions that exist between the parts, miss any emergent properties, and in turbulent situations become quickly out of date and lose any relevance they might have. Furthermore, it is often difficult to define precise objectives on which all stakeholders can agree. In these circumstances, methodologies demanding a predefined goal cannot get started because they offer no means of bringing about any consensus or accommodation around a particular goal to be pursued. In the 1960s and 1970s, it became clear to many that hard systems thinking floundered in problem contexts that turned out to be more complicated, complex, pluralist, and/or coercive. It is to the credit of applied systems thinking that it was able to move beyond its original simple–unitary assumptions and explore beyond the bottom left hand corner of the grid. The last 50 years have seen serious attempts made to extend the area of successful application of systems ideas by developing methodologies that assume that problem contexts are more complicated, complex, pluralist, and/or coercive in nature. This is the progress in applied systems thinking that we now seek to chart.

We begin with the vertical axis of the ideal-type grid of problem contexts; with those systems practitioners who sought to develop competence "further up" the axis because they assumed that "systems" were more complicated or complex than hard systems thinkers believed.

Following Kurtz and Snowden's description, complicated systems are those in which cause–effect relationships are difficult to understand because they are often separated in space and time and interlinked in tangled positive and negative feedback loops. Jay W. Forrester, working at MIT, believed that his "system dynamics" could take advantage of the power of the new digital computers and provide the expertise necessary to gain insight into such contexts. The solution he proposed was to use system dynamics to uncover the structural relationships between the feedback loops that govern system behavior. Once these are understood, it becomes possible to determine what leverage points decision-makers can exploit to change the balance of the feedback loops to produce more desirable outcomes.

At around the same time, in the 1960s, the sociotechnical systems approach was being developed by action researchers based at the Tavistock Institute of Human Relations in London. Much influenced by von Bertalanffy's theory of "open systems," they evolved principles for the design of complex organizations that had to meet changing human, technological, and economic needs while seeking to flourish in increasingly unpredictable, "turbulent field" environments. A similar endeavor, but from within the cybernetics tradition, was being pursued by Stafford Beer. Building upon the concepts of black box, feedback, and variety, and using the human body and nervous system as an exemplar, he constructed his "viable system model." The viable system model sets out the organizational features that any system must possess if it is to be capable of responding to unforeseen environmental changes. The model is designed to be perfectly general and is therefore applicable to organizations of all shapes and sizes. It seeks to explain how systems can survive and be effective over time by continually self-organizing in the face of internally and externally generated change. Both sociotechnical systems thinking and Beer's "organizational cybernetics" can be seen as responses to the nature of complex systems. These we defined as having a large number of interconnected subsystems and as existing in turbulent environments. Such systems must adapt and evolve over time as they are affected by disturbances emanating from their own purposeful parts and their environments.

In making progress up the vertical axis of the grid, the systems approaches mentioned have not entirely ignored the "stakeholders" axis. We will consider this when we come to look at them in detail. Nevertheless, it has been the "systems" dimension that has been their primary focus. If we want to see how progress has been made by applied systems thinking along the horizontal axis, we need to turn to "soft" and "emancipatory" systems approaches.

Soft systems thinkers, such as Churchman, Ackoff, and Checkland, developed their methodologies in the 1970s and 1980s. They concentrated on the complexity that arises from "stakeholders" having a plurality of different values and beliefs and acting according to multiple interpretations of reality. In these circumstances, practitioners cannot assume there is agreement about goals or pretend to produce an "objective" model of the system of concern. Instead, they must try to generate a systemic learning process in which the stakeholders come to appreciate alternative worldviews and the possibilities for change they offer, and reach a consensus or, at least, an accommodation about what needs to be done. The primary interest of soft systems thinkers is in exploring the culture and politics of the problem situation to see what change is desirable and feasible and in gaining commitment from participants to agreed courses of action. An appropriate methodology for bringing about change is developed, used again and again, and gradually improved.

If we shift further along the horizontal axis of the grid of problem contexts, the issue arises of how to intervene in problem situations that are regarded as coercive. Soft systems thinking cannot respond appropriately because it presupposes that a genuine consensus, or at least accommodation, between different stakeholders can be achieved. We have defined coercive contexts as those where significant differences of interest among the stakeholders exist and power operates in support of sectional interests. Systems practitioners who identify such contexts as common have sought to formulate "emancipatory" systems approaches to try to address the issues to which they give rise. Beer's "team syntegrity" seeks to specify an arena and procedures that enable all stakeholders to debate openly and democratically the situation they are confronting. Ulrich's "critical systems heuristics" allows questions to be asked about who benefits from particular systems designs and seeks to empower those affected by policy decisions but not involved in them.

In making progress along the horizontal axis of the grid, soft and emancipatory systems approaches do not entirely ignore the "systems" axis, as we shall see. Nevertheless, it has been the "stakeholders" dimension that has been their primary focus.

One important systems approach, "The Vanguard Method," escapes this historical timeline. It is a later development that combines aspects of the systems approach with lessons learned from Toyota's "lean manufacturing" process adapted for service organizations. It has had a significant impact, especially in the public sector, and we shall devote a chapter to it. In terms of the grid of problem contexts, it concentrates on improving processes as systems but with some acknowledgement of the importance of the issues raised by the "stakeholders" dimension.

In short, the argument of this section is that applied systems thinking has matured sufficiently over the past 50 years to take into account the characteristics of a much wider range of the ideal-type problem contexts represented in the grid. It has progressed up the vertical dimension to take greater account of complicated and complex systems. It has progressed along the horizontal dimension, recognizing that problem contexts can be defined as pluralist and coercive. The SOSM makes it possible to see a healthy future for systems thinking using the range of methodologies now available in informed combination. As Jackson and Keys argued, when first formulating the SOSM:

> OR [operational research] is regarded by many as being in crisis. If OR is taken to be 'classical OR', this is indisputable .... If, however, the definition of OR is widened to embrace other systems-based methodologies for problem solving, then a diversity of approaches may herald not crisis, but increased competence and effectiveness in a variety of different problem contexts.
>
> *(1984, p. 484)*

8.5   Systems Thinking and the Management of Complexity

Each of the 10 systems approaches considered, in Chapters 9–18, makes use of the systems theory presented and draws on the systems concepts introduced in Parts I and II of the book. They are all systemic. However, the manner in which they use systems ideas and the range of systems concepts employed is different. It will help to close the loop with our theme of the "management of complexity" if we order the 10 methodologies in terms of where each of them sees the main source of complexity arising for decision-makers and, linked to this, what they regard as the most difficult aspect of a leader's or manager's task. To this end, the 10 systems approaches are classified into 6 types according to the form of complexity to which they appear to give most attention:

- *Type A*: Systems approaches for "technical complexity"
- *Type B*: Systems approaches for "process complexity"
- *Type C*: Systems approaches for "structural complexity"
- *Type D*: Systems approaches for "organizational complexity"
- *Type E*: Systems approaches for "people complexity"
- *Type F*: Systems approaches for "power complexity"

The rest of Part III of the book is divided into six sections based on these initial thoughts. The argument will be elaborated in the case of each systems methodology in the individual chapters.

8.6   Conclusion

Time has been taken to explain the need for and to develop a suitable vehicle for helping us undertake a thorough critical systems review, or second-order analysis, of systems methodologies in the chapters that follow. The revised SOSM has already helped us to understand the development of contemporary applied systems thinking. For the purpose of organizing the material, we will divide the 10 methodologies under review into six types according to how they perceive and treat complexity. They can now be considered in detail.

Before passing on, however, I want to present something to meditate on. While writing this chapter, I became fascinated by the drawing shown in Figure 8.4. It illustrates so many aspects of complexity.

The waterfall is shown "with all its molecular logic" (Beer, quoted in Blohm et al. 1986), manifesting the laws of nature but always changing according to the rules of "chaos," the weather, and the seasons. The waterfall is actually High Force in the Yorkshire Dales. I know it well and carry a visual image of it in my mind. Here, however, it is seen through the eyes of a Chinese "silent traveler" in the Dales. I struggle to conceive how he could see the waterfall this way. The silent traveler is there in the scene, both creating it and impacted by it. And, so am I.

By the end of Part IV, we may have a better understanding of how we can appreciate such complexity and act appropriately. For the moment, here are some words from the

Figure 8.4 High Force and myself. *Source:* From Yee (1941).

famous "Ten Oxherding Pictures" of Zen Buddhism. The seeker begins the journey "searching for the ox":

> Alone in the wilderness, lost in the jungle, the boy is searching, searching!
> The swelling waters, the far-away mountains, and the unending path;
> Exhausted and in despair, he knows not where to go,
> He only hears the evening cicadas singing in the maple-woods.
>
> *(Suzuki 1973, p. 371)*

Type A

Systems Approaches for Technical Complexity

In certain circumstances ... a mechanical mode of organization can provide the basis for effective operation. But in others it can have many unfortunate consequences. It is thus important to understand how and why we are engaging in mechanistic thinking; and how so many popular theories and taken-for-granted ideas about organization support this thinking

(Morgan 1986, p. 22)

Technical complexity, as I define it here, occurs when you seek to design a complex system to achieve a predefined purpose by organizing the various components and subsystems (of machines, materials, money, and people) in the most efficient way possible. A certain kind of systems approach, which has come to be known as "hard systems thinking," addresses the issues arising from technical complexity by providing the specialist knowledge and techniques which will enable the predefined purpose to be achieved in an optimum manner. It includes individual methodologies such as:

- Operational Research, Systems Analysis, Systems Engineering (Hard Systems Thinking)

Critical Systems Thinking and the Management of Complexity, First Edition. Michael C. Jackson.
© 2019 John Wiley & Sons Ltd. Published 2019 by John Wiley & Sons Ltd.

9

Operational Research, Systems Analysis, Systems Engineering (Hard Systems Thinking)

Many elements of such [socio-technical] systems exhibit forms of regular behaviour, and scientific scrutiny has yielded much knowledge about these regularities. Thus, many of the problems that arise in sociotechnical systems can be addressed by focusing such knowledge in appropriate ways by means of the logical, quantitative, and structural tools of modern science and technology

(Quade and Miser 1980, p. 2)

9.1 Prologue

In 2012, Pauline and I bought a house close to the Segura River, in a town called Blanca in the Murcia region of Spain. It was a year after my original cancer diagnosis. We had decided to "keep on truckin" as the Grateful Dead, and others before them, advised. It was a quick decision, which I soon thought we would regret when I came across an article comparing the Segura to an open sewer. It was apparently contaminated to 10 times the legal limit and emitted foul smells in towns through which it passed. The state of the river had provoked a demonstration of 12 000 people in the city of Murcia. Panicking, I started doing some research on Google and found, fortunately, that the article I was reading was out-of-date. By 2011, the Segura had become the least polluted of all the great Spanish rivers, with contamination imperceptible over its entire 350 km length. As Bill McCann writes:

> It must be a rare, if not unique, experience for a nationally important European river to be transformed within ten years from the most polluted to the best quality in its homeland, but that can be said of the Segura in south–eastern Spain.
>
> *(2012, p. 28)*

This transformation happened as a result of a systems engineering (SE) project led by Miguel-Angel Rodenas, now President of the Segura River Basin Authority. In 2011, the project was awarded the *Acueducto de Segovia de Obra Civil y Medio Ambiente* prize,

Critical Systems Thinking and the Management of Complexity, First Edition. Michael C. Jackson.
© 2019 John Wiley & Sons Ltd. Published 2019 by John Wiley & Sons Ltd.

for Civil Works and Environmental Projects, by the Spanish Society of Civil Engineers, the citation concluding:

> The **Integrated Urban Water Reclamation and Reuse System in the Murcia Region** represents an example of the modern engineering approach: **Systems Engineering**. It is the right approach for solving problems of such complexity as the environmental restoration of a great river within a structural water deficit scenario, or the integrated management of water scarcity. According to these criteria, Systems Engineering does not only tackle the construction of single hydraulic works or a group of them, but also takes into account all necessary actions for achieving multiple objectives, amongst which technical and economic sustainability are of paramount importance.
>
> *(2011, p. 45)*

2016 brought a further accolade with the Segura River Project awarded the European River*prize* from the International RiverFoundation. That same year, Miguel-Angel and his team were kind enough to meet Pauline and I for discussions on the project and what it had achieved. The following brief account is based on our talks, and on Rodenas and Albacete (2014), Spanish Society of Civil Engineers (2011), and McCann (2012).

The Segura River Basin has the lowest annual rainfall of all European Basin Districts. Thanks to the sun, and expertise in irrigation techniques dating back to the Moors, it has nevertheless become a prime producer of fruit and vegetables, accounting for around 20% of Spanish exports in that sector. The Segura River traverses the entire Murcia region and is the main source for irrigation. In the 1980s, the region's historical problems of water shortage were made worse by a significant increase in need for water for irrigation. Further, the river was becoming badly polluted, especially at times of low flow, due to an increase in inadequately treated wastewater discharges from both domestic sources and the expanding fruit and vegetable canning industry. In the early 1990s, the problems were exacerbated by severe drought. All these factors combined and resulted in further stress on water supplies, little water reaching the river mouth, the river banks degrading, fauna and flora dying, and social discontent. In 1995, the Regional Government instructed the "General Directorate for Water" (GDW), headed by Miguel-Angel Rodenas, to commence planning for a regional scheme that would address the issues. Action was necessary on a number of fronts. Engineers, expert in wastewater treatment plant design and construction, reviewed the relevant literature and visited various advanced treatment plants in other countries. Manuel Albacete, another key figure in the planning group, focused on the management and operation of the future system and on the associated legal and tax issues. Public consultations took place and an environmental impact assessment was prepared. In 2001, enough preparatory work had been done for a 10-year "General Plan for Wastewater Reclamation" to pass through the Murcian parliament. The plan aimed to ensure the environmental recovery of the river at the same time as increasing water availability by treating and reusing wastewater. Four objectives were specified:

- Construction of a region-wide wastewater collection and treatment system
- Setting up of an independent regional agency to maintain and operate the new infrastructure

- Institution of a robust system for monitoring of discharges to sewers
- Facilitation and encouragement toward widespread take-up of industrial wastewater treatment at source

Crucially, the plan formalized the transfer of responsibility for wastewater from local authorities to the GDW and enabled the introduction of a levy on all dischargers on the "polluter pays" principle. The total investment in the plan was eventually around € 645 million, of which 75–80% came from European funds.

By 2011, 46 large and 51 smaller wastewater treatment plants had been built, with careful phasing of primary, secondary, and tertiary facilities, and all are achieving processing results above European Community standards. The new collection network, made possible by crossing local authority boundaries, takes in 99% of the urban population of the region. A new unitary authority ESAMUR had been established to administer the discharge levy and significant progress was being made in encouraging industry toward on-site treatment. The overall effect has been a 13% increase in the amount of reclaimed water available in the river basin, all used for irrigation and thus benefitting agriculture. Pollution levels have fallen drastically and river flow has improved significantly, especially in its lower reaches. During the course of the project, perceptions shifted from seeing the Segura as a "water canal" to viewing it as a "natural system." Attention turned to the flora and fauna, which regenerated as the water quality improved. The river now presents a pleasing aspect and is enjoyed by fishermen, canoeists, and other water sports enthusiasts. Otters have been spotted in the river in the City of Murcia. Several large lagoons were created for the storage of reclaimed water and these have become favored stopping-off places for migratory birds, including several endangered species. Two of these are now recognized within the Ramsar Convention of Wetlands of International Importance.

The particular circumstances that made this "systems engineering" project so successful, and the challenges it had to overcome, will be reviewed later in the chapter. For the moment, I can assure readers that I have myself experienced some extremely exciting rafting adventures on the stretch of the Segura between Cieza and Blanca.

9.2 Description of Hard Systems Thinking

9.2.1 Historical Development

Hard systems thinking (HST) was a name given by Checkland (1981) to a group of systems approaches for solving real-world problems that were developed during and immediately after the Second World War. The approaches most commonly associated with this label, and the ones examined below, are operational research (operations research in the United States), systems analysis (SA), and systems engineering. They share their common philosophy, theory, and methodology with many later variants, such as decision science, cost–benefit analysis, planning–programming–budgeting systems (PPBS), and policy analysis.

The term "operational research" (OR) originated in 1937 during a project, at the Bawdsey Research Station, in which scientists sought to assist military leaders to improve the workings of the United Kingdom's early warning radar system. What justified the new name was that this was scientific research carried out into operational processes rather than into natural phenomena. The use of a systems approach can be discerned in that the radar devices were viewed as an integrated set, rather than as individual units, and the project also embraced the behavior of operating personnel. According to Royston, the system developed

> ... has been estimated to have *doubled* the efficacy of UK fighter command in the Battle of Britain. National survival was then hanging in the balance. A small reduction in air power would almost certainly have led to a very different result.
>
> *(2013, p. 793)*

Largely due to the efforts of P.M.S. Blackett, often referred to as the "father" of operational research, the use of OR was soon extended from the RAF to the army and navy. It is reckoned that, during the Second World War, close to 1000 men and women in Britain were engaged in OR. The approach was used successfully in many projects, for example, to improve bombing and artillery effectiveness and to reduce losses suffered by convoys (Cummings 2011, pp. 22–25). OR also developed in other countries, including Canada, the United States, France, and Australia. In the United States, it was first used in the Naval Ordnance group dealing with mine warfare. After the war, OR found civilian application in government departments and, particularly, in the newly nationalized industries of the United Kingdom, including coal, gas, steel, and transport. The 1950s saw professional societies formed to promote OR, scholarly journals launched, and the beginnings of the academic study of the subject. In 1959, the International Federation of Operational Research Societies (IFORS) was founded and its purpose declared to be "the development of operational research as a unified science and its advancement in all nations of the world." Good accounts of the early years of OR, in war and peace, are provided in Keys (1991) and Kirby (2003).

Systems analysis (SA) was developed in the United States out of wartime military operations planning. The name was first applied to research done for the US Air Force on future weapon systems in the late 1940s. In the 1950s and 1960s, the approach was promoted by the influential RAND (an acronym for "Research ANd Development") Corporation, a nonprofit body in the advice-giving business and, in Hammond's (2003, p. 54) words, "the archetypal brain trust of the Kennedy/McNamara era." Thanks to official endorsement, the use of SA became widespread in the defence and aerospace industries. In 1965, President Johnson gave the approach, under the label PPBS, a further boost by ordering its adoption in all departments and agencies of the US federal government. SA soon began to be applied to problems with significant social dimensions and, as will be seen, was far less effective in this milieu. In 1972, the International Institute for Applied Systems Analysis (IIASA) was established in Austria, on the initiative of the academies of science (or equivalent) of 12 nations. Founded to promote East-West scientific co-operation during the Cold War, IIASA sought to apply RAND-style systems analysis to problems of common interest in areas such as energy, food supply, and the environment. Its aim was to promote interdisciplinary research into complex world problems with a view to providing valuable policy options to decision-makers involved in shaping the modern world. Since that time, IIASA has become the official guardian of the development of SA as a discipline and profession.

Many applied systems approaches, such as cybernetics and system dynamics, owe a profound debt to the discipline of engineering. The same is true, as its name suggests, with systems engineering (SE). In the 1940s and 1950s, engineering began to extend its scope from individual components to the design of complex systems involving many interacting components. This led to the development of a new branch of the discipline, systems engineering. The leaders were the Bell Telephone Laboratories, who pioneered the approach in the United States to deal with the networking challenges they faced in the communications industry. A good definition of this new branch of engineering is provided by Jenkins:

> *Systems Engineering is the science of designing complex systems in their totality* to ensure that the component subsystems making up the system are designed, fitted together, checked and operated in the most efficient way.
>
> *(1972, p. 82)*

SE spread rapidly to the defense, space (e.g. the Apollo program), aerospace, energy, and manufacturing industries. In the 1960s, a range of guidelines and standards were established for the use of the approach and, following the lead of the US Department of Defense, it was taken up by Boeing, Lockheed Martin, British Aerospace, Marconi, and many others (Open University 2016). Paul Gartz, of Boeing, came to regard SE as "a problem-solving discipline for the modern world" and describes how it was tested to its limits in the development of the Boeing 777 in a way that led to many advances in its practice (Gartz 1997). The origins of the International Council on Systems Engineering (INCOSE) can be dated to meetings that took place in 1990. INCOSE sees SE as applicable to a wide range of systems in product, service, enterprise, and "systems of systems" contexts. Certainly, it has come to dominate modern-day complex project management.

The three main strands of HST continue to support their own separate professional associations, research communities, journals, and conferences. The differences between them are, however, often blurred in practice. Ormerod (2011) sees systems analysis in IIASA as simply OR applied to (usually) large problems with policy implications. Systems analysis jobs in Network Rail are advertised in OR magazines and promise training in systems engineering.

9.2.2 Philosophy and Theory

The pioneers of HST were extremely proud of the fact that they were applying scientific methods to problems of immediate significance to leaders and managers. They were not the first to do this. Isaac Newton, in 1696, used "time-and motion" study to improve the efficiency of the Royal Mint:

> Two mills with 4 millers, 12 horses, two horse keepers, 3 cutters, 2 flatters, 8 sizers, one nealer, three blanchers, two markers, two presses with fourteen labourers to pull at them can coin after the rate of a thousand weight or 3000 *lib* [pounds] of money per diem.
>
> *(Newton, quoted in White 1997, p. 261)*

In the early years of the twentieth century, Frederick Taylor sought to apply science to work processes and management through his influential doctrine of scientific management. Hard systems thinkers, however, were the first to recognize that one of the main tenets of the scientific method – reductionism – had to be questioned if science was to make headway in tackling real-world problems. It is from this realization that much of the originality of the approach derives.

The philosophy and theory underpinning HST are rarely declared openly. It is taken for granted that the traditional scientific method provides the example that must be followed. Discussion does, however, sometimes focus on the adjustments that have to be made to that method to make it applicable to the real-world problems of interest to hard systems thinkers. We can tease out these adjustments by looking at the common features of well-known definitions of OR, Systems Analysis (SA), and Systems Engineering (SE). The British Operational Research Society stuck, for many years, to a definition of OR first propounded in 1962:

> Operational research is the application of the methods of science to complex problems arising in the direction and management of large systems of men, machines, materials and money in industry, business, government and defence. The distinctive approach is to develop a scientific model of the system, incorporating measurements of factors such as chance and risk, with which to predict and compare the outcomes of alternative decisions, strategies, or controls. The purpose is to help management determine its policy and actions scientifically.

Miser and Quade, in the first *Handbook of Systems Analysis*, state that:

> The central purpose of systems analysis is to help public and private decision and policy-makers to solve the problems and resolve the policy issues that they face. It does this by improving the basis for their judgement by generating information and marshalling evidence bearing on their problems and, in particular, on possible actions that may be suggested to alleviate them. Thus commonly, a systems analysis focuses on a problem arising from the operations of a sociotechnical system, considers various responses to this problem and supplies evidence about the costs, benefits, and other consequences of these responses.
>
> *(1985)*

INCOSE regards Systems Engineering as an

> ... interdisciplinary approach and means to enable the realization of successful systems. It focuses on defining customer needs and required functionality early in the development cycle, documenting requirements, then proceeding with design synthesis and system validation while considering the complete problem.
>
> *(2017)*

Although a commitment to aspects of the scientific method is obvious in all these definitions, it is also clear that the purpose the method is to serve is different from that normally found in pure science. The primary aim in HST is to serve "management," "public and private decision, and policy-makers" and "customers," and not to bring about the advancement of knowledge for its own sake. For Keys (1995), this shift in emphasis away from seeking truth and toward an investigation of the usefulness of methods and techniques from the point of view of "clients," means that OR (and, it follows, other hard approaches) should be understood as a technology rather than a science.

An important consequence follows. In order to tackle the real-world problems of interest to their clients, hard systems thinkers had to abandon the laboratory as the place where they tested their hypotheses and invent a different way of proceeding. Real-world problems resist reduction to simple cause–effect relations that can be explored under controlled laboratory conditions. Furthermore, it is often too costly or just unethical to carry out experiments on large systems involving people. Finally, intervening in sociotechnical systems inevitably changes their character and prevents experiments being repeated. To progress, an alternative to the laboratory had to be found. The response of all versions of HST was to propose that models, primarily mathematical models, can perform, in the management sciences, the role that the laboratory plays in the natural sciences. Models need to be built that capture as accurately as possible the workings of the system that gives rise to the problem under investigation. Once constructed, a model can be used to simulate the behavior of the system without taking any action that might alter the real world or affect the people involved. In particular, different possible ways of improving the system from the client's perspective can be tested using the model.

Finally, it is recognized in the definitions that no one field of science is likely to be able to deal with real-world problems. Such problems usually do not fit into the domains of the established disciplines. The hard systems thinker, being problem rather than discipline-centered, will therefore have to draw on a range of disciplinary areas or be interdisciplinary in his or her approach. As Keys notes, understanding OR as a technology means it is

> ... free to draw upon and influence those sciences which are concerned in any way with the situations in which OR is involved at any time.
>
> *(1991, p. 159)*

9.2.3 Methodology

Methodology in applied systems thinking, the reader will recall, refers to the guidance given to practitioners about how to translate the philosophy and theory of an approach into practical application. Looking at the methodologies proposed by hard systems thinkers, Checkland (1981) concluded that they all take the same form. They assume that they must define an objective for the system they are seeking to improve and all see their task as the systematic pursuit of the most efficient means of achieving that objective. We can now review this conclusion in relation to the specific methodologies of OR, SA, and SE.

An early expression of the classical OR methodology appeared in an influential textbook by Churchman, Ackoff, and Arnoff published in 1957. OR, the authors argue, is the application of the most advanced scientific techniques by interdisciplinary teams to the overall problems of complex organizations. This requires a systems approach. They then set out a six-stage methodology:

- Formulating the problem
- Constructing a mathematical model to represent the system under study
- Deriving a solution from the model
- Testing the model and the solution derived from it
- Establishing controls over the solution
- Putting the solution to work (implementation)

Emphasis is placed on problem formulation (specifying the decision-makers' objectives and identifying the relevant system of concern), on a modeling phase and an implementation stage.

It is reasonable to take the three IIASA handbooks, edited by Miser and Quade (1985, 1988) and Miser (1995), as representing the state of the art as far as systems analysis methodology is concerned. The handbooks make clear that systems analysis always starts with the recognition by someone involved with a sociotechnical system that a problem exists. This problem requires proper formulation. Once that is achieved, the methodology prescribes a research phase during which a scientific approach is brought to bear on the problem. The research should be multidisciplinary. It requires identifying alternative ways of tackling the problem and building models that can be used to test the alternatives. The alternative means are then evaluated and ranked according to the decision-makers' preferences, bearing in mind costs, benefits, and other consequences. Finally, assistance is given with implementation and with evaluation of outcomes. The Figure 9.1 representation of this systems analysis methodology appears in all three of the handbooks.

For A.D. Hall, reflecting on his experiences with the Bell Telephone Laboratories (see Keys 1991), systems exist in hierarchies and should be engineered with this in mind to efficiently achieve their purposes. The systems engineer is charged with co-ordinating a multidisciplinary team that must clarify the objectives and then ensure the optimum integration and consistency of system and subsystems in pursuit of those objectives. Jenkins (1972), a British systems engineer, provides a detailed elaboration of the steps required:

1. Systems analysis
   1.1 formulation of the problem
   1.2 organization of the project
   1.3 definition of the system
   1.4 definition of the wider system
   1.5 objectives of the wider system
   1.6 objectives of the system
   1.7 definition of an overall economic criterion
   1.8 information and data collection
2. Systems design
   2.1 forecasting
   2.2 model building and simulation
   2.3 optimization

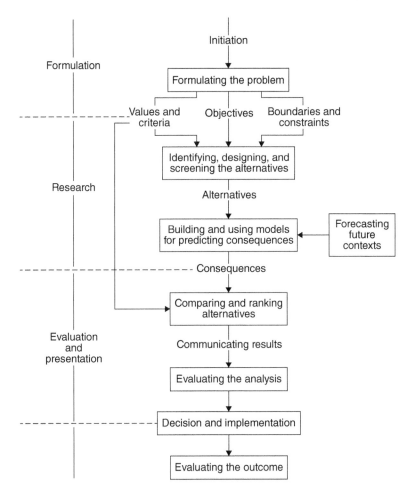

Figure 9.1 The systems analysis methodology. *Source:* From Miser and Quade (1988). Reproduced with permission of John Wiley & Sons.

    2.4 control
    2.5 reliability
3. Implementation
    3.1 documentation and sanction approval
    3.2 construction
4. Operation
    4.1 initial operation
    4.2 retrospective appraisal
    4.3 improved operation

At the beginning of this subsection, we stated that Checkland recognized a commonality in the form of all types of hard systems methodology. In essence, he argued, they take what is required (the ends and objectives) as being easy to ascertain and see their task as undertaking a systematic investigation to discover the most efficient "how" that will realize the predefined objectives. Hard systems methodologies presuppose that real-world problems can be tackled on the basis of the following assumptions:

- There is a desired state of the system S_1, which is known
- There is a present state of the system S_0
- There are alternative ways of getting from S_0 to S_1
- It is the role of the systems person to find the most efficient means of getting from S_0 to S_1

9.2.4 Methods

It is not entirely correct to say that HST has concentrated on methods of model building at the expense of methods to support other stages or phases of the methodological process. The three IIASA handbooks on systems analysis contain a good deal on the craft skills necessary to support problem formulation, communication with decision-makers, and implementation. The rapid development of "Soft-OR" in the 1980s and 1990s in the United Kingdom (see Section 9.2.5) is evidence of a tradition of work that has paid attention to the process of operational research. Successful practitioners of all strands of HST have of necessity to develop well-tuned social and political skills. Nevertheless, it is true that the mainstream academic research in the field shows an overwhelming bias in the direction of perfecting methods of modeling. In these circumstances, it is no surprise that this is the area in which HST has most to offer.

Models are explicit, simplifying interpretations of aspects of reality relevant to the purpose at hand. They seek to capture the most important variables and interactions giving rise to system behavior. They are used as surrogates for real-world systems for the purpose of carrying out experiments. The literature of HST identifies various types of models: iconic, analogic, analytic, simulations, gaming, judgmental, and conceptual.

Iconic models are simply scale (usually reduced scale) representations of what is being modeled, such as an aircraft model used in wind tunnel testing or an architect's three-dimensional model of a new building. Analogue models are very different in appearance to the reality but, nevertheless, seek to mimic the behavior of what they represent. An example would be an electrical network used to represent water flowing through pipes. Analytic models are mathematical models that are used to represent the logical relationships that are believed to govern the behavior of the system being investigated. They are widely used in OR. Wilson (1990, p. 20) helpfully provides a matrix (adapted as Figure 9.2) that divides analytic models into four classes, depending on whether they represent behavior at one point in time (steady state) or over time (dynamic) and whether the behavior is described by fixed rules (deterministic) or statistical distribution (nondeterministic).

Referring to Figure 9.2, algebraic equations can be used to formulate, for example, problems about the most appropriate way to allocate productive resources in order to maximize profit when many alternatives exist and resources are limited. The linear programming technique has been developed to provide an optimal solution in this type of situation. Statistical and probability relationships can be employed to determine the

	Steady state	Dynamic
Deterministic	Algebraic equations	Differential equations
Non-deterministic	Statistical and probability relationships	Discrete-event simulation

Figure 9.2 Types of analytic model. *Source:* Adapted from Wilson 1990.

degree of dependence of one variable on another in the absence of complete knowledge about the interrelationships in a system. Wilson (1990) shows, for example, how the linear regression technique can provide a model of how total electricity sales are related to the level of industrial production. Differential equations provide a modeling language for the dynamic deterministic category. Wilson demonstrates how the problem of designing a suspension system for a vehicle can be tackled on the basis of differential equations solved by computer. Dynamic nondeterministic systems require simulation, which we are treating here as another type of modeling.

Simulation of quite complex systems, in which the relationships between variables are not well understood and change over time, has been made possible with the advance in computing power. It refers to the process of mapping item by item and step by step the essential features of the system of interest. The model produced is subject to a series of experiments and the outcomes documented. The likely behavior of the system can then be predicted using statistical analysis. For example, the essential features of a traffic flow system can be represented in a computer simulation as long as the key factors impacting the traffic flow can be identified. Computer-generated random numbers determine amount of traffic, numbers turning left, etc., so that the system's behavior can be monitored under different conditions. A sophisticated type of simulation modeling, system dynamics, is described in Chapter 11. Gaming is a kind of modeling in which "actors" adopt and play out the roles of significant decision-makers in a system. They are supposed to behave as would their real-world counterparts in order that matters of choice, judgment, values, and politics can be investigated. Judgmental models usually rest on group opinion of the likelihood of particular events taking place. Techniques such as "Delphi" (developed at the RAND Corporation) and "scenario writing" are used in systems analysis to develop the best group models from the individual mental models of members of multidisciplinary teams. Delphi employs an anonymous, iterative process to guide experts and other knowledgeable individuals toward a reasonable consensus about an issue. Scenario writing explores the likelihood of particular future states of affairs coming about. Conceptual models, as the name suggests, are qualitative models used to make explicit the particular mental models held by parties interested in a decision. They are more frequently employed in soft systems thinking than in HST.

9.2.5   Developments in Hard Systems Thinking

The most important development that could occur in HST would be to extend the scope of the approach beyond technical problem situations, in which it has proved successful, to situations of greater system complexity and/or in which people and politics play a central role.

We begin with SA. The IIASA handbooks declare that systems analysis needs to be extended beyond problems that "are relatively easy to structure and in which some important aspect is dominated by technology" to more "people-dominated" problems (Miser and Quade 1988, p. xiv). As mentioned, the handbooks actually devote considerable effort to documenting the "craft skills" needed to cope with people and politics. However, this remains a long way from developing the concepts that would allow these matters to be discussed theoretically and a methodology that would enable any learning gained to be utilized in practice. In a helpful article, Leen Hordijk (2007), IIASA Director from 2002 to 2008, sets out to "explain" systems analysis using a nine-step framework:

- Marshall all the information and scientific knowledge available on the problem in question
- Determine the stakeholders' goals
- Explore different ways of achieving those goals
- Reconsider the problem in the light of the knowledge accumulated
- Estimate the impacts of the possible courses of action
- Compare the alternatives
- Present the results to the stakeholders
- Provide follow-up assistance
- Evaluate the results

It is clear that little has changed since Miser and Quade's statement of SA methodology and that it still corresponds to Checkland's schematic of a hard systems approach. The result as Hordijk states, honestly enough, is that it can be helpful in "addressing issues dominated by science and engineering" and for "budgetary decisions," but it is

> ... more problematical where political, organizational, and social factors predominate and where goals may be obscure and authority diffuse and overlapping.
>
> *(Hordijk 2007, p. 2)*

IIASA, now with 24 nations contributing, continues to set up research programs and initiatives using systems analysis to consider critical twenty-first century issues such as global, environmental, economic, technological, and social change. But it is largely ignored by those it deems its stakeholders.

INCOSE has long declared its intention to expand systems engineering into nontraditional domains, such as transportation, housing, infrastructure renewal, and environmental systems. A typical proposal to do this, made by Hipel et al., seeks to employ SE's "system of systems"(SoS) approach to investigate the global food system. This is clearly an extremely complex system but is to be tackled using a framework that demands a detailed, scientific grasp of its exact nature. In other words, it is treated as if it were simple:

> Current and future generations are faced with complex global challenges in their journey towards achieving general well-being and security for all. The interconnectedness of natural, societal and technological systems means that decision makers and policy makers must understand the SoS which they influence, as a whole, in order to achieve their desired outcomes and avoid unintended consquences.
>
> *(Hipel et al. 2010, p. 15)*

Kalawsky, setting out the "next generation of grand challenges for systems engineering research," similarly demands "modeling and simulation environments that allow a full system and its environment to be modeled." However, this is difficult when complex systems involve human beings:

> The major weakness today is that we don't yet have comprehensive models of human behaviour and cognitive performance. It would be a great asset if we could reliably predict human behaviour under all circumstances.
>
> *(Kalawsky 2013, p. 839)*

In the absence of this predictive capability, a proposed solution is to replace human-based systems by deploying "autonomous systems" with "little if any reliance on human operator intervention to ensure they act responsibly when performing tasks" (*Kalawsky* 2013, p. 842).

A much healthier approach to developing the competence of SE is revealed in the March 2017 version of the "Systems Engineering Book of Knowledge" (SEBoK). This online, updateable compendium of key knowledge sources and references clearly recognizes that SE has to be pursued, and its results implemented, in complex political, budgetary, and institutional contexts. It recommends complementing the technical-systems focus, that has been dominant in SE, with the learning-systems focus of soft systems approaches such as Checkland's. Discussions with the International Society for the Systems Sciences (ISSS – representing the "general systems theory" persuasion) have led to a recognition that it is necessary to integrate theory and practice from different sources and to the development of a framework for unifying "systems praxis." It seeks to provide a common language for systems practitioners and to combine hard and soft methods "in a pragmatic, pluralist, or critical multi-methodology." This nod in the direction of "critical systems thinking" is to be applauded but other parts of the compendium suggest that the "integration" needed is currently conceived of as being on SE's own, imperialistic terms. Other approaches seem to be welcomed to the degree that they can contribute to a largely traditional version of SE methodology:

- Transform stakeholder needs into a common set of requirements
- Develop requirements
- Select life-cycle model
- Architect a system of systems

SE will have to move quickly to bring its approach to tackling problems into line with its new "critical multi-methodology" mode of thinking. Senior figures in project management, a discipline long dominated by SE, have become fed up with waiting and have established the "International Centre for Complex Project Management" (ICCPM) to develop, harness, and propagate the knowledge and skills necessary to ensure successful delivery on complex projects and programs. These include SE but also an understanding of the world from the point of view of complexity theory, and a "critical systems" perspective on how other systems approaches can help in working with diverse and competing stakeholder views and in organizing viable project systems. Masters programs and executive courses have made use of the earlier version of this book (Jackson 2003) to teach a critical systems approach to complex projects.

Operational Research, at least in the United Kingdom, has been the strand of HST most willing to promote and embrace change. The reason frequently given for this (e.g. Rosenhead 2009) is that it has remained close to the world of practice while in the United States, for example, it retreated into the universities to engage in, using Ackoff's (1977) telling phrase, too much "mathematical masturbation." In the attempt to improve its practice, OR in the United Kingdom has been willing to engage with and learn from social scientists (see the conference proceedings edited by Lawrence 1966; and Jackson et al. 1989) and systems thinkers (see Paucar-Caceres 2011). It has responded to new developments that have threatened its client base, for example, its recent embrace of "business analytics." It has pursued initiatives around "Community OR" (see Midgley and Ochoa-Arias 2004) and "Behavioral OR," designed both to extend its range of clients and improve its methodology. It has been willing to take on board criticism from such as Ackoff (1979a,b) that suggested the profession was losing its way and failing to establish itself at the strategic level in organizations. Most importantly of all, propelled by this inventiveness, a group of operational researchers has made a paradigm shift and established "Soft-OR" as a complementary practice to the traditional hard approach (see Rosenhead 1989b; Rosenhead and Mingers 2001b). All these things are reflected in the official definition of OR employed by the UK Operational Research Society today:

> **Operational Research (O.R.)** is the discipline of applying appropriate analytical methods to help those who run organisations make better decisions. It's a "real-world" discipline with a focus on *improving* the complex systems and processes that underpin everybody's daily lives – **O.R.** is the "science of better."

It is clear that this represents a considerable "softening" compared to the original 1962 version.

The beginnings of the Soft-OR approach can be found in the work conducted by the former *Institute for Operational Research* (IOR), which was established in London in 1963. The IOR was a joint venture of the UK Operational Research Society and the Tavistock Institute of Human Relations, negotiated with the help of Russ Ackoff. The IOR quickly became engaged in significant action research projects and, according to John Friend:

> It is striking to see how soon the then orthodox [1962] definition of O.R. ... was overtaken within IOR by a recognition of the subtleties of the more diffuse

inter-organisational decision processes that were widely encountered in our early projects.

(2012, p. 21)

One project was a study of policy-making in local government, involving Coventry City Council. Friend muses:

> I found that there wasn't much data of the conventional form that I could analyse and gain insights from, but I did have colleagues who were social scientists and I sat beside them in committee meetings and all kinds of meetings including political meetings where people were arguing out how to deal with difficult strategic problems .... My colleagues and I realised we had to become facilitators of other people, drawing on the information they carried with them in their heads, rather than analysts of any kind of data.

(2017, p. 20)

Work at the IOR gave birth to the "strategic choice approach," which, refined over 50 years, now consists of various tools that can help decision-makers manage the uncertainty they face and make incremental progress toward decisions (Friend and Hickling 2004). During the 1970s and 1980s, this and other, independently developed Soft-OR methods came to prominence and Rosenhead (1989b) was able to put together a collection of papers describing and illustrating six such approaches: "strategic options development analysis" (Eden), "soft systems methodology" (Checkland), "strategic choice" (Friend, Hickling), "robustness analysis" (Rosenhead), "metagame analysis" (Howard), and "hypergame analysis" (Bennett, Huxham, Cropper). All share a common concern with unstructured problems recognizable, say Mingers and Rosenhead (2004), from the existence of:

- Multiple actors
- Multiple perspectives
- Incommensurable and/or conflicting interests
- Important intangibles
- Key uncertainties

The Soft-OR methods necessary to manage unstructured problems are characterized by (Rosenhead 1989a; Mingers and Rosenhead 2004; Mingers 2011):

- An emphasis on structuring problems rather than "solving" them
- The use of qualitative (nonmathematical) modeling
- Enabling alternative perspectives to be expressed and explored
- Encouraging participation of stakeholders
- Being transparent to users in the way that they operate
- Being able to operate in situations of uncertainty and lack of reliable data
- Operating iteratively so that the problem representation keeps pace with the discussion between actors
- Aiming for exploration of the problem situation, learning, and commitment to improvements

Thus, by the end of the 1980s, the OR community in the United Kingdom had responded comprehensively and decisively to the critique of traditional OR (launched by such as Ackoff) by developing Soft-OR as an approach to unstructured problem situations to set alongside OR's traditional competence in dealing with well-structured problems. This was achieved by making a "paradigm shift" in the discipline (Rosenhead and Mingers 2001a; Kirby 2007). If traditional OR is "functionalist" (in terms of the social theory of Chapter 4), using the methods of science to discover how "objective" aspects of reality work in order to improve their functioning, then Soft-OR is "interpretive." Its primary area of concern is people and their values, beliefs, and interests. Soft OR accepts that multiple perceptions of reality exist, and often come into conflict, and it aims to help managers and decision-makers operate successfully in pluralist contexts of this kind.

Use of the three most popular Soft-OR approaches, "strategic options development analysis" (SODA – including "cognitive mapping" and "journey making"), "strategic choice"(SC) and "soft systems methodology" (SSM) has been significant in the United Kingdom, as Ranyard et al. (2015) show. If this book was covering all management science, and not just systems approaches, there would have to be a chapter on SODA and on SC (SSM is, of course, covered). As it is, we can rest content that Rosenhead and Mingers (2001b) is an excellent complementary text to this volume. In Europe, there has been considerable interest in Soft-OR or "problem structuring methods" (PSMs), as these approaches are sometimes called. In the United States, however, despite letters to *OR/MS Today* (Ackermann et al. 2009) and the *Harvard Business Review* (Georgiou 2013) telling them what they are missing, Soft-OR has made hardly any progress among either academics or practitioners. The paradigm shift achieved in Soft-OR has proved a step too far for those in thrall to the traditional positivist/quantitative version of the discipline. It only remains to say that once OR, in the United Kingdom, accepted Soft-OR as a complementary approach to traditional OR, it was inevitable that an interest in "multi-methodology," using methodologies in combination, would also develop (see Mingers and Gill 1997). Multi-methodological theory and practice, however, had its origins, and has been most fully explored, as part of "critical systems thinking." It is fully considered in Part IV of the book.

9.3 Hard Systems Thinking in Action

Given that this chapter covers three strands of HST, it is difficult to provide one representative example of the approach in action. The problem is tackled by showing different aspects of use for each of OR, SE, and SA.

Once the original pioneering spirit had faded, operational researchers, or at least those of an academic persuasion, began to concentrate their efforts on developing mathematical models to apply to what they saw as frequently occurring types of problems. Each problem type was assumed to have a particular form and structure, which determined its nature and how it could be tackled regardless of the context in which it was found – military, manufacturing industry, service sector, etc. Fortuin et al. (1996) present 15 case studies of OR at work in application areas as diverse as transport and logistics, product and process design, maintenance and financial services, health care, and environmental decision-making. Keys (1991) and Cavaleri and Obloj (1993) provide good introductory material on the most common OR problems, a typical list being:

- Queuing problems
- Inventory problems

- Allocation problems
- Replacement problems
- Co-ordination problems
- Routing problems
- Competitive problems
- Search problems

Mathematical models relevant to queuing problems seek an optimum trade-off between the costs of providing service capacity and keeping customers happy, e.g. how many service points should a supermarket keep open at particular times. Inventory models aim to establish the optimum reorder point for stocks of resources so that production flow can be maintained while the costs of holding excess inventory are minimized. Allocation models seek to apportion scarce resources in the most efficient manner, maximizing output or minimizing costs, while achieving overall objectives. Keys comments on an example involving a farming enterprise that both reared cattle for beef and produced crops that could themselves be sold or, alternatively, used to feed the cattle. A linear programming-type model was constructed containing 640 constraints and 1801 variables. A solution that maximized profit was discovered in 34 seconds of computer time. Replacement models help to minimize costs by identifying the point at which acquisition of new assets is justifiable. Co-ordination techniques, such as PERT (Programme Evaluation and Review Technique) and critical path analysis, set out how tasks must be sequenced in a project to ensure completion in the least time and at minimum cost. The goal of routing models is to determine the most efficient route between different locations in a network. Competitive problems are conceptualized in terms of games, the aim being to maximize outcomes for one or more participants. Search models try to optimize the outcome of a search (say, for a location for a new factory) by minimizing both costs and the risks of error.

If the academic textbooks are to be believed, OR mainly offers a "cookbook" of approaches to narrow, well-structured, quite specific problems. But this is grossly unfair representation of the work of many highly skilled operational researchers who continue to achieve significant gains for their organizations by using the OR approach flexibly and with a sharp eye on context. In 2016, General Motors (GM) was awarded the INFORMS Prize in recognition of its integration of OR into its business. According to Owen and Inman (2017), today no area of GM is untouched by analytical methods. A scientific approach to decision-making and problem-solving is deemed indispensable in the modern world and OR and management science is used to tackle the most complex issues the company faces – prognostic algorithms, vehicle content optimization, optimizing new vehicle inventory, and optimizing sales incentives. Even in the dynamic environment of the high street fashion chain Zara, a company renowned for favoring human intuition, vision, and judgment, there are some decision areas where OR can deliver significant benefits, including "initial merchandize shipments to stores, store inventory replenishment, clearance pricing, and supplier purchase quantities" (Clegg 2015, p. 8). Typically, these areas involve large amounts of quantitative data, frequent and quick decisions, and a decision process capable of high automation.

INCOSE has established "systems engineering profiles" for 18 different application domains: agriculture, commercial aircraft, commercial avionics, criminal justice system and legal processes, emergency services, energy systems, environmental restoration, facilities systems engineering, geographic information systems, health care, highway transportation systems, information systems, manufacturing, medical devices, motor vehicles, natural resources management, space systems, and telecommunications. Additionally, there are profiles for seven cross-application domains: e-commerce, high-performance computing, human factors engineering, Internet-based applications, Internet banking, logistics, and modeling and simulation. Not surprisingly, a number of these profiles are rudimentary, with the most extensive being in areas of traditional systems engineering practice, such as the design and development of commercial aircraft. The commercial aircraft industry operates in a very competitive environment and relies on complex manufacturing processes dependent upon highly integrated subsystems, advanced technologies, use of advanced materials, detailed specifications, and very rigorous testing. The systems engineering specifications for this domain insist on the principle that commercial aircraft are considered as wholes and not as collections of parts. Both customer and regulatory requirements are first identified. Aircraft architecture is then seen as a hierarchy in which the functions and constraints operating at the top level, the aircraft system itself, flow down into requirements for the subsystems. A typical decomposition of the aircraft system into parts would identify the mechanical, propulsion, environmental, airframe, avionics, interiors, electrical, and auxiliary subsystems. These subsystems are then further decomposed into subordinate components with their own requirements deriving from those of the subsystems. Thorough monitoring and control is essential at all stages of design and construction to ensure that requirements at the different levels are verified and validated by testing.

The IIASA handbooks provide some comprehensive descriptions of SA applications, which are referred to and analyzed throughout the three volumes. The main illustrations are of improving blood availability and utilization (summarized in Jackson 2000), improving fire protection, protecting an estuary from flooding, achieving adequate amounts of energy for the long-range future, providing housing for low-income families, and controlling a forest pest in Canada. Of these examples, the fire protection case is regarded as one that closely follows the prescribed systems analysis methodology.

The fire protection study began in 1973 in Wilmington, Delaware, and was conducted by a local project team with technical assistance from the New York City – RAND Institute (Miser and Quade 1988). The eight existing firehouses in Wilmington were getting old and the mayor wanted to find out if they offered adequate protection, whether they were located in the right places, and whether any new firehouses needed building. The main objectives of fire protection were to protect lives and safeguard property while, at the same time, keeping costs low. Unfortunately, there was no reliable way of evaluating how different deployment strategies related directly to these objectives. Three "proxy" measures were therefore developed: approximate travel time to individual locations, average travel time in a region, and company workload. The consequences of changes in locations and numbers of firehouses were then considered against these. The next stage required the analysts to build models that could be used to test various deployment alternatives. The primary tools employed to encapsulate the data were a parametric allocation model, based on a mathematical formula for determining fire-company allocations to different regions, and a more descriptive, simulation model, known as the firehouse site evaluation model. The transparency of this latter model was crucial as it enabled city officials to be involved in suggesting alternatives. The recommendations to close one of the firehouses and reposition most of the remainder provoked a long battle with the firefighters union before they were eventually implemented. When the results were finally evaluated, it was found that the fire protection service was just as effective as before, but with costs significantly reduced.

9.4 Critique of Hard Systems Thinking

According to Quade and Miser, in the quotation at the start of this chapter, systems analysis is designed to tackle problems that arise in sociotechnical systems. A reasonable critical question to ask, therefore (applying the logic of such as Kant, Bhaskar, and Churchman), is: "What must socio-technical systems be like for hard systems thinking to help improve their performance?" The circumstances that enabled the Segura River project to be successful are illuminating. As Rodenas and Albacete insist, the period of intensive preplanning was decisive:

> This work also shows the advantages of comprehensive pre-planning as a rational, effective method that helps to solve complex problems associated with water management.
>
> *(2014, p. 56)*

This preplanning involved gaining the necessary technical expertise and also, crucially, addressing the main social and political barriers to success. Strong political support had to be marshaled to overcome the resistance of the local authorities to the transfer of decision-making power over wastewater to the GDW and to pass the discharge levy into law. Even then, not all the wastewater treatment plants could be sited exactly to plan. In spite of attempts to minimize the environmental and social impacts of their construction, local politics or "not in my backyard" attitudes sometimes prevailed. This led to the overall plan being implemented in a somewhat less than optimal manner and remains a regret of Miguel-Angel Rodenas. Nevertheless, transferring power to the GDW extended the locus of responsible decision-making to correspond to the boundary of the problem situation and transformed the project into a largely technical one within the scope of good systems engineering practice. Although the Segura River rises outside the Murcia region, the issues faced all lay within. It made it possible to address them comprehensively. That said, the new Segura River Basin Authority is still confronted by a substantial water shortfall in the region of Murcia. In October 2017, following three years of extreme drought in the Segura basin, water allocations to many farmers had to be slashed. This, and other national issues of water supply, can only be overcome, says Rodenas, by further inter-basin transfers. For this to happen, the nine River Basin Authorities in Spain must be brought into closer collaboration.

In the Segura River case, the political will was available to ensure that the social, economic, and environmental objectives largely coincided. Complexity was reduced to pursuing these objectives in a technically optimal manner by designing efficient and effective parts (the individual hydraulic works) and managing the relatively simple relationships between them. HST was able to bring scientific rigor to the solution of the problem and to present results it could regard as being "objective," free from the taint of personality or vested interest. Assumptions, data, and calculations were made clear and outcomes validated in a way that could inform the work of scientists engaged on similar projects.

The great achievement of HST, therefore, has been to take the scientific method and learn how to apply it successfully to practical problems of management. Hard systems thinkers do research aimed at serving the interests of clients, decision-makers, and problem owners. This shift to valuing knowledge directly relevant to application rather than simply to the advancement of a scientific discipline was revolutionary and enabled management scientists to steal a march on other disciplinary areas that, to this day, are still struggling to put in place the conceptual apparatus that would enable them to make their findings more relevant (see Tranfield and Starkey 1998, on the struggle to establish more application-oriented management research). As a result, HST is able to offer a systematic approach to decision-making, inspired by the scientific method, which constitutes a significant advance over ad hoc thinking about the management task. This is particularly the case in the domain of public spending where no natural control mechanism, such as the market, exists. The careful setting of objectives, the search for alternative means of reaching those objectives, and the evaluation of the alternatives in terms of measures of performance, make the efficient step-by-step control of projects feasible. Applying the scientific method to management problems required hard systems thinkers to come up with significant innovations in their practice. They pioneered the use of multidisciplinary teams of researchers and became advocates of an interdisciplinary or trans-disciplinary approach. Particularly valuable to them, in this respect, was the existence of various systems ideas and concepts. Reductionism was useless because of the complexity and unbounded character of real-world problems and because of the interactive nature of their parts. What was required was a more holistic, integrating approach that sought to be comprehensive by drawing the boundaries of the system of concern more widely. The systems language, employing concepts such as system, subsystem, hierarchy, boundary, and control, was perfect for this purpose. Another problem that hard systems thinkers had to overcome was how to test the hypotheses they developed. They could not carry out experiments directly on the systems they were hoping to improve because it was too costly and/or unethical. Unlike natural scientists, the problems they faced were too interconnected to be broken up and taken into the laboratory for analysis. The solution was to construct a model or models that accurately captured the behavior of the real-world system and to run tests on those. Considerable progress had to be made by hard systems thinkers on the techniques of mathematical and computer-based modeling if this approach was to succeed.

The apparent success of systems analysis and systems engineering in the 1950s and 1960s, in the fields of aerospace, defense, and space exploration, invested these approaches with a kind of "man on the moon" magic: They became "bathed in moonglow" (Hoos 1974, p. 87). Calls came from President Johnson, Vice-President Humphrey, senators, and government officials to transfer these systems approaches from the technological domain to the management of social problems – the state of the inner cities being a favorite target. One of the first to be caught by this mood was Governor Brown of California. In 1964, the State of California called for proposals from the aerospace industry for systems designs to be applied to information, crime, welfare, transportation, and waste management. As Hoos carefully documents, this attempt at "technological transfer" was unsuccessful, sometimes disastrously so. She concludes:

> The question we have asked in this research study is, 'Are the techniques of systems analysis appropriate when we are dealing with problems which are

essentially human and social?' The findings indicate that in their present condition they are not. And the direction in which they are developing promises little improvement. Refinement of methodology has led only to greater preoccupation with abstraction while the mythology that social problems can be solved remains unchallenged.

(1974, p. 247)

The reasons, set out below, for the difficulties encountered in trying to use HST in the social realm are derived from the work of Hoos (1974), Ackoff (1979a,b), Checkland (1981), Churchman (1979a), Boguslaw (1981), Lilienfeld (1978), and Rosenhead (1981). They suggest that it has a limited domain of applicability, which extends to social systems only in special cases. In summary, this is because of its difficulties in coming to terms with extreme complexity, an inability to appreciate "subjectivity," and its innate conservatism.

Hard systems thinkers recognize that the systems with which they deal exhibit aspects of complexity but still believe they are simple enough to be represented in mathematical models. When highly complex systems are involved, however, the building of a quantitative model is inevitably a highly selective process and will reflect the limitations of vision and biases of its creator(s). Far from recognizing this, hard systems thinkers tend to treat the model too readily as synonymous with the reality. The model, which is of course far more easily manipulated than the real world, becomes the focus of attention and the generator of "optimum" solutions. It is convenient and cozy to play with the model, but the result is solutions that are out-of-date answers (since the model soon becomes an out-of-date representation) to the wrong questions. Confronted with the complexity of California's "solid waste management system," Hoos shows how system designers simplified the issues involved by basing their models on interviews with just 39 people, concentrating on their attitudes to 13 categories of "bad effects" of waste – flies, water pollution, air pollution, rodents, etc.:

> The "hard data" were nothing but a crystallizing of the hastily contrived catalogue of unpleasantries into arbitrary and overlapping categories through an artificial weighting procedure in which three individuals, no matter how expert, were taken to represent the total community attitude.
>
> *(Hoos 1974, p. 140)*

Another consequence of the demand for quantification and optimization is the tendency to ignore those factors in the problem situation that are not amenable to quantification or, perhaps even more seriously, to distort them in the quest for quantification. Different aspirations or matters subject to differing value interpretations are forgotten or ground down on the wheel of optimization. The Bay Area Rapid Transport study failed to take any account of issues of environmental degradation.

HST is further accused of not paying sufficient attention to issues of "subjectivity" that usually dominate in debates about social systems and their objectives. Hard approaches demand that objectives be clearly defined at the very beginning of the methodological process. This may be fine for engineering-type problems when ends are easy to specify and attention can be concentrated on means. In the vast majority of managerial situations, however, the very definition of objectives will constitute a major part of the problem faced. Involved parties are likely to see the problem situation differently and to define objectives according to their own worldviews, values, and interests. This will give rise to many possible accounts of what the objectives of a particular system are, some of which might well conflict. In these "softer" problem situations, involving multiple perceptions of reality, it is not clear how hard systems methodologies can get started, since they lack the procedures for bringing about an accommodation between alternative definitions of what the objectives might be. Unfortunately, a common response to this difficulty from proponents of HST is to distort the nature of the problem situation in order to make it fit the requirements of the preferred methodology. One objective or set of objectives will be privileged over others on the basis of the "expert" understanding of the system achieved by the systems analysts. Hoos shows how a "police mentality" came to dominate the systems design of criminal justice in California:

> Law enforcement at the street level is stressed and serves as *raison d'etre* for more and more comprehensive information networks, designed to catch persons who rightfully or wrongfully have been tabbed "potential offenders." When detection and control are construed as the prime objectives of a criminal justice system, there is an instant 'rational' justification for the acquisition of "hard" technology, in the form of helicopters for surveillance and crowd control … such systems leave out such crucial considerations as the social environment, police activities, the state of the court calendar, and the functions and functioning of penal institutions.
>
> *(Hoos 1974, pp. 135–136)*

A related criticism highlights the failure of hard systems approaches to pay proper attention to the "purposefulness" of the human component in the sociotechnical systems with which they aspire to deal. People are often treated as components to be engineered just like other mechanical parts of the system. The fact that human beings possess understanding, and are only motivated to support change and perform well if they attach favorable meanings to the situation in which they find themselves, is ignored. This deterministic perspective in HST, which puts the system before people and their perceptions, extends to the ability of humans to intervene in their own destiny. Hard systems thinkers take the future to be determined by factors outside the control of organizational actors. It is the job of the systems consultant to predict what will happen in the future and help managers cope. Thus, the opportunity to mobilize people to design their own future is missed (Ackoff 1979a,b).

The innate conservatism of HST, the way it offers succor to the *status quo* and to the already powerful, is frequently noted. As we have just seen, in order to get going in softer problem situations, hard methodologies require the privileging of one objective or set of objectives over others. It goes without saying that the best way to ensure the continuation of a consultancy project, and the implementation of the proposals, is to privilege the objectives of powerful stakeholders. Forced inevitably into making such political choices, hard approaches seek to cover their tracks by encouraging "depoliticization" and "scientization" (Rosenhead 1981). The complicated mathematical modeling discourages ordinary people from believing that they might have anything useful to contribute to decision-making. It also suggests that differences of opinion and interest can be rationally dissolved by experts using the latest tools and techniques. Thus conflict is hidden. Conclusions emerge from a computer model programmed by white-collar scientists and take on an air of objectivity that is entirely spurious. Lilienfeld (1978) argues that systems theory of this ilk should be regarded as an "ideology." It flourishes because of the service it renders to scientific and technocratic élites. Presenting, as it does, a view of systems as entities to be manipulated from the outside on the basis of expertise, HST justifies the position and privileges of these élites. Boguslaw, himself a former RAND man, came to see hard systems thinkers as "the new utopians":

> Our own utopian renaissance receives its impetus from a desire to extend the mastery of man over nature. Its greatest vigor stems from a dissatisfaction with the limitations of man's existing control over his physical environment. Its greatest threat consists precisely in its potential as a means for extending the control of man over man.
>
> *(1981, p. 204)*

If we now dig deeper, using the philosophy and theory outlined in Part I of the book, it is obvious that hard systems thinkers take their lead from a "realist" version of the physical sciences that ignores the work of Kant. They believe that there is a "real-world" existing independently of both observers and their theories. This orientation is reinforced by the roots of much of HST in the engineering tradition. According to Hoos (1974, p. 26), engineers have a "trained incapacity" to see systems as anything but things governed by predictable laws. Their interest is in ensuring the efficient engineering of systems to achieve known goals. The behavior of systems has to be predicted and regulated in pursuit of their controllers' objectives. In social theory terms, HST is governed by the functionalist paradigm. The concerns of the interpretive paradigm in bringing about mutual understanding among those with different values and beliefs, of the "radical change" paradigm in alleviating disadvantage, and the postmodern paradigm in unpredictability and diversity, do not get a look-in. Again, a reservation to this conclusion needs inserting on behalf of sophisticated practitioners of HST who know, only too well, that social and political factors can intervene and determine how "rational" models are received and made use of. Here, as Ormerod (2006) insists, a more "practical, commonsense, scientific approach," akin to pragmatism, may prevail. Models will be judged in terms of their practical outcomes rather than by strict adherence to the scientific method. Hard systems thinkers can be pragmatists even if their methodologies are thoroughly functionalist.

The critique conducted makes it easy to relate HST to the system of systems methodologies (SOSM). The hard systems approach makes "simple" and "unitary" assumptions about the problem situations it seeks to tackle. The system of concern is deemed simple enough to capture in a quantitative model that can be used to test possible solutions. The world views of stakeholders are taken to converge to the extent that the problem becomes one of finding the most efficient means of arriving at agreed-upon objectives. Problem situations that yield to a systematic approach of this kind are meat and drink to HST. On the other hand, hard systems approaches will struggle to make progress, and achieve improvement, when systems are more complicated and complex, and when stakeholders are in a pluralist or coercive relationship with one another. This positioning on the SOSM is shown in Figure 9.3.

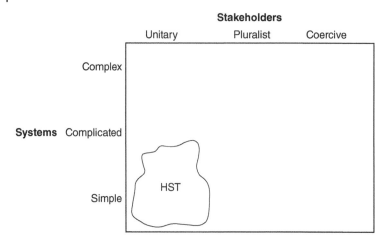

Figure 9.3 The positioning of hard systems thinking on the SOSM.

The grid of problem-contexts, of course, only offers guidance as to where we should look to find significant differences between systems methodologies. Because it seeks to provide a classification of systems methodologies, its meaning only becomes clear as we interrogate the different methodologies. Having now considered "hard systems thinking," we will be somewhat clearer about what is meant by "simple" and "unitary" assumptions. It also needs stating that proficient hard systems thinkers are well aware that they operate in problem situations that are more complex and pluralistic than their methodologies can hope to tackle on their own. They will often bring into play their own hard-earned craft skills to manage the reception and implementation of their "rational" solutions in these cases. This is to their credit. The only claim being made here is that there are alternative, well-documented and thoroughly tested systems approaches available, which may provide a greater guarantee of success and better enable the accumulation and dissemination of knowledge about how to proceed in more complicated/complex and pluralist/coercive problem situations. Finally, the SOSM draws no conclusions about how common "simple–unitary" problem contexts are and, therefore, about the relative value of the hard against other kinds of systems approach. It does, of course, insist that unless the other perspectives it highlights are taken seriously managers will have little option but to see all problem situations as simple–unitary.

9.5 Comments

In November 2017, just last week as I write, I was honored to receive The Beale Medal of The Operational Research Society "for an outstanding and sustained contribution by one person to operational research in the UK." Having spent a career looking at how other systems approaches can "enhance" the contribution of OR, I was somewhat disarmed by this award. I could have been dismissed as an enemy. Instead I was being embraced as a friend. I take this as a recognition by the OR community that a clearer understanding of the theoretical and methodological underpinnings of their approach, and the limitations these impose in practice, can actually make a community of thinkers and practitioners stronger. If the systems movement as a whole is to prosper, then advocates of its various strands will need to show the same kind of maturity as OR and recognize the important role critical systems thinking, as "second-order" critique, can play in augmenting what they are seeking to achieve.

9.6 The Value of Hard Systems Thinking to Managers

In the case of each of the 10 systems methodologies covered in Part III of the book, we will highlight five obvious benefits that they bring to management. Although the word "managers" is used in these summary sections, it must be interpreted broadly to include leaders and decision-makers of all kinds. With HST, although managers need to employ expert practitioners to get the most value out of the approach, the following five lessons can be learned that will benefit them in their everyday work:

- Scientific expertise can help managers in dealing with a significant set of the technical issues that they confront
- The insistence of the systems approach on holism rather than reductionism, in tackling real-world problems, can assist managers to obtain comprehensive and integrated solutions, and avoid suboptimization
- The use of a systematic methodology to tackle problems is usually preferable to an ad hoc approach based on the manager's common sense
- It is extremely helpful, in seeking to improve problem situations, if managers can clearly set objectives, seek alternative means of achieving those objectives, and evaluate those alternatives on the basis of precise measures of performance. This is especially the case in the public and voluntary sectors where the market does not operate to ensure efficiency
- Faced with the complexity of problem situations, managers inevitably use models. Whether these models are mathematical or not, it can assist learning if managers are explicit about the models they are using and the assumptions on which they are based

9.7 Conclusion

Hard systems thinkers established many of the tenets of applied systems thinking that have since served it well in dealing with the increased complexity of real-world problem situations. The approach has limitations, however, and these became obvious to both theoreticians and practitioners in the 1970s and 1980s. In general terms, despite its many strengths and achievements, the hard systems approach is today thought of as having a limited domain of application. Away from the realm of engineering-type problems, hard methodologies only prosper in those circumstances where worldviews converge and significant consensus is achieved about the need to maximize the performance of some relatively simple and easily separable system. New thinking was needed to enable the systems approach to extend its scope to more complex and pluralistic contexts.

Type B

Systems Approaches for Process Complexity

Most people imagine that the present style of management has always existed, and is a fixture. Actually, it is a modern invention – a prison created by the way in which people interact

(Deming 1994, p. 15)

Another form of systems thinking, which evinces a primary concern with process complexity, is introduced in Chapter 10. Process complexity arises when we have to put together a series of interdependent actions to achieve a purpose. The aim of systems thinking, in this context, is to tame complexity so that the process achieves the right things in terms of customer purpose and, by getting rid of "waste" in the system, is as efficient as possible. There are similarities with "hard systems thinking," but the emphasis upon "actions" rather than "components" implies more engagement with the people involved. An excellent example is:

- The Vanguard Method

Critical Systems Thinking and the Management of Complexity, First Edition. Michael C. Jackson.
© 2019 John Wiley & Sons Ltd. Published 2019 by John Wiley & Sons Ltd.

10

The Vanguard Method

The Vanguard approach to systems thinking is a methodology for change and improvement that engages the organisation. Any change is based on an under-standing of demand from an "outside-in" or customer perspective, identification of the value work, adoption of relevant measures and then designing out the waste within key processes. People who do the work must be engaged in these activities. The results: better service, reduced costs and improved morale

(Seddon 2005, p. 179)

10.1   Prologue

In 2006, the housing department of Portsmouth City Council owned and was responsible for the management of over 17 000 tenanted and leasehold properties in which around 51 000 people lived (O'Donovan and Zokaei 2011). The department's operational budget was £80 million and it delivered its services through a directly employed staff of approximately 600 and a variety of private sector contractors. The housing service was meeting the Council's own Key Performance Indicators (KPIs) and the Audit Commission rated it highly. Despite these favorable results, the Head of Housing Management, Owen Buckwell, was constantly having his ears burned by councilors whose surgeries were full of residents complaining of poor service. Buckwell decided to find out what was really going on and began by applying Seddon's Vanguard Method (VM) to "reactive repairs," where residents were complaining of having to wait for ages to get repairs completed satisfactorily (see O'Donovan and Zokaei 2011, for details of this project). The team chosen to conduct the intervention studied customer demand and determined that the purpose of the system, from the customer's point of view, was "Do the right repair at the right time." Checking the performance of the system revealed that this purpose was not being met. It was predictable that some repairs could take 98 days to finish, with a mean time of 24 days (see Figure 10.1). Sixty percent of calls received by the service constituted "failure demand." They were chasing up work not done or not completed properly first time. How could things be so bad? Analyzing the "flow" of work revealed that it was the imposition of "command and control" management practices, designed to reduce costs and meet targets set by local managers and government inspectors, which led to the poor performance of the system from the customer's perspective. For example, one customer

Critical Systems Thinking and the Management of Complexity, First Edition. Michael C. Jackson.
© 2019 John Wiley & Sons Ltd. Published 2019 by John Wiley & Sons Ltd.

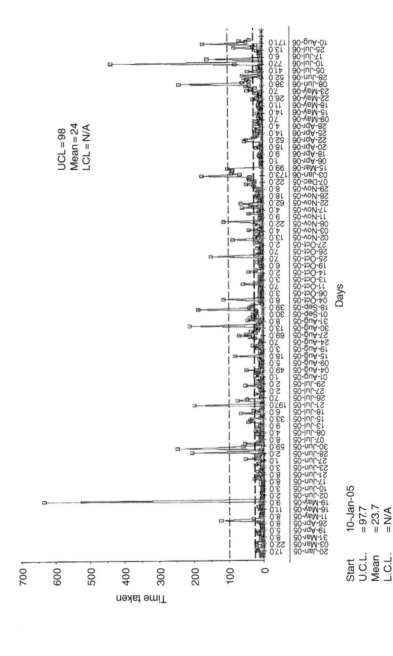

Figure 10.1 End-to-end times taken to make repairs. *Source:* From O'Donovan and Zokaei (2011). Reproduced with permission of SNCSC.

job was often turned into multiple jobs on the system so all could be completed on time and budget to meet targets, even though the result was disruption and distress to the customer. A plan to redesign the system, concentrating on those aspects of end-to-end flow crucial to meeting customer purpose, was produced and was gradually rolled out in the "do" stage of the approach. The type and frequency of demand were studied so that resources could be used to best effect. Timed appointments were introduced to improve access to properties to do jobs. Operatives were freed up to use their professional expertise to diagnose the exact repair needed, decide how long the repair would take and, if possible, undertake any other repairs the householder requested on the same visit. They were trusted with the materials necessary to undertake the kinds of repairs likely to arise and supplied with any additional materials needed at the place where the repair was being carried out. IT systems were redesigned to support the new way of working and private contractors were required to work according to systems thinking principles ensuring that the whole supply chain was transformed.

This comprehensive systems thinking intervention led to a housing service in Portsmouth that could deliver repairs on the day and time requested by the customer. The original 24 day mean time to complete repairs was reduced to 6.9 days to fix the originally reported repair and 11.2 days to fix all repairs identified at the property. Failure demand fell from 60% of calls to 14%. A satisfaction rating of 9.9 out of 10, based on follow-up phone calls immediately a job was completed, was recorded for the redesigned service over six months period. Overall, the housing repairs service reduced its expenditure from £35 million in 2007/2008 to a budgeted £32 million in 2009/2010. Employees were empowered by the new system to be more creative and their morale improved. Private contractor companies that adopted the VM managed to halve their costs per repair. In spite of these remarkable outcomes, the redesigned housing service started to draw criticism from the Audit Commission and the government's Department for Communities and Local Government. Doing what customers wanted did not seemingly translate into success in meeting bureaucratic targets. Notwithstanding these external reservations, the housing department's systems team extended its interventions in housing services beyond "reactive repairs" and its success led to it being commissioned to redesign the Council's corporate human resources and recruitment systems. In 2010, Portsmouth's housing department was honored as an M-Prize winner by US management expert Gary Hamel's Management Innovation eXchange (MIX). The commendation for the award concludes:

> Finally, the housing council's culture has shifted from one of 'learned helplessness to meet targets' to one that encourages employees to show up with their initiative and imagination fully engaged; a new ethos that emphasises action has taken hold. 'Owen always had a sneaking suspicion that people go to work to do a good job', says Seddon. 'It turns out he was right'.
>
> *(Hamel and LaBarre 2010, pp. 1–3)*

10.2 Description of the Vanguard Method

10.2.1 Historical Development

The VM was developed by John Seddon, originally an occupational psychologist, who had become interested in change programs and why they often failed. Its origins were in consultancy activities undertaken by Seddon and his colleagues in the mid-1980s. Seddon's approach is widely seen as an adaptation of the principles of the Toyota Manufacturing System to the design and management of service organizations. It has, as a result, commonly been associated with the "lean" movement. In fact, while adhering to the basic philosophy of lean, serving customer demand, Seddon is keen to distance himself from its "codification" and the use to which the tools and techniques of lean are often put. He objects to practitioners who seek to transfer lean principles unthinkingly from the manufacturing to the service sector. And, even more forcibly, to the "toolheads" who subsume its tools and techniques into traditional "command and control" thinking and use them to contribute to the further standardization of work (Seddon et al. 2011). As a former employee of Her Majesty's Revenue and Customs (HMRC), who continues to have a friend who works there, I can attest to the disasters that followed from introducing this Taylorist version of lean into that department. As Seddon argues:

> What HMRC is practising is 'factory lean'. It has been led down this path by people who think the tools developed to solve problems in manufacturing can be applied without thought with equal success in service organizations.
>
> *(2008, p. 159)*

Nowadays, Seddon and his colleagues promote VM as a type of systems thinking (O'Donovan 2014). Exactly what type will become clear in what follows. Unfortunately, devotees of the method, "Seddonistas" they have been called, have a tendency to speak of VM as if it stands for the whole of systems thinking. Such "imperialistic" maneuvers are, unfortunately, common in the systems movement – see, for example, Senge's claim for "system dynamics" in the next chapter. They need to be called out because they misrepresent the state of affairs, cause confusion, and encourage the use of a particular approach in all circumstances, even when other systems methodologies might be more appropriate. They risk discrediting the favored method and offer a disservice to the rest of systems thinking.

Seddon provided the first full account of VM in his 2003 book *Freedom from Command and Control* (2nd edn, 2005). In the book, he argues against the predominant "command and control" form of management, with its roots in Taylorism, and in favor of a "systems thinking" approach, based primarily on the work of Deming and Taiichi Ohno. VM is presented as capturing the spirit of the Toyota Manufacturing System, and the insights that can be gained from it, but suitably modified to apply to service organizations. Various, then popular, "management fads" such as ISO 9000, the "excellence" model, Investors in People, the Charter Mark, the balanced scorecard, and codified versions of lean, are dismissed, from a systems thinking perspective, as simply reinforcing a command and control mentality. In *Systems Thinking in the Public Sector* (2008), Seddon broadens his analysis to unmask the control freakery he sees as dominating government attempts to reform the public sector. The reform regime is said to be "preoccupied with creating more factories for mass-producing public services" (p. vii). There has been considerable investment:

> But our experience is that things haven't changed much. This is because governments have invested in the wrong things. Belief in targets, incentives and

inspection; belief in economies of scale and shared back-office services; belief in "deliverology" ... these are all wrong-headed ideas and yet they have underpinned this government's attempts to reform the public sector.

(Seddon 2008, back-cover)

Governments are urged to adopt the alternative offered by systems thinking. In 2014, the battle is escalated in *The Whitehall Effect*. Seddon's introduction begins: "Politicians don't know much about management" and the book proceeds to provide a damning account of the "industrialization of public services" in the United Kingdom by the Thatcher, Major, Blair, Brown, and Coalition governments. The main components of this industrialization are call centers, back offices, shared services, outsourcing, and IT systems. Rather than improve things, they make public services worse while forcing costs up. Whitehall's interventions in public services are seen as driven by ideology, often grounded in misplaced economic theory, fashions, and fads, rather than by evidence. Examples are "choice," personal budgets, commissioning, managing demand, "nudge," procurement, risk management, lean, and many large-scale IT initiatives. According to Seddon, Whitehall is "systemically incapable of doing evidence" (p. 167) and simply ignores the lessons provided by numerous studies using VM that "targets and standards make performance worse," "inspection can't improve performance," "regulation is a disease," "it's the system, not the people," and "incentives always get you less." Seddon adduces the many published examples of how VM has profoundly improved public services to argue for a better way based on changing the system – "the way the work works." This demands "a better philosophy," a recognition that "effective change starts with study," "better thinking, better design," "locality working," and "IT as pull, not push."

The roll-out of evidence that VM can drastically improve services has been impressive. Middleton (2010) details six case studies from different public sector organizations in the United Kingdom and New Zealand. Zokaei et al. (2011) provide four private sector examples of service improvement and six from the public sector, featuring disabled facilities, adult social care and housing. Pell (2012) draws attention to six more successful interventions in a wide range of public services, including the police, National Health Service (NHS), and fire and rescue. In 2013, Simon Guilfoyle, drawing on his own experience as a police officer, published *Intelligent Policing* with a view to showing how "systems thinking," understood as VM, could lead to greater efficiency, improve service delivery, reduce costs, and enhance morale in the police service. Simon Caulkin (2010) reflects on a "Vanguard Leaders Summit" at which nine leaders from the private, public, and voluntary sectors reported to an audience of 400 how they had used VM to transform their organizations and achieve significantly improved outcomes. Although this evidence was not gathered according to scientific principles, and so does not constitute unambiguous proof of the superiority of VM over other approaches, it is far more than any government has been able to muster in support of its change initiatives. It has given Seddon the confidence to continuously and vigorously berate ministers, civil servants, and regulators about their "factory management methods" and obsession with micro-management, targets, and inspections. For example, he points to the failures associated with "Universal Credit" – seven announcements of delays and implementation costs increasing to £16 billion even while focusing on claimants whose circumstances are "the least complex." His evidence to the Works and Pensions Select

Committee, in March 2017, argues that the scheme can never work because it seeks to employ computer systems, relying on rules, to deal with extremely high-variety demand, inevitably driving even more "failure demand" back into the system (Seddon 2017a).

By 2018, the Vanguard Consultancy had offices in over 12 countries. Its website declares that:

> The Vanguard Method is a propriety means by which we help leaders of service organizations change their organization from a command-and control design to a systems design; resulting in dramatic improvements in service, revenue, efficiency and morale.

The website outlines Vanguard's consultancy services, provides comprehensive on-line resources and software to support users of VM, offers popular courses on the method, and publishes a newsletter containing reports from those frustrated with command-and-control regimes and advocates of VM. Aviva has made VM a "signature" for how it works and has mandated its use world wide. Further international recognition of the approach can be found in an OECD document, *Systems Approaches to Public Sector Challenges: Working with Change* (2017), which offers a Vanguard intervention, designed to improve child protection in Amsterdam, as one of its major case studies. In 2008, The Child and Youth Protection Service in the Amsterdam area was in crisis and put under special supervision. Following the intervention, it was elected, in 2015, the best public sector organization in the Netherlands. The OECD document does, however, have some interesting things to say about the circumstances in which VM is likely to prove successful.

10.2.2 Philosophy and Theory

The "systems thinking" that underpins VM is derived primarily from the work of Deming (1982) and Taiichi Ohno (1988), figures outside the mainstream of the trans-discipline. Nevertheless, the insights gained are clearly systemic and Seddon is right to see Deming and Ohno as important systems thinkers in their own right.

Seddon insists on Deming's principle that the flow of work in organizations must be geared to what customers want. It is necessary to try to predict customer needs and then to design the work accordingly. This requires that concentration be focused on the horizontal flow of work. This is the thing that ultimately yields profit even though it is often forgotten about in organizations, which are arranged hierarchically and divided into functional silos. VM, therefore, always sees the purpose of a system in terms of its customer; what matters is what matters to the customer. Once the customer's purpose has been established, attention is given to how the parts or tasks must be fitted together in order best to achieve that purpose. Here, in line with systems thinking, it is the interactions between the parts that are viewed as crucial and as, possibly, giving rise to unexpected outcomes. VM requires that this customer focus remains central throughout an intervention. The design of support systems, such as IT systems, should follow design of the primary customer serving system. Evaluation must be in terms of overall system performance in pursuit of customer purpose. Inappropriate targets will distort the behavior of the system in ways that are not beneficial to customer purpose.

VM also honors Deming's principle that the best way to think about an organization, and manage it to achieve its purpose, is to treat it as a "system." On this basis, it is possible to create innovative, adaptive, and energized organizations that behave and learn according to what their customers want. This way of thinking about organizations must replace the traditional, "command-and-control" approach, which emphasizes hierarchy, the separation of decision-making from the work, and measures based on budgets, standards, and targets. The distinction between these two management philosophies is made clear by Seddon in Table 10.1. An important theoretical justification for this shift to systems thinking, as Seddon recognizes, comes from Ashby's "law of requisite variety" – the need for a system to match the variety thrown at it by its environment (see Section 6.3). Reducing environmental variety by trying to predict demand will help, but is unlikely to be enough in turbulent, rapidly changing situations. In these circumstances, it is also essential to improve the capacity of the organization itself to handle uncertainty. This is done, according to systems principles, by increasing its learning capacity and flexibility. VM is opposed to mechanistic thinking and the use of command-and-control management practices, which suppress creativity and the ability of a system to respond to its environment. There is a particular need for service organizations to possess high variety, and so be responsive, because of the more intangible nature of the services they deliver, the crucial role the customer plays in defining satisfactory service, and the difficulty of maintaining quality control over service transactions. Attempts to "standardize" service delivery, often through computerization, reduce the capacity of the system to provide the appropriate service to the customer first time. This is something we have all experienced when preprogrammed telephone responses fail to relate to the problem we have, we cannot get through to speak to a human being, or we are diverted to call centers, which can only offer us solutions that are irrelevant to our problem. Customers inevitably get frustrated as they have to contact the service organization again and again to get what they want, thus significantly increasing demand on the service. One of Seddon's most powerful distinctions is between "value demand," the kind of demand the system is there to serve, and "failure

Table 10.1 Command-and-control versus systems thinking.

Command and control thinking		Systems thinking
Top-down, hierarchy	Perspective	Outside-in, system
Functional	Design	Demand, value, and flow
Separated from work	Decision-making	Integrated with work
Output, targets, standards: related to budget	Measurement	Capability, variation: related to purpose
Contractual	Attitude to customers	What matters?
Contractual	Attitude to suppliers	Co-operative
Manage people and budgets	Role of management	Act on the system
Control	Ethos	Learning
Reactive, projects	Change	Adaptive, integral
Extrinsic	Motivation	Intrinsic

Source: From Seddon (2005). Reproduced with permission of Vanguard Press.

demand," which arises from a failure to do something for the customer or do it right first time. The latter simply clogs up the system and increases costs:

> Behind the idea of a crisis in our public services caused by mounting demand is the reality that the crisis is not generated by 'demanding customers' but by the failure of services to deal effectively with the customer first time round. This creates the need for more contacts, more explanation and more activity – all forms of failure demand that are of 'negative value' to the customer, the service and certainly the tax payer.
>
> *(Seddon 2014, p. 8)*

Another lesson from Deming, embraced by Seddon (2014, chapter 15), is that performance is 95% a result of the system and only 5% of the people who work in it. Take the case of an operative responsible for housing repairs in Portsmouth prior to the redesign described in Section 10.1. That person could be sent out to a wrongly classified job, with limited prospects of actually gaining access to the property (because the householder did not know when they were coming), without the right resources to complete the job, no way of immediately calling for the right resources and, quite possibly, without proper training. Clearly, the system is at fault but managers were concentrating on the 5%, measuring whether the operative completed the task in a fixed time. As Seddon argues, Deming's conclusion has radical implications for the various "people management" activities with which organizations can become obsessed:

> Culture change is probably the most absurd of all the ideas in the people management lexicon. People's behavior is governed by the system, so when you change the system the culture changes (for free!).
>
> *(Seddon 2014, p. 104)*

Both influencing and influenced by Deming was Taiichi Ohno, the "father" of the Toyota Production System (TPS). Ohno's work, suitably adapted for service organizations, provided the second major "systems thinking" inspiration for VM. His conception of "lean manufacturing" turned on its head the "make and sell" philosophy dominant in the United States and Europe. In Toyota, the work is based on customer demand. This "demand pull" approach means that something is only made when it is needed, creating a huge saving on inventory. When it is needed, it is essential that all the right resources are at the right place in the production process at the right time. At Toyota, the notion of "Just in Time" describes how material is processed and moved in order to arrive just in time for the next operation. In the West, at the time, it was assumed that costs could be driven down by "factory designs" that maximized scale. The success of lean manufacturing, by contrast, demonstrated that economies came from managing flow through the whole system, from end to end. Efficiency would follow inevitably from effectively meeting customer demand. In VM, therefore, the aim became to "clean-stream" the work, removing those activities that contribute nothing to the purpose of the customer and redesigning the system around the most vital tasks necessary in terms of that purpose. To ensure continuous improvement in the flow of work, Ohno was convinced that those who were doing the job had to have the responsibility to act to solve problems as they arose and to improve the system. Variety could be built into the system by making intelligent use of intelligent people. He wanted workers to be "constructively unreasonable," questioning the way the work is done and looking for better ways. In order to do this successfully, they should be supplied with continuous feedback on how the system is performing – another crucial systems thinking concept. The role of managers is to ensure that the system allows people freedom to think and take decisions and to support them in implementing change.

These insights from Deming's and Ohno's "systems thinking" are combined, in VM, with Seddon's own take on the kind of "intervention theory" likely to see them implemented in practice. Systems ideas are often counterintuitive, he argues, so their force is best demonstrated in practice. The preference is for the consultants to work with an in-house "systems team" selected to learn the approach by carrying it out. The systems team should be small and should have a bias toward those directly familiar with the work flow. Members of the team must be willing to challenge the status quo, be respected in and beyond their own work position, and be committed to service and quality. The systems team is supported throughout by an "organizational sponsor" who acts as a project manager. This person should be of sufficient seniority to clear any obstacles the team encounters. VM encourages those who work in the system to diagnose its faults and to lead the redesign process in order that their learning and self-development are promoted. Managers themselves must "walk the work" and be exposed to the system's failings and the improvements that can stem from redesign. Thus, both those directly involved in the work and their managers are brought to a recognition of the limitations of command and control thinking and make the leap to systems thinking by seeing its value in practice rather than just through theory. The system becomes a "learning organization" and problems that appear intractable given the current design of the system are, in Ackoff's terms, simply "dissolved" in a redesign oriented to serving customer purpose (O'Donovan 2014).

10.2.3 Methodology

The philosophy and theory enunciated above are translated into action by a three-stage methodology: *check–plan–do*. This change to Deming's well-known "plan–do–check–act" cycle is made because Seddon believes that performance can only be improved if managers change the way they think about their role. Putting "check" first ensures that they are confronted immediately and starkly with the failings of the current system and, therefore, become more willing to embrace alternatives. A summary of the methodology is provided in Table 10.2.

"Check" is an analysis of how the current system performs from the customer's perspective. The team involved in the intervention begins by trying to discover the real purpose the system serves at present. Often this will bear little relation to what the customer wants, being more influenced, for example, by targets. The next step is to establish the type and frequency of customer demand on the system. This often proves to be more predictable than expected. It is important to distinguish between "value demand" and "failure demand." In the case of repairs to properties, value demand will consist of people making first-time calls requesting repairs. Failure demand will be calls caused by a failure to do something or to do it right for the customer – for instance, tenants progress-chasing their repair or complaining that the repair has not been completed to their satisfaction. High levels of failure demand indicate something is seriously wrong with the system and, of course, put further pressure upon it. In the case of

Table 10.2 Description of "check," "plan," "do."

Stages in process	What is it?	What does it do?
"Check"	An analysis of the what and why of the current system	Provides a sound understanding of the system as it is and identifies waste and the causes of waste. "Check" asks: • What, in reality, is the purpose of this system? • What is the nature of customer demand? • What is the system achieving? • How does the work flow? • What is value work and what is waste? • Why does the system behave like this?
"Plan"	Exploration of potential solutions to eliminate waste	Provides a framework to establish what the purpose of the system should be and how the flow of work can be improved to meet it. "Plan" asks: • What is the purpose of the system from the customer's perspective? • What needs to change to improve performance against purpose? • What measures are necessary in order to gauge improvement?
"Do"	Implementation of solutions incrementally and by experiment	Allows for the testing and gradual introduction of changes whilst still considering further improvement. • Develop redesigns with those doing the work. • Experiment gradually. • Continue to review changes. • Work with managers on their changing role.

"reactive repairs" in Portsmouth, failure demand made up 60% of all demands (O'Donovan and Zokaei 2011, p. 90). For those types of demand that occur frequently, the team will assess how the system currently responds. It will consider average response times and the system's overall "capability to respond," based on upper and lower limits. The data are captured on a "capability chart" – one of the major tools employed in VM and described in the next subsection. Once the systems team understands demand and the system's response capability, it can start to map how the work flows through the system. It is vital that the mapping corresponds to the actual work and the team, therefore, will follow the set of tasks as they are carried out. Discussions center on issues that stop the work flowing smoothly and the causes of delay and frustration experienced by the workers and customers. The systems team produces flow charts of the work that they validate with those involved in doing it. The next step is to distinguish "value work," activity required to deliver what matters to the customer, from "waste." Waste takes three forms in VM. Some waste can be eliminated immediately as it serves no purpose.

Other waste may take some time to get rid of because it has been designed into the work. A final type of waste can be difficult to remove because, although it involves tasks of no direct value to this system's purpose, it is wanted by another system. The final stage in "check" involves trying to document why the system behaves as it does. A "systems picture" and a "logic picture" (both described later) are constructed on the basis of the observations of the work flow made by the team and discussions with staff, managers, and customers. The results of "check" are then presented to the organizational sponsor and other stakeholders, such as senior managers and boards. If permission to proceed is obtained, the redesign phase, incorporating "plan" and "do," can commence.

"Plan" is concerned with identifying levers for change based on systems thinking. "Do" is about taking direct action to redesign the system, eliminating waste wherever possible. The first step in "plan" is to rethink the purpose of the system from the customer's perspective. The next step is to consider principles for redesign, which will enable the system to achieve its new purpose. Common principles are "design against demand," "only do value work," "pull on expertise as needed," "IT should support the work." What is suggested in the case of "pull on expertise as needed," for example, is that it will reduce failure demand if appropriate expertise is placed at the front end of the process. A service organization, possessing knowledge of likely demand, should place an employee qualified to respond to customer requests as the first point of contact – only in exceptional circumstances will the customer then have to be referred on to someone with more specialist expertise. The final step requires performance measures to be established, which reflect the revised purpose, are relevant to improving the system, and are directly useable both by the people doing the work and their managers. "Plan" is undertaken primarily by the systems team working with the organizational sponsor and consultants. Once the "do" stage is entered, however, front-line staff and customers again become central to success. "Do" begins with possible redesigns being discussed with those doing the work. Changes are introduced incrementally, working closely with the people operating in new ways and taking careful account of direct customer feedback. Redesign is, therefore, a much slower process than "check;" the ethos being "do it right rather than do it quick." Nor can the results of the "do" stage be predicted in advance; unpredictable, emergent outcomes are always possible. A final important element is to help managers think differently. Special sessions may be necessary to help them abandon a command and control philosophy, think in a systems way, and adopt a role supportive of the staff carrying out the work.

In the VM methodology, "check," "plan," and "do" constitute a never-ending cycle designed to promote continuous improvement.

10.2.4 Methods

VM has a number of methods at its disposal to assist the methodology just described. Chief among these are "demand analysis," "capability charts," "flow charts," "systems pictures," and "logic pictures."

Demand analysis is employed to identify type and frequency of demand and to distinguish between value and failure demand. The amount of failure demand is often shocking for both staff and managers and can help galvanize a determination to change. An example of demand analysis is provided in the next section. Capability charts show how well the system responds to customer demands in customer terms. Figure 10.1 is an example, showing the end-to-end times for a sample of 451 repairs undertaken by housing services in Portsmouth between January 2005 and August 2006. The chart reveals a mean time to complete repairs of 23.7 days and the likely maximum time it could take of 97.7 days – the upper control limit (UCL).

Flow charts show how the work flows through the system. They can be produced at "keystroke" level, where every small step is included and charted, or in a more summarized form in which minor tasks are grouped together. They are employed to distinguish value work from waste. Systems pictures demonstrate how potentially simple flows become distorted and complicated by internal constraints and controls and external influences. They allow the causes of waste to be identified and classified. Logic pictures capture the manner in which the dominant mode of thinking in an organization influences system design and can contribute to it not serving its customers. There are examples of both types of picture in the 2005 report for the Office of the Deputy Prime Minister (ODPM) evaluating the use of VM in housing.

10.3   The Vanguard Method in Action

The Northern Housing Consortium (NHC) was interested in the use of systems thinking to improve housing services and had witnessed the improvements achieved by members employing the Vanguard Method. In January 2004, the NHC hosted a three-day VM awareness session, supported by Vanguard Consulting, and attended by representatives from the ODPM, the Audit Commission, and social housing organizations from across the North of England. Expressions of interest for involvement in a pilot were sought and members of the ODPM and NHC selected three pilot sites. The aim was to provide analysis across as wide a range of organizations and services as possible, covering work representative of the bulk of the mainstream business of housing landlords, and offering the most potential impact on resource use and customer service. The organizations selected were Tees Valley Housing Group, Leeds South East Homes, and Preston City Council. Tees Valley operated from a single site in Middlesbrough and was a traditional regional housing association managing almost 4000 homes. Its performance was very good but it believed there was still room for improvement in its responsive repairs service. Leeds South East Homes was an "Arms Length Management Organization" (ALMO) with delegated responsibility, from Leeds City Council, for providing housing management and maintenance services for approximately 7600 homes. Its performance was improving from a low base but there remained significant problems, according to the most recent Audit Commission inspection report, with the time taken to repair properties that had become vacant (voids) and in re-letting those properties. Preston City Council was, at the beginning of the research, a stock-owning local authority with a Housing Department managing around 6700 properties. Despite some recent improvement, major concerns persisted about service delivery. An obvious service area to review was rent collection and debt recovery where Best Value Performance Indicators remained in the lower quartile. The ODPM commissioned the NHC to undertake a project to evaluate VM in housing and work started on the pilots during 2004, continuing in 2005. The overall project was managed by Ian Wright of the NHC. At each pilot site, an in-house "systems team" was established to work with the Vanguard consultants. At the same time, the ODPM appointed an Evaluation Panel to monitor the interventions and to evaluate success. This panel was made up of key stake-holders and an academic advisor (myself). During the summer of 2005, a report, setting out the findings, was prepared for the ODPM by the NHC, based on advice from the Evaluation Panel. This report, "A Systematic Approach to Service Improvement: Evaluating Systems Thinking in Housing," was published by the ODPM in September 2005. The report was launched on 22nd September in London and 26th September in Leeds, with speakers including Dawn Eastmead (ODPM), Roy Irwin (Audit Commission), and representatives from each of the three pilots. A follow-up report, evaluating the sustainability of the initial improvements, was published by the NHC in October 2006 and titled "A Systematic Approach to Service Improvement – an update." The following account (drawn from Jackson et al. 2008) aims to illustrate the VM methodology and methods in action, taking examples as appropriate rather than offering a detailed account of what happened in each pilot.

10.3.1 Check

In all three cases, the current "purpose" of the system was found to reflect adherence to targets rather than pursuit of customer purpose. In Tees Valley, it was:

> *To do repairs within the target time set and maximize use of the in-house team*

In Leeds South East, it was:

> *To re-let empty properties in line with the Lettings Standard*

and

> *To repair the house to Lettings Standard within 28 days*

In Preston, it was:

> *To meet ICPIs in respect of rent arrears, that is, to reduce current rent arrears as a percentage of the debit*

The next step was to discover the nature of customer demand. At Tees Valley and Preston, this information was collected by closely observing the work of call center and reception staff. Tees Valley wanted to be involved in the pilot because of a concern they had about customer satisfaction with their responsive repairs service. The suspicion that all was not well was borne out in the demand analysis, which revealed 45% of the calls received as generated by failure demand. Table 10.3 illustrates this.

Once demand was understood, it became possible to assess the current system's performance on a capability chart. As an example, the mean re-let time for voids in Leeds South East was 82.2 days, with a likely maximum (UCL) of 179 days.

It was then necessary to track how the work flowed through the system and to distinguish value work from waste. Systems team members literally followed the work step by step and covered all areas that the work "touched," both within the organization and outside. As an illustration, this involved working with maintenance staff and

Table 10.3 Value/failure demand at Tees Valley.

Nature of call	% of total	Summary
Can I report/can you fix	49	55% value
Can I tell/give you information	3	
Can you confirm/can I check	3	
I'm still waiting for .../when will you be coming	17	45% failure
Someone has been out to fix and it's still not working	9	
Someone has been out but it's not finished/completed	9	
Can you give me more information	4	
I was out when you came	3	
Others	3	

contractors (Tees Valley – repairs), with the homelessness team (Leeds – voids), and speaking with tenants and staff at the Post Office (Preston – rents). In discussing "waste," it was necessary to convince staff that they were not doing themselves out of a job and that gains in efficiency would be used to provide better services. Once the flow charts were completed, they were validated with those involved. In each of the pilots, the flow charts demonstrated that value work was minimal and that there were significant areas of waste. In Tees Valley, flow charts were produced for each stage of the process from "repair reporting" to "invoicing." Of the 285 steps documented, only 6 could be said to represent value work; the rest being "waste" from the customer's point of view.

Some of the waste encountered was of the kind that could be easily eliminated. Maintenance assistants at Tees Valley completed a timesheet (taking 20 minutes a day), which was not used for any meaningful purpose. This was scrapped. A situation had evolved at Leeds where people, in order to remain on the homeless register, had to make three bids per week on the properties available under the Choice Based Lettings Scheme. The effect was that people bid for properties they did not want in order to continue to be considered homeless. This requirement was scrapped by agreement with Leeds City Council. Much of the waste, however, was firmly designed into the system and would take some effort to design out again. In Tees Valley, procedures existed whereby the diagnosis and time allocated to a repairs job were determined by staff in the customer services center and jobs allocated to operatives on the basis of one hour time slots. If a job could not be done owing to lack of time, materials, or access, it was passed back for rebook. This process involved canceling the previous job, entering a new one, and starting the whole process again. In Preston, the Housing Department relied on a council-wide IT system, which would have to be changed before any new finance system could be implemented. The third type of "waste" recognized in VM is that generated by another system. Some of this is fair enough since it clearly has value to that other system and indirect value to customers if, for example, it reduces the prospects of fraud and misuse of resources. In other cases, however, it seemed to be related to targets that bore no relation to customer service. In Tees Valley, for example, a target to respond to requests for repairs within 10 days meant that a tenant going on holiday for a fortnight would not be booked in but told to ring back again when they returned. The final stage of "check" is to use systems and logic pictures to try to ascertain why the system behaves the way it does. They indicated management thinking dominated by targets and performance indicators.

In general, the systems team members enjoyed doing "check." A Preston team member remarked:

> We presented our findings and our directorate was astounded. All our hard work was worthwhile and we were over the moon to be told that we will be going ahead with the 'redesign' phase …

It also assisted their self-development. They examined areas of work in which they had not previously been involved and grew the confidence to deliver presentations to senior staff.

One of the noticeable features of "check" is the speed with which it can be accomplished. With the systems teams working on the pilots for three days a week in Tees Valley and Preston, and full-time in Leeds, the time from setup to analysis of the organization's thinking was approximately six weeks. This speed led to impatience for change and, in Preston, some disillusionment later when the IT system caused delays in implementation. All the presentations went well and agreement was obtained in all cases to continue to the redesign stage.

10.3.2 Plan

In Tees Valley, the revised purpose, bringing it in line with customer purpose, became:

> *To do the repair right, first time and achieve what matters to the customer*

In Leeds South East, it became:

> *To repair and re-let homes and create sustainable communities*

In Preston, it became:

> *Right amount, right time so customer knows what to pay, when to pay and how to pay it*

Rethinking the purpose is important work in its own right, as well as being the basis for enacting change and measuring performance. It allowed the Preston team, for example, to recognize how the previous system had been "setting up customers to fail." The new purpose concentrated minds on ensuring they succeeded. The principles of redesign, established to guide future change, turned out to be remarkably similar in all three pilots. This is hardly surprising. The consultants clearly had a strong influence and these principles derive directly from the systems ideas they embrace. Measures were then established to evaluate performance against the revised purposes. In the case of Tees Valley, the measures introduced were:

- Measurement of true end-to-end time for the whole repair
- Number of repairs completed right and first time
- Customer satisfaction
- Reduction in failure demand

10.3.3 Do

"Do" begins when suggestions for redesign are considered in consultation with those carrying out the work. Using the flow charts, it is relatively straightforward to distinguish between value work and waste. At Tees Valley, the team recognized that considerable savings could be made by giving more responsibility to the in-house maintenance assistants and the contractors. In the redesign, the customer services center began to take brief details of the repair and contact details for the tenant. This information was sent via e-mail to the operative's handheld unit or the contractors. Ownership then passed to them until the job was fully completed. This included arranging an appointment with the tenant and obtaining all necessary materials. In Leeds South East, the proposed redesign halved the number of steps involved in letting a property from 64 to 32. The creation of a dedicated Lettings Support Team, to deal with all voids management and allocations, allowed many duplicated practices at area offices and protracted hand-offs between the offices and the contractor to be removed. A close working relationship developed between the dedicated team and the partnering repairs contractor. In Preston, it was clear that if customers were to be set up to succeed, they needed more information and proper communication had to be established between the allocations, arrears, and Housing Benefit teams. Prospective tenants should be told at the point of sign up what, when, and how to pay their rent. The tenancy start date should be agreed and, where applicable, a Housing Benefit claim form completed. If it could not be established how the rent would be paid, the organization refused to sign the tenant to the property.

The instruction in redesign is to experiment incrementally and to discuss issues arising with those doing the work and the customers. The systems teams at Leeds and Preston agreed with their senior managements to pilot redesign geographically, in two housing areas, before roll-out to the whole organization. At Tees Valley, initially, only one maintenance assistant kept his own diary and arranged his own work. Once any teething problems were sorted out, this procedure was taken up by a second assistant and this continued until all were working to the new system.

The next step in "do" is to continuously review the changes made. At Tees Valley, a capability chart revealed end-to-end repair times decreasing from an average of 46 to 5.9 days under the new way of working. The follow-up study (NHC 2006) showed them remaining at this low level. Customer satisfaction, evaluated using follow-up phone calls, also improved, with 75% scoring the service 10/10. Failure demand decreased from 45% to 23%. In Leeds South East, a capability chart showed void re-let times reducing from an average of over 50 days to around 25 days. Although these increased again to 34 days after all six offices went live, the number of empty properties and amount of consequent rent lost continued to fall. In August 2004, when the work started, the number of empty properties was 240. After April 2005, voids averaged 144 and in February 2006 stood at a new low of 118 empty properties. In Preston, in one remarkable pilot, the time for the first payment to hit the account reduced from an average of 34 days to an average of 3 days. As the redesign was extended to all offices, however, performance leveled off, with payments reaching the account within 20 days on average. The predictability achieved, as represented on the capability chart, was still, in the words of one Preston team member, "a Himalayas to Holland phenomenon." Furthermore, a study of 360 new tenancies, with 180 starting before the redesign work and 180 post redesign, showed that only 18% of new tenants had fallen into arrears after redesign, compared to 43% previously. The work had a negative impact, however, on re-let times in Preston, which rose on average from 32 to 40 days. Ensuring that tenants "succeeded" was taken too literally in some cases, with tenants able to delay their start date by up to six weeks after receiving the offer.

The final stage of "do" involves working with managers on their changing role. Throughout the interventions they were encouraged to become engaged, but few became directly involved and more general sessions proved difficult to arrange.

Systems team members continued to enjoy working on the projects during redesign. In Tees Valley and Preston, however, tension increased between working on the project and doing the "day job." In Leeds, where the day job essentially became working as the Lettings Team created by the project, the problem dissolved. Other staff and front-line workers who became engaged in redesign were also enthusiastic, with maintenance assistants at Tees Valley talking of "getting my brain back" and "getting into the rhythm of the new way of working." Other staff not directly involved had mixed reactions. The full-time secondment of the team in Leeds lent an air of "exclusivity" to their work. In the new system, the team assumed full responsibility for the allocation of properties. Area managers and housing officers, who had previously been involved, thought that this would create future problems as they were not meeting tenants before the tenancy began. Tenants were losing the benefit of the area staff's local knowledge, which, it was argued, was important in sustaining tenancies. The system was modified giving area offices two weeks to select from the shortlist. This added, on average, two days to the re-let time. In Preston, where the redesign eventually touched on four of the council's five directorates, there were instances of progress being hampered, making the sponsor's job very challenging. There were concerns among staff that despite the assurances given they were, by identifying and removing waste, designing themselves out of a job. Others grew frustrated as improvements they thought that they had identified did not come to fruition. The employees' concerns and disillusionment led to a fragmented approach to implementation.

It is clear that the attitudes and expectations of staff not directly involved need to be carefully managed. The same is true of more senior managers. There were occasions where investigation of long-standing systems was treated with suspicion. Senior managers at Preston became concerned that the work focused on rent arrears in isolation and did not pay due regard to other service areas affected, in particular the management of voids. In Tees Valley and Leeds South East, however, senior managers were happy with the results, supported the work, and have championed the cause of systems thinking in their boards and committees.

With work on the three pilots concluded, the Evaluation Panel was able to review the results of the interventions. It concluded that VM has the capacity to deliver significant efficiencies in service delivery. All the pilot organizations demonstrated potential annual six-figure gains from removing waste and redeploying resources more effectively. In Table 10.4, each system change is considered and monitored in terms of efficiency. The potential gains are also split between cash savings (e.g. reduced void loss bringing more rent into the organization) and efficiencies allowing for improved service (e.g. removal of timesheets allowing maintenance staff to spend more time maintaining properties). A more detailed analysis, including how the cost savings were calculated, is provided in the ODPM report (2005).

The follow-up report (NHC 2006) reveals that two of the pilot organizations, Tees Valley and Leeds South East, were succeeding in realizing most of the gains predicted. At Tees Valley, the main anticipated areas of gain were reduced processing time and an increase in work done in-house. The reduction in works orders in 2005/06 was 3500

Table 10.4 Potential annual efficiency gains in the three pilots.

Process removed	Actual cost savings if applicable	Efficiency gain to be reinvested
Leeds SE Homes		
Administration		£11 250
Removal of pre tenancy termination visit		£4 000
Transfer of reletting process into dedicated Lettings Support Team		£22 500
Reduction in relet time	£95 000	
Unnecessary bidding by homeless applicants removed; improved accompanied viewing	£1 900	£900
More focused sign up procedure		£4 500
Preston C.C.		
Personal contact improvements	£277	£2 271
First week's rent taken for all non-HB payers at sign-up and payment profile set up for each new tenant	£450	£10 824
Correct completion of claim forms at sign-up	£206	£3 382
Rigid tenancy start date removed	£300	£12 573
Connection of gas/electric improved		£1 189
ICT, HB, and administrative improvements	£7 503	£89 286
Tees Valley		
Removal of timesheets		£3 700
Reduction in failure demand		£4 000
Fewer works orders arising out of end-to-end completion (estimates based on 7000 fewer orders) leading to more work in-house (estimates based on 2000 fewer orders to contractor)	£3 500 postage £70 000 on contractor costs	£70 000 arising out of reduced processing
Working with contractors	£2 000	£3 330
Invoicing		£6 500

per annum rather than 7000, which, at £10 per works order, meant a gain of £35 000 rather than £70 000. On the other hand, 2200 fewer jobs were passed to contractors than in the previous year, yielding an efficiency gain (at £37 per repair) of over £80 000 – £10 000 more than predicted. At Leeds South East, the predicted efficiency gain from reducing the re-let time in the original report was £95 000, based on a four-week reduction. As the figures showed only a three-week reduction, the efficiency arising became £71 000. However, as we saw, the number of empty properties and amount of consequent rent lost continued to fall. This equated to a reduction in rent lost of over £360 000 since the work started. In Preston, the fragmented nature of implementation, the failure to spread the intervention to the arrears built up by existing tenants (because of problems with the IT system), and the matter of the unanticipated impact of the new system on re-let times, made it impossible to accurately calculate efficiency gains. Nevertheless, it is reasonable to surmise that, in this instance, the potential gains did not materialize.

Turning to qualitative indicators, improving the capacity of staff to learn is probably the best means of ensuring long-term gains in efficiency and in efficacy in achieving customer purpose. VM pays particular attention to this and the benefits were visible in the three pilots. The systems team members became more critical of existing ways of working, able to recognize how what they were doing fitted into the overall system, and better able to communicate their insights. This learning was extended to front-line workers engaged with the flow charts and redesign experiments, and to other groups drawn into the project, such as the Housing Benefit teams in Preston and contractors in Tees Valley and Leeds. Of course, not all staff impacted by the pilots felt as happy about what was happening. Some were suspicious and, in Preston, even hostile. In this regard, VM proved its own point that systems ideas are best learned through involvement in practice. Further difficulties can be encountered if a systems study starts to come into contact with other systems that are not being redesigned at the same time, as was the case with voids and re-let system in Preston. There is evidence, moreover, that the sustainability of improved performance is an issue. In Leeds and Preston, the excellent results obtained immediately following redesign were not fully maintained in the later time period. All sorts of factors were involved including, in some instances, the success of the pilot itself. In Leeds South East, for example, properties that had once seemed impossible to let began to get tenants and this affected the statistics. There is no doubt, however, that maintaining success means continually working at it, and here the commitment of managers, especially senior managers, is crucial. Where they are fully supportive, become "converted" to the new, systems way of thinking, and are willing to extend projects to new areas, the chances of long-term success are excellent. In Leeds South East, additional work was undertaken on rent collection. This work dovetailed with the voids and re-housing pilot and widened the knowledge of systems thinking within the organization. The sponsor at Leeds summed up their experience thus:

> We all recognized in the early days that systems thinking gave staff the opportunity to develop and to influence how services develop. Getting the culture to change to allow this is not easy and needs constant effort to get it embedded. It is my view that developing staff and releasing their creativity gave us the impetus to reduce void numbers as people thought of new ways to attract and keep

customers. Once embedded, the culture change feeds into other service streams and the capacity for improvement goes up exponentially. It is like sitting on the tarmac at the end of the runway – you can feel the surge as the pilot turns the power on, and you know you are off. It takes time and effort but the rewards are huge.

(NHC 2006)

Further VM work was also carried out at Tees Valley and was not ruled out in Preston.

The Evaluation Panel commented on the possibility that those organizations adopting VM might experience a continuing tension between the new customer-focused, "whole system" measures of performance and more traditional indicators imposed on them. This occurred, for example, in Tees Valley when the number of "urgent" jobs completed on target, according to the usual measure, fell from 98% to 72% following redesign. A "job" had traditionally been measured by the completion of a works order and this bore little relation to the actual end-to-end time needed to finish the full repair. The deteriorating figures were the result of the change to measuring "whole repair" times – which is the thing the customer is interested in. The same thing had happened in the example of "reactive repairs" in Portsmouth. Unfortunately, the problem continues even after a redesign is complete because "inappropriate" measures may still be applied by a higher level system. Tees Valley had to go back to measuring "completion" of urgent jobs in the old way in order to conform to the requirements of the traditional indicators. Portsmouth ploughed on regardless. For this reason, although VM is able to deliver efficiency savings, evaluating it only on that basis is, in a sense, to miss the point. The area in which VM seeks to make its main contribution is in improving efficacy – doing things right from the customer's point of view. This is why there is such an emphasis on revising the purpose of the system to bring it into line with the customer's purpose. Doing the right thing, in VM, is doing what the customer wants. Inappropriate measures can distort behavior in ways that are not beneficial to customer purpose. They can lead to people finding better ways to do the "wrong things righter" (Ackoff 1999a). To achieve the goal of effective redesign, VM ensures that service users' requirements are paramount throughout. It is from the customer's perspective that the service is analyzed, redesigned, and evaluated – even if VM then runs into trouble in terms of conventional targets and performance indicators.

10.4   Critique of the Vanguard Method

The critique begins by looking for clues in the case studies about where VM might succeed and where it might fail. The ODPM report mentions a number of conditions that seem to assist success. Among these are organizations willing to challenge the status quo; in which employees are not exhausted by previous change initiatives; with good relations with other bodies likely to become involved; and with sufficient capacity to release the resource necessary to run the initiative. In the Portsmouth case, another factor mentioned was the degree of independence of housing services from other council functions (O'Donovan and Zokaei 2011). The 2017 OECD report on the use of VM to improve child protection in the Netherlands emphasizes the need for top management support.

This is reinforced by Keith Bennett's account of his unsuccessful attempt to put what he had learned as a Vanguard consultant into practice as an employee of a bank:

> If leaders won't spend time truly understanding the links between their thinking about the design and management of work and performance – then don't start with a team. Once people have learned to see, they can't unsee. There is nothing worse than helping frontline staff to 'get' what's wrong and why and then have leaders not change the system because they don't understand that it is their own thinking about the design and management of work that is the invisible barrier to change.
>
> *(Bennett 2016)*

The OECD report also mentions requirements for the people involved in the flow of work to understand why change is needed and for recruitment policies that look for "a new type of person" suited to the new system. Indeed, it criticizes VM for ignoring the people element in design and the effect of organizational culture. Is it really the case that 95% of performance is down to "the system?"

The ODPM report ultimately sees the key to understanding the domain of competence of VM as resting on issues surrounding the unity of purpose of stakeholders and the degree of independence of the system under review from other systems. On the first point, VM requires a clear definition of the purpose of the system from the customer's point of view. It will obviously work best when such a definition is easily obtained. This may come about because the relevant stakeholders share a common purpose or one purpose is enforced by a powerful sponsor with strong senior management support. Both these possibilities are mentioned as avenues to success in the ODPM report. VM will encounter difficulties when there are different stakeholders, all who have a claim to the role of customer, and they have different aims and objectives. For example, one can imagine different definitions of purpose for a local authority housing system being suggested by the tenants, by representatives of other local authority departments, by private householders, by council tax payers, by employees of the system, by central government, and by sections of society who see themselves as discriminated against in the housing stakes. In these circumstances, it can be argued, it is the role of systems thinking not to espouse one purpose but to express the implications of the different purposes, perhaps in a variety of models as Checkland's soft systems methodology does (see Chapter 16), so that discussion and the political process are better informed. From this perspective, VM fails to pay due attention to the variety of possible purposes. Tees Valley was a self-contained organization, with a strong corporate identity, operating out of a single site, and all employees could see the work progressing on a day-to-day basis. This helped to create the unitary climate in which VM can flourish. In Leeds South East and Preston (see NHC 2006), pluralism raised its head when areas not initially involved in the projects became affected by them. In Leeds, concerns were expressed in a staff satisfaction survey, which reported:

> There are some internal issues that mainly reflect local offices feeling disconnected from the Property Letting Team allocations process. These concerns spread over the whole process and range from concerns about the

introduction to the house/area at the accompanied viewing, to undertaking a new tenancy 'blind'.

The tension was between the purpose of filling voids as quickly as possible and the purpose of creating "sustainable communities." In Preston, there was resistance in some area offices and other directorates to the new way of working. In these circumstances, when there is a pluralism of purposes, VM is much less well equipped to succeed. At the least, it can require a lot of effort devoted to keeping on side those who, for whatever reason, do not share the formal purpose articulated. Seddon's view, expressed in O'Donovan (2014), that other stakeholder interests can be accommodated in better system design, only reveals his essentially unitary viewpoint. Finally, the neglect of multiple possible purposes by VM can see it closing down interesting and creative possibilities for rethinking purpose.

On the second issue, there is a well-known systems principle that we should plan simultaneously and interdependently for as many parts and levels of a system as possible (Ackoff 1999b). VM seems willing, however, to redesign subsystems with little reference to other parts or levels. Unfortunately this bottom-up approach makes it impossible to address the danger of "suboptimization." Suboptimization refers to the possibility that the improvement of one subsystem might make the performance of the whole system worse. This danger arises because of the importance of the interactions between the parts in complex systems. It is impossible to evaluate whether improving one subsystem in isolation makes the whole better. This criticism is not met by O'Donovan's (2014) response that, while VM may initially accept accommodation with higher level systems, it returns to them once the micro-level is improved and stabilized. By that time, the damage will be done. VM, it seems, recognizes the possibility of suboptimization at the system level at which it operates but pays less attention to the issue at the wider system level. As was mentioned, in regard to Preston, improving the speed with which tenants' first payments hit the Council's bank account directly led to an increase in re-let times. Senior management were of the view that the work focused on rent arrears in isolation and did not give due attention to other service areas affected, in particular the management of voids (NHC 2006). There are other systems approaches, such as Beer's "organizational cybernetics" (see Chapter 13), that require a more comprehensive design brief up-front and have more to say about how sets of processes can be coordinated and controlled to ensure that each contributes to, rather than endangers, the viability and effectiveness of other parts and the whole system. On a related point, VM can encourage legitimate "auditing" requirements of higher level systems to be seen purely as bureaucratic annoyances. In the OECD-reported child protection example, a questionnaire required by law was ignored by most case workers. Measures designed by other systems to ensure financial probity, fire safety, fairness, etc. can come under threat from VM's all consuming "customer" focus. There is little wonder, in our litigious age, that the UK NHS has so many procedural checks. Of course, there is a danger in pushing these points too far. In a certain sense everything is "interconnected" and you can never be absolutely certain that an intervention will bring comprehensive, system-wide, sustained improvement. You have to get started and take chances even if some stakeholders are reluctant to commit. It is also reasonable to point to the significant dangers involved in attempting to undertake whole system improvement all at once.

A balance has to be struck between the risk of suboptimization and the often desperate need to take some appropriate action. As O'Donovan also argues, attempting whole system improvement risks working without knowledge of the true nature of customer demand and how the system responds. Nevertheless, these are important matters and ones with which VM has not yet successfully come to terms. Caulkin (2010) sees the next stage in the development of VM as "moving beyond incremental change." At present, however, it is reasonable to argue that VM is more appropriate for intervening in subsystems which are relatively independent rather than more complex problem situations where subsystems share close interrelationships and co-exist in a turbulent environment.

As "second-order" observers, seeking to deepen the critique, we are bound to recognize that VM privileges just one among the variety of social theories expounded in Chapter 4. It is essentially functionalist. On the basis of a definable customer purpose, a boundary is drawn and the aim becomes to predict and control the system to deliver that purpose. Social reality is seen as possessing a hard, objective existence and of consisting of systems and structures that determine the behavior of individuals. Ninety-five percent of performance is deemed to be due to the system and only 5% to the individuals who are assumed to behave in unison once the system is redesigned around customer purpose. Social reality is rendered predictable and manageable by VM on the basis of its assumptions about a unified purpose, regularities in demand, and the uniformity of people. Interpretive sociologists, to take one example of a possible alternative theoretical perspective, require attention to be directed instead to multiple purposes, the social construction of reality, and to people and cultural issues.

The ODPM report did deepen its analysis of VM by relating it to the assumptions about the nature of systems and stakeholders highlighted by the SOSM. We can follow suit using the redesigned SOSM explained in Section 8.3.5. VM copes with some aspects of complexity by considering the interrelationships between the elements in the systems it defines as being of concern and by employing the concept of "variety" to try and balance system and environmental complexity. It is limited, however, in its ability to manage interactions at the higher system level and in those circumstances when the environment is turbulent and demand becomes so unpredictable that whole system adjustments must be made. VM insists that those who do the work are involved in redesign and so caters for certain aspects of pluralism. On the other hand, it makes a unitary assumption in believing that all those involved in a service will naturally align their purposes to that of the customer simply because the system is well designed. Even more significantly, no account is taken of alternative purposes arising from different stakeholders and the need to manage these. The positioning of VM on the SOSM is, therefore, as in Figure 10.2.

It is important to insist, again, that the positioning of any systems methodology on the grid does not make it any "better" or "worse" than other methodologies placed elsewhere. It simply points out differences. Something will always be lost as well as gained by adopting particular assumptions. Nor does it say anything about how useful a methodology is. It might well be that most of the important problems lie within VM's domain. This is the stance that Seddon takes. He is more than willing to accept that VM is akin to hard systems approaches in assuming a single unifying purpose. However, he argues that setting boundaries on the basis of customer purpose is the only way to cut through

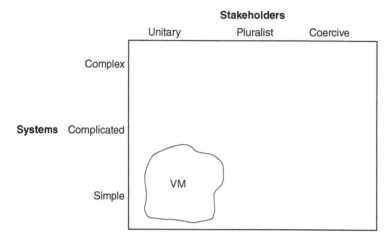

Figure 10.2 The positioning of The Vanguard Method on the SOSM.

the vast complexity of entities such as the British healthcare system and to start making changes in the right direction:

> The Vanguard approach sets a boundary. Other approaches to systems thinking might be interesting but not, in my experience, profitable. The Vanguard approach is *interesting* and *profitable.* Interesting, because it provides a method for developing relevant knowledge and, consequently, achieving the ideals all managers would aspire to: a learning, improving, innovative, adaptive and ener-gized organization. Profitable, because it provides the means to develop a cus-tomer-driven, adaptive organization; an organization that behaves and learns according to what matters to customers. If the system is to have viable econom-ics, it could only be understood and developed from this point of view.
>
> *(Seddon 2005, p. 179)*

10.5 Comments

John Seddon is sometimes portrayed as arrogant and rude. Calling lean practitioners "toolheads" is regarded as typical. It is certainly difficult to shift him and his followers from their fixed position – as I can attest having spent a day trying to convince a large group of "Seddonistas" of the value of alternative systems perspectives. On a personal level, however, John is a very pleasant individual and good company. Before condemn-ing the rhetoric, we need to ask just how much is at stake in the arguments. As Caulkin says, referring to VM,

> ... the least that can be said is that in any organization whose operations are untouched by systems thinking, the potential for improvement is great. In an entity as large as the NHS or public services as a whole it is simply dizzying.
>
> *(2010)*

In other words, there is something worth fighting for and his arguments need to be seen in that light. They have to be pursued forcefully and repeatedly against those who ignore the burgeoning evidence that VM can bring significant improvements in efficacy and efficiency. It is clear to Seddon (2014) that, despite massive investment, public sector reform is not delivering the goods for customers. In his view, the public sector faces not a resource but a design crisis. This is a difficult argument to make in the current consensus, but he puts it well:

> I'm astonished to see reports of whole teams of people (experts, thus super-costly) being stood down from operating on patients simply because there is no bed in which to put patients post operation. It is complete madness; what kind of mind would make beds a constraint in a health system? Of course the answer is a command-and-control mind. The problem with the NHS is the industrial design; it is ineffective. The management factory is focused on efficiency and this undermines effectiveness. It's a classic example of my aphorism: managing costs causes costs.
>
> *(Seddon 2017b)*

Anyone with any doubt about the frustrations this type of thing causes in staff should read *Do No Harm* (2014), the heartfelt reminiscences of the respected neurosurgeon, Henry Marsh. As a patient, I have myself suffered from the impact of the "factory design" that Seddon criticizes: worrying about operations going ahead because of lack of beds; kept in the recovery ward because of lack of beds in the high dependency unit; unable to get out of hospital (and free up my bed) because I couldn't get the necessary medicines; spillages not cleaned up because that doesn't come under anyone's job description; centralized appointment systems booking me for appointments with my consultant before the scans that would make the appointment meaningful; being refused pain-killing medication because some computer has got the timing wrong. And, I declare that I wholeheartedly support the rationale behind the NHS and have the utmost respect for all those who do their best and often succeed in making a broken system work for patients.

10.6 The Value of the Vanguard Method to Managers

It imposes a discipline if we restrict ourselves to highlighting just five ways in which each systems methodology can help managers. We stick to that rule with VM even though, in this as in all other cases, the relevant chapter reveals many other helpful aspects of the approach. Here are five important lessons that can be learned from VM:

- Not all problems are resource problems. Many stem from poor system design and dealing with them depends on changing the way managers think about their role
- Start an intervention by learning about customer demand and the ability of the organization to meet it. This helps provoke action. Redesign "outside-in" on the basis of customer purpose. The clarity this brings can help "dissolve" problems
- Efficiency follows efficacy. Savings come from smoothing the end-to-end flow, thus reducing failure demand, not from cutting unit costs

- The system must make the best possible use of the brains of those people doing the work. They are best equipped to understand it and to redesign it. They can provide the enhanced variety necessary to cope with environmental uncertainty
- Ensure that the system immediately captures feedback from customers on how well it is performing and that this is used both by managers and those doing the work to improve the system. In this way the system becomes self-evaluating

10.7 Conclusion

VM provides a well-specified methodology embodying many important aspects of systems thinking. It is able, as the ODPM (2005) investigation found, to contribute significantly to improving service delivery. Since that study, further solid evidence of its worth has continued to mount from projects conducted in financial services, local government, the NHS, IT services, universities, and more. It can only be helpful to the approach if, while acknowledging this, we are clear that there are some instances in which VM on its own fails to bring about the expected improvements in performance. We can then consider whether using it in combination with other systems methodologies can help broaden the range of circumstances in which it can contribute the strengths that it does have to offer. Soft systems methodology, which can assist accommodation between different purposes, is an obvious potential partner. The viable system model is another strong contender. It concentrates on the co-ordination and control of interacting processes toward an overall purpose while minimizing hierarchical constraints on the individuals who are engaged in those processes.

Type C

Systems Approaches for Structural Complexity

Over the course of the last 30 years there has evolved at the Massachusetts Institute of Technology a new method for understanding the dynamic behavior of complex systems. The method is called System Dynamics. The basis of the method is the recognition that the structure *of any system – the many circular, interlocking, sometimes time-delayed relationships among its components – is often just as important in determining its behavior as the individual components themselves*

(Meadows et al. 1972, p. 31)

A third type of systems approach is primarily concerned with structural complexity. Structural complexity stems from the arrangement of and dynamic interrelationships between the many elements that make up a complicated system. The nature and quantity of the interactions makes it difficult to understand the system's behavior. Systems thinking can help by identifying the significant variables and interactions that impact what happens and then modeling these interactions, on the basis of positive and negative feedback loops and lags, in order to grasp the underlying causes of system behavior. The systems methodology that has this as its main area of concern is:

- System Dynamics

Critical Systems Thinking and the Management of Complexity, First Edition. Michael C. Jackson.
© 2019 John Wiley & Sons Ltd. Published 2019 by John Wiley & Sons Ltd.

11

System Dynamics

Systems thinking is a discipline for seeing the 'structures' that underlie complex situations, and for discerning high from low leverage change .... Ultimately, it simplifies life by helping us to see the deeper patterns lying beneath the events and the details

(Senge 1990, p. 69)

11.1 Prologue

In 1968, a group of prominent international politicians, government officials, diplomats, scientists, economists, and business leaders met in Rome to discuss the present and future "predicament of mankind." Their conviction was that the problems facing mankind were so complex and interrelated that they were beyond the scope of traditional institutions and individual policies and needed to be studied and addressed as "a whole." This led them to find a multidisciplinary think-tank that became known as the "Club of Rome." One of the Club's earliest initiatives was to commission an international team of researchers, centered on the Massachusetts Institute of Technology (MIT) and funded by the Volkswagen Foundation, to produce a model that described the behavior of the different but interrelated components that make up "the global system." To achieve this, the 17 researchers employed the global model and system dynamics (SD) approach set out by Jay Forrester in his book *World Dynamics* (1971a). Their report, known as *The Limits to Growth* (1972), sent shock waves around the world. Its foremost conclusion stated starkly that:

> If the present growth trends in world population, industrialization, pollution, food production, and resource depletion continue unchanged, the limits to growth on this planet will be reached sometime within the next one hundred years. The most probable result will be a rather sudden and uncontrollable decline in both population and industrial capacity.
>
> *(Meadows et al. 1972, p. 24)*

To many readers and commentators, it seemed that Doomsday was just around the corner.

Critical Systems Thinking and the Management of Complexity, First Edition. Michael C. Jackson.
© 2019 John Wiley & Sons Ltd. Published 2019 by John Wiley & Sons Ltd.

The researchers took the five key factors influencing the behavior of the global system to be population growth, industrialization, food production, consumption of nonrenewable natural resources, and pollution. Looking at the statistics from 1900 onward, they all seemed to be growing exponentially. But the world was finite. If the growth of population and industrialization continued at their current rates, then a point would come when there was no new arable land to provide increased supplies of food, nonrenewable resources became severely depleted, and pollution began to endanger the natural environment through the production of CO_2, global warming, and radioactive wastes. But what were the exact limits to growth and when would they bite? To find out, the researchers elaborated on their intuitive conclusions using a computer model, *World 3*, based on the principles of system dynamics set out by Forrester. Experts were consulted for their opinion on the interrelationships between the five elements. They pointed to around 100 causal links and helped quantify the force of the interactions. Positive and negative feedback loops and lags were identified. Data about the five components could now be plugged in and the model run to produce an output that showed the effect of the simultaneous operation of all the relationships. The model made it possible to identify the critical determinants of behavior and to look at what happened if the numerical values of the initial assumptions were changed. Different policy proposals could also be tried out and their effects were monitored over time.

The "standard run" of the model, based on there being no significant change in the trends, showed that well before 2100 resource depletion would lead to a collapse of economic development and a decrease in the global population. Changing the assumptions on which the model was built could extend the period before collapse but, however drastic those changes were, there was always collapse before 2100 – if not due to resource depletion then because of food shortage or pollution. For example, if resource depletion and growth in pollution were reduced to a quarter of their current levels, starting in 1975, then collapse still occurred because of food shortages. Even if it was assumed that there were significant technological innovations, unlimited resources, and more effective birth control initiatives, the system collapsed due to pollution. The fact that significant changes in input data did not alter the overall pattern of behavior convinced the researchers that system dynamics was correct to assume that the feedback structure of the model was a more important determinant of behavior than the actual numbers that are used to quantify the feedback loops. Thus, while they were clear that the model could not produce exact predictions, they were able to insist that:

> The basic behavior mode of the world system is exponential growth of population and capital, followed by collapse.
>
> *(Meadows et al. 1972, p. 142)*

All was not lost. The model did not say that the future was determined. Social factors were excluded from the model because they were much too complex to deal with. The authors of the 1972 report now reintroduce them into their thinking and argue that new forms of social behavior are needed to avoid catastrophe. If we can learn from the simulation and take action to weaken the positive feedback loops, by imposing our own limits on growth of population and industrial output, then a sustainable future is possible in which the basic material needs of all people are met. We must start immediately to make an ordered transition from growth to global equilibrium. This was going to require a huge shift in human values and the Executive Committee of the Club of Rome end their commentary on the report with:

> The last thought we wish to offer is that man must explore himself – his goals and values – as much as the world he seeks to change. The dedication to both tasks must be unending.
>
> *(Meadows et al. 1972, p. 197)*

The publication of *Limits to Growth* provoked uproar and brought forth a tide of criticism. Much of this related to the authors giving too little weight to possible scientific and technological advances. They had, in fact, allowed for these but calculated that they would come too late to prevent collapse. Some of the criticism had merit, as when the Science Policy Research Unit, at the University of Sussex, suggested that the simulations were more sensitive to a few key assumptions than the researchers recognized (SPRU 1973). Some of it was sensible, e.g. asking how developing countries would react to calls for a curb on growth. Some of it, however, was blatantly ideological. Whatever the facts, the Catholic Church had to resist arguments in favor of birth control; politicians had to champion growth; and classical economists had to believe in equilibrium models in which prices would always ensure that a correction took place. The report's conclusions were dismissed, even ridiculed.

By the early years of the twenty-first century, however, attitudes to *The Limits to Growth* began to shift drastically. Follow-up research by Meadows et al. (2004) and other independent investigators, based on data from 1970 to 2000, showed that the results of the "standard run" of the model gave good approximations of what had happened for almost all the reported outputs. The global system was on track to collapse sometime in the second half of the century. A recent report of a UK All-Party Parliamentary Group reviewing the latest research concludes that

> … there is unsettling evidence that society is still following the 'standard run' of the original study – in which overshoot leads to an eventual collapse of production and living standards.
>
> *(Jackson and Webster 2016)*

In 2018, Earth Overshoot Day – the point each year at which consumption of the planet's resources exceeds the capacity of nature to regenerate – falls on August 1st. A year's worth of carbon, food, water, fiber, land, and timber has been consumed in a record 212 days. Thirty years ago, the overshoot was on October 15th; 20 years ago September 30th; and 10 years ago August 15th. Coming on 50 years of opportunity to do something about it has been wasted.

11.2 Description of System Dynamics

11.2.1 Historical Development

Jay Forrester developed his lifelong concern to find practical solutions to real-world problems as part of growing up on a remote ranch in Nebraska (Lane and Sterman

2018). He was bemused when others did not share his fascination with how things worked. His obituary in *The New York Times* (Hafner 2016) recounts how he once asked students in an engineering class at MIT if they had ever taken the lid off a toilet tank to see how the feedback device operated to maintain the water level. When he discovered that none of them had, he remarked: "How do you get to MIT without having ever looked inside a toilet tank?" He had a point.

Forrester studied for a batchelor's degree in electrical engineering at the University of Nebraska before moving to MIT for graduate studies and, from 1940 to 1951, acting as Associate Director of its Servomechanism Laboratory. During World War 2, as with a number of the pioneers in operational research (OR) and cybernetics, his expertise was put to good use in supporting the military effort. One project saw him designing and building a novel feedback controlled mechanism to stabilize radar antennae on naval ships (Lane and Sterman 2018). It is interesting, in the light of later developments, that this device also incorporated a positive feedback element in order to continuously move the heavy equipment. In the aftermath of the war, he made significant contributions to improving computer memory, to the development of aircraft flight simulators, and to defense information systems. In 1956, he joined the newly formed Sloan School of Management, at MIT, on a mission to apply what he had learned about feedback and the capabilities of modern digital computers to the broader domain of managing industrial organizations and economic systems. It was in this context that he established and developed the field of "system dynamics" (SD).

A foundational experience was a project he took on at an appliance plant of General Electric. The company was struggling to control large and costly fluctuations in stock levels and workforce numbers. Plant managers regarded the fluctuations as caused by external factors. Forrester, however, took an endogenous point of view and looked at the business as a system of interacting elements. He was able to demonstrate that the fluctuations resulted from managers at various points in the supply chain relying on the knowledge available to them locally and adopting policies which interacted within a system of multiple, nonlinear feedback loops (Lane and Sterman 2018). Computer simulations were created that accurately modeled the dynamic behavior giving rise to the unexpected fluctuations. The approach he was pioneering, at first called "industrial dynamics," was announced to the world in 1958 in a seminal article published in the *Harvard Business Review* called "Industrial Dynamics: A Major Breakthrough for Decision Makers" and, at book length, in *Industrial Dynamics* (1961). In the latter, the four foundations of the new field were stated as being:

- The theory of information-feedback systems
- A knowledge of decision-making processes
- The experimental model approach to complex systems
- The digital computer as a means to simulate realistic mathematical models

Although the book emphasizes the crucial role of rigorous computer-based simulation models, the success of industrial dynamics is also seen as dependent on the analysts maintaining a close engagement with managers and other stakeholders so that the models contribute to their learning. Those who understand how complex systems work, it is assumed, are more likely to act on them to bring about improvement.

Experiments continued throughout the 1960s and, as Forrester's ambitions increased and the scope of the work widened, a more general name, System Dynamics, was adopted for the approach he was using. *Principles of Systems* and *Urban Dynamics* were published in 1968 and 1969, respectively. *Urban Dynamics* marked the extension of SD from the corporate sector to public policy issues. It was particularly controversial because it argued that certain policies designed to alleviate urban problems actually made them worse. The reason for such unintended consequences is that the complexity of the structure of many systems is too difficult for the human mind to grasp. As far as humans are concerned, the dynamic behavior of social systems is frequently counterintuitive (Forrester 1971b). The solution, according to Forrester, is to harness the growing power of digital computers that are capable of tracing the interactions of many variables and thus of taking complexity in their stride. Even greater controversy, as we have seen, followed the 1971 publication of *World Dynamics* and the work on the "limits to growth" that derived from it. Forrester directed the System Dynamics Program he had established at MIT until his retirement in 1989 and, even after that, his work continued apace. A particular project he championed in retirement was the introduction of the teaching of systems principles and dynamic modeling in schools throughout the United States.

By the time of his "retirement," Forrester had done all the necessary groundwork to establish SD as a rigorous and respected applied systems approach. Others had become engaged in helping to develop the field; some concentrating on the building and validation of computer-based simulation models, others on qualitative issues surrounding implementation and the use of the approach to change managers' "mental models." Still, SD was hardly taking management theory and practice by storm. That changed, in 1990, with the publication of Peter Senge's book, *The Fifth Discipline*. Senge, also based at MIT, dropped any pretense to rigorous modeling and promoted a qualitative form of SD (the "fifth discipline" of the title) as the key to creating "learning organizations." His book hit the best-seller lists worldwide and his version of SD – under the label systems thinking – became an essential component of the management tool-kit.

Today, the SD community is the third largest, after systems engineering and complexity science, of the groups formally championing systems thinking. The System Dynamics Society goes from strength to strength, has around 15 national "chapters," runs an academic journal *System Dynamics Review*, and organizes annual conferences. The journal *Systems Research and Behavioral Science* publishes the proceedings of the bi-annual European System Dynamics Workshops and now receives more papers that employ system dynamics than any other type of systems approach. SD is well-established in universities worldwide and finds a place in many business school curricula. It boasts some excellent websites, such as Gene Bellinger's *Systems Thinking World*, hosting pedagogical material, news, and views. Perhaps most important of all, it has been and is being applied in a wide range of contexts, including conflict, defense and security, the economy, energy, project management, public policy, industry, education, health, policing, the environment, information systems.

11.2.2 Philosophy and Theory

Forrester was more interested in developing the methodology and methods of SD than he was in elucidating its philosophy and theory. One result, as Lane (1999) shows, is that he came out with apparently contradictory statements about the status of its models. Sometimes he argued that they are underpinned by principles that allow them to capture the actual nature of physical and social reality, elsewhere he suggests that their purpose is simply to illuminate a mental model in a precise way. Despite this ambiguity, which continues to bedevil the field, the manner in which SD actually proceeds to construct its explanations of phenomena reveals a clear epistemological commitment. Lane puts it like this:

> The only universal law/theory on offer is a grand methodological, or structural theory, associated with a representation scheme. System dynamics offers a new structure for thinking about causality but it does not specify the content of that structure.
>
> *(2000, p. 12)*

SD would not undertake the kind of modeling it does if it did not assume that, at a certain level of aggregation, causal laws can be found that provide insight into the dynamic behavior of systems. According to the theory, the multitude of variables that constitute complex systems become causally related in feedback loops that interact among themselves. As Forrester explains:

> The structure of a complex system is not a simple feedback loop where one system state dominates the behavior. The complex system has a multiplicity of interacting feedback loops. Its internal rates of flow are controlled by non-linear relationships. The complex system is of high order, meaning that there are many system states (or levels). It usually contains positive-feedback loops describing growth processes as well as negative, goal-seeking loops.
>
> *(1969, p. 9)*

In such complex systems, causes are not easily identifiable. They cannot usually be found in immediately preceding events. They result from the structure of the system and are often distant in both space and time. It is essential, therefore, to secure a deep understanding of the system's structure if we are going to make sense of its behavior and implement policies that bring favorable outcomes for stakeholders.

An example can help provide some clarity. In 2010, the Department of Education commissioned Professor Eileen Munro to undertake a review of the child protection system in England. Munro wanted to bring a holistic approach to understanding why previous policy initiatives had failed to bring the desired results and to suggest reforms. David Lane, a system dynamics specialist then at the London School of Economics, was appointed as an advisor. The OECD report on "Systems Approaches to Public Sector Challenges" describes the "Munro Review of Child Protection" as one of the best known examples of the use of systems thinking in the public sector (2017, p. 27). Other system concepts, such as "requisite variety" and "double-loop learning," were employed as part of the study, but it was SD that was dominant in providing the structure for the review process and the eventual recommendations. As Lane et al. remark:

> Here, the focus on causal mechanisms and behavior over time, combined with the wish to consider anticipated and unanticipated consequences of policy initiatives, indicated a central role for system dynamics modelling.
>
> *(2016, p. 614)*

In a preliminary "systems analysis" phase, examining the child protection system and how it had got into its present state, a set of "causal loop diagrams" (CLDs) were constructed. These were designed to flesh out an initial causal hypothesis that the system had become overly bureaucratic and that this had given rise to a compliance culture in which adherence to procedures trumped the attention given to the welfare and safety of children (Munro 2010). These CLDs were supported by evidence gathered from published research and information gleaned from experts and relevant professionals. A plausible feedback loop was identified in which policies based on the belief that a prescriptive approach to the work was needed had indeed led to compliance enforcement, then to reduced opportunities to use professional judgment, and then to further compliance with prescriptions. However, the damaging impact of the prescriptive regime on the system could only be understood if its unintended consequences, or "ripple effects," were also mapped. The reduced ability to exercise professional judgment decreased job satisfaction, leading to higher staff turnover, a reduced level of experience among staff, and lower public regard for child protection workers. Even more significantly, given the increasing variety of the work, the reduced ability to exercise professional judgment inevitably impacted on the quality of help that could be provided and led to errors. This should have set in motion a learning loop providing feedback that, on the basis of the errors arising, questioned the effectiveness of the prescriptive approach. In reality, compliance cultures always produce another feedback loop following a "we followed procedures" logic. The best defence against failure is to appeal to the rules and show that correct procedures have been followed. Thus, the potential learning loop is nullified because participants find it impossible to acknowledge errors and these are not properly investigated. This then reinforces the conviction that following procedures eliminates errors and that a prescriptive approach is good. The organization becomes locked in an "addiction to prescription." The various explanatory CLPDs were combined into a final diagram, which illustrates the full dynamics of the situation. It is shown as Figure 11.1. The diagrammatic conventions are explained in Section 11.2.4.

The "systems analysis" struck a chord in the profession, and among other stakeholders, and a second phase of work began. This was conducted in a much more participative manner as various experts and professionals engaged in group model building, over a number of sessions, to create a complex system dynamics map that captured their perception of current operations. The map, of over 60 variables, revealed nine "reinforcing" loops contributing to the "addiction to prescription" and demonstrated why possible balancing loops were not working or had further unanticipated effects. "Lags" in the system, especially the time involved in recognizing decreased quality of outcomes for clients, further complicated the picture. In this case, the map was sufficient to provide stakeholders with an understanding of what was going on and an insight into the impact of possible actions. It was not necessary to develop a computer-based simulation of the system. In the words of Lane et al.:

> The work around the complex systems map supported a concentration on causal mechanisms. This enabled poor system responses to be diagnosed as the unanticipated effects of previous policies as well as identification of the drivers of the sector. Understanding the feedback mechanisms in play then allowed experimentation with possible future policies and the creation of a coherent and mutually supporting package of recommendations for change.
>
> *(2016, p. 621)*

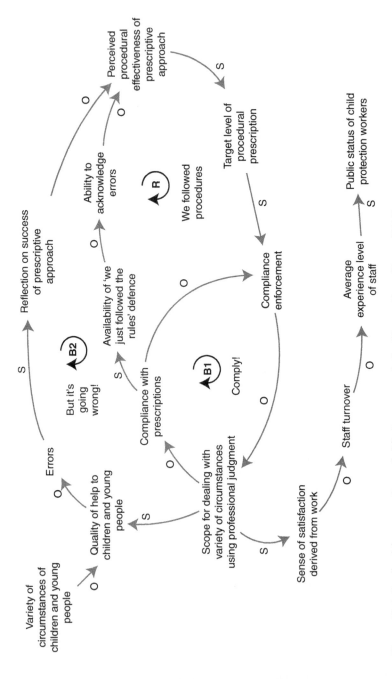

Figure 11.1 Feedback loops creating an "addiction to prescription." *Source:* From Munro (2010).

Only a simplified CLD can be found in the final report (Munro 2011), but the SD analysis governs the text and provides the framework for the recommendations. As the report states:

> Although only a few 'ripple effects' are illustrated here, they are indicative of a range of unintended consequences resulting from an overly-prescriptive approach to child and family social work. This collection of reinforcing loops has restricted the capabilities of the profession, increasingly reducing its effectiveness.
>
> *(Appendix A, p. 137)*

Of the 15 recommendations, the Government accepted 10 in full and 5 in principle. Considerable progress was made in implementing a more "child-centered system," although, as the 2017 OECD report says, public sector cuts and changing government priorities prevented the full rollout of the recommendations. The final word can be left to a senior civil servant who commented on the work:

> The 'causal loop' thinking that was introduced early in the review helped civil servants, ministers and critically, leaders and frontline practitioners to think about reform of a whole system in which single actions would create chains of causality and ripple effects.
>
> *(Quoted in Lane et al. 2016, p. 622)*

Forrester, while himself concentrating on methodology and method, did recognize the need for a fuller and deeper examination of the philosophy and theory of SD. Fortunately, this is exactly what David Lane has undertaken in a series of excellent articles (Lane 1999, 2000, 2001a, 2001b; Lane and Husemann 2008). Since this reflection comes from inside the system dynamics fold, it will be summarized here. A broader critique, from outside the field, follows in Section 11.4.

Lane's task is to make explicit the model of social reality that is implied by the methodology and methods of SD. He is obliged as part of this to respond to ill-informed critics (including me) who have labeled it a "deterministic" approach. Lane points out that SD has rarely sought to make prophecies of what will happen but simply predictions of what is likely to occur based on some declared assumptions. *Limits to Growth* provides a perfect example. Humans can change the assumptions built into the model by imposing their own limits on growth of population and industrial output. The model demonstrates that if they can do this, then a sustainable future is possible in which the basic material needs of all people can be met. Thus, SD, Lane convincingly argues, takes the view that human behavior is only partially governed by system structure. Further, SD is uniquely capable of suggesting to stakeholders what must change, including aspects of their own thinking, in order that a more desirable future can be attained. This is because SD modeling enables us to understand and influence, through our actions, the "vicious circles" of reinforcing feedback loops that appear to drive us toward some inevitable future. Lane and Husemann examine the phenomenon of "globalization" in this light. There are advocates and critics of globalization but what they appear to share in common is a belief that it is inevitable. Even a figure as powerful as US President Bill Clinton could declare, in 1998:

> I do not believe that you can defy the rules of the road in today's global economy, any more than I could defy the laws of gravity …
> *(Quoted in Lane and Husemann 2008, p. 50)*

SD points out that we have choices in situations where there may appear to be none.

Lane pursues his unearthing of SD's social theory using Burrell and Morgan's (1979) framework identifying the sociological paradigms that underpin organizational analysis. The Burrell and Morgan framework maps in most respects onto the account of social theory in Chapter 4 of this book. It is possible, therefore, to link Lane's conclusions directly to that account. In broad terms, and using our own terminology, Lane identifies three strands in SD which we will call "hard," "mainstream," and "soft." In his view, the first two of these offer variants of "functionalism," while the soft version flirts with "interpretive" social theory. He then proceeds to develop a "bold conjecture" of his own – that SD is most compatible with the "integrationist" social theory that seeks to dissolve the agency/structure divide.

"Hard" SD corresponds rigidly to the functionalist paradigm (see Section 4.2). It looks for the causal relations that give rise to structures which condition the behavior of human beings. These structural relationships exist below the surface level of society and are often hidden from humans. As the quote from Senge that heads this chapter makes clear, SD resembles "structuralism" in believing in and seeking to uncover these deeper patterns of relationships. Some of Forrester's pronouncements support this orientation. It clearly underpins the work of those "austere" SD practitioners who emphasize the role of experts who produce computer simulation models that have to be validated scientifically on the basis of what actually happens in the real-world. The role of managers and other stakeholders is reduced to grasping the implications of the models and pulling whatever levers are suggested in order to bring about desirable change.

Over time and respecting Forrester's more reflective comments, SD practice has become much more participative in character. SD models still seek to capture real-world causal relationships, and so computer models remain extremely useful, but it is recognized that, to achieve its purposes, SD must involve stakeholders in model construction. This can help clarify and overcome initial differences in opinion on the main causal factors and relationships involved, provide greater insight, and ensure implementation of recommendations because the models and their implications are fully understood. The models developed with managers and stakeholders are regarded as "negotiative." They provide a vehicle through which opinions can be articulated more precisely. The aim is to promote experiential learning so that participants develop "mental models" more accurately attuned to the dynamics of system behavior. A battery of "craft" knowledge, designed to achieve these purposes, has become incorporated into mainstream SD practice and more user-friendly software has been created to facilitate model construction. "Management flight simulators" and "microworlds" have been developed with which participants can engage to enhance their understanding of counterintuitive behavior and test the impact of potential interventions. It is still useful to verify models against real-world events, but it is equally important that they are validated in terms of acceptance and use by stakeholders. This "mainstream" version of SD has, according to Lane, become its "enduring heartland" (2001a, p. 105). It is delineated in Sterman's (2000) *Business Dynamics: Systems Thinking and Modeling for a Complex World.* "Mainstream" SD remains functionalist but its encouragement of significant stakeholder involvement moves it in a "subjectivist" direction within that paradigm.

The key proponents of "Soft-SD" are Senge and Vennix. Senge (1990) regards SD, which he calls the "the fifth discipline," as the most important tool that organizations must master to become "learning organizations." It helps reveal the systemic structures that govern their behavior. Nevertheless, it is essential to support study of the fifth discipline with research on four other "disciplines" that are also significant in the creation of learning organizations. They are "personal mastery," "managing mental models," "building shared vision," and "team learning." These four disciplines rest more easily on "subjectivist" assumptions that seem to undermine the principle of causal efficacy that is central to SD. The least we can say is that the theoretical relationship between them and SD is not well explained. Vennix's (1996) influential work on "group model building" represents an even more extreme case. He seems more interested in using SD models to create "shared meaning" than to capture any "causal laws" governing human behavior. His view of SD practice is that:

> It focuses on building system dynamics models with teams in order to enhance team learning, to foster consensus and to create commitment with a resulting decision .... System dynamics can be helpful to elicit and integrate mental models into a more holistic view of the problem and to explore the dynamics of this holistic view .... It must be understood that the ultimate goal of the intervention is *not* to build a system dynamics model. The system dynamics model is a *means* to achieve other ends ... putting people in a position to learn about a messy problem ... create a shared social reality ... a shared understanding of the problem and potential solutions ... to foster consensus within the team ...
>
> *(Vennix 1996, pp. 3–5)*

This approach depends heavily on the skills of the facilitator who helps the group to elaborate their initial mental models into a system dynamics model that reflects a shared social reality and a consensus around the nature of the problem and possible solutions. The whole group is involved throughout and this enhances team learning and creates commitment to the resulting decisions. All this sounds very "subjectivist" in orientation. We should, however, note Smagt's (2006) observation. In his view, Vennix, and others in the "group model building" tradition, while accepting that there will be multiple interpretations of messy problem situations, see it as their role to bring about a consensus around one representation of reality as expressed in an SD model. They still appear to believe that there is an actual reality, which can be captured in a causal model once the different interpretations have been reconciled, i.e. they remain essentially "functionalist." There is certainly some ambiguity.

In Lane's view, "Soft-SD" moves so far in a subjectivist direction that it risks succumbing to the assumptions of the "interpretive" sociological paradigm. According to interpretivism, social reality is the product of the actions and interactions of human beings and should be studied by getting as close as possible to social actors so that their perceptions and motivations can be understood. Lane is worried, and one can see why. Within this paradigm of social theory, there are no "causal laws" influencing human behavior and SD loses its *raison d'etre*. He asks how Vennix, with his understanding of SD, can justify using an SD model as against any other modeling approach (1999, p. 517). Vennix argues that "people have a strong tendency to think in terms of causal processes" (1996, p. 3) but ultimately has to admit that he cannot impose SD on a group that doesn't fancy it. Lane is unhappy with this:

> It will indeed be a main task for the field to establish the status of causal reasoning and of its various 'theories'. However, system dynamics need not be apologetic about its attachment to causal models .... [While] acknowledging and responding to the contribution of subjectivism, system dynamics cannot move too far in this direction without losing most of what is distinctive and – more to the point – effective about the approach. Indeed, if placation of subjectivists involves the denial of the relevance of causal laws, causal explanations and the grand structural claim of system dynamics then the field should stop placating and start declaiming. While it is useful to clarify exactly what the system dynamics position is, there comes a point where criticisms must be turned on their head and worn as badges of pride.
>
> *(2000, p. 15)*

Lane argues that a more natural theoretical home for SD is with "integrationist" social theory (2001b). SD maintains a role for social structure as partially determining human choice but recognizes that human "mental models" themselves contribute to the creation of social structure. There is a kind of dialectical connection between society and the individual, between structure and agency, which SD is in a good position to understand and exploit because of its emphasis on feedback relationships. The "integrationist" theorists Buckley, Archer, Bourdieu, and Giddens are mentioned as sources upon which SD might draw. The attractiveness of the link between integrationist theory and SD is obvious from a quote that Lane takes from Giddens. Giddens describes his "structuration theory" as:

> A hermeneutically informed social theory ... [which] recognise[s] the need for connecting an adequate account of meaningful 'action' ... with the analysis of its unanticipated conditions and unintended consequence.
>
> *(Quoted in Lane 2001b, p. 297)*

SD can offer those interested in the agency/structure debate in the social sciences a well-honed expertise in formal modeling.

For the sake of completeness, we should mention Lane's view on how SD relates to the "sociology of radical change" (1999, pp. 518–519). He is able to offer Forrester's thoughts that making the insights of SD open to all can enhance debate and could have democratizing effects. And he suggests that an engagement with critical theory could produce "exciting and effective results." Unfortunately, however, he sees no current efforts in SD to seriously address issues of "power, ideology, coercion, and communication."

11.2.3 Methodology

Forrester is clear on the steps necessary to translate the insights of SD into practice. First, an adequate representation of the system and its important elements is obtained. Then, it must be modeled:

> To model the dynamic behavior of a system, four hierarchies of structure should be recognized: closed boundary around the system; feedback loops as the basic structural elements within the boundary; level variables representing accumulations within the feedback loops; rate variables representing activity within the feedback loops.
>
> *(Forrester 1969, p. 12)*

A final stage involves experimenting with the model and agreeing on appropriate changes that will improve the behavior of the system from the point of view of the stakeholders.

Forrester (1971a) argues that the SD approach must combine the power of the human mind with the capabilities of today's computers. The human mind is best in the early stages and the final stage of the methodology. The early stages consist of structuring the problem and identifying the relevant variables and possible feedback loops. The final stage consists of deciding what action to take. These steps require the creativity and sensitivity of the human mind because:

> The human is best able to perceive the pressures, fears, goals, habits, prejudices, delays, resistance to change, dedication, good will, greed, and other human characteristics that control the individual facets of our social systems.
>
> *(Forrester 1971a, p. 15)*

The middle stages, however, need the computer because the mind is pretty useless for anticipating the dynamic behavior of the system to which the interacting feedback loops give rise.

Wolstenholme (1990) similarly likes to think in terms of "qualitative" and "quantitative" phases in SD methodology and insists that both get equal attention. Other exponents, while largely following Forrester, bring additional insight to different aspects of the approach. Maani and Cavana (2000) pay attention to developing its potential for encouraging learning. Meadows (2008) wants to take SD out of the realm of "computers and equations" and to concentrate on how it can enhance the systems thinking skills of policy-makers. Probst and Bassi (2014) seek to align SD to the natural decision-making cycle employed by managers. Stroh (2015) presents SD in terms of a "four-stage change process" based on the work of Senge. Maani (2017) presents SD as an approach to multi-issue, multistakeholder decision-making. In attempting to popularize the insights of SD, it is the qualitative aspects of the methodology that tend to be emphasized most. To maintain a balanced overview, in the "mainstream" of the field, this account, slightly modifying Morecroft's (2010), pays attention to each of five distinct stages:

- Problem articulation (boundary selection)
- Dynamic hypothesis

- Feedback modeling
- Computer simulation and testing
- Learning and policy formulation

The methodology is described in outline in this section, with the associated methods and techniques covered in the next. It should also be mentioned that it is supposed to be used iteratively, as learning from later stages is fed back to improve the outcome of earlier phases.

Before the modeling itself can begin, it is necessary to undertake problem articulation, achieving clarity about the issue of concern, the timeframe over which it will be studied and the important factors involved. This stage draws on available data, existing theories about the area of interest, and the capacity of the human mind to determine what is relevant to the system's behavior. SD takes an "endogenous" perspective and so success depends on including within the boundary of the study all the significant elements that are interacting to produce the behavior under investigation. If this is achieved, then it can be safely assumed that the dynamic behavior observed is generated by interacting feedback loops within the system itself.

The "dynamic hypothesis" stage refers to the use of common "system archetypes" to gain an initial appreciation of what dynamics might be governing the system's behavior. System archetypes are classifications of frequently occurring feedback structures, which produce generic patterns of behavior within organizations. Their identification, as a result of years of SD modeling, can often provide a shortcut to understanding what is happening in any apparently new set of circumstances. According to Senge:

> One of the most important, and potentially most empowering, insights to come from the young field of systems thinking is that certain patterns of structure recur again and again. These 'system archetypes' or 'generic structures' ... suggest that not all management problems are unique .... Once a systems archetype is identified, it will always suggest areas of high-and-low-leverage change.
>
> *(1990, pp. 94–95)*

Senge (1994, ed.) has now identified about a dozen system archetypes, illustrated using CLDs, and his is the best-known offering in this area. Others, however, have been making important contributions. Wolstenholme (2003), for example, argues that all the system archetypes currently discovered can be condensed down to four totally generic types arising from "the four ways of ordering the two basic types of feedback loops (balancing and reinforcing)." These he names "underachievement," "out of control," "relative achievement," and "relative control."

To understand the third stage of the SD methodology, and indeed the second, it is necessary to introduce a little more history. Richardson's (1991) excellent account of feedback thought in the social sciences and systems theory is our guide. Richardson distinguishes two distinct ways in which feedback thinking has been used in these fields. In the "cybernetics thread" (and despite Maruyama's efforts, see Section 6.2), most emphasis is placed on the negative feedback mechanism as a means of ensuring control when a system drifts off course, often as a result of changing external circumstances. In the servomechanism thread, by contrast, positive feedback links gain equal recognition and their interrelationships with negative feedback mechanisms come to the fore. Forrester, of course, is the preeminent representative of the servomechanism thread and, as a result, feedback modeling in SD has always concentrated on the interplay between the positive and negative feedback loops within a system and how this generates dynamic behavior. The SD approach to describing complicated systemic interrelationships and patterns of change is now easy to grasp. It relies on three concepts: "reinforcing" (or amplifying) feedback, "balancing" (or stabilizing) feedback, and "delay." Reinforcing feedback processes are engines of growth. With such processes, whatever change occurs in one variable is amplified in others to produce more movement in the same direction. A small change builds on itself as does the increasing size of a snowball as it rolls down a hill. Balancing feedback processes encourages stability. They operate as self-correcting mechanisms and prevent a system running out of control. SD sees dynamic behavior as resulting from the interaction of these two kinds of feedbacks together with the impact of "delays." Delays occur when feedback processes take a long time to show their effects. They are present in almost all feedback loops and always "come back to haunt you in the long term" (Senge 1990). In SD, "CLDs" are used to provide an outline description of the networks of reinforcing and balancing feedback loops that constitute the structure of a system and drive its behavior over time. The use of CLDs has grown in popularity in SD because they are relatively easy to construct and understand and so can be used to engage managers and other stakeholders in the modeling process. Indeed, in the more qualitative reaches of SD, CLDs are nowadays portrayed as providing sufficient insight into system behavior for the purpose of guiding discussion and taking decisions.

In the "mainstream" of SD, however, it is usually regarded as essential to go through the fourth stage of the methodology, "computer simulation and testing," in order to attain the deeper level of learning required before properly informed policy decisions can be taken. Lane (2008, p. 6) regards computer simulation as necessary in all but the simplest of cases. Sterman argues that "simulation is essential":

> Eliciting and mapping the participant's mental models, while necessary, is far from sufficient ... the result of the elicitation and mapping process is never more than a set of causal attributions, initial hypotheses about the structure of a system, which must then be tested. Simulation is the only practical way to test these models. The complexity of the cognitive maps produced in an elicitation workshop vastly exceeds our capacity to understand their implications. Qualitative maps are simply too ambiguous and too difficult to simulate mentally to provide much useful information on the adequacy of the model structure or guidance about the future development of the system or the effects of policies.
>
> *(2003, p. 357)*

To prepare the ground for computer simulation, SD makes use of its second major diagramming technique – stock/flow diagrams (SFDs). This, in fact, was the original diagramming method of SD, pioneered by Forrester before CLDs became popular (Lane 2008). SFDs specify the actual processes that lie behind the feedback processes in terms of "stocks" and "flows" and so allow for a more detailed exposition of the causal network.

A "stock" is a quantity of some element that has accumulated in the system. It can change over time. "Flows" are relationships between elements that lead to changes in stocks. In a simple inventory system, for example, manufacturing rate and delivery rate together determine whether the stock level increases or decreases. Flows also change their values, often as a result of management decisions, but the impact on stocks can take some time. Because stocks and flows are potentially measurable, SFDs act as a stepping stone on the way to the construction of a computer simulation. In fact, there is now an abundance of SD modeling software (STELLA, VENSIM, POWERSIM, ANYLOGIC) designed to help in the process of building SFDs and creating from them simulation models. Advocates of computer simulation regard such models as going well beyond CLDs in their ability to reveal, in a rigorous manner, the behavior of systems governed by complicated interacting feedback loops. Once a plausible model is built, it can be tested by comparing its behavior with the relevant real-world activity. According to Morecroft, it can take experienced modelers weeks or months, depending on the size of the model, "to formulate equations and then create a calibrated and fully tested simulator suitable for evaluating new policies and strategies" (2010, p. 83).

It may be difficult to keep decision-makers involved during the construction of the computer simulation, but every attempt must be made to do so. The purpose, after all, is not to create a perfect representation of the real-world system but to initiate a learning process in which participants come to challenge their prevailing "mental models" as they become sensitized to the often counterintuitive behavior that complex systems exhibit. Equipped with this learning they can, in the final stage of the methodology, experiment with the model to gauge how alternative policies might impact and improve system behavior from their point of view. They will be looking for "leverage" points – those areas of the system at which they can direct action in order to achieve maximum payback in terms of their objectives.

11.2.4 Methods

The five methods most frequently used to support the various stages of the SD methodology are "CLDs," "system archetypes," "SFDs," "microworlds," and "leverage points."

The identification of causal relations and significant feedback loops is central, in SD, to understanding system behavior. A CLD can help because it can be annotated to show the direction of feedback and the relationship between different feedback loops. The example described in Section 11.2.2, from the Munro Review's preliminary report, is shown in Figure 11.1 in CLD form.

In Figure 11.1, the arrows indicate that a change in the value of one variable (always described in a noun or "noun-phrase") produces a change in one or more others. For example, a change in "scope for dealing with variety of circumstances" impacts both "quality of help" and "sense of satisfaction derived from work." When a change in a variable produces a change in the "same" direction in another, this is shown by an "S" on the arrow. When a change in a variable produces a change in the "opposite" direction, this is shown by an "O" on the arrow. Thus, following one of the chains of causality, a decrease in "scope for dealing with variety of circumstances" leads to a decrease in "sense of satisfaction derived from work," this leads to an increase in "staff turnover," then a decrease in "average experience of staff," and finally a decrease in "public status of child protection workers." Two types of feedback loops are recognized in CLDs – "reinforcing (positive) loops," marked with an "R"; and "balancing (negative) loops," marked with a "B." In isolation, reinforcing loops amplify changes leading to exponential growth (or decline) and creating "virtuous" or "vicious" circles depending on whether the outcome is viewed as desirable or not. Balancing loops help systems settle down and stabilize around their goals or limits. We noted the existence of the balancing loop "B1" that seeks to ensure the system reaches its goal of increased prescription. Another balancing loop "B2" should enable the system to learn from errors made and reappraise its goal but, in this case, it is counteracted by the operation of the strong reinforcing loop "R," a result of the compliance culture that champions the logic of "we followed procedures." The result of the interaction of the three loops is system behavior that demonstrates an "addiction to prescription."

It was noted that, over time, the study of the possible relationships between reinforcing loops and balancing loops, together with the phenomenon of "delays," has led to the identification of certain frequently occurring "system archetypes." They give rise to many problems but, because they show regular patterns of behavior, they are easy to identify if not always to manage. Once mastered by managers, according to Senge (1990), they open the door to systems thinking. One such is the "limits to growth" archetype, when reinforcing growth loops inadvertently set off balancing, negative loops that slow down growth or even send it into reverse. Other well-known archetypes are "shifting the burden," "balancing process with delay," "accidental adversaries," "eroding goals," "escalation," "success to the successful," "tragedy of the commons," "fixes that fail," and "growth and underinvestment." The "shifting the burden" archetype can be illustrated if we consider a developing country wishing to increase the standard of living for its people. It may need to make some fundamental adjustments to its economy in order to achieve this. In the interim, it seeks aid to ensure a reasonable standard of living is maintained. The danger is that the country becomes "addicted" to the aid before it sees any benefits coming through from the changes to its economy. There is a "delay" built into the latter. Once that happens, it loses its capacity for self-reliance, the economy is weakened rather than strengthened, and the country becomes overly dependent upon aid. All of these archetypes are easily represented using CLDs. Our example is shown in Figure 11.2. This figure uses an alternative form of annotation for its CLD, with reinforcing relationships indicated with a + (plus) sign and negative with the – (minus) sign.

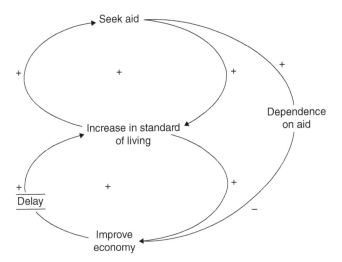

Figure 11.2 Increasing dependence on aid: an example of the "shifting the burden" archetype.

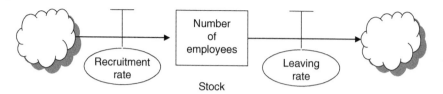

Figure 11.3 Stock-and-flow diagram.

If managers can learn to recognize system archetypes, they can save themselves a lot of wasted effort and target their interventions at points of maximum leverage.

Once the feedback structure of a system is understood and captured in a CLD, it is usually necessary to elaborate further by building a computer simulation designed to represent its dynamic behavior. Building the simulation involves plotting the systemic impacts of the various feedback loops on the "levels" and "rates" that exist within the system. SFDs (where the stocks are the levels and the flows are the rates) are used for this purpose. In Figure 11.3, we see some of the typical symbols used in stock-and-flow diagrams. The stock level of interest in this case is the number of employees in a company. This is shown as a rectangular "vessel." The recruitment rate is a flow, shown as a kind of valve, which adds to the number of employees. There is another flow of staff leaving, which subtracts from the stock of employees. Employees are recruited from a "source," shown as a cloud, and return to a "sink," similarly represented, when they leave.

Nowadays, SD modeling software makes the transition from SFDs to computer simulations relatively straightforward.

The software also facilitates the creation of microworlds or management flight simulators (see Morecroft and Sterman 1994). These are interactive and provide managers with a user-friendly interface that allows them to experiment with SD models in something much more like a gaming environment. Managers can try out different decisions on a simulator representing the situation they are facing and see what consequences ensue. A "learning laboratory" is designed to provide for the structured learning of groups of managers engaging with a microworld, and this is then employed to facilitate and diffuse learning throughout the organization. The aim is to get groups of managers to question their existing mental models. In particular, they should come to recognize the interdependence of the issues they deal with and replace superficial explanations of problems with a more systemic understanding. The detailed model structure and equation formulations can be made available to users at the press of a button.

In SD, the behavior of complex systems is counterintuitive; almost impossible for the human mind to understand. For the same reason, it is difficult to know exactly how to change their structure to get what we want. Often we make things worse. Nevertheless, Meadows offers a tentative list of "leverage points" – "places in the system where a small change could lead to a large shift in behavior" (2008, p. 145) and, if we are careful, in the direction we desire. Leverage can be gained by paying attention to such factors as stocks (as buffers), delays, and introducing new information flows. At a somewhat more sophisticated level, we can seek to strengthen balancing feedback loops (e.g. preventative medicine, monitoring systems for environmental damage, protection for whistle-blowers), and weaken reinforcing loops (e.g. population and economic growth in the "limits to growth" model). At a higher level still, we need to address our goals and the shared ideas and paradigms, which are "the sources of systems."

11.3   System Dynamics in Action

One of the interventions most frequently used to illustrate the power of SD is that conducted by the MIT System Dynamics Group with Hanover Insurance (see Cavaleri and Obloj 1993; Senge and Sterman 1994; Maani and Cavana 2000). It can, therefore, be regarded as being in the "mainstream" of SD practice. The case reveals that apparently rational action taken by management to reduce the costs of settling claims and to maintain customer satisfaction actually led to an erosion of quality of service and increased settlement costs. The SD study demonstrated, through an analysis of the interacting feedback loops, exactly why this happened, suggested less obvious but more efficacious ways of tackling the problems faced and led to the development of a microworld and learning laboratory to spread the learning obtained throughout the company.

Hanover Insurance had undergone an amazing transformation in the 1970s and early 1980s pulling itself from the bottom of the industry to become a leader in the property and liability field. During this period, it grew 50% faster than the industry as a whole. Nevertheless, it could not escape the many problems and resulting runaway costs that impacted on the industry during the 1980s. Automobile insurance premiums doubled causing a public backlash, the number of product liability cases increased massively, and the average size of claims settled in court increased fivefold. It was easy to blame dishonest policy holders, biased juries, greedy lawyers, and the increased litigiousness of society. Senior managers in Hanover, however, determined to look at how their own management practices were contributing to the problem situation. A good starting point was the claims management operation responsible, because of increasing numbers and complexity of claims, for more than 67% of total company expenses.

The project began with a team from Hanover, consisting of the senior vice-president for claims and two of his direct subordinates, meeting regularly with some MIT researchers. A vision statement expressed the desire to be preeminent among claims organizations and to provide a "fair, fast, and friendly" service. From this, it was possible to derive an image of the ideal claims adjuster and the performance measures he or she would be required to meet. The problem was finding a coherent path from the reality to the ideal. There were lots of candidate strategies, but these seemed disjointed. A more systemic solution was required. STELLA was used with the Hanover team to build computer-based simulation models. These were subject to basic reality checks and employed to test the results of current strategies and to seek improvements in management practices. Expert judgment was used, alongside whatever quantitative data were available, to bring to the fore the many "soft variables" involved and to estimate their effects. The final model was both sophisticated in its treatment of problem dynamics and fully owned by the Hanover team. Figure 11.4 is a causal loop diagram that expresses the problem dynamics.

The existing implicit strategy of Hanover, in the face of pressure from increased claims, is captured in the productivity loop and the work week loop of Figure 11.4. The operating norm for claims adjusters was simply to work faster and work harder. Working faster helped reduce pending claims because less time was spent per claim and therefore more claims could be settled. Working harder helped reduce pending claims because of the effect on productivity of working longer hours and taking shorter and

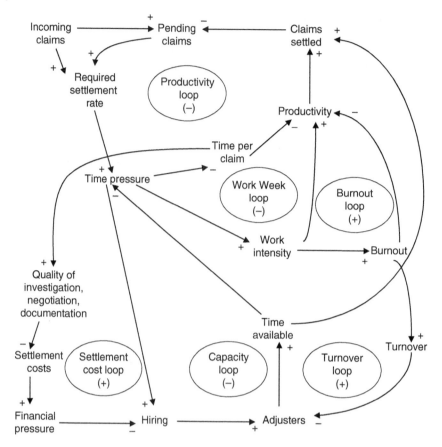

Figure 11.4 System dynamics of claims processing in the insurance industry. Feedback loops controlling claims settlement, with processes causing self-reinforcing erosion of quality and increasing settlement costs. Arrows indicate the direction of causality. Signs ("+" or "−") at arrow heads indicate the polarity of relationships: a "+" denotes that an increase in the independent variable causes the dependent variable to increase, ceteris paribus (and a decrease causes a decrease). Similarly "−" indicates that an increase in the independent variable causes the dependent variable to decrease. Positive loop polarity (denoted by (+) in the loop identifier) indicates a self-reinforcing (positive feedback) process. Negative (−) loop polarity indicates a self-regulating (negative feedback) process. *Source:* From Senge and Sterman (1994). Reproduced with permission of Productivity Press.

fewer breaks. In the short term, these "fixes" might appear to work. In the longer term, however, other relationships bring about unintended consequences (initially hidden because of the existence of "delays") that make matters worse rather than better. They are examples of the "fixes that fail" archetype. The unintended consequences of working faster are shown in the settlement cost loop of Figure 11.4. Spending less time per claim reduces the quality of settlements, and this leads to increased settlement costs. Less time to investigate and negotiate claims means that inflated settlements are agreed. Other customers become dissatisfied with the small amount of time devoted to them by the adjuster and are more ready to resort to law. Where litigation results, inadequate documentation means longer preparation time and less successful outcomes. The unintended consequences of working harder are shown in the burnout and turnover loops of Figure 11.4. Working harder increases fatigue, ill health, and staff burnout, impacting adversely on productivity. Burnout also increases staff turnover, which means fewer assessors and even greater time pressure on those that remain. Through the settlement cost, burnout, and turnover loops, therefore, the initial fixes provoke longer term unintended consequences that reduce quality, increase time pressures, and increase costs. Because this feedback is delayed and its causes are not easy to trace, management will tend to react by relying further on the original fixes. In an insurance industry facing runaway costs, the temptation to require claims adjusters to work even faster and harder was virtually irresistible. The idea of addressing the problem by hiring new claims adjusters seemed ridiculous. In archetype terms, therefore, the burden was shifted from capacity expansion to quality erosion.

Hanover, at the time of the study, had the highest number of assessors per claim in the industry. Nevertheless, because the Hanover team had been involved in building the model themselves and, through the model, had come to appreciate the interconnections that produced the counterintuitive behavior of the claims adjustment system, they were prepared to embrace the only fundamental solution to the problem. This, as is shown by the capacity loop in Figure 11.4, was increasing adjuster capacity. However well it compared with the rest of the industry, only by hiring new adjusters and training them properly could Hanover genuinely address issues of service quality and increased costs.

The next challenge for the project was to extend the learning gained to the entire company. As implied earlier, there was a tendency to blame outside factors, and not internal management practices, for the travails of the insurance industry. Furthermore, in Hanover responsibility for decision making is widely distributed. If there was to be a real change in the way the company behaved, it was essential that all those managers with influence experienced for themselves the counterintuitive behavior of the claims processing system. To achieve this aim, a claims learning laboratory, incorporating a computer simulation game (or management flight simulator), was developed. Managers were familiarized with CLDs and helped to think through the variables and relationships associated with the claims system. The design of the game ensured that they were forced to make explicit their mental models and to challenge them when the results of the strategies they championed defied their expectations. As a result, "double-loop" learning was facilitated.

11.4 Critique of System Dynamics

Forrester set out to create a revolution in management studies, bringing scientific rigor to a field which he saw at the time as "a very skilled art generated by a *melee* of empirical observations" (Lane 2006, p. 486). In Forrester's words:

> People would never send a space ship to the moon without first testing prototype models and making computer simulations of anticipated trajectories. No company would put a new household appliance or airplane into production without first making laboratory tests. Such models and laboratory tests do not guarantee against failure, but they do identify many weaknesses which can be corrected before they cause full-scale disasters. Social systems are far more complex and harder to understand than technological systems. Why then do we not use the

> same approach of making models of social systems and conducting laboratory
> experiments before adopting new laws and government programs?
>
> *(1971b, p. 53)*

Feedback ideas could, in his view, provide the theoretical armory, and computer simulation the means to understand social systems and improve their performance.

Lane (2016) is in no doubt of the value of SD modeling and has provided many examples of the insightful use of its key concepts – for example, in overseeing labor costs (unintended consequences and feedback), fishery management (stocks and non-linearity), and child protection (feedback and mental models). A report from the "Government Office for Science" (2018) discusses how computational modeling can be better used, in both the public and private sectors, and commends the SD modeling employed in "The Limits to Growth" investigation and "The Munro Review of Child Protection." We read in the press every day about policies that exhibit "destructive perversity" and would not, one hopes, have been entertained given some understanding of SD. Monbiot (2014) in *The Guardian*, commenting on increased flooding in the UK and the demand for more spending on flood defences, notes that the European Union's unbreakable rule for farmers wanting to receive their "single farm payment" is that the relevant land should be free of "unwanted vegetation." This when it is known that, because of the channels provided by tree roots, water sinks into the soil under trees at 67 times the rate it does under grass. Even policy-makers who have "had enough of experts" would be interested in the ReThink Health (2018) system dynamics model, which suggests that the combined effect of President Trump's policies on health, the economy, the environment and crime, if implemented, could lead to a cumulative total of almost 1 million unnecessary deaths by 2029, an extra 19.1 million person-years of chronic illness, and a loss of $3.2 trillion in productivity. Right or wrong, if it was your field, you would want to understand the logic and interrogate the assumptions made in the model.

If the claim is granted that structure, in the form of interacting feedback loops, is the main determinant of system behavior, and that it can be modeled on computers, then SD provides a unifying transdisciplinary approach capable of seeing beyond the surface detail presented by individual disciplines to the deeper patterns that are really responsible for generating behavior. If that is the case, it is clear that it has wide applicability and can help managers in diverse areas to penetrate structural complexity and get their hands on the real levers of change in social systems. This is even more the case because SD models help to pinpoint key decision points. The beliefs and actions of decision-makers can be incorporated in the models, and it is therefore easy to follow the consequences of present policies and to explore alternatives. This leads to the emphasis on learning and changing mental models. Managers, unaware of the systemic relationships to which their actions contribute, are prone to act in ways that exacerbate existing problems. If they are involved in building SD models, however, they become aware of the underlying structures at work. They are more willing to question their own mental models, which might be contributing to the consolidation of damaging archetypes. In Schwaninger's (2009) opinion, one of the great strengths of SD is that its specific features make it "an exceptionally effective tool for *conveying systemic thinking* to anybody." Combine SD's powerful theoretical insights with the advances made in facilitation, group model building, computer simulation and software packages, and there is much to support the argument that it has gone far in achieving Forrester's ambition of providing well-grounded scientific and practical help to managers and policy-makers.

The two interrelated criticisms of SD that seem to me to carry most weight are that its models are incapable of capturing the complexity of social reality and that they are the wrong kinds of model for social systems in the first place. We will try to keep these separate for analytical purposes.

SD's endogenous perspective requires that we include within the boundary of the study the key variables involved in the interactions that take place in the system of concern. Later, these elements and the strengths of the interactions between them have to be plausibly quantified. Cavaleri and Obloj provide, as an example of a CLD, a map of "Russian Society in Transition" with 29 variables (1993, p. 351). How can we possibly know if these actually are the significant variables and how can we arrive at an accurate figure for their current settings and the force involved in their interactions? If SD models are inevitably imprecise in complex situations of this type, then it becomes difficult to see how they can generate accounts of the future state of the system convincing enough to engage policy-makers. This criticism is reinforced if we take seriously the findings of chaos theory (see Section 7.2). Chaos theory insists on the "butterfly" effect – the idea that small changes in a system's initial conditions can radically alter its long-term behavior. If SD has no precise grasp on the relevant variables, on initial conditions, and on the impact the different variables have on one another, then the claim to make useful forecasts of behavior is surely preposterous. In response, as in the case of the "limits to growth" model, advocates of SD would argue that it is the structure of the model that is more important in governing overall system behavior than the actual numbers used to quantify it. And chaos theory offers some support here because it highlights the frequent existence of order underlying apparently chaotic situations. Boulton et al. (2015, p. 94) recognize this and argue that when little is changing in the environment and in the system itself, a system can "run smoothly" to an attractor basin. In these circumstances, SD models can provide a useful representation of system behavior and can reveal patterns and time-delayed outcomes difficult for the human mind to comprehend. Such models can also help judge the probable effects of an intervention "assuming that no structural changes are provoked." But there is a big sting in the tail, because Boulton et al. simply do not regard social systems as operating in this way. As we saw in Section 7.3, they see Prigogine's account of "dissipative structures" as considerably more revealing of the actual behavior of social systems. According to this theory, social systems such as organizations are continually subject to internal and external fluctuations, including attempts to intervene in them, which due to positive feedback can drive them far from equilibrium. They can change radically over time, evolving to higher levels of complexity as they spontaneously self-organize into new unpredictable forms. If this is a reasonable picture of how social systems behave, then SD falls well short of what is necessary:

> Yet, system dynamics models are still deterministic, they still only allow for one solution or path from a particular starting point. It is the path into the future traced by average elements interacting through average events and is only reasonable if nonaverage elements and nonaverage events have no systemic effect, e.g. if there is no learning or interaction. Such models can function but not evolve
> *(Boulton et al. 2015, p. 94).*

Proponents of SD would no doubt argue that all models are simplifications. SD models can be tested and modified and, an important consideration, policy-makers find them useful. To critics, however, the compromises required in constructing an SD model mean that an approach that sets out to handle complexity is in fact guilty of gross reductionism. It is easy to show, using simple feedback diagrams, that providing aid to a developing country is not a long-term solution to its economic problems; that increasing police recruitment may not be the best way of reducing crime; that building more roads may not ease road congestion; that reducing costs might not be the best solution to declining profits. But these conclusions are hardly enlightening or useful to decision-makers confronted by the myriad of cultural, political, ethical, and coercive factors that prevent them acting in the apparently rational way that SD would prescribe. None of these problems are as simple as SD makes them out to be. Few management problems are as straightforward as Senge's archetypes suggest. Perhaps a saving grace is, as Forrester says, that SD models "can be far more comprehensive than the mental models that would otherwise be used" (1971b, p. 53). Even this, however, does not hold water against a second type of criticism of SD models.

This line of criticism asserts that SD models are just not appropriate for understanding and helping us to change social systems. The argument goes that systems in which human beings are central do not lend themselves to causal explanations and, as a consequence, SD is unsuitable to the subject matter of its concern. To these critics, human motivations, intentions, and actions shape social reality. If we want to learn about social systems, we need to understand the subjective interpretations of the world that social actors and groups employ. Social structure emerges through a process of negotiation and renegotiation of meaning. SD misses the point when it tries to study social systems "objectively," from the outside. Stacey's work on "complex responsive processes of relating," discussed in Section 7.5, takes the argument forward from the "interpretive" perspective. In his view, SD is part of the traditional paradigm that pictures managers as overseeing and improving the interactions between the parts of their organizations, something they are able to accomplish because organizations develop in ways that can be anticipated. By contrast, Stacey sees organizations as being in a process of perpetual construction stemming from the microinteractions between human agents relating to one another through symbols. From this process perspective, the social world is mutually created, sustained, and changed through the interactions of human agents. Radical changes in organizational and individual identity can occur. It is the agency of individuals and their capacity to create novelty through self-organization on which we should concentrate if we wish to understand the dynamic behavior of social systems. If we want to change them, the appropriate course of action is to participate in the construction of meaning by engaging in the ongoing conversations. It is hopeless to try to steer organizations from the outside in the direction of particular purposes. That said, in his later work, with Mowles, Stacey accepts that communicative interaction can become patterned and particular "narrative themes" can become influential in organizations. Perhaps these could be incorporated in the causal models at the heart of SD? Lane, after all, argues that SD "is concerned with aggregate social phenomena, not individual meaningful actions" (2000, p. 13). But this argument cuts no ice with Smagt, who points out that aggregate social phenomena are themselves constituted as a result of compromises emerging from "interest-based political processes." Their definition is unstable and can easily change. Consequently, Smagt argues that

> ... causal models and particularly System Dynamics models, are of limited use when studying social phenomena .... This is because causal models require modelled phenomena to be stable and determinate .... If the constitution of social phenomena is inherently indeterminable and constantly subject to change, and if causal models are unable to represent this constitutional change, then causal models can easily produce invalid results.
>
> *(2006, pp. 513–514)*

One example he provides is the notion of "proven reserves" in an SD model created by Sterman. Smagt argues that the meaning of this term is taken to be stable and unproblematic whereas, in actuality, it results from an ongoing negotiation process between stakeholders with different interests. Depending on the outcome of the negotiations, say environmentalists gain the upperhand, the meaning applied to "proven reserves" could change without any easily identifiable causal mechanism coming into play. Another example suggests that an apparent reduction in the number of disabled people in the Netherlands can best be understood as due to changes in the constitutive domain, how disability is defined, rather than from addressing the actual causes of disability:

> This intervention is unmistakably not a causal intervention, since not a single condition that is causally related to disability (in its original definition) has changed. The only thing that has changed is the constituting rule, which tells us what counts as disabled and what does not. To put it differently, the social kind 'disability' has changed.
>
> *(Smagt 2006, p. 520)*

A similar argument can be made around the way "employment" is defined in order to "massage" unemployment rates. Smagt resists the argument that processes of constitution can themselves be seen as "rational designs" and included in a causal model. He sees them as emanating from processes of meaning construction with indeterminable outcomes. We have, in studying social theories in Chapter 4, come across the same debate. While it seemed to Durkheim that suicide rates were "social facts" that could be related to other social facts in a causal model, they were, from the "interpretive" viewpoint, socially constructed – the outcome of negotiations of meaning between social actors, principally family members and officials.

Vriens and Achterbergh discuss further problems that arise from SD's impoverished understanding of social theory. First they argue that, in SD interventions *for* social systems, it is useful to distinguish three groups – "modeling group," "problem owner," and "affected social system":

> For instance, in the case of the problem of waiting lists, the modelling group consists of consultants and people from the university, the problem owner is a group within the Department of Health, and the affected social system is Dutch society.
>
> *(Vriens and Achterbergh 2006, p. 556)*

This gives rise to problems of communication and coordination, especially as the categories of "problem owner" and "affected social system" will themselves consist of multiple stakeholder groups. Inevitably there will be different values, agendas, and interests involved – the situation will be pluralistic – and it will be difficult to find a consensus about problem variables and possible solutions. Second, SD purports to make models *of* social systems but, in the absence of an appropriate social theory, lacks understanding of the relevant factors that contribute to their production and reproduction. Vriens and Achterbergh hint that Luhmann's work might be able to assist the SD modeling process. In Section 4.7, we saw how, according to Luhmann, the different "function systems" of society construct reality in different ways and that this makes it virtually impossible to mount a coordinated response to global issues such as climate change. This is surely insightful in helping us understand the difficulties Meadows and her colleagues encountered in creating a group of stakeholders who could own and act upon the conclusions of the "limits to growth" work. Finally, SD models are made "in" social systems. Here attention must be given to how to make teams more productive and to eliminating "social biases and deficiencies" that might interfere with the group modeling process. To be fair, SD has given more consideration, in its "group model building" literature, to this issue than has almost any other systems approach.

We can begin a critical systems thinking interrogation of SD by considering its relationship to science and philosophy. Many of Forrester's pronouncements suggest that his primary concern was to bring the rigor of classical science to practical problems of management. Scientific models of "reality" were to be constructed and validated using traditional scientific criteria. Barton (1999), however, notes some very close parallels between SD and the philosophy of pragmatism (see Section 1.4). Pierce's notion of the "predictive success" of theories is relevant but so is Dewey's idea that knowledge is confirmed when it is recognized by the community as successful in transforming practice to overcome problems. A closer engagement with the literature of pragmatism can only benefit SD.

Turning to the social theory that lies behind SD methodology and methods, we have already indicated, in Section 11.2.2, our agreement with many of Lane's conclusions. "Hard" and "mainstream" SD, as we defined them, find their home in the "functionalist" paradigm. They see system structure as governing system behavior and seek to model that structure in terms of causal relationships. The aim is to improve goal-seeking. The former is clearly on the "objectivist" wing of the paradigm; the latter, with its attention to group model building, mental models and learning, has been pushed by a desire to ensure implementation toward the more subjectivist edge. The functionalism that underpins SD differs from the functionalism of hard systems thinking in embracing a "structuralist" rather than "positivist" epistemology. In other words, its claim to knowledge is that it gives access to underlying structures that exert a significant influence on system behavior. It can see beneath the jumble of surface detail with which hard systems thinkers have to contend. The ability to discover deep structural relationships provides SD with its capacity to cut through apparent complexity. It can then present itself as a transdisciplinary approach of very wide applicability. At home in the functionalist paradigm, SD can make a significant contribution to understanding and improving problem situations. This is particularly the case when mental models become widely shared and "fixed" and give rise to apparent "social facts." The drive for continuous economic growth highlighted in the "limits to growth" model and the embrace of "command and control" in the public sector, underscored in the realm of child protection by the Munro Review, are cases in point.

The work of Senge and Vennix, demonstrating a primary concern with ensuring mutual understanding through group problem solving, flirts with the "interpretive" paradigm. This type of "Soft-SD" threatens the credibility of the field, as Lane warns. If SD leans too far in this direction, it risks jettisoning the claim to our attention derived from its functionalist presumption that it can discover structural arrangements that govern system behavior. If human beings are free to construct social systems as they wish, what determining influence does system structure have? And what confidence can we have in the power of feedback analysis to uncover likely system behavior? SD, to be worthy of our attention, must maintain its functionalist aspirations. Otherwise, its models become simply addenda to a debate orchestrated between stakeholders in the manner of soft systems thinking. This is exactly what Lane and Oliva (1998) risked reducing them to in their failed attempt to ignore the issue of paradigm incommensurability and "synthesize" SD and soft systems methodology – and Lane learned his lesson (1999, p. 518). Ultimately Lane concludes that, while there is much to learn from the "interpretive" paradigm, it is "probably a blind alley for the field" of SD (2001b, p. 304).

From the point of view of "the sociology of radical change," SD possesses some very worrying characteristics. As Vriens and Achterbergh suggest, SD interventions often impact a wide variety of stakeholders in the "affected social system." These stakeholders will in all likelihood represent many possible viewpoints on the nature and purpose of the system of concern. In these circumstances, SD models appear, from the radical change paradigm, to be selective accounts of "reality" constructed from one point of view. Furthermore, SD provides no means of questioning the rationale underlying any model, for example, by comparing it to another model built on the basis of an alternative worldview. The criteria for deciding what is the optimal behavior of a system will similarly be bound up with one model and one worldview, and so go unquestioned. In "coercive" problem situations, SD modelers will almost inevitably privilege the perspectives and purposes of powerful groups. They serve the purposes of élites even more effectively when they present themselves as "élite technicians," using their expertise to provide decision-makers with objective and neutral advice.

Finally, what of Lane's assertion that SD is uniquely suitable for articulating the agency/structure dialectic theorized in "integrationist" social theories? The reader will remember from Section 4.6 that it is difficult to maintain an integrationist position that avoids privileging either the agency or structure side of the divide. Giddens' "structuration theory," for example, collapses structure into agency, as Lane recognizes. This makes for a difficult marriage with SD:

> Squaring this with the nature of the causal 'laws' that underlie system dynamics is problematic. Compare the system dynamics notion of causal 'laws' with Giddens's work .... [They] do not particularly feel like mere hermeneutical generalisations!
>
> *(Lane 2001b, pp. 300–301)*

If SD accepts Archer's argument for "analytic dualism," however, and seeks to serve our understanding of how the structure side of the equation works, then a rapprochement with the ideas of such as Bourdieu and Bhaskar might be feasible. Bourdieu, for example, while allowing a space for agency, argues that class stratification continues to dominate society. It would certainly be interesting to see SD entertain his social theory and seek a causal account of how certain classes maintain their position of privilege. Attempts to unmask the effects of power relationships have been sadly lacking in SD to date.

To conclude this second-order, critical systems review of SD, it needs to be positioned in terms of the system of systems methodologies (SOSM). This positioning is not in doubt because in Chapter 8, based on Snowden's distinction between "complicated" and "complex" contexts, we modified the SOSM to accommodate the special features of SD. SD, therefore, defines exactly what is meant by the SOSM in referring to "complicated-unitary" situations. In Snowden's terms, complicated domains exhibit cause–effect relationships but these are often separated in time and space and linked in chains that are difficult to comprehend. Policy-makers can make use of experts to help them discern the relationships and so understand better the behavior of the system. This is a reasonable short-hand description of SD and its appropriate place in the SOSM is therefore as in Figure 11.5. There is some bleeding into the pluralist section of the grid to reflect the endeavors of those who emphasize that SD models must be formulated as part of a "group model building" exercise.

This positioning of SD can be clarified if we consider what it would have to be like if it occupied any other location. Commenting on traditional management science and operations research, in 1956, Forrester acknowledged that it "did pay its way" but concluded that it did not deal with "major [problems] that made the difference between the companies that succeed and those that stagnate or fail" (quoted in Lane 2006, p. 484). His ambition in developing SD was to extend the range of applied systems thinking to more strategic problems. OR was, he believed, losing touch with the important issues confronting managers as it concentrated more and more on specific tactical problems, amenable to mathematical modeling because they involve just a few variables in linear relationships with one another. SD, by contrast, would employ the science of feedback, harnessed to the power of the modern digital computer, to unlock the secrets of complex, multiple-loop nonlinear systems. Lane presents a more detailed argument as to

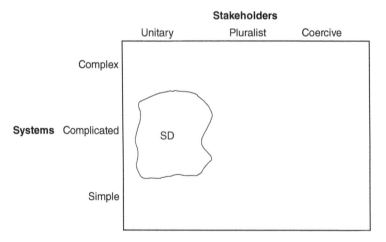

Figure 11.5 The positioning of system dynamics on the SOSM.

why SD cannot be regarded as a form of hard systems thinking or systems engineering, concluding:

> The aim of hard/SE approaches is to design and implement an improved system .... As indicated ... specific policy directions and new 'rules' for control are not the aim of system dynamics. It is the case that a mathematical model is created, and that the results of changing policies are experimented with using the model. But the model is never seen as the system then to be implemented. Instead, the purpose is to promote individual and organizational experiential learning ...
>
> *(2000, p. 17)*

The argument has been won that SD does not lie with "hard systems thinking" in the simple-unitary region of the SOSM. Its assumptions and practice place it higher along the vertical dimension.

On the other hand, SD does not reach up into the realm of what are defined by the SOSM as "complex" contexts. These domains, according to Snowden, consist of so many agents and relationships that it is impossible to chart their interactions and anticipate what the outcomes will be. They are in a state of constant flux with any patterns that do emerge only discernible in retrospect. The only way to proceed is to open up discussion and encourage diversity until favorable attractors eventually hold sway. This is social reality as described by Stacey in his work on "complex responsive processes of relating," it is not a world in which SD methodology and methods are relevant. Another characteristic of complex contexts, according to the SOSM, is environmental turbulence. This again is not a feature that SD, with its endogenous approach, is well equipped to handle. To include relevant variables in its models, SD has to know about them in advance. If the environment is turbulent, it will be constantly throwing up unpredictable fluctuations. Schwaninger puts it like this:

> [SD] does not provide a formal procedure for an organization to cope with the external complexity it faces, namely, for designing a structure which can absorb that complexity. In contrast, OC [Organizational Cybernetics] and LST [Living Systems Theory] offer elaborate models to enable the absorption of variety, in the case of the VSM [Viable System Model] based explicitly on Ashby's *Law of Requisite Variety*.
>
> *(2009, p. 8981)*

This is in tune with Beer's assertion that "exceedingly complex" contexts, where systems and their environments cannot be described in a precise and detailed fashion, are the province of cybernetics. As we shall see in Chapters 12 and 13, socio-technical systems thinking and Beer's VSM are more closely matched to complex domains because they offer designs for keeping systems viable in the face of endogenous and exogenous fluctuations that cannot be predicted in advance.

Turning to the "stakeholders" dimension of the SOSM, it has been shown that SD recognizes an important role for human agency and usually accepts that it must pay attention to addressing a pluralism of perspectives, among involved participants, if it is to secure a smooth implementation of proposals for change. The extensive work undertaken on "group model building" aims to illuminate how SD models themselves can be used to reveal pluralism and help move a group toward consensus. That said, Lane admits that SD itself "lacks any structured way of eliciting and handling such diversity" (2000, p. 15). He suggests that the problem-structuring methods of "Soft-OR" might be able to help. This is the point. SD itself makes an essentially "unitary" assumption that a group of people, working with an SD model, can come to agreement about the nature of the problems they are facing in the external world, the key variables bearing on those problems, and some possible solutions. Those "Soft-SD" thinkers such as Senge and Vennix, who believe that pluralism is a rather more difficult issue to resolve, are forced to call upon other approaches – four "disciplines" in the case of the former, Soft-OR methods for the latter – in support of their SD practice. An example of the use of group model building, to find ways of addressing the problem of the declining size of the Dutch-registered merchant marine, sees Vennix having to go to extreme lengths to make his approach work in the context of genuinely divergent views about the causes of the problem and the likely future consequences. Nominal Group Technique and a variety of well developed consulting skills have to be employed to keep the show on the road (1996, pp. 174–186). Finally, SD has virtually nothing to say about "power" and "coercion" – matters that deeply trouble observers who identify coercive contexts as pervasive.

To reiterate, the process of positioning SD, and any other applied systems approach, on the SOSM says nothing about how often it is useful. An advocate of SD will tend to see most problem-contexts as complicated-unitary and, therefore, SD as a solution to the majority of problems. Somebody equipped with a critical systems sensibility, on the other hand, will have SD as an important methodology in their systems tool-kit but will judge how helpful it is according to the traction it brings to problem resolving in a particular set of circumstances.

11.5 Comments

The relationship between SD and other strands of the systems movement has not been as healthy and productive as it should. System dynamicists have a reasonable claim that other systems thinkers have misrepresented their approach as "hard" and "deterministic." The greatest transgression of academic propriety, however, has come from those in the SD community who have sought to seize the label "systems thinking" for their version of the systems approach and to ignore all the other strands. SD has a particular competence that it would do well to consolidate and cultivate, but it is simply misleading for it to lay claim to the whole of the systems thinking territory. This imperialistic land-grab started in earnest with Senge's *The Fifth Discipline* but a number of others, who should know better, have followed in his wake. It does no good to SD or the rest of the systems movement.

11.6 The Value of System Dynamics to Managers

As with hard systems thinking, managers usually need the help of experts in order to get the most out of SD. Nevertheless, it is worth following the pattern begun in previous chapters and picking out five lessons that managers can easily learn from SD that will certainly benefit their practice:

- It is often helpful to look beyond the apparent mess presented by surface appearances to see if there are any underlying patterns of feedback loops that are determining system behavior. Sometimes, computer simulation can help to reveal and clarify counterintuitive outcomes that the relationships between variables and loops produce
- An understanding of how feedback loops interact to cause system behavior can inform the way managers work. For example, they become much more aware of the dangers of unintended consequences, of treating symptoms rather than causes, and of the importance of "delays." Senge's (1990, 1994) 11 counterintuitive "laws of the fifth discipline" are genuinely thought provoking
- Rather than jumping to what appear to be obvious solutions to problems, managers need to appreciate that complex systems often behave in subtle and unexpected ways. It is worth spending time looking for smaller interventions that may provide levers that can bring about substantial changes
- SD supports the conclusion that "no man is an island." It is no good, therefore, just blaming the environment or other people for our problems. We are all in it together. Our decisions feed into the set of relationships that give rise to the difficulties that we face
- SD models, management flight simulators, and microworlds can assist managers to appreciate the systemic relationships in which they are involved and to which their decisions contribute. They teach managers that they often need to radically change their thinking before improvement becomes possible. The double-loop learning involved in changing mental models is crucial to successful management practice

11.7 Conclusion

Donella Meadows, one of the most astute thinkers to apply SD in environmental and social analysis, argues that, as no paradigm is right, "you can choose whatever one will help to achieve your purpose." That conclusion, she believes, provides us with the preeminent leverage point for bringing about desirable change:

> That is to keep oneself unattached in the arena of paradigms, to stay flexible, to realize that *no* paradigm is 'true', that everyone, including the one that sweetly shapes your own worldview, is a tremendously limited understanding of an immense and amazing universe that is far beyond human comprehension.
> *(Meadows 2008, p. 164)*

To embrace this insight and to play a full part in addressing the complexity of real-world problem situations, SD practitioners will have to learn to use their own powerful approach in informed combination with other versions of systems thinking.

Type D

Systems Approaches for Organizational Complexity

Antifragility is beyond resilience or robustness. The resilient resists shocks and stays the same; the antifragile gets better

(Taleb 2013, loc. 436).

Some systems approaches are primarily interested in managing organizational complexity. Organizational complexity is seen as driven both by the internal interactions of the parts of a system and by the interactions between the system and its turbulent environment. These, often unpredictable interactions make it very difficult to steer the system toward goals. The viability of the system becomes of central concern and also its ability to reconfigure itself to take advantage of new opportunities. Provision must be made for appropriate arrangements for the different parts and levels of the system in order that the relationships between the subsystems, the system itself and its environment can best be managed. The aim is to ensure that the organization can survive and thrive over time. Two systems approaches are worthy of particular attention under this heading:

- Socio-Technical Systems Thinking
- Organizational Cybernetics and the Viable System Model

Critical Systems Thinking and the Management of Complexity, First Edition. Michael C. Jackson.
© 2019 John Wiley & Sons Ltd. Published 2019 by John Wiley & Sons Ltd.

12

Socio-Technical Systems Thinking

From the beginning the sociotechnical concept has developed in terms of systems, since it is concerned with interdependencies. It has also developed in terms of open system theory, since it is concerned with the environment in which an organization must actively maintain a steady state

(Trist 1981, p. 12)

12.1 Prologue

In Norway, in the early 1960s, the state of the economy was of growing concern to both employers and trade unions. Industry had failed to modernize, growth had declined, multinationals were taking over Norwegian companies, and the largest paper and pulp firm went bankrupt. There were increasing calls for workers' control. The Confederation of Employers and the Confederation of Trade Unions asked Einar Thorsrud, director of the Institute for Social Research at the Technical University of Norway, to carry out an investigation. Thorsrud invited Emery and Trist from the Tavistock Institute of Human Relations, in London, to become involved and, together with representatives from the two Confederations, they became part of the planning committee tasked to develop the research design. In 1962, the committee initiated "The Norwegian Industrial Democracy Project," the most ambitious attempt to humanize work and democratize industry ever proposed in the West.

The first stage of the work consisted of an inquiry into the role of workers' directors. It was found that, however well they performed, allowing workers' representatives onto the board had, on its own, little effect on levels of commitment and perceived involvement among the mass of workers. Nor did it help improve performance. Emery and Thorsrud (1969) concluded that the attempt to extend industrial democracy should instead begin at the level of the shop floor, with control over the task itself. The socio-technical systems thinking developed at the Tavistock, and particularly the idea of semi-autonomous work groups, showed how the process of democratization could be started at the bottom of an organization, where the benefits would be immediately registered, and then proceed upward. This conclusion was accepted and an ambitious scheme formulated, involving employers, unions, and gradually the government, to democratize Norwegian industry using ideas derived from socio-technical systems

Critical Systems Thinking and the Management of Complexity, First Edition. Michael C. Jackson.
© 2019 John Wiley & Sons Ltd. Published 2019 by John Wiley & Sons Ltd.

thinking. There were even hopes that the lessons learned in industry would diffuse to other sectors of Norwegian society. Four major pilots were set up, positioned in key industries, as demonstration projects from which the ideas could be spread outward (reported in Emery and Thorsrud 1976).

The pilot projects were successfully launched and, in three cases, sustained. They were deemed a "moderate success" (Bolweg 1976) and, although the expected widespread diffusion in Norway did not take place, a law was passed in 1976 granting workers the right to demand fulfilling jobs conforming to socio-technical principles and a program launched to increase trade union knowledge about technology (Mumford 2006). The main payoff actually occurred in Sweden where industry was expanding and firms were having difficulty recruiting and keeping employees. By 1973, as a direct result of studying the Norwegian results, nearly 1000 work-improvement programs had started in a whole variety of different sectors. The socio-technical design of Volvo's car plant at Kalmar became world famous. Mumford (2006) details further diffusion of the ideas, although on a smaller scale and with varying degrees of success, in Denmark, France, Italy, Germany, The Netherlands, the United Kingdom, and the United States.

12.2 Description of Socio-Technical Systems Thinking

12.2.1 Historical Development

Engineers and mathematicians were not the only scientists called to help with real-world problems during the Second World War. Social scientists, from the Tavistock Clinic in London, were engaged by the British Army to work on projects such as officer selection and the readjustment of psychologically damaged soldiers to civilian life. After the war, when the Tavistock Clinic became part of the National Health Service, a group of those involved in the wartime work, helped by a grant from the Rockefeller Foundation, were determined to continue with the practical application of social science and, in 1947, established the Tavistock Institute of Human Relations as a not-for-profit organization with charitable purpose. They had developed a particular fascination with working with groups to improve work situations from the human perspective. In their view, because of the influence of Taylorism, lower-level tasks in organizations were often degrading and offered no opportunities for personal fulfillment or development. Once the initial money ran out, the Tavistock Institute became dependent for its survival on research grants, contract work, and course fees. As a result, the assignments undertaken had to bring direct value to clients. The Institute succeeded in being useful and, as a result, has been long-lived and successful. It gave birth to the influential socio-technical systems approach which, associated with names such as Eric Trist, Fred Emery, and A.K. Rice, found application worldwide. In 1963, as mentioned in Section 9.2.5, it joined with the United Kingdom Operational Research Society to establish, within its orbit, the Institute for Operational Research – from where "Soft-OR" took flight. The journal *Human Relations*, founded in 1947 with Kurt Lewin's group at the University of Michigan, remains important. The Tavistock Institute continues in existence to this day as a not-for-profit organization "which applies social science to contemporary issues and problems."

One of the earliest projects carried out by members of the Tavistock Institute gave rise to the field of inquiry, which became known as socio-technical systems thinking (STS) thinking. The project was based in the British coal mining industry, shortly after nationalization, and studied the impact of the mechanization of the work process and its outcomes. Trist and Bamforth (an ex-miner), writing in *Human Relations* in 1951, give an account of how work was done before and after mechanization. In the traditional "hand-got method" of coal getting, prior to mechanization, small groups of skilled men worked in an essentially self-regulating and autonomous way on their own part of the coal face. They could choose who to work with and were responsible for their own pace of work. Supervision was internal and individual workers developed multiple skills. Each group made its own contract with management. The kind of social system that evolved with this form of work organization seemed appropriate to the tasks that needed doing in the dangerous and unpredictable environment at the coal-face. With the advent of mechanization, however, the traditional form of work organization was abandoned and a "conventional long-wall" method of coal getting adopted. This was a factory-like system with 40 or 50 specialists, on different pay rates, all working on a single long face. Furthermore, a three-shift system was introduced, with each shift doing a different part of the overall task. The whole system required the constant involvement of management to ensure coordination and control. This system was introduced with a view to getting the most out of the new technology. In reality, however, productivity was disappointing, absenteeism and employee turnover increased, and there were constant problems for management, especially in handling the changeovers between shifts. Trist and Bamforth concluded that the premechanized equilibrium between "interactive technological and sociological patterns" had broken down. Introducing the new form of work organization, without a thought for the social system, had resulted in extremely dysfunctional social and psychological consequences. They expressed interest in some recent innovations in work organization that had sprung up in a colliery in South Yorkshire, apparently in reaction to the deleterious impact of mechanization on the work situation. There, autonomous group working had been reintroduced but in a different form and this seemed to suggest the possibility of a new type of equilibrium between the social and technical systems, getting the best out of both. Productivity was up and there were impressive improvements "in the social quality of the work life of face-teams."

In a later study, Trist et al. (1963) were able to use the new concept of a "production unit as a *socio-technical system*" to chart, in detail and over a four-year period, more extensive developments of the same kind in the Durham coalfield. Some miners, unable to tolerate the conventional long-wall system, had originated and won acceptance by management for what was called a "composite long-wall" system:

> Here, the research team found what the conventional wisdom had held to be impossible: the working of the conventional, semi-mechanized, three-shift long-wall cycle by a set of autonomous work groups (locally known as composite).
>
> *(Trist 1981, p. 16)*

This form of work organization was able to operate the new technology efficiently and also paid attention to the needs of the social system. Demarcation between shifts disappeared and, on each shift, self-selected groups of 40 or 50 men took on responsibility for the whole task. These groups allocated work, allowed individuals to become multiskilled and were self-regulating. They were paid on a group bonus system that seemed equitable to them. Where the composite long-wall form of work organization had been introduced, productivity was 25% higher and with lower costs. Accidents and sickness decreased, and absenteeism was cut in half. The miners were generally more satisfied with their work (Trist 1981, p. 16). As the title of the book by Trist et al. proclaimed, there can be considerable "organizational choice." Given a particular technology, alternative forms of work organization are possible. If attention is paid to the social system as well as the technical system, then a choice can be made that leads to optimum performance of the system as a whole even if that requires a less than optimum state for the separate dimensions (Trist et al. 1963, p. 7).

Meanwhile, Rice (1958) was demonstrating the wide applicability of STS in his consultancy at the Calico textile mills in Ahmedabad, India. One project, in an automated weaving shed, reinforced the benefits of autonomous group working. The overall process had been subdivided into small, specialized tasks and workers allocated to the tasks according to scientific management principles. The aggregate of individuals that resulted was confused as to relationships and responsibilities, with chaos ensuing if any sort of unexpected event occurred. Rice suggested a reorganization of the work based around those groups of tasks and roles necessary to keep the looms running. This would permit internally structured and led groups of seven workers to take collective responsibility for 64 looms as well as carrying out the necessary ancillary tasks. The results were spectacular:

> This reorganization was spontaneously and enthusiastically accepted by the men and despite some setbacks led to an increase in efficiency and less damaged cloth. In addition, management of the shed was considerably simplified by the existence of team leaders responsible to the supervisor.
>
> *(Brown 1967, p. 48)*

The later extension of his work in India to whole organizations led Rice to develop existing socio-technical thinking about enterprises as "open systems" related to their environments by their "primary task," and to produce significant original contributions around leadership as a boundary function, subsystems with their own primary tasks, and relationships between the principle "operating systems" of an organization and service and control systems (Rice 1963).

The high-water mark of STS, in both scale and spread of application, occurred in the 1960s and 1970s. The chapter began with a glimpse at the ambitious Norwegian Industrial Democracy Project. In countries where the economic and labor relations conditions were favorable, such as Sweden, numerous smaller experiments followed that example. Famously, Volvo's car factory at Kalmar, designed according to socio-technical principles, became a major tourist attraction for managers. In this factory, opened in 1974, there was no assembly line. At a time of very high employment, it was built as a factory where people would want to work and stay working. It was organized to allow car manufacture by semi-autonomous groups of around 20 people (Gyllenhammer 1977). There were around 30 such groups. Design changes made it possible for each to be given responsibility for an identifiable part of the car. A self-propelled vehicle – the "Kalmar carrier" – transported vehicles around the various areas of the factory, controlled by the different work groups, as they were being built. Each group could pace and organize its own work, was responsible for its own inspection, and for making its own contract with management. Kalmar was about 10% more expensive to build than an equivalent ordinary factory, but Volvo regarded this as worthwhile given the better productivity and lower staff turnover and absenteeism that resulted.

The projects in Scandinavia saw established socio-technical principles consolidated and also new ideas coming to the fore. Socio-technical theory became a vehicle for promoting industrial democracy. With Kalmar, joint optimization of the social and technical systems embraced significant changes in technology for the first time. As will be seen later, a lot of thought was also given to the detail of job design and to the nature of the environments in which organizations were operating. Finally, a nine-step methodology for putting socio-technical thinking into practice emerged.

The 1960s also witnessed the initiation of Enid Mumford's important research program using STS in information-systems design and implementation (Mumford 2006, p. 318). This was research that contributed significantly to the development of information systems as an academic discipline. Computers were being introduced into the office environment to handle sales accounts, stock control, invoicing, and other simple information processing tasks. The impact on the working lives of lower-level white-collar workers was considerable. If the staff involved were going to welcome the technology and use it effectively, Mumford reasoned, they had to be involved in its introduction and in the reorganization of work required. This would ensure that proper account was taken of their local knowledge and help convince them that the technology would bring improvements in their working lives and not simply result in deskilling. Mumford developed a close relationship with the Tavistock Institute and took on board STS ideas concerning participative design, giving equal weight to the social system, job design, and autonomous group working. Her skill was in adapting these ideas and applying them, using her ETHICS (Effective Technical and Human Implementation of Computer-based Systems) methodology, to the new context. ETHICS was not just a neat mnemonic, it signaled that ethical and social issues were central to her work. For Mumford:

> Socio-technical systems design provides a new worldview of what constitutes quality of working life and humanism at work. It facilitates organizational innovation by recommending the removal of many elite groups and substituting flatter hierarchies, multi-skilling and group decision-taking. It wants to replace tight controls, bureaucracy and stress with an organization and technology that enhances human freedom, democracy and creativity.
>
> *(Mumford 2003, p. 262)*

In the 1980s, Mumford undertook projects involving the introduction of information technology in hospitals, banks, and major manufacturing firms such as ICI and Rolls Royce. A project with the Digital Equipment Corporation in the United States used socio-technical ideas to launch expert systems in manufacturing plants and sales offices.

Despite the successes reported for STS interventions in improving productivity and employee satisfaction, and the continued efforts of Mumford and others, the 1980s and 1990s were a disappointing time for exponents of the approach (Mumford 2006). Challenging international markets encouraged employers to take the easy option of cutting costs rather than embarking on expensive and risky socio-technical experiments. It was an era of "lean-production," "downsizing," and "computer-aided neo-Taylorism." STS morphed into the "quality of working life" movement, with its much less radical aspirations, and reappeared in a very watered-down form in the idea of "quality circles."

Partly as a result of reflecting on the failure of STS initiatives to sustain themselves and to proliferate as widely as expected, Emery and Trist had moved on as well. In the 1970s, they began to concentrate on developing "change strategies" that would help organizations accept the "new paradigm" of thinking needed to cope in increasingly "turbulent environments." Their major concern became organizational purposes, values, and philosophies (Trist 1981, p. 45). The theoretical groundwork was laid in the book *On Purposeful Systems* (Ackoff and Emery 1972). Emery had already been developing the "Search Conference" as a methodology for operating at this level. Trist joined Ackoff at the Wharton School, University of Pennsylvania, in 1969 and worked with him on the influential "Social Systems Sciences" (S-cubed) program until his nominal retirement in 1978. In the 1990s, Emery and his wife Merrelyn, working in Australia, continued to run Search Conferences and added the "Participative Design Workshop" to the armory of STS methods. Merrelyn Emery has stayed with these endeavors since Fred's death in 1997. The fact that Trist and Emery worked so closely with Ackoff (see Chapter 15) indicates a radical reorientation in their thinking which will be reviewed in what follows. The continued relevance of the traditional version of STS will be considered in Section 12.5.

The Tavistock Institute has carried out some very important work but has not been quite as influential, I fancy, as some conspiracy theorists would have us believe. Coleman's (2006) *The Tavistock Institute of Human Relations: Shaping the Moral, Spiritual, Cultural, Political and Economic Decline of the USA* has a title that speaks for itself. The Institute is held to be in the frontline of an attack on the United States and its constitution, to have led the country into wars, and employed rock music and drugs – having launched the Beatles – to undermine education and the family unit. Estulin's (2015) *Tavistock Institute: Social Engineering the Masses* (2015) is in a similar vein, accusing the Tavistock of mass brainwashing.

12.2.2 Philosophy and Theory

The early Tavistock researchers were much influenced by Bion's work on leaderless groups and Lewin's experiments on the superior performance of democratic groups. As Trist later put it:

> Both emphasized the capacity of the small group for self-regulation, an aspect of systems theory which received increasing attention as cybernetics developed.
>
> *(1981, p. 14)*

Psychoanalysis and *Gestalt* psychology also impinged on their thinking. Of huge significance, as will be seen, was Lewin's development of and commitment to "action research" as the most appropriate means to intervene in and gain knowledge of social systems. Overall, these influences pushed the Tavistock thinkers toward a multidisciplinary and holistic approach to human problems. Holism, of course, can assume a number of forms and STS thinking has taken on mechanical-equilibrium, organismic, and purposeful systems orientations over its history.

One variant of holism, as we saw in Chapter 4, is the mechanical-equilibrium model associated with Pareto, Henderson, and Roethlisberger and Dickson. Trist and Bamforth (1951) were working within this tradition in the early coal-mining studies. They argued that the "premechanized equilibrium" had been disturbed by the long-wall method of coal getting as a result of "the magnification of local disturbances," and that a new equilibrium was needed that respected the social system as well as the technical system. However, as Trist commented in 1981:

> In our action research projects at that time, we and our organizational clients were baffled by the extent to which the wider societal environment was moving in on their more immediate concerns, upsetting plans, preventing the achievement of operational goals, and causing additional stress and severe internal conflict.
>
> *(Quoted in Scott 1987, p. 108)*

It was von Bertalanffy's concept of "open systems" (see Section 3.2), popularized in his 1950 article *The Theory of Open Systems in Physics and Biology*, which resolved this conundrum for them. Work organizations were not "closed socio-technical systems" but "open socio-technical systems" in constant matter-energy exchanges with their environments. This organismic analogy soon began to dominate. By 1960, Emery and Trist were rethinking all the major concepts of STS in open systems terms and stated:

> Considering enterprises as 'open socio-technical systems' helps to provide a more realistic picture of how they are both influenced by and able to act back on their environment.
>
> *(In Emery 1969, p. 293)*

If the 1963 report, on the later coal-mining studies, shows vestiges of the old "mechanical equilibrium" language, the authors are nevertheless clear that

> ... the socio-technical concept requires to be developed in terms of open rather than closed system theory, especially as regards the enterprise-environment relation and the elucidation of the conditions under which a steady state may be attained...
>
> *(Trist et al. 1963, p. 6)*

Rice's book of the same year manages a clean break and uses von Bertalanffy's open systems theory to build a complete picture of enterprises as living organisms interacting with their physical and social environments in order to survive:

> This book seeks to establish a series of concepts and a theory of organization that treats enterprises – industrial, commercial, government, academic, or social – as living organisms. An enterprise, like an organism, must work to live .... The more complex the enterprise, the more complex the organization that is necessary to relate the parts to the whole, and the whole to its environment. The theory

outlined here is based on organic concepts. Its language is the language of process and growth rather than that of mechanics.

(Rice 1963, p. 179)

Many commentators continue to identify STS with the classic case studies of the 1960s and 1970s, which were dominated by organismic thinking. It is not unreasonable to see this phase as generating the most exciting intellectual developments and the most rewarding interventions. Nevertheless, there is a later stage to which we need to give attention.

The exact date when Emery and Trist began to shift from "organismic" to "purposeful" systems thinking is not easy to pin down. Their 1965 article "The Causal Texture of Organizational Environments" (1969a) was certainly a breakthrough moment. Systems thinking had always understood internal or L 1–1 interdependencies, where "L" indicates some potentially lawful interconnection and "1" the organization. von Bertalanffy had conceptualized organization – environment interactions in both directions, called L 1–2 and L 2–1 transactional interdependencies, where "2" refers to the environment. But, Emery and Trist insisted, a comprehensive understanding of organizational behavior also requires an understanding of the L 2–2 interdependencies that become established in the environment itself. This extra concept, "the causal texture of the environment," is clearly necessary because L 2–2 interactions are among the determining conditions of L 1–2 and L 2–1 exchanges. If environmental interdependencies become complex, it is impossible for individual organizations to predict the consequences of their actions. They can try to heighten their own adaptive capabilities, but this may not be enough. Ultimately, they will need to seek joint solutions with other organizations. The aim became to change "turbulent field" environments, produced by L 2–2 interactions, into a more manageable form through

> ... *adaptive cultural change* at all societal levels. This requires a shift from a society based on hierarchy to a *participative democracy* where **all** systems want to be and are purposeful and responsible.
>
> *(Emery M. 2000, p. 629)*

Somehow, collaborating enterprises would have to develop shared values and purposes. Thinking through how this might be achieved requires a "purposeful" systems perspective that can generate an "ideal seeking" capability (Emery 1981a).

Another significant step beyond von Bertalanffy's general systems theory (GST), toward a completely conceptualized "open systems theory" (OST) of a purposeful type, was Fred Emery's discovery, in 1967, of the "genotypical organizational design principles":

> There were only three choices of organization: responsibility for coordination and control is located with the actors, the second design principle (DP2) yielding equality; responsibility for coordination and control is not located with the actors, the first design principle (DP1) yielding inequality; or there is no coordination and control (laissez-faire), that is, no organization at all.
>
> *(Emery M. 2011, p. 403)*

This insight allowed certain anomalies to be explained and, according to Merrelyn Emery, it became possible for STS to abandon its current "normal science," in the Kuhnian sense. As Fred Emery put it:

> When, in Norway, we brought forward **our** studiously thought-out plans for sociotechnical redesign and allowed the workers to vote on them *we* made the question of trust the foremost issue. In each instance, our designs were patently fairer and more liberating than existing management practices but we had done the same thing that management was always doing – we scorned their knowledge. We learnt the hard way.
>
> *(Quoted in Emery M. 2010, p. 709)*

Clarity about its commitment to DP2 changed everything and allowed a paradigmatic break in STS from "organicism" to "contextualism," a world hypothesis that allowed for true novelty and emergence (Emery M. 2010, p. 701). The first purpose of OST could now be stated as being to

> ... promote and create toward a world that is consciously designed by people, and for people, living harmoniously with their ecological systems, both physical and social. Included within this is the concept of open, jointly optimized, sociotechnical (and sociopsychological) systems, optimizing human purposefulness and creativity, and the best options afforded by changing technologies. Again these organizational systems are designed by people themselves.
>
> *(Emery M. 2010, p. 623)*

These two milestones on the road to purposeful systems thinking were consolidated theoretically in Ackoff and Emery's (1972) book *On Purposeful Systems.* The aim was to lay a new foundation for looking at human and social behavior in terms of purposeful systems that could change or select both their own goals and the means of attaining those goals. This required an examination of perception, consciousness, memory and belief, and the model or representation of a situation to which they give rise. That representation can lead to dissatisfaction with the current situation and the identification of a problem. Problems can only be resolved by ideal-seeking systems working in cooperation with other ideal-seeking systems. This means that conflict between purposeful systems must be eliminated. Trist, drawing upon Ackoff's thinking, sums up nicely what is involved in the shift to purposeful or open systems theory:

> Traditional organizations serve only their own ends. They are, and indeed are supposed to be selfish. The new paradigm imposes the additional task on them of aligning their own purposes with the purposes of the wider society and also with the purposes of their members. By so doing, organizations become both 'environmentalized' and 'humanized' – and thus more truly purposeful – rather than remaining impersonal and mindless forces that increase environmental turbulence.
>
> *(1981, p. 43)*

We are now in a position to consider the main theoretical concepts used in STS, bearing in mind these developments in the underlying philosophy of the approach.

The first, foundational principle came to the fore in the early coal mining studies. Taylorists had emphasized the technical elements of work systems; human relations theorists the social. Socio-technical thinkers concluded that work groups or organizations should be regarded as *interdependent socio-technical systems*. They have interactive technological and social aspects, and in designing the structure of the group or organization both of these should be considered. If the structure of the work organization is designed with only the technology in mind, then it may be disruptive of the social system. If it is designed with only the social system in mind, it is unlikely to make very good use of the technology. In either case, the overall outputs will not be optimum. It is possible to give proper attention to the social system because, as Trist et al. (1963) argued, there is "organizational choice." The miners themselves had demonstrated that there was no "technological imperative" by developing a form of work organization that defied Tayloristic and bureaucratic principles. They showed that it is possible, within the same technological constraints, to operate with different forms of work organization producing different social and psychological consequences. So, given the constraints, managers should exercise choice over the type of work organization to adopt taking the social system into account. They would then achieve positive economic as well as human results. According to Trist, Emery was able, at the time of the Norwegian Industrial Democracy project, to use the more advanced systems theory that had become available to conceptualize these ideas appropriately as the *joint optimization* of the social and technical systems:

> The technical and social systems are *independent* of each other in the sense that the former follows the laws of the natural sciences while the latter follows the laws of the human sciences and is a purposeful system. Yet they are *correlative* in that one requires the other for the *transformation* of an input into an output, which comprises the functional task of a work system. Their relationship represents a *coupling* of dissimilars which can only be jointly optimized. Attempts to optimize for either the technical or social system alone will result in the suboptimization of the socio-technical whole.
>
> *(Trist 1981, p. 24)*

In the early socio-technical studies, in spite of these notions, it was usually the case that the existing technology was accepted, with the real adjustments being made to the work organization and the social system. In the later work there were genuine changes to technology in order to realize the true meaning of joint optimization. In building Volvo's car factory at Kalmar, for example, the technology was designed to facilitate group working.

The technical and social systems had to be jointly optimized to achieve the *primary task* of the group or organization – another key concept of STS. If this is done effectively, the "economic dimension" will also be optimized. The primary task, therefore, integrates "the technological, economic and socio-psychological aspects of a production system" (Trist et al. 1963, p. 20). The meaning of the idea of primary task changed as STS evolved. Rice, in line with traditional management thinking, defined it in 1958 as "the task an institution had been created to perform" (Rice 1963, p. 185). Once the organismic analogy took hold, inspired by von Bertalanffy, the primary task became the "key transaction" relating an operating system to its environment and allowing it to maintain a "steady state" (Trist et al. 1963, p. 20). Rice recognized that complex organizations were differentiated into subsystems each of which would have its own primary task. The mission, or overall objective, of the enterprise would then consist of all the tasks necessary for its survival. In the face of incompatible tasks, a compromise would have to be reached if the overall organization was to achieve its mission. This could be difficult for organizations such as the prison service where at least three apparently appropriate primary tasks could be identified – punishment, confinement and rehabilitation. Anticipating Checkland's "soft systems methodology" (see Chapter 16), he argued that, in such cases, a model needed to be constructed for each task (Rice 1963, pp. 185–191). Once Emery and Trist began to move STS onto purposeful systems foundations, it was quickly established that the main problem was reconciling the multiple "tasks" legitimated by the different purposes that could arise at subsystem, system and supra-system levels.

Trist and Bamforth (1951) were convinced that working in small groups in the underground situation of the mining industry produced greater adaptability and enhanced job satisfaction. These groups, which the later study (Trist et al., 1963) showed could be 40 or 50 strong, were able to tackle whole tasks in a more flexible manner and yielded more meaningful work for individuals. The establishment of *semi-autonomous groups* as the basic unit of work organization became, therefore, another important idea associated with STS. Trist (1981, pp. 9, 34) suggests that the efficacy of autonomous work groups is based on the cybernetic concepts of *variety* and *self-regulation*. Within the groups, job rotation is encouraged and workers become multi-skilled. Such "redundancy of functions" is "organic" and allows for considerable adaptive flexibility. Individuals and the group as a whole can display an elaborate array of responses to changing circumstances. This contrasts with the "mechanistic" approach, based upon "redundancy of parts," in which jobs are narrowly defined and closely monitored by management; an approach that can only be efficient if internal and external changes are predictable in advance. In short, "redundancy of functions" increases variety, while the traditional, bureaucratic mode of organization decreases it. With "redundancy of functions," groups gain *equifinality*, the ability to reach a steady state from differing initial conditions and in differing ways (Emery and Trist 1969b, p. 293). Further, no individual is ever without a job – if you finish your own, you can always help with someone else's. When semi-autonomous groups are in place, decision-making and control are exercised internally by the groups not externally by management. There will be instruction about what to do but not how to do it. This capacity for self-regulation enables problems, or "variances" to be dealt with quickly, as they arise, and by those as near as possible to the point where they occur. Better results and increased job satisfaction ensue. Groups negotiate contracts with their supply and user departments, and some of the details of their own labor contract with management. Semi-autonomous groups, exhibiting high variety and self-regulation, become self-developing social systems, capable of maintaining themselves in a steady state of high productivity. In "healthy and maturing" systems of this kind there is constant growth in capability to achieve the mission. In Trist's words:

> Autonomous groups are learning systems. As their capabilities increase, they extend their decision space. In production units they tend to absorb certain maintenance and control functions. They become able to set their own machines.

> The problem-solving capability increases on day-to-day issues.... As time goes on, more of their members acquire more of the relevant skills.... The overall gain in flexibility can become very considerable, and this can be used to enhance performance and also to accommodate personal needs as regards time off, shifts, vacations, etc.
>
> *(1981, p. 34)*

The next important STS concept is directly related to the setting up of semi-autonomous work groups. Because these groups coordinate and control themselves, they do not have to be regulated from the outside. This frees up managers to undertake the essential role of *boundary management*. The task of relating the system to its environment is crucial to the survival and effectiveness of open systems. Managers can now devote their energies to the exchanges occurring across the boundary between the system and its environment, ensuring that the group doing the work is supplied with the necessary inputs and that its outputs are those required by its customers. In a complex organization, of course, boundary management will be necessary to serve the subsystems carrying out discrete parts of the primary task as well as to ensure the effective performance of the system in its overall environment. Rice (1963) discusses how "line" subsystems supporting the primary task can be differentiated according to the dimensions of technology, territory and time, and then suggests how various specialist control and service functions should be designed to manage the relationships between these operating systems and supply them with their needs. It is the job of leadership, located on the boundary between the enterprise as a whole and the external environment, to take responsibility for these matters. Commenting in 1981, however, Trist is still not certain that STS has got this right:

> Self-standing primary work systems exemplify a holographic principle of organization in which the whole is represented in the part. The forms through which holographic primary work systems may best become linked to the overall organization constitute an area requiring further research.
>
> *(Trist 1981, p. 37)*

Certainly a dialogue with organizational cybernetics (see Chapter 13) would be beneficial to both these strands of the systems movement.

About the time of the Norwegian Industrial Democracy project, it became clear that achieving joint-optimization of technical and social systems demanded a clearer conception of the psychological requirements individuals have of their work. Concentrating on the "intrinsic" characteristics, rather than the "extrinsic" characteristics included in the employment contract, and drawing on the work of Lewin and Bion, Emery identified six principles of *job-design* that became central to STS. The first three relate to job content and the second three to the social climate of the workplace. Emery's six principles, as set out in Trist (1981, pp. 29–30), are as follows:

- For the content of a job to be reasonably demanding in terms other than sheer endurance and to provide some variety (not necessarily novelty)
- To be able to learn on the job and go on learning. Again it is a question of neither too much nor too little

- For an area of decision-making that the individual can call his own
- For a certain degree of social support and recognition in the work place for the value of what he does
- To be able to relate what he does and what he produces to his social life, for it to have meaning and to afford dignity
- To feel that the job leads to some sort of desirable future (not necessarily promotion)

A little later, Fred Emery and Merrelyn Emery (1974) placed these principles at the center of their "Participative Design Workshop" approach to shifting organizations from bureaucratic structures, based on the DP1 principle, to "ideal-seeking" systems expressing the DP2 principle. In broad terms, these workshops consist of three phases. First, participants work in groups to determine how their job is currently done and the extent to which it falls short of the six job-design principles. Second, having been informed about the nature of democratic structures, they redesign their way of working to better accord with the six principles. Finally, they work on implementation using a participative learning process.

Our final key STS concept is the *causal texture of organizational environments*. As we saw earlier, Emery and Trist (1969a) accepted von Bertalanffy's open systems formulation but felt that it neglected to deal with processes in the environment that are, themselves, among the determining conditions of organization-environment relations. They therefore added an additional concept – the causal texture of the environment. This refers to the degree of system-connectedness that exists in the environment itself. Emery and Trist isolate four ideal types of causal texture, forming a series in which the degree of causal texturing increases. First, there are "placid-randomized" environments in which there is no connection between the parts of the environment and the environment is homogeneous in character. Second are "placid-clustered" environments in which there is still no connection between environmental parts, but the environment is diverse with certain resources in certain places (so the organization must know its environment). Third are "disturbed-reactive" environments. These are dynamic environments in which a number of organizations of the same type compete. There is connection between environmental parts and each organization has to take account of the others. Finally, there are "turbulent fields." With the increasing interaction of organizations and interconnectedness of the environment, powerful dynamic properties arise not only from the interaction of the component organizations but also from the environment itself. The environment takes on its own momentum. For example, timber enterprises, in the course of competing with one another, may overexploit the available timber, encouraging soil wash and erosion and making regeneration of timber resources impossible. Emery and Trist argue that environments are becoming increasingly to resemble turbulent fields. This makes management extremely difficult since uncertainty for organizations is increased as the consequences of their actions become unpredictable. Organizations must abandon bureaucracy and adopt flexible structures to increase their adaptive capabilities. Even this may not be enough. Individual organizations will have to enter into joint collaboration with other organizations to seek solutions. The development of a set of values that can be shared by organizations will be important in this.

As these last points make clear, the arguments of Emery and Trist's paper took socio-technical thinkers beyond the mere reconsideration of organization-environment relations, and on to looking at how they could help organizations adopt goals and values suitable for turbulent field situations. It becomes essential, says Trist referencing Vickers (see Chapter 16), "to secure an *appreciation* of the issues at the highest level of the corporation or agency" (1981, p. 45). The new thinking had immediate application because, as soon as the 1965 article (Emery and Trist 1969a) was completed, the Tavistock researchers became involved in shaping a "new philosophy of management" for Shell UK. This intervention is described at length as our example of "STS in Action" below.

12.2.3 Methodology

From the start the Tavistock researchers were committed to "Mode 2" research, carried out in the context of application and produced to satisfy the demands of users, rather than "Mode 1," the "pure science" model (see the introduction to Part III). Partly this was because they needed to satisfy clients in order to pay their way. But it was also based on their belief that the social sciences had not advanced sufficiently to boast a set of generally accepted theories that could be taken off the shelf and applied in practice. In these circumstances it was better to follow the "professional" model, adopted in medicine, engineering and architecture, beginning with practice, working back to theory, and then using the theory developed to further improve practice (Brown 1967, p. 40). To this end, they could draw on the work of Kurt Lewin with which they were already very familiar. As was seen in Section 4.8, Lewin had developed an approach called "action research" designed to serve at the same time the needs of organizations and of scientific rigor. This research process requires the involvement throughout of those actually in the social situation that is being addressed. The Tavistock researchers can be regarded as Lewin's heirs in pursuing and developing action research. Sometimes the work began by emphasizing research, as in the coal-mining studies, sometimes action, as in Rice's consultancy at Ahmedabad, but always attention was eventually directed to both. It was a prolonged action research program on which they became engaged.

It is now possible to see how the key concepts of STS were incorporated into the action research framework to produce methodologies. Three examples are provided – the "nine-step model," Mumford's "ETHICS" methodology, and the "Search Conference" approach.

The nine-step model, specifically for the socio-technical analysis of production systems, emerged during the Norwegian Industrial Democracy project when, in one of the pilots, an "action" group of workers, technicians, and supervisors was created to do the diagnosis of problems. It was first used to train departmental managers in the Shell intervention. Trist (1981, p. 33) provides a summary:

1. An initial scanning is made of all the main aspects – technical and social – of the *selected target system* – that is, department or plant to be studied
2. The *unit operations* – that is, the transformations (changes of state) of the material or product that take place in the target system – are then identified, whether carried out by men or machines
3. An attempt is made to discover the *key variances* and their interrelations. A variance is key if it significantly affects (i) either the quantity or quality of production, and (ii) either the operating or social costs of production
4. A table of variance control is then drawn up to ascertain *how far the key variances are controlled by the social system* – the workers, supervisors, and managers concerned.

Investigation is made of what variances are imported or exported across the social-system boundary

5. A separate inquiry is made into *social-system members' perception* of their roles and of role possibilities as well as constraining factors
6. Attention then shifts to *neighboring systems*, beginning with the support or maintenance system
7. Attention continues to the *boundary-crossing systems* on the input and output side – that is, supplier and user systems
8. The target system and its immediate neighbors are then considered in the context of the *general management system* of the organization as regards the effects of policies or development plans of either a technical or social nature
9. Recycling occurs at any stage, eventually culminating in *design proposals* for the target and/or neighboring systems

Enid Mumford's ETHICS methodology was developed in the late 1960s as an attempt to influence the course of the impending information technology revolution. The methodology demands the participation of all stakeholders in a design process based upon socio-technical principles. If followed, she believed, it yields the greatest returns in terms of productivity as well as ensuring that new technologies are introduced in an ethical manner and used to increase rather than diminish human freedom and self-determination. Mumford has presented ETHICS as a 7, 15, and 22 stage methodology at various times. The common starting point is the establishment of a steering committee and design group, both representing all major interests. In the 15 stage version (Mumford 1983), the design group proceeds through the following steps:

1. Consider why change is necessary
2. Identify the boundaries of the system to be designed and its interfaces with other systems
3. Description of the existing system
4. 5 & 6. Definition of key objectives and tasks, along with their key information needs
7. Diagnosis of efficiency needs
8. Diagnosis of job satisfaction needs
9. Future analysis to determine a better and more flexible version of the existing system
10. Specifying and weighting efficiency and job satisfaction needs and objectives
11. Evaluation of the organizational design options according to efficiency, job satisfaction, and future change objectives. This step embraces the socio-technical principles of good job design, multi-skilling, and self-managing groups
12. Evaluation of the technical options against efficiency, job satisfaction, and future change objectives. The organizational and technical options are now merged to ensure compatibility and are evaluated against the primary objectives and the one that best meets the objectives is selected. This selection is performed by the design group with input from the steering committee and other interested constituencies
13. The preparation of a detailed work design
14. Implementation
15. Evaluation

The Search Conference, first developed by Emery and Trist as a participative strategic planning method, came into its own following the publication of their 1965 article on the causal texture of organizational environments (Emery and Trist 1969a). Emery and Trist argued that, in unpredictable "turbulent field" environments, it becomes necessary for organizations to change their philosophies and adopt strategies which reduce uncertainty either by facilitating their own coevolution with their environments or by developing shared values with other enterprises giving rise to common responses to turbulence. The Search Conference seemed to offer a methodology that could be used at the highest levels in organizations to address these most important issues. As Emery put it:

> Values, ideals and broad encompassing social processes are the main fare of search conferences, as they are the main features of social change.
>
> *(1981b, p. 466)*

The methodology has been refined over the years by Fred and Merrelyn Emery as they developed "open systems theory," beyond the work of von Bertalanffy, as part of STS's "purposeful systems" turn. From the 1970s to the present day, there have been hundreds of applications worldwide. Examples include work with Hewlett-Packard, with mental health professionals in Nebraska, with citizens and community leaders in Michigan, with water engineers in Colorado, and to help plan community health promotion actions in Norway. The following brief account of the approach is derived from Rehm and Cebula (1996), Emery and Purser (1996), Cabana et al. (1997), and Magnus et al. (2016).

The Search Conference is a participative planning methodology that can be used by single enterprises, groups of organizations, and whole communities to plan for a desirable future. In line with open systems theory, it focuses attention on the relationships any system has with its wider environment. In the case of an organization, it helps it to understand its current environment and develop action plans that enable it to integrate with its environment in an active adaptive relationship. With multiple organizations, the approach seeks to bring about a shared appreciation of the uncertainty faced and helps to establish sustainable relationships between the organizations themselves and with their global environment. The aim, following purposeful systems principles, is always to enable those involved to express their own goals and values and then agree upon collective human ideals, which they can pursue together to create a future worth living in.

Successful search conferences require considerable preplanning. This is usually undertaken by "search conference managers," consultants from outside the system of concern, together with several individuals from the system itself. This planning group is responsible for the extremely important task of selecting who is to participate in the event. The recommended size for a search conference is 20–50 people. In the case of a single organization, the chosen will be those who have responsibility for the overall direction of the enterprise. In the case of a community, a "community reference system" is employed which allows the community itself to decide who should participate. The planning group must also collect and distribute to participants any reports, statistics, expert information, results of interviews with stakeholders, etc. relevant to the system and its environment. They may also commission new research. Participants are fully briefed in advance about the nature of the search conference. Most search conferences are planned to last two days and two nights (Cherns 1976), although Magnus et al. (2016) report on a successful one-day event with a modified structure. During the conference itself, the search conference managers do not engage in the content work. Their role is to ensure that an open and safe atmosphere exists in which a fully demo-cratic dialogue can take place, disagreements can be freely expressed, and out-of-the box thinking can occur. Considerable advice exists in the literature about how this can be achieved and how a self-managing structure can be created. Given an appropriate event environment, the search conference process consists of three stages:

1. Learning about the turbulent environment; taking account all perceptions of the global environment and how it is likely to play out in the future
2. Learning about our system, its history, its current functioning, and agreeing upon its most desirable future
3. Action planning in which all those involved develop and agree upon specific strate-gies, action plans, and steps to achieve the desirable future by building a proactive and flexible relationship with the environment

Sometimes the work is conducted in small groups, sometimes as a whole conference community. After the Search Conference, all those who have participated must ensure that what they have agreed is implemented. Participative design sessions should be organized to involve and spread the learning to other individuals, groups and, if neces-sary, organizations. Participants must continue to scan the environment so that an adaptive relationship with it is maintained.

12.2.4 Methods

Cherns' (1976) sets out a nine-principle checklist for socio-technical design, which acts as a reminder of the key "methods" employed in the approach:

- *Compatibility*: The process of design must be compatible with its objectives (so if the aim is a democratic organization, the design process must be participative)
- *Minimal critical specification*: Of the way in which the work is actually carried out and who should carry it out; only the essentials are decided a priori
- *The socio-technical criterion*: Variances from specifications are to be controlled as near to the point where they arise as possible
- *The multifunction principle*: To provide for flexibility and equifinality, each individual should be able to perform more than one function
- *Boundary location*: Control of activities in a department should become the respon-sibility of the members, with the supervisor concentrating on boundary activities
- *Information flow*: Information systems should be designed to provide information, in the first place, to the work teams who need it for task performance
- *Support congruence*: Systems of social support should reinforce the organizational structure (so, if it is based on group working, payment should be by group bonus, etc.)
- *Design and human values*: High-quality jobs based on the six design characteristics
- *Incompletion*: Design as an iterative process (once at the end, one must go back to the beginning again)

Cherns added a tenth principle in revising these guidelines in 1987. This principle, "power and authority," states that those who need resources to carry out their tasks should have the authority to command them and must accept responsibility for their proper use.

12.3   Socio-Technical Systems Thinking in Action

This case study documents a STS project of the 1960s, Shell UK's "New Philosophy of Management," which was ahead of its time and has considerable contemporary relevance. It demonstrates the full panoply of the newly developed socio-technical concepts in action, making use of ideas primarily from the organismic but, to some extent, also from the purposeful systems phase of thinking. It draws upon the accounts of the experiment provided by Hill (1971) and Blackler and Brown (1980).

In the early 1960s, Shell UK was a company confronted with some very pressing external and internal problems. Externally it was faced with the beginnings of OPEC, rapid technological change, and the birth of the ecological lobby. A rethink of its traditional values was necessary. Internally, the main problems were of an industrial-relations nature, with a multiplicity of unions complicating wage negotiations, frequent demarcation disputes, over-manning, excessive overtime working and supervisors who felt they were losing control to the shop stewards. A 1964 "rundown" of the company, including some dismissals, had created bad feeling among some employees. Shell was, therefore, hampered in adapting to the turbulent environment it faced. In 1965, a special study group headed by Hill proposed to top management some radical solutions for getting the company out of the difficulties it faced. These involved, firstly, promulgating a new philosophy of management throughout the company that would help change attitudes; and, secondly, attending to the conditions of work of employees, improving these as part of productivity deals with the unions that would require greater flexibility and less demarcation on the employees' side. As soon as the recommendations were accepted by top management, socio-technical researchers from the Tavistock Institute were brought in to help develop the philosophy and to consider how it could best be diffused. To these researchers, the Shell situation was ideal for trying out their ideas on how organizations should be managed to deal with turbulent field environments. They were committed to encouraging a new way of thinking, introducing flexible structures, and using the methods of implementation developed during the Norwegian Industrial Democracy Project.

A statement of "objectives and philosophy" (issued in May 1966) saw the primary objective of the company, maximizing its contribution to the long-term profitability of the Shell Group, as subject to certain social objectives. For example, the statement declared that the resources the company used were community resources and should be employed to contribute to the satisfaction of the community's need for products and services. In addition, employee potential had to be enhanced, the safety of employees and the public given high priority, and pollution of the environment minimized. The document then went on to spell out the socio-technical requirement for joint optimization of the technical and social systems, although certain features of the technical system were regarded as fixed for the foreseeable future, and to detail the psychological requirements that related to the content of jobs. The philosophy was diffused throughout the organization in a series of conferences held between 1965 and 1967 – first for top management, then for senior staff, and finally for lower staff levels: foremen, supervisors, and union officials. To give some indication of the effort that went into this, the philosophy conferences for senior staff were two-and-a-half-day events involving around 20 people. At the time, Hill believed that the conferences were getting across the message of the philosophy and its implications.

To complement and reinforce the lessons of the conferences, four channels of implementation were opened. Demonstration projects were set up at three locations aiming to show the power of STS in action. Second, departmental managers were charged with the task of acting as change agents and taught the nine-step method of socio-technical analysis. They were required to motivate their staff with the philosophy and to encourage spontaneous job-design experiments. Early reports gave the impression of great activity in this area and new projects did start, but it seems that enthusiasm soon waned. One initiative, in the wax department at the Stanlow refinery, was played up as a great success story but came later, unfortunately, to be seen as something of a disaster and got the philosophy a bad name in certain quarters. The third leg of implementation was the productivity deals with the unions, which offered improvements in working conditions in tune with the philosophy (including staff status for all workers) in exchange for increased efficiency, greater flexibility, and less demarcation. These were successfully negotiated in 1968, the philosophy playing a very significant part in ensuring agreement and easing introduction. Finally, the design of a new refinery at Teesport was heavily influenced by STS. At this green-field site, genuine joint optimization of the social and technical systems was possible, the job design criteria for satisfying work were observed, and single-status employment was introduced.

Reaching a balanced assessment of the Shell experiment is not easy, as we will see when we consider Blackler and Brown's criticisms in the next section. There can be no doubt, however, that it was remarkable for its time – for the sophistication of the philosophy developed, the socio-technical ideas employed, and the considerable and genuine enthusiasm generated among many who were involved. In Hill's view, the philosophy brought many positive changes. In the short term, it did something to restore the legitimacy of the company in the eyes of its employees, and helped the firm get through a period of industrial relations difficulties and successfully negotiate new productivity deals. In the longer term, it provided great learning opportunities from which other projects benefited. For example, Shell Canada's Sarnia plant in Western Ontario was built, in 1978, on STS principles after a "collective contract" had been agreed with the unions.

12.4 Critique of Socio-Technical Systems Thinking

An analysis by Pasmore et al. (1982) of 134 socio-technical studies shows the remarkable success that can be achieved using STS. For example, 53% of the studies employed "autonomous work groups" and, of these, 100% reported success in terms of attitudes (the social dimension) and 89% in terms of productivity (the technical dimension). Forty percent of the studies encouraged "skill development," of which 94% witnessed improved attitudes and 91% improved productivity. In the 9% of studies where "minimal critical specification" was emphasized, 100% saw both improved attitudes and productivity. Overall, 18 "common socio-technical measures" were reviewed with a reported average success rate of 94% for attitudes and 87% for productivity.

The measures reviewed by Pasmore et al. were all developed during the classical, organismic phase of STS. It seems clear that viewing organizations as open, socio-technical systems provides a much richer picture than that supplied by the traditional and human relations models. The traditional, Tayloristic approach emphasizes organizational goals, structure, and technology but largely ignores the human subsystem. It is also based on a closed-system view, saying nothing about organization – environment relations. The human relations model emphasizes the human subsystem but neglects the others. And it, too, offers a closed-system perspective. STS embraces both the technical and social aspects of organizations and their interrelationship, and seeks a form of work organization that jointly optimizes them. This has made it possible to challenge technological determinism. Socio-technical theory also brings to the fore the interactions between the organization and its environment. This has encouraged the development of organizational structures that are flexible enough to respond to environmental changes. The practical results of the improved theory are evident in the results reported by Pasmore et al.

One of the admirable features of the socio-technical thinkers has been their willingness to admit the problems and failures of their approach. Trist reports on the resistance of the Divisional Board to the extension of autonomous group working in the coal mines. Their priority was intensifying managerial controls in order to push through full mechanization:

> The Divisional Board's reaction suggested that any attempt to reverse the prevailing mode would be met with very serious resistance. To move in the opposite direction meant going against the grain of a macrosocial trend of institution-building in terms of the model of the technocratic bureaucracy, which had yet to reach its peak or disclose its dysfunctionality.
>
> *(Trist 1981, p. 14)*

The Communist Party of India, fearing a loss of influence, agitated against the Ahmedabad experiments (Trist 1981, p. 18). During the Norwegian Industrial Democracy project, 500 middle managers, "sensing all loss and no gain" so far as they were concerned, said "no" to diffusing socio-technical change throughout Norshydro, the largest enterprise in the country at the time (Trist 1981, p. 48). Mumford (2006) puts the success of STS in Scandinavia down to the common set of values shared across society, which had produced explicit legislation requiring employers and trade unions to cooperate. In Italy, by contrast, management feared the unions and resisted changes in work design because of the additional power they might afford to workers. In France, the main resistance came from the trade unions, which saw the socio-technical approach as just another, more sophisticated way to exploit workers. Even in Norway, similar thoughts occurred to workers, while "engineers and technologists saw some of the changes as threatening to their positions and status" (Mumford 2006, p. 325).

Not noticing the positives, or that Emery and Trist were learning from their mistakes and further developing their thinking, the critics went for the jugular. To them, the organismic model that lay at the heart of classical STS meant that it was clearly "functionalist" and suffered from all the faults of that social theory. Readers of Chapter 4 of this book will not be surprised that the assaults came from sociologists of both an "interpretive" and "radical change" persuasion.

From the interpretive perspective, Poggi has argued that analysis in STS terms tends

> ... to impute functions to systems and sub-systems rather than register actions which are the result of choices made in the face of a range of action alternatives.
>
> *(Quoted in Brown 1967, p. 46)*

Silverman (1970), from his "action frame of reference," similarly accuses STS of "reifying organizations," ascribing to them a life and purposes of their own, and downplaying purposeful action. In his view, employing the organismic analogy leads to explanations of organizational behavior in terms of an organization's requirement to meet functional imperatives and to adapt to its environment. Individuals are seen as subject to forces that are beyond their control and which they do not always understand. The conscious reasons people give for their actions are no substitute for a scientific, functional explanation of what is really occurring in the organization. While the criticism may seem a bit harsh in relation to STS, it is fundamentally true, at least of the organismic variant. Attention is given to human beings, but there is a tendency to see people as predictable if their psychological requirements are met. All humans are supposed to respond favorably to appropriately designed jobs, group working, and an organization that provides them with a clear sense of purpose. People are not treated as self-conscious, autonomous actors capable of reading different meanings into the situations they face. Socio-technical theorists did not believe that all workers in all circumstances would immediately respond enthusiastically to enriched jobs, and take more responsibility, but they did think they would naturally grow to appreciate these things.

From the point of view of the sociology of radical change, STS exhibits a managerial bias and is incapable of providing a proper analysis of conflict and change. The Tavistock researchers, according to critics, were always in thrall to the powerful because they had to find clients who could pay and who, therefore, imposed their own definition of the problem. The use of the organismic analogy only reinforced this managerial bias. Organismic thinking represents the organization as an integrated whole, the survival of which is in the interests of everyone. Such thinking appeals to managers because it presents all stakeholders as benefiting equally from company success. Conflict is a dysfunctional threat to the system. Managers and consultants are, therefore, seen as acting paternalistically, for the good of all, using their expert knowledge to adjust the organization in ways that will ensure its survival. STS, according to the radical change logic, simply relieves managers of one onerous chore. It gets the workers to control themselves by convincing them that they have genuine influence over their working conditions. According to Blackler and Brown (1980), Shell's "new philosophy of management" was ideologically manipulative, whatever the best intentions of those involved, because it tried to cover real disputes with a gloss of common interest. In their view, the conferences were top-down events at which the philosophy was presented as a *fait accompli* rather than subjected to thorough debate and discussion. The guiding ethos was indoctrination not mutual understanding. The philosophy helped passage of the productivity agreements, but the idea of collaboration gradually gave way to political manevering as negotiations proceeded. Management drove a hard bargain as the deals took on the character of ordinary haggling over cash. In reply, it should be said, Foster and Hill (in Blackler and Brown 1980) defend the philosophy experiment. Foster regards it as having made a positive and worthwhile contribution. Hill sees Blackler and Brown's interpretation as negatively biased, narrow, and academic. To sociologists who adopt the radical change orientation, however, the conclusion is clear. STS neglects the actual source of radical change – conflicts of real interest – and fails to recognize the power of some groups over others. It is misguided in believing that its techniques can be other than manipulative in conflict situations.

Stahl (2007) notes that there are overlaps between Mumford's work and "critical research" in information systems inspired by the thinking of the Frankfurt School sociologists. As was seen, Mumford was on a "humanistic mission" to ensure that technology is designed in an ethically acceptable manner and used to enhance "human freedom, democracy, and creativity." Stahl is forced to conclude, however, that there is an overriding commitment to the status quo in her work, which makes it subject to the same criticisms leveled at other forms of STS. She seems to believe that "liberation" can occur within the bounds of the capitalist system because the interests of employers and workers can be reconciled in the course of participative action. This is anathema to "radical change" theorists who regard society as characterized by irreconcilable class conflict and participation as being, as Land paraphrases the critique, no more than allowing the worker to "determine the direction of the stripes on his prison uniform" (Land, quoted in Stahl 2007, p. 483). Stahl also ponders the status of "ethics" in Mumford's ETHICS methodology. Is its role intrinsic or instrumental – of primary importance or just there to facilitate systems design?

There is an important question around whether the criticisms leveled against the organismic version of STS also carry weight against the later "purposeful systems" variant. An interesting debate in the pages of *Systems Research and Behavioral Science* illuminates but does not entirely settle the issue. Kira and van Eijnatten (2008) started the ball rolling by claiming that their "chaordic systems approach" offers a superior way of achieving socially sustainable work organizations than traditional STS. We will restrict ourselves to their criticisms of STS. STS is labeled as "functionalist"; specifically as being positivist/objectivist in methodological terms and seeing causality as simple to untangle. It is then portrayed as having four major weaknesses. First, STS is guilty of "shaping work and organization toward a predefined goal through predefined paths" (p. 746) and largely ignoring the spontaneous way work organization develops in the course of stakeholder interaction. Second, it is inconsistent because it falls into the "Kantian split" that Stacey sees as bedeviling all systems approaches. Managers are perceived, using a rationalist teleological framework, as capable of making intentional choices to improve organizations, while workers are viewed, using a formative teleological framework, as objects to be controlled for the greater good of the whole enterprise (see Section 7.6). The essential role employees can play is organizational development is therefore neglected. Third, STS is accused of wanting to hold organizations in a steady state, using feedback control, rather than in the "multiple, more complex equilibrium states" (p. 746) necessary to promote sustainability. Finally, STS forgets that the process of achieving sustainability goals is just as important as the goals themselves. It is essential to continually engage *en route* with stakeholder values and worldviews to encourage learning and the emergence of novelty.

Merrelyn Emery is, quite reasonably in my view, aggrieved (Emery M. 2010) and is not placated by Kira and van Eijnatten's (2010) claim that their criticisms are aimed more at the earlier type of STS than the Emerys' later "open-systems thinking" (OST) (Emery M. 2011). In her view, any plausible critique has to take into account the significant developments made by the "purposeful systems" type of STS pioneered by Trist, her husband Fred Emery and herself. Emery argues that OST, embracing the work on the "causal texture of organizational environments," the "genotypical organizational design principles," "participative design workshops" (PDW), and "search conferences," learned from past mistakes and situates STS in a different paradigm. OST, Emery argues, does not impose predetermined designs, treats all people, including workers, as open, purposeful systems, facilitates debates about human values, and encourages the coevolution of the organization with its environment as stakeholders develop their ideal-seeking capabilities and translate them into strategic plans. According to Emery M. (2010, p. 701), a fundamental shift in paradigm has taken place that makes the old sort of STS "obsolete," and Kira and van Eijnatten's criticisms of STS, based on that version, misplaced. The change, using Pepper's theory of different "root metaphors," is from the "world hypothesis of organicism," which champions the "integration" of parts in whole systems, to the "contextualist world hypothesis," which deals with the historic act in its context, accommodates purposeful behavior, and can produce and explain "novelty and emergence." As Fred Emery (1969, p. 15) remarked, all Pepper's "root metaphors" are in operation in different systems theories and the result is much mutual incomprehension. Kira and van Eijnatten's critique certainly struggles to find its mark when aimed at the "purposeful systems" version of STS.

One of the aims of critical systems thinking is to eliminate incomprehension and encourage mutual understanding and respect among advocates of the different branches of the systems approach. To that end, let us continue this second-order analysis of STS by relating it to the system of systems methodologies (SOSM). Early attempts to do this (Jackson and Keys 1984; Jackson 2000), as Baburoglu (1992) points out, made the same mistake as Kira and van Eijnatten by treating STS as a unified paradigm and concentrating on the earlier organismic formulation. I plead guilty and will seek to provide a richer appreciation of the approach in what follows. Baburoglu identifies four tracks within STS, or what he calls the "Emery-Trist Systems Paradigm," on the basis of "a liberation theme." While we will stick with the simpler categorization into "organismic" and "purposeful systems" variants, we will use his work to add further depth to the analysis.

In terms of the reworked SOSM, the organismic version of STS makes "complex" assumptions about systems and their environments. It certainly does not regard causality, *pace* Kira and van Eijnatten, as simple and easy to grasp. Rather it sees internal interrelationships as potentially giving rise to emergent properties and the environment as changing and unpredictable. The recommendations made concern how systems should be designed to be responsive and capable of learning so that they can evolve appropriately over time as they are affected by their own parts and by the turbulent environments in which they exist. This last point distinguishes STS from system dynamics, which we labeled as making "complicated" assumptions because it takes an endogenous point of view and seeks to map the myriad interrelationships between all the variables impacting system behavior. In respect of the "stakeholders" dimension of the SOSM, the organismic form of STS clearly makes "unitary" assumptions. This is explicitly the case when the "primary task" of the organization is defined as the task it was set up to perform or must undertake in order to survive. Even when it becomes accepted that multiple primary tasks can exist, there is an implicit assumption that all areas of the organization can reach an agreement on what the overarching objective should be. Little thought is given, in this version of STS, to how this agreement can be reached. As Fred Emery stated, in the Norwegian Industrial Democracy project, the workers were required to "trust" the consultants. Mumford speaks honestly when she says of socio-technical design that:

> It can be used to contribute to most problem-solving in work situations, providing that both the innovators and recipients are willing to use a democratic

approach. It will be difficult to use successfully if the parties involved are hostile to each other, disinterested in developing strategy and unwilling or unable to co-operate.

(2006, p. 317)

The approach shows very little interest in organizational politics, conflict, or issues of power and coercion. Nevertheless this "first developmental track" of STS, as Baburoglu calls it, does offer to employees "liberation from the domination of the machine" and "liberation from single and meaningless tasks and from external control" (1992, p. 287). In summary, the organismic version of STS can reasonably be represented as making "complex-unitary" assumptions and positioned on the SOSM as in Figure 12.1.

It should now be asked whether the "purposeful systems" version of STS requires a different positioning on the SOSM. Emery argues that OST marks a paradigm break with the earlier formulation. In Baburoglu's account it opens up two more "developmental tracks" for STS – one offering "liberation of the futures" and "liberation from the expert and expertise"; and another promising "liberation from single social system referential design" and "liberation from institutional and organizational constraints on individuals" (1992, p. 287). Both seem to demand that this version should be recognized as fully appreciating the challenges posed by pluralistic contexts. A couple of points give me pause however. Without digging too far into the detail, I think OST retains a "whiff" of functionalism about it. If the first purpose of OST (see Section 12.2.2) is to "promote and create toward a world that is consciously designed by people, and for people," the second purpose is

> ... to develop an internally consistent conceptual framework or social science, within which each component is operationally defined and hypotheses are testable so that the knowledge to support the first purpose is created.
>
> *(Emery M. 2000, p. 623)*

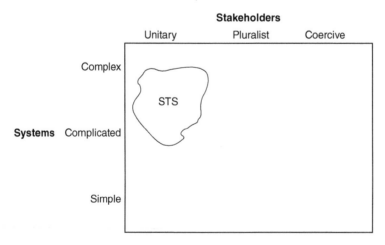

Figure 12.1 The positioning of Socio-Technical Systems Thinking on the SOSM.

Not necessarily a bad thing, of course, but it does suggest that a "complex-pluralistic" positioning on the SOSM is problematic. I am influenced here by Kira and van Eijnatten's critique which, in one respect at least, finds the mark. They argue that STS is guilty of "shaping work and organization toward a predefined goal through predefined paths." If one looks closely at how PDWs and Search Conferences are meant to run, and the kinds of outcomes to which they are intended to give rise, then one finds many "socio-techni-cal" principles accepted without question. In PDWs, we find the job-design principles and the desirability of democratic structures; in Search Conferences, it is easy to iden-tify participative planning, self-managed groups, and particular dissemination prac-tices. What are we to make, for example, of this:

> There are no remaining theoretical or practical challenges in producing *socially* and *ecologically* sustainable organizations. The theory, methods and outcomes are clear. We know what preparations must be made, what conditions must be in place and what issues must be addressed before employees walk into the PDW workshops to change the design principle. All of these are documented from diverse organizations in different countries.... *We know how to produce sustain-able organizations.*
>
> *(Emery M. 2011, p. 408)*

If these socio-technical "certainties" are smuggled in, and have to be accepted, then are we really achieving the sorts of "liberation" Baburoglu claims for the new develop-mental tracks? It is true that OST has much that is new to offer, especially with regard to the implementation of STS. But it does not seem that a clean break with "organicism" has been made and that "contextualism" has been fully embraced. Barton and Selsky (2000) concur that there is an "essentialist stance" in OST in which environments are concrete and people can perceive them directly without the need for abstract concepts. The same conclusion can be reached from a different direction. If we remove the organ-ismic principles embedded in the PDW and Search Conference methodologies, what do we have left? I would suggest that we have significantly underdeveloped forms of the kind of soft systems approach that will be studied later. It was left to Ackoff (see Chapter 15), who had worked closely with Emery and Trist, to make the clean paradigm break that eluded STS. This is, of course, for better or worse depending on the observer.

For completeness, we should state that neither Merrylyn Emery nor Baburoglu argue that OST entertains "coercive" assumptions. Search conferences seek cohesion around "collective human ideals." Perhaps the absence of such ideals explains why, although "we know how to produce sustainable organizations," we have great difficulty in successfully developing them. Postmodern and Luhmanian, as well as radical-change, social theo-rists would have a field-day here. Baburoglu (1992), in response to OST's failure to address coercive contexts, is intent on establishing an enhanced developmental track for STS which will ensure "liberation from harmony and consensus orientation" and "liberation from trying to control and self-assertiveness."

It may be necessary to repeat that there is no shame in any particular positioning on the SOSM. STS's main contribution is in conducting a rich and comprehensive explora-tion of the organismic analogy. This provides it with explanatory power and produces recommendations for action that can yield significant benefits for all stakeholders. Intelligently working out the implications of making "complex-unitary" assumptions may be just what society requires in the current era. But, to be sure about that, we have to review all the alternatives and test them in practice.

12.5 Comments

If the heyday of STS was in the 1960s and 1970s, it is reasonable to ask whether it is still relevant in the twenty-first century. In fact, many regard its concepts and conclusions as even more germane today. Netland et al. (2009), engaging in a four-year project in Norway, "Ideal Factory," call for STS to be reborn to cope with the issues posed by modern production systems. Competitiveness in high-technology manufacturing systems depends on managing complex interactions between state-of-the-art technologies and "social factors such as knowledge, learning, and innovation." If new high-speed information and communication systems and modern technologies, such as automation, robotics, and AI, are implemented without thought for the social system, the result can be increased bureaucratization and alienation. The original version of STS, in their view, still has much to offer in helping to build manufacturing systems that are more productive and profitable while, at the same time, improving the quality of working life for employees. Robertson et al. (2015) summarize investigations by a group of international researchers into how STS can be applied to improve worker safety. If, as is argued, occupational safety is an emergent property of the dynamic interactions between the social and technical components of work environments, then the concepts developed by STS will be more powerful than traditional approaches to workplace safety in generating insights that can lead to appropriate action. Walker (2015) hopes that STS can be brought back into prominence and used to improve the resilience of civil engineering projects. All engineered systems involve humans at one level or another. STS suggests how the human capacity to be flexible and responsive can be jointly optimized with engineering, in an open systems approach, to ensure that the overall system can adapt to any unpredictable, nonlinear behavior. The recent catalogue of failures associated with the introduction of complex IT systems – in the Ministry of Defence, Her Majesty's Revenue and Customs, and the National Health Service (in the United Kingdom alone) – contributes to the feeling that Mumford's STS-based ETHICS methodology, for information system design, must offer improvements on current practice.

In terms of its foundational recommendation to jointly optimize the social and technical aspects of work organizations, we can reiterate Fox's conclusion that the STS approach

> ... has more relevance today than ever before, as organizational personnel seek more fruitful means of empowerment and as their organizations strive for greater productivity and viability in increasingly turbulent environments.
>
> *(1995, p. 91)*

And, STS keeps on the agenda the even more important matter of whether people and society really have to surrender to technological determinism. Can't we go back to considering technology as a means to achieving human well-being rather than abandoning ourselves to it and accepting it as an end in itself? Heidegger posed this "question concerning technology" in its most fundamental form when asking whether "enframing," the purely technical mode of thought, is overwhelming human beings and cutting them off from a more authentic way of experiencing and being in the world. The technological imperative that insists everything is "ordered" must be kept within its own bounds, he argues, if we are not to be blinded to the truth:

> But enframing does not simply endanger man in his relationship to himself and to everything that is. As a destining, it banishes man into that kind of revealing that is an ordering. Where this ordering holds sway, it drives out every other possibility of revealing.
>
> *(Heidegger 1978, p. 309)*

STS reminds us of a concern that will, hopefully, never become out-of-date.

12.6 The Value of Socio-Technical Systems Thinking to Managers

Here are five lessons from STS which managers would do well to observe:

- When designing work organizations, it is essential to pay attention to both social and technical aspects; the two should be jointly optimized to achieve the best overall outcome in employee satisfaction and productivity
- Work groups and organizations are best able to manage the variety of their changing environments if they have the autonomy to control variances, operate on the multifunctional principle, and are provided with the information that will enable them to improve their performance
- Managers should control boundary exchanges not interfere in the internal affairs of work groups and departments. Their role is to bring the external stability necessary for effective group working, by securing the inputs necessary for task completion and ensuring that the outputs are valuable to other stakeholders
- Ensure that human values are honored in the socio-technical design by encouraging participation and creating high-quality jobs
- As environments themselves become turbulent, and so unpredictable, it is necessary to pay increased attention to an organization's philosophy and values so that it can be shared among all stakeholders and reconciled with those of other organizations

12.7 Conclusion

Nassim Taleb (2013) pictures a world in which complex interdependences between systems give rise to nonlinear consequences, which are impossible to predict. "Black swan" events proliferate. Organizations that rely on traditional scientific "rationality," and continue to manage using classical strategic management methods and hierarchical structures, will make things worse rather than better. Successful enterprises cannot just be resilient and robust in the face of turbulence, they must be "antifragile" – thriving on uncertainty and using it to continually remake themselves; learning and improving all the time. This requires them to abandon what they have learned from machine thinking and to become more "organic." Studying living systems in nature, we learn that they grow and develop as a result of disorder and flourish by becoming more complex, increasingly differentiated and decentralized, exhibiting redundancy, and reacting positively to randomness by continually regenerating themselves. STS anticipated by decades much of what Taleb has to say. It has developed, through its thorough exploration of the organismic analogy, and hundreds of project experiences, a richer, more comprehensive, and more fully tested set of directives for achieving antifragility than he is able to offer. Further, having entered a "purposeful systems" phase, STS tends to suggest that more may be needed to ensure antifragility for organizations, over the long term, than the thinking yielded by the organismic analogy in which Taleb appears stuck. We will explore exactly what this "more" is when we consider "soft" and "emancipatory" systems thinking in later chapters. For the moment, let us note that socio-technical systems thinkers made considerable use of cybernetic concepts, such as variety, redundancy, and feedback, in progressing their theory. The next chapter considers what further advances can be made, toward Taleb's goal of antifragile organizations, if all the various cybernetic ideas are brought together to produce a coherent model of what any "viable system" must be like.

13

Organizational Cybernetics and the Viable System Model

The viable system is a system that survives. It coheres; it is integral. It is homeo-statically balanced both internally and externally, but it has none the less mechanisms and opportunities to grow and to learn, to evolve and to adapt – to become more and more potent in its environment

(Beer 1981, p. 239)

13.1 Prologue

On 12 November 1971, Stafford Beer met the democratically elected Marxist president of Chile, Salvador Allende, to explain a plan for the cybernetic regulation of the economy of the country. Taking Allende through his model of any viable system, which was to provide the theoretical foundation for the project, Beer eventually reached "System 5," the ultimate policy-making entity. He thought, naturally enough, that Allende would identify this with himself and his government. Instead, the President threw himself back in his chair and declared: "at last, *el pueblo*" (Beer 1981, p. 258). There was a meeting of minds and what became Project Cybersyn ("cybernetic synergy") began in earnest. In Eden Medina's assessment:

> Project Cybersyn was short-lived, but it was one of the most ambitious applications of cybernetic ideas in history because of its national scope and because it formed part of a larger project for economic, social, and political transformation.
>
> *(2014, p. 219)*

In Allenna Leonard's view:

> No country since has been willing to try such an innovative approach although the less than satisfactory performance of most governments would lead a reasonable person to wonder why something new should not be tried.
>
> *(2009, p. 232)*

Eaton, writing in the *New Statesman*, sees the use of technology in the project as having anticipated both the Internet and the era of "big data," and states that

> ... in its ambition and its noble ideas, it pioneered a glimpse of a daringly alternative order: one in which humans are the masters, rather than the slaves, of machines.
>
> *(2018)*

Beer had been invited to Chile by Fernando Flores, the Technical General Manager of the State Development Corporation (CORFO). The first edition of his book *Brain of the Firm* (1972) had just gone to the publishers. This book set out, for the first time, Beer's model of any viable system (VSM) derived from neurophysiology. Allende's government was engaged in a program of mass nationalization of the banks and major industries. Inevitably, such a complete reorganization of the economy threw up significant problems of oversight, coordination, and regulation, especially when there was a crucial need to increase productivity. However, there was no thought of replicating the Soviet model. Everything was to be done with respect for Chile's democratic institutions and in a manner that encouraged "bottom-up" initiatives. Allende insisted on this point during his meeting with Beer. The plan had to be "decentralizing, worker-participative, and antibureaucratic" (Beer 1981, p. 257). The commonalities between what Allende hoped to achieve and the lessons of organizational cybernetics were clear to Beer. He put it like this:

> In Chile, I know that I am making the maximum effort towards the devolution of power. The government made their revolution about it; I find it good cybernetics.
>
> *(1975, p. 428)*

Cohesion, effectiveness, and freedom could all be enhanced in tandem using the VSM.

Beer conceived of the Chilean industrial economy as one of the key functions of government alongside health, education, finance, etc. The government was nested in the nation of Chile, and that in the world of nations. Looking at the subsystems of the industrial economy, it was possible to identify "branches," such as heavy and light industry and, heading down through the logical levels, sectors within branches, enterprises within sectors, plants within enterprises, and so on. If the industrial economy was to flourish and play its proper part in the wider system of the nation, it was essential, Beer's cybernetic theory held, for it to become a "viable system" – capable of responding to environmental change and evolving in a cohesive manner. That in turn required that all the levels of possible agency embedded within it should themselves become viable systems. If this could be achieved, then the economy would not need to be planned from the center as in the Soviet system. It would be constituted in a way that maintained coherence and, at the same time, ensured improvements in performance on the basis of the learning and adaptation that was possible at all levels. The VSM was the perfect model for the job at hand because of its generality. It could be used to model and improve all the levels of potential viable systems embedded in the industrial economy.

There were four components of Project Cybersyn, all designed to support the learning and adaptation required by the VSM. The first, nicknamed *Cybernet*, was a communications network, which was required to collect and transmit data on performance to decision-makers at all levels of the Chilean industrial economy. In theory, every factory within the nationalized sector would need to be in communication with a computer capable of analyzing its own data. In practice, there were only around 50 computers in the whole of Chile, and most of these were antiquated. The solution adopted was to requisition every available telex machine and to use the existing telex network to send the information to the machines available in Santiago, for processing, before sending the results back to the factories. Beer insisted that the whole system operate in "real-time" with indices of performance submitted every day, evaluated, and the outcomes communicated to managers and workers' committees the next day. This would avoid the "lags" that bedeviled economic management even in the developed countries where decision-makers still had to act on information that was months out-of-date. Beer had already developed indices of performance – based on measures of actuality, capability, and potentiality – general enough to operate in different settings and at different levels of the economy. But the question remained of what activities to measure in the plants and also at the enterprise, sector, and branch levels. Operational Research (OR) teams were sent out to plants to construct quantitative flow charts that highlighted the most important activities as agreed with the local workers' committees. These would include, for example, measures of input stocks of raw materials, outputs of finished products, processes that could generate bottlenecks, and absenteeism as a proxy for morale. Around 10 or 12 activities were adequate for most plants. The OR teams were also expected to agree with the workers' committees, or whoever was managing the plant, timescales for correcting poor performance locally before a signal, indicating a problem, was automatically sent up to the next higher level in the industrial economy. That level would then have an agreed time to put right any problems impacting its own indices before a still higher level was alerted. Beer describes this as follows:

> In principle, then, it is possible for an algedonic [pain or pleasure] signal originating at plant level to reach the Minister himself. If that were ever to happen, it would be a disgrace: the management at plant, enterprise, sector, and rama [branch] levels … would all have failed. This is why the principle is precious: it is clearly an instrument of *cohesiveness* in the nest of viable systems. But, again, it offers the maximum decentralization that is consistent with cohesion – since, if all concerned do their agreed jobs properly, algedonic signals will rarely be fired.
> *(1981, p. 276)*

As early as July 1972, Medina states, Cybernet had connected CORFO with various central agencies, eight sector committees, and 49 plants (2014, p. 135).

For Cybernet to provide real-time information of maximum benefit to decision-makers, it needed the capacity to monitor and evaluate thousands of performance indicators daily, send the results to the plants or other appropriate level, and trigger signals to higher levels if appropriate. To prevent bureaucracy and information overload, it was essential that it transmitted only the most vital information. Beer wanted a predictive program suite that, from the inputs it received, was capable of recognizing when an unexpected change was about to occur. The development of this *Cyberstride* program, the second vital tool of Project Cybersyn, was subcontracted to consultants from Arthur Anderson and Co. based in London. Describing the outcome, Beer says:

> Briefly, the method uses Bayesian probability theory to quantify a multi-state data-generating process. The filter can automatically recognize changes in the stream of input indices, and determine whether they represent transient errors, step functions, or changes in time trend and slope.
>
> *(1981, p. 263)*

Provided with this information in real-time, managers could anticipate what was going to happen and act in advance to forestall problems. By July 1972, a temporary suite running the Cyberstride program was running. By the end of Project Cybersyn, Beer estimates that some 400 entities, accounting for seventy percent of the socio-industrial economy, were operating the Cyberstride system linked through Cybernet (1981, p. 264).

Cybernet and Cyberstride provided the industrial economy with the information necessary to gain the maximum benefits from decentralization, both economically and in terms of the political agenda, while seeking to guarantee cohesion overall. But the VSM insists that external as well as internal information is vital to any viable system. The third component of Project Cybersyn, the *Checo* (CHilean ECOnomy) programs, was designed to ensure that all the viable systems identified at various levels in the economy could understand what was happening in their environments and be responsive to opportunities and threats. At the national level, what was required was a simulation model capable of charting the effects of the interactions between the main factors that had an impact on the economy as a whole. This could potentially provide government with an experimental laboratory to try out different policy options. Beer suggested system dynamics models of the type developed by Jay W. Forrester. The available DYNAMO compiler could be used to construct the models and Beer sought out a systems engineer, Ron Anderton, to assist with the task. By September 1972 a macro-economic model of the economy had been built, together with more micro-level models to serve the light industry "branch" and the automotive "sector" (Beer 1981, p. 267). Unfortunately, the economic turmoil caused by internal and external opposition to the government, including the economic blockade, meant that the models could not be supplied with accurate and up-to-date information. Although Flores, who had become Minister of Economics, took ownership of the Checo project, there was too much uncertainty about the data fed into the models to have faith in the three to five year simulations they produced.

The final element of Project Cybersyn was the *Operations Room*. This was an "environment for decision" in which the information provided by the other components of Project Cybersyn was brought together. The Operations Room was designed by Gui Bonsiepe, of the Ulm School of Design that also inspired Steve Jobs and the Apple designer Jonathan Ive (Morozov 2014). It was constructed in Santiago in a building that had previously housed *Readers Digest*, and was in experimental working order by January 1973 (Beer 1981, p. 270). The room consisted of a large hexagonal space, 33 ft in diameter, with information displayed on screens on the six walls. Its design was extremely futuristic and is frequently compared to the style found in Stanley Kubrick's film *2001: A Space Odyssey* (Medina 2014, p. 121). Three of the walls were used for a diagram of the VSM and to show production trends and any algedonic signals, indicated by flashing red lights. Two walls were employed for animated versions of the Checo simulations with which participants could interact. The sixth wall was available to show electronic feedback from those watching the discussions taking place in the room. The decision-makers sat in seven swivel chairs in a circle with no obstructions between them. They used buttons on their chair arms to take control of the equipment and to call up information about any part of the economy that was of interest. The ultimate objective was to have operations rooms in all the factories and at all the other levels identified as viable systems in the industrial economy.

Although Project Cybersyn was never completed, and so its overall success is not easy to evaluate, it was able to demonstrate its worth in an unexpected way. In October 1972, the *gremios*, business associations of lorry-owners, retailers, owners of local shops, and small distribution centers, called a strike. They were able to sustain their action with resources made available from outside the country. Faced with this serious attempt to bring the government down, Flores immediately saw the potential of the Cybernet system. A central emergency operations center and similar, regional centers were established which employed the telex system to register shortages as they occurred and to track and make the best use of the distribution resources still under government control. Soon, according to Beer, 2000 telex messages were flowing every day. They were sorted at the emergency centers into crucial and less important. The ability to use this information to take instant decisions and support decentralized action, together with the high degree of redundancy in the existing distribution system, was a major factor in helping the government defeat the strike. There had always been government support for Cybersyn:

> But it was not until the top officials and the socially responsible ministers were plunged into the traumatic experience of the *gremio* battle, lived with the problems non-stop, used the tools provided however makeshift, and mastered the revolt, that they fully and deeply understood. We really had been talking about a managerial revolution, and not about the introduction of some rather slick administrative tricks.
>
> *(Beer 1981, p. 314)*

They were right to be impressed. Raul Espejo, Operations Director of Cybersyn, has calculated that during a second *gremio* action, in August 1973, supplies of fuel and essential food were maintained at normal levels despite there being only between 10% and 30% of the lorry fleet in operation (Beer 1981, p. 346).

Beer and the team in Chile were greatly taken by Allende's insistence that "System 5," the policy-making function in the VSM, should be "the people." Another project, *Cyberfolk*, was therefore proposed to provide cybernetic support to "the interactions of the people and policy makers" (Espejo 2017, p. 42). As Beer expressed it in 1973:

> But the tools of science are not anywhere regarded as the people's tools; and people everywhere become alienated from that very science which is their own. Hence we are studying all these matters with the workers. Hence the systems I have to tell you about so far are designed for workers as well as ministers to use.

Hence we are working on feedback systems to link the people to their government.

(Beer 1975, pp. 428–429)

Project Cyberfolk proposed "algedonic participation," whereby people could express their satisfaction or dissatisfaction with policies discussed on TV. They were to do this through individual electrical meters, which they could manipulate by turning a knob to register happiness or unhappiness with what was being suggested. As was noted, one of the walls of the Operations Room was available for displaying the sum of public reaction. Artists, poets, and musicians were also encouraged to amplify the people's will to politicians and to explain the value of the scientific tools to ordinary people. The famous folk singer, Angel Parra, wrote a song named, in English, "Litany for a Computer and a Baby about to be Born." "So let us heap all science together," it went, "before we reach the end of our tether" (Beer 1981, p. 290). It sounds a lot better in Spanish.

As the crisis in Chile escalated, Beer set out the need for a "people project" and an "allocation project" to run alongside and support Project Cybersyn under the Minister of Economics. The former was a more focused version of Cyberfolk, which would assist the "economic community" that was taking management into its own hands in the regions. The latter was a reaction to the October strike of the *gremios*, which had demonstrated the importance of "commerce" to the economy. It was to use cybernetics to help the local volunteer bodies that had self-organized to supervise distribution. But it was too late. On 11 September 1973 came the military coup that the internal opposition and external enemies, particularly the US, had long wished and worked for. Allende shot himself rather than surrender to Pinochet's soldiers. His last order to the Cybersyn project team, on the 8th September, was to move the Operations Room into La Moneda, the presidential palace (Beer 1981, p. 346).

The Pinochet regime imprisoned and tortured around 40 000 people and killed or "disappeared" over 3000. Flores was imprisoned. Espejo fled abroad. Beer was in England at the time of the coup, seeking support for the Chilean government, and found out about Allende's demise from a newspaper placard. Though he did not suffer in the manner of the Chileans, the whole experience changed him radically. He had gone to Chile a rich and successful consultant who owned a fine house and a Rolls Royce. Soon after his engagement there ended, he gave up many of his material possessions and relocated to a small cottage in Wales which, until quite late in his life, had no running water, central heating, or telephone. Beer was frequently questioned about what more could have been done to avoid the coup in Chile. A reply:

It is like complaining that man, who is supposed to be an adaptive biological system, cannot adapt to a bullet through the heart.

(Beer 1981, p. 346)

13.2 Description of Organizational Cybernetics

13.2.1 Historical Development

There are two main ways of applying cybernetic insights to management depending on whether the machine or organism analogy is privileged. I call these "management cybernetics" and "organizational cybernetics." Management cybernetics, the earlier version, consists essentially of "bolting-on" some cybernetic ideas to the mechanical model. The goals of organizations are still regarded as unambiguous and organizations seen as input-transformation-output systems designed to achieve the goals efficiently. Because of inevitable internal and external disturbances, however, managers are now recommended to employ the negative feedback mechanism to ensure that their organizations remain on course. In the same style, the feedback concept was added to traditional "hard systems methodologies," such as systems analysis and systems engineering, in order to ensure proper monitoring and control of outputs. While this is useful, it is hardly radical. By contrast organizational cybernetics, drawing upon analogies with "brains" and "organisms," is capable of transforming the way we look at the management task. Suddenly the notion of regulating an organization in pursuit of a known, predetermined goal becomes problematic. Attention needs to be devoted to integrating differentiated subsystems possessing disparate priorities. And, an organization confronting a turbulent environment will usually, as with an organism, emphasize its own survival at the expense of achieving particular goals. The previous chapter, on socio-technical systems thinking, demonstrated the value of cybernetic concepts used in the context of organismic formulations. For example, the concepts of variety and feedback were employed to think through how organizational systems can be made more adaptable to the rapidly changing and unpredictable environments with which they increasingly have to cope. In this chapter, we consider in detail a fully developed version of organizational cybernetics. Stafford Beer took advantage of almost all the cybernetic ideas introduced in Chapter 6 and converted them into a model, the "viable system model" (VSM), which applies them coherently, directly, and with great acuity to the management of all forms of organization. A little more history before the detail of that is explained.

We must give due recognition to two strands of organizational cybernetics – the "St. Gallen Management Model" as well as Beer's VSM. The St. Gallen Model (SGM) was developed at the University of St. Gallen, Switzerland, initially by Professor Hans Ulrich. According to Schwaninger (2001), on which this account is based, Ulrich was convinced that a generalist, integrative, and pragmatic approach to management research and education was needed and that this could only be provided by systems theory and cybernetics. In the mid-1960s, a task force was set up to study the work of Wiener, von Bertalanffy, Beer, Pask, Churchman, Buckley, and others, and its findings, together with work by Ulrich, led to the first elaboration of the SGM in a book by Ulrich and Krieg in 1972. A highlight of the model was its in-depth treatment of the environment of organizations. It employed a stakeholder perspective and, unusually for the time, saw managers as having responsibilities toward the natural environment. The SGM was taught to thousands of students and became very influential in German-speaking countries having, in Schwaninger's opinion, a significant impact on the attitude of businesses to environmental concerns. The SGM has been continually developed over the years, by the likes of Bleicher, Schwaninger, and Ruegg-Sturm, as attempts have been made to increase its "variety" as a general management model and to address new issues as they emerged. One well-known adaptation, made by Bleicher, was a focus on the three "logical levels" of management originally identified by Parsons – the operative, strategic, and normative levels. At the operative level, the concern is with efficiency, at the strategic with effectiveness, and at the normative with legitimacy. The demands of the three levels can conflict and it is the job of systemic management to meet all three requirements over the long term. As we shall see, these three levels correspond, respectively, to Systems 3, 4, and 5 in Beer's VSM.

What follows concentrates on the VSM rather than the SGM. They have much in common and there in no point in duplicating the argument by considering both. The "New SGM" (Ruegg-Sturm 2005) mirrors the VSM with its organization-environment focus, in seeing management as primarily about mastering complexity, in emphasizing the three logical levels, and in adopting a process perspective that distinguishes between management, business, and support processes. And many of those originally associated with the SGM, such as Markus Schwaninger and Fredmund Malik, have also become leading advocates and exponents of the VSM. Of the two related approaches, I have chosen the VSM because I agree with Huhn (2012) that it makes its radical break with traditional management thinking more explicit, is more fully corroborated, and has retained its applied focus and relevance to practitioners while the SGM has tended to become more descriptive in character.

In Chapter 6, we left Beer ruling out mechanical analogies and searching for "naturally occurring systems" that might contribute inspiration for building a cybernetic factory capable of "regeneration, growth, and adaptation." Eventually, he concluded that the human body, controlled by the nervous system, was the richest and most flexible viable system of all. This "known-to-be-viable" system, therefore, was the most appropriate model to use. In his book, *Brain of the Firm* (1972), Beer delves into neurophysiology and discovers that there are five essential subsystems, which he labels Systems 1–5, which can be identified in the brain and body in line with major functional requirements. This is the basis of the VSM, which he then seeks to show is equally relevant to social organizations. It was this model that Allende, who himself had trained as a pathologist, immediately understood and which served as the basis for Project Cybersyn. In a later book, *The Heart of Enterprise* (1979), Beer derived the same model, consisting of five subsystems and appropriate feedback loops and information flows, from cybernetic first principles. This provided the evidence that the VSM was not just an organismic metaphor applied to organizations. The cybernetic laws it espoused were relevant to all systems, including enterprises whatever their size. Indeed, in a one-person business all five functions will need to be performed by that one individual if viability is to be ensured. The year 1985 saw the publication of *Diagnosing the System for Organizations* in which Beer introduces the VSM and provides advice on how to use it in the form of a handbook or manager's guide. A second edition of *Brain of the Firm* (1981) provided the details of the Chilean experiment. Beer continued to publish articles about the VSM and to use it in consultancy projects until his death in 2002. Particularly significant were commissions for the Presidential Offices of Mexico, Uruquay, and Venezuela. Medina (2014, pp. 225–226) charts how the bureaucracy in Mexico, financial difficulties and lack of a powerful local champion in Uruguay, and political unrest in Venezuela contributed to the lack of success of these endeavors. He also went on to develop a complementary systemic methodology known as "team syntegrity," which is the subject of Chapter 17. Following his death, the *Cwarel Isaf Institute* was established, named after the cottage in Wales where Beer lived for more than 30 years. It is dedicated to preserving his legacy and continuing his work. It is possible to visit the cottage and stay and study there. The "Metaphorum" community organizes conferences on Beer's ideas and continues to develop them.

Other accounts of the VSM, and developments in the theory and practice of organizational cybernetics, can be found in books by Espejo and Harnden (1989b); Espejo and Schwaninger (1993); Espejo et al. (1996); Espejo and Reyes (2011); Schwaninger (2006); Türke, another luminary from St. Gallen (2008); and Espinosa and Walker (2011, 2017). In what follows, we will see that Raul Espejo has made important contributions to methodology, Angela Espinoza to the theory and application of the VSM in the context of sustainability, and Jon Walker to the use of the VSM with workers' cooperatives. Beer's own books, and all these secondary sources, contain numerous case studies. Outside the academic domain, the VSM has been the mainstay of a number of consultancy and training firms. Malik Management founded by Fredmund Malik, previously a professor at St. Gallen University, is the largest with a significant presence in the German-speaking world. In the United Kingdom, Patrick Hoverstadt's SCIO (Systems and Cybernetics in Organisation) is doing important work promoting and applying the VSM. Hoverstadt's *The Fractal Organization* (2008) is the most accessible introduction to the VSM for managers.

13.2.2   Philosophy and Theory

In *Decision and Control* (1966), Beer presents his view of how scientific models are produced. The scientist will likely proceed by metaphor and analogy to build a conceptual model that appears to have utility in relation to the part of the world they are interested in. An attempt will then be made, using a neutral scientific language, to produce a rigorous model, which is structurally identical, or "homomorphic," to the conceptual model. This must capture its essential features and ignore those that are irrelevant. Successful homomorphic mapping requires

> ... many elements in the system that is conceptually modelled [to] map on to one element in a rigorous model.
>
> *(Beer 1989a, p. 14)*

A model of this type can be generalized and, therefore, compared with a homomorphic version of the real-world situation. If the two are "isomorphic," corresponding in form or relations, then you have a valid scientific model that can be tested for its predictive value. For Beer cybernetics, as "the science of control and communication" provided exactly the kind of neutral scientific language that was required in management:

> Now if cybernetics is the science of control, management is the profession of control – in a certain type of system.
>
> *(Beer 1966, p. 239)*

Thus, the cybernetic armory of concepts – feedback, variety, black-box, self-organization – could be employed to transform the sorts of flabby conceptual models in use in management into a homomorphic model that would be formulated rigorously enough to be falsifiable, and so scientific in Popper's terms. It must have occurred to Beer, at some stage that, since cybernetics was transdisciplinary and relevant to all "viable systems" whether physical, biological, social, or economic, he could shortcut the task of building a homomorphic model in management by borrowing one from another domain. This is exactly what he does in *Brain of the Firm* (1972). Already committed to organismic rather than machine analogies, he takes a rigorous model, developed in neurophysiology to explain the workings of the human body and nervous system, to inform the development of a homomorphic model for management. He ends up with the VSM, which he regards as a valid scientific model because, on the basis of cybernetics, it captures the structural correspondence between the neurophysiological and the organizational models. The two are isomorphic. In *The Heart of Enterprise* (1979), having established the VSM as a scientific model, Beer moves on from neurophysiology, explains the cybernetic laws in depth, and seeks through examples to demonstrate how they underpin the organization of all complex systems. In his view, the model is yet to be falsified:

> As experience of the VSM grew, as its format was made tidier, and as others became involved, more and more viable systems were mapped on to the model: the invariances held .... Other scientists around the world have confirmed the VSM in various modes and situations, most but not all of them managerial.
>
> *(Beer 1989a, pp. 15–16)*

We will now describe the model, with hardly any reference to neurophysiology, drawing on Beer's trilogy – the two books mentioned above and *Diagnosing the System for Organizations* (1985).

Beer's VSM specifies the criteria that any enterprise must meet if it is to be *viable*, i.e. capable of surviving and maintaining its identity in an often unpredictable and turbulent environment. These criteria are based on the laws of cybernetics, which Beer defines as the "science of effective organization." The starting point for understanding the VSM is Ashby's law of requisite variety: "only variety can destroy variety" (see Section 6.3). Indeed, with its emphasis on "variety engineering," the VSM can legitimately be seen as a sophisticated working out of the implications of Ashby's law for management. Figure 13.1 pictures the management (shown as a square) of a set of operations (drawn as a circle) confronted by an environment (represented by the amoeba shape). The problem faced by management is that the operations can exhibit greater variety – more system states – than can the management and the environment can display much greater variety than the operations. However, as Ashby's law tells us, the varieties must be balanced if management is to be in control of the operations. And the same is true for the operations seeking to prosper in their environment. As Figure 13.1 suggests, this requires variety engineering – the attenuation of the variety of higher-variety systems and the amplification of the variety of lower-variety systems.

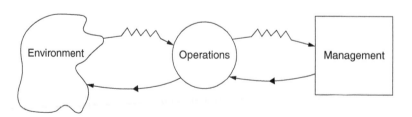

Figure 13.1 Variety engineering.

This, however, confronts managers of organizations with a difficult conundrum. They will be keen to have tight control over their operations so that the organization is successful in realizing whatever goals it is pursuing. At the same time, if they restrict the autonomy of the operations too much then the operations will lack the variety to respond to changes in the environment. The dilemma presents itself in management theory and practice as the perennial problem of centralization versus decentralization. As Hoverstadt says:

> Organisations that are too centralised are too rigid, do find it difficult to adapt to changes in their environment and do die as a result. Organisations that have no centralizing decision making structures are incapable of acting as coherent wholes and do fall apart.
>
> *(2010, p. 92)*

The VSM claims to provide a solution by balancing the variety equations in a way that grants the maximum autonomy possible to the operations that is consistent with overall systemic cohesion.

The VSM is shown in Figure 13.2. It is made up of five elements, Systems 1–5, which we label, following Beer's usual nomenclature, as *implementation, co-ordination, operational control, development, and policy.* The functions signaled by these five elements must, the model dictates, be adequately performed in any system that wishes to remain viable. They are now described in turn together with the information flows that connect the various parts of the model.

System 1 consists of the various parts of the organization concerned with *implementation* – with carrying out the tasks directly related to the organization's purpose. This means, for example, that, in a university, the System 1 elements are likely to include teaching, research, and outreach. But marketing the university, however important, is not part of System 1 because it is not what universities are usually seen as being about. On the other hand, "marketing" would be part of System 1 in a firm offering marketing services to other organizations. In Figure 13.2, the operations directly contributing to the purpose have been broken down into four parts, labeled A, B, C, and D. There could be more or less depending on the enterprise. Each of these has its own localized management 1A, 1B, 1C, and 1D, and its own relations with the relevant part of the outside world. The parts may interact (shown by the wavy lines) in various ways, perhaps by passing on subassemblies, sharing facilities, or simply competing for resources. Dividing up the operations in an appropriate way, according to the organization's purpose, reduces the variety of the operations for management. It can treat the parts as "black-boxes," concerning itself only with performance reports on how well they are achieving their subobjectives. It is also good cybernetics, in terms of the whole organization, if senior management doesn't interfere in the internal affairs of the parts of System 1 because, in variety terms, they need to be as autonomous as possible so that they have the variety necessary to respond to changes in their own localized environments. Localized management 1B, for example, having agreed its goals with higher level management, will interpret these for its own operations B, receive feedback information on performance, and takes corrective action as necessary. The autonomy of the parts is the basis for spreading control through the architecture of the whole system and matching environmental variety.

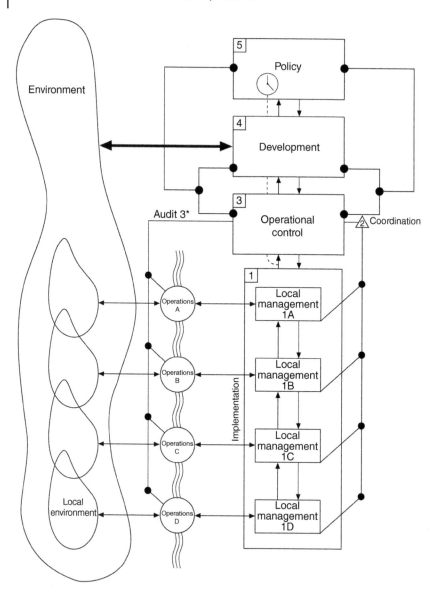

Figure 13.2 The Viable System Model (VSM).

System 2 fulfills a *co-ordination* function designed to prevent tensions arising between the parts of System 1. It will also dampen conflict if, as inevitably happens, things do not go exactly to plan; for example, one part does not deliver its subassemblies on time. Much of its effort will go into overseeing the development and maintenance of the various rules and regulations that ensure the System 1 parts act cohesively and do not get in each other's way. It will also embed in the organization any legal requirements that must be obeyed. In a University, there are likely to be System 2 regulations relating to governance, finance, human resources and quality, among others. If one unit, say the faculty of science, employs a member of staff on different terms and conditions than usually used by others, this might set up reverberations around the whole of System 1 and have a destabilizing effect. Hoverstadt provides a memorable example of what can happen when System 2 doesn't do its job properly:

> In 1999, NASA had the embarrassing and expensive experience of crashing a probe into Mars. It emerged that the problem had been that two teams were using different measuring standards, one metric and one imperial.
>
> *(2010, p. 106)*

The role of System 2, if it is conducted appropriately, will feel helpful rather than constraining to the parts of System 1. Other examples of System 2 in action are a time-table in a school or production scheduling in a manufacturing concern. Leonard's list of services that might come under System 2 includes access for disabled persons, accounts payable/receivable, certifications, computer/ICT services, diversity promotion, docu-mentation, employee assistance and benefit programs, energy efficiency, food service, insurance coverage, purchasing, scheduling of common facilities, safety and security, tax compliance, training in existing practices, and travel (2009, p. 227). System 2 is there to ensure harmony between the elements of System 1. As Hoverstadt states, it becomes increasingly important when the number of operational activities increases, when their interdependencies are greater, and the more they disturb the same parts of the environ-ment (2008, p. 92).

The role of System 3 is to exercise *operational control* over System 1 and, in so doing, optimize its performance. Hoverstadt calls it "delivery." Espejo uses the word "cohesion" for System 3, which perhaps best captures its role in ensuring synergy between the System 1 elements. System 3 has overall responsibility for the day-to-day running of the enterprise, making sure that policy is implemented appropriately. It sits on the vertical command axis and must take the overall strategy and produce a coordinated opera-tional plan, passing it down the line to System 1. It engages in a "resource bargain" with the parts of System 1 during which targets are agreed together with the resources to achieve them. Operational management, finance, human resources, sales, etc. all typi-cally have roles in System 3. As soon as the "resource bargain" has been struck, System 3 should, as far as possible, operate indirectly through coordination and audit. The autonomy of the System 1 elements is best protected, the VSM asserts, if senior man-agement exercises influence through coordination (System 2) and audit (System 3* – see below) rather than acting authoritatively down the vertical command channel. Occasionally, however, on the basis of information it receives from System 4, 3*, or 2, it will need to employ more hierarchical control measures. It is, after all, in possession of information relevant to the whole organization while the System 1 parts have only local facts to go on. System 3 also has to report upward through normal channels any infor-mation needed by the policy function, System 5.

System 3* is a servant of System 3, fulfilling an auditing role to ensure that targets specified by System 3 and rules and regulations promulgated by System 2 are being adhered to. This channel gives System 3 direct access, on a periodic basis, to the state of affairs in the operational elements. Its existence should assure senior management that it can remain in control of what is important without micromanaging. Through it, System 3 can check more immediately on things such as quality and adherence to finan-cial and maintenance standards. Leonard lists financial, energy, security, and

IT-compatibility audits, studies of customer complaints, and sporadic employee satisfaction and needs surveys, as examples of System 3* activity (2009, p. 228).

Systems 1, 2, and 3 (and 3*) make up what Beer calls the "autonomic management" of the organization. They can maintain internal stability and optimize performance, within an established framework, without reference to higher management. Autonomic management does not however possess an overall view of the organization's environment, and it is therefore incapable of responding to threats and opportunities by modifying corporate strategy. It is capable of "single-loop learning," changing how it does things in pursuit of pregiven goals, but does not possess the capacity for "double-loop learning," reviewing and changing those goals. This is why Systems 4 and 5 are necessary.

While Systems 1–3 concentrate on the "inside and now" System 4, *development*, focuses on the "outside and future." It makes the case for innovation and change in response to what is happening externally. System 4, therefore, must capture all relevant information about the organization's total environment, registering significant changes and identifying the opportunities and threats they present. It must then communicate relevant information to System 3 if quick action is required or to System 5 if it has important long-term implications. System 4 is charged with bringing together information about the overall external environment with the information it receives from System 3 about the internal state of the system. And it must present it to managers in a form that facilitates decision making. Beer proposes it become the "operations room" of the enterprise, a real "environment for decision." According to the Conant–Ashby theorem, every good regulator must contain a good model of whatever is regulated. Consequently, a viable system must have at its disposal high-variety models of its environment and of itself. It is the role of System 4 to provide these models. The reader will recall the *Operations Room* in the Chile project drawing on external information from the *Checo* simulations and internal information from *Cybernet* and *Cyberstride*. System 4 is also home to all those activities that seek to amplify the whole organization's variety in relation to its environment and attenuate the environment's variety viz-a-viz the organization. These will include marketing, research and development, public relations, lobbying, long-term financial planning, staff development, strategic planning, OR, and market research.

System 5, *policy*, has to articulate the identity of the whole system to the wider system of which it is part. It must ensure good governance. It is also responsible for the direction of the whole enterprise. It will frequently be asking "what business are we in?" It therefore has a particularly close association with System 1, which must implement policy. If the organization changes its overall purpose, the System 1 units will have to change to reflect that. One or more units may have to be sacrificed to make way for new units embodying the new purpose. In reverse, a System 1 unit at a low level of recursion, or even a service system, might strike up a favorable relationship with its environment, grow in importance, and lead the overall system to ask questions about its present identity. I witnessed the IT department of a supermarket chain in Hull become extremely successful at selling its own products and confront its parent organization with just this dilemma. The organization did not change its identity to reflect the success of this element. Almost inevitably the IT department broke away and became a business in its own right, thus proving that it was indeed a viable system. System 5 formulates policy in the operations room provided by System 4 and communicates this downward to System 3 for implementation by the divisions. An essential task is balancing the often conflicting internal and external demands placed on the organization. Here it needs to adjudicate between System 3, championing stability so that it can optimize the efficiency of ongoing operations, and System 4, which will constantly push for change to keep up with what is happening in the environment. System 5 has to ensure that the organization adapts to the external environment as and when necessary but otherwise reaps the benefits to be gained from internal stability. Beer recommends that System 5 increase its variety by employing integrated teamwork and organizing itself as an elaborate, interactive assemblage of managers – a "multinode." Decision making needs to be formalized, and the effects of decisions monitored without threatening the freedom and flexibility of interaction in the multinode. System 5 might also seek to enhance its variety by recruiting experts or employing consultants.

It is easy to see the VSM as hierarchical and the various restrictions on the freedom of the System 1 parts as "constraints." This is, however, a misrepresentation. The VSM, as represented in Figure 13.2, can just as easily be turned upside down. All of Systems 1–5 are interdependent in a viable system and, if anything, it is more accurate to say that System 1 is the most important and that Systems 2–5 facilitate its functioning in pursuit of the purpose of the whole. Element B cannot realize its goals if it is destabilized by the actions of A, C, or D – hence the need for coordination or System 2. There is no point in B continuing to pursue its existing goals, to take another example, if a massive change in the overall environment makes these irrelevant – hence the need for development driven by System 4. Systems 2–5, therefore, are all meant to be facilitative of System 1 as it implements the organization's purpose. The only restrictions on the autonomy of the System 1 elements stem from the requirement that they continue to function appropriately as parts of the whole organization. They are granted the maximum freedom consistent with the survival of the overall system.

Also crucial to the successful use of the VSM is the proper design of the information flows it prescribes. Given the importance of negative feedback for control, much of this information is about how the organization and its divisions are doing in relation to their goals. According to cybernetic logic, the information flows should be designed to reduce the variety managers have to handle. They should convey only variances from planned performance to avoid managers suffering from information overload. This is especially true in the case of System 5, which might otherwise be overwhelmed by irrelevant data. Finally, Figure 13.2 shows a "hatched line" leaving the normal System 1 to System 3 reporting channel and rushing vital communications through System 4 directly to System 5. This "algedonic," or pleasure–pain, signal "wakes up" (the alarm clock) System 5 to any potential disaster further down to which it really must pay attention. This is the pain felt by the bladder of a person who has been out on a night's drinking and is now unconscious. It is hopeless trying to coordinate other organs to relieve the strain and catastrophe would be inevitable if the issue was dealt with in the usual leisurely manner. The only way the danger can be averted is to bypass all normal channels and rapidly rouse the policy-making function. With the help of System 4, the now conscious mind can steer the person toward the normal facilities and deal with the problem.

One of the most important features of the VSM is that it can be used to model all levels of organization. This was the case with the industrial economy in Chile. It is possible because organizational cybernetics regards complex systems as possessing a "recursive" character. Recursion means that the characteristics of higher level systems are repeated in their parts: "All viable systems are contained within viable systems" (Beer 1975, p. 441). Complex systems are seen as consisting of viable systems nested in viable systems nested in viable systems. They are like Mandelbrot's fractals in complexity theory (see Section 7.2). As with a series of Russian dolls, they retain the same basic organizational form even though they may differ a little in the detail. It follows that the same principles that apply to any viable system can be applied at all the different levels of recursion. For example, the VSM can be used to model a subsystem (a division) in an organization, that organization and its supra-system (the system of which the organization is a part). Using the VSM, lower level systems, which inevitably appear as "black boxes" when the organization as a whole is being observed, can become the focus of interest with only a slight adjustment of attention. In Figure 13.2, each of the operational elements, A–D, is a viable system in its own right, with its own policy, development, operational control, coordination, and implementation functions. A "blown-up" version of element B, with its localized management 1B and own environment, would resemble Figure 13.2 as a whole. By dropping a level of recursion, it can be examined in detail using the same model. This allows elegant representations of complex organizations to be constructed and acts as a great variety reducer for managers and management scientists. Again it is necessary to insist that the recursive nature of the VSM does not make it a hierarchical model. In a hierarchy, those at the top are assumed to have the knowledge to make all the important decisions and are given the power to do so. A fractal arrangement, as found in the VSM, is not a means of concentrating power and influence. Rather, it is a way of managing complexity. According to Espejo:

> Recursion is the most powerful strategy to distribute complexity, as autonomous units take responsibility for larger chunks of the environmental complexity.
>
> *(2017, p. 51)*

Its success depends on different managers throughout the organization taking responsibility for those decisions they are best placed to make.

Toward the end of his time in Chile, Beer had meetings with Maturana and Varela and became familiar with their concept of *autopoiesis* (see Section 3.2). As a result, and despite Maturana and Varela's reservations about extending their ideas to social systems, two additional pieces of theory became central to the VSM. The first arose from Beer's identification of viable systems as autopoietic and his working out of the implications. He argued that the autopoietic faculty for a viable system "is embodied in its totality and in its Systems 1, and nowhere else" (Beer 1981, p. 338). It is the system as a whole and its key operational elements that must be designed as self-producing systems on the basis of the VSM. In a company, that means the company itself and its business units. Further,

> ... any viable system developing autopoiesis in any of its Systems Two, Three, Four, or Five is *pathologically autopoietic*; and that entails a threat to its viability.
>
> *(Beer 1981, p. 338)*

Systems 2–5 should see themselves as support services and not viable systems in their own right. If they start to prioritize their own survival, they inevitably turn bureaucratic and get in the way of what System 1 needs to achieve to realize the organization's identity and purpose. In Beer's view, this was happening in Chile, and it was hampering his project. Whether or not Beer is technically correct to identify viable systems as autopoietic systems (Brocklesby and Mingers 2005, argue he isn't) the analogy is certainly insightful. It provides, as Medina says, a powerful conceptual vocabulary "for understanding and critiquing" bureaucracy (2014, p. 219). Anyone with experience of being a System I manager will have encountered finance and human resources departments more interested in maintaining their own comfortable lives than serving the business. I have seen "international offices" in universities grow fat on preventing academic departments from developing international links.

Another insight derived by Beer from *autopoiesis* rests on the theory's distinction between "organization" and "structure." "Organization" describes the fundamental relations that must exist between the components of a system if it is to maintain its identity. These relations have to remain invariant if the system is to preserve itself in existence. In Beer's view, it is these essential organizational features that the VSM captures and describes. "Structure" concerns itself with the particular form the components and their relations take in order to "realize" the necessary "organization" at a point in time. Structure will change, as the system coevolves with its environment, but the system will retain its identity as long as the new structure remains capable of supporting the underlying "organization." Thus, a variety of structural arrangements are possible in organizations as long as they all pay attention to the requirements of the VSM. Its focus on "organization" helps explain the applicability of the model to such a wide range of situations.

We can now turn to the philosophy underpinning the VSM and ask what kind of a model it is. As long ago as 1992, I pointed to a battle for "the soul of the VSM" between advocates of positivist, structuralist, and interpretive theoretical positions (Jackson 1992a). It is clear that, as a model, the VSM can be put to different uses according to the philosophical inclinations of the user. It is nevertheless still worth asking a question about which philosophical and theoretical position it is best able to serve. We pursue this here by considering how it might contribute to the three traditions of cybernetics identified in Chapter 6 – "first-order," "British," and "second-order." The argument is broadened in the critique section of the chapter.

First-order cybernetics is the "cybernetics of observed systems." It sought, we argued, to extend classical science by offering an explanation of the purposive behavior of complex systems. In chapter 1 of *The Heart of Enterprise* (1979), Beer insists that systems, their boundaries, and purposes are all observer-dependent. However, if we can agree on a "convention" about the nature of a system, its boundaries and purposes, then we can begin to communicate about the system and, indeed, "essay scientific statements" about it. Systems are what "we declare them to be" but, once we have a consensus, "there are a great many things that can be scientifically said about systems," not least about the systemic relationships that keep them viable:

> What truly is given in nature is a collection of principles that systems cannot violate if they are to maintain their identity. These principles are the content of the science of cybernetics, which may conveniently be defined as the science of effective organization.
>
> *(Beer 1979, p. 24)*

And Beer is furious with those who fail to acknowledge the existence of a science of effective organization:

> Our institutions are failing because they are disobeying laws of effective organization which their administrators do not know about, to which indeed their cultural mind is closed, because they contend that there exists and can exist no science competent to discover those laws.
>
> *(Beer 1974, p. 19)*

To Beer, Ashby's law of requisite variety stands to management science as Newton's laws do to physics (1989a, p. 18). Flouting it is like trying to ignore the law of gravity.

Beer declared himself to be a scientist, and the VSM, it seems, can contribute to first-order cybernetics by providing objective knowledge about the laws governing the viability of systems. It is certainly traditional science that many of Beer's followers think they are getting. Flores welcomed cybernetics because it offered "scientific views on management and organization" that could be applied on a national scale in Chile. Hoverstadt confirms Beer's view that the VSM has stood up to rigorous testing as a scientific instrument:

> So far, we have not found any sort of organization to which it does not apply. Just as all types of flying things from airliners to Frisbees are subject to the same laws of thermodynamics, so, what it means to be a viable organization seems to be universal.
>
> *(Hoverstadt 2008, p. 6)*

Huhn states that the VSM "has been thoroughly corroborated" and "never falsified" in tests. Further, he says, Beer proposed at least one axiom, recursivity, which was only finally proved mathematically 20 years later, by Mandelbrot in his theory of fractals:

> Epistemologically, this feat may even be on a par with Newton's Law of Gravity, which was vindicated only by the discovery of Pluto 130 years after Newton proposed its existence.
>
> *(Huhn 2012, p. 18)*

Situating Beer's work in the tradition of "British cybernetics," Pickering (2009) presents a very different picture of what the VSM has to offer. According to him, the unique feature of British cybernetics is that it abandons the search for objective knowledge in the traditional sense and, instead, champions a process of finding out what is possible when systems engage with the world. The most significant precursor of the VSM then becomes Ashby's model of the brain as an "embodied organ" interested in performance rather than representation. The brain cares about its own and the body's purposes and not about building a scientific model of the world. Its primary concern is with responding successfully to unexpected fluctuations and changes. For Ashby, it follows that the way to find out what the brain is capable of is to expose it to perturbations and to see how it reacts. Beer's cybernetics is seen, by Pickering, as an extension of this philosophy of "becoming" and "revealing" to organizations and their environments. Beer accepts that exceedingly complex probabilistic systems, such as brains and economies, are unknowable. The most important thing for any viable system, therefore, is not that it possesses a scientific model of its environment but that it is set up appropriately to respond to things it has not encountered before. Successful adaptation becomes the most significant goal. Beer's VSM is important because it shows how a "dance" of "becoming" can be established between an organization and its environment during which they explore each others' possibilities, establishing mutually satisfactory relationships along the way. They evolve together continuously, and it is impossible to predict what will happen. We simply have to watch where the performance staged by the VSM takes us. Knowledge is not "fixed" but emerges as system and environment find out what each other are capable of. Using Heidegger's terms, the world is "revealed" rather than "enframed." It is, to Pickering, confirmatory of his thesis that Beer shortcuts the traditional scientific method to arrive at his model of any viable system. He turns to what the natural world has to offer in the form of a "known-to-be-viable" system, the human body and nervous system, and draws his insights from that. We are back to cybernetics taking advantage of what already exists in nature although, as Pickering remarks, not quite as literally as when Beer was experimenting with ponds controlling factories. None of this is surprising when we remember that what really matters is adaptability rather than a scientific model of either the details of the firm or its environment. Admitting that Beer does sometimes seem to talk in terms of representational models – system dynamics models of the economy and internal models of what is happening in System 1– Pickering argues that this is just a hangover from an old type of cybernetics and that these models are, in any case, not fixed but constantly updated and "geared directly into performance."

I find Pickering's account of the philosophy behind the VSM convincing in relation to Beer's work in Chile. The VSM provided a way of organizing the different levels of the industrial economy in a manner that would allow them to learn their way to increased performance, under workers' control, while adapting to unpredictable disturbances. There was no detailed, scientific plan setting out how the industrial economy should achieve its goals. More importantly, because he was there, Espejo also agrees with Pickering, seeing Allende's "third way" as "performative" (Espejo and Reyes 2011). Medina, who has studied the case in detail, concurs:

> Indeed Beer's work bears the hallmarks of British cybernetics as described by Pickering …. Beer was more interested in studying how systems behaved in the real world than in creating exact representations of how they functioned …. He stressed that cybernetics and operations research should drive action, not create mathematical models of increasing complexity and exactitude …. Instead of using science to dominate the outside world, scientists should focus on identifying the equilibrium conditions among subsystems and developing regulators to help the overall system reach its natural state of stability.
>
> *(2014, pp. 25, 28)*

However, the evidence remains that Beer regarded his cybernetics as more than just "geared directly into performance." For him, they are fixed features of the world, which have been discovered by the science of cybernetics. The VSM itself seems not to be revisable. We have apparently, as Pickering suggests, "an ontology of becoming" within a fixed framework.

There is a possible way of reconciling the view from "first-order" cybernetics with that from "British cybernetics." This involves taking advantage of the distinction between "organization" and "structure" explained earlier. It could be argued that the VSM is scientific in the traditional sense that it is based on laws that explain how viable systems can maintain themselves in existence. These concern the system's "organization." However, these laws do not relate to features of the world in the normal sense. Rather, they are laws that explain what a system must be like in order to be performative. In other words, they make possible the adaptation, through change of "structure", the "becoming" and "revealing" that is central to the perspective of British cybernetics. The VSM provides the necessary initial conditions for the coevolution of system and environment. Once these are in place, the system is enabled to explore the possibilities open to it – possibilities that are sanctioned but not determined by the laws that maintain the system's "organization." This argument is appealing but it requires working on. Beer is not really talking about laws of "conservation of organization" in Maturana's sense. His emphasis is on laws of "conservation of adaptation." Tantalizingly, Maturana mentions such laws, without elaborating on how they are linked to "conservation of organization":

> Indeed, I could speak of the laws of conservation of organization and adaptation as ontological conditions for the existence of any structure-determined system in the same manner as physicists speak of the laws of conservation in physics as ontological conditions for the occurrence of physical phenomena.
>
> *(Maturana 1987, p. 346)*

Perhaps Beer's organizational cybernetics and the theory of autopoiesis address complementary rather than identical aspects of viability.

Second-order cybernetics is the "cybernetics of observing systems." How is the VSM related to this third strand of cybernetics? Second-order cybernetics, as we saw in Section 6.4, was nurtured by von Foerster, drawing on the phenomenological aspects of autopoietic theory, von Glasersfeld's "radical constructivism," and Spencer-Brown's work on "distinctions." All the thinkers involved, at a minimum, argue for "bracketing" objectivity. Attention shifts to observers and the distinctions that they make. These distinctions bring forth a "reality." The role the VSM must adopt if it is to contribute to second-order cybernetics is that of a "hermeneutic enabler." We encountered hermeneutics briefly, in Section 4.3, as the theory and methodology of interpretations. It is Roger Harnden (1989, 1990) who makes the strongest case for the VSM as a hermeneutic enabler. He questions "representational" accounts of the VSM which see it as expressing fundamental laws governing the organization of complex systems. In his opinion, models such as the VSM are best regarded as aids to structuring and orienting ongoing conversations about complex social issues rather than attempts to describe objective reality.

There are, in fact, two possible ways that the VSM can act as a hermeneutic enabler and these tend to get conflated in the early discussions. The first is by supplying a set of concepts that allow a rich discourse to unfold about possible organizational forms. Here it provides an "umbrella of intersection" for different perspectives, allowing a "consensual domain" to emerge with subsequent coordination of actions (Espejo and Harnden

1989a,b). This is the route pursued by Espinosa (Espinosa and Walker 2011, 2017) who sees the VSM

> ... as a meta-language, as a hermeneutical tool, to enable people to engage in a structured conversation about organizational viability and adaptability.
>
> *(Espinosa et al. 2015, p. 205)*

The second possibility, formalized by Espejo as part of his *Viplan* methodology, has the VSM acting as a hermeneutic enabler by setting out the structural context in which good conversations and problem solving can take place (Espejo and Reyes 2011). The VSM is used to uncover relational issues that are restricting proper communication and to suggest other arrangements, based on the logic of the model, that will ensure that the right people are participating in discussing the right issues. The VSM prescribes the conditions most conducive to appropriate conversations leading to effective action.

Returning to the main argument there are, according to second-order cybernetics, as many "truths" as observers can bring forth in their distinctions. All are equally valid. If then, the VSM is regarded as just one means of orientating discussion or one possible way of structuring an organization to permit conversations then, as Harnden says, "the Viable System Model has no incompatibility with second-order cybernetics" (1989, p. 402). Its value, however, is immediately much reduced. It is one possible model among many. Why should we prefer it to classical management theory, socio-technical systems thinking, soft systems thinking, or numerous other points of view, including those derived from common sense? Shackling the VSM to second-order cybernetics, in my view, does it a severe disservice. In fact, the way Espinosa and Espejo use the VSM makes it clear that they do think that it captures something special. They believe it has some knowledge to impart and, whatever they sometimes write, in practice resist the relativism of second-order cybernetics.

The VSM cannot be squared with second-order cybernetics for anyone who regards it as being of significant value for managing organizations. Upholding its worth is incompatible with the philosophy of that approach. Second-order cybernetics can provide no justification for an assertion that the VSM supports a particularly rich discourse about organizational forms or that it can define the right structural arrangements for the best discussions to take place. Any claim that the VSM is an excellent "hermeneutic enabler" is, therefore, just as much a demand for "obedience to knowledge" as is a claim that it offers an objective account of the nature of viability. It closes off alternative perspectives and other options. It is also circular because it can only be made by those who already believe in the value of the VSM. The VSM is an excellent hermeneutic enabler because it leads to appropriate and rich conversations as judged by the VSM. I suppose that a case can be made that it privileges the "bringing forth" of organizations that are more participative, democratic, and fair. But that is not a case that cybernetics itself can make. It must be left to a second-order critique based on critical systems thinking.

The following discussion on methodology throws further light on these arguments.

13.2.3 Methodology

The VSM, Beer claims, embodies in a highly usable form the various cybernetic laws and principles essential to ensuring viability and improving the performance of organizations. It can be used to reveal whether actual or proposed enterprises obey cybernetic laws or flout them. If employed in a "design" mode, its gaze is focused on plans for new organizations, and its aim is to ensure that they are constructed according to good cybernetic principles. The model can also be used in a "diagnostic" mode. Here it acts as an exemplar of good organization against which the structures and processes of an actually existing system can be checked. In this mode, the organizational pathologist can "X-ray" the actual system and judge what is going wrong on the basis of his or her knowledge of what a healthy organization should look like. In either case, little methodological sophistication is required in translating VSM theory into practice. In Beer's view, we simply need to apply its principles in a logical order. Hoverstadt agrees, stating that "the basic methodology for the VSM is so simple that it hardly warrants the term" (2010, p. 128).

The VSM can, however, appear to managers as Utopian, rationalistic, and overly prescriptive when presented to them by experts. Not surprisingly, therefore, recent methodological developments have sought to "soften" the way the approach is used as a means of easing the problems often associated with its implementation. This is the case with Espejo's Viplan methodology (Espejo and Reyes 2011) and Espinosa's "methodology to support self-transformation" (Espinosa and Walker 2017).

Viplan consists of interacting "cybernetic" and "learning" loops. The cybernetic loop encourages stakeholders to question structures that inhibit effective relationships and to transform them so that they facilitate conversations that will, according to the VSM, improve effectiveness:

> The VSM is above all about enabling connectivity and structuring the system to facilitate the healthy development of relationships and ultimately effective performance.
>
> *(Espejo and Reyes 2011, p. 88)*

In the learning loop, the stakeholders take advantage of the enabling structures to engage in a process of continual learning about their situation. They engage in discussions about problematic issues, agree on any changes they want to make, and implement changes that are feasible in their context. The methodology is participative throughout and the two loops feed off one another. For example, the cybernetic loop might be revisited once to help clarify the implications of proposed changes and again to see how they can be sustained. Viplan is said to rest on "constructivist" foundations because it helps stakeholders to align their perspectives and translate their aspirations into structures which can realize them and keep them under review:

> In summary, the Viplan Methodology is a methodology to increase the chances of co-ordinated actions in an organization, with the purpose of aligning participant's actions with the requirement for organizational viability.
>
> *(Espejo and Reyes 2011, p. 213)*

In their book, Espejo and Reyes illuminate this methodology using as an example a large project conducted for the National Audit Office in Colombia. Harwood (2012) provides a clear account of Viplan and discusses its use with small- and medium-sized enterprises.

Espinosa's "methodology to support self-transformation" is, as its name suggests, similarly supportive of involving participants in change processes at all levels of complex systems:

> In particular, when the members have identified an opportunity for self-transformation towards more sustainable and adaptive structures, processes and practices, this methodology contributes to guide their collective learning process. The focus on this learning is to enable people to identify key barriers to organizational viability and sustainability; and to design and develop the required organizational arrangements, to improve their own chances of long-term viability (and/or sustainability).
>
> *(Espinosa and Walker 2017, p. 67)*

The role of the analyst is to facilitate the debate among stakeholders so that a shared diagnosis is arrived at and suggestions for improvement are agreed. Espinosa (Espinosa et al. 2015; Espinosa and Duque 2017; Espinosa and Walker 2017) has provided many examples of the use of this methodology in government projects, and with companies, workers' cooperatives, an Irish eco-community, and indigenous people.

Taking all this into account, it is generally agreed that there are three main phases to be considered when using the VSM. The following account draws primarily on Beer (1979, 1981, 1985) while speaking to particular innovations suggested by Espejo, Espinosa, and Hoverstadt. The first stage seeks to bring about agreement on the identity of the organization. System 5 should express and represent this identity but cannot be its sole repository. It should reflect purposes that emerge from and are accepted by both internal and external stakeholders. Equally, the identity needs to take into account the state of the organization's environment and the opportunities and threats that exist. Proper management of this first step attenuates environmental variety massively by identifying those aspects that are relevant to the system. For Beer, as we saw, the first phase is complete when a "convention" is agreed about the nature of the system, its boundaries and purposes. It then becomes possible for stakeholders to communicate about the system and for cyberneticians to start their scientific work using the VSM. The first step of Espinosa's methodology is called "understanding organizational identity" and should involve representatives from many different stakeholders. It makes use of "rich pictures" and "CATWOE analysis," from Checkland's "soft systems methodology" (see Chapter 16), with a view to expressing different perspectives on the organization, exploring its interactions with its environment, collectively agreeing its identity, establishing a boundary for the study, and clarifying purposes and tasks. The first two activities of Viplan similarly require stakeholders to create a "rich picture" of the problem situation and to examine and clarify their different viewpoints by "naming" and defining systems relevant to those viewpoints. CATWOE, or Espejo's variant "TASCOI" (specifying the Transformation, Actors, Suppliers, Customers, Owners, and Interveners implied by each viewpoint), can be employed to express the different perspectives in a precise manner:

> In methodological terms we need a tool that helps stakeholders to articulate their viewpoints in order to reach agreements and align their purposes to co-ordinate their actions.
>
> *(Espejo and Reyes 2011, p. 121)*

The different systems named are debated in "identity workshops" until stability is achieved around one or more possible identities. These activities are an essential precursor to both the cybernetic and learning loops of Espejo's methodology. Hoverstadt (2011) suggests that it is better to see an organization's identity as emerging from its "structural coupling" with the environment rather than as a reflection of the conscious purpose of senior management. The VSM incorporates many purposes and different groups and levels in the system will seek to adapt to their environments in different ways. The identity of the system as a whole will, therefore, be created, defined, and redefined over time as a result of all these different interactions. Identity emerges at an "unconscious level" in the sense that it is not determined by strategic decision making. The role of senior management is to monitor the "structural coupling" and handle intelligently the changes that occur as a result. They want to understand the significance of the changes for the organization's identity and its future relations with its environment. They can then manoeuvre their organization to ensure a productive strategic fit. Hoverstadt and Loh (2017) argue that this requires a different approach to strategic decision making. It should be based on choosing between available futures as they present themselves rather than slavishly following some mission statement. This is very much in line with the "performative" perspective of British cybernetics.

The second phase of VSM methodology involves "unfolding" the complexity of the organization through different levels of recursion and deciding on the primary activities necessary for it to achieve its purposes at each of those levels. In the Viplan approach, this is about exploring the possible structural implications of the named systems. Primary activities produce the organization's products or services. According to Espinosa and Walker (2011, Appendix 1), primary activities normally have defined tasks, resources allocated to them, and some managerial capacity. Often they are capable of being viable systems in their own right outside of the organization in which they are currently embedded. HR, marketing, sales, finance, etc. are support rather than primary activities unless the enterprise's purpose is actually to offer these services to the market. The logic of the model requires that the business units responsible for the primary activities at each level of recursion are as autonomous as possible within the constraints of overall systemic cohesion. They must have power to act. The strategy of unfolding complexity, therefore, amplifies the variety of the organization with respect to its environment. It also attenuates the overwhelming variety that senior management in the organization would otherwise have to face because most of this is "absorbed" when decisions can be taken at lower levels.

There will always be a choice of "dimensions" along which an organization can unfold its complexity. A university, for example, might decide that its primary activities should first be decomposed into arts, social sciences, sciences, business, and medicine or, alternatively, undergraduate programs, postgraduate programs, research, and consultancy or, possibly, UK, European and international students. Hoverstadt suggests that the usual drivers are technology, geography, customers, and time (2008, p. 63). It is in making this decision that the VSM requires the greatest creativity from its users. The key is to find the best expression of identity and purpose given what the nature of the organization and its environment will allow. So, a university that has chosen to be teaching-centered, or been forced by history and circumstances in that direction, will not want to devote resources and management time to research and reach-out. As we go down to lower levels of recursion activities will again be divided, perhaps now on the basis of a different driver. Once a decision has been made about primary activities and how to divide them, at the different levels of recursion, it is usual to focus the VSM analysis on three levels. At recursion level 1 resides the system with which we are currently most concerned, called the "system in focus." In the university example, this might be the university itself. At level 0 is the wider system of which the university is part, which might be universities in the north of England, or universities in the United Kingdom. At level 2 are the primary activities that have now been determined – perhaps arts, social sciences, sciences, business, and medicine. This can be seen in Figure 13.3 where University X is our system in focus. As Figure 13.3 suggests, if the management of the university wishes to delve into what for it is normally the "black box" of the business school, it could go down to what would be recursion level 3 and use the VSM to ask questions about the various activities into which the business school has been divided, for example, how well is it doing in research?

The third phase of the methodology applies the model to diagnose problems in existing systems or to test the viability of proposed designs. Beer's *Diagnosing the System for Organizations* (1985) goes through the process in detail. In Espinosa's "methodology to

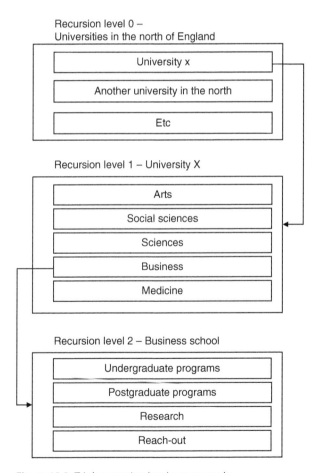

Figure 13.3 Triple recursion levels: an example.

support self-transformation," this stage is called "VSM analysis" or "structural diagnosis" and consists of analyzing the primary activities and checking on the mechanisms for coordination, control, and adaptation. The Viplan methodology is assisted by a Viplan method, which guides inspection of the different structural levels to ensure that the primary activities, at all levels, possess the necessary resources, support functions, and information to be viable systems. Following Beer, and repeating in summary form what we have already said about system identification and unfolding complexity, this third phase runs as follows:

(i) System identification
- Identify the purpose(s) to be pursued (using some appropriate participatory approach)
- Determine the relevant system for achieving the purpose(s) (this is called the "system in focus" and is said to be at recursion level 1)
- Specify the system of which the system in focus is a part (wider systems, environments) (this is at recursion level 0)
- Specify the viable parts of System 1 of the system in focus ("unfolding complexity") – these are the parts that "produce" the system in focus (they are at recursion level 2)

(ii) System diagnosis
Study System 1 of the system in focus:
- For each part of System 1 detail its environment, operations, and localized management
- Ensure that each part of System 1 has the capacity to be viable in its own right
- Study what constraints are imposed upon the parts of System 1 by higher management
- Ask how accountability is exercised for each part and what indicators of performance are used
- Model System 1 according to the VSM diagram

Study System 2 of the system in focus:
- List possible sources of disturbance or conflict in the organization
- Identify the various System 2 elements that are needed to ensure harmonization and coordination
- Ask how System 2 is perceived in the organization – as threatening or facilitating

Study System 3 of the system in focus:
- List the System 3 activities of the system in focus
- Ask how System 3 exercises authority – is this seen as autocratic or democratic by System 1 and how much freedom do System 1 elements possess?
- How good is System 3 at translating overall policy into operational plans?
- How is the "resource bargain" with the parts of System 1 carried out?
- Who oversees the performance of the parts of System 1?
- What audit, or System 3*, enquiries into aspects of System 1 does System 3 conduct and are these appropriate?
- Are all control activities clearly facilitating the achievement of purpose?
- How is the performance of System 3 elements in enabling achievement of purpose measured?

Study System 4 of the system in focus:
- List all the System 4 activities of the system in focus
- How far ahead do these activities consider?
- Do these activities guarantee adaptation to the future?
- Is System 4 monitoring what is happening in the environment and assessing trends?
- Is System 4 open to novelty?
- Does System 4 provide a management center/operations room, bringing together external and internal information and providing an "environment for decision"?
- Does System 4 adequately process, filter, and distribute relevant information?
- Are all development activities clearly facilitating the achievement of purpose?
- How is the performance of System 4 elements in enabling achievement of purpose measured?

Study System 5 of the system in focus:
- Who is responsible for policy (e.g. on the "board") and how do they act?
- Does System 5 provide a suitable identity and convey clear purposes for the system in focus?
- How does the "ethos" set by System 5 affect the perception of System 4?
- How does the "ethos" set by System 5 affect the relationship between System 3 and System 4 – is stability or change emphasized?
- Is System 5 organized to behave creatively?
- Does System 5 share an identity with System 1 or claim to be something different?

Finally, check that all information channels and control loops are properly designed.

Espinosa and Walker (2017, Appendix 1) add "aligning strategy and structure," "agreeing and implementing changes" and "monitoring performance" to their version of VSM methodology.

13.2.4   Methods

It will be useful to set out three "methods" that have been developed to help with the application of the VSM. Shorthand designations for these might be "measures of performance," "frequent faults," and "guidelines for decentralizing support units."

The success of any organization designed on the basis of the VSM depends on the rapid transmission of useful information to where it is required. Beer argues that this is best done on the basis of some general "measures of performance" that allow for appropriate filtration to ensure that only that which is most useful is received. The general measure of achievement used in most organizations is money – the criterion of success being the extent to which profits are maximized. This is not regarded as satisfactory by Beer. It ignores how well the organization is doing in terms of preparing for the future by investing in research and development or employee training. It fails to reveal the cost-cutting manager who, in search of immediate profits, is damaging the organization's long-term future. Instead, Beer advises adopting three levels of achievement (actuality, capability, and potentiality) that can be combined to give three indices (productivity, latency, and performance) expressed in ordinary numbers. These can be used as comprehensive measures of performance in relation to all types of resources throughout the organization.

Defining more clearly the three levels of achievement:

- *actuality* is what we manage to do now, with existing resources, under existing constraints
- *capability* is what we could achieve now, if we really worked at it, with existing resources and under existing constraints
- *potentiality* is what we might be doing by developing our resources and removing constraints, although still operating within the bounds of what is already known to be feasible

The three indices are then:

- productivity: the ratio of actuality and capability
- latency: the ratio of capability and potentiality
- performance: the ratio of actuality and potentiality and also the product of latency and productivity

This is shown diagrammatically in Figure 13.4. These measures are able to detect the irresponsible cost-cutter. The cost-cutting manager will increase productivity not by increasing actual achievement but by lowering capability (e.g. by neglecting research and development and employee morale). This will show as an increase in productivity and, no doubt, profits will rise in the short term. But the manager's latency index, under this scheme, will deteriorate. This should signal that a careful watch be kept on their overall performance. Future profits are being threatened. It was these indices, filtered by Cyberstride and transmitted over Cybernet in real-time, that were the backbone of Project Cybersyn in Chile.

A list of the "frequent faults" discovered in organizations when the VSM is used can help to direct investigations quickly into fruitful areas. Such faults are often called "archetypes" following Senge's use of that term for the problem-types most commonly

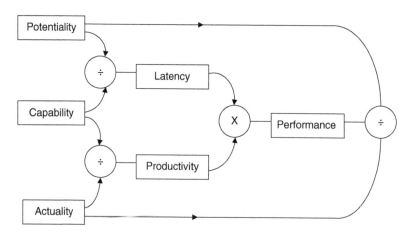

Figure 13.4 Measures of performance.

found by system dynamics. Beer (1989a) identifies four "diagnostic points." Jackson (2000) recognizes six "common threats to viability." Espejo (2008) distinguishes between "identity" and "structural" archetypes and named about 30 or 40 as part of his work with the National Audit Office in Colombia. The former relate to discrepancies between discourse and behavior; the latter – embracing "unfolding of complexity," "cohesion," "stretching," and "policy" variants – to inadequate relations between deployed resources. Hoverstadt (2008) details 21 "pathological archetypes." Espinosa and Walker (2017), drawing on work by Beer, Espejo and Reyes, and Perez-Rios, compile a list of around 50 "organizational archetypes." An example of an important archetype that appears in all the lists is what Espejo calls "the control dilemma." This rears its head when senior management, in an effort to exert more control, continually asks for reports on performance from lower levels. As a result, the resources of the primary activity are exhausted on satisfying requests from higher management rather than dealing with environmental complexity. Here are some more of the frequent faults in organizations that are revealed by the VSM:

- Organizational identity is not adequately defined, represented externally, and spread throughout the system
- Mistakes in articulating the different levels of recursion so, for example, an operational element crucial to overall success is "hidden" lower down and does not get the attention it deserves
- Failure to grant autonomy, with appropriate resources, to System 1 elements
- Failure to ensure adequate localized management exists at the System 1 level
- Systems 2, 3, 4, or 5 seeking to become viable in their own right (pathological autopoiesis) rather than serving the whole system by promoting implementation – leads to "red tape"
- Any of Systems 1–5 being absent or not working properly, but particularly –
  - System 2 is too weak, so coordination is jeopardized
  - System 3 fails to ensure cohesion among the System 1 elements
  - System 4 is too weak so System 5 "collapses" into System 3 and becomes overly concerned with day-to-day affairs
  - System 5 does not represent the essential qualities of the whole system to the wider systems of which it is part
- Failure to achieve a proper balance between the requirement of System 3 for stability and the demands of System 4 for change
- Information flows do not correspond to those necessary in any viable system, e.g. feedback loops not closed

A significant issue for many organizations is whether support activities, such as finance, HR, and marketing (usually located in Systems 2–4), should be managed centrally or distributed to the business units. The VSM is not decentralizing in the sense that it always insists that these are located physically with the operational elements. While viable systems at all levels of recursion need to have these functions available to them, they do not have to own them. They can be performed for them by centralized units. I therefore find particularly valuable Espejo and Reyes' (2011) "guidelines for decentralizing support units." These favor decentralization when the support function is a critical success factor for the primary activity; when it needs characteristics particular to the primary activity it supports; when demand for its services within the primary activity is high; where resources to carry out the support function are plentiful in the overall organization; and when those resources are distributable. Business schools in universities often argue for their own support services on the grounds that they must deal with different types of customers, for example, overseas students and clients for consultancy activities.

13.3 Organizational Cybernetics in Action

Perhaps because of the overlap in meaning between "viability" and "sustainability" there has been considerable interest in how the VSM can be employed to promote sustainability research and practice. Leonard (2008) has contributed by showing how the model can be used for integrating sustainability practices at three levels of recursion – the household, neighborhood, and city levels. The most ambitious attempt to develop the VSM in this direction has, however, been made by Espinosa and Walker (2011, 2017; Espinosa et al. 2008). They argue that the ability to maintain viability, promoted by the VSM, is a preequisite for sustainability. This must be complemented, in progressing toward a sustainable society, with a commitment to ensuring that organizations achieve their purposes in a way that is respectful to all stakeholders while maintaining a balance with their ecological niches. Moreover, the capacity for viability and autonomous action, together with these values, must be spread throughout the system. The traditional hierarchical approach to organization will not do. To achieve sustainability, we must not only change our attitude toward the environment but also the governmental structures we employ to organize ourselves. The only way of dealing with the massive variety exhibited by the rapidly changing environment is to promote self-governance at all levels of society. For Espinosa and Walker, the VSM provides the perfect model for realizing this ambition. It emphasizes self-organization at the local level in a way that amplifies the capacity of the whole system to respond:

> From the VSM perspective, participation is so fundamental that it often escapes notice. Variety balancing between all operations at all levels requires empowered, engaged individuals/communities/organizations, and the only way effective organization can be articulated is to devolve power to the level that things get done .... From the environmental point of view, an empowered community would be in a position to actually DO something about polluted rivers or choking fumes rather than complain that nothing ever gets done by some "higher authority".
> *(Espinosa et al. 2008, pp. 642, 645)*

Further, the VSM offers mechanisms for maintaining maximum local autonomy while ensuring a coherent response from the whole system in the face of environmental challenges. This is essential:

> When dealing with environmental issues, it is not enough to understand individual, institutional, or local issues in isolation: the key is to find a way of thinking about the interrelationships between all these levels and the global ecological systems which sustain all life on the planet.
> *(Espinosa and Walker 2011, p. 11)*

Chief among the mechanisms provided by the VSM to ensure overall systemic cohesion is the notion of recursion – viable systems within viable systems. To achieve a learning society that can coevolve sustainably with its environment, Espinosa and Walker suggest that eight levels of recursion be recognized and integrated using the logic of the VSM. These are individual, family/household, neighborhood/community (urban and rural), town/municipality, eco-regional, national, continental, and global. They then go on to articulate each level of recursion in terms of the issues that need to be resolved and what is needed to make it work – asking what is involved in an individual, family unit, etc. attempting to live sustainably.

To implement these ambitious proposals requires, as Espinosa and Walker acknowledge, smashing through traditional administrative, economic, and political boundaries. An example of what can be achieved is provided in the following case outlining a VSM project in Colombia to create an eco-regional approach to sustainable development. I am grateful to Angela Espinosa, who worked on the project, for sharing with me the details (see also Espinosa 2002).

The Colombian Constitution of 1991 recognized a national environmental system (SINA) consisting of the Ministry of the Environment, a set of regional environmental corporations (CARs), and various environmental research institutes, including the Environmental Development Institute (IDEAM). SINA was seen as directly responsible for developing environmental policies, programs, and instruments that would lead to responsible environmental practices and the protection and regeneration of environmental resources at the local level. One of the main difficulties in the early years was the lack of coordination experienced at the level of the ecosystem, where the most critical problems arose and required action. The agencies charged with dealing with the problems were organized according to the existing political and administrative structures, whereas problems such as pollution and flooding require a coordinated approach based on an ecoregion. Another problem was a failure to follow through on the development of a National Environmental Information System (NEIS) which was to provide the data and models to support environmental management by SINA. This was despite several attempts, sponsored by agencies such as the International Development Bank, to design a centralized system to hold information on the main environmental issues as defined by academic disciplines – flora and fauna, water resources, forests, mineral resources, oceans, etc.

In 1999, the project described here began when the Ministry and IDEAM initiated a new approach to the NEIS using cybernetic thinking based on Beer's VSM. By 2001, a project team consisting of representatives from the Ministry, the CARs, IDEAM, other research institutes, and other national institutions with an interest in the environment were using the VSM to radically reformulate their understanding of SINA and NEIS. The identity of SINA was redefined as follows:

> The National Environmental System [SINA] is a network of recurrent fluid interactions involving individuals, communities and institutions aimed at achieving a sustainable way of living and of interacting with each other and with nature.

Workshops were held at which agreement was reached about the levels of recursion in the system. The basic level of recursion, where environmental action happens, was seen as the community (individuals, families, local industries, and organizations inhabiting a particular settlement) interacting with its natural environment. The organizational purpose at this level was seen as being sustainable development. At the next higher level of recursion, the System 1 would consist of networks of communities with an interest in a particular subecoregion. They would need to interact to solve environmental problems stemming from, for example, a specific ecosystem like a river basin, a natural reserve, or a forest region. A higher level of recursion, concerned with managing environmental issues shared by several subecoregions, was recognized as an ecoregion. At this level, more than 10 ecoregions were identified in Colombia. Finally, at an even higher level of recursion, there was the natural environment of the whole nation. Figure 13.5 represents these recursion levels.

With this analysis in place, it was possible to use the VSM to understand what had gone wrong with SINA and the NEIS and to propose alternative organizational and structural arrangements. The starting point was to ensure that the parts of System 1 at every level of recursion (such as the communities and groups of communities) had the capacity to develop the practices and relationships necessary to achieve sustainable development (i.e. they could be autonomous). This meant that, for each level of recursion, Systems 2, 3, 4, and 5 would have to be put in place to support the System 1 parts. All the levels of recursion would also have to cohere as a whole capable of realizing sustainable development as a national priority. We can now consider, as illustrative examples, work under way at the level of the ecoregion and the community, and aimed at the redesign of the NEIS.

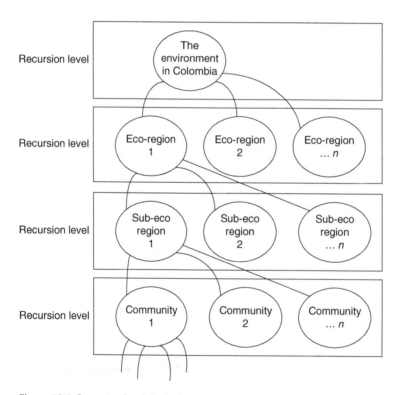

Figure 13.5 Recursion levels in SINA.

Figure 13.6 can help us understand developments at the ecoregion level. It shows the VSM applied to SINA with an ecoregion as the system in focus at recursion level 1. Recursion level 0 is then the natural environment of Colombia and the operational System 1 elements (at recursion level 2) are the networks of communities concerned with environmental problems at the level of the subecoregion. The task for the project team was to assist in establishing a suitable metasystem (Systems 2, 3, 4, and 5) to manage and coordinate these networks of communities so that they could act on relevant environmental problems. Previous attempts had failed because the CARs, which might have done the job, were administratively related to existing governmental structures rather than to ecoregions and subecoregions. The way forward, it seemed, was to redirect the CARs and get them to cooperate with other relevant agencies to produce the kind of metasystemic function required by the VSM diagnosis in order to deal with key environmental issues at the ecoregion level.

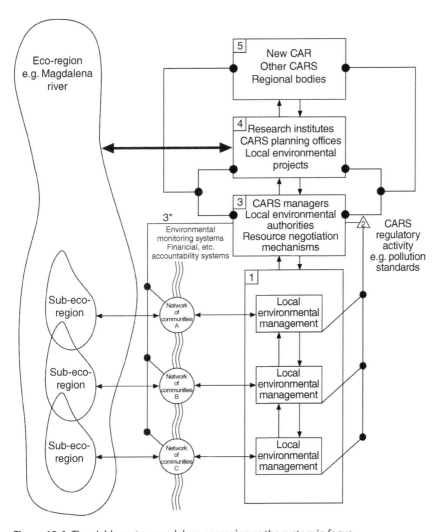

Figure 13.6 The viable system model: an ecoregion as the system in focus.

Aware of this, and determined to reorientate the CARs, the Ministry of Environment focused the main investment resources, from both national and international sources, on what they felt were important strategic ecoregions. The CARs, and other relevant institutions, had to put their effort into developing goals and programs relevant to these ecoregions. In the process they learned cooperatively to overcome the divisions between them created by the existing governmental and administrative arrangements and, gradually, appropriate Systems 2, 3, 4, and 5 were established (see Figure 13.6) in some of the most strategic ecoregions in the country.

An example of this happening was the coordination of projects on the Magdalena River. This is the second most important river in Colombia and provides water to dozens of communities, towns, and cities. The health of this water source is critical for national development. There were significant problems of pollution that required coordinated management because no one would take on a commitment to clean the river unless they were convinced those upstream would do the same. A new CAR was established by the Ministry specifically to address the problems of the Magdalena River. System 2 was established through negotiations with other CARs from areas crossed by the river, which led to agreements on such matters as pollution standards. System 3 emerged from discussions between CAR managers and various local environmental authorities. It began to plan longer term policies and programs to clean the river and to get agreement on clearer norms and the sanctions to impose if these were broken. Efforts began to pool resources to set up a proper System 4. This would need to collect and structure required knowledge on the main ecosystems bordering the river, as well as about the socioeconomic and cultural characteristics of the industries and neighborhoods using and sometimes abusing it. The main challenge for System 5, in the medium term, was establishing a common culture of sustainable development among relevant regional bodies. This would require a significant educational effort to convince the appropriate people of the necessity, especially as it could mean loss of short-term profit.

Of course, success at the levels of the ecoregion and subecoregion depends critically on success at the community level of recursion. Everything ultimately rests on communities of individuals, families, local industries, and organizations. Previous SINA efforts at this level had emphasized the establishment of mechanisms for controlling the use of resources, managing information, and meeting legal requirements. The VSM diagnosis, however, focused on the need for communities to develop self-organizing skills in managing their local environments. Attention was paid to sharing and improving local knowledge on environmental issues and their management, on employing local skills for environmental development, on conservation programs, and on utilizing local information to predict and respond to emergencies.

Success at all levels of recursion depended on rethinking the NEIS. The NEIS had previously concentrated on collecting information based on the various relevant academic disciplines. It was unable to help at all if, for example, a local agency required information useful for dealing with high levels of pollution in a particular environment. Following the VSM diagnosis, it was asked to devote its attention toward information systems that would support decision making relevant to the achievement of environmental goals at each level of recursion. Clear examples of such a reorientation have been some pilot projects, conducted for NEIS, in which CARs and the research institutes have sought to design shared information networks providing information on the type, current state of and critical issues surrounding the main environmental resources characterizing an ecoregion. To finish the job, it would be necessary to combine appropriate geographic information systems, databases, and environmental indices in a manner suitable for supporting the main decision areas in the ecoregion.

Although the SINA project required considerable cybernetic expertise, it also recognized the need to cascade the VSM language downward. Long-term change in any organization cannot be achieved on the basis of remote, expert diagnosis by consultants and recommendations endorsed only by higher management. Participation at all levels is essential.

13.4   Critique of Organizational Cybernetics and the Viable System Model

A reasonable reaction to my account of Beer's work in Chile is that the example is now out of date. Computing power has grown exponentially, connectivity has been enhanced massively through the internet and social media, real-time information systems are common in many organizations (although not government), and use of the Internet to discuss and approve or reject legislative proposals is well established, for example, in Italy's populist "Five Star Movement." Reasonable but wrong, because although technology marches on, the cybernetic principles underpinning effective organization remain the same. Beer's VSM captures those principles in a model of great generality that can be employed to understand and redesign all types of system. It can be applied to systems at different levels in the organization and also to integrate multi-agency arrangements – as Brocklesby (2012) seeks to demonstrate using the example of combating transnational organized crime. It can do this because it focuses on those organizational functions and relations that are essential to viability rather than getting lost describing the many possible variants of organizational structure that act in support, or otherwise, of that viability. The generality of the model acts as a great "variety attenuator" for managers and management scientists. There is only one VSM.

I have argued elsewhere (Jackson 1988a, 2000, pp. 172–174) that the VSM provides an extremely rich account of organizations and how they should be managed. It integrates seamlessly the findings of around 70 years of work in the systems-based version of organization theory. The contributions of the great and good of that persuasion – Burns and Stalker, Lawrence and Lorsch, Parsons, Thompson, Emery and Trist, J. R. Galbraith, etc. – can be subsumed within the model and integrated. And the VSM goes beyond organization theory by assimilating its findings into an applicable management tool that can recommend very specific improvements to the design and consequent effectiveness of enterprises. The model is insightful in the way it treats organization-environment and subsystem–subenvironment relations. The organization is seen as capable of influencing its environment as well as adapting to it. In essence, it needs to set up a sustainable balance with its milieu. It is sophisticated in the manner it understands the tension between the requirements for stability and change, the vertical embedding of different levels within an enterprise, and the horizontal interdependence of elements integrated by coordination and control. The VSM proposes a convincing solution to the perennial problem of centralization versus decentralization, suggesting how cohesion can be maintained at the same time as guaranteeing maximum autonomy to the parts. Control can then be spread throughout the architecture of the system, permitting decisions to be taken and problems addressed as closely as possible to the point where they manifest themselves and leaving senior management free to concentrate on strategic issues. The VSM provides a scientific justification for empowerment and democracy in organizations. The parts must be granted autonomy so that they can absorb some of the massive environmental variety that would otherwise overwhelm higher management levels. Finally, it offers a particularly suitable starting point for planning information systems. Indeed, it puts information processing first when making recommendations for design based on the law of requisite variety. It is not farfetched to see the whole history of systems-based organization theory as an empirical commentary upon the cybernetic principles underlying viability as unearthed deductively by Beer.

On a more contemporary note, the VSM provides a foundation from which the outpourings of many contemporary would-be "management gurus" can be viewed and evaluated. Overwhelmingly, they call for less hierarchy, a process orientation, more organic and adaptable structures, and employee empowerment. Gary Hamel, in *The Future of Management*, states:

> So here we are; still working on Taylor-type puzzles and living in Weber-type organizations …[when]… to thrive in an increasingly disruptive world, companies must become as strategically adaptable as they are operationally efficient … they must learn how to inspire their employees to give the very best of themselves every day.
>
> *(2007, p. 14)*

Wolfe (2011), in *The Living Organization*, demands a shift from viewing organizations as machines to thinking about them as "living systems." Morieux and Tollman (2014) see leveraging the intelligence of employees as the most successful strategy for managing complexity. Laloux (2014), in his amazingly successful book *Reinventing Organizations*, calls for enlightened "Teal Organizations" that empower front-line staff, employ structures that allow workers to self-manage processes, combine pluralism with integration, strive for wholeness, and consciously seek to listen to and unfold their evolutionary purpose within a "paradigm of collective thought." Robertson (2016) wants to abolish hierarchy and replace it with "holacracy" – a model of organizations as decentralized, nested self-organizing teams in which everyone can be a leader. Taleb (2013), having discovered that the world is made up of exceedingly complex, probabilistic systems, wants to create "anti-fragile" organizations that are not just resilient but actually improve in the face of unexpected shocks. Mother Nature, he thinks, is the best guide on how to do this. Biological systems have survived and developed over millions of years without much command and control. He therefore favors decentralized systems that can continuously regenerate themselves following random events, if necessary by sacrificing certain subunits. They will possess redundancy and evolve, learning from their mistakes on the basis of trial and error. In an unpredictable world, Taleb believes, strategic plans which cut down a system's options make no sense. All these authors make a case and often do so using very fine words. However, what they say can appear arbitrary and shallow because it is not underpinned by sound theory. The VSM has sought to get across the same messages for decades and has done so on the basis of the science of cybernetics and numerous practical applications. It has substance and coherence and can actually explain why systems must be organized in particular ways if they are to be viable and effective.

All advocates of the VSM present compelling case studies in support of the claims they make for it. Schwaninger provides a survey of VSM applications and adds two more of his own, in a health services organization and in a small corporation in the chemical industry. Systematic follow-up in these two cases showed that the organizations prospered after the intervention. This leads him to conclude that the VSM conveys "durable and reliable knowledge" and that "high confidence in the model is justified" (2018). Walker (2017) offers further support to the VSM, demonstrating that a number of positive transformations in English local government owe their success to following, although implicitly it must be said, the organizing principles espoused by the VSM. Schwaninger and Scheef have supplied some empirical evidence for the value of the model based on a relatively large sample of firms. The researchers initially set out eight hypotheses based on VSM theory, for example, that an organization's strength of System 4 is positively associated with its viability. These were tested using a questionnaire presented to senior managers. About 261 questionnaires were completed in a satisfactory form of which 70% related to organizations employing 50 or more persons. The results largely supported the hypotheses. In the case of System 4, its strength showed "a significant effect in the direction of a culture of viability." In the researchers' opinion:

> This implies that the VSM is a reliable orientation device for the diagnosis and design of organizations to strengthen their vitality, resilience, and development potential.
>
> *(Schwaninger and Scheef 2016, p. 544)*

They are, nevertheless, keen to point out that all they can definitively claim is not to have falsified the VSM.

Despite its wide applicability and its strengths, other commentators highlight significant problems with its use. In particular, perhaps, the VSM is accused of failing to understand the peculiarities of human and social systems and of failing to gain traction in highly politicized environments. I was a member of a commission looking at different models for the devolution of decision-making from central government to the Hull and Humberside region. The VSM seemed the perfect model to help with the debate about these matters. History, politics, and power crowded and befuddled the agenda to such an extent that it was impossible to make a space for any such "rational" discussion. Jopling (2012) worries about how realistic Espinosa and Walker's model of "a recursive world of embedded viable systems" can ever possibly be. How could we get everyone to agree on what the autonomous units should be and what would constitute the meta-system (Systems 2–5) at each level? The attempt to fit Project Cybersyn into the realities of the Chilean political landscape, in order that it could become more relevant to what was really needed, is a constant theme of Medina's study. She summarizes the issues:

> In the climate of political and economic collapse in which the Cybersyn technologists were working, it was impossible to make the organizational changes Beer wanted or convince factory managers to give serious attention to a high-risk technological prototype .... But perhaps the most important shortcoming of the project, and why it was not adopted more broadly, was that it did not connect to

the political, economic, and social processes that consumed the country ... such problems as runaway inflation, lack of foreign credit, falling copper prices, and black-market hoarding. The system also did not connect to the changes that were taking place on the factory floor ... to make Cybersyn participative in the way he [Beer] desired would require a massive training program. Such a program would have diverted attention from the daily crises taking place in the factories as workers and managers struggled to maintain operations.

(2014, p. 216)

Beer made desperate attempts to align the technology with the government's political aspirations for decentralization and workers' control through the Cyberfolk initiative and, later, the "people" and "allocation" projects. But he could not take other stakeholders with him. Some of the engineers associated with Cybersyn saw it, or for good reason sought to represent it, as apolitical and technocratic. The teams sent out to the factories to model the key processes often preferred to talk to the senior managers rather than workers. With their scientific background, the way in which they operated could seem like the imposition of Taylorism, for example, through the employment of time analysis techniques. The industrial managers, overwhelmed by daily emergencies, were reluctant to use Cybersyn. They had to be chased for the statistics and became further disillusioned by the failure of the system to deliver reports back to them in a timely manner. The different parties in the Popular Unity coalition had different views on how workers' control should be brought about. The Communist Party, with their power-base in the unions, preferred top-down initiatives. In general, Cybersyn failed to empower the workers in the way that was hoped for and they played a limited role in its implementation. Medina also makes the point that the project tended to concentrate on male groups of industrial workers and, to a lesser extent, rural peasants, while largely ignoring other professions and forms of work. Female workers were marginalized as a result. The design of the Operations Room itself, Medina suggests, encouraged masculine forms of expression. Keyboards were deliberately not incorporated in the chairs meant for the senior decision-makers because they were associated with female clerical work and not the serious business.

The internal opposition in Chile and external commentators opposed to the government found it easy to represent Project Cybersyn as an instrument of authoritarian control. Beer pointed to its decentralizing intent, and the mechanisms embodied in the technology to prevent central interference unless it was clear that a lower level of recursion could not cope on its own. This was to no avail. In January 1973, *The Observer*, the UK's oldest Sunday newspaper, ran a story under the headline "Chile Run by Computer." In February, the *New Scientist* published an editorial and article presenting Beer as a super-technocrat running a project that would create a more centralized state and benefit technocrats rather than managers and workers. In April, *Science for People*, an organ representing left-wing opinion in the scientific community in the United Kingdom, published a piece under the headline "Chile: Everything under Control," portraying Cybersyn as a device to increase government control and turn workers into parts of a well-oiled machine (Adams 1973). Medina concludes:

Because Cybersyn never reached completion, I cannot say with absolute certainty whether the system would have empowered Chilean workers or whether it would have increased the influence of a small group of government technologists.

(2014, p. 184)

While exhibiting much sympathy for the project, she accepts that others might regard it as "quixotic" and marginal to what was actually happening in Chile. The journal *Policy Sciences* criticizes an account of the project given by Schwember, who had been closely involved, as displaying a

> ... high level of naivity about processes of social change, political institutions, and organizational behavior on the part of the Cybersyn ... team.
>
> *(Medina 2014, p. 228)*

A second-order analysis of the VSM, in terms of social theory, can throw more light on its strengths and shortcomings. In essence, organizational cybernetics, with the VSM as its main vehicle, is about helping organizations to become viable and effective. If they want to be successful in pursuit of goals, it is argued, they need to be extremely adaptive to their environments and also efficient in their use of resources. The VSM, therefore, is functionalist in orientation. It seeks to provide knowledge, based upon cybernetic principles, that supports regulation in the social domain. Its aim is to increase our ability to "steer" organizations and other social systems. Having said that, it is not a normal manifestation of functionalism we are dealing with here. Beer's organizational cybernetics embraces a "structuralist" rather than "positivist" epistemology to learn about the laws governing system behavior. As was explained in Chapter 4, structuralists believe that the behavior we observe at the surface of social systems is generated by underlying structures or systems of relationships that need to be uncovered and understood. In this respect, the VSM has similarities with system dynamics (see Chapter 11). With system dynamics, it is the relationships between feedback processes operating at the deep structural level that give rise to system behavior at the surface level. In the case of organizational cybernetics, it is cybernetic laws and principles at work below the surface that generate the phenomena we observe. Thus, a violation of cybernetic laws at the deep structural level in an organization will show itself in various pathologies at the surface level. Knowledge of cybernetic principles permits a trained analyst to pinpoint the underlying problems and suggest how they can be corrected. The symptoms will then disappear.

The convergence between cybernetics and structuralism has long been apparent. Levi-Strauss (1968) regards Wiener as having made an outstanding contribution to structural studies. Jackson and Carter (1984) demonstrate a correspondence between the function of myth in Levi-Strauss's structural anthropology and the way variety attenuation works in Beer's organizational cybernetics. Piaget (1973) is very complimentary about the achievement of cybernetics in synthesizing information and communication theories with guiding and regulatory theories. Molloy and Best (1980) argue that Beer's VSM can be used as an "iconic" model to reveal underlying mechanisms giving rise to surface system behavior and to provide explanations of observable phenomena. Beer himself, in *Decision and Control* (1966), is explicit that management science should not content itself simply with discovering the facts but should seek to know what they mean and how they fit together and should seek to uncover the "mechanisms" that underlie them.

Beer's structuralism seeks, as was suggested above, the general "laws of conservation of adaptation." It is these that give the VSM its wide applicability and explanatory power. We are making use of general scientific principles, not proceeding as positivists do to try to find regularities through empirical observation, analysis, and classification of countless surface phenomena. However, from the point of view of the positivist version of functionalism, it is certainly worth asking if we lose something as a result. Beer's criterion of viability is very broad. Viable systems can just "muddle along" – like the coelacanth (Beer 1989a, p. 239). If we want to say more about what makes one viable system more "effective" than another, don't we need to say more about the "structures" that support viability and their usefulness in different situations? Pickering (2010) would say "no." Having set an organization up to be adaptable, it should be left to pursue its performative dance with the environment, each discovering what potential exists in the other. Facilitating this process of "becoming" and "revealing" is what is truly unique and compelling in British cybernetics. Achterbergh and Vriens, by contrast, bemoan the lack of research in this area. In their view, at least two additional types of knowledge are needed to complement the VSM:

> (1) knowledge about designing organizational infrastructures needed to realize the VSM-functions, and (2) knowledge about organizations as *a particular kind* of (viable) system.
>
> *(Achterbergh and Vriens 2011, p. 20)*

The first type concerns guidelines for the effective design of a proper division of work, HR-systems, related technologies, etc. (and their mix) in support of the functions for viability. This would require the development of an empirical research agenda around the VSM. We might then know what structures are beneficial for viability and which impair it. From the functionalist perspective this is certainly essential, and it is disappointing that so little has been done. The second requires a greater recognition of the social dimension of organizations. Social organizations are a special type of viable system. To understand them, we must draw upon other knowledge from social systems theory. Achterbergh and Vriens suggest that Luhmann's work is a good starting point.

It is certainly valuable to consider how the VSM appears to social theorists operating from other sociological paradigms. For Checkland (1980), looking at things from the perspective of interpretive social theory, the VSM offers only one among many possible ways of viewing organizations. It seems to him as legitimate to regard an organization as a social grouping, an appreciative system, or a power struggle, as it is to see it through the eyes of the VSM. The VSM viewpoint, moreover, misses the most important feature of organizations – the fact that their component parts are human beings who attribute meaning to their situation and can ascribe to organizations whatever purposes they wish and make of them what they will. This links to perhaps the most frequent criticism of the model: that it underplays the purposeful role of individuals in organizations. Morris (1983), while not agreeing with this criticism, captures its flavor nicely with his phrase "the big toe also thinks!" Organizations are different to organisms – consciousness and purposefulness are found at all levels. In Ulrich's (1981b) view, cybernetic models leave out the most important feature of socio-cultural systems – human purposefulness and self-reflection. Beer (1989a, pp. 20–21) tries to counter this criticism by insisting that what matters in a viable system are the roles undertaken by individuals not the identity of the individuals themselves. The roles are subject to modeling because they are invariant. Individuals are not and so, however important, cannot feature in a model like the VSM. This answer is logical in cybernetic terms but will hardly satisfy the critics. Interpretive theorists will see it as simply reinforcing their judgment that the VSM places too much emphasis on organizational design and too little on people. If this is correct, managers seeking to promote the effectiveness of their enterprises by concentrating effort on their logical design, following the VSM, are misplacing their energies. Social organizations can exist and perform well while employing a host of apparently illogical structures. The emphasis placed on organizational design may preclude proper attention being given to the generation of shared perceptions and values – to organizational culture. The point can be overdone. The VSM caters for the purposeful role of individuals performing different functions at different levels of recursion. Indeed, it suggests that it is to the advantage of organizations to grant maximum autonomy to individuals within minimal constraints imposed to ensure overall systemic cohesion. Nevertheless, once the identity of the viable system has been established, the emphasis remains overwhelmingly on systemic/structural design to the neglect of the need to manage processes of negotiation between different viewpoints and value positions.

From the interpretive perspective, a major problem facing the VSM is how to arrive at a clear identity for the system, so that the modeling can begin, and then to sustain an agreed identity over time. The problem is exacerbated because the VSM, as Beer formulates it, offers no substantive assistance to managers who must seek to bring about a consensus or accommodation between people with different worldviews and interests. Little attention is paid to methods that might help, at the level of conscious meaning, to achieve and sustain shared understanding about purposes. Ulrich's (1981b) distinction between "syntactic" and "semantic-pragmatic" levels of communication helps to further establish this argument. The syntactic level is solely concerned with whether a message is well formed or not, in the sense of whether it can be "read." This matter can be dealt with by information-processing machines. The semantic and pragmatic levels are concerned, respectively, with the meaning and the significance of messages for the receiver – they inevitably involve people. Ulrich argues that the concept of variety, which underpins the VSM, operates only at the syntactic level. More significant in organizations is the management of processes of negotiation between different viewpoints and value positions. This demands methodologies that function at the semantic and pragmatic levels. We can see how restrictive the VSM is, in this respect, by considering the criterion of "good" management that it implies. Good management is, apparently, management that establishes requisite variety between itself and the operations managed, and between the organization as a whole and its environment. The lesson, from the interpretive perspective, is that good management is also about the meaning and significance of purposes for participants in an enterprise and the creation of intersubjective agreement to pursue a set of purposes. Churchman makes the argument well:

> And this brings us back to the same theme: The Ethics of Whole Systems. We are talking not about techniques of improving performance *given* the goals that certain people wish to attain. We are also talking about whether the goals themselves are proper ones, and we are asking how the scientist can possibly come to answer this question. The underlying theme is that if he fails to answer the question, he fails to 'apply knowledge to system improvement'.
>
> *(1968, p. 15)*

Flores makes a similar point in reflecting back on Project Cybersyn, in which he was a central player:

> My problem [in Allende's cabinet] was not variety; my problem was the configuration of reality, persuading other people.
>
> *(Quoted in Medina 2014, p. 229)*

Following his experiences on the project, Flores began to turn away from the kind of cybernetics he saw as represented by the VSM toward the more subjectivist stance found in the second-order cybernetics of Maturana and the philosophy of Heidegger. Medina summarizes his rationale:

> Flores found that management through variety control did not allow intuitive forms of decision making, nor did it account for the previous experiences and cultural situation of decision makers or accommodate the importance of communicating effectively and with intention.
>
> *(2014, p. 231)*

The methodologies developed by Espejo and by Espinosa do, as we witnessed, pay more attention to the different perceptions and purposes of individuals and seek to engage them in a learning process using the VSM as a guide. These thinkers seek to free the VSM from its functionalist origins and embed it in the interpretive paradigm as a "hermeneutic enabler." At the same time, and in my view wisely, they refuse to abandon their commitment to the VSM as something that possesses an "objective" value. In Espejo's Viplan, its significance is that it can tell us what organizational structures will enable the "right" people to engage in the "right" conversations to ensure effectiveness. In Espinosa's "methodology to support self-transformation," its importance stems from its ability to facilitate a rich discourse about effective organizational forms. From an interpretive perspective, this is like inviting stakeholders to play any sport they want but only providing them with one pitch of a particular size and shape (Espejo) or with just a football (Espinosa). Even while attempting to ground the VSM on constructivist philosophy, Espejo and Reyes refuse to sell out the VSM completely:

> Everything we have said about managing complexity suggests that in the end constructing a systemic world is a *learning process* where the transformations we want to produce are adjusted and modified as we hit walls that make apparent that the cost and consequences of pursuing them are unacceptable. The Law of Requisite Variety asserts itself in all situations but systemic thinking can help us anticipate these walls or regulatory failures to avoid unnecessary pains.
>
> *(2011, p. 255)*

The commitment of Espejo, Espinosa, and others to the "scientific" significance of the VSM makes it impossible for them to make the transition to an interpretive position. Ultimately, they believe that the model does have something unique to offer beyond facilitating conversations. If an enterprise does not respect the law of requisite variety, for example, it will not be as effective as one that does. And it doesn't matter how long stakeholders debate the matter.

For sociologists of a "radical change" persuasion, what organizational cybernetics fails to take into account is that conflict, contradiction, and power play a significant part in social systems. Because of this, it is argued, there are potentially autocratic implications when cybernetic models are used in practice. This is an old criticism. Lilienfeld (1978, p. 73) comments on a 1948 review of Wiener's *Cybernetics* in which a Dominican friar, Pére Dubarle, expresses his fear that cybernetics can help some humans to increase their power over others. The same objection is made against the VSM by Adams (1973), in the *Science for People* article already mentioned, and by Checkland (1980). Beer is aware of the issue, of course, and discusses some of the problems that certain social arrangements present to the proper operation of the VSM, which he insists is in essence decentralizing and democratic. At the top of the list are the existence of power relationships and our acquiescence in the concept of hierarchy (Beer 1985). In Chile, he sought to "immunize" the system against authoritarian usage by designing the technology to allow autonomous units at lower levels to put things right before higher levels were alerted to problems. To critics, however, acknowledgement of the issue and minor corrective adjustments are scarcely enough in relation to such a pervasive aspect of organizational life. In an organization disfigured by the operation of power, many of the features of the VSM that Beer sees as promoting decentralization and autonomy instead offer to the powerful means for increasing control and consolidating their own positions. Even the granting of maximum autonomy to the parts can be interpreted as the imposition of a more sophisticated but equally compelling, management control technique. Workers are encouraged to believe that they possess freedom but this is only the limited freedom to control their own contribution in the service of someone else's interest.

For radical change theorists, an emancipatory approach must actively seek to redress power imbalances and do away with hierarchical forms of organization. The VSM is not equipped to take on this activist role. As Jopling remarks about the many examples Espinosa and Walker (2011) give to show how the VSM can promote the cause of sustainability:

> It is noteworthy that all the case studies concern projects undertaken at the invitation of the relevant government or organization or someone closely involved in it. They do not convince me of the value of the VSM in bringing about change within systems currently dominated by values inconsistent with sustainability, such as most of today's nation-states, or that are embedded in larger systems, such as the global economy, dominated by such values.
>
> *(2011, p. 255)*

To fulfill the liberating role that Beer intended, the model actually requires a milieu that is already democratic and participatory – and ideally, perhaps, a leader who, when System 5 is explained, can exclaim: "At last, *el pueblo*." In Beer's view the System 5 of a firm, as well as looking after the shareholders, should also embody

> … the power of its workforce and its managers, of its customers and of the society that sustains it. The board metabolizes the power of all such participants in the enterprise in order to survive.
>
> *(Beer 1985, p. 12)*

If the stakeholders of a system are agreed about its purposes, and those purposes are embodied in System 5, then the VSM can fully deliver on what it promises and provide the means for pursuing those shared purposes effectively and efficiently, and with only those constraints on individual autonomy necessary for overall cohesion. It is not surprising therefore that the VSM has been inspirational to those promoting communes and workers' cooperatives. These organizations are committed to democratic values and structures. However, if they also want high performance, there can seem little choice but to introduce more hierarchical arrangements. The VSM provides a choice that allows them to combine liberty with operational effectiveness. Seymour and Gunton (see Coates 2013) suggest that the VSM offers communes a means of allowing freedom to the individual without dissolving into chaos. Walker (2018) has produced a guide to the VSM for cooperatives and federations designed to allow them "to function with increased efficiency without compromising democratic principles." His own extensive work aiming to achieve this at the SUMA cooperative is described in Espinosa and Walker, who conclude from the experience:

> A co-operative provides the perfect environment to apply the [VSM] principles described in this book. SUMA survived its crisis, restructured, avoided command-and-control and created a structure which enabled it to compete successfully, and to create a more rewarding working environment. Departmental autonomy has ensured that every part of SUMA has found the most environmentally appropriate way to go about its business.
>
> *(2011, p. 100)*

This compatibility between the VSM and democratic values and structures is what led me to argue (Jackson 1990) that it has a "critical kernel" at its heart. It demands a participatory environment in order to function as well as it can.

It would be possible to continue this second-order critique of the VSM from the perspectives of postmodernism, poststructuralism, and Luhmann's systems theory. Its life would then become even more difficult. A Foucauldian analysis of the work in Chile, for example, would reveal the nature and influence of the power relations involved in the discourse surrounding the VSM. Luhmann's account of the way differentiated "function systems" interpret and construct the world, according to their own logics, should give pause to those who believe that agreement can ever be reached about the autonomous units and recursive levels required to adequately manage a sustainable future.

Instead, the conclusion to this critical systems thinking interrogation of the VSM will take the form of considering its positioning on the "system of systems methodologies" (SOSM). The SOSM grid is meant to suggest what sort of "ideal-type" problem context the VSM is best set up to address. It says nothing about how frequently problem situations resembling this type are found in the world. We can assume that advocates of the VSM will see them all the time. With that proviso, I would argue that the VSM is most at home with "complex-unitary" contexts, as shown in Figure 13.7.

To date, we have argued that "hard systems thinking" sees problem contexts as "simple-unitary" and that system dynamics operates with "complicated-unitary" assumptions, according to the meaning of those terms established in Chapter 8. Beer's VSM was designed for "exceedingly complex probabilistic systems," meaning that

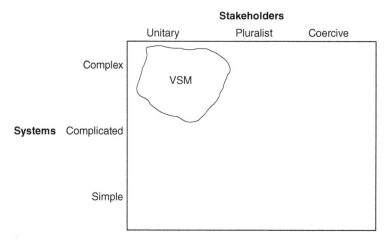

Figure 13.7 The positioning of the Viable System Model on the SOSM.

they cannot be described in precise detail and their behavior cannot be predicted. In my view, Beer has succeeded in this respect and the VSM is, indeed, suitable for "complex" problem contexts in SOSM parlance. The difference between system dynamics and Beer's approach makes this clear. System dynamics believes that it can capture all the important variables impacting system behavior in its models and then describe how they are interconnected in a way that provides insight into that behavior. The VSM, by contrast, assumes that systems and their environments are far too complex for this. They involve so many agents and relationships that it is impossible to identify them all and pointless to try to anticipate the outcome of their interactions. What is required is a design for organizations that enables them to be "antifragile" in the face of unexpected disturbances. This is what the VSM provides. Although complexity theorists do not acknowledge it, therefore, there is already a model in existence that recognizes the issues posed by the most complex type of system conceptualized in Snowden's Cynefin framework and can guide action in those circumstances. Complexity theory has no equivalent and still seems a long way from developing one. As Espinosa and Walker argue:

> It is our experience that there simply isn't enough time to sit back and wait for the breakthrough to occur. CAS [Complex Adaptive System] theory offers insightful ways of understanding societies as complex adaptive systems, but the authors have been unable to identify a coherent set of tools from within this approach, which would support an organisation's journey through the chaos point .... The VSM does exactly this, and can be used to enable a set of sub-organisations, struggling at a chaos-point to break through into new forms of self-organisation and self-management.
>
> *(2011, p. 132)*

Looking at the "stakeholders" dimension of the SOSM, I have allowed some slippage in a "pluralist" direction to accommodate the fact that the VSM encourages a degree of diversity of purpose, within its recursive structures, in order that the system as a whole can be responsive to the variety of its environment. Primarily, however, the VSM occupies the "complex-unitary" space. As we have argued, it struggles with "people complexity" and the "pluralism" of values and interests that this entails. An overall identity for the system has to be agreed, at some stage, before serious modeling can begin. The VSM, on its own, lacks the means to bring about and maintain such an agreement. We therefore find aspects of "soft systems thinking" being corralled into service to help out. Both Espejo and Espinosa make use of parts of Checkland's soft systems approach in their methodologies. Zlatanovic (2016) suggests that "strategic assumption surfacing and testing" (see Chapter 14) can be employed alongside the VSM to manage pluralism and help establish a system's identity. Espinosa et al. explain the need for additional systems tools:

> The VSM provides a meta-language, a template to map the complexity of real life organisations. Additional tools will be required to analyse cultural, political and technological aspects relevant to sustainability. As shown before, other systemic approaches offer complementary tools and methodologies to support sustainability agendas and related complex societal decisions.
>
> *(2008, pp. 648–649)*

While this strategy may help with establishing an initial identity for the system, it does little to meet the objections of interpretive theorists. To them, identity is something that is continually created and recreated through the interactions of organizational stakeholders. This process needs constant attention using methodologies that express the different interpretations held by the various parties and which facilitate debate so that temporary accommodations between them can be achieved and agreed action taken. The last thing interpretivists would support is the use of soft approaches to help consolidate one particular identity in structures that cannot easily be changed. For me the problems posed by this need to establish an "identity," and manage the "unfolding of complexity" accordingly, are well illustrated in Espinosa and Duque's (2018) thoughtful account of the application of the VSM with an Amazonian indigenous association to help with governance issues. The study, as the author's state, is necessarily more "expert driven" than would normally be the case with Espinosa's "self-transformation methodology." But that is not the primary issue. The main point is the consequences that follow from prioritizing one identity. The identity agreed on for the socio-ecological system involved was:

> The association of ethnic communities living there since ancestral times, who have responsibly used their ecosystem services, guided by their clearly ecological life philosophy, their traditional authorities (World Orderer, Kubus, Maloqueros), and their "cultural and ecological calendar".
>
> *(Espinosa and Duque 2018, p. 1011)*

This "identity" seems to have emerged from the priorities of an agency of the Colombian Home Office, for whom the project was undertaken, the authors' own views on what makes for "a viable and sustainable community" and field research, with participative elements, conducted with the communities involved. Once the identity is established, the values it enshrines become embedded in the VSM analysis that follows. The VSM diagnosis revealed:

> The main dilemma they need to resolve involves the interplay of traditional and western values; they need to ensure the new generations are inspired by the traditional knowledge and made aware that their culture has evolved mechanisms far superior to western culture in terms of sustainability, while being exposed to the westernised knowledge they get through public schools.
>
> *(Espinosa and Duque 2018, p. 1015)*

Possible ways forward include the need "to re-establish respect for the Kubus, their traditions and rituals among the youth"; "to limit the use of the digital kiosk and Internet to moments of leisure and to ban them when there are other community activities that need to be undertaken"; to "restore the social order" by creating collective spaces for decision making "where captains and traditional authorities are equally represented and empowered"; and to develop "new and clear mechanisms for social coexistence ... i.e. punitive or social control mechanisms for theft, blackmail, violation, lack of values" (Espinosa and Duque 2018). The consequences of adopting this particular identity and then reinforcing it with semi-permanent supportive structures, using the VSM, seem to me more likely to contribute to preserving a culture in aspic than to help the community coevolve with its changing environment. Espinosa and Duque are well aware of the issues and deal with them sensitively. I would be encouraging the young to revolt, for the same reasons as they do against the norms and mythology of the rigid society described in Chris Beckett's (2012) wonderful science-fiction novel *Dark Eden*.

Moving along the horizontal axis of the SOSM, we encounter "coercive" contexts manifesting "power complexity." Enough has been said to indicate why the VSM is unsuitable for managing the issues that arise in this domain. Beer himself came to the conclusion that a complementary approach to the VSM was required that made possible creative discussion and debate in large groups free from the distorting effects of power relations. Accordingly, in the 1990s, he developed "team syntegrity" as his own response to coercive contexts, as we shall see in Chapter 17.

13.5   Comments

Stafford Beer cut an imposing figure, especially following his experiences in Chile when he turned "hippie," as Maturana put it, and allowed his beard to grow to Tolstoyan proportions. Jonathan Rosenhead describes him as "a charismatic, even flamboyant character" with a "larger-than-lifesize personality," "tall, broad, and brimful of energy, and in later years bearded like an Old Testament prophet" (2006). His long-term partner, fellow cybernetician Allenna Leonard, writes this about him:

> The first thing to say about Stafford was that he was a big man – both in stature and in the scope of his ideas. He was also a polymath – a scientist who painted, wrote poetry, taught yoga and cooked a delicious Yorkshire pudding.
>
> *(2009, p. 224)*

His poetry was published, and he exhibited his paintings in the Roman Catholic Cathedral in Liverpool. Examples of both can be found in the book *Think before you Think* (Beer 2009). You can make up your own mind. He was an accomplished practitioner and teacher of yoga. He was a lover of fine cigars and a prodigious drinker – bottles of whisky, brandy, and sherry in his prime and white wine mixed with water later in life. He was extremely generous of his time with individuals, possibly too generous. In group situations, he did like to command the status of a "guru." I was fortunate enough to meet him a number of times and three occasions particularly stay in the memory.

In January 1986, around the time of the Challenger disaster, which is why I know the date, a workshop on the VSM was organized at Manchester Business School with Stafford presiding. The night before the workshop Stafford could be found holding court with a number of disciples all of whom were describing how the VSM had changed their lives. I was scheduled to deliver a paper on the afternoon of the workshop. It described a respectable VSM intervention that I later turned into an exercise in cybernetic diagnosis, XY Entertainments, which thousands of Hull University MBA students have had to struggle with. During lunch at the workshop, I was told that my presentation had been canceled. Large speakers were erected on stands in the conference room, and we were all informed that the musician, composer, and ex-Roxy Music star, Brian Eno was about to arrive to showcase some new music composed according to the principles enunciated in Stafford's *Brain of the Firm*. I suppose that I shouldn't have been surprised but, at the time, it was news to me just how widely his ideas had spread. Eno was not the first composer to be inspired by cybernetics. Roland Kayn specialized in "cybernetic music" and Bebe and Louis Barron, who wrote the score for the classic science-fiction film *Forbidden Planet,* frequently cited Wiener. But he is certainly the best known in popular circles. Eno was introduced to Beer's work by his mother-in-law who borrowed *Brain of the Firm* from Swiss Cottage Library:

> Stafford's book hooked me with one sentence: *Instead of specifying it in full detail, you simply ride on the dynamics of the system to where you want to go ….* And the lovely thrill of it was that the system could produce beautiful things which you had never even conceived of in advance – which hadn't ever existed in any mind.
>
> *(Eno 2009, p. 8)*

Beer's thinking fed into Eno's experiments with "self-generating music":

> Actually everything I did from 1975 was to some extent informed by him. *Music for Airports* is a pure example of a systems record: set a few things in motion and record the result.
>
> *(Quoted in Whittaker 2003, p. 58)*

Generative music seeks to liberate sound from the tyranny of the composer. It is composed "bottom-up" with only a few basic rules specified. Regulation is built into the system rather than externally imposed. Taking off from the initial settings, the feedback dynamics of the system can produce something surprising, even more so since accidents are encouraged. Complexity and intelligence emerge from simplicity. In many interviews, Eno uses a gardening analogy to explain the difference between traditional and generative forms of composition. The classical style, based in the engineering/design paradigm, tries to specify everything in advance. It is like engineering a tree. Generative composition follows a biological/evolutionary paradigm. A gardener tries to create the right conditions for selected seeds to grow but cannot predict exactly what will occur, as the system develops with its environment, and so is always likely to be surprised by the outcome. There are real insights here into the nature of British cybernetics and the VSM. At some stage in their friendship, Beer invited Eno to his cottage in Wales. He explained that the "torch" of cybernetics had been passed from Wiener to McCulloch, from McCulloch to Ashby, and then to himself. Beer now wanted to pass it on to Eno. Eno was flattered but declined, fearing that he knew too little about cybernetics (Eno 2009, p. 10). He did, however, introduce other musicians such as Robert Fripp and David Bowie to Beer's ideas and it is said, though I cannot find direct evidence for this, that Bowie mentioned *Brain of the Firm* on the radio program *Desert Island Discs* as one of his favorite books.

The second occasion was at the International Federation of Operational Research Societies (IFORS) specialized conference on "Operational Research and the Social Sciences," held at Queens College, Cambridge, in April 1989. Beer delivered the after-dinner speech at a sumptuous banquet in College, and it was my job, as one of the conference organizers, to keep him topped up with white wine. His speech, over an hour and a half long, contained, among much else, an absolutely coruscating attack on Mrs. Thatcher and her government's policies. He was genuinely worried that he would get arrested that night and dragged to a prison cell. His life had indeed been in danger in Chile. Nevertheless, this seemed to me to be a complete misreading of the situation. The British have other ways of dealing with dangerous dissidents – not least by emasculating them with sumptuous banquets in Oxbridge colleges. His wonderful speech, "I am the Emperor – and I Want Dumplings," was published in *Systems Practice* (1989b).

The final memorable encounter was at a systems conference in New Delhi in January 1993. I had traveled to India with Peter Dudley also from the Centre for Systems Studies at Hull University. We had a number of conversations and many drinks with Stafford and he told us, and it was announced at the conference, that his presentation was going to contain something earth-shattering. I had been conned this way before by Gordon Pask at a conference in Vienna and then not understood anything he said. Perhaps it is a cybernetic ploy. Peter and I decided that we would rather go to see India versus England in a cricket match in Jaipur. Stafford gave one of the earliest public presentations of "team syntegrity." His book, *Beyond Dispute*, on the subject was published the following year. It would have been nice to be there but, then again, England did win on the last ball of their innings. Stafford found reason to remark on our behavior in a letter to David Whittaker (2003, p. 370) and was, apparently, prone to bringing it up in conversation to the end of his days.

13.6 The Value of Organizational Cybernetics to Managers

A little knowledge of the VSM can take managers a long way toward achieving a better understanding of their task. It can also save them a lot of time – it incorporates and explains the findings of traditional organization and management theory and makes the pronouncements of most contemporary management gurus appear trivial. That is not to say that managers do not need expert help to master the approach and use it successfully. The VSM has many levels of sophistication, and we have certainly not been able to touch on them all in this chapter. Let us outline the five main insights that managers can gain from organizational cybernetics:

- It is essential to establish a clear identity for an organization that embodies purposes achievable in the environment and is understood and agreed upon throughout the enterprise. If the environment or that identity changes, the organization will need to reinvent and reconfigure itself
- The VSM can provide for a shared understanding of organizational complexity and provides for a rich discourse on issues of organizational design and structure, stability and change, control and coordination, centralization, and decentralization, etc
- It demonstrates that there is a solution to the perennial problem of centralization versus decentralization. The parts can be given autonomy and empowered without any threat to managerial control and organizational cohesion. Indeed, freedom and control are complementary rather than in opposition
- Once an identity and purposes have been developed, the VSM enables essential business units and their necessary support services to be determined. It is a vehicle for design or diagnosis that tells managers which structures and processes are essential and which can be dispensed with
- Because the VSM spreads decision making and control throughout the "architecture" of the system, it makes sense of the idea of leadership at all levels

13.7 Conclusion

Beer was determined to break with traditional management thinking. He looked at company organization charts and despaired. To him they were totally unsatisfactory as a model for complex enterprises. They suggested that the person at the top needed a brain weighing half a ton – since all information flowed up to him and all decisions appeared to be his responsibility. As Beer commented, peoples' heads do not get bigger toward the top of an organization, except perhaps in a metaphorical sense. Starting with neurophysiological analogies and drawing upon cybernetics, the "science of effective organization," he set out to construct a more useful model. The result was his influential VSM: a model of the key features that any viable system must exhibit. This provides a radical alternative to "mechanistic" thinking and suggests organizational forms that are more appropriate in today's turbulent environments. Furthermore, the VSM conveys its insights in a relatively useable form. Managers just need to apply the model. They have only themselves to blame if they don't try to understand it and the lessons it imparts. They will face difficulties. All systems models and methodologies struggle with the complexity of today's problem situations. But they can rest assured that they are better and more relevant to practice than those produced by any other discipline.

Type E

Systems Approaches for People Complexity

For some people, when you say 'Timbuktu' it is like the end of the world, but that is not true. I am from Timbuktu, and I can tell you we are right at the heart of the world

(Farka Touré 1994)

Many systems thinkers argue that systems involving people are unique and that, as a result, a radically different kind of approach is required if they are to be successfully managed. They have developed a strand of the systems approach called "soft systems thinking," which shows an overriding concern with people complexity. This form of complexity is seen as emerging from human self-consciousness and free will, and is reflected in the very different ways individuals and groups see and respond to the world. The plurality of perspectives that exists has been enhanced, in contemporary society, by an increase in education among the populace and by a loss of belief in many of the old guarantors of consensus (religion, capitalism, communism, etc.). Soft systems approaches seek to improve organizational performance by exploring different perspectives and ensuring that enough agreement is obtained among stakeholders about the purposes they wish to pursue to enable them to take action. Their primary orientation is to review different aims and objectives, promote mutual understanding, ensure an accommodation is reached, and gain commitment to purposes. It is argued that only when an agreement has been reached about "doing the right things" can other types of systems approach begin to contribute to "doing things right." Three soft systems approaches are presented and reviewed in this section:

- Strategic Assumption Surfacing and Testing
- Interactive Planning
- Soft Systems Methodology

Critical Systems Thinking and the Management of Complexity, First Edition. Michael C. Jackson.
© 2019 John Wiley & Sons Ltd. Published 2019 by John Wiley & Sons Ltd.

14

Strategic Assumption Surfacing and Testing

The systems approach begins when first you see the world through the eyes of another

(Churchman 1979b, p. 231)

14.1 Prologue

Two important Strategic Assumption Surfacing and Testing (SAST) interventions have involved the US Bureau of Census. Both sought to help resolve messy public policy issues. The first in 1979, documented by Mason and Mitroff (1981) and Barabba and Mitroff (2014), was concerned with the possible adjustment of the inevitable undercount in the 1980 census. The second conducted in 2005, and reported in Barabba and Mitroff (2014), was around the very contemporary issue of the rights of individuals to privacy.

The problem of whether the Bureau should adjust its 1980 population count or not was highlighted on January 14th of that year in *Time* magazine:

> Counting Americans every ten years would seem to be a noncontroversial enterprise, but the 1980 US census has become immersed in politics up to its last decimal point. A growing number of people are worried about the accuracy of the tally because so much – political power as well as the distribution of billion dollars in federal funds – is riding on the outcome.

Census data are used in the United States for, among other things, reapportioning seats in the House of Representatives and allocating back to states, cities, and communities, for various mandated programs and grants, billions of dollars collected by the Federal government through general taxation. The issue was that the census always produced an undercount because people failed to fill in the documents and could not be traced on follow-up. More importantly, the degree of undercount varied significantly according to, for example, ethnicity. Studies suggested that Whites were undercounted by 1.9%, Blacks by 7.7%, and Hispanics by even more. Certain ethnic groups, and those regions and municipalities where they were over-represented, stood to lose out both in terms of political representation and cash. The Bureau was coming under pressure to adjust its raw data and law suits were being filed in case it failed to do so.

Critical Systems Thinking and the Management of Complexity, First Edition. Michael C. Jackson.
© 2019 John Wiley & Sons Ltd. Published 2019 by John Wiley & Sons Ltd.

The policy problem facing the Bureau was whether to adjust its direct numbers or not and, if so, what adjustment method to use. To ensure that all the relevant issues and assumptions were identified and considered Vincent Barabba, the Census Director, called for a SAST workshop that took place over five days in the first week of September 1979. There were around 30 participants consisting of key staff from the Bureau and pertinent external stakeholders, including representatives of those preparing to sue. Following the SAST methodology, the participants were divided into groups – in this instance, they were allocated randomly into four groups. Each group was tasked with making the strongest possible case for a particular policy position. The first group was to argue for sticking with the raw count; the second for following the 1970 model of making minor adjustments on the basis of "imputation" when supporting information was available about households; the third for making significant changes to the direct figures using estimates based on age, race, and sex; and the fourth for making radical adjustments taking into account any information held on the critical variables leading to undercount, including age, sex, ethnicity, relationships, income, race, and language. Working independently, the groups listed the stakeholders who they felt were important to the success or otherwise of their preferred strategy. Stakeholders identified included the Census Bureau, the Administration, Congress, other governmental agencies, radical and ethnic groups, other vested interest groups, the legal community, media, the statistical community, and other data users. There were many other sub-groups within these main categories. Each group then identified the critical assumptions that needed to hold about the different stakeholders if their strategy was to be successful, classifying them according to their importance and certainty. For example, the first group was certain that the professional staff in the Census Bureau wanted to avoid setting a precedent for the ongoing politicization of the census and felt that this was very important. The third group thought it is important that the statistical methods used to make adjustments were defensible but were not certain that they could be made watertight. Each group was thus empowered to make a coherent case for its position, emphasizing its strong points and taking into account possible weaknesses. On the following days, the groups came together to present their preferred strategies, clarify their positions in the light of questioning, and openly debate the arguments and assumptions for and against the different strategies. Each group took some time on its own to consider which of the assumptions of the other groups, if true, would be most damaging to its position. It was not possible to agree a strategy to recommend to the Bureau, even of a "hybrid" kind, but the discussion did reveal seven issues that continually surfaced during the debate and continued to divide the groups: politicization, statistical defensibility, legal defensibility, "does accuracy lead to equity," timeliness of data availability, public demand for adjustment, and congressional demand for adjustment. Further analysis of these issues was deemed essential before taking a final decision, and information sources were identified that could help reduce the uncertainty that continued to surround them.

In the knowledge that the different possible strategies had been thoroughly examined, the crucial issues identified, and the necessary further research done, a decision was taken by the Bureau to run with the unadjusted, actual census count. Prior to this, however, it released a full account of the SAST workshop to justify its decision. The workshop made it possible for the Bureau to be completely transparent about the assumptions it was making in taking this decision and rejecting other possibilities. Over 50 lawsuits followed, and it was only in December 1987 that the US District Judge, John E. Sprizzo, reversed a lower court's decision and upheld the Bureau's determination not to adjust. The Bureau's case was considerably strengthened because it could show in what respects and to what degree it was failing to get the actual numbers right and had critically examined, before rejecting as flawed, the possible ways of adjusting the figures. In a footnote to his position, the Judge commented:

> The Court finds as a matter of fact that ... the Bureau's decision was primarily based on its determination that it was not feasible to develop and implement an adjustment methodology which would be more accurate than the census itself, a determination supported and confirmed by the evidence at trial.
>
> *(Quoted in Barabba and Mitroff 2014, p. 41)*

A second significant SAST workshop held by the Census Bureau, on 2–4 March 2005, was around the issue of privacy. On two occasions, in 2002, the Department of Homeland Security asked the Bureau to provide information on Middle-eastern ethnic communities and it obliged. When this became known, even though the information could have been easily obtained from the Bureau's website, serious concerns were raised about its practices. The Director, Louis Kincannon, asked Barabba to conduct another SAST workshop on how the Bureau could reconcile the rights of individuals to privacy with its commitment to providing vital information to inform the decision-making of individuals, organizations, and government. The participants included executives from the Bureau, experts in privacy issues and the use of data, and representatives from the Arab American League and the Civil Liberties Union. On this occasion, three teams representing diverse perspectives on the issues were formed. Group A was asked to support and defend the position that the value of having information available that could contribute to knowledge overwhelmed the concerns of individuals for privacy. Group B was to make the best case it could for the efficient and effective collection of accurate information while recognizing that the Bureau, whatever it did, was still likely to be accused of breaching confidentiality. Group C took the position that the current laws adequately protected the identity of individuals and that the Bureau needed to work harder to convince individuals that their fears were groundless. Each group identified relevant stakeholders and the assumptions they needed to make about them to support their arguments. These assumptions were rated as to their importance and certainty. The best cases that could be made for the different positions were presented and clarified in a plenary session. The groups went away to consider the assumptions made by the others that were most troubling to them and then reconvened for further discussion and debate. Six issues arose continually and were targeted for further research:

- What is the process by which the Census Bureau exercises discretion in determining how and what to collect and to distribute it?
- Does the Census Bureau need to address inferential harm? For example, where information collected about a group could result in harm to an individual member of that group
- If the Census Bureau collects information, is it thereby compelled to report it?
- How can the dialogue between the Census Bureau and its stakeholders be enhanced?
- How can we ensure that technology has positive and not negative effects?

- How does the Census Bureau design robustness into operations to address future conditions and unintended consequences?

According to Barabba and Mitroff:

> The Privacy and Data Use Workshop succeeded in surfacing assumptions that had been implicit prior to it. As a result of the exercise, the Census Bureau achieved a better understanding of the issues, assumptions, and concerns related to privacy and data use.
>
> *(2014 p. 76)*

14.2   Description of Strategic Assumption Surfacing and Testing

14.2.1   Historical Development

The reader will recall the name of Churchman from Chapter 9. He was one of the most influential pioneers of operations research (OR) in the United States and, in 1957, coauthored an influential textbook on OR methodology. During the 1960s and 1970s, he became increasingly disappointed in OR and turned his considerable talents to the development of what he came to call "social systems design." Churchman's disillusionment (see 1979a) stemmed from what he saw as the betrayal of the original intention of OR. In his view it set out to be a holistic, interdisciplinary, experimental science addressed to problems in social systems. It had ended up obsessed with perfecting mathematical tools and techniques of relevance only to a narrow range of tactical problems. As a result, he established his own educational program in "social systems design," at Berkeley, to keep alive the original vision he had for OR. Both Dick Mason and Ian Mitroff, the key figures in the formulation of SAST, studied with Churchman at Berkeley and were hugely influenced by his thinking. They regarded the systems approach they were learning about as philosophically enlightening and relevant to broad strategic issues and wanted to develop it further in tandem with a methodology that would allow its insights to be put into practice:

> Over the past fifteen years [since 1966] the research of the authors has proceeded on two parallel but closely related fronts: the methodological and the philosophical .... It should be acknowledged that the constant interplay between these two fronts is itself a partial expression of our own philosophical stance. As such, it derives from, among many, Ackoff, Churchman and Dewey with whom we are in deep sympathy. In a word, we do not believe that philosophy, or any other discipline for that matter, can proceed in isolation from the insights and knowledge of other disciplines or from contact with practice .... To twist an old saying, theory without practice is sterile and irrelevant; practice without theory is unreflective and hackneyed.
>
> *(Mitroff and Mason 1987, p. 137)*

They initiated an interdisciplinary research program into "dialectical pragmatism" and credit Mason (see 1969) with conducting the first case study of the dialectic inquirer "both as a naturally occurring phenomenon within an organization and as a tool for strategic planning" (Mitroff and Mason 1987, p. 138). It was this research program that culminated in the formulation of the SAST methodology.

A major breakthrough came during Mitroff's year on a visiting appointment at the Wharton School, University of Pennsylvania. He took this up because it enabled him to work closely with Russ Ackoff, another student and then close collaborator of Churchman. While at Wharton, Mitroff became involved, together with Jim Emshoff, on an important consultancy with a Philadelphia-based pharmaceutical company (see Barabba and Mitroff 2014, pp. 23–28). The company's most important product was a painkiller which, at the time, was only available with a doctor's prescription. Unfortunately for the company, its profitability was being threatened by the ready avail-ability of generic alternatives that were just as effective. The President convened a meet-ing of the company's 12 most senior executives to seek a way forward. The executives divided themselves spontaneously into three groups, each championing a different solution to the problem. One group wanted to lower the price of the product in order to "out generic the generics." A second favored raising its price further to indicate to cus-tomers that the company's pill was a premium product that was better than the gener-ics. The third group argued for concentrating all efforts on this product, adopting a half-way house in terms of price, and increasing overall profits by abandoning any new and expensive research and development. Unable to agree, they set off to collect the data they thought would enable them to prove which of the strategies was best. This approach failed miserably because each group collected data that supported their exist-ing perspective and interpreted it in line with their preferred solution. Barabba and Mitroff explain it thus:

> Ever since the great German philosopher Immanuel Kant, it has been known that one can't collect data without presupposing some theory as to the nature of the phenomenon one is studying. Otherwise, one just wanders around collecting everything and ultimately nothing. In other words, data are not theory and value free. Unbeknownst to them, each group's pet proposal functioned as a hidden theory that steered it to collect data that insidiously supported the particular theory each group was in favor of! Data are anything but neutral.
>
> *(2014, p. 25)*

To make any progress it was necessary to make explicit and challenge the assump-tions underpinning the different theories the participants were bringing unconsciously to the discussion. A flash of insight led Emshoff to propose that "assumptions are the presumed properties of stakeholders." Emshoff and Mitroff got the groups to list as many stakeholders as they could relevant to their preferred strategy. They were then asked to consider what assumptions they were making about these stakeholders in arguing for their strategy and to plot them on a grid composed of the two dimensions of importance and certainty. Patients were a prime stakeholder and all the groups thought it important and certain that they wanted low-cost, high-quality drugs. Physicians were clearly key stakeholders and what was assumed about them was important because they had to prescribe the company's painkiller. However, as debate proceeded, it became clear that there was high uncertainty about whether they were price-sensitive or not. The group that wanted to reduce the price thought that physicians were becoming more price sensitive because of concerns about the rising cost of health care. The group that wanted to raise the price of the product implicitly believed they were price-insensitive – they would prescribe the drug they thought effective regardless of cost. The assumptions relating to physicians were of very high importance but were the most uncertain. At last, the company knew exactly what information it needed before deciding on its strategy. Eventually, it was concluded that if the company raised the price and then, for market reasons, had to lower it again, physicians would still prescribe it. The opposite was not true. The company was *McNeil Pharmaceuticals* and the product *Tylenol*. It decided to raise the price.

Since its original formulation SAST has been employed in a variety of settings; from the Census Bureau to companies such as Xerox and in public health agencies. Barabba (2004) details significant applications in Eastman Kodak and General Motors. Mitroff has taken the ideas into his influential writings, workshops, and projects on "crisis management." Crises occur when our taken-for-granted assumptions about how things ought to work prove to be badly wrong. They happen more frequently as things become more interconnected, but can sometimes be avoided if our operating assumptions are revealed and challenged and we take appropriate action in advance. Mitroff insists that there are members from different industries in his crisis management workshops who can challenge the "normal thinking" of those inside a company or industry (Barabba and Mitroff 2014, p. 33). I have had some success using SAST with "Thornton Printing Company" (Ho and Jackson 1987), with a Co-operative Development Agency (see later in this chapter), at the University of Hull and, recently, in developing and refining a five-year strategic plan for Hull and East Yorkshire Children's University. Chowdhury and Nobbs (2008) describe a SAST study that looks into the relative merits of top-down versus bottom-up IT strategy in the UK National Health Service.

14.2.2 Philosophy and Theory

Churchman, whose work inspired SAST, is frequently referred to as the major philosopher of the systems approach and management. He obtained his undergraduate, master's, and PhD degrees in philosophy from the University of Pennsylvania, where he was taught and profoundly influenced by the pragmatist philosopher Edgar A. Singer. Singer had been a student of William James at Harvard. It is, therefore, on a philosophy heavily influenced by pragmatism that Churchman builds his systems thinking. Churchman takes from pragmatism (see Section 1.4) the belief that it is mind that imposes a structure on the world and not vice versa. According to this perspective, reality is in the making and people have a crucial role to play in creating their own future. The "truth" of ideas should be established according to their practical outcomes. In other words, those beliefs that are effective in helping people achieve their goals in the long-run, as judged by the relevant community, can be accepted as true. Churchman saw his task as applying Singer's philosophy to practical problems. His overriding concern, throughout his career, was with how we can produce knowledge that can improve the quality of human life. Hence, his early interest in OR, which he saw as an attempt "to secure improvement in the human condition by means of the scientific method" (quoted in Mason and Mitroff 2014, p. 38) and, once that had failed him, in social systems design. In pursuing this endeavor, he faced one massive problem. Because of the interconnections between the parts of systems, attempts to address problems in one part often lead to unforeseen consequences. To be sure of improving things overall, we need to understand the whole system. As Churchman put it:

> The problem is very simple: How can we design improvement in large systems without understanding the whole system, and if the answer is that we cannot, how is it possible to understand the whole system?
>
> *(1968, p. 2)*

In trying to answer this question, and in formulating social systems design, he drew upon the whole of the Western philosophical tradition. His greatest work, *The Design of Inquiring Systems* (1971), is a review of how five philosophers, Liebniz, Locke, Kant, Hegel, and Singer, suggest we go about acquiring knowledge that can be useful to us. According to Churchman, the Singerian inquiring system is the most comprehensive, because it is capable of incorporating the others. In practical terms, Churchman's approach translates into saying that while we must try to be comprehensive we should also recognize that all attempts to grasp the whole system will inevitably fail. All we can do is to take into account as many viewpoints as possible when trying to improve social systems. Social systems design is about "unfolding" the implications of partial perspectives in terms of their restricted "boundary judgments" and "sweeping-in" as many viewpoints as we can. This must include the perspectives of "enemies" of the systems approach (Churchman 1979c), for example, "aspects of humanity" such as religion, politics, ethics, and esthetics, which remind the systems approach that its own best efforts to be rational and comprehensive are themselves flawed. If we ignore them, we will never get implementation. In Churchman's view, it was the failure to take account of these four aspects of humanity that prevented the recommendations of the "limits to growth" model, considered in Chapter 11, from being implemented (Churchman 1979b, p. 131).

If Churchman is the philosopher of the systems approach, he is also its moral conscience. In his view, if failing to take account of the whole system is bad science it is also bad ethics. Any science of management must concern itself first and foremost with helping humans to manage their affairs ethically, i.e. in the best interests of all. We should feel moral outrage when our systems designs fail to deliver for all their stakeholders:

> It would be a good thing if the systems planner's germination was moral outrage and not just a mild felt need. In other words, I do not think we should view the major problems of the world today with calm objectivity. We shouldn't first ask ourselves for a precise and operational definition of malnutrition. We should begin with 'kids are starving in great numbers, damn it all'!
>
> *(Churchman 1982, p. 17)*

Churchman sometimes refers to his thinking as "a deception-perception" approach to systems. We have to take all points of view seriously while also recognizing their limitations. What is required, therefore, is a continual review of problem situations conducted by constructing debate between different perspectives. At the close of *The Systems*

Approach (1979b), he presents four principles of this approach. We shall take each of these in turn and expand on its meaning.

> The systems approach begins when first you see the world through the eyes of another.

This principle draws upon the work of Kant and Hegel. We are reminded that it is the mind, through its sensibilities and categories, which constructs and integrates our experiences (Kant). Following Hegel, we also need to recognize that whatever world-view, in German *Weltanschauung* (W), we hold is historically and culturally conditioned. There are an infinite number of alternative worldviews. Systems designers, therefore, must accept that completely different evaluations of social systems, their purposes, and their performance will exist. The only way we can get near to a view of the whole system is to look at it from as many perspectives as possible. Subjectivity should be embraced by the systems approach. It follows, for Churchman, that the systems approach begins with philosophy, which pictures the world, through radically different lenses, as "modeled," experienced, dialectical, etc. The systems thinker must learn to inhabit and believe in the "rightness" of each of these positions.

> The systems approach goes on to discovering that every world view is terribly restricted.

In *The Design of Inquiring Systems* (1971), Churchman shows that each of the five designs for acquiring useful knowledge is incomplete in itself, resting upon assumptions that cannot be proved using its own logic. This has profound implications for our understanding of objectivity in systems thinking. In the hard, system dynamics, and management cybernetics traditions, objectivity is seen to rest on how well the model constructed explains and, preferably, predicts the behavior of the system of concern. Churchman is arguing that any model can give only a partial view of a social system. The manner in which any worldview is "terribly restricted" can be revealed if we "unfold" the account it provides of a purposeful system using 12 categories based on the roles and responsibilities people must undertake for a system to be teleological, as shown in Figure 14.1.

Client
    Purpose of system
    Measure of performance of system
Decision-maker
    Environment of system
    Components of system
Planner
    Implementation of plans
    Guarantor-of-design
Systems philosopher
    Significance of systems approach
    Enemies of the systems approach

Figure 14.1 Churchman's 12 social systems categories. *Source:* Adapted from Churchman (1979c).

It follows that a different understanding of objectivity is necessary – one that embraces subjectivity rather than trying to exclude it. Only by "sweeping in" different subjectivities, through stakeholder participation, can the restricted nature of any one W be overcome. Churchman recommends employing a dialectical approach to objectivity, which he derives from Hegel. Faced by a set of proposals emanating from one particular, inevitably restricted W, we first need to understand the nature of that worldview and why it makes the proposals meaningful. This first W (or *thesis*) should then be challenged by another "deadly enemy" W, based on entirely different assumptions that give rise to alternative proposals (*antithesis*). Whatever facts are available can then be considered in the light of both worldviews. This should help to bring about a richer (i.e. more objective) appreciation of the situation, expressing elements of both positions but going beyond them as well (*synthesis*). We know that Ws are very resistant to change. They interpret facts, even apparently contrary ones, in a way that fits their existing assumptions. The dialectical method enables us to get at the foundations of Ws and to examine them systemically.

From an ethical standpoint, Churchman argues that it is the role of the planner (or systems designer), no doubt aided by the systems philosopher, to challenge the mental models of powerful decision-makers so that they come to act according to the purposes of the "clients" of the system – those people supposed to benefit from it. This is only likely to occur if a deep "mutual understanding" can be established between systems designer and decision-maker (Churchman and Schainblatt 1965).

> There are no experts in the systems approach.

This principle challenges adherents of any particular branch of science who claim special expertise in resolving problems in social systems. They all offer partial viewpoints and need to be engaged in a dialectical inquiry process with others. It applies particularly to systems thinkers themselves. Systems designers, because they seek to address systems as wholes, may become arrogant in the face of opposition from apparently sectional interests. They need to listen carefully to all "enemies" of the systems approach, such as religion, politics, ethics, and esthetics, since these enemies reflect the very failure of the systems approach to be comprehensive – to draw the boundary wide enough (see Churchman 1979c). Churchman insists that when it comes to matters of aims and objectives (and appropriate means), which inevitably involve ethical considerations and moral judgments, there can be no experts.

> The systems approach is not a bad idea.

The attempt to take on the whole system remains a worthwhile ideal even if it cannot be realized in practice. Systems designers must pursue their profession in the "heroic mood." This is the spirit advocated by Churchman's mentor, the pragmatist philosopher E. A. Singer. Increasing purposefulness and participation in systems design, by dialectically developing worldviews, is a never-ending process:

> Hence, the Singerian inquirer pushes teleology to the ultimate, by a theory of increasing or developing purpose in human society; man becomes more and more deeply involved in seeking goals.
>
> *(Churchman 1971, p. 254)*

There is a need to achieve some consensus around a particular perspective so that decisions can be made and action taken. Before this worldview can solidify into the status quo, however, it should itself be subject to attack from forceful alternative perspectives.

The radical reorientation in the systems approach demanded by Churchman's philosophy comes through in a story of the engagement of his research group with NASA during the period of the Apollo space programme. NASA wanted an evaluation of the innovative, hard systems methods it was using to manage the project. Churchman's group drew the boundary much wider and began asking challenging questions about the purpose of the Apollo programme, which from *their* systems perspective did not obviously contribute to the betterment of humanity. Reviewing the performance of Churchman's group, NASA awarded it the highest mark available for interdisciplinarity and the lowest for relevance to their mission.

A phrase was coined to describe the influence of Churchman and Mitroff on the Swedish systems movement: "Churchman is Churchman and Mitroff is his prophet" (Agrell and Leonarz 2006). It was not just in Sweden that this was true. Surveying the field of management science in the late 1970s, Mason and Mitroff (1981) concluded that the methodologies and tools available could only be successful if used on relatively simple problems. This was selling managers short because most of the policy, planning, and decision-making problems they had to deal with were "messes," ill-structured problem situations made up of highly interdependent problems. They determined to design an approach specifically to deal with messes or, as they preferred to describe them, "wicked problems." The result was SAST.

Most organizations fail to deal well with wicked problems, Mason and Mitroff argue, because they find it difficult to challenge accepted ways of doing things. Policy options that diverge from current practice are dismissed. An approach to planning and problem resolving is needed which ensures that alternatives are fully considered. The presentation of new data will not do the job. These will simply be interpreted in terms of existing theory and will not lead an organization to change its existing practice. What is needed is the generation of radically different policies firmly based on alternative worldviews. An organization only really begins to learn when its most cherished assumptions are challenged. Assumptions underpinning existing policies and procedures should therefore be unearthed and alternatives put forward based on counter-assumptions. A variety of policy perspectives, each interpreting the data available in its own way, can then be evaluated systemically. Furthermore, because problem situations are increasingly complex, dynamic, and thorny, this type of organizational learning needs to be ongoing. Organizations must constantly challenge the conscious and unconscious assumptions and arguments that guide their strategies:

> Examining key arguments is no longer a luxury. It's literally a matter of life and death.
>
> *(Barabba and Mitroff 2014, p. 21)*

Mason and Mitroff recognize that tensions will inevitably arise during the process of continuously challenging assumptions – not least because its success depends, initially at least, on different groups being strongly committed to particular policy options. However, they regard it as naive to believe that wicked problems can be tackled in the absence of such tensions. Organizations are arenas of conflict between groups holding to and expressing alternative worldviews. This offers great potential for developing and examining alternative strategies but, clearly, the manner in which this is done needs to be carefully managed. The SAST methodology attempts to surface conflicts and to direct them productively as the only way eventually of achieving a productive synthesis of perspectives.

14.2.3 Methodology

SAST is a systems methodology that is meant to be employed when managers and their advisers are confronted by "messes" or "wicked problems." In Mason and Mitroff's opinion, most policy, planning and strategy problems are "wicked problems of organized complexity" (1981 p. 12). They are characterized by interconnectedness, complicatedness, uncertainty, ambiguity, conflict, and societal constraints. In tackling such issues, problem formulation assumes greater importance than problem-solving using conventional techniques. If problem formulation is ignored or badly handled, managers will end up solving, very thoroughly and precisely, the wrong problem. Another interesting way of looking at this is provided by Barabba and Mitroff (2014). They argue that organizations display "Jungian personality types" in that they often perform better or worse in each of four quadrants – in terms of hard measures (e.g. production records), strategic measures (e.g. innovation), personal measures (e.g. good HR), and humane measures (e.g. flat organizational structures). High performing organizations do well in all four. Wicked problems occur when issues cross boundaries, developing and exhibiting themselves in all the quadrants.

Four principles are highlighted by Mason and Mitroff (1981) as underpinning the SAST methodology. It is:

- *Participative*: Based on the belief that different stakeholders should be involved, because the knowledge and resources required to resolve wicked problems will be spread among different parts and levels in an organization and different groups outside the organization. And also because individuals only give their full commitment to implementing solutions if they have also been involved in problem definition. Mitroff and Mason argue that:

  > As problems have become increasingly more complex in their definition, let alone solution, widespread participation is no longer a luxury. It is necessity.
  >
  > *(1987, p. 144)*

- *Adversarial*: Based on the belief that different stakeholders perceive wicked problems very differently and that judgments about how to tackle such problems are best made after full consideration of opposing perspectives
- *Integrative*: Based on the belief that the different options thrown up by the participative and adversarial principles must eventually be brought together again in a higher-order synthesis, so that an action plan can be produced and implemented
- *Managerial mind-supporting*: Based on the belief that managers exposed to different assumptions that highlight the complex nature of wicked problems will gain a deeper insight into the difficulties facing an organization and appropriate strategies that will enable it to move forward

It is not obvious that an approach can be both adversarial and integrative. That it can was brought home to me while conducting an intervention, with Joseph Ho, in a printing company that was in the process of adopting a quality management program (Ho and Jackson 1987). In that firm there was an apparent consensus around the need for a particular type of quality program. In fact, this consensus was founded on very different interpretations of the key concepts underpinning the program. Only through a process of adversarial debate could the significant differences be highlighted and the ground prepared for a more soundly based consensus built on a genuine common understanding.

The four principles are employed throughout the stages of the SAST methodology. The following account is drawn from a variety of sources (Mason, 1969; Mitroff et al. 1977, 1979; Mason and Mitroff 1981; Barabba and Mitroff 2014). It represents the approach as having four main stages:

- Group formation
- Assumption surfacing
- Dialectical debate
- Synthesis

We consider these in turn.

The SAST process should involve as wide a cross section of individuals as possible who have an interest in the policy or problem being investigated. They are first carefully divided into groups. Within each group, the aim is to maximize convergence of perspectives so as to minimize interpersonal conflict and achieve constructive group processes. Between groups, the aim is to maximize divergence of perspectives so as to get the most out of the totality of groups by taking advantage of their differences. A number of techniques can be used to accomplish this – grouping according to functional area, organizational level, or time orientation (short- or long-term perspective), on the basis of personality type, using the Myers-Briggs instrument or Jungian theory, vested interest, or advocates of particular strategies. The method chosen will depend on the problem and organization. Each group should have or develop a preferred strategy or solution, and each group's viewpoint should be clearly challenged by at least one other group.

The aim of the second, assumption surfacing, stage is to help each group uncover, map, and analyze the key assumptions on which their preferred strategy or solution rests. It is important to provide a supportive environment and good facilitation so that people can be as imaginative and creative as possible. Barabba and Mitroff argue:

> We cannot overstate the importance of this particular aspect of SAST. Very rarely do people have the ability or opportunity to examine their key assumptions systematically (methodically) and systemically (comprehensively as a system). They almost never have the ability or opportunity to see a map of their assumptions, in effect, a map of their fundamental beliefs about an important issue or problem. Such maps allow one to go beyond seeing assumptions individually and in isolation. They allow one to see the potential interactions between sets of assumptions.
>
> *(2014, p. 28)*

Three techniques are recommended to help with this stage and they are dealt with in Section 14.2.4.

The groups are then brought together and encouraged to enter into a dialectical debate. A spokesperson for each group will present the best possible case for its preferred strategy or solution, being careful to identify the key assumptions on which it is based. During this presentation, other groups are allowed to ask questions but just for clarification. It is important that each group understands each other's viewpoint and assumptions before debating and challenging them. Open, dialectical debate begins only after each group has presented its case clearly. The debate may be guided by asking the following questions:

- How are the assumptions of the groups different?
- Which stakeholders feature most strongly in giving rise to the significant assumptions made by each group?
- Do groups rate assumptions differently (e.g. as to their importance for the success of a strategy)?
- What assumptions of the other groups does each group find the most troubling with respect to its own proposals?

The debate should take the form of "constructive conflict" in which differing and opposing alternatives are ranged against one another and discussed as strongly as possible:

> In other words, we don't merely assume, but we claim unequivocally that crucial decisions always involve more than one 'good' or 'best' alternative. If a decision involves only one, then the 'first problem' is to create more viable alternatives.
>
> *(Barabba and Mitroff 2014, p. 42)*

After the debate has proceeded for a time, the groups separate again to consider independently whether they wish to modify their assumptions. This process of debate and "assumption modification" can continue for as long as progress is being made.

The aim of the final, synthesis stage is to achieve a compromise on assumptions from which a new, higher level of strategy or solution can be derived. Assumptions continue to be negotiated and modified, and a list of agreed assumptions is drawn up. If this list is sufficiently long, an implied strategy can be worked out. This new strategy should bridge the gap between the different strategies of the groups and go beyond them as well. If no synthesis can be achieved, points of disagreement are noted and research undertaken to resolve the remaining differences. At the least, any strategy adopted will be fully understood and the assumptions on which it is based can be tested as it is put into effect.

Mason and Mitroff (1981, pp. 80–83) provide a schedule for a typical four-day SAST workshop.

14.2.4 Methods

The three methods most closely associated with SAST are stakeholder analysis, assumption specification and assumption rating. They all serve the assumption surfacing stage of the methodology.

Mason and Mitroff believe that a strategy or solution can be thought of as a set of assumptions about the current and future behavior of an organization's stakeholders. In surfacing the assumptions underlying a particular strategy or solution, therefore, it is useful to decide who the relevant stakeholders are taken to be. Stakeholder analysis recommends that each group putting forward a strategy be asked to identify the key individuals, parties, or groups on which the success or failure of their preferred strategy would depend were it adopted. These are the people who have a "stake" in the strategy. The process can be helped by asking questions like:

- Who is affected by the strategy?
- Who has an interest in it?
- Who can affect its adoption or implementation?
- Who cares about it?

Stakeholder analysis was used to help decide whether to undertake substantial capital expenditure on a swimming pool for the social club of a hospital in the Middle East (see Flood 1995). In this example, it yielded the following list of stakeholders:

- Western nurses
- Developing countries' male staff
- Arab families
- Matron
- Residential medical staff
- Government liaison staff
- Recruiters
- Swimming pool manufacturers
- Financial controller
- Hospital administrator
- Support services manager
- Hotels in the city
- Religious fundamentalist groups

The output of stakeholder analysis is a consolidated list of stakeholders relevant to the set of proposed strategies.

The second technique is assumption specification. This asks each group to list what assumptions it is making about each of the stakeholders identified in believing that its preferred strategy will succeed. Two or three assumptions should be unearthed for each stakeholder. Assumptions can be generated and tested for relevance by using the following procedure (Mason and Mitroff 1981, p. 101):

- Ask what must we be assuming about this stakeholder and its future behavior in order for the plan to be successful
- For every assumption, formulate a counter-assumption – an assumption that is opposite from and a "deadly enemy" to the stated assumption. For example, if the assumption is that religious fundamentalist groups wouldn't care about the planned swimming pool, then the counter-assumption would be that they would care very much
- If the counter-assumption's truth would have no significant impact on the plan, then the assumption is not very relevant and should be discarded. The counter-assumption concerning the religious fundamentalist groups could have a significant impact on the plan and so the original assumption makes the cut

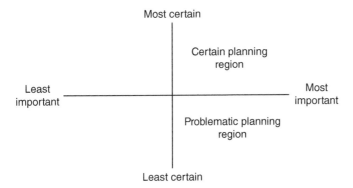

Figure 14.2 Assumption rating chart.

We now have a list of the assumptions on which the success of the group's preferred strategy or solution depends.

Assumption rating, our third method, requires each group to rank each of the assumptions, it is making according to two criteria:

- How important is this assumption in terms of its influence on the success or failure of the strategy?
- How certain are we about the likelihood of occurrence or the truth of the statement of events contained in the assumption?

The results are recorded on a chart such as that shown in Figure 14.2.

Assumptions falling on the extreme left of the chart are of little significance in terms of the success or failure of the proposed strategy. Those falling in the top right (certain planning region) are important but at least we are reasonably certain about them and can plan accordingly. It is those in the lower right-hand quadrant (problematic planning region) that are most critical. They are important to the strategy, but we are uncertain about them. They deserve close attention.

14.3 Strategic Assumption Surfacing and Testing in Action

This case study is taken from an intervention I was involved in, with Ellis Chung, in the Humberside Cooperative Development Agency (CDA).

The CDA was established to serve its region by fostering, encouraging, developing, and promoting industrial and commercial activity through the formation of cooperative enterprises – enterprises owned by the people who work in them and usually also managed by those same people. The particular focus of the SAST exercise was a disagreement in the CDA over the relative merits of "top-down" as opposed to "bottom-up" cooperative development work. The top-down approach, which involves identifying business opportunities and then recruiting individuals to form workers' cooperatives in these fields, is usually viewed with great distrust in cooperative circles. The preferred approach is bottom-up, essentially encouraging and assisting groups already thinking about starting co-operatives in particular areas of work. This description of the case follows the four stages of the SAST methodology.

Within the CDA the idea of trying a top-down strategy had some support, although there was also vehement opposition from other development workers. The development workers were therefore divided into two groups, one consisting of those with some sympathy for the top-down approach and the other of those opposed. The opposed group was asked to make the best case it could against top-down. It was felt that this, rather than asking them directly to make the case for bottom-up, would lead to the most fruitful debate.

The separated groups then went through the assumption surfacing phase, using the stakeholder analysis, assumption specification, and assumption rating techniques. The groups came up with widely different lists of stakeholders, obviously influenced by initial perceptions about which individuals or groups might or might not support the case for the top-down strategy. The stakeholders listed by each group are shown in Table 14.1.

The lists of stakeholders were combined and each group was asked, following the logic of assumption specification, what it was assuming about each stakeholder in believing that its arguments for or against the top-down strategy were correct. This facilitated the emergence of numerous assumptions supporting/against the top-down strategy. These were ranked as to their importance and certainty by each group and the results recorded on assumption rating charts. Table 14.2 contains lists of those assumptions rated most significant by the two groups (i.e. those appearing in the right-hand quadrants of Figure 14.2). The particular stakeholder generating each assumption is noted in parentheses.

The groups were then brought back together to engage in a dialectical debate. During the presentations, it became clear that the groups were emphasizing assumptions derived from consideration of different stakeholders as the main props for their arguments. Group 1 (for top-down) drew heavily on the stakeholders "funding bodies" (increase in credibility, ensures continuous support, carries out expectations) and

Table 14.1 Stakeholders listed in the study for the Humberside Cooperative Development Agency.

Group 1 (for top-down)	Group 2 (against top-down)
The development workers	The development workers
The unemployed	Potential clients
Local authorities	The ideologically motivated
Business improvement schemes	Local authorities
Existing cooperatives	Department of trade and industry
Funding bodies	Existing cooperatives
Other CDAs	People already in work
Marketing agencies	
Trade unions	
General public	
Other businesses	

Table 14.2 Significant assumptions concerning the stakeholders listed in Table 14.1.

Group 1 (for top-down)	Group 2 (against top-down)
Another way to set up workers' co-operatives (potential clients)	Mixed feelings of the development workers toward the strategy (development workers)
Increases the CDA's credibility in job creation (funding bodies)	Lack of group cohesion among the cooperators (potential clients)
Ensures continuous support to the CDA (funding bodies)	Lack of willingness to cooperate among the cooperators (potential clients)
Carries out the expectations of the funding bodies (funding bodies)	Getting people who are not motivated (the unemployed)
Strengthens the co-operative sector (existing co-operatives)	Less development workers' time on helping existing co-operatives (existing co-operatives)
Provides employment (the unemployed)	Lack of knowledge of business opportunities hinders "top-down" (development workers)
Provides the unemployed with a solution in a package (the unemployed)	Lack of experience of the development workers in this area of activity (development workers)
A more effective way of starting workers' co-operatives (development workers)	Lack of commitment to business idea among the new co-operators (potential clients)
Establishes a successful precedent (other CDAs)	Against principle of self-determination (ideologically motivated)
Increases numbers working in co-operatives (existing co-operatives)	Could be criticized as a waste of development workers' time (funding bodies)
Increase in industrial democracy (trade unions)	Very dangerous if failed (funding bodies)
	Suspicions of other co-operatives, fear of hierarchy and getting co-operatives a bad name (existing co-operatives)
	Too risky a venture for them (funding bodies)
	No previous association of co-operative members (potential clients)
	May have nothing in common with other co-operators (potential clients)

"unemployed" (provides employment, gives unemployed a solution in a package). Group 2 (against top-down) concentrated on assumptions generated by the stakeholders "development workers" (mixed feelings, lack of knowledge about business opportunities, lack of experience in the area), "potential clients" (lack of group cohesion, lack of willingness to cooperate, lack of commitment to business idea, etc.), and "existing co-operatives" (less development workers' time for them, suspicion). This analysis helped the participants to clarify the nature and basis of the arguments for and against top-down and contributed to a very productive debate.

As debate continued, other interesting results emerged. The two groups interpreted the reaction of the stakeholder "funding bodies" from entirely different perspectives. Group 1 insisted that top-down would assist the CDA's credibility in job creation and fulfill the expectations of funders, so ensuring continued support. Group 2 believed that top-down might be seen as a waste of development workers' time on risky ventures, and this dangerous experiment could lose the CDA credibility with the funders if it failed. On the issue of whether top-down promoted industrial democracy, Group 1 argued that more people in workers' cooperatives would inevitably bring this effect; Group 2 argued that the very idea of top-down took choice away from the individuals concerned. Group 1 argued back again that many of these were unemployed and had few choices anyway, so work in a cooperative could only increase their options. The most troubling assumptions of the other side for Group 1 (for top-down) were the divisions among the development workers themselves and the possible lack of commitment from those brought together in a top-down scheme. Group 2 (against top-down) worried that, if no top-down work took place, a genuine opportunity to set up more cooperatives would be wasted, and chances to improve the lot of the unemployed and to gain credibility with funders would be missed.

Following the dialectical debate, attempts were made at assumption, negotiation and modification, but it proved impossible to arrive at any overall synthesis. Consensus was however reached on particular matters, such as the need to seek out sources of information about business opportunities, to research other top-down experiences elsewhere and to carry out some experiments with a modified top-down approach.

The intervention using SAST in Humberside CDA was useful in assisting creativity, in helping to clarify where differences of opinion lay, and in generating a very full and rich discussion. Overall synthesis proved impossible to achieve, but agreement around specific issues was obtained and this brought benefits. The inclusion of the items mentioned above in an action plan would not have been possible without the changes in perception and culture brought about through the use of SAST.

14.4 Critique of Strategic Assumption Surfacing and Testing

Churchman's social systems design provides a response to one of the most common criticisms of systems thinking. Critics argue that the systems approach is hopelessly idealistic and impractical because it requires us to understand "the whole system" and that is something only God could do. Churchman agrees that such an understanding is necessary and also that it is impossible. But he points out that this is the fate of all the applied sciences that have no option but to live with the prospect that localized actions based on limited information can have unforeseen, potentially deleterious consequences in terms of whole system improvement. As Ulrich (1981a) puts it, all the critics are therefore doing is "blaming the messenger for the bad news." The important question is how to proceed given that what is necessary for our designs to be foolproof – that we understand the whole system – is impossible. Churchman's answer is that we should use the theoretical indispensability of comprehensive system design as an ideal standard to force us to reflect critically on the inevitable lack of comprehensiveness of our actual designs. We need to make the lack of comprehensiveness of our designs transparent so that we can easily reflect on their limitations. A good way of doing this is to expose our designs to the "enemies" of the systems approach and learn from what they have to say. Churchman's insistence that we constantly reflect upon the lack of comprehensiveness of our systems designs, and particularly at who is excluded from consideration, links to another great strength of his thinking. In a way that is not seen in the systems methodologies previously considered, ethical concerns lie at the very heart of his approach. For Churchman (1982), "wisdom" only emerges when we combine thought with a concern for ethics. He is emphatic that the systems approach should address serious problems such as hunger, poverty, and war, and systems designers should take responsibility for the social consequences of their work.

Mason, Mitroff, and their collaborators work hard to show how Churchman's thinking can be put into practice at the level of companies, agencies, and social programs. SAST is a systems methodology designed to deal with "messes" – strategic, wicked problems rather than narrow technical ones. There is actually little empirical evidence in favor of the superiority of SAST over conventional planning approaches (see Jackson 1989), but this does not surprise its advocates. In the context of wicked problems, the concern is with clarifying purposes and finding elegant ways forward, rather than with producing the "best" solution that can be compared with outcomes derived from other methodologies. We are drawn back, therefore, as with all soft systems approaches, to the coherence of the underlying philosophy and the way it is operationalized in the methodology as the only possible guarantees for the benefits said to be associated with the approach. Mason and Mitroff recognize this well enough. In their view, the "guarantor" for SAST, validating the process and its outputs, rests on its capacity to manage "controlled conflict between different viewpoints," "to inspect and challenge assumptions," and on "managerial judgment itself" (1981, p. 122).

In the absence of empirical evidence, Mason (1969) tries a "thought experiment" to highlight SAST's advantages over the "expert" and "devil's advocate" approaches to planning. In the expert approach, organizations set up special planning departments or obtain the services of outside experts and require them to produce a plan based on the best available evidence. However, the planners' or experts' own worldviews and strategic assumptions often remain hidden. Moreover, experts usually provide plans from a limited perspective and ignore the wide range of additional perspectives that policymakers and managers might usefully take into account. The lack of transparency over assumptions and the failure to test assumptions leave the decision-makers handicapped at crucial stages of the formulation and implementation process. The devil's advocate approach does allow the surfacing and testing of some assumptions when the planners present their proposals for scrutiny by senior management. However, this approach often encourages the top management to be hypercritical, with the added problem that, if they are too destructive, the suggested plan disintegrates with no alternative to replace it. In these circumstances, planners may be tempted to produce "safe" plans to protect themselves from severe criticism. Again, with the devil's advocate approach, the chance is lost to develop alternative plans constructed on the basis of different worldviews. Mason argues that the dialectical philosophy embedded in SAST overcomes the weaknesses of the other two methods.

Starting our second-order analysis, using social theory, it is clear that SAST operates from a different sociological paradigm to any methodology so far considered. SAST is an "interpretive" systems approach which embraces subjectivism rather than the objectivism that underpins functionalist methodologies. This shift in paradigm arises from following Churchman's philosophy and the type of pragmatism, derived from

William James and Singer, from which his social systems design emanates (see Barton 1999). Singer, having attended a reading of James' essay *The Will to Believe*, goes on to describe pragmatism as

> ... a moment in the swing of thought from realism to idealism, and how for it the most vital, that is to say, the moral and religious, aspects of our world are things to work and fight for, to make and to mould, not just to find and come across.
>
> *(Quoted in Barton 1999)*

According to Singer's version of pragmatism (see Britton and McCallion 1994), there are no fundamental truths that can be taken as a sure starting point for inquiry, but a process of learning can be put in train that pursues the ideals of humankind as a whole. Churchman interprets this in terms of his systems perspective. For him, systems exist in the minds of observers not in the "real world." A model captures only one possible perception of the nature of "the whole." It is in this context that purposeful endeavor can create the future. To do so ethically, and to gain an appreciation of the whole, we have to engage with multiple perspectives. The primary "guarantor" of the results of a systems study is the maximum participation of different stakeholders, holding to various Ws, in the design process. "Objectivity," it turns out, can only emerge from open debate among holders of many different perspectives. Churchman posited that the main "challenge to reason" is how to design improvement in large systems without understanding the whole system. He concludes:

> The thinking man's reply to the challenges to reason has been to construct a debate; whether the debate is an adequate response to the challenge remains to be seen.
>
> *(1968, p. 197)*

To interpretive social theorists it is. They are concerned with understanding and articulating the way people view the world with the aim of allowing them to arrive at a consensus, accommodation, or synthesis of their positions. If people can change their perceptions, then it is assumed that they can also bring about the changes they desire in social systems. Thus, Mason and Mitroff (1981, p. 10) support Rittle's conclusion that formulating wicked problems is synonymous with solving them. From this social theory perspective, therefore, SAST is an excellent vehicle for promoting intersubjective communication and understanding in organizations.

To those who do not share SAST's commitment to the interpretive paradigm, however, the approach passes the main problems by. Churchman's question about whether debate is an adequate response to the challenges we face is answered in the negative. For functionalists, who see "complicatedness" and "complexity" as originating primarily from objective features of systems, SAST has only a marginal role to play – perhaps ensuring that agreement is reached about goals at the very beginning of a project. Advocates of socio-technical systems thinking and organizational cybernetics, for example, will regard it as skipping over the many daunting issues associated with structuring large-scale organizational systems to be responsive to their turbulent environments. SAST does not concern itself with whatever "systemic principles" they think must be followed to achieve success in this task.

From the point of view of social theorists of a "radical change" orientation, there are other problems that frustrate SAST's ambitions. They will be encouraged, initially, by Churchman's insistence that a systems designer's first obligation is not to the decision-makers, even if they are paying the bills, but to the "clients," "customers," or "beneficiaries" of the system – Churchman uses various words to describe this group. He is referring to all those people who have an interest in a system and whose objectives should, in view of this, be served by the system. In the case of an industrial firm, it will include employees, shareholders, customers, suppliers, the local community, and other interested sections of the public. The aim of social systems design is to identify the interests of these "customers" and to influence decision-makers to realize those changes that benefit the customers. If a systems designer is convinced that the decision-makers are serving the wrong customers, then she has a professional obligation to change the decision-making process (Churchman 1970). So far, so good. The difficulty for "radical change" theorists comes with the means that Churchman advocates for bringing about this transformation. He hopes to engage the decision-makers, along with other stakeholders, in a dialectical debate during which their most cherished assumptions are challenged by counter-assumptions and a synthesis between different positions emerges. To those who see the social world as characterized by asymmetry of power, structural conflict and contradiction, this is a pious hope.

The problem radical change theorists have with Churchman's social systems design is that they see it as underpinned by a consensus worldview. He ends *The Design of Inquiring Systems* with the question: "What kind of a world must it be in which inquiry becomes possible?" In other words, what must social reality be like for social systems design to work. And he recognizes that the world can confound his own position. In the conclusion of *The Systems Approach*, he writes:

> Hence, I, too, am biased and deceived. It's naive to think that one can really open up for full discussion the various approaches to systems. People are not apt to wish to explore problems in depth with their antagonists. Above all, they are not apt to take on the burden of really believing that their antagonist may be right. That's simply not in the nature of the human being.
>
> *(Churchman 1979b, p. 231)*

To his critics (Jackson 1982a), he makes no progress in addressing these issues. They do not lead him to fundamentally question his own worldview. In this he is not helped, radical change theorists of an "objectivist" orientation argue, by his subjectivism. The highly structured, coercive social world studied by "radical structuralists" (see Burrell and Morgan 1979) is foreign to Churchman. His book on *The Systems Approach and Its Enemies* surprises one reviewer (Sica 1981) because in a book supposedly about social systems we are told remarkably little about what they are actually like. For Churchman, there are no "objective" aspects of social systems to worry about. Bringing about change simply means changing the way people think about the world – changing their *Weltanschauungen*. The difficulty, from the point of view of objectivist social science, is that Ws are not so easily changed. They are closely linked to other social facts in the social totality. Changing Ws may depend crucially on first of all changing these other social facts. If we really wish to bring about change, we need some understanding of the laws that govern the transformation of the social totality. Only then can the real blockages, which may not be in the world of ideas, be located and pressure applied. To radical structuralists, therefore, Churchman misses the real leverage points for bringing about change. While he confines himself to the world of ideas, all that he can guarantee through the process of dialectical inquiry is a continual readjustment of the ideological consensus.

These criticisms of Churchman, from a radical change perspective, easily transfer to SAST. The success of SAST may depend on encouraging conflict, but it is a type of conflict that can be controlled and, ultimately, "integrated:"

> SAST is based on the premise that a unified set of assumptions and action plans are needed to guide decision-making and that what comes out of the adversarial and participative elements of the process can be integrated.
>
> *(Barabba and Mitroff 2014, p. 29)*

In the face of real conflicts of interest, SAST appears helpless as viewed from the radical change perspective. Even if it can get dialectical debate started, it will be constrained by power relationships from addressing issues of real concern to disadvantaged stakeholders. It is little more than a kind of multigroup brainstorming that tinkers with the ideological status quo in ways which are only likely to further benefit the powerful. No doubt a review of SAST from the point of view of postmodern and poststructuralist theory would raise additional issues around the deceptiveness of language, the pervasiveness and role of micropolitics, and the relationship between power and discourse. And Luhmann's work would suggest why, in a social world made up of functionally differentiated autopoietic systems, a cohesive, holistic approach to hunger, poverty, and war is immensely difficult. But we will leave it at that.

The positioning of SAST on the system of systems methodologies (SOSM) is, I think, an easy task. It is shown, in Figure 14.3, as responding to simple-pluralist contexts. SAST appreciates the usefulness of formal analytic modeling in simple-unitary situations. Churchman, after all, was one of the pioneers of OR. But it regards it as being of "limited ability" for handling messy, ill-structured problems (Mason and Mitroff 1981, p. 301). SAST does not see value in charting, in the manner of system dynamics, the

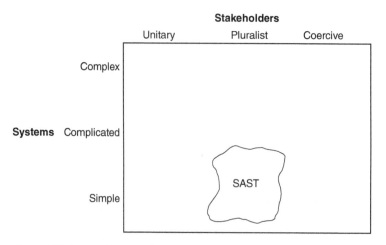

Figure 14.3 The positioning of strategic assumption surfacing and testing on the SOSM.

interdependencies between feedback loops as a means of understanding the "complicated" behavior of systems. Nor does it believe that there are laws, cybernetic or otherwise, which we can uncover and use to help manage complex systems seeking to be viable in rapidly changing environments. The focus of SAST is squarely on "pluralism." As Mason and Mitroff argue:

> Unlike other processes, which regard conflict as bad, signifying the breakdown of natural discourse, SAST regards such conflict as necessary to the emergence of a natural, synthetic policy .... Frequently we have instructed [participants] to carry their positions to a position more extreme than the one they believe in. Unless this is done, the positions become 'so reasonable' that they are acceptable without adequate debate by all parties. In effect, suppressing the extreme differences inhibits the entire examination of real alternatives.
>
> *(1981, p. 122)*

SAST sees its primary role as encouraging productive intersubjective communication. In the language of the SOSM, SAST seeks to develop systems thinking along the horizontal dimension of problem contexts. It is primarily concerned with philosophies, perceptions, values, beliefs, and interests. As such, it concentrates managers' minds on the diverse worldviews, multiple perspectives, and different assumptions of stakeholders and seeks to achieve a greater degree of mutual understanding. To this end, it works well. It manages pluralism, first making use of it, generating it if necessary, to encourage creative discussion, and then seeking to resolve it in a higher-order synthesis. SAST is well equipped to assist in structuring the exploration of different perceptions and values and to help in bringing about a synthesis, or at least accommodation, among participants so that action can be taken. It is an example of a successful soft systems methodology.

Even advocates of SAST recognize that it depends for its success on the willingness of participants to have their assumptions exposed and challenged. Mason and Mitroff see its main weakness as being its dependence upon the "willingness of participants to lay bare their assumptions" (1981, p. 301). Kilmann (1983) points out that assumptional analysis takes it for granted that stakeholders want their assumptions exposed. This will be less of a problem in situations where some basic compatibility of values and interests exists and compromise is possible. There will be other instances, however, when there are barriers to the extension of the participative principle and, in these circumstances, many of the benefits of SAST will be lost. The powerful are likely to need some persuading to have their assumptions revealed and disputed. In "coercive" contexts, by definition, it is not going to happen, and any employment of SAST will get distorted and provide benefit only to those who already hold power. Its journey along the horizontal dimension of the SOSM takes SAST far enough to highlight and thrive in pluralist contexts but not far enough to encounter and address coercive problem situations.

14.5 Comments

This is a chapter about SAST and not C West Churchman. Those who are intrigued by Churchman's many contributions should consult his books; the *festschrifts* of *Systems Research and Behavioral Science* (20.6) and *Systems Practice* (1.4) dedicated to his work;

and the three volumes in the *C West Churchman and Related Works Series* (van Gigch, 2006; McIntyre-Mills, 2006a,b). It is worth, however, paying some attention to his legacy as a teacher and mentor. As the philosopher and moral conscience of the systems approach, he had many followers. Mason and Mitroff state that

> He wanted his students and audience to listen to his ideas and then go out and use them on their own. He had thousands of followers but never demanded that they form a tight rigid 'Churchman-school' or to necessarily pursue his personal research agenda. Rather, he released a diaspora of systems thinkers who spread out widely through academia, industry, and society.
>
> *(2014, p. 43)*

As well as influencing the originators of SAST, his work impacted many of the systems thinkers that we shall be considering in detail in later chapters: Ackoff, Checkland, Ulrich, Midgley, and Jackson. Mitroff and Linstone (1993) dedicate their book on "unbounded systems thinking" to Churchman. This important work draws upon his philosophy to integrate elements of SAST with Linstone's (1984) "multiple perspectives" systems thinking. Linstone had argued for the benefits of examining any problem situation through three lenses, which offer radically different worldviews. Harold Nelson (2003), whose influential coauthored book *The Design Way* presents design as "a culture of inquiry and action," describes the huge impact Churchman's thinking had on his work. Wijnhoven (2009) demonstrates the relevance of Churchman's ideas to information and knowledge management. Peter Checkland tells a nice story about visiting Churchman in the old people's home where he had ended up residing. Rather than discussing past achievements, he insisted on telling Peter all about the "philosophy study group" he had established at the home.

14.6   The Value of Strategic Assumption Surfacing and Testing to Managers

We are committed to setting down five lessons that managers can learn from each systems approach. Having gained an appreciation of SAST, it is reasonable to suggest that the following list captures the main learning it has to impart:

- SAST demonstrates how important it is for every organization to have a map of the most critical assumptions it holds about its stakeholders and to monitor, critique, and update these assumptions regularly
- In the right circumstances, SAST is a methodology that can encourage and orientate a participative style of problem management. The involvement of many stakeholders brings a large spread of opinion to bear on a problem situation and eases the implementation of proposed courses of action
- It is often argued that the best and most creative debate occurs when there is strong opposition to a preferred set of proposals. SAST takes advantage of "conflict" to encourage debate while managing the tensions that arise. It shows that an approach can be both adversarial and integrative. The understanding gained by participants of the assumptions underlying their favored options, and of the deeply held convictions of other parties, prepares the ground for a more soundly based consensus

- The use of dialectical debate helps managers overcome the tyranny of the "either–or" – proposals and counterproposals seen as exclusive alternatives. There is the possibility of the "and:" combining two opposites as part of a new and grander synthesis
- The methods associated with SAST – stakeholder analysis, assumption specification, and assumption rating – are profoundly "managerial mind supporting." They are excellent means of feeding a comprehensive debate on planning or problem resolution and throw up many previously unconsidered issues

14.7 Conclusion

During the 1960s and early 1970s, systems thinking had begun to lose its way. OR and management science had become obsessed with perfecting mathematical solutions to a small range of tactical problems. Meanwhile, most of the pressing problems managers faced were of the ill-structured variety, strategic in nature, and set in social systems. SAST was specifically developed by Mason and Mitroff to deal with "wicked" problems of significance to senior managers of organizations. As we have seen, it has much to offer. SAST is one of three soft systems approaches considered in this book. In looking at Ackoff's "interactive planning" and Checkland's "soft systems methodology," in the next two chapters, we will gain a deeper appreciation of what it is like to pursue systems thinking and practice from a "soft systems" perspective. As Vickers (1983) proclaims, "human systems are different." With "people complexity," new issues emerge that need addressing in distinctive ways. The three approaches detailed emphasize that effective problem resolution depends on recognizing the variety of stakeholder values and purposes and the formulation of intelligent and effective ways forward that can command their commitment.

15

Interactive Planning

I must Create a System, or be enslaved by another Man's;
I will not Reason and Compare: my business is to Create

(William Blake 1815)

15.1 Prologue

Perhaps the defining moment in Russ Ackoff's development of "interactive planning" (IP) was his involvement in a Bell Telephone Laboratories project in 1951. This account draws on Ackoff et al. 2006, and on Ackoff's "Bell Lab. Lecture" (2018). The project saw Bell Labs imagining and creating the telephone system of the future.

Ackoff became involved almost by accident. On a consulting trip to New York, he called in to see his friend Peter Meyers, who worked at Bell Labs in Murray Hill, New Jersey. Meyers was in a bit of a state as he and other section heads had been called into an emergency meeting that morning by the VP of Bell Labs. Probably not knowing what to do with Ackoff, and convinced that another participant would not be noticed, he asked him if he wanted to come along. Arriving at the meeting, Meyers and Ackoff found themselves in a classroom with about 40 other people. When the VP eventually arrived he seemed in a state of shock. He was visibly upset, looking down at the floor. Finally, he declared:

> Gentlemen, the telephone system of the United States was destroyed last night.

Despite protestations from those who had used their phones that morning, he persisted with his claim. The audience started to conclude that he was "off his rocker." But just before they called for help, and to the relief of all present, he straightened up, returned to normal, and let out a hearty laugh. Now that he had the full attention of the group, he went on:

> Well, in the last issue of the 'Scientific American' there was an article that said that these laboratories are the best industrially based R&D laboratories in the world. I agreed, but it got me thinking. I've made a list of those contributions to the development of telephonic communications that I believe have earned us this

reputation …. I'd like your opinions. What do you think are the most important contributions we have ever made to this development?

The first suggestion was the *dial*. The VP agreed that this was one of the most important but had to tell them that it had been developed before 1900. The next was *multiplexing*, which allowed the simultaneous transmission of multiple conversations over one wire and, as a result, massively increased the capacity of the network. The VP revealed that this also had been invented before 1900. Someone mentioned the *coaxial cable* across the Atlantic. This had been laid in 1882. The VP turned to the group:

> Doesn't it strike you as odd that the three most important contributions this laboratory has ever made to telephonic communications were made before any of you were born? What have you been doing? I'll tell you, you have been improving the parts of the system taken separately, but you have not significantly improved the system as a whole …. We have got to restart by focusing on designing the whole and then designing parts that fit it rather than vice versa. Therefore, gentlemen, we are going to begin by designing the system with which we would replace the existing system right now if we were free to replace it with whatever system we wanted, subject to only two not-very-restrictive constraints.

The constraints were that the redesign should be "technologically feasible," i.e. not science fiction, and "operationally viable," i.e. able to function and survive in the current environment.

The VP then divided the group into six subgroups of six or seven individuals, each responsible for a subsystem of the overall telephone system, e.g. intercity communication, in-city communication, switching, and the telephone set. He was acutely aware that this could lead to interactions between the subsystems being missed and insisted that each group appoint a representative to meet with other representatives, at least once a week, to discuss these interactions:

> At the end of the year, I want to see one completely integrated system design, not six subsystem designs. I don't even want to know what the individual teams came up with.

Ackoff found himself in the telephone set team and was invited by that group to participate even though it was now revealed that he was not an employee of Bell Labs. They began work the same day. An initial suggestion that they list what was wrong with the current telephone was dismissed as exactly not what was required. It was this approach that had led them astray in the past. Instead a list was drawn up of the properties they wanted it to have. The first four were:

- No wrong numbers – every call I receive is intended for me
- I want to know who is calling before I answer the phone so I can decide whether to pick-up
- A phone I can use with no hands
- A phone that comes with me wherever I am – not one I have to go to in a fixed location

Over several weeks the list expanded to over 90 desirable properties.

Work then began on the design of a phone that would possess the first property – eliminating wrong numbers. Ackoff consulted the head of the psychology department at Bell Labs who was able to tell him that four out of five wrong numbers resulted from the caller dialing incorrectly the right number in their head. Within an hour the subgroup came up with a design to drastically reduce such errors. It replaced the dial with 10 buttons, 1 for each digit, a register, and a red key in the lower-right corner. The caller was to leave the phone on the hook while pressing the appropriate buttons for the number, check the number was correct on the register, and if correct, lift the receiver sending the whole number through at once. If the number was wrong, the red button could be used to clear the phone. The subgroup then asked for technical help to assure themselves that the device was technologically feasible. After some weeks, the R&D department had built a push-button phone, tested it on over 2000 people, and determined that it was indeed feasible. They found that, as expected, it massively reduced wrong numbers. Moreover, it saved 12 seconds on the time usually taken to dial a seven-digit number and more time because the line was not occupied until after the number was put in and the receiver picked up – yielding a 20% increase in the carrying capacity of the telephone system, worth millions of dollars to the company. The R&D department had already begun a project to develop the phone under the code name "Touch tone." Before the year was out, the technological feasibility of all the 90 plus desirable properties had been established. Overall, the six subgroups anticipated almost all the changes that took place in the telephone system in the twentieth century – touch-tone phones, consumer ownership of phones, mobile phones, call waiting, call forwarding, voice mail, caller ID, conference calls, speaker phones, and speed dialing of numbers in memory. Ackoff comments:

> The impact of the design we produced was greater than the impact of any other effort to change a system that I have ever seen .... This experience is a convincing example of how "dealized design" can literally move mountains of change. However, applying the process involves not only discarding old mindsets that inhibit creative thinking but knowing the steps that we have learned work best in applying it.

A huge advantage of idealized design for Ackoff is that everyone in an organization can participate in it. You don't have to be an expert to come up with useful answers to the question "what should the system be like?"

15.2 Description of Interactive Planning

15.2.1 Historical Development

Russ Ackoff took his undergraduate degree in architecture at the University of Pennsylvania and, during that time, became a student and friend of Churchman's and attended classes provided by his mentor, E.A. Singer. He developed a keen interest in philosophy and, following service in the US Army during World War II, returned to the University to become Churchman's first doctoral student, graduating in 1947. Ackoff was a very practical philosopher and recounts (Ackoff 1999a) that he was dismissed from his first full-time appointment in philosophy, at Wayne State University, Detroit, because he wanted to create a center for applied philosophy and dared to put on a conference on philosophy and city planning. It was difficult to find another position in philosophy but posts came up in operations research at the Case Institute of Technology, in Cleveland, and both Ackoff and Churchman were appointed:

> Now West and I had only a vague idea as to what operations research was ... [we] had noted the similarity of intent between operations research and what we were trying to do, putting interdisciplinary teams together to work on real problems in real organizations. We did not care much about the name we went under when the Case Institute offer came up. By then we knew that we could not do what we wanted to do within philosophy departments.
>
> *(Ackoff 1988, p. 354)*

Together at Case, they formed the first Operational Research (OR) group in the United States to work on industrial and governmental problems and established the first doctoral program in OR. In 1957, as we know, they published (with Arnoff) *Introduction to Operations Research* described, by Kirby and Rosenhead (2005, p. 130), as "the most influential early OR textbook." The group broke up when Churchman left for Berkeley in 1958 and Ackoff returned to the University of Pennsylvania, as a professor in the Wharton School, in 1964.

Ackoff entered OR, and became one of its most important pioneers, because of its willingness to engage with practical problems. He became disillusioned with it when he judged that it had become wedded to "mathematization" ("mathematical masturbation" he sometimes called it) and lost touch with the real issues that concern managers. OR abandoned him, he claimed, rather than the other way round. In his view (Ackoff 1977), those who promote OR as a set of mathematical techniques, committed to optimization and objectivity, are out of touch with the requirements of the modern era and inevitably opt out of tackling the important issues of the day. To cling to optimization in a world of multiple values and rapid change is to lose your grip on reality. The emphasis has to be on adapting and learning. Objectivity in the conventional sense is also a myth. Purposeful behavior cannot be value-free. Ackoff's "apostasy" from OR reached its climax with two incisive papers delivered at the annual conference of the UK Operational Research Society, in 1978, and published in the *Journal of the Operational Research Society* the following year (Ackoff 1979a,b). In the first, he mounted a scathing attack on the way OR had developed and was continuing to develop, particularly in the United States, and called for it to adopt a new paradigm based on a holistic approach. The second outlined the new program in Social Systems Sciences (S^3) that he had established and which he held up as being what OR should have become.

At the University of Pennsylvania, Ackoff established himself as a leader in the systems movement. In a series of brilliantly conceived and written books (Ackoff 1974b, 1981, 1999b) he outlined his vision for systems thinking and set out "interactive planning" as a methodology for putting systems ideas into effect. The details follow in later sections of this chapter. The postgraduate educational program, S^3, went from strength to strength, with large numbers undertaking a highly innovative educational experience. It combined theory and practice, involving students in projects and driving them toward independent thought and action. It was unambiguously interdisciplinary and was supported by visiting luminaries such as Churchman, Emery, and Trist. Meanwhile Ackoff, often together with John Pourdehnad, continued his wide-ranging consultancy activities with industry and government.

In 1986, Ackoff retired from the Wharton School and founded a consultancy firm, *Interact*: the Institute for Interactive Management, which continued to undertake numerous projects. In 2000, the Ackoff Center for Advancement of Systems Approaches (ACASA) was established at the University of Pennsylvania in his honor. It provides a base from which Ackoff's type of systems thinking can be further refined and applied to the many organizational, societal, and world problems to which it is relevant. His friends and colleagues have not let him down. Jamshid Gharajedaghi's book *Systems Thinking: Managing Chaos and Complexity*, now in its third edition (2011), argues that Ackoff's systems approach can help us meet the challenges to society revealed by complexity theory, shows how Forrester's system dynamics can support the interactive planning methodology, especially in the "formulating the mess" stage, and provides new case studies. John Pourdehnad remains extremely active as a consultant and educator. His latest book, written with Peter Smith (2018), sees interactive planning as a means of creating the effective decision-making procedures that stakeholders of enterprises need in order to deal with the dynamic complexity unleashed by the newly emerging digital technologies of the "fourth industrial revolution."

In later life, three motifs start to dominate in Ackoff's writings. The first is an attack on management "fads." He felt obliged (Ackoff 1999a) to denounce the "panaceas, fads and quick fixes" with which managers are assailed and to which they are, unfortunately, prone. He shows that approaches such as quality management, business process re-engineering, and the balanced scorecard are always likely to fail because they are fundamentally antisystemic. They treat the whole as an aggregation of parts that can be improved independently of one another. As systems thinking explains, this often does not lead to any improvement in the performance of the whole because it is the interactions between the parts that are fundamental and need managing.

The second sees him extending systems thinking to broader, societal issues. A book with Sheldon Rovin (2003) addresses the "societal messes" of "Governance," "The City, Housing, and Transport," "Health Care," "Education," "Welfare," and "Crime and Punishment." Three core ideas are offered to guide the "redesign of society" in a way that eliminates the problems that arise from these messes. First is a distinction, due to Drucker, between doing things right, i.e. efficiency, and doing the right thing, i.e. effectiveness. The criminal justice system, in order to do the right thing, needs to focus at least as much on the correction of society, which produces criminality, as on criminals. A second idea follows. We should focus on what we want. This requires an "idealized design" approach. Idealized design can "dissolve" messes by focusing attention on a desirable future. The health care system should be redesigned to keep people healthy rather than tinkered with as a "sickness and disability care" system. The third concept is "thinking systemically". Societies are systems in which the interactions between the parts are more important for overall performance than the way the parts function separately:

> For example, a welfare system cannot succeed without some interactions with the education and health systems.

> *(Ackoff and Rovin 2003, p. 6)*

The education system is a huge "mess" and this is disastrous because of its interrelationships with all the other subsystems in the societal whole.

Ackoff (1978) had long demonstrated a predilection for capturing his insights in "fables" and stories. This only intensified in his later years – our third motif. *Beating the System* (with Rovin 2005) tells a number of stories designed to assist creative managers "to outsmart bureaucracies." *Memories* (2010), published the year after his death, presents "a set of discrete bits and pieces of the past ... that readers might find useful and, if not useful, entertaining." Ackoff is determined to retain the humanity in the stories while indicating what he learned from each incident. Then there are the two volumes of "management *f-LAWS*" (Ackoff and Addison 2007; Ackoff 2008) setting out, with wit and wisdom, the many flaws of management that get in the way of change and development. Stefan Stern (2009), in his obituary in the *Financial Times*, picks out some of his favorites:

- All of our problems arise out of doing the wrong things righter. The more efficient you are at doing the wrong thing, the wronger you become. It is much better to do the right thing wronger than the wrong thing righter. If you do the right thing wrong and correct it, you get better
- Business Schools are high security prisons of the mind
- An organization that cannot accommodate nonconformity will not be able to retain creative people
- Organizations fail more often because of what they have not done (errors of omission) than because of what they have done (errors of commission)
- The less managers expect of their subordinates, the less they get
- The only problems that have simple solutions are simple problems. The only managers that have simple problems have simple minds. Problems that arise in organizations are almost always the product of interactions of parts, never the action of a single part. Complex problems do not have simple solutions

Simon Caulkin, commenting in *The Observer* on the first volume of *f-laws*, writes:

> Reading Russell Ackoff's slim new volume, *Management f-Laws* ... is like being pricked by a series of delayed electric shocks .... The first shock comes as the implication sinks in, followed by a chain-reaction of secondary ones as the first implication interacts with subsequent ones, until shocks are going off all over and you are left in no doubt that you are in the presence of one of the profoundest and wittiest brains ever to engage with the bizarre human activity called management.
>
> *(2007)*

Ackoff has been called the "Einstein of Problem Solving" (Brant 2009) and the impact of his work in the management field gives some credence to this. Kirby and Rosenhead (2005) describe his influence on the early development of OR, both in the United States and in Britain, as "hard to over-estimate." Turning to his effect on the "systems movement," they state that his writings have been "enormously influential." For evidence they point to the compilation of the four volume set of readings on systems thinking edited by Midgley (2002). Forty-seven international experts in the field were asked to nominate papers for inclusion and more papers by Ackoff were suggested than by any other author. In the general management area, he was a collossus and counted Deming and

Drucker as among his friends. The influence of his writings in the specific fields of corporate planning (1970a), applied social science (with Emery 1972), management information systems (1967), and management education (1968, 1979b), has also been considerable. His impact has had a wide geographical as well as disciplinary spread. Carvajal (1983) has discussed the influence of Ackoff and his work on the development of the management sciences in Mexico. A book in tribute to his achievements there has been published (Jimenez 2006). We can take the United Kingdom as another example. Patrick Rivett spent six months with Ackoff at the Case Institute and went on to become the first professor of OR in the United Kingdom, modeling the department at Lancaster University on the arrangements at Case. Ackoff was "marriage broker" between the Tavistock Institute and the Operational Research Society when they worked together to form the Institute for Operational Research (later the Center for Organizational and Operational Research – see Friend et al. 1988). This initiative helped to extend the influence of OR in the public sector and to establish links between OR and the social sciences. He was prominent at the influential "OR and the Social Sciences" conferences held at Cambridge in 1964 and 1989 (Lawrence 1966; Jackson et al. 1989). The famous, or infamous, "Ackoff papers" (1979a,b), delivered at the OR conference in York in 1978, helped to revolutionize the theory and practice of OR in the United Kingdom and encouraged the development of "Soft-OR" or "problem structuring methods," one of the strongest strands of work in the field today. In September 1992, he opened the "Centre for Systems Studies" at the University of Hull and freely gave of his time to support the Center. He was an excellent public speaker – erudite, engaging, and humorous. Finally, we should pay attention to his efforts as a consultant. During his working life, he was involved with over 250 companies, in a broad range of industries. His books report on many successful interventions and his work with Alcoa, General Electric, Super Fresh, and Anheuser-Busch is particularly well known. According to Pourdehnad (in Ackoff 2010, p. vi), his relationship with Anheuser-Busch lasted from the late 1950s until August Busch III retired in 2006 – a period during which the company's share of the US beer market grew from 8% to 52%. Ackoff was heavily engaged in its expansion and marketing strategies. Pourdehnad cites it as an example of Ackoff's "uncompromising ethics" that he never owned a single share in the company during that time. Interactive planning is equally at home in nonbusiness settings and Ackoff worked with more than 50 government agencies including, during the Clinton and Bush administrations, the White House Communications Agency and the White House Military Office. Ackoff's work with the black ghetto in Mantua (1970b) gave birth to early attempts at "Community OR" and, later, inspired the full-scale Community OR initiative in the United Kingdom. Pourdehnad and Hebb (2002) provide another fine example of the use of interactive planning in the nonprofit sector to transform the fortunes of the Academy of Vocal Arts. One could go on, but it is time to turn to what it is in the nature and detail of the work that gives it its power.

15.2.2 Philosophy and Theory

Ackoff's (1974b, 1999a) philosophical orientation endorses that of Churchman and Singer. In particular, he has contributed to the new understanding of "objectivity" that is embraced in soft systems thinking. For him, the conventional view that it results from constructing value-free models that are then verified or falsified against some real world "out there" is misguided. In the natural sciences, objectivity can be approximated by science seen as a system but not by individual scientists. Objectivity in social systems science can only be approached through the interaction of groups of individuals with diverse values. It is "value full," not value-free. These conclusions give birth to some of Ackoff's most distinctive themes. One is that planning and design must be based on wide participation and involvement. Another is the idea that people must be allowed to plan for themselves. People's own ideals and values should be paramount in the planning process, although operationalizing the process may require assistance from professional planners. This sidesteps one of the major paradoxes of conventional planning – how to quantify quality of life – because this is only a problem if you are trying to plan for somebody else. The need to respect multiple worldviews also suggests that improvement needs to be sought on the basis of the client's own criteria. It may well be that the analyst's model of reality differs markedly from that of the client. Nevertheless, if you want to serve that client, you are better off granting rationality to them than rationality to yourself as the analyst.

This philosophy takes on a more precise form when it is applied to the management of organizations in the "systems age." Ackoff (1974a) argues that, about the time of the Second World War, the "machine age" associated with the industrial revolution began to give way to the "systems age." The systems age is characterized by increasingly rapid change, by interdependence, and by complex purposeful systems. It demands that much greater emphasis is placed on learning and adaptation if any kind of stability is to be achieved. This, in turn, requires a radical reorientation of worldview. Machine-age thinking, based on analysis, reductionism, a search for cause–effect relations, and determinism, must be complemented by systems-age thinking, which proceeds by synthesis and expansionism, tries to grasp multiple causality, and accepts the existence of free will and choice.

The systems age requires systems thinkers to be very careful about how they classify systems. Ackoff (1999a) identifies four different types of system and four different types of systems model. "Deterministic" systems have no purposes and neither do their parts, although they can serve the purposes of other purposeful systems. "Animated" systems have purposes of their own, but their parts do not. "Social" systems have purposes of their own, contain purposeful parts, and are usually parts of larger purposeful systems. "Ecological" systems contain interacting mechanistic, organismic, and social systems, but unlike social systems have no purposes of their own. They serve the purposes of the biological and social systems that are their parts. Problems arise if a model of one type of system is applied to a system of a different type. A particular issue has been the tendency to apply deterministic or animate models to social systems:

> The effectiveness of any model used to describe and understand behavior of a particular system as a whole ultimately depends on the degree to which that model accurately represents that system. Nevertheless, there have been and are situations in which application of deterministic or animate models to social systems have produced useful results for *a short period of time*. However, in a longer run, such mismatches usually result in less than desirable results because critical aspects of the social systems were omitted in the less complex model that was used.
>
> *(Ackoff 1999a, p. 34)*

Similarly, those who manage corporations, in the systems age, need to alter the way they think about them (Ackoff 1981). In the past it has been tempting to regard corporations either as machines serving the purposes of their creators or owners, or as organisms serving their own purposes. Today, organizations must be viewed as social systems serving three sets of purposes: their own, those of their parts, and those of the wider systems of which they are part. It follows that corporations have responsibilities to themselves (control problem), to their parts (humanization problem), and to those wider systems (environmentalization problem). Managers should seek to serve purposes at all these three levels, developing their organization's various stakeholders and removing any apparent conflict between them. If this is achieved, internal and external stakeholders will continue to pursue their interests through the organization and ensure that it remains viable and effective.

Drawing upon his philosophy, and reflecting upon these changing conceptions of the world and of the corporation, Ackoff sets out a new approach to planning that, he believes, is more appropriate to our current predicament. This is "interactive planning" (IP), which is the main operating tool of Ackoff's S^3 and brings his philosophy to bear on "messes" – systems of interdependent problems. It is best understood if it is compared with three other ideal types of planning: reactivist, inactivist, and preactivist. Reactivist planners want to return to some "golden age" they believe existed in the past. They treat problems in a piecemeal fashion and fail to grasp current realities. Inactivists want to keep things as they are. They too treat problems separately as they muddle through, trying to avoid real change. Their approach is to satisfice as they try to resolve day-to-day difficulties. Preactivist planners are future-orientated and seek to predict what is going to happen in order that they can prepare for it. Their aim is optimization on the basis of forecasting techniques and quantitative models that allow problems to be solved. To Ackoff, this "predict and prepare" approach is illogical since if the future was so determined that we could accurately predict it, there would be nothing we could do about changing it. Ackoff's preferred approach is IP (see Ackoff 1974b, 1981, 1999a,b). Interactivists do not want to return to the past, keep things as they are, or accept some inevitable future. They take into account the past, the present, and predictions about the future, but use these only as partial inputs into a methodology of planning aimed at designing a desirable future and inventing ways of bringing it about. Interactivists believe that the future can be affected by what the stakeholders of an organization do now, especially if they are motivated to reach out for ideals. In the process problems simply "dissolve" because the system and/or environment giving rise to them are changed so radically. Ackoff (1981) recounts how severe inventory and customer satisfaction problems plagued General Electric's Appliance Division because of uncertainty about how many left-hinged and right-hinged refrigerators to build. The problem failed to respond to either resolving (salesmen-generated forecasts) or solving (statistically based forecasts) approaches. It was eventually dissolved by designing refrigerators with doors that could be mounted on either side and thus could be made to open either way. As well as eliminating the inventory mix problem, this proved an attractive marketing feature as customers were not faced with the possibility of having to replace their fridge when they moved.

We have noted the depth and breadth of Ackoff's influence in the management sciences. An important explanation is the inspiring vision he presents for the discipline, deriving from his philosophy and theory. The job of the systems practitioner is no longer just to build mathematical models in order to enable key decision-makers to "predict and prepare" their enterprises for an inevitable future. Rather, it is to assist all the stakeholders to design a desirable future and to invent the means of realizing it. While carrying out development work with leaders of the Mantua ghetto in Philadelphia, Ackoff was delighted to find many of the lessons he was trying to impart to management scientists captured in the motto of the Mantua Community Planners: "plan or be planned for." The project is described in an article by Ackoff (1970b) titled *A Black Ghetto's Research on a University*. In 1968, Ackoff's group at the University of Pennsylvania received a request for assistance from the Mantua Community Planners, a coalition of neighborhood groups. The Mantua ghetto in Philadelphia had, at the time, a population of about 22 000, which was 98% black. It was an area of critical underdevelopment and its population suffered from considerable poverty and disadvantage. The approach adopted by Ackoff's group was to insist that, while help could be provided, the ghetto community had to solve its problems in its own way. To get the benefits of planning, the ghetto community had to plan for itself. Three people from the ghetto community, soon to be joined by a fourth, were employed at the university to work on the development of their community, taking advantage of university facilities (office space, secretarial aid, a graduate student assistant) but using these only as *they* saw fit. Soon requests for assistance from the ghetto community were flooding in, and a full-time senior member of the university staff had to be appointed to coordinate the requests for aid and the many offers of help from university personnel. After six months, further funding was received from the Anheuser-Busch Charitable Trust and additional money secured from the Ford Foundation to guarantee the project for two years.

The remarkable achievements that stemmed from the project are set out in the paper. Reflecting on this success, Ackoff makes five points. First, the methodology adopted in the Mantua project offers a better way of carrying out research. University staff learned far more about the ghetto by being directly involved in it, under the guidance of its members, than they could have done using traditional research methods. Second, the consultancy relationship was enhanced by not specifying to the client what skills the university faculty could offer. The ghetto leaders had to carry out research on the university to see how it might be useful. Third, because the ghetto could not rely on receiving resources to fund its planning proposals, its planning efforts constantly had to consider and respond to the wider system of which it was part in order to generate new resources. It had to develop in tandem with the wider community to secure support for ghetto activities. Fourth, planning had to be participative. Plans could not be imposed from above because of the dependence of the ghetto leaders upon their constituents. The ghetto leaders had continually to respond to the wishes of the "subsystems" that made up the systems they were trying to manage. Finally, the approach to planning, which simply predicts change in the environment and attempts to respond to it (predict and prepare), has to be abandoned. Most of the trends in the larger systems of which the black ghetto was part were detrimental to it. The future could not, therefore, be allowed to run its course. Active intervention in the wider environment was required in order to change the trends – what Ackoff describes, elsewhere, as designing a desirable future and finding ways of bringing it about. It is worth mentioning that he was still attending meetings of "Mantua Cares," in pain and using a walker, just weeks before his death: "In the streets of Mantua he was plain Russ Ackoff, a man who cared" (Morrison 2009).

The lessons of the Mantua research were given full expression in the book *Creating the Corporate Future* (1981), which Ackoff subtitled "Plan or Be Planned For." These sentiments, in turn, bring to mind the words of the English poet William Blake, which head up this chapter. It is certainly the case that the spirit of Blake's words is well captured in Russell Ackoff's work. He has shown why they are apposite to systems thinking and why they are even more relevant to the systems age than to the time when Blake wrote them. Ackoff's achievement goes beyond this, however. In his books, he sets out a detailed methodology that can actively be used by stakeholders to plan and pursue a desirable future.

15.2.3 Methodology

Interactive planning is of particular relevance to us in this book because it was specifically designed to cope with the "messes" that arise from the increased complexity of the modern era. Three principles underpin the methodology (Ackoff 1981, 1999a). The first is the *participative* principle. If possible all stakeholders should participate in the various phases of the planning process. This is the only way of ensuring "objectivity." It also secures the main benefit of planning – the involvement of members of the organization in the process. This is more important than the actual plan produced. It is by being involved in the process that stakeholders come to understand the role they can play in the organization. It follows, of course, that no one can plan for anyone else. The role of professional planners is not to do the planning, but to help others plan for themselves. The second principle is that of *continuity*. Because values change and unexpected events occur, plans need to be constantly revised. The third is the *holistic* principle. Because of the importance of the interactions between the parts of a system, we should plan simultaneously for as many parts and levels of the organization as possible.

With these principles in mind, we can now consider the five phases of the IP approach:

- Formulating the mess
- Ends planning
- Means planning
- Resource planning
- Design of implementation and control

These phases may be started in any order and none of them, let alone the whole set, can ever be regarded as completed. They constitute a systemic process.

Formulating the "mess" involves determining the future an organization will be in if it continues its current plans, policies, and practices and if its environment changes only in ways that are expected. Three types of study are necessary:

- *Systems analysis*: Giving a detailed picture of the organization, what it does, its stakeholders and relationships with its environment
- *Obstruction analysis*: Setting out any obstacles to corporate development
- *Reference projections*: Extrapolating on the organization's present performance in order to predict future performance if nothing is done and trends in the environment continue in entirely predictable ways

Synthesizing the results of these studies yields a "reference scenario," which is a formulation of the mess in which the organization currently finds itself. It reveals the seeds of self-destruction inherent in its current policies, practices, plans, and external expectations.

Phase 2, ends planning, is about "where to go" and involves specifying the purposes to be pursued in terms of ideals, objectives, and goals. It has five steps. First, a mission statement is prepared. This should outline the organization's ultimate ends (its "ideals"), incorporate the organization's responsibilities to its environment and stakeholders, and aim to generate widespread commitment. Second, planners should help the stakeholders to prepare a comprehensive list of the desired properties stakeholders agree should be built into the system. Third, an "idealized design" of the organization should be prepared. The fourth step requires formulation of the closest approximation to this design that is believed to be attainable. Finally, the gaps between the approximation and the current state of the system should be identified.

Idealized design is both the unique and most essential feature of Ackoff's approach. To make it happen requires leaders with an inspiring vision who can guide others in pursuit of ends using means that they approve (Ackoff 2005). It must capture the purposes that the stakeholders invest in the organization. An idealized design is the design for the enterprise that the stakeholders would replace the existing system with today if they were free to do so. The idea that the system of concern was destroyed yesterday, no longer exists and must now be designed as a whole from scratch, is meant to generate maximum creativity among those involved. Imagining an ideal future, and then working back to where you are today, ensures that you do not erect imaginary obstacles before you even start planning. It is recommended that idealized design should be repeated twice; once to produce a "bounded" design, assuming no changes in the wider system, and once to produce an "unbounded" design, assuming that changes in the containing system can be made. In Ackoff's view, designers will find that most organizations can be considerably improved just within the context of the bounded design. This is because the barriers to change are usually in the decision-makers' own minds and in the organization itself.

To ensure that creativity is not hindered during idealized design, only three constraints on the design are admissible. First, it must be technologically feasible and not a work of science fiction. It can be based on likely technological developments but not, for example, on telepathy. Second, it must be operationally viable, capable of working and surviving if it were implemented in what would be its environment now. Third, the design must be capable of being continuously improved. The aim of idealized design is not to produce a fixed "Utopia," but an "ideal-seeking system" that will be in constant flux as it responds to changing values, new knowledge and information, and buffeting from external forces. Beyond these three constraints, everything is open. Constraints of a political, financial, or similar kind are not allowed to restrict the creativity of the design.

An idealized design should cover all aspects of an organization and Ackoff (1999a) provides the following as a typical list:

- Products and services to be offered
- Markets to be served
- Distribution system
- Organizational structure
- Internal financial structure

- Management style
- Internal functions, such as:
  - purchasing
  - manufacturing
  - maintenance
  - engineering
  - marketing and sales
  - research and development
  - finance
  - accounting
  - human resources
  - buildings and grounds
  - communications, internal and external
  - legal
  - planning
  - organizational development
  - computing and data processing
- Administrative services (e.g. mail and duplicating)
- Facilities
- Industry, government, and community affairs

The remaining three phases of interactive planning are directed at realizing the idealized design as closely as possible – at closing the gaps that have been identified in the previous stage. "Means planning" is concerned with "how to get there." Policies and procedures are generated and examined to decide whether they are capable of helping to close the gap between the desired future and the future the organization is currently locked into according to the reference scenario. Creative thought is needed to discover appropriate means of bringing the organization toward the approximation of the idealized design favored by the stakeholders. Alternative means must be carefully evaluated and a selection made. "Resource planning" involves working out "what's needed to get there." Five types of resource should be taken into account:

- Money
- Plant and equipment (capital goods)
- People
- Materials, supplies, energy, and services (consumables)
- Data, information, knowledge, understanding, and wisdom

Each of the chosen means will require appropriate resources. It is essential to determine how much of each resource is wanted, when it is wanted, and how it can be obtained if not already held. It almost always turns out that there is an excess or a shortage in terms of the initial means plan. This, in turn, leads to a planning cycle in which either a productive use for excesses is found or they have to be disposed of, or shortages have to be overcome or plans changed. "Implementation and control" is about "doing it" and learning from what happens. Procedures must be established for ensuring that all the decisions made hitherto are carried out. Who is to do what, when, where, and how is decided. Once implementation is achieved, the results need to be monitored to ensure that plans are being realized. The outcome is fed back into the planning process so that learning is possible and improvements can be devised.

Before we leave this section on methodology, it is worth saying something about the importance of facilitation to the success of the IP process. Ackoff was naturally brilliant at it and saw skilled facilitation as an art rather than a science. For lesser mortals, Donna Lumbo's (2007) study of what is required to be a successful IP practitioner is useful. She concludes that it is necessary to have:

- Excellent communication skills
- A high degree of analytical competence
- Considerable people skills
- An understanding of personality characteristics and group dynamics, and an ability to guide groups in productive ways
- An ability to establish an atmosphere conducive to sharing and developing ideas
- The capacity to maintain energy levels and focus
- An ability to connect previous experience to the current situation
- The capacity to be creative, express unusual thoughts, and be interesting and stimulating

15.2.4 Methods

Readers of Ackoff's books and articles will find him using many methods, tools, and techniques in support of the interactive planning process. Here we concentrate on four models that he believes should be employed to help shift from a "mechanistic" or "organismic" to a social–systemic form of organization, and so are essential to the success of the methodology. In considering these four models – of a "democratic hierarchy," a "learning and adaptation support system," an "internal market economy" and a "multidimensional organizational structure" – we also need to bear in mind Ackoff's injunction that transformational leadership is essential in putting them into effect.

The need for a democratic hierarchy, or "circular organization," arises from the fact that managers are best employed focusing on the interactions of the parts rather than on controlling the parts directly; from the rising level of educational attainment among the workforce; and from dissatisfaction, in democratic societies, with working in organizations structured along Stalinist lines (Ackoff 1999a). The proposed democratic organization strongly supports the participative principle that underpins interactive planning. In a circular organization, every manager is provided with a "board." At the top level, this will involve external stakeholders. Each manager's board should minimally consist of the manager whose board it is, the immediate superior of this manager and the immediate subordinates of this manager, and any others invited to participate in a way defined by the board. This design is shown in Figure 15.1 (from Ackoff 1999b). The functions of the board are planning for the unit whose board it is; setting policy for that unit; co-ordination; integration; quality of work life; performance improvement; and approving, and if necessary dismissing, the head of the unit. Although this arrangement may seem unwieldy and time-consuming, Ackoff's experience is that the benefits in terms of synergy and motivation considerably outweigh the disadvantages.

An ideal-seeking system obviously requires a very particular kind of organizational design – one that encourages rapid and effective learning and adaptation. To this end,

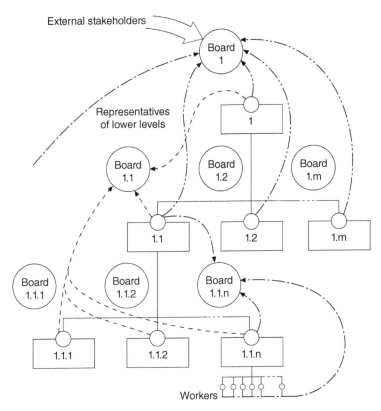

Figure 15.1 A circular organization. *Source:* From Ackoff (1999b). Reproduced with permission of John Wiley & Sons.

Ackoff supplies a suggested design for "a learning and adaptation support system." It contains five essential functions (Ackoff 1999a,b):

- Identification and formulation of threats, opportunities, and problems
- Decision-making – determining what to do about these
- Implementation
- Control – monitoring performance and modifying actions
- Acquisition or generation, and distribution of the information necessary to carry out other functions

This model is built on an array of feedback controls and takes account of the work of Argyris and Schön on organizational learning, especially the importance of double-loop learning. It also insists that an organization monitors errors of omission as well as errors of commission. Errors of omission, which occur when something is not done that should have been done, are usually ignored in bureaucracies, which tend to be much more interested in punishing people who do things.

Ackoff (1999a) has found that many organizational problems involve internal finance. These can be dissolved if the typical centrally planned and controlled corporate economy is replaced with an "internal market economy." Every unit within an organization, including the executive office, must become a profit center, or a cost center for which some profit center is responsible. It is then permitted to purchase goods and services from any internal and external supplier it chooses and to sell its output to any buyer it wishes. Higher authorities can override these decisions, but if they do so are required to compensate the unit for any loss of income or increased costs that arise due to the intervention. Ackoff (1999b, pp. 209–210) gives the example of Mobil's corporate computing center. This center was budgeted for and subsidized from above and, because none of the internal users paid for its services, there was no measure of their value. All that was known was that the computer center was overloaded and its customers complained of the service they received. A new CEO agreed with Ackoff that it be made a profit center, charging what it wanted and selling externally if it wished. Internal users of the center were permitted to seek similar services outside. The costs of the computer center were reduced dramatically because, now they had to pay for its services, others used it more sparingly. At the same time, the center became profitable by improving its service and gaining a substantial amount of external business.

A "multidimensional organizational structure" (Ackoff 1999a) is recommended because it increases flexibility and eliminates the need for continual restructuring. Organizations divide their labor in three ways and, in so doing, create three types of unit:

- Functionally defined units whose output is primarily consumed internally
- Production- or service-defined units whose output is primarily consumed externally
- Market- or user-defined units defined by type or location of customers

Restructuring occurs when the relative importance of these three ways of dividing labor changes and the system adjusts the level in the hierarchy at which different types of unit are manifested. Ackoff argues that time and effort can be saved if units of each type are placed permanently at every level. Reorganization can then be replaced by reallocating resources.

15.3 Interactive Planning in Action

The account that follows picks some highlights from an interactive planning project conducted in the DuPont Specialty Chemicals Safety, Health and Environment (SHE) Function. Further details can be found in Leeman (2002), who was central throughout. Ackoff was involved at various points as an advisor and has endorsed the project as a good example by including a version in *Re-creating the Corporation* (1999b). My account is drawn from these two sources. During the period that the work was ongoing, 1995–1998, DuPont was undergoing major corporate-wide transformation and downsizing. The circumstances were not necessarily favorable, therefore, for a successful application of IP.

Throughout its long history, in the chemical production industry, DuPont has prided itself on the attention it pays to the health and safety of its employees. More recently, it has evinced similar concern about its impact on the environment. The year 1994 saw a further step forward in all these areas when its SHE policy was revised and released as "The DuPont Commitment – Safety, Health and the Environment." From this commitment, the corporation derived a new slogan, "The Goal is ZERO." Its business units were to aim for zero injuries and illnesses, zero wastes and emissions, and zero environmental, process, and transportation accidents and incidents. In the SHE function, it was recognized that it would need to go about its work in a radically different fashion. Leeman realized that bringing about such a transformation would require a methodology that allowed purposes to be rethought, new goals systemically pursued, and encouraged widespread participation. Following a chance meeting with Ackoff, and an analysis using the system of systems methodologies (SOSM), it became apparent that interactive planning would be suitable.

Mess formulation was curtailed as Ackoff pointed out that DuPont was already a leader in safety, occupational health, and environmental protection, and the real effort, therefore, should go into improving its leadership position by gaining competitive advantage by further developing the excellence of its SHE function. Nevertheless, a brief systems analysis was conducted to ensure that SHE professionals were fully acquainted with Specialty Chemicals' businesses. And a brief obstruction analysis helped to reveal certain weaknesses in the way SHE performed its current role. It was structured in a centralized–hierarchical manner with a regulative rather than facilitative orientation. As a result, its expertise was not integrated into business decision-making. SHE professionals were caught in a situation of having to perform a multitude of mundane tasks and operating in a reactive mode to crises. Knowledge management was poor.

Ends planning, featuring idealized design, was divided into two parts. First, a group of consumers of SHE information and knowledge was invited to specify the properties of an ideal SHE system. Second, a designer group was asked to redesign the SHE system according to those specifications. The consumer group consisted of individuals chosen on the basis of six criteria:

- Use SHE information
- Are responsible for its implementation
- Are capable of specifying what they need from a SHE system
- Represent diversity in thought and in gender and race
- Are capable of thinking "out of the box"
- Understand the need for SHE and its role in the business

A number of SHE professionals were irritated by the role given to the consumer group, believing that only they had the expertise to contribute sensibly to the redesign.

A first session with the consumer group concentrated on identifying positive and negative outputs from the current SHE system. Some of Senge's ideas (see Section 11.2.4) helped the group to recognize the main structural problem as being the failure of the SHE function to connect to the business needs of the Strategic Business Units (SBUs). The next step was to specify the properties for an ideal SHE system based on the presumption that the existing system had been destroyed the night before and the new system could be designed unhindered by traditional constraints. The consumer group identified 58 specifications for the ideal system; later narrowed to 19 and categorized in 9 major arenas.

The designer group was then put together, consisting of an even number of SHE professionals and other managers from within Specialty Chemicals with detailed knowledge of SHE. The group was tasked with designing an ideal SHE system, replacing the one destroyed the previous night, using all the specifications from the consumer group, and being sure to *dissolve* all the negative outputs that had been identified. The process began with the identification of nine stakeholders crucial to SHE's success: customers; employees; representatives from plant sites; from business functions; from, government agencies; from SBUs; from local communities; from the DuPont Company; and from the corporate SHE function. The expectations of each of these groups with regard to the SHE system were considered. John Pourdehnad, from ACASA, then suggested following a version of idealized design that required three iterations of four major steps:

- Creating a mission statement
- Identifying the functions of the SHE system
- Formulating the processes for doing the SHE work
- Organizing the SHE structure to do the SHE work

The final version of the mission statement read as follows:

> A seamless SHE system that integrates, enables, and installs the core DuPont SHE competency to successfully make chemicals, win in business, and sustain our communities.

Paying attention to stakeholder expectations, the designer group then identified the key functions that the redesigned SHE system needed to offer. The eventual list was:

- Performance auditing and analysis
- Related project front-end loading guidance
- Training and education
- Personnel development
- Knowledge and learning
- Risk assessment and recommendations
- Federal, state, and local regulatory advocacy
- Community interactions
- Methodology and technology development
- Management and decision-making (planning)
- Core competency management
- Information management

For each function, the group then designed the necessary work processes for getting the work done. Finally, an organizational structure was proposed for SHE that provided for appropriate relationships between units and flows of responsibility, authority, communications, and resources, in order to deliver the functions to the business units. At the end of the three iterations, the designer group was convinced that their idealized design would meet stakeholder expectations, match consumer specifications, and dissolve all the output issues. The SHE ideal system is shown compared with the SHE current state, in terms of mission, function, process, and structure, in Table 15.1.

The next phase of interactive planning, means planning, involves determining how the gaps between the idealized design and the current state are going to be filled. The team chosen for this task consisted of SHE professionals from within Specialty

Table 15.1 Safety, Health and Environment (SHE) fundamental changes.

SHE current state	SHE ideal state
Mission	
1) Compliance-driven	1) Stakeholder-driven
2) Reactive/Intervention	2) Proactive/Prevention
3) Not aligned with business	3) Fully integrated within the business
4) Cost of doing business	4) Revenue enhancer/Value adder
Function	
1) SHE is operations support	1) SHE is business, operations, and customer support
2) Regulation tracking and interpretation	2) Regulation knowledge: • shaping regulations • quick access to regulation interpretations • shaping business plans
3) SHE training	3) SHE education
4) Data/information generation	4) Knowledge/understanding generation
Process	
1) Policing through auditing	1) Risk assessment and loss prevention
2) Government report preparation	2) Automated/electronic reporting
3) Manual data collection/documentation	3) Automated data collection/documentation
4) Classroom training	4) Online learn–teach–learn SHE system
Structure	
1) SHE is "centralized"	1) SHE is leveraged and distributed
2) "Stovepiped"	2) SHE is on cross-functional business teams
3) Hierarchical	3) "Lowerarchical"
4) SHE personnel confined to plant	4) SHE personnel on transunit teams
5) Line accountable for safety	5) Business and line accountable for SHE
6) SHE reports to operations	6) SHE reports to vice-president/general manager

Chemicals, most of who had also been in the designer group. The critical gaps identified were as follows:

• Current SHE system does not adjust to changing needs
• SHE professionals do not have time to deliver the "high-value" functions – prevention versus intervention
• We currently do not deliver many of the ideal-state primary functions
• We do not know where the "required inputs" reside
• SHE is focused on operations, not on increasing the business competitive advantage

For each of these, ways of closing the gap were proposed. For example, "SHE is focused on operations …" could be addressed by:

• Clearly defined connections between SHE and business leaders

- Clearly defined and supported SHE functions for business teams and customers
- Make SHE part of the SBU staff to increase status

Leeman emphasizes the efforts needed during this phase to keep the team focused, integrated, and committed and to ensure continued high-level management support.

The resource plan, aimed at identifying and providing the resources necessary to bridge the gaps and realize the idealized design, was then put in place. This consisted of personnel planning, financial planning, facilities and equipment planning, and materials, suppliers, and services planning. In personnel planning, for example, the need to hire a full-time project manager and new facilitator, and to set up eight SHE knowledge networks, a core team and a steering team were identified. Implementation and control concerns who is going to do what, when, where, and how. It was achieved successfully by paying particular attention to "the human factor," "the organizational factor," and "the commitment factor" and by ensuring that controls over implementation were designed that allowed tracking of progress.

At the end of the interactive planning project in DuPont, it was possible to identify the following clear benefits:

- A step-change improvement in SHE performance, shown by the most significant improvement in operational SHE performance metrics in its history
- SHE work was aligned with business goals and objectives to such an extent that relationships between SHE professionals and people in the business were transformed into a partnership
- SHE began to be perceived as a value-adding profit center rather than a "cost-of-doing-business" unit, and its services became highly valued by the wider organization and recognized as a powerful differentiator by external customers
- Organizational learning flourished among SHE professionals who were enabled to do more higher value-adding work, while their knowledge was made available to all to aid decision-making and routine SHE tasks were carried out by line employees
- A wider range of creative and less expensive solutions to SHE issues and problems were explored and implemented

Leeman is convinced that the project's success was due to the principles of participation, continuity, and holism embraced and operationalized through interactive planning. The idealized design process was essential for unleashing creativity, and participation ensured that energy and commitment levels were maintained during the hard work of implementation.

15.4   Critique of Interactive Planning

Ackoff's development of interactive planning, with its commitment to dissolving problems by designing a desirable future and inventing ways of bringing it about, has taken him a long way from some of his erstwhile colleagues stuck in the predict-and-prepare paradigm of OR. In the systems age, he believes, it has become necessary to shift the emphasis of management science toward exploring purposes and institutionalizing agreed ideals, goals, and objectives in a manner that allows for continuous learning and adaptation. Idealized design seeks to unleash creativity while harnessing the diverse purposes of different stakeholders by focusing their attention away from petty differences onto the ends they would all like to see their organization pursue. The process is meant to generate consensus, mobilize the stakeholders with a crusading zeal, and reveal that only the participants' limited imaginations prevent them approximating the future they most desire "right here, right now." Participation at the different stages of the planning process allows stakeholders to incorporate their esthetic values in the idealized design and the means necessary to achieve it.

Ackoff has demonstrated the effectiveness of interactive planning as a practical systems approach in hundreds of projects with organizations of all types in the United States and elsewhere. Much of its success is due, he believes, to the fact that it is based on an appropriate kind of systems model. As we saw, he grants some utility to deterministic and organismic models in some circumstances and in the short term. He is clear, however, that, in the systems age, their usefulness has diminished drastically:

> Our society and the principal private and public organizations that it contains have reached a level of maturity that eliminates whatever effectiveness applying deterministic and animalistic models to social systems may once have had.
> *(Ackoff and Gharajedaghi 1996, p. 22)*

Ackoff builds interactive planning on the back of a social-systemic model. This is more appropriate, he believes because modern organizations are best seen as purposeful systems containing other purposeful systems and as part of wider purposeful systems. Those who continue to apply deterministic or organismic models to social systems will get less desirable results because these models ignore the most essential characteristics of social systems.

In social theory terms, Ackoff is signaling a break with the functionalism that underpins hard systems thinking and much of system dynamics, sociotechnical systems theory, and cybernetics. An emphasis on understanding how systems function, so that we can control their behavior, gives way to an interpretive rationale emphasizing the need for mutually agreed purposes among stakeholders. In Ackoff's work, a subjectivist orientation holds sway. He does countenance an analysis of the systemic characteristics of messes, "formulating the mess," but the main purpose of this is usually as a "shock tactic," to wake people up. Thereafter interactive planning concentrates on people's perceptions and purposes. He hopes, through idealized design, to engage all the stakeholders of a system in the design of a desirable future for that system. In doing so, he assumes that they have free will and that their behavior is not determined by external circumstances. It is an axiom of Ackoff's approach that the stakeholders of a system do not have to accept the future as inevitable. Instead, they can plan a desirable future for themselves and seek to bring it about. Ackoff's work, along with that of Churchman's and Checkland's, represents an "epistemological break" between paradigms, from the functionalist to the interpretive, that opens up a completely different perspective on the way systems ideas can be used to help managers. Advocates of "soft systems thinking" would argue that it brings within the scope of systemic treatment all those wicked, messy, and ill-structured problems that escape or are distorted by functionalist methodologies because of the strict prerequisites that have to be met before those can be employed.

Given the severity of his assault on what he saw as the "mathematization" of OR, it is perhaps surprising that the main criticisms of Ackoff's work do not come from hard systems thinkers but from advocates of the "radical change" theoretical position. In Dando and Bennett's (1981) opinion, proponents of the "official" position in OR and systems thinking have been unable to defend themselves against "reformist" soft systems thinkers and have been intellectually routed. Instead it was a group of "revolutionary-minded" thinkers who took up the reformist challenge and sought to advance the debate further. There has always been a tension in social theory between those who emphasize the consensual and those who concentrate on the conflictual aspects of social systems. To his critics, the nature of interactive planning suggests that Ackoff's orientation is toward consensus. He seems to believe that there is a basic community of interests among stakeholders, which will make them willing to enter into interactive planning and to participate freely and openly in idealized design. Apparently there are no fundamental conflicts of interest between and within the system, wider system, and subsystem that cannot be dissolved by appealing to this basic community of interests. The critics believe that this does not give serious enough attention to the deep-seated conflict and coercion they see as endemic in organizations and society (Chesterton et al. 1975; Rosenhead 1976; Jackson 1982a). If irreconcilable conflict between stakeholders is frequent, as the critics assert, then Ackoff's approach is impotent because no agreement can be reached in such cases concerning the idealized future. Ackoff, it is argued, tries to fix the argument in his favor by defining irreconcilable conflicts as those that involve apparently logically incompatible ends. Since his methodology is oriented to the world of ideas, to expanding individual conceptions of the feasible, it is always going to be open to him to claim that a conflict is resolvable at a higher level of desirability. As Bryer (1979) argues, there is always going to be a "higher" system in which interests can be reconciled because the only limits to systems boundaries are subjective conceptualizations. In the real world, however, it is argued, a social structure can operate such that it is impossible for all different groups to achieve their ends. Rosenhead (1976) argues that "only by abolishing the sweat-shop owner as a social category can his interest and those of his laborers be made compatible." From this perspective, we need to talk about the social incompatibility of ends not their logical incompatibility.

Second, it is argued by critics of a "radical change" persuasion, Ackoff's one-sided understanding of social reality leads him to take the possibility of participation for granted and to overestimate it as a remedy for organizational problems. Participation is essential to interactive planning, philosophically because it provides the justification for the objectivity of the results and practically because it generates creativity and helps ensure implementation. Perhaps because of its significance, Ackoff plays down the obstacles to full and effective participation. To get started, his interactive planning depends on all the stakeholders being prepared to enter into participative planning about the future. But will the powerful be willing to forgo their dominant position and submit their privileges to the vagaries of idealized design? Even if interactive planning can be started, another problem will be encountered. The methodology depends for the objectivity of its results on free and open discussion between stakeholders, but planning is complex and time-consuming. We cannot realistically expect that less privileged stakeholders will be able to participate equally in the IP process. Whatever assistance the analyst can give to less fortunate groups, the various stakeholders will enter the process with widely divergent informational, political, and economic resources. The less privileged may additionally feel threatened by the massive resources that can be mobilized by the powerful and limit their demands to what is "realistic" (Rosenhead

1984). The organization will already represent a "mobilization of bias" against them in a way that requires no representation or advocacy (Bevan 1980). The less privileged may find themselves under the sway of a "dominant ideology" through the mists of which they fail altogether to recognize their own true interests. Any discussion or debate among stakeholders can only be, from this perspective, exceptionally constrained.

Third, it is said that Ackoff's belief in a consensual social world, and in the efficacy of participation, is only sustained because he artificially limits the scope of his projects so as not to challenge his clients' or sponsors' fundamental interests. In his research, it is argued, he only goes so far as "circumstances permit." No matter, as Rosenhead (1976) insists, that these circumstances are not forces of nature but the result of particular social arrangements, which could be changed. If Ackoff were to truly challenge the sponsors' interests, he would, it is claimed, soon provoke conflicts that revealed deep status, economic, and other inequalities in organizations that could not be spirited away by participation in idealized design. As a result, all the methodology is likely to achieve are acts of social engineering, which re-adjust the ideological status quo. Chowdhury (2015) documents a use of interactive planning with an NGO, The Universal Team for Social Action and Help (UTSAH), as the client. Once the idealized design was finalized, it was not shared with the community it was designed to serve. Chowdhury comments:

> In a way, this model portrays itself as an administrative mechanism that will legitimately direct the ways of life in the communities partnered with. This situation, according to Habermas lends a kind of statutory authorization to the party in the position of power .... The above understanding can lead us to believe that the notion of participation was a false notion, and the range of deliberations held with the community members and children involved participation that was fore-designed by the privileged parties, and hence, incomplete!
>
> *(2015, pp. 571–572)*

Finally, for some "radical change" theorists, such as Mingers (1984), the real problem with soft approaches such as interactive planning is their "subjectivism" or "idealism," which makes it impossible for them to recognize let alone come to terms with structural features of social reality such as conflict and power. To Rosenhead (1984), the fact that Ackoff ignores these aspects of social systems is attributable to the fact that he ascribes "prime motive power to the force of ideas." For Ackoff, the critics argue, conflict is always at the ideological level and is essentially dealt with by manipulating world views. Perhaps it is possible to alleviate conflict temporarily at the ideological level by getting people to believe they have interests in common. But the subjective beliefs of groups about their interests do not necessarily coincide with their objective interests. Permanent reconciliation of conflicts between stakeholders needs to be in terms of objective as well as subjective interests. Another consequence of Ackoff's idealism, it is claimed, is that it limits his ability to understand how change comes about and hence his ability to promote it.

To "radical change" thinkers, therefore, Ackoff's approach remains as "regulative" as the hard systems thinking that he criticizes. Ackoff (1975, 1982, 2003) has responded vigorously to this charge. He does not think much of any of the critics' arguments. If his work appears consensual to the critics, he believes, this is because they are obsessed with the notion of irresolvable conflicts – as they would be from their radical change perspective.

Ackoff has never encountered one of these in more than 300 projects on which he has worked. He argues that in every case in which conflicting parties have been willing to meet him, face to face, he has been able to find a solution to the conflict. He suspects that the critics merely assert that such conflicts exist. If they went out and tried to use IP on conflicts they presuppose are irresolvable, they might find out differently. If participation is an issue, Ackoff argues, it is possible to find ways around this, such as by introducing stakeholders first as consultants and then gradually increasing their role. The ability of low-level stakeholders to participate can, of course, be aided by professional planners. The idea that such stakeholders might not recognize their own true interests is élitist. What do the critics expect? It is surely better to work with stakeholders to see what changes are possible in the circumstances prevailing than to wait for the arrival of a Utopia in which no inequalities exist. Nor does Ackoff feel constrained because he spends much of his time working with managers as his sponsors. He finds that they are one of the most enlightened social groups and can see that benefiting other stakeholders will also benefit themselves. If he did come across a genuinely irresolvable conflict, and could not bring the parties together, Ackoff is quite prepared to work with the willing, usually disadvantaged, party to help them get what they want and what he feels is just. He will not work for clients whose values he cannot endorse. Finally, Ackoff simply does not accept the existence of the structural aspects of social reality that the critics emphasize. For him, the chief obstruction between people and the future they most desire is the people themselves and their limited ability to think creatively and imaginatively. Idealized designs are not meant as fixed utopias but they do represent desirable futures that can be worked toward. They appear challenging because they require people to rid their minds of self-imposed constraints. Provide people with a mission, with a mobilizing idea, and the constraints on their and their organization's development will largely disappear. The critics (e.g. Jackson 1983) respond that Ackoff underestimates the frequency of irresolvable conflicts. Getting the disadvantaged to believe they have interests in common with the powerful is likely just to lock them further into "false-consciousness" and deprive them of the ability to represent properly their real interests.

We are witnessing, of course, a battle between sociological paradigms. This becomes even clearer by bringing in the perspective of postmodernism. Thinking back to Section 4.5, it is clear that postmodernism would identify IP as a rather underdeveloped form of "critical modernism," based on Kant's program of enlightenment and seeking the progressive liberation of humanity from constraints. Ackoff (1974b) declares that he wants man to take over God's work of creating the future and offers interactive planning as the vehicle for achieving this. However, because it is so underdeveloped a version of critical modernism, for all the reasons we have been detailing, it is particularly prone to slipping back into becoming no more than an adjunct of "systemic modernism," readjusting the ideological status quo by engineering human hopes and aspirations in a manner that responds to the system's needs, thus ensuring its smoother functioning. These disputes between paradigms are impossible to resolve because of "paradigm incommensurability" – the fact that different paradigms rest on incompatible theoretical assumptions. The interpretation of an attempt by Haftor to evaluate interactive planning will, therefore, be left to the reader. On the basis of a case study, Haftor concludes:

> The results revealed that most of the propositions provided by IP [interactive planning] were actually realized during its application process, and also that the targeted

organization did develop successfully, chiefly in terms of its operational efficiency and quality. However, some of the IP-proposals could not be realized; these included securing the participation of all stakeholders in the process of IP execution and successfully managing power structures in the social setting addressed. These two latter findings conform well to the previously delivered critique of IP.

(2011, pp. 374–375)

In considering IP's positioning on the SOSM, it should be remembered that Ackoff views systems age thinking as complementing rather than replacing machine age thinking. He regards the purposeful systems orientation as the most sophisticated but accepts that the machine and organism models can be useful in the right circumstances and is willing to make use of what the earlier systems thinking has to offer. For example, his model of a "learning and adaptation support system" has much in common with Beer's neurocybernetic "viable system model." It is interesting that his chapter on this subject, in *Ackoff's Best* (1999a), is headed by a quotation from Stafford Beer. Beer was a friend but someone with whom Ackoff profoundly disagreed about the fundamental nature of social systems. IP's embrace of earlier systems models means that it can be allowed to extend along the vertical, "systems" dimension of the SOSM to embrace "complicated" contexts. However, Ackoff's work clearly prioritizes the complexity arising from people and their different perspectives rather than that stemming from the design of "complex adaptive systems." This was the source of his disagreement with Beer. IP cannot, therefore, be said to adopt "complex" assumptions in the way those are defined by the SOSM.

Looking at the horizontal axis of the SOSM, it is without question that IP emphasizes the need to manage pluralism by incorporating the variety of purposes that arise in systems, their subsystems, and their wider systems, into an idealized design that all stakeholders can embrace. His social-systemic model brings pluralism sharply into focus. It says little directly, however, about the characteristics of social systems that "radical change" theorists regard as most significant; things such as irreconcilable conflict, forms of domination, power, and false-consciousness. IP does not, therefore, extend as far as coercive contexts. John Pourdehnad disagrees with this conclusion and points to examples such as Ackoff's intervention in Alcoa (Ackoff and Deane 1984). Here IP was used to transform traditionally adversarial relationships between management and labor into a program of change, based upon co-operation, which apparently yielded significant benefits to all stakeholders. IP, it is suggested, can be of use in "coercive" contexts. However, a radical change theorist would simply point out that such co-operation, in the context of the capitalist system, can be read very differently, as making a small adjustment to how things are done in order to ensure the continuance of existing, unfair ownership structures. The grid is a theoretical device that gives an airing to a wide range of theoretical positions. The reader knows that there is no implication in saying this that Ackoff is wrong or that his approach is misguided. He may be right about the nature of social reality in which case IP is a useful approach in all conceivable problem contexts. The SOSM simply points out that there are alternative readings that point to the existence of circumstances in which IP will either struggle or reinforce the status quo. Ackoff had a rare appreciation of the different insights provided by the full range of systems models and this gave him access to almost the full dictionary of significant systems ideas and concepts. He (1999a) even took to talking of his "multidimensional organizational structure" as a fractal design. It is hardly out of

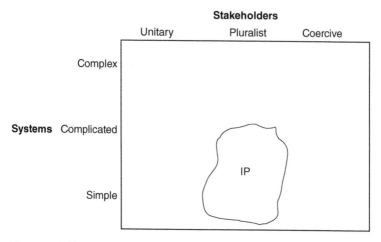

Figure 15.2 The positioning of interactive planning on the SOSM.

keeping with his message that we should explore the possible limitations, as well as the strengths, of the different systems models, including his own. According to SOSM logic, therefore, the positioning of interactive planning is as in Figure 15.2.

15.5 Comments

Jamshid Gharajedaghi, Ackoff's coworker at Wharton and Interact, has written that

> ... to be accepted as a friend by Russ, is to become the subject of his incisive and unrelenting critique ... at the same time, he is equally tireless in promoting you as nothing but the very best.
>
> *(Quoted in Ramage and Shipp 2009, p. 140)*

This is exactly right. My own friendship with him got off to a less than auspicious start. Having published my early critique of his work (Jackson 1982a), I approached the great man at a conference in Washington DC to ask what he thought of it. His response was immediate: "I don't think you've understood one thing I've ever written." Things did improve. He never sought to play the role of a "management guru" but regarded himself as an "educator" who helped others to work things out for themselves. I find the following words of encouragement inscribed in the various books he had written and that he gave to me over the years:

> 'To Mike
> ... who has given me a lot to think about and I am grateful'
> ... more than a colleague, a friend'
> ... our level of aspiration'
> ... whose work keeps me feeling young'
> ... with great affection and admiration'
> ... what a joy you and your work are to Russ Ackoff'
> ... from one of the flaws'

Ackoff lived modestly. I was lucky enough to stay with him and Helen Ackoff in Philadelphia, where he told me his stories, explained baseball, and introduced me to Philly Cheese Steak. He enjoyed visiting Hull. He kept an ancient dinner suit at my house. It was the only one he owned and, apparently, he had no use for it anywhere but the United Kingdom. We went to see Hull City play soccer and enjoyed watching the "1919 New Orleans Jazz Band," who played around Beverley at the time and were coincidentally named after the year in which he was born. On his last couple of visits, he took to our dog, Molly, and "trained" her to sit with him on our best furniture. This is a part of his legacy that we still live with every day.

15.6   The Value of Interactive Planning to Managers

The advantages for managers of adopting interactive planning are many and these are well documented in Ackoff's work. The five most frequently cited and evidenced are the following:

- The approach facilitates the participation of all stakeholders in the planning process and therefore secures the main benefit of planning – continuous engagement in the process itself rather than the production of some final document
- Allowing stakeholders to be dominant in the planning process and to incorporate their own esthetic values into planning relieves professional planners of the impossible task of measuring "quality of life" on behalf of others
- Idealized design releases large amounts of suppressed creativity and harnesses it to organizational and personal development
- Interactive planning expands participants' conception of what is possible and reveals that the biggest obstructions to achieving the future we most desire are often self-imposed constraints
- The participative principle helps generate consensus and commitment and eases the implementation of the outcomes of planning

15.7   Conclusion

Interactive planning was developed by Ackoff to assist stakeholders design a desirable future for themselves, their organizations, and their organizations' environments and to help them find ways of bringing that future about. This is a powerful vision of the role of management science and explains in large part why Ackoff's work has had such a huge impact on the OR and systems communities. Examined in detail, and in terms of what it is really able to deliver, interactive planning does not disappoint. It is arguable, however, that no approach can be comprehensive enough to achieve what Ackoff wants from interactive planning and that, in representing itself as being comprehensive, it inevitably disguises its own particular biases. Critical systems thinking and practice shares Ackoff's vision, as we shall see in Part IV, but tries to pursue it in a rather different way.

16

Soft Systems Methodology

Soft systems methodology (SSM) is an approach for tackling problematical, messy situations of all kinds. It is an action-oriented process of inquiry into problematic situations in which users learn their way from finding out about the situation, to taking action to improve it. The learning emerges via an organized process in which the situation is explored using a set of models of purposeful action (each built to encapsulate a single worldview) as intellectual devices, or tools, to inform and structure discussion about a situation and how it might be improved

(Checkland and Poulter 2010, p. 191)

16.1   Prologue

In the late 1980s, Peter Checkland became involved in a project with the "manufacturing function" (MF) department of the Shell Group. It employed around 600 people at the company's headquarters in the Hague. Shell, of course, is in the business of obtaining raw materials, principally oil, from around the world and transforming them into profitable products. MF had a variety of tasks: keeping abreast of the latest technology and informing the Board accordingly, advising on appropriate research and development programs, monitoring the performance of production units, and providing assistance to production plants as necessary. Its staff consisted principally of engineers and other scientific and technological experts. Support functions of this type are constantly questioned about what exactly they contribute to the business and so it is not surprising, perhaps, that the new head of MF, Rob de Vos, decided that it would be prudent to conduct a thorough investigation into its role, structure, and procedures. Two Shell managers, Jaap Leemhuis and Kees van der Heiden, persuaded de Vos that a systems study, with Checkland as adviser and using his "soft systems methodology" (SSM), should be undertaken. SSM, which had already been refined over nearly two decades of action research, was undergoing one of its more drastic revisions and Checkland saw this as an excellent opportunity to investigate how the new "two streams" version would work in an important and complex problem situation. This account of what happened is drawn from Checkland and Scholes (1990) and Checkland and Poulter (2006).

Before the SSM study began, Leemhuis and van der Heiden had been engaged in a significant process of "finding-out," consisting of 80 interviews with stakeholders of MF

Critical Systems Thinking and the Management of Complexity, First Edition. Michael C. Jackson.
© 2019 John Wiley & Sons Ltd. Published 2019 by John Wiley & Sons Ltd.

about how they perceived its role and issues. The results were documented and revealed the variety of conflicting demands that were made on MF. One example was the requirement to focus on technology development while serving a great variety of internal customers. Another, was to maintain staff continuity in MF so that managers could fully understand MF's core activities while rotating them regularly to ensure they kept up-to-date with what was actually going on in the production units. In planning the study, therefore, the team of Leemhuis, van der Heiden, and Checkland already had a considerable amount of information that they could share with participants. It was decided to involve as many managers as possible – Shell professionals and "customers" of MF as well as those directly employed in MF – and to proceed on the basis of the common Shell practice of two-day workshops involving 15–20 managers. Any new "vision" for MF, it was felt, should come from the department as a whole. The decision was also taken to conduct the workshops in the everyday language of the Shell managers, explaining SSM only if requested to do so by the participants. The first two workshops involved the range of stakeholders mentioned and focused respectively on the themes of "technology development" and "a service business." Both followed the same format of discussion and model building on the first day and a comparison of the models with what was actually happening on the second day. Documents, which included all the models built and the comparisons, were prepared and circulated for comment and suggestions after each workshop. There followed three further workshops, attended only by de Vos, his senior management team, and the consultants, which used SSM to evaluate what had been produced so far and to agree a way forward.

SSM was employed to serve a number of different purposes during the project. Checkland produced a "conceptual model" of the process necessary to prepare and conduct the seminars. This enabled him to recall what exactly needed doing at each stage. It included activities such as "define theme for next seminar," "conduct two-day seminar in Shell language," "record outcome of seminar," "send seminar outcome to audience." The "two-streams" version of SSM requires the analyst to continuously reflect on and record the roles, norms, values, and power relationships observed in the situation so that they are better able to judge what change is likely to be "feasible." Checkland was reminded by the "conceptual model" to update the "Analysis 2 and 3" files he was keeping on these matters. SSM was used in the workshops not to impose a formal structure but rather to intervene to prevent discussion going round in circles, to provide greater clarity to what was being said, and to encourage progress toward the intended outcome of reorientating MF. For example, following a particularly intense morning of discussion, during the first workshop, order was restored by dividing the participants into groups and asking them to write on flip-charts some statements about MF's "core purpose." Checkland then demonstrated how each of these statements could be "tightened up" to the extent that it became possible to work out the structure of activities necessary to realize the particular view of the "core purpose" enshrined in each statement. In SSM terms, this is expressed as moving from "relevant systems" to "root definitions" to "conceptual models." The participants were now encouraged to have a go at constructing their own models from particular worldviews and began to appreciate the role these could play in enabling a better, more focused debate to take place. This, in turn, made them more willing to engage with the preprepared models the facilitators would sometimes introduce to further enrich and bring more rigor to the ongoing discussions.

Finally, the comparison of what was taking place in MF with the activities implied by the models was guided by a standard SSM template, which asked whether the activity in the model existed now, how it was done, who did it, whether it was done well or badly, and in what alternative ways it might be done.

The first workshop highlighted the tension that had been noted by Leemuis and van der Heiden between MF's concern for the longer term development of technology and the demands on it as a service provider. These two requirements competed for time and resources. Checkland, reviewing the evidence as part of Analysis 2, concluded that MF exhibited an "engineering culture" with the greatest emphasis placed on technical excellence. This only confirmed the need to orientate the second workshop around the theme of service provision. In that workshop, the participants, aided by a specialist service management specialist, looked at MF in a variety of service roles – giving technical help to operations, advising on the construction of plants, and as an enabler of strategy. Ten "root definitions" and their corresponding "conceptual models" were reviewed, five preprepared by the facilitators and five that emerged from the groups. On the second day, comparison with the way that MF actually operated revealed just how much structural, procedural, and attitudinal change would be necessary for MF to develop as a service business. It had become clear to the participants that much could be gained by taking forward a new vision for MF as an internal provider of service within the Shell Group. And this is exactly what de Vos did in the three workshops with his senior management team. The first of these reviewed the work done so far and sought to express the challenges facing MF in a precise manner using "root definitions" and "conceptual models." It became clear that current practice paid insufficient attention to the competitive position of MF's customers and to the range of information they needed. The second workshop concentrated on the detail of what was necessary for MF to become a service business. The service management consultant was again present and insisted that a key question was "who is my customer's customer?." A successful service business had to consider what the recipients of its services required if they were to meet their customers' needs. It was apparent that, to serve its clients, MF would need to manage simultaneously activities related to developing the technology base, supporting operations, and developing manufacturing strategy. Eventually, after a number of iterations, a "root definition" and "conceptual model" emerged, which were later recognized, by Checkland, Leemhuis, and van der Heiden, as the most crucial of all in reorientating thinking about MF's role. These, reproduced as Figure 16.1, highlight the need for the coproduction of added value by MF and its customers and for that value to relate directly to the competitive position of the customers.

With this vision now firmly implanted in their minds, senior managers focused, in the final workshop, on the structures, processes, and attitudes that would be necessary in MF to bring it about, always bearing in mind what was culturally and politically feasible in the actual Shell context. It was regarded as essential to continue the participative process that had already taken the department so far in its thinking. Presentations, information sessions, and seminars were held at which all members of MF were invited to contribute their ideas and suggestions. Thus the open and transparent process, that had begun with the widest possible distribution of documents following each workshop, was maintained. From the initial "finding-out" stage to the point when the new

A system owned and staffed by service supplier and client – within an environment of access, open communication and transparency – which, by joint examination, seeks to match the capabilities of party A (supplier) with the potential for opportunities of party B (recipient) to maximize B's success, thereby simultaneously increasing the experience and income of A.

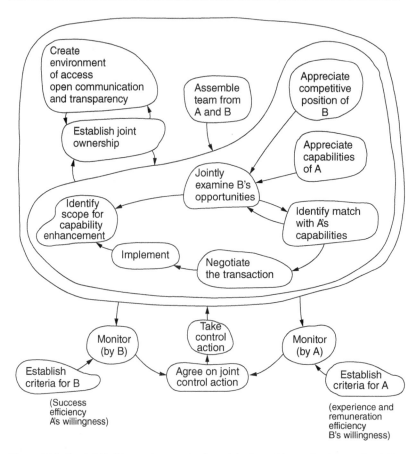

Figure 16.1 Root definition and conceptual model relevant to service transactions. *Source:* From Checkland and Scholes (1990). Reproduced with permission of John Wiley & Sons.

MF was in operation, with new organizational units serving named client groups, took 15 months. One senior manager in MF, commenting on the whole intervention, said:

> Though we could not have articulated it at the start, we were at that stage thinking of MF as a creator and preserver of technical skill pools. Now we are thinking of it as an internal service business.
>
> *(Quoted in Checkland and Poulter 2006, pp. 80–81)*

Checkland and Poulter see this as a "very neat summary of the change of worldview." It was that stark; and made the manager in question so uncomfortable that he asked for a transfer out of MF.

16.2   Description of Soft Systems Methodology

16.2.1   Historical Development

Peter Checkland, the key figure in the development of SSM, studied chemistry and did research on spectroscopy at Oxford before spending 15 years with ICI. Initially he worked in research and development in fibers but eventually, after a number of promotions, found himself manager of a 100-strong group of other people doing research and development (Checkland 2011). As his managerial responsibilities increased, he began to look for assistance to the literature of management science dominated, at that time, by "hard systems thinking." He was shocked and disappointed to find that most of what he read was completely irrelevant to his job. It concerned itself with the "logic of situations," with a few regularities, rather than with the unique events that seemed to dominate his life as a manager. A growing fascination with such matters, together with an interest in the application of systems ideas, led Checkland to leave ICI in 1969 to join the postgraduate Department of Systems Engineering, which had been established at Lancaster University by Professor Gwilym Jenkins with a grant from ICI. Brian Wilson, an electrical and control engineer with whom Checkland collaborated in developing SSM, was already there having joined the department as a founder member in 1966.

The systems approach favored in the new department at Lancaster was inevitably Jenkins' systems engineering, as described in Section 9.2.3. This demanded well-structured problems and clearly defined objectives and measures of performance – it was a typical hard systems methodology. Jenkins was determined, however, to interpret the word "engineering" broadly and to see whether the methodology could be developed to deal with a wider range of organizational issues than had traditionally been the case. The department was later renamed simply a "Department of Systems" to signal this broad conception. Crucially, he also believed that systems engineering could only advance if it constantly engaged with "real-world" problems. It would stagnate if absorbed by academia. He therefore initiated an action research program in which systems engineering was "tested" in organizations outside the University. A one-year Masters course in "Systems in Management" was launched and this included a five-month project. The average age of the students recruited was around 30; mature enough to take part in projects in a wide variety of organizations. Later, Jenkins created a University-owned consultancy company, called ISCOL, which could take on projects at all times and of all lengths, and this was run out of Lancaster for 21 years. A virtuous circle of interaction between ideas and experience was now possible, which was fully exploited by Checkland and his colleagues.

Jenkins believed that the action research program would lead to the discovery of powerful generic models of functional systems, such as production, marketing, and distribution, which would have very wide applicability. Wilson, as we shall see, continued to pursue this agenda. Checkland was less certain, but willing nevertheless to follow the research strategy of trying to use Jenkins' systems engineering approach to tackle management problems and learning from the results. Unlike Jenkins, however, he rapidly concluded that the methodology itself would have to be radically changed to make it appropriate for dealing with the greater complexity and ambiguity of the social as opposed to engineering context. It was of little use, as Checkland soon discovered during work on the Anglo-French Concorde project, to come up with a structured set of means to create a supersonic jet aircraft if the political differences between the British and French about the purpose of the project could not be reconciled. What finally emerged, after considerable project work and reflection on the experience gained, was an entirely different kind of approach – SSM.

In the first full account of this methodology, published in 1972, Checkland (1976) describes three of the most significant early project experiences that led to the break from systems engineering and the formulation of SSM. In all three, it was clear that serious problems existed in the organizations of interest, but the clients simply could not say what they were in precise terms. Each of the problem situations was vague and unstructured. One of the projects, in a textile firm, gave rise to at least a dozen candidates for the role of "the problem." Generalizing from these three projects, Checkland was able to specify how SSM needed to differentiate itself from hard approaches. First, in confronting "softer" problems, the analysis phase of a methodology should not be pursued in systems terms. In the absence of agreed goals and objectives, or an obvious hierarchy of systems to be engineered, using systems ideas too early can lead to a distortion of the problem situation and to jumping to premature conclusions. Analysis, therefore, should consist of building up the richest possible picture of the problem situation rather than trying to capture it in systems models. Second, given that it is not obvious, which if any system needs to be engineered, it is more appropriate for the analysis to define a range of systems possibly relevant to improving the problem situation, each expressing a particular world view (or *Weltanschauung*). These notional systems can be named in "root definitions" and developed more fully in "conceptual models." The use of SSM will therefore lead to the construction of a number of models to be compared with the real world, rather than just one as in hard methodologies. These models are of possible "human activity systems" and Checkland came to recognize the delineation of this type of model as one of the most important breakthroughs in the development of SSM. Previous systems thinkers, he said, had sought to model physical systems, designed systems, even social systems, but they had not treated purposeful human activity systemically. (Actually Caws (see 2015), with his notion of "relations as components of intentional structures," had at least anticipated the idea). A human activity system is a model of a notional system containing the activities people need to undertake in order to pursue a particular purpose. Finally, while the models produced by hard approaches are meant to be models of the real world or blueprints for design, human activity system models are contributions to a debate about change. They explicitly set out what activities are necessary to achieve a purpose meaningful from a particular point of view. On the basis of such models, participants in the problem situation aim to learn their way to what changes are systemically desirable and culturally feasible. The models are thus epistemological devices used to find out about the real world.

The action research program that yielded these early breakthroughs was rigorously pursued in the several hundred projects that followed. It has ensured that the lessons learned from the project experiences can be considered and used to improve SSM, that refinements can be made to supportive methods and techniques, and that reflection can take place on the philosophical underpinnings of the methodology. The developments that have occurred can be traced in a series of books written by Checkland about SSM. *Systems Thinking, Systems Practice* (1981) details the emergence of systems thinking as a response to the complexity with which the traditional scientific method struggles, its key characteristics, the parting of SSM from systems engineering, the seven-stage version of the methodology, and the philosophical underpinnings of this approach. *Soft Systems Methodology in Action* (1990, with Scholes) provides further discussion on the nature of "methodology," introduces the "two-streams" version of SSM, a "Mode 2" way of using it, and provides many examples of the methodology employed in both the public and private sectors. *Information, Systems and Information Systems* (1998, with Holwell) sets out to demonstrate how SSM can be used to reconfigure thinking and practice in the important field of information systems. In 1999, Checkland added a 30-year retrospective to new editions of *Systems Thinking, Systems Practice* and *Soft Systems Methodology in Action*. This revisits the emergence of SSM, reconsiders the methodology as a whole, and in terms of its parts, and argues that "four key thoughts" dictated its development. *Learning for Action* (2006, with Poulter) followed. This is described as "a short definitive account of soft systems methodology and its use for practitioners, teachers, and students" and largely fulfills its aim. It is also important for students and potential users of SSM to consider the accounts of the approach offered by Brian Wilson and to recognize his contributions to the development of the methodology and its associated methods and tools. These can be found in *Systems: Concepts, Methodologies, and Applications* (1990, originally 1984), *Soft Systems Methodology: Conceptual Model Building and Its Contribution* (2001), and *Soft Systems Thinking, Methodology and the Management of Change* (2015, with Van Haperen).

As Checkland sees it, SSM has now become a mature methodology capable of intervening in soft, unstructured problematic situations where answering questions about "what" we should do is as significant as determining "how" to do it. It is also capable of sharpening up to tackle more structured problems if circumstances permit. SSM has been central to a revolution, which has released systems thinking from the intellectual straightjacket in which it was locked and, at the same time, has made it much more relevant to managers. In Mingers' view:

> SSM has reoriented an entire discipline and touched the lives of literally thousands of people ... [soft] thinking is now completely taken for granted within the systems discipline.
>
> *(Quoted in Ramage and Shipp 2009, p. 149)*

SSM has been readily accepted, by academics and practitioners, as a successful approach to "wicked" problems and has spread its influence to many countries outside the United Kingdom. Ledington and Donaldson (1997) chart how widely it has been adopted and used in Australia. A number of papers (Mingers and Taylor 1992; van de Water et al. 2007; Hanafizadeh and Mehrabioun 2018) demonstrate the impact it has had in many different fields – such as strategy development, information systems, knowledge management, general problem solving, education, health care, agriculture, project management, performance management, and sustainable development.

In examining the methodology, and the thinking behind it, it is helpful to bear in mind those "four key thoughts" that Checkland (1999, 2000) says helped move SSM research forward:

- Human situations are characterized by people trying to act *purposefully*. It is therefore useful to model the activities that, when appropriately structured, give rise to the emergent property of being *purposeful*

- In any situation, there will be many possible interpretations of the purposeful activity. So build a number of relevant models and declare the worldview behind each
- Conduct the inquiry as a *learning* process, using the models as devices to structure a debate/discussion about change that is both feasible and desirable
- Activity models can be turned into information models and so SSM can play an important role in development of information systems

16.2.2 Philosophy and Theory

Checkland (1981, pp. 260–261) states that the work that led to the development of SSM was not based on any explicit philosophy or theory. Nevertheless, he soon displayed an interest in relating the methodology to theory and it is clear that its later manifestations benefit from being theoretically informed, at the very least, in terms of their articulation. For example, his readings of Vickers, Weber, and Husserl enable him to bring greater theoretical clarity to his break with hard systems thinking. His writings do not betray the "tensions" between the hard and soft positions that, as we shall see, he has identified in Ackoff's and Churchman's work. Checkland is the most theoretically pure of the soft systems thinkers because he recognized the direction that the approach was taking, made this explicit, and began to consciously base SSM on different philosophical and theoretical foundations. It is because of this, I would argue, as much as the methodology itself, that his writings have had such an impact. Checkland and Poulter (2006, p. 171) believe that knowing the theory that underpins SSM will "more than likely" improve a practitioner's use of the methodology. I would suggest that it is indispensable for the beginner to prevent them falling into the trap of employing SSM simply as an adjunct to hard systems thinking.

Checkland (2011, p. 500) describes his introduction to the work of C West Churchman, just six weeks after he had started his academic career, as "one of the most important weeks of my life." Churchman was delivering a one-week course on his version of "the systems approach" on behalf of the Operational Research Department at Lancaster. From the course, Checkland took away the concept of *Weltanschauung* or worldview. He quotes Churchman from *Challenge to Reason*:

> For the applied scientist, scientific method must include a philosophy …. It is what the Germans call a Weltanschauung, a perception of what reality is like … the applied scientist is not merely applying the results of pure research; he is applying his Weltanshauung.
>
> *(Checkland 2011, p. 500)*

For Checkland, it is not just applied scientists but everyday people who carry around a Weltanschauung (W) – a relatively stable tendency to interpret the situation in which they find themselves in a particular way. Further, individual Ws can differ completely such that one person's "terrorism" is another person's "freedom fighting." To make matters even more difficult for managers, Ws can change over time and sometimes in a radical manner.

Following this first encounter, Checkland became an admirer of the work of Churchman and his collaborator Ackoff. Nevertheless, one of their influences on him was unpredictable. He noticed (1981, 1988) that in their writings they still used notions such as "the design of systems" (Churchman) and "idealized design" (Ackoff) that were redolent of hard systems thinking. To me, while this is true, it is looking too closely at the words and not at the way Churchman and Ackoff actually employ them as part of their broader conceptual schemas. It is to miss the essence of what they are about as I have described it in Chapters 14 and 15. Whatever, the fact that Checkland believed them, in certain respects, to be still in thrall to the "goal-seeking paradigm," with all the limitations that implies, made him even more determined to distance himself from that way of thinking, and had an important impact upon the development of SSM.

A big step forward occurred when Checkland became acquainted with the work of Sir Geoffrey Vickers, which, he says, "has had by far the greatest impact on the development of SSM" (Checkland 2011, p. 507). Vickers had a long and distinguished career as a lawyer, soldier, and public administrator and, in retirement, published a series of books reflecting deeply upon his years in the "world of affairs." He was appointed a visiting professor at Lancaster at the age of 84. Vickers (1965, 1970, 1983) found systems concepts and ideas of more use than those of any other discipline he came across for explaining what he had experienced in his long career as a manager and decision-maker. At the same time, he rejected the impoverished formulations, at least for management practice, of the goal-seeking and cybernetic systems paradigms dominant at the time. For Vickers, achieving stability in social systems involves much more than establishing a goal from outside the system, monitoring performance, and taking corrective action on the basis of feedback information. He wanted to extend systems thinking to embrace the "peculiarities" of human systems – what was different about them compared to other types. Of primary importance is the fact that multiple perspectives and multiple possible courses of action are generated from within social systems. Systems thinking therefore needs to embrace "meaning" and "judgment" and how these are linked to the culture and history of society. Societies and organizations can only be governed through a complex process in which shared norms and values are established and maintained. This depends on the negotiation of relationships between different participants.

Checkland recognized in Vickers' writings a rich description of the social reality he was discovering in the process of originating SSM. Of particular importance, in developing the methodology, was Vickers' concept of the "appreciative system." Here, Vickers was building on work by that renegade pioneer of general systems theory, Kenneth Boulding:

> Professor Boulding has described our inner view as 'the Image' and he has most usefully stressed its importance and its dimensions; but to picture the inner world we must look behind the image and ask what causes an individual or a society to see and value and respond to its situation in ways which are characteristic and enduring, yet capable of growth and change. A national ideology, a professional ethic, an individual personality, resides not in a particular set of images but in a set of *readinesses* to see and value and respond to its situation in particular ways. I will call this an appreciative system.
>
> *(Vickers 1970, p. 59)*

We therefore have a direct link from Boulding's proto-phenomenological systems thinking to what became Checkland's fully fledged version. Vickers argues that the components of human systems, active individuals using appreciative systems to attribute meaning to their situation, make it impossible to study them using the methods of the natural sciences. The only way to understand human systems is to grasp the different appreciative systems that people bring to bear on a situation. Vickers summarizes this aspect of his thinking as follows:

> To account for the appreciated world .... I postulate that experience, especially the experience of human communication, develops in each of us readinesses to notice particular aspects of our situation, to discriminate them in particular ways and to measure them against particular standards of comparison, which have been built up in similar ways. These readinesses in turn help to organize our further experience, which, as it develops, becomes less susceptible to radical change .... Since there are no facts, apart from some screen of "values" which discriminates, selects and relates them, just as there are no values except in relation to some configuration of fact, I use the word appreciation to describe the joint activity which we call knowing and which we sometimes suppose, I think mistakenly, to be a separable, cognitive activity which is "value-free." Since these readinesses are organized into a more or less coherent system, I speak of them as an appreciative system. I sometimes refer to their state at any point of time as their appreciative setting and to any act which expresses them as an appreciative judgement. The appreciative world is what our appreciative system enables us to know.
>
> *(1972, p. 102)*

An individual's appreciative system will determine the way they see ("reality judgment") and value ("value judgment") the situation they are confronted by and condition how they make "instrumental judgments" (what is to be done?) and take "executive action" – in short how they contribute to the construction of the social world. It follows, according to Vickers, that if human systems are to achieve stability and effectiveness, the appreciative systems of their participants need to be sufficiently shared to allow mutual understandings to be achieved. Social systems depend upon cultures that permit shared appreciations to flourish. Checkland, already acquainted with Churchman's related concept of *Weltanschauung*, immediately recognized the significance of Vickers' notion of the "appreciative system" for SSM. His methodology could be seen as a way of engaging with the social world as described by Vickers. It subjected current appreciative settings to close examination, brought them into contact with different appreciative systems, and discovered what change could occur based on whatever accommodations became possible as a result of the learning achieved:

> I see SSM as giving epistemological guidance to the exploration of appreciative settings in situations regarded as requiring action-to-improve. It explores current and potential settings, and the standards used to judge fact, value, and action. Its sought accommodations inevitably entail envisaging new or modified relationships among those seeking the action-to-improve.
>
> *(Checkland 2011, p. 509)*

Also related to Vickers' thought is the notion of "organization" that Checkland and Holwell (1998) see as underpinning SSM. As they describe it, SSM treats the notion of "organization" as problematical. It only arises because of the readiness of people, members and nonmembers alike, to talk and act as though they are engaging with a collective entity capable of purposeful action in its own right. On this basis, a degree of agreement on purposes may emerge, social processes to pursue those purposes, and criteria for evaluating performance. This, in turn, may lead to the definition of organizational "roles" and the establishment of norms and values. Despite the willingness of individuals to conform in this way, there will be many different conceptualizations of the nature and aims of the "organization," premised on the values and interests of individuals and subgroups, apart from any "official" version of its purpose. People constantly seek to renegotiate their roles, norms, and values and are capable of displaying considerable "cussedness and irrationality" in the face of official goals. Because the different values and interests will rarely coincide exactly, the "organization" depends for its existence on the establishment of temporary "accommodations" between individuals and subgroups. These also provide the basis for any action to bring about change. In summary:

> In the present analysis, an organization is clearly an abstraction: it is a social collectivity concerned with some collective action, and there are associated social practices which relate to this. But what causes it, as an entity, to exist? The answer can only be: the *readiness* of some people, usually large numbers of people, members and non-members alike, to talk and act as if there were a collective entity which could behave like a conscious being, with the ability to do things and then make them happen .... This way of thinking about an organization is rather abstract, but it is necessary to make sense of what we all know from observation and experience, namely that members of organizations are not necessarily simply quiescent contributors to the achievement of organizational goals, as the conventional model suggests.
>
> *(Checkland and Holwell 1998, pp. 80–81)*

A further insight provided by Vickers' is that, in social systems, goal-seeking is a special case of "relationship maintaining." In developing this insight, Vickers differentiates his position from that of Nobel Prize winner Herbert Simon. The idea had a big impact on Checkland. It became clear to him that hard systems methodologies are predicated on the goal-seeking model of human behavior as exemplified in the work of Simon, while SSM reflects a model of human behavior oriented to "relationship-maintaining" as set down in the writings of Vickers (Checkland 1985). Checkland (1983) carries this thought further in reflecting on the shift in philosophical perspective necessary to establish SSM. In particular, he suggests that whereas the hard systems approach is premised on a paradigm of optimization, his own methodology embraces a paradigm of learning. Hard systems methodologies assume the world contains systems. They then seek to optimize their performance by following systematic procedures. These procedures involve establishing clear objectives and then using generic models, based on systems logic, to establish how control of the real-world systems of concern can be obtained and their objectives realized with maximum efficiency. Unfortunately for hard systems thinking, at least in Checkland's experience, logic is usually much less significant in terms of what happens in organizations than is the history, culture, and politics of the situation as reflected in multiple worldviews and constant change. On this basis, Checkland regards real life as simply too complex and unpredictable to model. SSM accepts that reality is problematical and rejects any attempt to capture it in systems terms. Instead, it seeks to work with different perceptions of reality, facilitating a systemic process of learning in which different viewpoints are examined and discussed in a manner that can lead to purposeful action in pursuit of improvement. Participants use a systemic methodology to learn what changes are feasible and desirable given the peculiarities of their problem situation. Checkland puts this concisely in stating the change in theoretical viewpoint that occurred in moving from hard to soft systems thinking. SSM, he says, shifts systemicity from the world to the process of enquiry into the world:

> In order to incorporate the concept of worldview into the approach being developed, it was necessary to abandon the idea that the world is a set of systems. In SSM the (social) world is taken to be very complex, problematical, mysterious, characterized by clashes of worldview. It is continually being created and recreated by people thinking, talking and taking action. However, our coping with it, our process of inquiry into it, can itself be organized as a learning *system*. So the notion of systemicity ('systemness') appears in the process of inquiry into the world, rather than in the world itself.
>
> *(Checkland and Poulter 2006, pp. 21–22)*

Checkland and Holwell (2004) further capitalize on Vickers' insight, that goal-seeking is a special case of relationship maintaining, to argue that hard systems approaches should only be used in very particular circumstances as part of an SSM intervention. In their terms, the two are "asymmetrically complementary." During a soft systems study, those involved might make a conscious decision that, in their circumstances, it is useful to engineer some system to achieve a goal as efficiently as possible. Thus, hard approaches are not just thrown away in soft systems thinking. Under the "umbrella" of SSM, they can be called into action as a deliberate choice if the circumstances are propitious and the participants think it is desirable.

It is worth noting that Checkland's conclusions about the importance of "nonlogical" factors, in decision-making in organizations, are supported in recent work by the psychologist and "behavioral economics" guru, Daniel Kahneman (2011). Kahneman details the instinctive, emotional, unconscious ways of thinking we use to decide what to do which undermine the consciously deliberative, logic-driven side of thinking. We tend to employ short-cuts and heuristics associated with existing pasterns of thought rather than make the effort to create new patterns.

The debt to Vickers was significant. However, there was further theoretical development to come. During his reading of philosophy and social theory, in preparation for writing *Systems Thinking, Systems Practice* (1981), Checkland became aware that he was taking systems thinking in a much more radical direction than he had hitherto realized. He was initiating a paradigm break from one philosophy and sociology to another. He characterizes it in that book, which remains the best source for the relationship between SSM, philosophy, and social theory, as a shift from the positivism and functionalism that underpins hard systems thinking to an approach based on phenomenology and interpretive sociology. To make his case, Checkland argues that SSM has much more in common with the interpretive sociology of Weber than with the functionalism of Durkheim; and with the hermeneutics of Dilthey, and the phenomenology of Husserl and Schutz, than with the positivism of Comte. SSM is, indeed, more akin in its presuppositions to the "action frame of reference" of Silverman than to the function-alist organizations-as-systems approaches dominant in the organization theory that Silverman attacks. In SSM, as in phenomenology (see Section 1.5), it is observers' men-tal constructs, yielding different interpretations of the world, which are the focus of attention. In SSM, as in hermeneutics and interpretive sociology (see Section 4.3), it is subjective meaning that is privileged. SSM works with different descriptions of reality, based on different worldviews, and embodies them in "root definitions." These root definitions are turned into conceptual models that are explicitly one-sided representa-tions expressing a *Weltanschauung* – in other words, they are Weberian ideal types. A debate is then structured around the implications of these different perceptions of what might be done in which actors negotiate and renegotiate their interpretations of reality.

Checkland (1981) draws together his theoretical arguments by referring to Burrell and Morgan's (1979) categorization of sociological paradigms, according to whether they are objectivist or subjectivist and whether they support regulation or radical change. It is clear to him that hard systems thinking is objectivist and regulative in ori-entation and therefore, according to Burrell and Morgan's logic, that its implied social theory is functionalism. SSM, however, he argues, is more subjectivist in character and extends somewhat toward the radical change axis of their grid. As a result:

> The social theory implicit in soft systems methodology … would lie in the left-hand quadrants with hermeneutics and phenomenology, although the position would be not too far left of the centre line because the methodology will over a period of time yield a picture of the common structurings which characterize the social collectivities within which it works. Also, given the analyst's complete free-dom to select relevant systems which, when compared with the expression of the problem situation, embody either incremental or radical change, the area occu-pied must include some of the 'subjective/radical' quadrant.
>
> *(Checkland 1981, pp. 280–281)*

Checkland supports his claim that the implied social theory of SSM embraces aspects of radical humanism (the subjective/radical change quadrant) by providing a brief account of the sociology of the Frankfurt School. In particular, he draws on a paper by Mingers (1980) that reveals some similarities between SSM and the social theory of Habermas. Both see human action as reflecting the nature of the human animal but also as purposeful; both regard systems analysis as inadequate to deal with the complexities of the social world; and both try to reconcile rationality and values through communi-cative interaction:

> Habermas' communicative competence would enable social actors to perceive their social condition in new ways, enabling them to decide to alter it; Checkland's methodology aims at consensual debate which explores alternative world-views and has as criteria of success 'its usefulness to the actors and not its validity for the analyst' …. It is perfectly possible to see the latter eventually as a vehicle for what …. Habermas calls 'radical reformism', namely an attempt 'to challenge and to test the basic or kernel institutions' of present-day society.
>
> *(Checkland 1981, p. 283)*

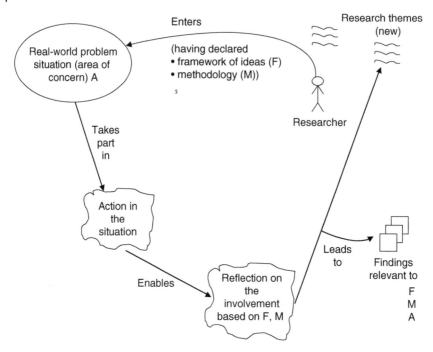

Figure 16.2 The cycle of action research. *Source:* From Checkland and Holwell (1998). Reproduced with permission of John Wiley & Sons.

It has to be said that, as well as seeking a liaison between SSM and Habermas, Checkland has also encouraged flirtations with postmodernism (Checkland and Scholes 1990, p. 235) and Luhmann's social theory (Checkland 2011, p. 510; 2012, p. 467). It seems that a lot of disentangling will have to be done in the "Critique" section below.

What has certainly remained consistent for Checkland has been his commitment to "action research." As was noted in Section 4.8, Lewin sees the action researcher as participating in a real-world exploration of social systems. He thought that learning about social systems while trying to change them could provide outcomes that were useful both for the advancement of social science and to the people in the situation. Checkland accepted this orientation but wanted to make the whole exercise more rigorous. The outcome is the process set out in Figure 16.2. As described by Checkland and Holwell (1998), the researcher first seeks out a real-world problem situation relevant to his or her research themes. They then negotiate entry into that area of concern (A), declaring in advance the framework of ideas (F) and methodology (M) they will use in trying to bring about improvements. They take part in action in the situation and reflect on what happens there using F and M. This yields findings that are relevant to F, M, and A and possibly some new research themes. In the case of work with SSM, F constitutes the set of systems ideas that inform SSM (M) as a learning system. In action research, no two social situations are ever the same, so the results cannot be justified on the basis of the natural scientific criterion of repeatability. However, Checkland and Holwell argue, rigor can be introduced by declaring the F and M in advance and keeping careful records of what exactly occurs in A. The outcomes can then be related back and described in terms of the F and M. This allows the *recovera-bility* of the whole research story by someone standing outside the process and enables coherent debate about the findings.

The action research approach endorsed by Checkland and Holwell is meant to yield improvements from the point of view of the participants in the problem situation and refinements to the F and M as they are enhanced over time, on the basis of what has been learned, to make them more fit for purpose. The experiences reported suggest that improvement to the "area of concern" is a frequent outcome. Anyone who reads the next section will not, I feel, be in any doubt that the methodology has moved on as a result of being employed in action research. What of the F? Checkland and Holwell argue that the failures that accumulated when trying to use systems engineering as the F discredit the functionalist account of social reality. In contrast, the hundreds of suc-cessful action research projects conducted using SSM as the F speak for the superiority of the phenomenological account of what the social world is like as elucidated in inter-pretive social theory.

16.2.3 Methodology

Checkland finds it necessary, in all his writings, to insist that SSM is a methodology and not a method. A method will always deliver a particular outcome if it is used properly. Methodology cannot guarantee this. It is the *logos of method*, a set of principles of method that guide how methods are used in a particular situation. A methodology is flexible enough to adjust to the requirements of the user and the situation in which it is employed. To explain this, Checkland has formulated the so-called LUMAS model – **L**earning for a **U**ser by a **M**ethodology-guided **A**pproach to a **S**ituation. He describes the model in this manner:

> Here a user, U, appreciates a methodology M as a coherent set of principles, and perceiving a problem situation S, asks himself (or herself): *What can I do?* He or she then tailors from M a specific approach, A, regarded as appropriate for S, and uses it to improve the situation. This generates learning L, which may both change U and his or her appreciations of the methodology: future versions of all the elements of LUMAS may be different as a result of each enactment of the process ...
>
> *(Checkland and Scholes 1990, p. A33)*

In Checkland's opinion, the failure of commentators to distinguish between a meth-odology and a method has led to significant misrepresentations of SSM in the second-ary literature. In particular, those who label it as "managerialist" or "conservative" or "radical" or "emancipatory" fail to recognize that, as a methodology, it can be all of these things depending on the user and the circumstances.

But if no two uses of SSM are ever the same, how can we be sure we are using the methodology? What is the thing that is being tailored by the user into different shapes according to the circumstances in which it is used? This was a pressing question for Naughton, who was trying to teach SSM coherently at the Open University. He answered it by providing some "constitutive rules," which anyone claiming to use SSM must adhere to (Naughton 1977). It was also a significant question for Checkland, who was seeking to distance his approach from hard systems thinking but was finding that the knowledge base of SSM was being polluted, as Yearworth and White (2014) put it, by declared uses, which, on closer examination, followed a hard systems logic. In *Systems Thinking, Systems Practice* (1981), Checkland was happy to endorse a modified version of Naughton's constitutive rules. As SSM developed, however, he came to feel that they needed rethinking. In *Soft Systems Methodology in Action* (1990), he and Scholes sought to rewrite the rules to embrace a wider range of SSM practice. The new rules consisted of five characteristics of the use of SSM together with a statement of its epistemology. By the time of his *30-Year Retrospective*, Checkland had moved on again and was welcoming a contribution of Holwell's as "a solid basis for definitive constitutive rules for SSM" (Checkland and Scholes 1990, p. A35). Holwell regarded the 1990 rules as too loose and not extensive enough and provided another attempt to answer the question "what is SSM?" framed at three levels: the "taken-as-given assumptions"; the "process of inquiry"; and the "elements" used in the process. She writes that there are three "taken-as-given assumptions":

(1) you must accept and act according to the assumption that social reality is socially constructed, continuously;
(2) you must use explicit intellectual devices consciously to explore, understand, and act in the situation in question; and
(3) you must include in the intellectual devices "holons" in the form of systems models of purposeful activity built on the basis of declared worldviews.
(Quoted in Checkland and Scholes 1990, p. A35)

The first of these "assumptions" underlines the point that you need to understand the philosophy and theory underlying the approach if you want to do SSM "properly." The "process of inquiry" should then be "cyclical and iterative" and should use activity models, informed by the history, culture, and politics of the situation, for the purpose of enabling a discussion and debate through which participants arrive at accommodations from which either "action to improve" or "sense making" is possible. The "elements" used in the process should include, but are not limited to, a selection from rich pictures, root definitions, CATWOE (**C**ustomers, **A**ctors, **T**ransformation process, **W**orld view, **O**wners, and **E**nvironmental constraints), etc.

With all this in mind, we can now chart the four stages of methodology development that Checkland (1999) identifies in his "Thirty-Year Retrospective."

The first full account of SSM was published, as we noted, in 1972 (Checkland 1976). Checkland calls this version "Blocks and Arrows." Looking back, he detects a "whiff of Systems Engineering" about it as it sought to escape its origins (2011, p. 503). The emerging methodology is presented somewhat mechanistically. There are references to "the problem," and the nine steps of the methodology include "design" and "implementation" stages. Conceptual models need checking against a "formal system" concept, which suggests they represent something that might be engineered. Nevertheless, as noted in the Section 16.2.1, this early paper does contain many of the key insights that remain fundamental to SSM.

Although Checkland no longer favors it, the representation of SSM as a "seven-stages" learning system, which appeared in 1981 in *Systems Thinking, Systems Practice*, is still frequently used today. It is shown in Figure 16.3.

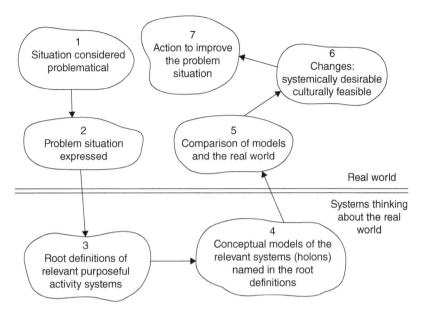

Figure 16.3 The learning cycle of Soft Systems Methodology.

In the first stage, a sense of unease felt by individuals leads to the identification of a problem situation that is considered problematical. The second stage requires that this problem situation is expressed, not in systems terms but in the form of a rich picture. The aim is to gain and disseminate creative understanding of the problem situation. The early guidelines emphasized the need for a "neutral" rich picture constructed by gathering information about the structures and processes at work and the relationship between the two – the "climate." Later, it became clear that a good way of doing the expression stage was to take the idea of rich pictures literally and to draw pictorial, cartoon-like representations of the problem situation that highlight significant and contentious aspects in a manner likely to lead to original thinking at stage 3 of SSM. The rich picture technique is one of the most successful and frequently used of the methods that have come to be associated with SSM and, as with other such methods, is explained below.

It is now time for some "below the line" (see Figure 16.3) systems thinking. In stage 3, some relevant purposeful activity systems, potentially offering insight into the problem situation, are selected and from these "root definitions" are built. A root definition should be well formulated to capture the essence of the relevant system and, to ensure that it is, must pay attention to the factors brought to mind by CATWOE. As the W indicates, each root definition reflects a different way of conceiving the problem situation. Checkland and Poulter (2006, p. 10) provide the example of the Olympic Games, which it could be insightful to define from the perspective of the International Olympic Committee, the host country, the host city, the athletes, the athletes' coaches, the spectators, hot dog sellers, commercial sponsors, those responsible for security, television companies, a terrorist group seeking publicity for their cause, etc. In stage 4, these root definitions are used to construct conceptual models. Conceptual models consist initially of seven or so activities, each governed by a significant verb, structured in a logical sequence, and representing the minimum that are necessary to achieve the transformation enshrined in the root definition. They are perhaps the key artifact in SSM and Checkland refers to them as "holons" to emphasize their artificial status and to distinguish them from the "systems" people loosely refer to as existing in the real world. Thus, using another favorite Checkland example, no prison *is* "a punishment system" or "a rehabilitation system" or "a system to protect society" or "a university of crime" – these are notional concepts relevant to exploring the realities of any actual prison.

The conceptual models, developed if necessary to a higher level of resolution, are then brought back above the line, see Figure 16.3, to be compared with what is perceived to exist in the problem situation according to the rich picture. This constitutes stage 5 of the methodology. The aim is to provide material for debate and for a dialogue about possible change among those interested in the problem situation. Thus SSM articulates a social process in which Ws are held up for examination and their implications, in terms of purposeful human activities, are made explicit and discussed. Stage 6 might then see an accommodation developing among concerned actors over what changes, if any, are both "desirable" in terms of the models and "feasible" given the history, culture, and politics prevailing. Checkland (2011, p. 501) provides the example of "the Concorde project" to explain the difference between desirable and feasible. It would have been desirable if BAC had been able to pursue "the Concorde project" using a project management approach but this was simply impossible because of the culture of dividing the engineers according to functional specialism – hydraulic, electrical, etc. All that was feasible was a process to define horizontal information flows between the vertically differentiated departments. When accommodations are found, action can be taken that alleviates some of the initial unease and, therefore, improves the problem situation. The conclusion of the methodological cycle is more likely to lead to the emergence of another, different problem situation than it is to provide a long-standing "solution." Ending a systems study is therefore, for Checkland, an arbitrary act. Problem resolving should be seen as a never-ending process in which participants' attitudes and perceptions are continually explored, tested, and changed, and they come to entertain new conceptions of desirability and feasibility.

One important point to note before we leave the "seven stages" version, and one that is clear from Figure 16.3, is that SSM is doubly systemic. First, the methodology describes a cyclic learning process. This draws upon Vickers' account of the process of appreciation and the way appreciative systems originate, develop, and change in organizations. SSM is said by Checkland (1985) to articulate in a formal way the process Vickers calls appreciation. Also explicit (Checkland 1981) is the connection with Churchman's work on inquiring systems. The methodology searches for a possible Lockean consensus through a Kantian and Hegelian route in which different assumptions about reality are set against one another. SSM is Singerian in that it accepts that learning is never ending and should be sought in the heroic mood. During the cyclic learning process, participants learn their way to new conceptions of desirability and feasibility as attitudes and perceptions are tested and changed. If it works well, changes that could not be conceived of when the study began, because of the culture and politics of the situation, can seem obvious by the time it has finished. The methodology is doubly systemic because it also uses systems models as part of the cyclic learning process just described. Checkland (1981) is prepared to make an "epistemological commitment" to systems models, as a means of learning about the world outside ourselves, because it gives every impression of being densely interconnected and seems to reveal a degree of coherence and interrelatedness. Systems models are constructed during stage 4 of SSM and carried into the real world to help structure a debate in which different perceptions of the facts of the situation and different value positions are revealed and discussed. An appropriate type of model had to be invented for this role, called a "human activity system." The idea is that pure models of purposeful activity can be built, each expressing explicitly a particular viewpoint relevant to the problem situation. These contain sets of logically linked activities that when combined together produce, as an emergent property, a purposeful whole. As we know, SSM seeks, at stage 5, to assist learning by making a comparison between these models and what is perceived to be taking place in the real world.

Another important point follows. In order that participants in an SSM study can learn their way to new conceptions of desirability and feasibility, reach an accommodation, and be willing to implement action likely to improve the situation, they must be involved as much as possible. SSM has to be participative and to engage as many interested parties as it can. Only then will stakeholders come to "own" the study and its outcomes. For this reason, soft systems practitioners are more concerned to "give away" their approach to interested parties than they are to conduct a study themselves and provide a set of recommendations.

Finally, we should mention that Checkland uses the word "accommodation" rather than "consensus" for the outcome SSM seeks from the debate informed by the systems models. This is because he regards consensus as rare in everyday life and not necessarily a good thing:

> Our experience is that genuine consensus is extremely rare in human groups and is probably unwelcome because differences of worldview generate energy and lead to fresh ideas. The norm is to find accommodations between conflicting world views, an accommodation yielding a version of a problematical situation which different people not in agreement can nevertheless live with.
>
> *(Checkland 2011, p. 505)*

Stowell (2016) still regards the "seven-stages" version as the best place to start when teaching SSM because it provides students with an "easy-to-remember" framework. Checkland accepts that that this point has legitimacy. Nevertheless, as experience of using SSM accumulated, he began to regard the "seven-stages" account as misleading and limiting. He had always stressed that the learning cycle could be commenced at any stage and that SSM should be used flexibly and iteratively, but the seven-stages model still seemed to contribute to a systematic, rather than systemic, understanding of the process and one, moreover, in which use of the methodology appeared cut off from the ordinary day-to-day activities of the organization. In an attempt to overcome this, and to demonstrate that SSM in use requires constant attention to and reflection on the historical, cultural, and political aspects of the problem situation, a third representation of the methodology was developed. This "two-streams" model, which first appeared in Checkland and Scholes' *Soft Systems Methodology in Action* (1990), is shown in Figure 16.4.

The "two-streams" version, as can be seen, gives the same attention to a "stream of cultural analysis" as to the logic-based stream of analysis that had tended to dominate

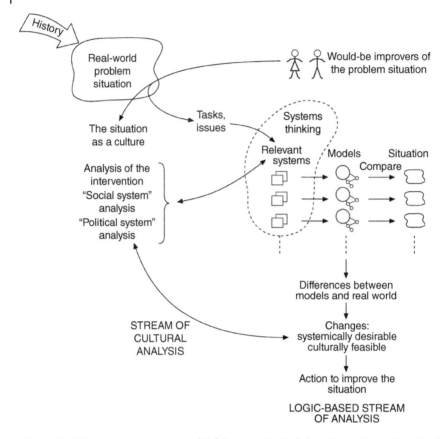

Figure 16.4 The two-streams version of Soft Systems Methodology. *Source:* From Checkland and Scholes (1990). Reproduced with permission of John Wiley & Sons.

the seven-stages version. SSM after all regards its task as being the management of the "myths and meanings" that are central to the functioning of organizations because they are the means by which individuals make sense of their situations. The enhanced cultural analysis takes the form of three types of inquiry, referred to as Analyses 1, 2, and 3.

Analysis 1 considers the intervention itself and the roles of client, problem-solver, and problem-owners, defined as follows:

- The client is the person(s) who causes the intervention to take place
- The problem-solver is the person(s) who wishes to do something about the problem situation
- The problem-owners are stakeholders with an interest in the problem situation

The way the intervention is planned needs to reflect the client's reasons for causing it to happen and take into account the problem-solver's knowledge, viewpoint, and resources. No one is intrinsically a problem-owner but, in order to be holistic, the problem-solver should consider a wide range of stakeholders as possible problem-owners. Looking at the problem situation from the various perspectives of many different problem-owners ensures a good source of relevant systems to feed the logic-based stream of analysis.

Analysis 2, social system analysis, looks at roles, norms, and values, defined as follows:

- Roles are social positions that can be institutionally defined, e.g. head of department, shop steward, or behaviorally defined, e.g. opinion leader, confidante
- Norms are the expected behaviors that go with a role
- Values are the standards by which performance in a role is judged

These three elements are assumed to be in continuous interaction with each other and constantly changing.

Analysis 3 examines the politics of the problem situation and how power is obtained and used. This can be overt or covert and rests on various "commodities" that bring influence in an organization, such as command over resources, professional skills, talent, and personality.

Analyses 1, 2, and 3 are not done once and then stored for reference. It is essential that they are continually updated and developed as the intervention progresses. Often it is helpful to incorporate them into an initial rich picture, which is then revisited and reworked. Recognition of the cultural and political aspects of a problem situation, and the ways they are changing, can significantly assist progress on the logic-based stream of analysis. It can guide the choice of insightful relevant systems, help an analyst secure more open discussion, and inform the process of arriving at feasible, as well as desirable, recommendations for change.

Another innovation made by Checkland and Scholes at this time, in line with the greater flexibility they wanted to encourage through the "two streams" version of the methodology, was establishing the distinction between Mode 1 and Mode 2 uses of SSM. They argue that much of early work on SSM concentrated on the methodology. This was inevitable because of the urgent need to move on from systems engineering and develop an approach more appropriate to managers. SSM was introduced as an external recipe to drive an intervention in a structured and sometimes sequential manner. This is now defined as Mode 1 usage. As experience and confidence grew, however, those using SSM began to internalize the approach. Once this happens they can remain much more situation-driven. The methodology ceases to dominate what is done and becomes instead the basis for reflecting on what is happening in the everyday flux of events. It is used flexibly and only occasionally breaks the surface to interact with ongoing ideas and occurrences. This is Mode 2 SSM. Although Mode 1 and Mode 2 are "ideal types," and any SSM study is likely to include elements of both, it seems clear that Mode 2 is more easily incorporated by managers into their daily working lives. Managers are absorbed by the concerns and pressures of their immediate environments. They act and react to events according to their personalities, knowledge, instincts, values, etc. and are unlikely, on an everyday basis, to operate according to the rules of a methodology. Inevitably, they are situation-driven. If, however, SSM has become sufficiently internalized, a manager or group of managers can use the approach naturally to help them think through the situation they are in and the possibilities it opens up without it appearing as a disruption to their proper jobs. They may wish to step outside the hurly-burly of ongoing events to make sense of what is happening using rich pictures, or apply some structured thinking to proposals for change using root definitions and conceptual models. In Mode 2 SSM, the distinction between the world of the problem situation and the world of systems thinking disappears. They are thoroughly intertwined. The "two-streams" representation of SSM drops the dividing line between the "real world" and "systems thinking about the real world" that is so prominent in the "seven-stages" depiction.

Already by 1990, however, Checkland was moving on from the two-streams version, which "was felt to carry a more formal air than mature practice was now suggesting characterized SSM use, at least by those who had internalized it" (Checkland 1999, p. A15). As a consequence, *Soft Systems Methodology in Action* also presented an alternative representation of "the basic shape of SSM" in terms of four activities. This "four main activities" version of the methodology has stood the test of time and is shown in Figure 16.5. Checkland attributes its longevity to the fact that there is now much less activity along the Learning (L) to Methodology (M) link in his LUMAS model. SSM has survived many tests and become a "mature" methodology. L continues, of course, and so there are refinements to M, but things have largely settled down. The only recent addition to the literature on the methodology, that Checkland and Poulter feel it necessary to mention, can be found in a paper published in 2006 by Checkland and Winter. This details twin uses of the approach. It had become clear that it can be employed to organize thinking about the *process* of carrying out a study (SSM "p") in addition to its usual use exploring the *content* of the perceived situation (SSM "c"). And this, they say, is just a case of the literature catching up with practice. Indeed, we saw in the Prologue, Checkland making use of a "conceptual model" of the process necessary to prepare and conduct the seminars in Shell.

We will now briefly describe the four activities shown in Figure 16.5, while being aware that we are considering a "learning cycle" that is never ending and that work will likely be happening on all four activities simultaneously. We should also note that the "stream of cultural analysis" has not been lost but has been subsumed into each of the activities. This account follows that of Checkland and Poulter (2006).

The first activity involves *finding out* about the initial situation deemed to be problematical. It is supported by making "rich pictures" and by carrying out Analyses 1, 2, and 3. The second consists of making some *purposeful activity models*. The models built should be relevant to the situation and are "intellectual devices" constructed on the basis of a particular worldview. They are built with the help of "root definitions," which

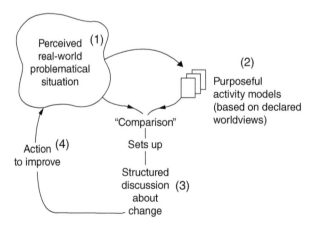

Figure 16.5 The iconic representation of Soft Systems Methodology's learning cycle. *Source:* From Checkland and Poulter (2006). Reproduced with permission of John Wiley & Sons.

have been shaped according to the PQR formula, enriched by the elements captured in the CATWOE mnemonic, and monitored by criteria for efficacy, efficiency, and effectiveness. These technical terms are explained later. The third activity uses the models to question the problematical situation and to structure a *discussion/debate* about its improvement. Various means of conducting this comparison between what is perceived to exist in the problematical situation and the relevant models have emerged. If an accommodation can be reached on changes that are arguably desirable, and also culturally feasible, then *action to improve* the situation, the fourth activity, can be defined and implemented. There is a fifth action, of course, which is at a meta-level to the other four. It demands critical reflection on the whole process and the insights gained. This then feeds into the learning loops of the LUMAS model and the action research process.

As well as settling upon a particular way of representing SSM, Checkland has also found some concise words for describing what it has revealed to him about the nature of the social reality in which it engages to bring about improvement. In short (Checkland and Poulter 2006, 2010), he sees it as made up of real-world problematical situations, which people perceive as needing attention and action. When they think and talk about these problematical situations, they inevitably do so in terms of their own *Weltanschauungen*. However, in all such situations, there will be people trying to get things done, i.e. acting purposefully. This provides the basis for using purposeful systems models as intellectual devices for exploring the situation, enabling discussion and debate, and seeking accommodations on action to be taken that is both desirable and feasible. Improvement can only emerge from a learning cycle that is never ending. To facilitate group learning, it is best conducted by those in the problematical situation although facilitators, with a sound knowledge of SSM, can be helpful. In the same 2006 book, Checkland reflects on the "mind-set" needed in approaching an SSM study and on some of the "craft skills" that can ease its use. He rightly insists that the best way of learning about SSM is to get out and try it. These reflections are meant to provide guidance and confidence to those who want to do just that. Looking to summarize the result of the journey that led to SSM, Checkland (2011, p. 511) turns to a comment made by J. M. Gvishiani, head of the Research Institute for Systems Analysis in Moscow in the Gorbachev era: "I see your approach as a rigorous approach to the subjective."

It remains to consider Brian Wilson's contributions. In his early works (1990, 2001), he pretty much towed the party line on SSM, and the philosophy behind it, while alarming some Checkland purists with the size of his conceptual models, usually decomposed into many subsystems, which seemed more appropriate for engineering than for promoting learning. The originality in these books lies in enhancements made to the methods supporting the approach. In a more recent publication, with Van Haperen (2015), some fundamental differences with Checkland have become clear. In Wilson's hands, SSM is an "expert-driven" approach, rather than one that is best "given away" and used by those directly involved in the problem situation. This manifests itself in three ways. The first is his further development of the notion of a "consensus primary-task model" based on the "Enterprise" concept. This model rests on the argument that any organization must have a core transformation system, supporting systems, linking systems, and adaptive elements. With this in mind, systems experts can construct a "consensus primary-task model," often from a number of "root definitions" reflecting different Ws, and seek "buy-in" from stakeholders to use it as a "reference model" during the course of an investigation. Wilson and Van Haperen regard such models as essential in, for example, the development of information systems requirements when multiple models are, in their view, inappropriate. They are likely to be fairly complex models. One designed for Tameside Children's Services (Wilson and Van Haperen 2015, p. 56) contains 23 subsystems and over 200 activities. This takes us on to the second major divergence from Checkland's approach. Whereas Checkland suggests that conceptual models should consist of a maximum of 7 ± 2 activities, because that is the limit of the human mind's processing capacity, Wilson and Van Haperen argue that much richer models are necessary for tackling the complexity faced by many organizations. Their models, consisting of many subsystems and activities, are developed as analysis tools by SSM experts away from the stakeholders and only introduced to them using a walk-through with slides that gradually unfolds their complexity. The availability of such rich models is said to yield significant benefits at the comparison stage. The third difference is the promotion by Wilson and Van Haperen of generic models that they regard as useful in a range of organizational settings. Thus, they present generic models for ISO 9001, Programme Realisation, Vision Realisation, Logistics, Change Management, Benefits Management, A Generic Company Model, and "An Issue-based model relevant to respond to an Information-To-Tender" (2015, p. 84). This is a fulfillment of Jenkins' original hope that the action research program would lead to the discovery of powerful generic models of very wide applicability, and runs against Checkland's injunction that conceptual models should always be specific to particular situations. Wilson and Van Haperen (2015, p. lxxviii) accept that, in all these respects, their approach shows a "proximity" to systems engineering but argue that in many instances in the real world, away from the focus of academics, this is exactly what is required. In practical application, in areas such as organizational design and change, procurement, information systems, and engineering, they find a need for more elaborate models and more advanced modeling skills. It seems that these two successful consultants have been led by SSM to a rather different understanding of the nature of social reality than that of Checkland. They even suggest that a form of groupthink in the academic community, encouraged by the proponents and gatekeepers of Checkland's version of SSM, has played a part in their experiences being ignored and their voices silenced.

16.2.4 Methods

We will introduce the most important methods developed to support SSM in an order that corresponds to the stages of the "four main activities" version of the methodology: "finding out," "purposeful activity models," "discussion/debate," and "action to improve."

Alongside Analyses 1, 2, and 3, discussed earlier, the major method employed in the "finding out" stage of SSM is the drawing of "rich pictures." Rich pictures are actual drawings that allow the various features of a perceived problematic situation to be set down pictorially for all to see. The value of drawings is well captured by Isabel Seligman in the British Museum catalogue for the exhibition "Lines of Thought":

> Artists often use drawing to provide arbitrary or haphazard spurs to invention. The process of drawing is particularly conducive to associative thinking, complementing a more rational enquiry with the space in which to explore and dream. Such reveries often include the recontextualization or repurposing of images ....

Answers often appear in the places we least expect them, and neuroscientific studies suggest that these insights (the technical term, often described as "aha!" moments), are produced by brain mechanisms very different from, and yet equally important to, those we use to work out a problem methodically.

(2016, p. 87)

Specifically in relation to our topic, Bell and Morse comment:

How to allow groups to discourse, problem solve and review their own issues and concerns? Diagrams in general and Rich Pictures in particular can be great means to allow groups to explore their subconscious, their occult sentiments and conflicted understandings.

(2013, p. 331)

There are no fixed rules for drawing rich pictures. Some end up looking quite formalized while others are very cartoon-like in nature. If facilitation is offered then that will have an impact (Berg 2015). But the norm is just to provide groups with flip-charts and colored pens and to let them get on with it. Much will depend on the skill and enthusiasm of the person(s) doing the drawing and those supplying the ideas. Rich pictures are obviously selective, and it is an art to know which issues, conflicts, and other problematic and interesting aspects to accentuate. If done well, they can assist creativity, express the interrelationships in a problem situation better than linear prose, allow the easy sharing of ideas, catalyze discussion, act as an excellent memory aid, and help move on the thinking of those involved. Figure 16.6 is a rich picture produced by Superintendent P.J. Gaisford during a soft systems study of prostitution in London for the Metropolitan Police. Figure 16.7 provides another example, drawn by Giles Hindle, used with the Trussell Trust in a study in the foodbank sector.

Checkland's early guidance, that we should pay attention to "structure," "process," and "climate" (the relation between structure and process) when constructing rich pictures, still has followers. Armson (2011, loc. 1499 *et seq*.) makes some useful suggestions while insisting that all rules are there to be broken if necessary:

- Don't structure your rich picture in any way
- Don't use too many words
- Don't exclude relevant observations about culture, emotions, and values
- Include other points of view
- Include a representation of yourself … you are part of the messy situation
- Include a title and a date

The building of helpful "purposeful activity models" requires choosing "relevant systems," constructing rigorous "root definitions," and using these to derive "conceptual models." From the representation of the problematic situation in the rich picture(s), and drawing particularly on the perspectives of those listed as problem-owners during Analysis 1, various relevant systems are chosen for further analysis. Thus Concorde might be viewed as "an engineering project, a political project, an economic project, a system to pollute the environment" (Checkland 2000, p. S65). Relevant systems should offer insight into the problematic situation and free up thinking in a way that begins to

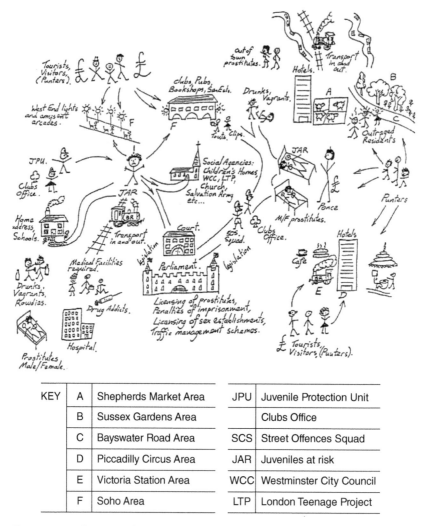

Figure 16.6 Rich Picture of vice in the West End of London. *Source:* From Flood and Carson (1988). Reproduced with permission of Plenum Press.

KEY	A	Shepherds Market Area	JPU	Juvenile Protection Unit
	B	Sussex Gardens Area		Clubs Office
	C	Bayswater Road Area	SCS	Street Offences Squad
	D	Piccadilly Circus Area	JAR	Juveniles at risk
	E	Victoria Station Area	WCC	Westminster City Council
	F	Soho Area	LTP	London Teenage Project

suggest actions that might yield improvement. I think Armson's analogy with a "log driver" captures neatly what the systems thinker is about at this point:

> Dealing with intractable situations is rather like my mental picture of the log drivers as they jump from log to log across the jam, freeing it up to get both lumber and river moving again. A skilled driver, I imagine, can spot which log to move and how to move it: freeing surrounding logs and creating movement .... Systems thinking allows me to spot the "logs" that, if moved, will improve the situation. I don't have to solve every bit of the puzzle created by the messy situation. I just have to set it on a path to continuing improvement.
>
> *(Armson 2011, loc. 225)*

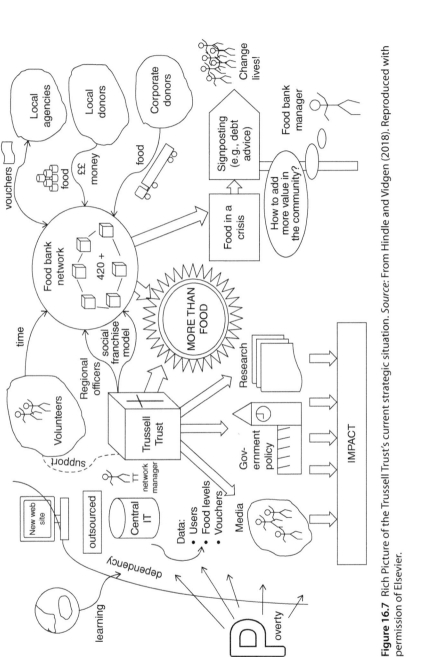

Figure 16.7 Rich Picture of the Trussell Trust's current strategic situation. *Source:* From Hindle and Vidgen (2018). Reproduced with permission of Elsevier.

The process of choosing and pursuing potentially insightful relevant systems will continue throughout an intervention, constantly suggesting new possibilities. It can assist in breaking up log-jams in thinking to take forward apparently contrary relevant systems. One of Checkland's favorite examples is to look at prisons as "universities of crime." Which, in one sense, they are. Hindle and Jackson confronted Humberside Training and Enterprise Council with a model based on the relevant system "to make the contracting process as frustrating as possible for suppliers" (1997). It has soothed my soul to view an "international office" at a university as "a system to do everything possible to prevent faculties make international links." Relevant systems are the creative bridges to what the "two streams" version of the methodology calls the logic-based stream of analysis.

Each relevant system is expanded into a "root definition" (RD), which is a concise statement of what that (notional) system is in its most fundamental form. RDs are an essential element in exploring the possibilities available for change in the problematic situation given its history, culture, and politics. To ensure that the exploration is thorough, it is always necessary to consider a number of different RDs. It is also useful to take forward two types of RD: "primary task" and "issue-based." Primary task RDs refer to officially declared tasks in the organization and give rise to models that reflect existing organizational structures. Issue-based RDs refer to current matters of concern, perhaps the need to be more innovative or to resolve a conflictual situation, that cross established boundaries.

Since RDs will be the basis for models of purposeful human activity, they need to have at their core a transformation process in which some input is changed into a new state or configuration, which then becomes the output. Checkland and Scholes (1990) suggest that it can help to conceive the core transformation in an RD as "a system to do P by Q in order to contribute to achieving R." This ensures that "what to do" (P), "how to do it" (Q), and "why do it" (R) are captured and draws attention to concerns at the system, subsystem, and wider system levels. The PQR "formula" allows the RD to be set out as a basic statement. It then requires further refinement using CATWOE as a guide. This is best done by first specifying the transformation intended and the *Weltanschauung* that makes the transformation meaningful in terms of the context. A well-formulated root definition is also likely to pay attention to who will benefit or suffer from the transformation, who will carry it out, who can stop the transformation from happening, and what environmental constraints limit the actions and activities. We now have six elements that an RD should make reference to unless justification can be provided for omitting any of them. CATWOE is the mnemonic that captures these:

- C = "customers" – the beneficiaries or victims of the transformation process
- A = "actors" – those who would undertake the transformation process
- T = "transformation" – the conversion of input to output
- W = "world view" – the world view that makes this transformation meaningful
- O = "owners" – those who could stop the transformation
- E = "environmental constraints" – elements outside the system that are taken as given

In a study for the Information and Library Services Department (ILSD) in ICI Organics (discussed in Checkland and Scholes 1990), Checkland produced the following RD of ILSD's role:

An ICI-owned and staffed system to operate wealth-generating operations supported by enabling support systems which tailor their support through development of particular relationships with the main operations

C = ICI
A = ICI people
T = need for supported wealth generation – need met via a structure of main operations and enabling support
W= a belief that this structure will generate wealth
O = ICI
E = structure of the main operations plus support; ICI ethos

Even when what seems a rigorous RD has been produced, it is often possible to sharpen it further by asking whether it makes apparent how the performance of the named system can be measured. Proper monitoring and control of performance depends minimally on specifying the criteria for "three Es" – efficiency ("are the minimum resources being used"), efficacy ("do the means work?"), and effectiveness ("is the transformation meeting the longer term aim?"). Checkland and Poulter suggest that other criteria such as "ethicality" ("is this a morally correct transformation?") and "elegance" ("is this a beautiful transformation?") can be added to the standard "three Es" in particular circumstances (2006, p. 43).

A root definition is a precise account of what a relevant system is. Once it has been formulated satisfactorily, a "conceptual model" (CM) can be built from it that sets out the purposeful activities that must be undertaken in order to carry out the transformation and fulfill the other requirements of the RD. CMs are not models of anything in the real world – they are purposeful "holons" constructed with the aim of facilitating structured debate about the problematic situation and any changes to it that might be desirable and feasible. They should therefore be developed from their relevant RDs alone, without reference to reality:

> To build such a model only requires logical thought. It requires assembling the activities relevant to acquiring the input, the activities to transform it, and the activities to do something with the input. Defining these activities, and linking them with arrows which show which activities depend upon which other activities, yields an operational whole. To that whole, we link a control (regulating) sub-system which monitors the activities as they occur and takes control action, or not, in light of defined measures of performance.
>
> *(Checkland 2011, p. 505)*

Since CMs are "purposeful activity models," they consist of verb statements describing actual activities that humans can undertake. They will normally, at the first stage of their development, consist of 7 ± 2 activities. They are models designed to promote learning and research suggests that the human brain can only cope with around seven concepts simultaneously. As Checkland says, a common feature is to have a number of verbs in one sub-holon concerned with operations and another set of linked activities that are responsible for monitoring and control. The verbs are logically ordered in terms of their interactions, showing how the activities depend on one another. The CM derived by Checkland from the earlier RD and CATWOE analysis is presented in Figure 16.8. Basic models of this type can, as is common in Wilson's approach, be expanded to higher levels of resolution by taking any activity within them as the source of a new RD and accompanying CM. Hindle (2011) identifies four types of conceptual

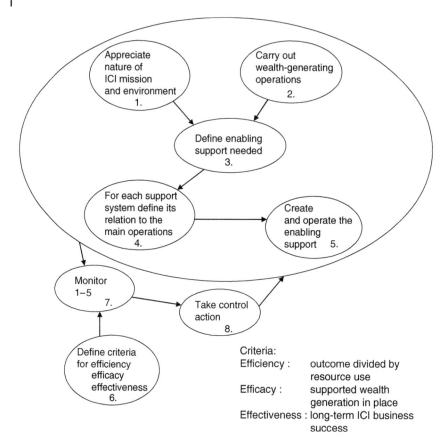

Figure 16.8 A conceptual model derived from the root definition presented earlier. *Source:* Adapted from Checkland and Scholes (1990).

modeling that he has found useful – "baseline" modeling of existing arrangements, "design" modeling of the kind preferred by Wilson, "theoretical" modeling articulating one W, and "temporary system" modeling. These last two are characteristic of Checkland's work.

The CMs are then employed as a source of questions to ask about the problematic situation. The purpose is to generate a structured "discussion/debate" about desirable and feasible change. Checkland (1981) mentions four ways in which this can be carried out:

- Informal discussion of the main differences between the models and what seems to be the case now
- A more formal questioning of the main differences, involving completing a matrix that asks of each activity and dependency in the CMs questions such as: Does it actually exist or not?, How is it done?, Who does it?, When is it done?, How is it judged?, Any other comments?
- Scenario writing based on notionally running a CM, in the mind or on paper, to see how it would be expected to behave in the future – this might be compared with how the actual system appeared to work in the past in similar circumstances

- Trying to model the real world using the same structure employed in a CM in order to highlight any significant differences that might provoke discussion

"Action to improve" involves reaching "accommodations" and putting into effect changes that everyone can live with. Analytically, these can be distinguished and talked about as being changes to "structures," "processes," and "attitudes."

16.3 Soft Systems Methodology in Action

As was mentioned, one of the "key thoughts" that moved SSM research forward was the recognition of its relevance to information systems (IS). Checkland and Holwell's book *Information, Systems and Information Systems* (1998) signals this innovation. The volume is an ambitious attempt to initiate "conceptual cleansing" in the field. This involves bringing intellectual clarity to confusions about concepts such as "data," "information," and "knowledge," and replacing the outdated model of the organization as a machine, traditionally used to underpin work in IS, with a "phenomenological" account that more adequately matches actual experience. This alternative model emphasizes values and meanings and the processes that occur as purposeful action is formulated. Once such action is agreed and represented in CMs, it becomes possible to provide appropriate ISs to support the operations. This is because CMs can form the basis for IS models showing the information needed to undertake each activity. SSM is seen as a perfect vehicle to guide the development of ISs that truly meet decision-makers requirements. To illustrate this, I have chosen an IS study as our example of SSM in action. It was conducted using the "two streams" version of the methodology.

This SSM intervention was conducted by Checkland and Holwell for an Information Department (ID) located in the central research and development laboratories of a multinational science-based group. ID had just over 100 staff and was linked to a library employing a further 25. It consisted of four sections; three concerned with technical aspects of IT and telecommunications, and one with the organization of ID itself and IS rather than IT issues. The role of ID was to serve the Research and Development laboratories (R&D), ensuring they were up to date with relevant knowledge and that any new knowledge they generated was appropriately managed and made available to those who needed and were entitled to it. R&D was itself divided into four sections: products research, process research, engineering research, and general administration (of which ID was a part).

The "Reorganization Project," as it was known (see Checkland and Holwell 1998), was led by Eva, a member of staff seconded from that part of ID (call it IDI) concerned with its organization and the more general information systems issues. The project was a response to the feeling that the service offered by the IS/IT professionals to R&D needed rethinking. Its justification was the rate of technical change overtaking IT. It was supposed to answer questions such as "what are the presentation requirements for information transfer?," "what are the costs of information access, storage and quality?" and "at what rate will changes occur within the laboratories?" SSM was called on because it was felt a more holistic approach to this broad "information support problem" would bring benefits.

In entering the problem situation, Checkland and Holwell noted that, although the project had been set up by the head of ID, its recommendations were supposed to be limited to the activities of IDI. It seemed inevitable that the project would impact on the whole of ID and its relationships with R&D. Perhaps the head of ID was worried about possible resistance to change from the other section heads within ID and about the reaction of clients in R&D. This was just one of the historical, cultural, and political factors they faced. Others were a lack of high regard for the work of ID and the difficulty of justifying expenditure on any aspect of R&D when the returns are indeterminate and often far into the future.

The "problem situation expressed" stage took the form of workshops involving Checkland and Holwell, Eva, other staff from ID, and around 20 clients of ID from R&D. A rich picture, reproduced as Figure 16.9, was developed, and this helped to bring about a shared understanding of ID's support function and to identify some important roles and processes worthy of further investigation. Another problem-structuring device

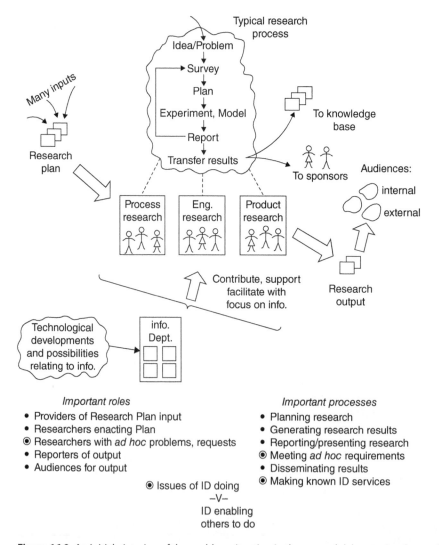

Important roles
- Providers of Research Plan input
- Researchers enacting Plan
- ◉ Researchers with *ad hoc* problems, requests
- Reporters of output
- Audiences for output

Important processes
- Planning research
- Generating research results
- Reporting/presenting research
- ◉ Meeting *ad hoc* requirements
- Disseminating results
- ◉ Making known ID services

◉ Issues of ID doing
–V–
ID enabling
others to do

Figure 16.9 An initial picturing of the problem situation in the research laboratories. *Source:* From Checkland and Holwell (1998). Reproduced with permission of John Wiley & Sons.

used was a CM derived by taking the ID's official mission statement as an RD. This had the added benefit of familiarizing participants with some of the most important techniques associated with SSM.

The logic-based stream of analysis was carried forward in two one-day workshops at which researchers and ID professionals were present. They began with the head of ID and Eva explaining the Reorganization Project, which, increasingly, was becoming associated with achieving greater "client orientation" and with determining how relationships between ID and R&D should develop going forward. Then some models, previously constructed and tried out on the researchers, were presented for discussion and modification in small groups. These were about the role of R&D in the company in the particular circumstances it faced. One example, and the root definition from which it came, is presented in Figure 16.10.

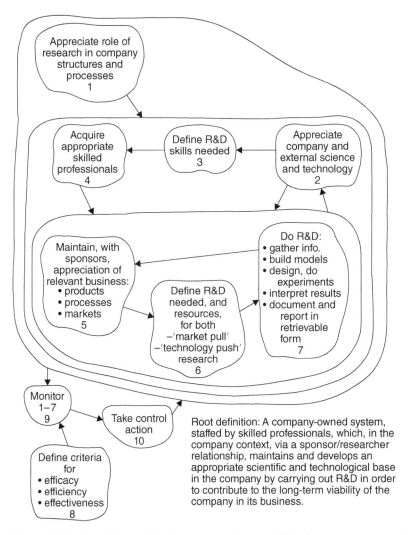

Figure 16.10 An activity model relevant to carrying out R&D in the company. *Source:* From Checkland and Holwell (1998). Reproduced with permission of John Wiley & Sons.

Later in the sessions small groups were again used, this time to discuss R&D–ID interaction. In the first workshop, for example, attention was focused on those activities in the models that were now or could, in the future, be supported by ID. The groups were then asked to consider, for each activity that might be supported, whether ID should do it, provide expertise relevant to it, offer appropriate education and training, or help manage it. Figure 16.11 reproduces one of the models used to assist the debates. It is an expanded version of activity 7 in the earlier model (Figure 16.10), with the thick arrows indicating activities to which ID could contribute support.

The learning from the two workshops was condensed into reports that highlighted a number of matters – such as whether ID should be more responsive or proactive. The most significant issue, however, was the need to improve mutual understanding and contact between ID and R&D. Unless this could be achieved, there was no chance of ID succeeding in its aim of becoming more client-centered. A conceptual model was therefore produced, setting out what activities would be necessary to ensure that ongoing dialogue between ID and R&D became institutionalized. Eva then argued for the Reorganization Project to continue: first, by making changes in IDI and then by spreading change to the whole of R&D in a manner that would realize productive ID–R&D relationships.

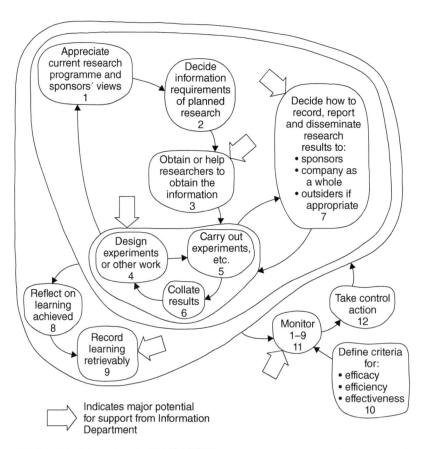

Figure 16.11 An activity model that expands activity 7 of Figure 16.10. *Source:* From Checkland and Holwell (1998). Reproduced with permission of John Wiley & Sons.

It is worth noting at this point that the "stream of cultural analysis" was proceeding alongside the logic-based stream throughout the project. It took the form of a very informal use of Analyses 1, 2, and 3. The cultural shift of ID toward a more client-driven perspective was monitored. Discussions at the workshops helped surface hidden agendas and possible political constraints on progress. There was growing realization that, despite its turbulent environment, things changed pretty slowly in this company. Checkland and Holwell were confronted with confirmation of their original concerns about the nervousness of the head of ID. Both the new conceptual model and the plan for continuing the Reorganization Project pointed to the need for change beyond the boundaries of IDI and, indeed, of the ID itself. His determination to proceed with extreme caution in these circumstances imposed a considerable delay on further progress.

About a year down the line, however, an idea emerged that allowed the institutionalization of dialogue between the ID and R&D in a natural rather than contrived manner. An "Information Market" was established in the headquarters building of R&D. This was a permanent exhibition space displaying the latest IT products. It acted as a magnet to researchers interested in the latest toys. It also became a forum for meetings and discussions with IS/IT professionals about the researchers' needs and how new technology could help address them.

Checkland and Holwell document the considerable learning they gained from this intervention. The idea that it is necessary to understand the purposeful activity that requires supporting before you can build effective IS systems was confirmed and refined. So was the notion that those delivering support services should be involved in the organizational discourses and interactions that give rise to higher level purposeful activity. Finally, they were able to make sense of the diffidence of the head of ID by reflecting on it as part of a more general problem. The IS function is potentially a very powerful one in organizations because of its position at the center of meaning generation. Those responsible for it have to play their cards carefully if they are not to be seen as a threat.

16.4 Critique of Soft Systems Methodology

At the beginning of his endeavor to develop a systems methodology more appropriate to the problem situations faced by managers, Checkland (1976) declared his intention to take systems thinking beyond the abstractions of general systems theory and the constraints of the specialized techniques then dominating approaches such as operational research. The story of SSM reveals the success of this enterprise.

The capacity to give structure to thinking about and intervening in problematic situations while being flexible in use is one of the major reasons for the popularity of SSM. Rather than requiring objectives to be agreed in advance, it sets in train an organized process through which participants learn their way to whatever "action to improve" is desirable and feasible. With SSM, therefore, it is always possible to get started on a study and to see where the interaction between the approach and the "real-world" situation takes you. Further, while it sets out some clear methodological principles, SSM does not constrain method use. It is capable of providing a different response in each situation depending on the user and the nature of the situation. Although it has some extremely powerful methods (rich pictures, root definitions, conceptual models, etc.)

associated with it, SSM does not require that they are all used, or that they are used in the same way, in each intervention. It is this flexibility that ensures its relevance in so many managerial situations. Finally, its use is not tied to any particular timescale. As Checkland and Poulter note, it can take an hour or so, or a year or two. Of course, in short investigations, the existing level of knowledge will have to be taken as given (2006, p. 197).

Another reason why SSM is so widely applicable and successful is that it directly relates to the all pervasive activity of people trying to act purposefully in organizations. As Checkland and Poulter have it, SSM "is not a tool or technique to be used occasionally but a way to think and act every day" (2006, p. ix). Managers trying to improve things in organizations have to worry about the present situation, try to get some handle on it, postulate alternative ways forward, and seek accommodations that allow change to happen. With all of these, SSM can help. There is nothing in SSM that is unnatural to managers. Indeed, once they have participated in a few studies, they should be able to internalize the methodology in a manner that will enable them to use it as a normal part of their work. In Mode 2 practice, as we saw, the methodology is placed in a secondary position to the demands of the problem situation. As Dean of Hull University Business School, it was my daily routine to open my e-mails, contemplate the mess, think of ways forward, and then consult with colleagues, over coffee, to see if those might work or if they had better ideas. Occasionally I would step back and employ SSM more formally to enrich particular steps in this process.

The empirical evidence that SSM can bring about improvement is inevitably, given the situations in which it is used, sketchy. Mingers and Taylor (1992) analyzed the responses of 90 UK users of the approach and found that 94% were at least reasonably happy with the results they got. Sixty-three percent thought they got good or very good results. The main benefits reported were that the methodology provided a structure for their interventions and that it generated an understanding of other perspectives. Ledington and Donaldson's (1997) review of the responses of around 130 Australian users found that 69% had success using SSM. The authors of both studies comment on the unavoidable subjectivity of the respondents. They also indicate the need for experienced practitioners to help with some of the stages of the methodology. Hindle and Jackson (1997) found that constructing RDs and building CMs caused particular difficulties for people new to SSM.

Checkland's own position on whether actual *proof* can be found that SSM works has remained consistent over the years. In his view this is impossible. There is no prospect of repeatable experiments because no two problematic situations are ever the same. And, even if they were, you are still left with the difficulty of refuting any claim that the methodology has or has not been successful:

> If someone says to me: "I have tried the methodology and it works," I have to reply on the lines: "How do you know that better results might not have been obtained by an *ad hoc* approach?" This is an undecidable question. If, on the other hand, the assertion is: "Your methodology does not work," I may reply, ungraciously but with logic, "How do you know the poor results were not due simply to your incompetence in using it?".
>
> *(Checkland 1981, p. 241)*

Checkland is clear that no methodology can guarantee "improvement." But its "guidelines" might help those using it to achieve better results than they might otherwise have done. Ultimately, therefore, only those concerned about a problematic situation can judge whether "learning [has] occurred, either explicitly or through implementation of changes" (Checkland 1981, p. 253) and "improvement" has resulted. Even here there are difficulties. It is not easy, for example, for people to recognize the extent to which working with the methodology has led them to alter their perceptions and change their attitudes. Nevertheless, on these criteria, Checkland does think that SSM has proved its worth.

We will now consider whether a "second-order" analysis, using social theory, can help evaluate the "guidelines" upon which SSM is based. We are interested in what social reality must be like for those guidelines to be appropriate and for SSM to work. As was seen, SSM embraces phenomenological philosophy and interpretive sociology. The "subjectivism" that Burrell and Morgan (1979) see as central to interpretive sociology is readily apparent in Checkland's work. Systemicity is transferred from the world to the process of inquiry into the world. The existence of different Ws is taken for granted and inquiry proceeds using "ideal-type" purposeful activity models to structure a participative debate/discussion. As a result of this process, participants may reach accommodations and initiate changes that they judge feasible and desirable. As summarized in Holwell's latest version of the constitutive rules of SSM, "you must accept and act according to the assumption that social reality is socially constructed, continuously" (Checkland and Scholes 1990, p. A35). Checkland's interpretive systems approach opens up a new perspective on the way systems ideas can be used to help with decision-making and problem alleviation. It arguably brings within the scope of systemic treatment all those wicked, messy, and ill-structured problems that either escape or are distorted by functionalist methodologies because of the strict prerequisites that need to be met before those methodologies can be employed. The result, it is claimed, has been a major extension of the area within which systems thinking can be used to help with real-world problem management.

But Checkland does not believe that the usefulness of SSM is entirely restricted to exploring the interpretive paradigm. He believes that it can also be employed to pursue Dilthey's concern with discovering common types of *Weltanschauung*. It has a role to play in revealing any recurrent Ws and opens up the prospect of discovering "the universal structures of subjective orientation in the world" (Luckmann, quoted in Checkland 1981, p. 279). These "common structurings" will relate to historically determined attitudes and behavior patterns stemming from "the emergent properties of social wholes which transcend individuals" (Checkland 1981, p. 279). These are what a functionalist thinker would take as the objective aspects of the social world. Furthermore, because the SSM practitioner has "complete freedom" to select relevant systems that promote sweeping change, the methodology can also reach into "radical change" territory. Here Checkland is encouraged by similarities between SSM and Habermas' critical social theory. The methodology can be seen as a means of increasing "communicative competence":

> Another way to describe the methodology, in fact, is as a formal means of achieving the 'communicative competence' in unrestricted discussion which Habermas seeks.
>
> *(Checkland 1981, p. 20)*

In another foray into social theory, Checkland and Scholes describe Mode 2 SSM, as used in the Shell study outlined in the Prologue, as "postmodern" in the sense of helping "an observer community which constructs interpretations of the world, these *interpretations* having no absolute or universal status" (Cooper and Burrell, quoted in Checkland and Scholes 1990, pp. 235–236). Finally, Luhmann is evinced as a theoretical fellow-traveler for the reason that he is one of the rare systems thinkers who make a clear distinction between objects themselves and the labels, like the concept *system*, that we find useful to describe them (Checkland 2011, p. 510).

It is now necessary to consider what adherents of other sociological paradigms make of SSM, and the interpretive social theory that underpins it, and to evaluate the argument that it can extend its usefulness by offering support to other theoretical positions. From the functionalist perspective, as Mingers suggests, Checkland's interpretive stance "reduces the force of systems thinking." Functionalists believe that concepts such as system, feedback, and equilibrium are

> ... genuine explanatory concepts in that the existence of such systemic processes in the world was necessary to explain the phenomena that were observed; to deny reality to systems concepts is to reduce them to an essentially arbitrary language game.
>
> *(Mingers, quoted in Ramage and Shipp 2009, p. 153)*

For Checkland, the models produced in hard systems thinking, system dynamics, and cybernetics, at best merit a place at the debating table. They cannot provide any objective truth about how organizations should be designed and managed. They are seen as appropriate only in "special cases," in those situations where SSM reveals that worldviews have coalesced to the extent that there is consensus about what system to design. Sadly, as we witnessed, this is a role that some in qualitative system dynamics and second-order cybernetics seem willing to accept. Checkland and Poulter (2006, p. 174) welcome them with open arms. Confirmed functionalists, by contrast, should argue that it is SSM that has little to offer. When it comes to important matters like managing technical, process, and structural complexity and designing complex adaptive systems, it is they who can provide the knowledge necessary to successfully guide practice. Their work is, therefore, relevant in large swathes of social and organizational life. Within these areas, it is the functionalist rationality, justified by its ability to increase prediction and control, which must hold sway. What is the best queuing system for a supermarket or what is a suitable organizational design for a corporation in a turbulent environment are, for functionalists, not primarily matters of inter-subjective agreement. To put it simply, they believe that there is something scientific about the various models produced by experts in management science that managers should take seriously.

Checkland thinks that the learning achieved in his action research program, using SSM, supports the conclusions of interpretive social theorists. He has not found evidence of significant "common structurings" that could provide a solid basis for functionalist approaches. Advocates of the hard, system dynamics and cybernetic strands of systems thinking, and there are many more of these than there are soft systems thinkers, will wonder whether he has really been looking. For them, there are abundant regularities in social reality for their approaches to seize upon. It is worth remembering that

Wilson, working with his version of SSM, believes that he has discovered and developed a number of generic models of social processes that have very wide applicability.

Radical change theorists argue that the social world is characterized by conflicts of real interest, asymmetry of power, and coercion and that the failure of the interpretive approach to come to terms with these factors means that methodologies based on its assumptions become distorted in use and prove ineffective in bringing about significant change for the better. If social reality is as they describe, most of the contexts in which SSM is used will be hierarchical and will feature powerful clients. As Thomas and Lockett (1979) point out, this is likely to lead to conservative or, at best, reformist recommendations for change. The client can restrict the boundaries of the study at an early stage. If the SSM practitioner wants to continue working with the client, they will have to abandon any radical root definition as not being "culturally feasible" given the history, culture, and politics of the problem situation. The choice of which changes to implement will be subject to existing decision-making processes in which the client is dominant. As long as powerful stakeholders are not threatened by the study, because they can keep significant issues off the agenda, they may well be willing to let other groups participate and it might seem that all stakeholders share common interests. If, however, SSM really challenged the hierarchical nature of organizations, the ultimate decision-making rights of the powerful, or the unequal distribution of organizational resources to different stakeholders, it would soon provoke a response that revealed the deep class, status, and other inequalities that radical change sociology sees as such fundamental aspects of social reality. Burrell (1983) is convinced that the reason why Checkland fails to notice anomalies that might lead him to question his interpretive position is that he primarily uses the methodology in support of a community of managers sharing similar interests. This is a community, moreover, that usually has the power to impose agreement on any other groups involved in the proceedings. This capitulation of SSM to powerful interests can go unnoticed because Checkland is "uncritical" about the results of SSM interventions. In Ulrich's view, good professional practice requires that the outcomes of SSM interventions must be subject to critical reflection and discourse. They will be selective in terms of the facts and values that have been taken into account and this will have consequences for stakeholders, including those affected but not involved (2005, p. 5). His own methodology of "critical systems heuristics" can, he argues, supply the critique that can prevent methodologies like SSM unquestionably accepting "a managerialist notion of good practice" and provide a justification for their results.

Much of the disagreement between Checkland and the "radical change" critics of SSM has focused on the agreement between stakeholders that the methodology hopes to bring about so that action can be taken. For example, Jackson has argued at length (1982a, 1983, 2000) that, if the "agreement" is to be deemed "fair," then the debate/discussion stage that gives rise to it must conform as far as possible to the model of communicative competence proposed by Habermas. Although Checkland has long insisted that SSM seeks "accommodations" rather than "consensus," the argument still holds. There can be "false" accommodations that advantage some stakeholders over others, just as there can be a "false" consensus. For the "action to improve" to reflect a genuine accommodation, it must result from a discourse between stakeholders that is conducted in conditions that approximate to Habermas' "ideal speech situation." They must have equal chances to select and employ speech acts and to assume dialogue roles. There must be unlimited discussion that is free from constraints or domination, whether the source of these is the behavior of other parties or communication barriers secured through ideology. The ability of powerful stakeholders to impose sanctions must not affect the outcome of the discussion. Only if such conditions are met will any accommodations, at the end of the debate, be based on a fair exchange of views and not on various constraints on discussion. Of course, the kind of unconstrained debate envisaged here cannot possibly take place if, as radical change theorists insist, organizations and societies are characterized by significant inequalities. The stakeholders will bring to the discussion unequal intellectual resources and will be more or less powerful. The result of the unequal intellectual resources is that the ideologies of the powerful are imposed upon other actors who lack the means of recognizing their own true interests. The result of the inequalities in power is that the existing social order from which power is drawn is reproduced. As Giddens writes:

> The use of power in interaction involves the application of facilities whereby participants are able to generate outcomes through affecting the conduct of others; the facilities are both drawn from an order of domination and at the same time as they are applied, reproduce that order of domination.
>
> *(1976, p. 122)*

SSM, to radical change critics, merely facilitates a social process in which the essential elements of the status quo are reproduced – perhaps on a firmer footing since differences of opinion will have been temporarily smoothed over. In doing so, it supports the interests of the dominant group or groups in the social system. Checkland (1981, p. 283) does seem to take the point that the discussion/debate stage of SSM can be crucially inhibited by society's structure but concludes rather weakly from this that "it is the nature of society that this will be so." This says nothing about the degree of constraint on discussion imposed by particular social arrangements or about the possibility of changing the arrangements in order to facilitate communicative competence. The social environment in which the methodology has to operate nullifies, therefore, its attempts to bring about changes based on a genuine accommodation between different opinions and interests. The methodology is culpable in that it is prepared to accept for implementation changes emerging from false accommodations produced by distorted communication.

In order to counter the arguments of the radical change critics, SSM would have to use Habermas' conceptualization of the ideal speech situation to unmask cases of "systematically distorted communication," brought about by unequal power relations, and would then have to challenge the social arrangements that produce distorted communication. Unfortunately, the critics argue, SSM's adherence to interpretive social theory makes this impossible. As Willmott has it:

> Its major shortcoming lies in its unnecessarily limited capacity to promote reflection upon the possibility that the content and negotiation of *Weltanschauung* are expressive of asymmetrical relations of power through which they are constructed and debated .... Phenomenology, and SSM in particular, simply lacks a social theory capable of accounting for why particular sets of perceptions of

reality emerge, and why some perceptions are found to be more plausible than others.

(1989, p. 76)

Mingers' (1984) shares Willmott's view that it is SSM's commitment to phenomenology and, therefore, subjectivism that stands in the way of it developing the necessary social theory. It is condemned by this to act only at the level of ideas, seeking to change things by changing worldviews. SSM is not in a position to understand how particular social structures give rise to worldviews and determine their influence. It cannot recognize that it is difficult to change Ws without first doing something about these structures. According to the sociology of radical change, a sophisticated social theory is needed, which is capable of unmasking ideologies and providing an understanding of how emancipation can be brought about by changing the social structure. It is true that in Analysis 3, in the "two streams" version of SSM, Checkland does give some consideration to power. This, however, is only in terms of "commodities" that attach to people not as something inextricably entwined with the structures of the social system. And he pays no attention to how it affects the supposed "neutrality" of his methodology. From the radical change perspective, the interpretive foundations of soft systems thinking condemn it to regulation and severely limit its ability to bring about change that is in the interests of disadvantaged groups.

This conclusion is reinforced if we consider Checkland's bold claim that the social theory to which his methodology corresponds is not simply interpretive (subjective/regulative) but must occupy as well some of the radical humanist (subjective/radical) quadrant of Burrell and Morgan's map. Checkland bases this on an article by Mingers (1980) that reveals some similarities between the social theory of Habermas and SSM. If these similarities are fundamental, Checkland's argument that his methodology can be used as a radical social instrument is greatly enhanced. Unfortunately, as has been argued elsewhere (Jackson 1982a), it is the differences rather than the similarities between the work of Habermas and Checkland that are more significant. The major difference is theoretical and lies in Habermas's willingness (at least in his early work) to accept the usefulness, on appropriate occasions, of "objective" social theory. He recognizes that although the social world is created by the interaction of people, it is not transparent to them. It can "escape" human beings, take on objective features, and constrain them. In these circumstances, interpretive inquiry cannot be the sole method appropriate to the social sciences. There must also be a moment in which the objective features of the social world – when people do appear to act as things – can be studied. There is a need, too, for a critical moment, corresponding to an emancipatory interest. The hope here is to reduce the area of social life where people act as things and to increase the realm of the hermeneutic, where rational intentions become realized in history. Though the major difference is theoretical, it *does* have a political result. Habermas's work opens up the possibility of political action to accomplish real change – it is potentially radical. Checkland's methodology confines itself to working within the constraints imposed by existing social arrangements – it is essentially regulative.

This second-order critique can be continued by briefly looking at SSM from the point of view of postmodern and poststructuralist social theory. Checkland and Scholes (1990) argue that there is an element of postmodernism in SSM because the interpretations of the world with which it works have "no absolute or universal status." This is one similarity but it hardly makes for a convincing argument. Taking Cooper and Burrell's (1988) distinction between "critical modernism" and "systemic modernism," as outlined in Section 4.5, a stronger case can be made for identifying SSM as an underdeveloped form of critical modernism, based upon Kant's program of enlightenment and seeking increased mutual understanding between human beings so that progress can be made in creating a better society. Checkland after all, in other writings, sees SSM as a means of achieving the "communicative competence" sought by Habermas, a "critical modernist" whose work is anathema to postmodernists. At the same time, because it is so underdeveloped a version of critical modernism, for all the reasons we have been detailing in this section, SSM is particularly prone to slipping back into becoming an adjunct of "systemic modernism," readjusting the ideological status quo by engineering human hopes and aspirations in a manner that responds to the system's needs and so ensures its smoother functioning. Poststructuralist theory would also have a field day with SSM. Foucault, for example, sees worldviews not as free-floating but as intimately linked to power relations. The battle over whether Pierre Riviere should be labeled as bad or mad (see Section 4.5) is understood by Foucault as a struggle between the discourses of the criminal justice system and psychiatry to gain power over others.

Finally, we should consider Checkland's complimentary allusion to Luhmann's sociological theory. Superficially this seems a promising point of reference. As we know from Chapter 4, Luhmann was significantly influenced by Maturana, who himself has traveled on a path that led to subjectivism. Brocklesby (2007) enhances the argument by suggesting that SSM could benefit by adopting Maturana's work as a theoretical underpinning to complement and enrich that of Vickers. Once again, however, we need to conclude that it is the differences that are more important than the similarities when we compare Checkland's account of social reality with that provided by Luhmann. For Checkland, it is human actors who attribute meaning and continuously construct social reality. If you can change their appreciative settings and their actions, then you change the social systems of which they are members. In Luhmann's sociology, however, communication replaces action as the basic component of social systems. Social systems are themselves "observers," constructing their own realities independently of humans who exist as "psychic systems" in their environments. People can only "irritate" social systems. They should not pretend that they can change them according to their own intentions.

Checkland's SSM successfully takes systems ideas into a new paradigm and, in doing so, significantly extends the scope of systems thinking and its ability to help manage problem situations. It is a rigorous and thoroughly tested methodology that can facilitate stakeholders in learning their way to taking purposeful action leading to feasible and desirable change in a world of multiple and often conflicting worldviews. Work on messes and ill-structured problems can be confidently undertaken using the approach. It does the methodology no favors, however, to claim too much. SSM, by the nature of its philosophy and theory, and the way it embeds these in its methodology, has an "ideal-type" kind of problem context for which it provides the most appropriate response, just as do all the other strands of the applied systems movement. By relating it to the SOSM, the aim is to highlight what it can and cannot achieve in relation to the full range of "ideal-type" problem contexts, and thereby point out potential strengths and weaknesses depending upon what social reality could be like. This, of course, does not define what the methodology can do in practice. If social reality is as SSM's theory assumes, its usefulness will not be limited at all and it will be able to realize the ambitions it has.

Let us begin our SOSM critique of SSM by recognizing that it makes pluralist assumptions about the nature of problem contexts and placing it as in Figure 16.12. It is ideally suited to those problem situations where there is a need to create some shared appreciation among stakeholders with different Ws about what action is required to bring about improvement. As can be seen, there is some spread in the "complicated" and "complexity" direction because Checkland does use purposeful systems models, constructed on "adaptive whole system" principles, to structure the discussion/debate about "action to improve." Not too much, however, because these models are meant purely to improve the quality of the debate on the road to achieving "accommodations." They do not signal any commitment to manage "complicatedness" and "complexity" using knowledge about how complicated systems behave or how complex systems should be designed to survive in turbulent environments. For example, the idea that there are cybernetic laws that must be obeyed if you want to improve the viability of all complex systems is not taken seriously by SSM.

Concentrating on the right-hand side of Figure 16.12, SSM is not suitable if the problem context is "coercive." The sociology of radical change does not receive much attention from Checkland and his methodology is not designed to respond to its conclusions. If there is fundamental conflict, or Ws refuse to shift, or power determines the outcome of debate, then SSM cannot deliver on what it promises. Given the importance of participation for the success and legitimacy of SSM interventions, it is surprising that Checkland has not given more attention to how it might be promoted, arranged, and facilitated. He simply does not address the issue of how far participation should run or offer ground rules for what is to count as "genuine" participation. In the absence of such rules, any debate that takes place can be constrained and distorted because particular individuals or agendas are excluded, because of hierarchy and the threat of sanctions, because of unequal intellectual resources and/or because powerful ideologies hold sway. It is all too easy for those with power and influence to dominate the discussion/

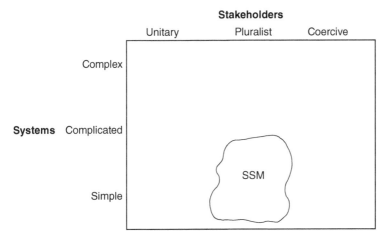

Figure 16.12 The positioning of Soft Systems Methodology on the SOSM.

debate and to have their own priorities reflected in the outcomes. To the critics, there-fore, SSM cannot be properly employed in coercive contexts.

It is only fair, at this stage, to point out that Checkland (1985, p. 766) has his own way of presenting the relationship between systems methodologies, which is very different to that of the SOSM. In his view, as was noted, SSM is relevant to all problematic situa-tions while other approaches are appropriate only in "special cases" identified by SSM. SSM is seen as a large circle containing smaller circles with other methodologies inside that become available in the service of SSM on special occasions. So, for example, a soft systems study may occasionally "condense" into a hard systems project if it reveals a consensus around designing an unproblematical system with defined objectives. According to Checkland, hard and soft methodologies relate to each other like "apples and fruit." In the SOSM, the relationship is akin to apples and pears. To my mind, the debate between paradigms staged above demonstrates that Checkland's argument doesn't hold water. No paradigm can absorb another. They are "incommensurable." It is possible to look out from the interpretive paradigm that houses SSM and see other methodologies as special cases, but it is equally legitimate to occupy an alternative paradigmatic position and see SSM as a special case of some methodology adhering to its theoretical assumptions. For example, from the perspective of the functionalist para-digm, the debates that SSM orchestrates will appear to be nonevents, superficially reflecting some more fundamental adjustments that the system is making to ensure its needs are met in a changing environment. An old adage makes the point:

> Man is like a fly riding on the trunk of an elephant who thinks he is steering the elephant. The elephant doesn't mind and it makes the ride more interesting.

It is not difficult to read both the extended SSM case studies we have considered, in MF (Shell) and ID, in this way. It is easy to inhabit all the different sociological para-digms and see SSM as a special case of their own associated methodologies. The reader will not want me to succumb to that temptation.

Do we have any evidence as to whose account of the nature of social reality is correct? Checkland believes that the results achieved during the decades of action research using SSM fully justify the interpretive perspective he adopts. He insists that if, during his experiences using SSM, he had found other theoretical perspectives and their asso-ciated systems methodologies useful, he would have employed them. He finds it disap-pointing that "most of the critique of SSM gives the impression of deriving from nothing more than a casual reading of some of the literature, rather than from real-world experi-ence – which could provide a legitimate grounding for criticism" (Checkland 2003). The trouble with this, as Churchman would put it, is that no data can ever destroy a *Weltanschauung*. Unless you are willing and able to view the world through alternative theoretical lenses, you are likely to go on finding confirmation for the one you favor, however much real-world experience you gain. Clarke and Lehaney (1999) and Connell have encountered significant problems using SSM in the face of prevailing power struc-tures. Connell ventures that:

> Within the stakeholder group meetings, the group occasionally displayed behav-ior which seemed to lend weight to arguments such as those proposed by Jackson

and Rosenhead about the effects of unequal power relationships on participative approaches.

(2001, p. 158)

Perhaps they took more cognizance of radical change theory than Checkland. Warfield looking back over many years using his own group problem-solving approach, interactive management, laments:

> While I recognized early in the work ... that organizational factors would be very important in getting acceptance of my work, I constantly underestimated and did not plan for the negative aspects of the great strength of the higher levels of power structures in organizations.
>
> *(2003, p. 555)*

16.5   Comments

I was born in Hull and adopted into a family there. I passed my 11+ but achieved only moderate "A" level results at a local grammar school. Nevertheless, benefiting from one of the social tinkering experiments that Oxbridge periodically engages in, I was awarded an Open Exhibition by University College, Oxford. Like Checkland, I had to learn Latin in six months, and pass an "O" level in the subject, to secure my place. My background and political views did not make for an easy fit with the institution but, in 1973, I graduated in Politics, Philosophy, and Economics. After a few wasted years, teacher training and as a Tax Inspector in Hull, I was still thinking about what to do with myself. I saw an advert for an MA in "Systems in Management" at Lancaster University and applied. I had read Parsons' work at Oxford and, with time on my hands as a Tax Inspector, immersed myself in Marx's *oeuvre*, Althusser, and early Foucault – all systems-orientated texts. When Checkland told me, at the interview, that "we've gone way beyond all that stuff," I was hooked and took the program in 1977–1978. I suppose that I was one of those "mature" students, rethinking their careers, who both Checkland and Wilson say they were looking for on the MA.

It was an exciting time to be at Lancaster. Burrell and Morgan, based in the Department of Organisational Behaviour, were just finishing *Sociological Paradigms and Organizational Analysis* and were happy to talk to students. In the Department of Systems, Checkland was working on *Systems Thinking, Systems Practice* and was the star lecturer. We also benefited from the presence of Brian Wilson and Ron Anderton; the latter an expert in system dynamics who had worked with Stafford Beer on the Chile project. Some wag put on the notice board an excellent drawing of Checkland as Don Quixote tilting at windmills, with Wilson as Sancho Panza trying to keep Checkland's feet on the ground. That pretty much captured their relationship at the time.

The MA was well designed and attracted excellent students. John Mingers had taken it the year before me. Bob Galliers, later a professor of information systems and Dean of Warwick Business School, was in my year. Its structure is described by Wilson (Wilson and Van Haperen 2015, pp. xxxiii–xxxix). I well remember the "examination" in which the staff role-played members of a company from whom students had to obtain information in order to conduct a soft systems study and write a report. The students conferred, of course, and I discovered I was missing some vital production figures. The staff member playing the production manager did not like me and, however I phrased my questions and however ingratiating I was, he would not give me the information. I learnt something. After the Easter break, most students went out on the projects that constituted the core of the SSM action-research program. I remained in the department and undertook a desk-based dissertation. This was both a good and a bad thing. Good in the sense that I had Checkland as my supervisor and he encouraged my early interest in the relationship between the social sciences and systems thinking. Bad in that I did not get to use SSM in the field and this certainly delayed my gaining a full understanding of the approach and what it can achieve. I agree with Checkland that you have to use SSM in anger to appreciate it. My contact with Brian Wilson was less intellectual. Just before Christmas, a roller-hockey match was arranged between staff and students. I was placed in goal as I could hardly stand up in roller skates. Brian Wilson was an aggressive sportsman and, at some stage, I looked up to see him bearing down on my goal at full speed. I raised my hockey stick in self-defense and Brian ran into the stick with the unfortunate consequence that it split and one slice lodged in his knee. He had to spend Christmas in hospital. Many years later, I heard a report that he was limping and was much relieved to hear that it was the other leg that was giving him problems. The Lancaster MA certainly achieved its purpose for me. It changed the way I thought and set me off on a new career.

I have kept in touch with Peter Checkland since and always enjoyed our meetings. As well as being an academic, he is a jazz *aficionado* and was a rock climber. Both jazz and rock climbing, as he says, require a combination of structure and passion – exactly what SSM also demands from skilled practitioners. He adopts a modest demeanor and lifestyle with no interest in fashion, good food, or fine wine. His sandals, universally popular among 1950s "radicals," are a trademark *accoutrement*. In the professional realm everything is different. It's not just that he doesn't suffer fools gladly but that he has little sympathy with anyone struggling to keep up with his intellectual gymnastics. Witness his ferocious attacks on the secondary sources on SSM – even in a book meant for "practitioners, teachers and students" (Checkland and Poulter 2006). Further, as Mingers (2000) points out, the SSM cycle of learning is completely self-contained. He cannot find a single example in Checkland's writings where any other methodology is used alongside SSM. Checkland displays the kind of supreme confidence in his own views that is necessary, I suppose, to successfully carry through a decades-long action research program such as gave SSM to the world.

16.6 The Value of Soft Systems Methodology to Managers

SSM argues that a philosophy of management based on relationship maintaining is more appropriate and will prove more useful to managers than the goal-seeking approach with which they have usually been inculcated. Using purposeful activity models, they can discover what accommodations are possible between concerned actors and learn their way to what changes to a problematic situation are both desirable and feasible. This is a process they can conduct formally guided by SSM or, even better, they can internalize the methodology and employ it naturally on an everyday basis without losing touch with the dynamics of their actual work situation. With this orientation, it is not surprising that it brings much of value to managers. Here are five major things:

- SSM does not require the establishment of clear goals before problem resolving can begin; rather, it maps onto the normal managerial tasks of considering a "mess," suggesting ways forward, and seeking agreements for action – thus it is easily absorbed into normal managerial processes
- SSM offers an excellent way of exploring purposes, using a range of relevant purposeful system models to find out what is possible given the history, culture, and politics of the problem situation
- SSM articulates a learning system that challenges existing ways of seeing and doing things and can lead to some surprising shifts in *Weltanschauungen*, opening up novel and elegant proposals for change
- SSM has shown that the effective design of support systems, such as information systems, depends on a clear understanding of the purposeful activity that is to be supported at a higher level
- Some powerful methods, such as rich pictures, root definitions, and conceptual models, have been developed and refined to assist with using SSM

16.7 Conclusion

Checkland's break with the predominant goal-seeking tradition was a revolution in systems thinking. His insight that he was engaged in making an "epistemological break," reframing systems thinking on new philosophical foundations, has helped bring clarity to the nature of that revolution and assisted in steering it through. SSM, developed and refined on the basis of an action research program directly linking ideas and experience, enables the alternative, relationship maintaining perspective to be followed in practice. With its associated principles and methods, SSM represents an achievement that has revitalized the systems approach and significantly increased its relevance to business and management. Any systems thinker would do well, therefore, to reflect on Checkland's "four conditions for serious systems thinking and action":

- Understand and argue for the real existence of emergent properties, for that is the core justification for systems ideas
- Make sure that [your] work embodies the "concept" of an adaptive whole, which can be used to explore the real world's complexity
- Attribute systemicity first to the process of enquiry into the perceived real world, that is, start from the stance of soft systems thinking
- In order truly to engage with perceived reality, adopt a sharply defined action research methodology, with recoverability of the research story as the best available validity criterion, allowing coherent discussion of both the course of the thinking during the research and its results

(Checkland 2003, p. 555)

Type F

Systems Approaches for Coercive Complexity

In general, it always comes back to not having gone as far as possible in my radicalism. Naturally, in the course of my life I have made lots of mistakes, large and small, for one reason or another, but at the heart of it all, every time I made a mistake it was because I was not radical enough

(Sartre 1975).

There are some systems approaches that can play a role in addressing the complexity that arises from "coercion" in organizations and society. Coercion is closely associated with the exercise of power, which can operate to ensure some individuals or groups have the capacity to control the behavior of others and benefit as a result. Power can be overt or operate in very subtle ways. It can be so overwhelming that it temporarily disguises the existence of different interests. Power is ubiquitous and can have beneficial (getting things done) as well as harmful outcomes. We become concerned when it results in structured inequalities and some groups consistently gaining at the expense of others. An important purpose of Type F systems approaches is to contribute to fairness by guaranteeing that those who are potentially or actually disadvantaged by coercion have a say in decisions and, if necessary, by working on their behalf. Improvement can only occur, it is argued, if decision-makers pay attention to the consequences of their actions for those affected, full and open participation takes place, and discrimination and disadvantage are eliminated. The two systems methodologies that currently pay most attention to the forms of complexity that originate from coercion are:

- Team Syntegrity
- Critical Systems Heuristics

Critical Systems Thinking and the Management of Complexity, First Edition. Michael C. Jackson.
© 2019 John Wiley & Sons Ltd. Published 2019 by John Wiley & Sons Ltd.

17

Team Syntegrity

When I started to construct physical polyhedra with my own hands, it was truly a revelation to follow Bucky's route. An unwholesome mess of wooden doweling, panel pins, rubber bands, string, and glue, strengthened with gratuitous contributions from skin and beard, quite suddenly transformed itself into a polyhedron so strong that I could actually stand on it

(Beer 1990, p. 122)

17.1 Prologue

In 2010, Fürth, a German city of 114 000 people, was Bavaria's most heavily indebted city. A further budget deficit of 25 million euros was expected in the 2011 financial year. Local efforts to put things right had demonstrably failed and the situation appeared hopeless. Inevitably, higher authorities became concerned, demanded permanent savings of 13 million euros by 2013, and threatened to take over direct control of the City's finances. It was at this point that Dr. Stefanie Ammon, city councilor for organization and financial affairs, contacted the "Malik Management" consultancy. She had heard of their *SuperSyntegration* approach, based on Stafford Beer's "team syntegrity" (TS) methodology, and hoped it might help bring about a "miracle cure." A *Malik SuperSyntegration* (MSS) was conducted over the weekend of 18 July to 20 July 2010, involving 32 participants. These included the mayor, employees from all branches of the city administration, and other stakeholders from private companies and public sector organizations. The results were spectacular, with Stefanie Ammon able to declare, in November 2010: "I am convinced that the SuperSyntegration is the optimal method for solving complex problems that deal with highly diverging interests." This account of what occurred is derived from "City of Fürth: Back to financial health after Malik Syntegration" (see Malik Management 2018).

To get an MSS started, a "key question" has to be declared. This was:

> Beginning from today, what has to be done so that the city of Fürth can reestablish its ability to function in all areas and, thereby, manage to effectively and durably eliminate the existing deficit and balance the administrative budget by 2013?

From this question, a list of 12 subissues was agreed by the participants that formed the agenda and would focus the discussion:

- Boosting income
- Mission statement, policy, and administration
- Sustainable cost-cutting
- Cutting social spending
- Legal directives contrasted against practical necessities
- Strengthening the business location
- Improving deployment of human resources
- Information for and involvement of citizens
- Improving the management culture
- Critical appraisal of activities
- Intensifying collaboration between municipal authorities
- Improving the organization

Because MSS allows discussion of a range of issues, participants were able to look beyond savings alone to new ways of functioning that would bring long-term effectiveness. Also, unlike most conventional methods, MSS enabled them to address all 12 issues simultaneously while also processing the complex interactions between them.

There followed three "syntegrative iterations" of Team Syntegrity's "sophisticated, mathematically optimized interconnectedness logic," working through the 12 subissues and ensuring that every bit of available knowledge was captured, interconnected, and employed in the search for innovative solutions. Hermann Schnitzer, Head of HR, for the city stated that:

> For the first time since I joined the company 25 years ago, we have succeeded in pooling the expertise and know-how of all executives to produce shared results.

The outcome was unanimous agreement on 50 comprehensive action packages, with more than 200 actions in total, aimed at reforming the City of Fürth.

The MSS identified a total savings potential of 24 million euros, considerably more than required by the state auditor and with enough left over to set up an investment reserve for future development. The City retained its financial autonomy. Implementation proceeded smoothly with over 70% of the agreed actions implemented in 7 months and 80% of supplementary measures, agreed in 2011, realized within another 12 months. Over 90% of the participants were both satisfied with the Syntegration and had their expectations fulfilled. Longer term socio-cultural benefits were increased enthusiasm and optimism, a greater awareness of dynamic complexity, and an appetite for cross-divisional collaboration to deal with the issues as a whole. The last word can be left to the mayor, Dr. Thomas Jung:

> The method brings issues to the fore. Especially on the second day, I felt a sense of group dynamics and experienced emotional highlights. I came to see many colleagues in an entirely new light. We now treat one another far more respectfully. This is a huge improvement.

17.2 Description of Team Syntegrity

17.2.1 Historical Development

Stafford Beer, the founder of organizational cybernetics and inventor of the "viable system model" (VSM – see Chapter 13), devoted his last years to the development and refinement of an approach to democratic decision-making called Team Syntegrity. TS provides a theory to underpin and a set of procedures (a "protocol") to support, nonhierarchical, equitable, and participatory discussions among substantial groups, representing a diversity of views on shared issues, so that they can arrive at and agree upon innovative solutions. Beer developed TS for two reasons. First, to correct what he saw as a misunderstanding of the VSM; second, to fulfill an important requirement of that model.

The misunderstanding is that the VSM is hierarchical. Beer believes that this arises from people taking at face value the diagram that depicts System 5 at the top. They assume that this is "the boss." In fact, Beer argues, it is a mere appearance. A more sophisticated grasp of the model reveals that the primary purpose of System 5, as with Systems 2–4, is to facilitate the functioning of the embedded viable System 1 elements. After all, do not "the lowest autonomic functions in the human body have representation in the cortex, and the leadership of a democracy supposedly embodies the will of the people" (Beer 1990, p. 121). However, he got past protesting and instead offered TS as a concrete rebuke to all those he felt had got him wrong. It is a complementary approach to the VSM, setting out the means of achieving the democratic milieu in which that model can operate as it is meant to.

TS also satisfies a significant demand highlighted by the model. In order to define and specify a resolution to policy, control, coordination, and monitoring issues, it is important to have conversational tools that can deal with the divergent, often conflicting viewpoints that participants bring to the discussions. In particular, as we saw in Section 13.2.3, it is essential to promote rich and purposeful debate at the point in an enterprise where information about its internal state, generated by System 3, is brought together with information about the external environment, generated by System 4. This is known as the "operations room" in VSM parlance. It must be given embodiment and constitute a productive "environment for decision" around issues such as strategic direction. Beer knew, based on Ashby's work on variety, that most organizational structures are variety inhibitors that constrain interaction and discussion because they impose barriers reflecting organizational rules and practices. If the conversations and debates are to take place at the high variety levels necessary and achieve the kinds of balances demanded throughout the architecture of the VSM, it is important to pay attention to the design of the negotiation spaces in which they occur and to create democratic conditions in which all relevant viewpoints can be freely expressed and taken into account. This is what TS does.

Beer's book *Beyond Dispute: The Invention of Team Syntegrity* (1994) describes in meticulous detail the work that went into developing the approach. This began around 1970, at a conference looking at the future of the UK Operational Research Society, and continued with experimental "syntegrations" held at Manchester Business School and in St. Gallen, California and Toronto. It took some doing to iron out all the various technical problems. The book also sets out the theoretical and practical aspects of TS. The back cover provides a neat summary:

> The underlying model is a regular icosahedron (20 sides). This has 30 edges, each of which represents a *person*. An internal network of interactions is created by a set of protocols. A group organized like this is an ultimate statement of *participatory democracy*, since each role is indistinguishable from any other. There is no hierarchy, no top, no bottom, no sideways.

Team Syntegrity Inc. (TSI) was founded in 1992 to find markets for TS and to continue to develop the methodology. It ran numerous syntegrations worldwide, and its success led to its incorporation as a core component into the Malik Management consultancy. To date, there have been roughly 600 applications of TS by TSI and the Malik organization, and these are said to "have proven successful without exception" (Malik Management 2018).

The value of such an approach is obvious for organizations that are promoting more democratic decision-making, and it is widely employed in such contexts. White (1994, 1998) has also demonstrated its usefulness in multiorganizational settings where, of necessity, the commitment of a variety of stakeholders to action has to be obtained. He describes TS sessions that debated the questions "How can we, sovereign world citizens, govern our world?" and "How should we run London?" But its use is by no means confined to such situations. In his opinion, it is an approach that answers to the "New Times" of the postindustrial age, where democracy and decentralization are valued more highly than hierarchy and centralization. In the modern age, there is likely to be an increasing need for TS to promote inclusiveness, participation, and self-management even in otherwise conventional organizations. The evidence seems to be there. Schwaninger (1997) is convinced of its worth in supporting teamwork, particularly in relation to problems of planning, innovation, and knowledge acquisition. Espinosa and Harnden (2007) chart applications to policy-making, group learning, organizational evaluation, and collaborative research. Malik Management (2018) see TS as relevant to "all segments of the economy and society and with all types of challenges": strategy development, restructuring, change of corporate culture, increase in earnings and innovations, maintaining and enhancing customer relations, postmerger integrations, organizational development, cost reductions, profit improvement, etc.

17.2.2 Philosophy and Theory

Beer acknowledges important debts to Warren McCulloch and W. Ross Ashby in developing TS. This directs us to "British cybernetics," as delineated in Section 6.3 with the help of Pickering (2010), in our search for the philosophy and theory underlying TS. Within this tradition, TS is another example of what Pickering describes as the "ontology of becoming although in a fixed framework." The fixed framework is supplied by the structure of the regular icosahedron. Pfiffner (2004) compares it to the genetic code that defines life. The icosahedron lays out the code that produces the properties required for effective communication in a large group of people. Malik Management (2018) sees it as embodying "the natural laws of modern communication theory and complex system cybernetics." Once set up, syntegrations guide conversations according to a "performative idiom." There is no attempt to arrive at a stable "representation" of the world.

The process not the product is most important. During the process, the agenda emerges and conversations take flight revealing creative possibilities. Ideas flow in all directions and become the property of the group as a whole rather than individuals. A quotation from Brian Eno, discussing the origins of his "self-generating music," is apposite:

> Without having the language [later supplied by Beer's *Brain of the Firm*] … what I was thinking about was self-organization, bottom up creation, a different conception about what it meant to be a 'composer' …. I'd become familiar with the work of composers like Steve Reich and Terry Riley and Philip Glass, who had been exploring ways of generating music from very simple initial conditions, and had started to think of the composer as someone who starts things rather than finishes them. I liked the idea of making what the composer Michael White called 'Machines' – intellectual constructions which produced music, and music you hadn't specified in advance.
>
> *(Eno 2009, p. 8)*

TS produces conversations of an innovative nature that cannot be anticipated advance. Espinosa and Harnden, drawing upon Maturana's thinking, observe another generative function performed by TS. The protocol patterns the "coordinations of actions" between participants in a manner likely to lead to follow-up activity after the event:

> What the formality of Beer's Syntegration model provides is an elegant routine for a dance of coordinations of actions in the course of which a particular quality of shared experience might emerge, which embraces the braiding of subjective and shared knowledge, *across all the parties* …. [T]his methodology as distinct from other systems approaches is to do with instituting a *pattern* of repeated coordinations of actions …. And … it is the recurrence of such a regular pattern of interactions that promotes the willingness for consequent coordinations of action (action plans, follow-up, etc.).
>
> *(Espinosa and Harnden 2007, p. 1062)*

Another important philosophical foundation of TS is Beer's "hylozoism." In a weak form, hylozoism claims that all living and nonliving systems obey the same laws – a doctrine common to many cyberneticians. Beer, as Pickering argues, seems to sail close to the stronger version that all matter is "infused with spirit." Mind is everywhere. Nature is an irreducible mystery. We cannot possibly represent it, but we can take inspiration from its performance. This is from his poem *Computers: The Irish Sea*, written in 1964 (quoted in Blohm et al. 1986):

> That green computer sea
> with all its molecular logic
> to the system's square inch,
> a bigger brain than mine,
> writes out foamy equations from the bow
> across the bland blackboard water.
>
> Accounting for variables
> which navigators cannot even list,

a bigger sum than theirs,
getting the answer continuously right
without fail and without anguish
integrals white on green.

We should learn from matter which "knows" so much more than us. This thinking always led Beer, as with the VSM, to want to bypass the traditional "representational method" of science and to borrow what he wanted from nature. In this he anticipates, by decades, Taleb's insight on how to gain knowledge appropriate to "antifragility":

> If there is something in nature you don't understand, odds are it makes sense in a deeper way that is beyond your understanding. So there is a logic to natural things that is much superior to our own. Just as there is a dichotomy in law: 'innocent until proven guilty', let me express my rule as follows: what Mother Nature does is rigorous until proven otherwise; what humans and science do is flawed until proven otherwise.
>
> *(Taleb 2013, loc. 2283)*

Not surprisingly, therefore, when Beer wanted a model of democratic, participatory decision-making, he did not follow the theoretical route prescribed in the social sciences. Habermas, as we saw in Section 4.4, took a prolonged "linguistic communicative turn" through speech-act philosophy, eventually arriving at prescriptions for an "ideal speech situation" that could be used to unmask "systematically distorted communication." Beer proceeded along a more direct cybernetic course, using metaphor and analogy until he found a rigorous model in another field that was "isomorphic," and subject to the falsifiability criterion, when applied to the social domain in which he was interested. He explored "holography" (the study of wholes that are manifest in their parts), experimented with lasers and toyed with various abstruse areas of mathematics. One day, in a completely different context, Beer remembered Buckminster Fuller's dictum: "All systems are polyhedra." At once he recognized that no structure could be less hierarchical than a regular polyhedron. Perhaps here could be found a model of perfect democracy. Fuller had been experimenting with icosahedra, as the most suitable form of regular polyhedron for architectural purposes, and had modeled his famous "geodesic domes" on the structural principles that they displayed. Beer was hooked and began to construct physical polyhedra himself in order to understand better their nature and properties. He found to his amazement that using wooden doweling, panel pins, rubber bands, string, and glue, together with, as he put it, "gratuitous contributions from skin and beard," he was able to build a polyhedron so strong that he, a man of large stature, could stand on it.

With an icosahedron the whole gains cohesive strength by achieving equilibrium between the forces of compression and tension:

> According to this, the wholeness, the INTEGRITY, of the structure is guaranteed not by the local compressive stresses where structural members are joined together, but by the overall tensile stresses of the entire system. Hence came the portmanteau term for Tensile Integrity: TENSEGRITY.
>
> *(Beer 1994, p. 13)*

For Beer, the geodesic dome was likely to be just one expression of a "natural invariance" that would be repeated in different domains:

> One matter in which (I argue) holistic thinkers ought to agree is that, when a good case for a natural invariant is presented, we are entitled to be excited by a possible advance in human understanding of the natural world and to seek out further examples.
>
> *(Beer 1994, p. 13)*

In fact, examples of icosahedra are relatively rare in nature, and the icosahedron is often cited as the first example of a geometrical object that was the free creation of human thought rather than the result of observation. The thought is credited to Theaetetus, working around 380 to 370 BCE, who may have been extrapolating from knowledge of the dodecahedron, found in crystals of pyrite which had been worked from early antiquity. Nevertheless, icosahedra do exist in nature, and Fuller was pressing on to produce designs for cars, ships, and houses based on their structural characteristics.

Beer determined, therefore, to investigate the phenomenon of *tensegrity* as a possible guide for democratic and effective dialogue. The analogy he develops is of a group of participants, an "Infoset," trying to express their integrity by compressing their ideas into an integrated statement, which benefits from the creative tensions that their discussions produce. The rigid structure of the icosahedron can be used as a model for organizing discussion in a manner that leads to a strong, cohesive outcome just because of the tensions generated by diverse viewpoints. Maximum tensegrity of communication is achieved. Beer adds to this the notion that, if the Infoset constitutes a closed system, ideas will inevitably "reverberate" around it and generate *synergy*. Thus, the word *syntegrity* was coined to express the full meaning of a Fullerian tensegrity balance which results, in the social context, in synergistic communication among teams of participants. The name Team Syntegrity was given to the whole approach.

Beer (1990, 1994) agreed with Fuller that the icosahedron is the most interesting of the regular polyhedra, exhibiting the most perfect tensegrity. It was this geometrical form that he took as a possible model for an ideal democracy. To understand the theory underlying TS, therefore, we will have to study the peculiar properties of the icosahedron and think through how he believed they could be transferred to the social domain to provide for democratic and effective conversations.

The icosahedron has 20 triangular sides described by 30 edges. It has 12 vertices each connecting 5 edges. Figure 17.1 shows its shape. Taking the icosahedron to represent a large group, the "Infoset," Beer asks us to regard each of the edges as a person. There are therefore 30 people. Each of the 12 vertices gathers 5 edges together. If a vertex is regarded as a discussion group, there are, therefore, 12 groups and 60 group members. As each edge (or person) has two ends, every person belongs to two teams and no two people belong to the same two teams. Again:

- There are 30 people, divided into 12 discussion groups
- Each person belongs to two groups, which might be called his or her left-hand and right-hand group. Thus, there are 60 team members. No person belongs to the same two teams as anyone else

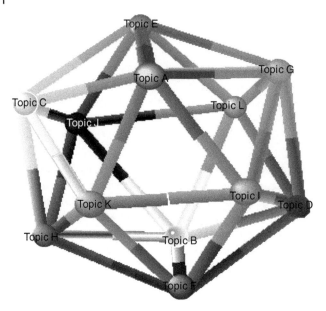

Figure 17.1 An icosahedron showing how 30 "people" (edges) are connected to 12 "topics" (vertices).

Beer regards this as a good start for a model of a perfect democracy because it exhibits no global hierarchy. He harks back to his experience in Chile remarking that:

> My idea was to replace the Marxist 'classes' (where the ruling class exploits the proletariat) with a richer and less tendentious categorization based on shared information.
>
> *(Beer 1994, p. 11)*

Moreover, the arrangement described demonstrates total closure – it is in logical terms self-referential. If the discussions are appropriately ordered, this should produce the phenomenon of "reverberation." Views emanating from one discussion group will reverberate around the structure, gaining and losing adherents, consolidating and subtly changing, and this should ensure maximum creativity among the groups and generate synergy.

The other benefit of the icosahedron, for Beer, lies in its strength. Fuller had reasoned that the icosahedron achieves its resilience from what he called "tensile integrity," or "tensegrity" for short. He defined tensegrity as the unity of the structure of any physical object as determined by its distributed tensile stresses. In structures like the geodesic dome, the forces of compression and tension achieve a natural, reinforced balance. The integrity of the structure then derives not from the local compressor stresses where elements conjoin, but from the overall tensile stresses governing the entire system. Fuller obtained cohesive strength in the geodesic dome by running struts between the centers of adjacent triangular faces, thus providing for a thick two-dimensional skin. Beer recognized that in his theoretical construct, of a group of 30 participants exhibiting tensegrity, even greater strength could be obtained by driving the "struts" straight through the central spaces that, in an architectural design, had to be kept open. He now had to work out what the correlate of a strut might be in social terms. He came up with the idea of "critics" appointed to each of the teams formed at the vertices from positions across the phase space. Critics were to offer constructive advice to the team on the basis that they were relatively disinterested members of the whole but possessed detailed specialized knowledge of one other team. The tension created by each critic was to be balanced by appointing a member of the receiving team as a critic to another distant team – in fact, the second team of which the original critic is also a member. Again:

- Each person is appointed a critic of a team of which he or she is not a member. The team to which he or she is appointed as a critic will appoint a critic to his or her right-hand team
- Similarly, each person becomes a critic in that team of which he or she is not a member, but of which a right-hand teammate is a member. And that team will appoint a critic to his or her left-hand team
- Since each person has two critical appointments, there are 60 critics spread with tensegrity over 12 teams. So, each five-member team has five more quasi-members who are its "critics"

We now have a model of an organization generating synergy out of perfect democracy and, at the same time, demonstrating great strength and cohesion because of its tensegrity. According to Pfiffner (2001), TS achieves this outcome by employing a reliable mathematical principle that organizes Infoset interactions "in the middle ground between absolute rigidity and chaos." It remains to specify how this theory is translated into practice through the "protocol" of TS.

17.2.3 Methodology

Essentially, TS is a process that guides nonhierarchical group decision-making for an "Infoset" of 30 people who share an interest in addressing an issue of particular concern to them and about which they have different opinions. The 30 individuals must agree to a communications protocol – a set of procedures designed to extract maximum advantage from the qualities of the icosahedron. The protocol establishes how these individuals share information about the issue, develop their conversations, and reach conclusions. It places the participants in roles of equal status so that every voice is heard and no individual is allowed to dominate. People are divided into teams, meetings are sequenced, and information is distributed in such a way as to ensure a highly interactive and democratic event offering the best opportunities for balancing tension and synergy as the teams negotiate different viewpoints. It will be clear that the protocol simply specifies the form of the interactions and conversations in the Infoset. It puts no restrictions and makes no comment on the content of what is said. That is left up to the judgment of the individuals working in teams.

Once the decision to use the methodology is made, the first step is to agree on an "opening" question that captures the issue on which all participants hope to reach agreement. We set out the "key question" used to launch the Fürth city syntegration in the Prologue to this chapter. Participants are then selected who should offer a rich variety of perspectives on that question. In a multiagency situation, for example, they will include representatives from different geographical regions, ethnic, age, gender, and status groups and from communities and institutions affected by the issue. In the syntegration on "How should we run London?," described by White (1998), the Infoset of 30 members included politicians, journalists, voluntary sector leaders, health experts, think-tank researchers, and other experts and citizens. In a business syntegration, there will likely be representatives from different departments and countries, along with suppliers, customers, and other stakeholders (Leonard 1996). Careful consideration will be given to the choice of an appropriate site for the meeting and to arranging the necessary amenities. Although Beer regards "icosahedral theory" as the main guarantor of a successful outcome, he also insists that skilled facilitation is necessary (1994, p. 113). Facilitators must therefore be appointed and any staff briefed as to their roles. The participants can then be invited to the opening of the event.

A classic TS exercise will last five days and follow a protocol that has three parts, as follows:

- Problem jostle
- Topic auction
- Outcome resolve

Beer sets out in some detail the requirements, the procedures to follow, and the timings for each of these parts in *Beyond Dispute* (1994, pp. 22–34). Included in the same book are alternative versions, with some additional recommendations, provided by Schwaninger (pp. 53–60) and Truss (pp. 281–299). Leonard (1996) offers another account.

During the "problem jostle" sessions, the participants concentrate on "generating an agenda" that will help them to address the opening question. Since they are going to be divided into 12 groups, this requires clustering the concerns of all 30 individuals around 12 main issues, or "topics," that will be the focus of discussion. The 30 participants and the facilitators assemble in a large room with extensive wall space, to which documents can be affixed, and containing 12 tables with a few chairs around each. The problem jostle begins with each player submitting to the facilitators at least one Statement of Importance (SI) that he or she feels is particularly relevant to the opening question. An SI should be a concise, one- or two-sentence assertion. The facilitators scrutinize the SIs, eliding any that say similar things, type them up and make them available to the "players." Any player who then regards a particular SI of extreme importance can move to one of the 12 tables (called "hours" because they are arranged in the pattern of a clock), name the SI and start a discussion group around that topic. Other players are free to champion alternative SIs in this way, join existing discussion groups or simply wander around the room. When enthusiasm develops about a topic at a particular table, its advocates can write the topic down and seek further adherents to their cause. Any topic that gains five signatories is classed as an Aggregated Statement of Importance (ASI). While this self-organizing process is taking place, the facilitators assist by pointing out similarities between certain ASIs and by seeking to position "polar opposite" statements at polar extreme "hours" (e.g. 6 and 12). The first session of the problem jostle should end with all 12 hours having ASIs beside them. Further sessions are designed to promote further reflection on existing ASIs, the emergence of new ASIs, and the reduction and refinement of these to just 12 Composite Statements of Importances (CSIs) arranged according to the prescribed pattern. This is known as "hexadic reduction." If necessary, a rating system can be employed to decide on the preferred CSIs. During the syntegration on "How should we run London?" the problem jostle produced 25 ASIs and, following a difficult hexadic reduction, these were reduced to 12 CSIs, as follows:

- Strategic planning
- Transport
- Children of the city
- Sustainability
- Exclusion and inclusion
- Money/finance
- Serious media coverage
- Arts and education
- Demographic diversity
- Representation
- The complexity of London
- Community involvement in governing London

The "topic auction" sees the 30 participants allocated to different groups according to their preferences among the topics and the logic of the icosahedron. Individuals are asked to rank their preferred topics, and an algorithm is employed to ensure that the highest level of satisfaction is obtained while respecting the constraints imposed by the structure of the icosahedron. Each of the 12 vertices is taken to represent one of the 12 topics for discussion (topics A–L in Figure 17.1) and, as we know, each edge represents one participant. Thus, every discussion group is allocated five people, corresponding to the vertex that is formed by five edges joining together. And each participant ends up assigned to two groups defined by the ends of their edge. The structure then determines which two other topics the person will be a critic of. There are, therefore, 12 teams each developing one of the topics and consisting of 5 members and 5 critics. Each of the 30 people involved plays two roles: a participant in the two groups defined by their edge and a critic in the two groups defined by the edge on the opposite side of the icosahedron. Once the groups are fixed in the topic auction, they remain the same for the remainder of the exercise. The topic auction ensures that maximum lateral communication is obtained and, in Beer's view, ensures that "reverberation" will happen. Participants feel ideas rebounding back to them in a different and enhanced form.

The third stage, the "outcome resolve," takes up the majority of the five days and consists of a series of meetings of the 12 groups. The meetings typically last between 45 and 75 minutes, and each group will meet three or four times depending on the time available. The sequence of meetings is determined by the protocol, which stems from the geometry of the icosahedron. As each participant is involved in developing two issues and as a critic of two others, it turns out that only two groups can meet at any one time. The five discussants in each group seek to work up their thinking on the topic into an insightful Final Statement of Importance (FSI). The five critics on the team must remain silent while this process is going on, but may then join in with 10 minutes of relevant questioning. The critics should act as devil's advocates by challenging the group to review any agreements reached and questioning the assumptions behind apparent agreements. Observers may also be present from other groups that are not meeting and, while they may not intervene, they can scrawl on any written outcomes ("graffiti") and make use of the information they garner from being present in the sessions when they are discussants or critics. Given that each participant belongs to four teams (two as participant, two as a critic), they act as information diffusion channels, rapidly spreading information around the different teams. This produces the desired reverberation or "echo effect," so that all the team members share information among all the others thanks to the closure inherent in the structure of the icosahedron. It has been demonstrated that after the third iteration around 90% of the information initially generated about the opening question has been distributed among all the participants. After each set of meetings, there is a plenary session at which the 12 teams present the current results of their conversations and raise questions and proposals concerning the content of the discussions or the methodology of TS.

At the end of the whole event, there is a closing plenary session where the teams present their FSIs. There are 12 of these, one for each team, which by now should meet with general assent in the Infoset. Teams are also invited to assess the event and methodology and their own learning and experience as participants. Plans for further meetings or actions can be made at this closing session.

In *Beyond Dispute*, Beer describes the traditional form of TS involving 30 people over five days. He was however, from the beginning, aware of the difficulties created by the strict demands this imposes. The corporate market wanted an approach that involved fewer people over a shorter time period. He therefore began to work with Joe Truss, originally with TSI, to develop alternative protocols that could yield the same benefits but with smaller numbers of people and less time. Delving into the geometry of the icosahedron, Truss discovered ways of adapting TS so that it could work with groups of virtually any size, over different time periods, without compromising the mathematical integrity of the methodology. So-called *SmallForms* and *ShortForms* of TS were developed for groups of 24, 18, 12, 6, even 3, lasting from one to a few days (Truss et al. 2000). At the same time, work took place to refine Beer's proposals so they could be used with very large groups. These might involve 900 people in 30 syntegrations, with one final Infoset made up of one representative from each. Such "mega-forms" could be useful for large organizations, including governments.

Reducing the numbers required to as few as three people also enabled Truss to suggest an answer to one of the main questions asked of TS: "What does it do to assist with implementation after a TS session has taken place?" His solution, "FACE planning" (Beer 1994, pp. 333–345; Truss et al. 2000, pp. 6–7), employs a version of the same methodology to coordinate and structure smaller meetings of willing and available Infoset members designed to take implementation forward after the syntegration.

17.2.4 Methods

TSI, the company that first brought the methodology to market, developed a number of special tools for supporting the planning and delivery of TS sessions. We have already mentioned the optimization software for deciding team membership. There are also various brainstorming, visual applause, and other creative and democratic techniques for sharing information and promoting enhanced group learning. Development of TS methods has continued under the auspices of the Malik Management consultancy.

It is worth saying a little more about the role of facilitators during TS sessions. They are important in helping participants to generate an agenda in the "problem jostle." They are also crucial in the "outcome resolve." During each team meeting, a facilitator is responsible for handling the variety of the conversations and integrating the contributions from all the participants and critics. They moderate the amount of time taken by each discussant in the conversation, making sure all have the opportunity to have an equal say. The critics too must be allocated time to make their contribution. It is not the role of the facilitators to influence the content of what is said but they should aim to give each participant the space necessary for them to have an impact. They need skills in hearing, understanding, and feeding back the participants' conversations. They are also responsible for making a record of the important issues addressed. In this they may have the support of technical assistants who record an electronic version of the development of team conversations so that they can be reproduced for the plenary sessions.

17.3 Team Syntegrity in Action

Readers of Chapter 13 will recall that the Colombian Constitution of 1991 established a Ministry of the Environment together with a set of organizations – the national environmental system (SINA) – and charged them, at the national, regional, and local levels, to deal with environmental issues in a more holistic and participatory way in order to promote sustainable development. The organizational design and policies of SINA sought to respect these goals. Nevertheless, by 1996, it was clear that many organizational problems remained at SINA, and its practices and results were actually far removed from its objectives. For this reason, the Ministry of the Environment and the National Auditing Office decided to sponsor a TS event to get agreement on the main strategic issues still plaguing SINA's development and operation. The event took place on the Colombian island of Gorgona, a beautiful and valuable ecological reserve in the Pacific Ocean. The principal organizer was Angela Espinosa and I owe the following description to her. Angela's accounts of the syntegration can be found in Espinosa (2003) and Espinosa and Walker (2017). She was assisted with facilitation at the event by Chris Cullen and Joe Truss of TSI.

Senior managers from the national institutions sponsoring the meeting and the consultants organizing it agreed that the opening question should be:

> How should public, private and voluntary sector organisations and institutions be co-ordinated in order to preserve the natural environment in Colombia?

They then drew up criteria for inviting participants with the aim of getting a varied and balanced representation of individuals from different bodies involved with strategic environmental matters around the country. Issues of ethnicity, gender, age, geographical spread, political, and professional background, etc. were taken into account. The final list of participants involved

(a) Senior managers from the following public sector institutions:
  - Ministry of Environment
  - National Auditing Office
  - Risaralda – the county auditing office
  - Ministry of Agriculture
  - National parks

- La Macarena – a regional environmental corporation (CAR)
- Environmental Development Institute (IDEAM)
- A Colombian petroleum enterprise (Ecopetrol)
- National Planning Department
- Van Humboldt Research Institute
- Cartagena's environmental agency (DAMA)
- The city of Cordoba (an ex-governor)
- The National University

(b) Individuals representing different communities, organizations, and institutions:
- A representative of black communities
- A senior manager from a private sector company – Antioquia
- A women's leader
- A researcher on environmental issues
- A postgraduate student studying environmental issues
- A countryside and peasants' representative
- The editor of *Environmental and Socio-Economic Review*
- Ecology Teachers' Association – Cundinamarca
- Colombian Environmental Funding Agency (Ecofondo)
- Environmental nongovernmental organization – Popayan
- Los Andes University
- Colombian Association of Small and Medium Sized Enterprises (ACOPI)

The organizers made the necessary travel and accommodation arrangements and the participants gathered on the first afternoon for the opening. At this session, the methodology and protocol were explained, as well as the roles of participants, facilitators and assistants, and the various facilities available were described. The rest of that afternoon and evening focused on generation of the agenda using the problem jostle technique. More than a 100 SIs were produced. With the help of the facilitators, these were then collapsed and prioritized, during hexadic reduction, into ASIs and, finally, 12 CSIs – the 12 issues relevant to the opening question which would guide later conversations. This whole process was stressful for the participants, already tired from traveling, and it was 9 p.m. before agreement was reached on the following 12 issues:

- Participatory mechanisms
- Search for peace – war, corruption, and environment
- Culture and education
- Equity and agrarian reform
- Institutionalization
- Territorial and environmental ordering
- Management and development
- Gender and environment
- Ethical values
- Environmental conservation
- Rationalization of environmental control
- Sustainable development and the international dimension

Once the CSIs were agreed, the topic auction stage could begin with the participants each expressing their preferences for discussing particular issues. The computer program did its best to satisfy these preferences within the constraints imposed by the icosahedron structure and the results were announced first thing on the second day. Not everyone was satisfied. Three participants in particular were unhappy with being allocated to the "gender and environment" group. Nevertheless, everyone eventually accepted membership of the groups to which they had been assigned. The outcome resolve could now commence. It lasted for three days, and the participants worked well in the 12 teams, as discussants, critics and observers, to address the 12 CSIs. They were supported throughout by the facilitators and technical assistants who recorded the conversations and their outcomes and fed them back to the whole Infoset during the plenary sessions. There were certain problems arising from faulty installation of computers and software, but these were overcome by good teamwork and everything ran to schedule. The closing session saw the groups sharing their conclusions on the 12 issues and the Infoset discussing the experience and the learning they had derived from it. A few individuals had been upset by conflict in their teams and were disappointed with the outcomes. The great majority, however, as shown by evaluation questionnaires, had a positive and pleasant learning experience and declared that they were satisfied both with their contributions and with the quality of the final results.

FACE planning that would have structured meetings of participants after the event, to ensure implementation of the outcomes, did not take place. However, some of the organizers and participants did meet again to work out a final version of the 12 FSIs and, some months later, the organizers published the *Gorgona Manifesto* setting out their conclusions. A very brief summary follows:

- *Participatory mechanisms*: The State should encourage communities to take advantage of the opportunities for participation open to them – unless they learned to participate sustainability could not be achieved
- *Search for peace – war, corruption, and environment*: War and corruption are social phenomena that destroy the environmental and social milieus
- *Culture and education*: Environmental education should be introduced at all educational levels and there is the need to value, conserve, and transmit traditional wisdom
- *Equity and agrarian reform*: It is necessary to agree on a new development model promoting agrarian reform on the basis of equity. Without this there cannot be democracy, peace, political stability, or sustainable development
- *Institutionalization*:
  - ○ SINA must grant autonomy to institutions involved in environmental management at the local level, but they must still act cohesively as a whole
  - ○ its components should cohere by region and theme so they can act jointly to tackle complex, transdisciplinary, and inter-regional issues
  - ○ there should be strategies promoting environmental investments aimed at key programs and institutions and encouraging participation and responsibility
  - ○ it is essential to involve universities and research institutes in developing environmental educational programs
  - ○ monitoring and control systems should be put in place to assess the impact of the development programs and industrial projects on sustainability
  - ○ SINA should review current environmental laws and design effective control and punishment mechanisms

- *Territorial and environmental ordering*: An integrated regional strategy was needed to manage the water basins of the main rivers and an integrated national strategy to ensure all possible benefits are obtained from environmental reserves
- *Management and development*: Knowledge, science, technology, and local wisdom form the basis for sustainable development, and these must be coordinated to support decision-making contributing to sustainable development at the local level in the context of a shared national ethos
- *Gender and environment*: The participation of women in environmental issues is fundamental because of their traditional roles in the family, education, agriculture, and the community, and relevant female values should be more highly valued and shared by males
- *Ethical values*: An environment of ethical responsibility should be collectively built, based on trust, throughout society – thus enabling cooperative behavior between individuals and between communities
- *Environmental conservation*: We need to monitor environmental damage, understand its causes, and be prepared organizationally and economically to prevent or respond to it
- *Rationalization of environmental control*: Communities must be involved participatively in defending their local environments
- *Sustainable development and the international dimension*: Colombia needs to improve its negotiating skills on environmental issues at international fora in order to protect its national interests

Following the publication of the *Gorgona Manifesto*, there were no more meetings of the organizers or participants. Nevertheless, many of those who took part in the syntegration continued to be involved with SINA and it does seem that the event and the *Manifesto* made an important contribution to the debate about organizational development going on within SINA at that time and into the future. For example, in the "environmental strategic plan" for the period of the 1998–2002 Colombian Government, the section on the "institutional development strategy" of SINA shows a precise alignment with the conclusions on "institutionalization" set out in the *Manifesto*. Three of the main action lines – "organizing actions by region and theme," "strategies for promoting participation and responsibility" and "developing monitoring and control systems" – echo the sentiments and even employ many of the same words used in the *Manifesto*. At a broader level, an evaluation report on SINA's performance during the same period of government, produced by the National Council of Socioeconomic Development (COMPES), highlights results that relate closely to the guidelines for action established at the Gorgona syntegration.

17.4 Critique of Team Syntegrity

In Allenna Leonard's words:

> The impetus behind the development of syntegration was to provide a structure for holding purposeful conversations which would be non-hierarchical and democratic but would be contained and not dissipate their energy or insights.
>
> *(1996, p. 408)*

The Malik Management consultancy believes it has shown that TS can deliver this and, at the same time, generate maximum participation and maximum consensus for the best solutions (2018). TS succeeds in making use of all the knowledge spread throughout an organization and optimizing it through integrating and enhancing the outputs that can be obtained from the fast pace and effectiveness of small teams.

Malik Management insists that TS enables an organization to take advantage of the cumulative knowledge of highly qualified persons from all its different units by engaging them in a participative search for good solutions. Others who have worked with TS (Pearson 1994; White 1994, 1998) attest to the fact that participants feel the process to be open, self-organizing, and nonhierarchical. There is also evidence that participants experience the power of "reverberation." In the London syntegrity exercise, this led to considerable cross-referencing between the 12 FSIs. Real progress seems to have been made in getting a diverse set of individuals to connect their goals and aspirations in a synergistic way. White concludes:

> The implications of such an approach are that it can help develop wider involvement in decision-making, improve the quality of decisions, build consensus around controversial (local) issues and encourage participation more generally.
>
> *(1998, p. 16)*

Based on the structure of the icosahedron, TS supports participative decision-making by ensuring that no individual or group dominates the conversations. Furthermore, it guarantees that the views of each participant are taken seriously and are "heard," through reverberation, even in groups that they do not belong to. The fact that opinions become detached in this process, from the individuals promoting them, helps to dissolve the force of hierarchical relationships. The approach is, therefore, an effective vehicle for achieving democratic dialogue. It can promote fairness in decision-making by neutralizing many of the baneful influences that power can have on discussion.

There is evidence from Malik Management's many applications to speak for the excellence of the results TS produces faced with complex problems. Apparently, they have "proven successful without exception." Even allowing for the hyberbole inevitable from a management consultancy firm, the repeat business must say something. Malik Management points to the "optimal, holistic, innovative, and super-fast" solutions syntegrations deliver. TS is said to be capable of addressing several levels of the problem simultaneously. It produces optimal solutions and creative strategies for implementation. It works on the level of organizational culture to generate maximum consensus around the solutions, energy, and perseverance to ensure the agreed changes happen. Pfiffner's (2004) experience tells him that the "implementation quota" from TS amounts to about 80% after 12 months. Only a few days are needed to get these excellent outcomes and so the approach more than pays for itself. Other, more independent, observers are also positive about the effectiveness of TS. Holmberg has sought to assess the methodology in an experiment with university students. He concludes that it makes the best possible use of participants' joint capacity, yields synergy, satisfies the participants, is easily modified to suit different circumstances, and is "straightforward to grasp, learn, and apply" (1997). Martin-Cruz et al. (2014) provide another assessment based on student groups undertaking a business simulation exercise with some employing TS and others not. They note a positive impact on teamwork, with groups using TS remaining more stable and involved. Those with experience of TS also performed better. TS sets out a design for democratic communication that maximizes the constructive engagement of participants and achieves the greatest possible learning from the resources brought to the event. Using the geometry of the icosahedron, as a guide to organizing the conversations, produces robust and effective dialogue because it allows a balance to be struck between integrating the knowledge and experience of the participants about an area of concern and creating a healthy tension from multiple viewpoints. The protocol ensures that information circulates freely, that people are exposed to the views of others and so can experience learning, and that new synergistic knowledge emerges because of the reverberation of the ideas produced around the 12 teams. The end result of a syntegration should be the agreement and commitment of the Infoset to a final document, structured around 12 issues and representing the best of the participants' understanding and experience brought to bear on the opening question. As we saw with the Gorgona syntegration, the event can produce a climate in favor of change that resonates through future action. It also creates a coterie of critical, active learners.

Critics find somewhat "far-fetched" the cybernetic logic that apparently connects icosahedra, geodesic domes, and participative decision processes. To them the constraints imposed by TS, which require 30 people to take five days out of their working lives to discuss exactly 12 topics, seem artificial. Efforts have been made, as was seen, to increase flexibility by offering alternative protocols – although these too reflect geometric designs and their effectiveness has not yet been fully investigated. Furthermore, Pearson (1994) argues that the process depends heavily on the participants "wanting to play" and being at the same level in terms of grasping meaning. White (1994, 1998) notes the lack of specific mechanisms to involve the silent or inarticulate and the danger that discussions can degenerate into "networking sessions" rather than being topic focused. Finally, critics argue, it may be true that participants gain a greater understanding of the complexities surrounding the opening question, but there is currently a lack of thorough research on how frequently this is translated into action. Good intentions delivered in the context of a democratically organized Infoset can soon dissolve when they encounter power relationships and hierarchy in the real world. Whatever the apparent success of a syntegration, and even if it is followed by FACE planning, it cannot guarantee that anything will actually get implemented in practice. Leonard (1996) puts it like this:

> The attitude of the sponsoring organization is important too. If the ethos of the organization is autocratic, it may not be able to utilize the results of an open outcome process. If a group inside an organization has neither decision making nor advisory power, holding a syntegration may lead to frustration or cynicism.
>
> *(1996, p. 411)*

In terms of a second-order critique based upon social theory, it is clear that TS steers clear of functionalism. It does not seek out procedures for managing and controlling the behavior of complex adaptive systems. That is, after all, the role assigned by Beer to his other major cybernetic contribution – the VSM. He designed TS as a complementary approach to the VSM and so it does not try to do the same job. The TS methodology actually has much more in common with soft systems approaches, such as SAST, IP, and SSM, which are based on interpretive thinking. It seeks to encourage the expression of different viewpoints and sees shared social expectations as emerging from the various viewpoints being negotiated and renegotiated in the process of social interaction. As Mejia and Espinosa (2007) point out, however, it differs from these methodologies by focusing only on the structure of the conversational processes and not at all on their content. It makes no attempt to increase the richness and rigor of the discussions by clarifying and redescribing different *Weltanshauungen* (Ws) using systems concepts and techniques. By contrast, soft systems methodologies deploy systems methods such as unearthing the assumptions underpinning different worldviews, encouraging pursuit of an "idealized design," or exploring the implications of different Ws by capturing them in "root definitions" and "conceptual models." These are designed to enhance mutual understanding and to bring to the fore matters that might not have otherwise been considered. For this reason, soft systems thinkers will see TS as offering an impoverished attempt to support the interpretive paradigm in comparison with their own efforts. It may prepare the ground for open discussion but contributes nothing to enhancing the content of the debate by employing systems models and methods to clarify and augment exactly what is being said and proposed.

However, an orientation that makes TS weak in supporting the interpretive paradigm, in the eyes of soft systems thinkers, can make it stronger when viewed through the eyes of others. From a radical change perspective, the ground rules it establishes for organizing discussion in the Infoset go a long way to ensuring that the debate is democratic and participatory. This is an issue that soft systems thinkers, with their emphasis on the content of the discussions, largely ignore. They give little thought to the kind of elaborate protocol established by TS to ensure that every voice is heard and no one can dominate the outcome. Mejia and Espinosa (2007) praise TS for the attempts it makes to overcome barriers to genuine communication and participation. It does much to ensure that power relations cannot impact the results achieved by guaranteeing everyone an equal say. The fact that ideas reverberate around the whole group means that they are more likely to be taken seriously whoever they emanated from. That said there remain doubts about just how thoroughly TS addresses the concerns of the radical change paradigm (Jackson 2003; Espinosa and Harnden 2007; Mejia and Espinosa 2007). A lot of power remains with the "owners" of the syntegration who are likely to decide on the opening statement and that, of course, will put boundaries around what can be discussed. They are also going to have the final say on who is invited to become a member of the Infoset. Furthermore, it has been argued that TS assumes that participants will have the same capacity to enter into dialogue and to grasp meaning. We have already noted White's concern about the lack of specific mechanisms to involve the silent or inarticulate. Mejia and Espinosa (2007, p. 34) suggest that some of these issues can be overcome by "adding or modifying particular elements in the Team Syntegrity protocol." Care should be taken to maximize the variety of backgrounds and the ideological orientations of those involved. The roles of "critic" and "facilitator" could be further specified to include representing the perspectives of any stakeholders who are not present at the event. They can also be more active in helping disadvantaged participants to get their points across.

A more fundamental line of criticism (Mejia 2001; Mejia and Espinosa 2007) builds on the argument that TS can ensure democratic interaction but it cannot impact on the actual content of the conversations that take place. It has to rely on the content brought to the sessions by the participants or developed as part of the learning process that occurs during the conversations. This content-less type of approach cannot guarantee that radical change concerns are addressed because it depends on the existing level of understanding and awareness of the participants. From a radical change perspective, they will enter the dialogue conditioned in their thinking by the "dominant ideologies" that act to maintain the positions of the powerful in the unequal society of which they are part. In short, they will suffer from "false-consciousness" and be unable to recognize their own true interests. As Mejia and Espinosa note, if participants enter the debate under the sway of ideologies that have been imposed on them in the past, then all a syntegration as currently envisaged can achieve, under the guise of democratic dialogue, is to further legitimize and reinforce those ideologies. The end result is that participants leave a TS event no better equipped and perhaps even less disposed to challenge the status quo. There may be action after a syntegration but it is unlikely to be of the radical change variety. Basing his thinking on Paolo Freire's "critical pedagogy," Mejia (2001; Mejia and Espinosa 2007) argues the need for "critical" facilitators in TS to do something about this. Freire, with the knowledge provided by Marxism and critical theory, believes that he has an understanding of the social, political, and economic conditions in society that give rise to exploitation and a "false perception" among the oppressed about the social world and the possibilities for its transformation. Dialogue in critical pedagogy is conducted between those aware of oppression and its causes and those who are actually oppressed. Through this process, the latter become aware of structures of domination and the need for political engagement. They become demystified and see social reality as it actually is. This is called "conscientization" by Freire. It is a kind of pedagogical politics of conversion.

> ... in which objects of history constitute themselves as active subjects of history ready to make a fundamental difference in the quality of the lives they individually and collectively live.
>
> *(McLaren and Leonard 1993, p. viii)*

Freire's approach is therefore, unlike TS, content-full rather than content-empty. It brings to the discussion a particular expert theorization of the nature of current social reality and the role of ideologies and is convinced of the kind of solutions that are needed to social problems. Its purpose is to get those with which it engages to throw off false consciousness and guide them to a similar understanding of the radical change required. They should then be committed to doing something about it. It goes without saying that this is far more than Beer intended for TS and way beyond what any sponsors of Malik Management SuperSyntegrations would be happy with. Nevertheless, Mejia and Espinosa believe that more critical forms of facilitation could go a long way to meeting this objection to TS by promoting "critically active learning" that can bring benefits to all stakeholders:

> ... a *critical* facilitator would have the responsibility for becoming aware of exclusions in the discussions within the teams. They could then try to compensate in some way for those exclusions, whenever possible, by means of either a direct inclusion of new ideas about unquestioned pre-suppositions, unexamined dimensions, and so on, or by means of the asking of questions that can help

participants do it …. A critical facilitation would also be alert to the presence of subtler mechanisms of exclusion that could appear despite the protocol's intention to prevent it.

(Mejia and Espinosa 2007, p. 34)

Another line of criticism emanates from postmodernist and poststructuralist thinking. TS appears to assume that people are willing to enter into discussion and debate because they want to reach an informed consensus. An alternative view is that dialogue is an arena for struggle and dissension in which speakers seek to defeat their opponents, often for the simple pleasure of the game. From this perspective, it is simply a different form of power struggle and whatever issues forth from "democratic debate" represents just another claim to power. Beer is relying on the participants wanting to play the language game that he prefers. Postmodernists believe that we live in a world of multiple truths that give rise to incommensurable interpretations of reality and think that we should promote difference rather than seek to subsume it in the quest for agreement. Espinosa and Harnden (2007) comment that TS does provide the opportunity for participants to express different, even conflictual, views and that it is not its aim to achieve a once-and-for-all consensus. An agreement to disagree is always a possible outcome.

In positioning TS on the system of systems methodologies (SOSM), we can take advantage of Beer's own classification of methodologies according to the "systems" to which they correspond, as described in Section 8.3.2. He sees operational research (OR) (we would add system dynamics [SD]) as appropriate for "complex probabilistic systems," which give rise to involved behavior but are describable. "Exceedingly complex, probabilistic systems," such as a brain, company or economy, can never be described in a way that can yield predictions about their behavior. They are the province of cybernetics, specifically of course his VSM. At the time of this classification, Beer was only interested in the type of "systems" complexity that constitutes the vertical dimension of the SOSM. With the invention of TS, however, he recognizes the need to give separate attention to the distinctive emergent properties of human and social systems that produce the complexity highlighted by the "stakeholders" dimension of the grid. It now seems worthwhile to him, at least for analytical purposes, to give separate attention to "systems" and "stakeholder" complexity. This is exactly what the "integrationist" social theorist, Archer, would recommend. TS is Beer's attempt to come to terms with the pluralism and coercion that can arise in social systems and we must place it on the SOSM accordingly.

Leonard says of TS that:

It is a particularly appropriate process to use when groups are characterized by high levels of diversity – either because they come from different countries … or because they come from different political, cultural or disciplinary backgrounds.

(1996, p. 407)

There is truth in this if she means that it offers a protocol that ensures that diverse viewpoints, if they are present, all get an airing in a democratic and participative debate. TS must, therefore, be allowed to occupy some part of the pluralist terrain of the grid. However, as we have seen, this is limited compared to soft systems approaches, because it offers no guidance on how the debate can be enhanced by using systems ideas to

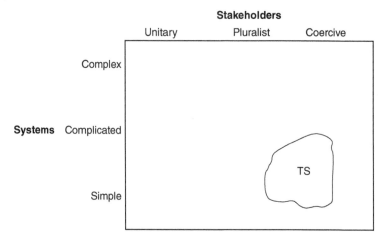

Figure 17.2 The positioning of team syntegrity on the SOSM.

express and interrogate different viewpoints. It stays clear of the content of the discussions. On the other hand, TS pays attention to issues of coercion, arising as we move along the horizontal dimension of the SOSM, that are largely ignored by soft systems thinkers. The protocol empowers participants by giving them the opportunity to take part in dialogue on an equal and fair basis. It, therefore, spreads into "coercive" territory that is foreign to soft systems thinking. Unfortunately, this spread is limited, from a radical change perspective, for the reasons we have seen. However, good TS is at ensuring full participation in the context of the syntegration itself, it cannot address the power relations that determine who takes part, how participants can become competent to contribute in a way that favors their own true interests, and what happens when the conclusions from the process come to implemented, or otherwise, in a society divided on a multitude of class, status, gender, and other inequalities. The positioning of TS is, therefore, as in Figure 17.2.

17.5 Comments

While engaged on Project Cybersyn, Beer was presented with his own personal mandala by a Buddhist monk from a mystical mission based in Arica. His yogic practice had already brought him into touch with such spiritual symbols which seek to capture, through balancing visual elements, the unity and harmony present in the universe. His background in "British cybernetics" (see Section 6.3) would also have predisposed him to taking such devices seriously. Mandalas do not try to convey knowledge by representing anything, but rather act as performative aids to meditation. They operate on an emotional and psychological level and offer a more direct route to enlightenment than can be achieved through ratiocination. Toward the top of his mandala, Beer immediately noticed an "enneagram" – a geometric figure of nine interconnected points. The enneagram is an ancient symbol that originated in the early days of the Sufi, Hindu, or Christian religions, depending on who you read. Beer knew that it had entered the Western mystical tradition by way of the works of Gurdjieff and Ouspensky. Today, the enneagram remains important in some people's eyes as a spiritual guide to higher states of consciousness. The nine points are seen as related to archetypical ways of being human. They can be seen as nine ways in which the human soul can be distorted by the ego and so offer nine mirrors for self-reflection on how we are flawed. Meditating on different versions of the symbol, and the connections between the points, can help us break the tyranny of the ego and reach ever higher levels of consciousness on the road to attaining enlightenment. In the Christian tradition, it can suggest what we must do to become more like Christ. Inevitably, it has been commandeered by consultants as a model of nine interconnected personality types that managers need to be aware of. Tests are available that allow you to learn your "enneagram type" in less than five minutes.

Beer became fascinated with the enneagram as he continued his meditations with his personal mandala. He had long harbored the ambition of giving a rendition of the Requiem Mass in paintings. The enneagram now provided a clue about how this could be done. If an enneagram was marked on the floor of a room, paintings could be mounted at each point of the star. The viewer would be guided around the exhibit of nine paintings according to the mathematical logic of the enneagram. They might then feel a mystical "reverberation" as the pictures, and the connections between them, impacted upon their consciousness. The work, with paintings 5 ft × 3.5 ft in size, was shown in 1992 and 1993 in the Catholic Cathedral in Liverpool.

Meanwhile, back at the Syntegrity project, Joe Truss suggested to Beer that the enneagram was deeply embedded in the structure of the icosahedron. It was not long before they produced the evidence to show that it was "ubiquitously present" in that geometrical form. Truss declared to Beer, in a mood of high excitement:

> Do you see what this means? The icosahedron is the actual *origin* of the enneagram, and the ancients knew it. Could it not be possible that the plane figure was coded esoteric knowledge?
>
> *(Beer 1994, p. 206)*

Beer's faith in the icosahedron as the basis for organizing effective participative debate was greatly reinforced. He already knew that, employed in TS, it could give rise, through the reverberations produced, to a kind of "group consciousness" in which the participants gained a powerful awareness of each other and the identity of the group as a whole. Given that the nine points of the enneagram reflect different manifestations of the human psyche, he now felt able to construct a stronger hypothesis:

> It is that there is such an entity as a *group* mind that operates at a different level from group consciousness as defined. It would operate, if it exists, as a self-conscious entity …. It is at least possible that the multiple enneagrammatic structure, reverberating as it does, provides a complex of linkages to constitute …[a] 'corporate brain' that would then give rise to its own consciousness …. If it exists it should make the infoset more competent than its component member consciousness, or even the group consciousness defined.
>
> *(Beer 1994, pp. 207–208)*

Those who want to continue their exploration of the scale of Beer's thinking should refer to two collections of his works (Harnden and Leonard 1994; Beer 2009). Anyone wishing to follow him on his spiritual journey can consult Blohm et al. (1986) and the *Chronicles of Wizard Prang* (Beer 1989c).

17.6 The Value of Team Syntegrity to Managers

TS's value to managers can be summarized in the following five points:

- It offers a "democratic" systems approach that emphasizes the benefits of the equal and participative involvement of stakeholders in defining and clarifying social and organizational concerns and seeking their resolution. TS promotes fairness
- TS produces robust and effective dialogue though a protocol that takes advantage of the tensions induced by multiple viewpoints, integrating these to strengthen the agreements reached among participants about an area of interest to them
- The end result of a syntegration, a record of the agreed responses to 12 topics, will reflect the very best of the participant's shared understanding and experience of the opening question, enhanced by their involvement in highly interactive and participative group processes
- Participants in a syntegration emerge with a shared understanding of a matter of concern and interest to them and with a commitment to ways of addressing it. These will serve later to help with implementation of particular actions in the real-world
- *Participants* in a syntegration gain an understanding and respect for participative processes and have their learning capacities enhanced

17.7 Conclusion

Beer's development of TS, and the associated methods and concepts, has added considerably to the armory of applied systems thinkers. It is essential, however, to recognize what a syntegration can and cannot achieve. The main purpose of a syntegration is to provide a proper context for developing democratic and insightful agreements about complex policy or strategic topics. It cannot guarantee that after the event the agreements reached will be respected or effectively implemented. A syntegration in isolation cannot ensure that long-term emancipatory changes are realized. Beer originally designed TS as a complementary approach to be employed alongside the VSM, and it is this combination that offers the best possibilities for profound organizational self-transformation. Concurrently with the Fürth SuperSyntegration, described in the Prologue, another Malik Management team was working with ideas from the VSM to ensure that the outcomes were holistically and sustainably implemented:

> This team developed solutions during the process for future management and regulatory systems of the city administration as well as support systems for an effective implementation of the many measures. In terms of end products, this team delivered interconnected steering and management models for a city that functions reliably.
>
> *(Malik Management 2018)*

18

Critical Systems Heuristics

Likewise, the meaning of critique threatens to be lost if we do not link it properly to the emancipatory interest, *that is, if we do not give the latter an adequate methodological status and place in CST [critical systems thinking] .... The question thus is, What is the proper role of the emancipatory interest in CST?*

(Ulrich 2003, p. 328)

18.1 Prologue

The main inspiration for Werner Ulrich's *critical systems heuristics* (CSH), as for SAST (see Chapter 14), is C. West Churchman's systems thinking. Churchman (1974) tells a story that illuminates some of the central concerns that Ulrich's approach is designed to address. My paraphrase of the story changes the names of the characters involved and simplifies their debate, but hopefully leaves the meaning intact.

There are three characters who find themselves together in the lounge of a Boeing 747. The plane was flying to New York but has been hijacked and is now bound for Cuba. One is a busy and successful executive. The second is a systems expert. The third is a professor of philosophy. They start a conversation primarily because the systems expert would like to convince the executive that the systems approach has value. A fourth character watches on while getting drunk. He utters the occasional expletive but otherwise remains silent.

The systems expert is explaining to the executive that social problems have to be seen as resulting from many different factors. The way to tackle such problems is not to concentrate on the individual parts but to consider the way they are interlinked. The executive responds that we should not forget that people are also involved and that successful managers know that their main role is leadership of those people. At this point, the philosopher feels obliged to intervene and ask an ontological question – how does anyone know a problem actually exists? The executive gets heated and states that he knows damn well that he has a problem. He was on his way to New York to try to settle a strike by garbage men. You could actually smell the problem on the streets. The way to tackle the problem is by surgery. You need to find the source of the problem and remove it as quickly as you can. The systems expert replies that any decent surgeon would have to consider the patient's general condition, blood pressure, heart, etc.,

Critical Systems Thinking and the Management of Complexity, First Edition. Michael C. Jackson.
© 2019 John Wiley & Sons Ltd. Published 2019 by John Wiley & Sons Ltd.

before operating. So it should be with the sanitation sector, which is inextricably linked to other sectors in the City. For example, an attempt to solve the dispute by giving way on wages could lead to others, such as policemen, questioning their remuneration and set off escalating wage demands as positive feedback loops took effect across the City. The philosopher retorts that they seem to be looking at problems in different ways. The executive searches for situations where he can use decisive leadership as a solution. The systems expert regards problems as resulting from interacting factors which he can model and then provide an answer. But what they both have in common is that they each conceptualize problems as something that they feel they can do something about and suggest the type of solution that pleases each most. He then backtracks a little and says he respects both their positions and is only really asking for a bit of humility. He doesn't want to come over as a relativist but asks them to accept that, just as they have different viewpoints, so there will be many other perspectives on any problem. For example, a black person from a ghetto in the City will likely possess a very different account of social reality, but it is just as refined, educated, and justified as theirs. There follows another argument between the executive and the systems expert about the relative merits of leadership, black or white, and the problems that leadership, in the absence of a full understanding of the context, can bring about. The executive insists that modeling can play an important role but that ultimately it is strong leadership that gets things done. The systems expert sees a role for leadership but only after systems models have revealed where it can best be applied.

Meanwhile, the philosopher has been thinking and wondering, as a good Hegelian would, whether any synthesis is possible between the different positions: "I think I see a new perspective of the systems approach," he says, "What both of you have been struggling to do is subsume the other into your own worldview." The executive, he argues, wants to make modeling an essential part of the total leadership function, so that he can "swallow" the systems approach into his own larger worldview. The systems expert wants to make leadership an essential element in the total system, so that he can "swallow" that into *his* larger worldview. Furthermore, a ghetto leader might see both leadership and the systems approach as subtle forms of racism and "swallow" both into *his* larger worldview. This is not easy to resolve, the philosopher goes on, because whatever evidence is produced will inevitably be interpreted according to whichever worldview is favored. No data can ever destroy a *Weltanschauung*. "So this may be the synthesis," he concludes, "the systems approach means enabling every man to appreciate as fully as possible his own view of social reality by listening seriously to other views."

At this point, the plane banks sharply and the pilot announces that he has been ordered by the hijackers to head for New York as originally scheduled. "Why?" cry the three protagonists in unison. The fourth man, now very drunk, explains that he is one of the hijackers and that it is he who has given the order. He explains that they had been hoping to capture some first rate talent from the United States but after hearing the gibberish they were talking he was now convinced they would do more harm than good. Revolution is the only thing that can liberate oppressed people he declares. And revolution will win while you lot are all busy debating worldviews. As the hijacker disappears a stewardess walks past. She tells them not to worry as the man was very drunk, but goes on to insist that he did have a point: "After all, none of you men ever once included a woman in your so elegant and comprehensive systems."

18.2 Description of Critical Systems Heuristics

18.2.1 Historical Development

Werner Ulrich had already earned a doctorate in Economics and Social Science, from the University of Fribourg (Switzerland), when he applied to join Churchman's PhD program at Berkeley in 1975. He had become fascinated with Churchman's thinking and the five years he spent with him did not disappoint. As well as becoming thoroughly grounded in Churchman's systems approach and the philosophy of Kant, he became familiar with the critical social theory of Jurgen Habermas, who he met at Berkeley in 1980. Habermas' discourse ethics became another significant source of inspiration. On returning to Switzerland, Ulrich continued his academic work at the University of Fribourg. He has also been a visiting professor at the Universities of Hull and Lincoln and the Open University, in the United Kingdom, and at the University of Canterbury in New Zealand. At the same time, he has been thoroughly engaged in professional practice as founder and head of the office of evaluation research within the Department of Public Health and Social Welfare in the Canton of Bern (see Ramage and Shipp 2009, pp. 159–160). His experience has, therefore, been gained in government rather than the private sector and his specific interests are in planning, policy analysis, evaluation, and social research into issues of health, poverty, citizenship, and civil society. Ulrich's thinking has, inevitably, also been influenced by spending most of his working life in Switzerland – a country of "citizens," cantons, and direct democracy. Now retired, he continues to be active in systems thinking as an *Ancien Professeur Titulaire*, at the University of Fribourg, and in many other capacities.

Ulrich's major work *Critical Heuristics of Social Planning*, published in 1983, was almost immediately recognized as a landmark in the development of systems thinking (Jackson 1985a). The book provided a methodology, "critical systems heuristics" (CSH), which can be used by designers and concerned citizens alike to reveal and challenge the normative content of actual and proposed systems designs. By normative content, Ulrich means both the underlying value assumptions that inevitably enter into planning and the social consequences for those at the receiving end. What was most radical was his insistence that the views of those affected by systems designs, but not involved in the decision-making, had to be taken into account before any design could be regarded as rational. CSH was hailed as establishing "emancipatory systems thinking" because of this and because it provided the means by which the affected but not involved could make their voices heard.

The chapter began with a quotation from Ulrich in which he embraces the "emancipatory interest" but asks for clarity about what this means. In his own work, he is clear that he sees it as a "methodological issue" (Ulrich 2003, p. 332). Following Habermas, he argues that extending rationality to the sphere of social interaction, where we debate "what we ought to do," requires that we secure the "possibility of discourse":

> The 'emancipatory interest' in this sense is without regard for persons; its only advocacy is in favour of a situation of undistorted communication in which all concerned parties have as equal a chance of articulating their concerns as possible.
> *(Ulrich 2003, p. 332)*

The emancipatory interest becomes a methodological necessity for CSH because Ulrich accepts Habermas' position that only an ethical foundation for discourse can secure the grounds for rationality in the practical domain. What it is not, for him, is a sweeping ideological commitment to some class or group seen as being in need of "emancipation." Indeed, Ulrich argues, such a predefined commitment would only serve to close down the surfacing of different ethical positions, and the debate about these, that CSH seeks to promote. The intent behind CSH's emancipatory interest is, therefore, critique. It offers a "discursive framework for promoting critical (reflective and emancipatory) practice" (Ulrich 2003, p. 327). We can obtain a fuller understanding of what Ulrich means by this if we consider the three ways in which CSH can be used.

The first purpose of CSH is to provide a basis for *reflective practice*. In Ulrich's view, planners are inevitably selective in the values they bring to a piece of work and what they take into account. This can lead to partiality in terms of who benefits from, and who is disadvantaged by, a particular system design. Sound professional practice, therefore, needs to be clear about the selectivity and partiality it is operating with in specific situations. CSH enables planners and decision-makers to critically reflect upon their own sources of bias and to make these explicit to others. This thinking applies, of course, to users of systems methodologies. Systems thinkers can use CSH to interrogate the designs emanating from their interventions to reveal the assumptions made and whose interests are being served in a particular case. A long section of *Critical Heuristics of Social Planning* is taken up with displaying the "sources of deception" inherent in Beer's work in Chile. Another example, provided by Ulrich and Reynolds (2010), sees CSH employed to evaluate existing practices of participatory rural appraisal used in a project in Botswana.

The second purpose to which CSH can be put is *dialogical*. CSH can help improve mutual understanding between those who take different perspectives on a situation. Often disagreements arise because people have different values and different opinions on what is relevant and "frame" situations differently. They end up "talking past" each other with very little idea of how their appreciations truly diverge. CSH is able to examine multiple perspectives systemically, assisting people to understand the nature of their differences more thoroughly. They become able to question their own assumptions and, perhaps, more willing to revise them. They are certainly in a position to deal more constructively with their differences if they wish to do so. Venter and Goede (2017) have successfully applied CSH to surface conflicting stakeholder perspectives on a new business intelligence system in the technology division of a large South African-based petro-chemical organization. The deeper understanding gained by the stakeholders allowed them to resolve their differences, easing implementation of the new system and ensuring it yielded the business benefits expected. More generally, Midgley and Pinzon (2011) have sought to demonstrate how an extended version of CSH, used in this way to increase mutual understanding, can inform mediation practice and help with conflict prevention.

The third purpose is to strengthen the position of stakeholders who are affected by a systems design but not involved in decision-making about it. These stakeholders are the ones most likely to be disadvantaged. Ulrich (2005) calls this third way of employing CSH "*emancipatory boundary critique*." He insists that there are no experts when it comes to issues of "what ought to be done" and that any rational debate about these matters must allow the voices of the affected to be heard. CSH is an approach that can be used to help ordinary people, and by ordinary people themselves, to refine their own perspectives in such a way that they can challenge the more powerful to reveal the undisclosed assumptions underpinning their proposals. It puts all the different perspectives on an equal footing by demonstrating that the views of experts on matters of practical reason are no more "objective" than those of other stakeholders. Ulrich's (1983) example relating to health systems planning in the Central Puget Sound, and Ulrich and Reynold's (2010) account of using CSH in the ECOSENSUS-Guyana study, suggest how marginalized stakeholders can be empowered to become more involved in the issues facing them. Donaires (2006) uses CSH to study stakeholders who are disadvantaged in software development projects. In this case, he identifies software developers as a group who are excluded from effective involvement. If the boundary was drawn wider to take into account their perspective, during the development process, they would feel they had more freedom and better performance might result. At the least some of the more interesting work, such as developing new technologies and tools, might be assigned to this group to ensure greater fairness. Algraini and McIntyre-Mills (2018) employ CSH to challenge the concept of human development embodied in policy-makers plans for the Saudi education system. In their practice, if not in the vision they propound, the policy-makers uphold the narrow instrumental view that the purpose of education is to promote economic growth. A more critically systemic approach, taking account of the perspectives of teachers and learners, and involving them in shaping the curriculum, would give equal attention to education as a means of developing the life-long learning skills, which would enable students to fulfill their human potential. Emancipatory boundary critique comes closest to the usual understanding of "emancipation" as being about liberating oppressed groups and offers something genuinely new to the systems approach. It chimes with Sartre's radical invocation that we ask how every situation looks from "the eyes of the least favored" and take their side (see Bakewell 2017). Because of these things, it has inevitably attracted most attention from commentators.

Since the publication of *Critical Heuristics of Social Planning*, Ulrich has continued to be very active in clarifying and developing his research agenda. A theoretical research program, on "Critical Pragmatism," concerns itself with further refining his philosophy of professional ethics by combining CSH with a critically revised version of pragmatism. We consider this in the next section. A related research program, "CST (critical systems thinking) for Professionals and Citizens" seeks to contribute to the revival of civil society by simplifying the key principles of CSH, particularly the idea of boundary critique, so that it can be used by ordinary citizens to help them participate fully in decisions over matters of public concern (e.g. Ulrich 1998). Recent contributions to these programs, and regular updates on progress, can be found on Ulrich's website (2018).

Before leaving this account of the historical development of CSH, we take the opportunity to briefly review the work of three other prominent systems thinkers who have incorporated many of Ulrich's conclusions in their thinking. Although not reliant solely on CSH, they all share a concern with boundary judgments that "marginalize" certain stakeholders and disadvantage them as a result.

Gerald Midgley, working at the Centre for Systems Studies at the University of Hull, draws on the work of Churchman, Ulrich, and the Hull school of CST (see Part IV), to formulate his approach to "systemic intervention." Midgley (2000) argues that conflict between groups arises when they hold different views about the facts and values that are relevant to the situation they find themselves in. He follows Ulrich in seeing the perspectives people take as governed by the "boundary judgments" they make and argues that the tendency to unreflectively accept existing boundary judgments, thereby potentially disadvantaging some stakeholders, should be addressed at the beginning of a systems study. Midgley's "systemic intervention" is considered further in Section 20.3.

Janet McIntyre-Mills also draws on insights from Churchman, Ulrich, and the Hull tradition to pursue her approach to "critical systemic praxis." She insists that the "governance" of public policy issues needs to be more systemic, taking into account all relevant historical, economic, political, sociocultural and environmental variables. The best way to do this is to ensure open and flexible dialogue among multiple stakeholders. As a result, it becomes possible to "sweep in" multiple meanings, "unfold" them using CSH, and reveal the full range of issues relevant for problem-solving. A variety of praxis tools can then be employed. Particular attention has to be given to ensuring that the narratives of disadvantaged groups are taken into account and that these groups are able to participate in finding solutions. Her study of indigenous people in Alice Springs leaves us with a powerful image of what can happen when a nonsystemic approach to social problems predominates. The technocratic "solution" to the issue of alcohol abuse translates, for the indigenous person, into "life on the machine," i.e. relying on dialysis for survival (McIntyre-Mills 2003). In work with others (Riswanda et al. 2016, 2017), she has argued for "systemic policy practice" to address issues associated with prostitution in Indonesia. This involves focusing on factors that contribute to disadvantage, listening to the voices of those with lived experience of poverty and prostitution, and ensuring their rights and dignity are respected in framing policy. As was noted above, she was engaged in a study in Saudi Arabia using CSH in a pure form to display the viewpoints of teachers and learners "marginalized" by the instrumental view of education embedded in the action of policy-makers (Algraini and McIntyre-Mills 2018).

Anne Stephens (2013) work, on how cultural ecofeminism and the critical systems approach can benefit from each other, is starting to get the recognition it deserves. She believes that ecofeminism can learn, from Churchman and Ulrich, the importance of boundary judgments and how they can be critiqued. In the other direction, Stephens notes that little account has been taken of gender-specific concerns and ecological issues in CST. In a recent paper (Stephens et al. 2019), she addresses specifically the need to provide "ecological justice for nature." CST has had an anthropological bias that has led it to marginalize nonhuman entities. We will look further at these issues, with the aid of Stephens thinking, in Section 21.2.2.

Although it is important in what follows to separate out the particular perspectives of Churchman, Ulrich, and the Hull school, it is encouraging to see ideas from all these versions of the critical systems approach used together to good effect by these authors.

18.2.2 Philosophy and Theory

Ulrich (1983) is careful to distance his systems thinking from the dominant use of the systems idea in what he calls "systems science," in which he includes OR, systems analysis, systems engineering, and cybernetics. Systems science, he argues, is based on restrictive mechanistic and organismic analogies – Beer's work in Chile being an example of the latter. In systems science, the systems idea is condemned to be used in the context of instrumental reason to help us decide *how to do things*. It refers to a set of variables to be controlled. Ulrich's purpose is to develop the systems idea as part of practical reason, to help us decide *what we ought to do*.

He calls his approach "critical systems heuristics" (CSH), interpreting each of these words according to the meaning given to them by Kant. To be *critical* means reflecting on the presuppositions that flow into both the search for knowledge and the pursuit of rational action. A critical approach to systems design means that planners must make transparent to themselves and others the normative content of their designs so that they can be subject to inspection and debate. They should not be presented as the only objective possibility. The *systems* idea in Kant refers to the totality of elements – ethical, political, ideological, and metaphysical – on which theoretical or practical judgments depend. In trying to grasp the whole system we inevitably fall short and produce limited accounts based on particular assumptions. Systems thinkers should accept and respond to this challenge. *Heuristics* describes a process of exploration and discovery whereby current assumptions about issues and how they can be addressed are kept under constant surveillance. Systems design should be conducted in a way that exposes objectivist deceptions and helps planners and concerned participants to unfold the inevitable partiality of their approaches through critical reflection.

Ulrich constructs the philosophy and theory of CSH upon his reading of two great traditions of thought. The first is European "critical philosophy" as represented, in very different ways, by the work of Kant, Popper, and Habermas. The second is American "philosophical pragmatism," primarily as mediated through the thinking of C. West Churchman. The label he applies to his "philosophy for professionals" is "critical pragmatism," reflecting both sources (Ulrich 2007). I seek to summarize what Ulrich takes from these two rich traditions.

For Popper critical thinking is applied to the logic underpinning theoretical reason to provide rules that we can use to test scientific hypotheses. The proper application of such thinking is in instrumental reason, which makes use of the results of science to determine how to do things. As far as social systems design is concerned, this means that critical reason can only help us with technical questions such as the most efficient means to achieve predetermined ends. Rational discussion about ends, and even about the value content of means, is apparently not possible. The central question of practical reason – "What ought we to do?" – is placed by Popper beyond the scope of critical reflection. It is left to "decision" and enacted without rational guidance. This same attitude pervades systems science. The goals it serves go unexamined as it puts all its efforts into finding scientific means for achieving prescribed ends. Ulrich, by contrast, wishes to make the question of "what ought we to do" subject to critical reflection.

Habermas's work is more useful to Ulrich's enterprise because it seeks to extend the application of critical thinking beyond instrumental reason. Practical reason and emancipatory reason are, as we saw in Section 4.4, equally important to Habermas and all three forms should be subjected to critical reflection. In order that questions arising in the domain of practical reason can be decided appropriately a process of rational argumentation must be established. All those affected by a decision in the social realm must be allowed to participate in the discussions that lead to the decision. The debate must be so arranged that all ideological and institutional constraints are eliminated and the force of the better argument persists. Through an analysis of the structure of actual speech situations, Habermas is able to determine what an "ideal speech situation" of this kind, free from systematic distortion, should be like. This underpins his theory of "communicative competence." Ulrich is willing to follow Habermas in grounding rationality in the social domain on discourse ethics and is convinced that his work provides a useful theoretical boundary experiment. However, he is concerned about its practical usefulness. In order to enter Habermas' debate it appears that speakers must already be able and willing to exhibit communicative competence. This presupposes the very rationality the discourse is designed to ensure. In attempting to ground critical reflection theoretically, Habermas cuts himself off from the real world in which personal and group interests exist, are powerful, and inevitably contaminate debate. Far better, Ulrich argues, to ground critical reflection heuristically by providing a methodology that enables practical judgments to be constantly reviewed and their partiality revealed by holding them up for review in the light of ordinary, everyday accounts of the nature of social experience. Ulrich, therefore, accepts Habermas' discursive approach but, in his search for a more practical form of critique, rejects the demands implied by his consensus-orientation and the ideal speech situation:

> For this reason, I do not equate a discursive approach with a consensus theory of truth and of normative validity but rather with a *discourse theory of critique*: while rationally defendable consensus is bound to remain an ideal, inter-subjectively compelling forms of critique are achievable.
>
> *(Ulrich 2003, p. 326)*

Where better for Ulrich to look for a practical, rather than theoretical, approach to the problem of grounding practical reason than in the pragmatism espoused by his mentor, Churchman. Churchman (see Section 14.1) had long struggled with the apparent need for social systems designers to understand the whole system in order to justify their interventions. To some critics this makes Churchman's work appear hopelessly idealistic and impractical. All that Churchman is doing, however, is pointing out the fate of all the applied sciences, which have no option but to live with the prospect that localized action, based on partial understanding, can lead to unexpected consequences in terms of whole system improvement. As Ulrich (1983) argues, Churchman's critics are blaming the messenger for the bad news. Furthermore, Churchman made good progress in thinking of a way through the conundrum. What he saw was that the theoretical indispensability of comprehensive systems design could be used as an ideal standard to force us to recognize the inevitable lack of comprehensiveness of our actual designs. This links to another of his greatest contributions to systems thinking. It was Churchman who established and developed the idea that our inability to operate comprehensively inevitably leads us to draw boundaries which are selective ("every worldview is terribly restricted") and which determine what aspects of the overall system we pay attention to. This notion, originally derived from Kant, is then employed by him to demonstrate that these boundaries are crucial in determining how systems designers define improvement and what action they recommend. He argues that, if this is the case, any rational justification for systems designs must require continually redrawing the boundaries to "sweep in" the *Weltanschauungen* (Ws) of any stakeholders excluded from consideration. In his story, retold in Section 18.1, the possible Ws of poor blacks, revolutionaries and women have to be brought to the attention of the executive and the systems expert. Ulrich is content to follow Churchman in seeing Kant's systems idea as an admonition to reflect critically on the inevitable lack of comprehensiveness and partiality of all systems designs. However he sees Churchman's solution, based on "sweeping in" all possible Ws, as a "heroic quest for comprehensive knowledge and understanding" which is simply unattainable (Ulrich 2003, p. 326).

For Ulrich (2003, 2007) the ideal forms of rationality championed by Habermas and Churchman are of little use in professional practice. Instead, we must make a "critical turn" in systems thinking and develop an inter-subjectively compelling form of critique which relies instead on making the very lack of comprehensiveness of our designs transparent and, with others, allows us to reflect critically on their limitations. Ulrich again draws on Kant and Churchman for clues as to how the selective judgments we make in planning and systems design can be subject to such critique. He argues that certain presuppositions, which he terms "boundary judgments," inevitably enter into any social systems design. These reflect the designer's "whole systems judgments" about what is relevant to the design task. They also represent "justification break-offs" since they reveal the scope of responsibility accepted by the designers in putting forward their proposals to other stakeholders. Thus boundary judgments provide an access point to the normative implications of systems designs. The task is to find an appropriate way of interrogating existing and proposed designs to reveal the boundary judgments that they make. Summarizing, Ulrich puts it like this:

> The strategy of CSH for dealing with the problem of practical reason thus consists in what it calls the *critical turn* of our notion of rational practice – practice is rational to the extent it is aware of its inbuilt selectivity and partiality and qualifies its claims accordingly. This is how CSH aims to support the quest for rational practice despite its unavoidable selectivity and partiality.
>
> *(2012a, p. 1237)*

18.2.3 Methodology

The philosophy and theory behind CSH requires a methodology that concentrates on supporting "systematic processes of boundary critique":

> Accordingly, CSH's methodological core principle is the *principle of boundary critique*: what a claim means and how valid it is depends on its reference system, that is, the boundary judgements that inform its view of relevant fact and values and thus its empirical and normative selectivity.
>
> *(Ulrich 2012a, p. 1237)*

Boundary critique, according to Ulrich (2005), faces five tasks:

- Identifying the sources of selectivity in a reference system by surfacing the boundary judgments it makes
- Examining these boundary judgments in terms of their practical and ethical implications
- Finding options to the reference system by considering alternative answers to some of the boundary questions – thus highlighting its selectivity

- Seeking some mutual understanding between stakeholders adhering to different reference systems to increase mutual tolerance and, perhaps, achieve a shared perspective
- Challenging the claims of stakeholders who take their own boundary judgments for granted or try to impose them on others

The detail of this can be understood if we start with the key concept of "reference system." This is the term that Ulrich uses for the assumptions, in terms of boundary judgments, that individuals use when they view a situation in the "real-world." It is akin to Boulding's "image," Vickers' "appreciative system" and Churchman's *Weltanschauung*. Reference systems are, of course, selective in what they highlight and what they hide, and they lead to particular "claims" about how problem situations should be seen and what should be done to improve them. Ulrich (2005) asserts that the "methodological core principle" of CSH involves thinking through the relationships between boundary judgments, facts, and values – the three elements that determine the nature of reference systems. He presents these three elements as making up an "eternal triangle," one at each point, and calls the process of examining the relationships between them "systemic triangulation." The elements mutually condition one another and a change in any one can lead to a significant reconfiguration of a reference system. For example, if new facts become available which we regard as significant they will broaden our perspective on a problem situation and lead to new values coming into play when we assess it and suggest improvements. Similarly, if someone or something pricks our moral conscience about an issue we will start to redraw our boundary judgments and new facts will become relevant. I am thinking, while writing this, of refugees coming to Europe and how different facts, values and boundary judgments interact to produce alternative reference systems for viewing the situation and managing it. Undertaking systemic triangulation on a continuous basis is, for Ulrich, a mark of good professional practice.

The next step is to find a way of reviewing the "merit" of reference systems. Kant, as we saw in Section 1.2, faced a similar problem when he sought to justify scientific knowledge about the physical world. He was particularly concerned about certain concepts which were deeply implicated in the production of knowledge but were little understood and difficult to justify. He proceeded by demonstrating the theoretical necessity of those concepts for thought and knowledge about the world – particularly the concepts of space and time and the 12 "categories." This is fine for Newtonian natural science, Ulrich (1983) argues, but social systems designers inevitably also come up against human intentionality (self-consciousness, self-reflexivity, and self-determination) as well as space, time and the categories. He reasons, therefore, that if we wish to understand and improve social systems, we must add an additional dimension of "purposefulness" and design social systems to become purposeful systems. Here, Ulrich again takes inspiration from Churchman who, himself following Kant's lead, had produced an "anatomy of system teleology" that "suggested an elementary list of nine conditions that must be fulfilled for a system to be purposeful" (Ulrich 1988a, p. 421). Churchman later extended this list to 12; the three new conditions pointing to the need to reflect on the lack of comprehensiveness of the systems approach. Each condition is designated by a systems category. These categories, in the four groups into which they were divided, were: client, purpose, measure of performance; decision-maker, environment, components; planner, implementation, guarantor; systems philosopher, enemies of the systems approach, significance (see Section 14.2.2 and Figure 14.1). In Ulrich's words:

> Churchman's idea is to employ the categories for tracing the different interpretations and valuations to which one and the same set of data about the problem situation lends itself, depending on the observer's world views and needs.
>
> *(1988a, p. 422)*

The way forward was now clear to Ulrich. Churchman had suggested 12 categories which underpinned the nature of purposeful systems. How these were viewed by different stakeholders would reveal their reference systems and the boundary judgments they were employing. By making them explicit it was possible to critically reflect on existing and planned social system designs. Different reference systems could be reviewed in a process through which the boundary judgments implicit in them were systematically "unfolded."

Ulrich's list of 12 boundary categories is a modified version of Churchman's. The main change is in the last three categories – replacing Churchman's "systems philosopher," "enemies of the systems approach" and "significance" with, respectively, "witness," "emancipation" and "worldview." The rationale is that the changed categories can be used to reveal the selectivity of particular designs rather than just point to the lack of comprehensiveness of the systems approach as a whole. A different understanding of "unfolding" is at work. The task is not the never-ending one of "sweeping in" the perspectives of more and more stakeholders but the more practical one of critically reflecting on existing and proposed reference systems. Ulrich's categories are shown in Figure 18.1. As with Churchman, they are divided into four groups. Three of these

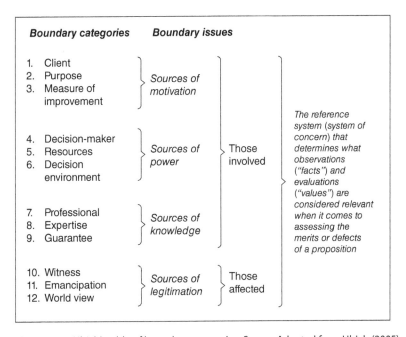

Figure 18.1 Ulrich's table of boundary categories. *Source:* Adapted from Ulrich (2005).

represent roles involved in any planning decision – client, decision-maker, professional. The boundary issues relevant to these three groups are, respectively, the system's "sources of motivation," "sources of power" and "sources of knowledge." The fourth, the role of witness, attests to the "sources of legitimation" that are employed. Together they express the different sources of influence the four groups can bring to bear in a purposeful system.

The 12 categories lead to 12 "boundary questions" which, Ulrich argues, need to be asked to reveal the boundary judgments underpinning a reference system. Three questions are asked of each of the four groups – client, decision-maker, professional, and witness. The first question is about their "social roles"; the second refers to "role-specific concerns"; and the third to "key problems" surrounding the determination of boundary judgments with respect to that group. The power of the 12 questions to reveal the normative content of systems designs is best seen if they are asked in an "is" mode and an "ought" mode, and the answers contrasted. For example, we could compare the answer to the question "who is the intended beneficiary of the system (S)?" with answers to the question "who ought to be the intended beneficiary of the system (S)?" Asking the questions in the "is" mode is called, by Ulrich, "actual mapping." It reveals what facts and values are considered by a reference system and which are left out. Asking them in the "ought" mode is called "ideal mapping" and reveals what facts and values ought to be considered relevant and which should be left out. Table 18.1 presents another version of the categories, and the sources of influence of the different groups, with the 12 boundary questions, asked in the "ought" and "is" mode, clearly shown.

Ulrich now has a methodology in place that can fulfill the five tasks of "boundary critique" that he identified and which, therefore, enables CSH to realize the three purposes set out in the historical development section of this chapter. This is achieved by using the 12 boundary questions in different combinations of actual and ideal mapping. A professional planner committed to *reflective practice* will begin by asking the questions in the "is" mode and then seek to expose the selectivity of their proposals by conducting experiments asking them in the "ought" mode. In a multiagency situation, the stakeholders might decide to use the methodology in a *dialogical* manner to develop a shared understanding of the situation they are confronting. To do this, they would each engage in ideal mapping and compare the results. If a group wants to engage in *emancipatory boundary critique*, it will answer the 12 questions in the "ought" mode, according to its own reference system, and use the results to show that the proposals of those undertaking a systems design are also based on boundary judgments, which they may be failing to make explicit. This last case will now be examined in more detail.

For Ulrich, no systems design can be regarded as rational unless it reveals its own normative content. This, however, is not the only criterion of rationality, and so it is not enough that the involved, making use of the boundary questions, be self-reflective about the partiality of their designs. The pragmatic criterion insists that any systems design must also be acceptable to those affected by its real-world consequences. It must be subject to a dialogue with the witnesses – in practice, representatives of those affected but not involved. The main obstacles in the way of this would seem to be the willingness of the involved to participate in the process and the lack of expertise of the affected. However, Ulrich argues, CSH helps to ensure that these are not insurmountable difficulties. All designs are based on boundary judgments incorporating justification break-offs and these are beyond the reach of expertise to justify. Anyone who understands the

Table 18.1 The boundary categories and questions of critical systems heuristics.

Sources of influence	Boundary judgments informing a system of interest (S)			
	Social roles (stakeholders)	Specific concerns (stakes)	Key problems (stakeholding issues)	
Sources of motivation	1) *Beneficiary* Who ought to be/is the intended beneficiary of the system (S)?	2) *Purpose* What ought to be/is the purpose of S?	3) *Measure of improvement* What ought to be/is S's measure of success	The involved
Sources of control	4) *Decision-maker* Who ought to be/is in control of the conditions of success of S?	5) *Resources* What conditions of success ought to be/are under the control of S?	6) *Decision environment* What conditions of success ought to be/are outside the control of the decision-maker?	
Sources of knowledge	7) *Expert* Who ought to be/is providing relevant knowledge and skills for S?	8) *Expertise* What ought to be/are relevant new knowledge and skills for S?	9) *Guarantor* What ought to be/are regarded as assurances of successful implementation?	
Sources of legitimacy	10) *Witness* Who ought to be/ is representing the interests of those negatively affected by but not involved with S?	11) *Emancipation* What ought to be/are the opportunities for the interests of those negatively affected to have expression and freedom from the worldview of S?	12) *Worldview* What space ought to be/is available for reconciling differing worldviews regarding S among those involved and affected?	The affected

Source: From Ulrich and Reynolds (2010). Reproduced with the permission of SNCSC.

concept of boundary judgments knows that planners who justify their proposals on the basis of expertise or "objective facts" are in fact employing boundary judgments, even if unreflectively or cynically. If they can be made to debate their boundary judgments, they are put in a position where they are no better off than ordinary affected citizens. It becomes a matter of trading value judgments about what assumptions should influence plans and what consequences are desirable or otherwise. In order to put recalcitrant planners into a position where they have to admit their boundary judgments, Ulrich advocates a method called the "polemical employment of boundary judgments." Affected citizens use different boundary judgments, reflecting alternative reference systems, purely with critical intent against the planners. This is quite good enough to shift the burden of proof onto the planners because it demonstrates:

- That the planners' proposals are governed by their boundary judgments

- That the knowledge and expertise of experts is insufficient to justify their boundary judgments or to falsify those of others
- That planners or experts who seek to justify designs on the basis of knowledge or expertise are, in fact, employing boundary judgments dogmatically or cynically, and so disqualify themselves from rational debate

The technique brings the systems rationality of the planners directly into contact with the "social rationality" of those who have to live in and experience the social systems designs. It should help secure a position in the dialogue for any ordinary citizens who care to employ it.

Finally, in this section on methodology, we should note Ulrich and Reynold's (2010, p. 288) suggestion that boundary critique works best when it functions in the background as a "reflexive framework" to inform our practice, rather than upfront to dominate it. Otherwise, there is a danger that the language will put people off, and CSH will appear to be using exactly the kind of expertise that it is at pains to show cannot resolve practical issues. This suggestion echoes Checkland's point (see Section 16.2.3) that experienced users of his methodology naturally employ it in a "Mode 2" rather than "Mode 1" style. A good example of a Mode 2 use of CSH, in this instance with emancipatory intent, is Buckle Henning and Thomas' (2006) employment of the methodology to consider the efforts, by the Project Management Institute, to develop the Project Management Body of Knowledge (PMBOK). They do not explicitly use the 12 questions, but the polemical employment of boundary judgments is always there in the background as they seek to show that PMBOK is severely constricted because it privileges masculine forms of cognition and action, which emphasize goal-seeking and performativity. Drawing the boundary so narrowly excludes feminine sense-making based more on relationship maintaining and improvisation. Setting the feminine perspective against the masculine enables Buckle Henning and Thomas to "emancipate" some very significant but neglected aspects of project management. The masculine view of project management is analytical and impersonal. It emphasizes mastery over the environment; advocates following a direct route to achieve predetermined goals regardless of the peculiarities of the situation; and uses objective criteria to measure success. The feminine perspective brings to the fore connectedness with others and is highly interpersonal. It emphasizes uncertainty about the end-state, multiple project realities, sharing power, participative decision-making and empathy. In the view of Buckle Henning and Thomas, successful project managers need to work with multiple reference systems.

18.2.4 Methods

The primary method developed to serve CSH is the 12 boundary questions. This is so intrinsic to the methodology that it was appropriate to cover it in the previous section. Nevertheless it is worth mentioning two additional areas in which work has taken place to specify how these questions should be used. The first concerns the order in which they are asked. While insisting that the questions are so interlinked that it is possible to start with any of them, Ulrich and Reynolds (2010, pp. 258–259) have set out a "standard sequence," which they suggest beginners use when unfolding the questions. This order is essentially as in Table 18.1. Their experience also tells them that it is often best to start by asking the questions in the "ought" mode since this motivates participants to think about their "hopes and visions" for the type of change they seek to bring about. The second area of work has been around the somewhat technical language in which the questions are phrased. As Ulrich and Reynolds (2010, pp. 275–276) demonstrate, it is relatively easy to adapt the questions according to the levels of literacy, knowledge of English, and familiarity with CSH of those participating. They describe four templates for applying boundary critique, corresponding to increased competence with the approach, which can be used to gradually train users. From a focus on "ideal mapping," participants are led to an appreciation of CSH's role in "evaluation," then "reframing," and finally "challenge." Possible questions, reflecting these levels of understanding, are as follows:

- Where do we want to go from here?
- How satisfied are we with the state of affairs?
- How else can we frame the picture?
- How can we rationally claim this is right?

18.3 Critical Systems Heuristics in Action

This example follows the use of CSH in a study of police strategy toward the carrying of offensive weapons. The background and an initiative meant to address the perceived problem are explained. The 12 boundary questions are then employed to look at possible interpretations of the initiative. In other words, to help reveal different reference systems through which it could be viewed. The contribution of Chief Superintendent T. Brydges to the study should be acknowledged. The project is described more fully in Flood and Jackson (1991b) from which this account is derived.

The study was centered on the Lambeth Borough in London which, for policing purposes, comes under the auspices of the Metropolitan Police Service. Since the approach proposed for tackling the issue of carrying offensive weapons was "multiagency," it is best to begin by describing the key stakeholders and, briefly, the various viewpoints on display. Twelve stakeholder groups were recognized – the Home Office, the police (both operational officers and community liaison officers), trade organizations, shops selling offensive weapons, the media, the Crown Prosecution Service (CPS), The Lambeth Consultative Group, the "Why Helpless Youngsters?" (WHY) group, carriers and assailants, youths, schools, and Magistrates. The Home Office is responsible for the administration of law and order and also for the rights of individuals. This can create problems, especially from a policing perspective. For example, the 1953 Prevention of Crime Act and the 1984 Police and Criminal Evidence Act (PACE) are often seen as inadequate for tackling crime. One particular gray area is the definition of "reasonable suspicion" to stop and search. The Association of Chief Police Officers was calling for various changes in legislation to combat crimes associated with carrying offensive weapons – as examples, lowering the standard of proof required for reasonable suspicion and updating of the offensive weapons act to include survivalist type devices. They also expressed the opinion that the courts were lenient, readily accepting that accused offenders were seeking to defend themselves. The Chief Superintendent of Kennington, in Lambeth Borough, was a key person from the policing angle. At the time Lambeth had a high crime rate and suffered from a high incidence of people carrying knives. Directives from the Chief Superintendent sought to encourage police officers to identify possible carriers and to confiscate their weapons without the need to arrest or charge. Operational police officers wanted to oblige but felt constrained because PACE seemed to equate a reasonable suspicion to stop and search with that necessary to arrest. From their point of view, there were three distinct reasons for the apparent increase in knife crime – availability and low cost, ease of concealment, and the ineffective law. When asked to describe typical knife offenders they saw them as male, between 14 and 30, unemployed, poorly educated and housed, and as having previous convictions. Another factor identified was race. Police officers recognized that there were a number of acute social factors that needed to be tackled to get to the root of the problem. Consequently, a consultation had been initiated and this suggested that they interact more closely with the community and gain access to children from an early age. The aim would be to promote honesty and instill trust in the police. Police community liaison officers argued that education that focused on traditional values was the key to a long-term solution. They suggested that information packs detailing the laws on offensive weapons should be made widely available. This approach would need support from the judiciary taking a harder line.

Shops selling weapons are often accused of being more commercially oriented than socially responsible, and there was a good deal of evidence of them selling to persons under the age of 18. There were no clear laws, however, barring the sale of many such weapons. The Home Office had drawn up voluntary guidelines, but these were considered as being not particularly influential. Working out effective legislation is difficult because certain vendors, such as iron-mongers, do need to sell knives. Another issue was the extent to which toy weapons mould aggressive attitudes in young adults. The South East Co-operative Society's decision to remove toys that glamorized violence from all its stores had met with a mixed response from other stakeholders. The British Toy Manufacturers Association claimed that it was an over-reaction. Other representatives of manufacturers felt that a blanket ban was unjustified. A Chief Inspector in the police, on the other hand, welcomed the move and Slough Council, in London, expressed their delight. The media are influential in forming attitudes but do so on the basis that they need a "good story" to generate sales. The CPS complained that the police too often provided insufficient evidence in support of accusations of "intent to use" offensive weapons. The Lambeth Consultative Group reckoned that there were five main issues relevant to the problem: the need for new legislation, a review of police "stop and search" powers, banning sales to juveniles, controlling magazine advertising, and waging an education campaign in consultation with the local police. "WHY" was a London pressure group campaigning against young people carrying offensive weapons. It was promoting three solutions: a relaxation of the reasonable suspicion criteria so that the police could react to public demands for tougher action; education programs beginning at the age of 8 or 9; and legislation to restrict sales. The views of carriers and assailants were sought but this did pose some obvious difficulties. The closest contact possible was via youth clubs, Afro-Caribbean clubs and youth custody centers. Various reasons for carrying knives emerged. There was a fear of being attacked, against which a knife carried some protection. There was a status attached to knife carrying. And knives could be used to extract money from "rich people" – a kind of wealth redistribution argument. During the study a request made to the Inner London Education Authority to hold discussions with children in schools was rejected. The following opinions were obtained from school children by less direct means. They offered three reasons for the frequency with which knives were carried – fashion, status, and ready availability. Schools were attempting to deal with the problem of knife carrying on their premises by confiscating weapons that were discovered, punishing and suspending offenders, and informing their parents. The School Curriculum Development Committee was considering promoting relevant rules in schools with the hope of engendering more social responsibility. Magistrates tended to feel that they had a balanced view of the issues. Most were drawn from the local community and thought they understood the social context of crime and the implications of their sentencing.

At the time of the study, the Metropolitan Police Service was becoming increasingly interested in adopting a "multiagency" approach to many of the problems it faced and the carrying of offensive weapons was thought to offer a suitable context for trying out the idea. Given the complexity of the situation just described, it is not surprising that the police in Lambeth were keen to experiment. Their particular description of such a multiagency approach was that it should be

> … designed to pursue an objective of identifying and locating key social agencies, with the aim of designing and implementing better community social crime prevention organization through decentralized community services, based on the principles of self-help and support neighbourhood watch schemes, geographically aligned to Ward boundaries, to be administered by the local council, for the purpose of providing a local ('ground-floor') multiagency forum within a consultative and accountably controlled and monitored hierarchical structure in respect of local crime and community affairs.

In this example, therefore, the system (S) we are concerned with is that part of police strategy in Lambeth Borough based on introducing a multiagency approach toward the carrying of offensive weapons. We concentrate on employing the 12 boundary questions of CSH is the "is" mode in order to unfold the reference system underpinning the proposal. We then deal, in less detail, with the "ought" mode from the perspective of an alternative client with different purposes and concerns.

1) *Who is the actual client of S's design?*
   The proposed design is a means of ensuring compliance with the law of the land. The direct beneficiaries are assumed to be those in Lambeth Borough who suffer when the laws governing offensive weapons are broken. The police will benefit if crime rates fall because their reputation will improve.
2) *What is the actual purpose of S's design?*
   The existing strategy was extremely frustrating for the police. It was difficult for them to ensure compliance with the law on carrying offensive weapons because of the constraints imposed by other legislation. In particular, reasonable suspicion to stop and search was equated with that needed to arrest. The multiagency design could potentially bring pressure to bear to change the law. The main purpose of S, however, was to address the issue in a different manner. It would seek to gain support from all the relevant stakeholders to take whatever action was appropriate to improving the problem situation.

3) *What is the design's built-in measure of success?*

   The traditional measure was standard crime statistics. However, it was accepted that these were inaccurate because of the way they were collected and because they left unreported incidents out of account. They also neglected important measures such as the level of satisfaction among the general public regarding how the law was enforced. The multiagency approach would seek to take into account a wider range of indicators – for example the number of crime prevention initiatives started and publicized.

4) *Who is actually the decision-taker, i.e. can change S's purpose and measure of success?*

   The key decision-taker is the police force in Lambeth Borough, which is taking forward the multiagency approach. If the approach takes off, then all the stakeholders previously mentioned – schools, trade organizations, pressure groups, the CPS, the media, etc. – will become involved as decision-takers. For example, the media could decide to change the way it reports police activity and the outcomes of that activity. When it comes to changing the measures of performance of S, to fit better with a multiagency approach, then the wider police force and the Home Office would need to become engaged.

5) *What conditions of successful planning and implementation of S are really controlled by the decision taker?*

   Although the multiagency approach suggests that the meetings of stakeholders will be administered by the local council, the police force will likely need to organize and run them. They will, therefore, have considerable control over who attends the meetings, the issues addressed, and the amount of time spent discussing different issues.

6) *What conditions are not controlled by the decision-taker, i.e. what represents the environment to him?*

   The currently favorable attitude of the Metropolitan Police Service to multiagency working could shift. The Home Office and wider police force may resist changing the measures of performance. The different stakeholders that S hopes to engage all have other priorities, and it may be difficult to gain and sustain their commitment. Additionally, many factors impacting this type of crime – unemployment, bad housing, unstable family lives – will be in the environment of the decision-takers.

7) *Who is actually involved as planner of S?*

   The initial organization of the scheme and the planning for meetings will fall to the police force. That said, the intention is to increase the role of other stakeholders in taking decisions about multiagency working as things progress.

8) *Who is involved as "expert" in S's design and what is the nature of their expertise?*

   The point of the multiagency approach was to involve as many stakeholders as possible in helping to tackle different aspects of the problem. They will all have a relevant contribution to make in terms of their own form of expertise, e.g. representatives from the schools can advise on the kinds of educational program most likely to succeed in bringing about a reduction in the carrying of offensive weapons.

9) *Where do the involved see the guarantee that their planning will be successful?*

   Traditionally, the police have been seen as the experts in reducing crime as measured by traditional crime statistics. With the multiagency approach, the guarantee is extended because it draws on a wider range of expertise and seeks a broader societal consensus on what constitutes improvement.

10) *Who will act as witnesses of S, representing the affected but not involved?*
    Ideally, in a multiagency situation, all those affected by S will be involved, at least in so far as they are represented by a relevant agency. It is likely to be difficult, however, to secure the involvement of offenders and potential offenders in multiagency discussions.

11) *Are the affected given an opportunity to emancipate themselves from the assumptions of the involved and take their fate into their own hands?*
    One group of the affected – the carriers and users of offensive weapons – seem to be given little chance to have their worldview taken into account. If a police mentality, constrained by traditional measures and statistics, remains dominant, then this group will continue to be seen as a means to an end. The overall thrust of the multiagency approach will still be control in pursuit of improving performance indicators.

12) *On what worldview of either the involved or affected is the design of S actually based?*
    Despite the good intentions of the designers, it is hard to see the worldview underlying S as other than that of the police force. The Lambeth police are subject to constraints imposed by the wider force and the Home Office. They have to respond to external measures of performance and their own concerns and responsibilities with respect to these are likely to determine who they involve in meetings and what is discussed.

The answers to the 12 questions in the "is" mode demonstrate the significant change in police strategy that would occur if the multiagency approach was adopted. However, responses to some of the questions, especially those about the "witnesses," suggest that S may still fall short of offering a fully multiagency approach because it represents the reference system of the involved, especially that of the police force, much more than that of the affected but not involved. The questions have revealed something about the boundary judgments, facts, and values feeding into the design of S. This "boundary critique" can now be deepened if we change the answers to some of the questions by asking them in the "ought" mode from an alternative perspective. Many such perspectives are possible but the worldview of potential offenders serves this purpose particularly well. Taking this line, the potential offenders become the client, and the purpose of S will be to remove the need that they may feel to offend. This means that S must address the mythology associated with carrying knives, the widespread fear this group has about being attacked, their particular vision of society and, where possible, the social conditions that lead them to conclude that carrying offensive weapons is a good means of improving their position in society. The decision takers of S will need to include representatives of the potential offenders. Their expertise, especially with regard to the reasons why people carry offensive weapons, should be recognized and made use of in the design of S. It is hard to argue with the proposition that taking account of the reference system of potential offenders early in the design of S would have led to significant improvements. As it stands, from the point of view of potential offenders, it is easy to portray S as simply removing their right to self-protection in a society that is massively biased against them on class and/or racial grounds. Of course, a design of S based on the reference system of potential offenders will appear bounded by the wrong facts and values by witnesses such as those who have been attacked or live in fear of being attacked. But further boundary critique from their perspectives could only further enrich reflective debate.

18.4   Critique of Critical Systems Heuristics

Ulrich's CSH provides a coherent philosophical foundation for the purposeful systems approach and a methodology and methods that enable it to be used to increase rational decision-making in society. In order for plans for social systems designs to be deemed rational they must be made transparent with regard to the "boundary judgments" they express; which facts and values they focus on and which they exclude from consideration. In order to ensure that a comprehensive "boundary critique" takes place, plans must be examined and debated with other stakeholders viewing them from alternative reference systems. A debate can then take place, which is considerably more informed about points of agreement and disagreement. Particular attention needs to be given to ensuring representation of those affected but not involved in the planning process. Groups that do feel excluded can call systems designers to account using the "polemical employment of boundary judgments" and so secure for themselves a position in the debate.

A great strength of CSH, according to Ulrich, is that it offers a way of improving rational decision-making that does not depend on achieving Habermas' "ideal speech situation" or engaging in Churchman's endless process of "sweeping in" new stakeholder perspectives. Communicative rationality is enhanced if we make clear to ourselves and others the selective nature of our perspectives in such a way that they can be systematically analyzed and their implications discussed. Consensus may emerge but is not necessary:

> It is already an essential gain in communicative rationality to understand people's different rationalities. We thus at least have a chance to agree that we do not agree and why this is so – and how we might handle the situation in a decent way. Mutual recognition, tolerance and co-operation can grow on this basis.
>
> *(Ulrich 2012b, p. 1318)*

Another strength claimed by Ulrich for CSH is that it can be used by all concerned citizens to help them get their voice heard. Boundary critique is not unnatural to us. Everyone practices it every day, although unconsciously – for example when we make decisions about working harder or trying to achieve a better work/life balance. And it is easy to hear boundary judgments at play when we listen to arguments on the TV. The protagonists talk past one another because they operate with different reference systems. One may privilege jobs and economic growth while the other puts forward arguments based on the protection of the environment and the quality of life of future generations. CSH builds on this primitive understanding of the importance of "boundaries" by enabling us to systematically explore the nature of boundary issues and to reflect on them ourselves and in discussion with others. CSH also assists citizens to use boundary critique to challenge those who claim to know the answer to problems because they possess some special expertise. Once we recognize that all plans are underpinned by particular boundary judgments, it becomes clear that it is impossible to sustain a claim to "objectivity" in seeking to improve social systems. There is no right or wrong answer. All plans for systems designs must be subject to legitimate processes of decision-making involving the relevant stakeholders. CSH ensures that these processes can take place in a more meaningful way, and that the perspectives of those affected but not involved are brought to the table and assume the same status as those of "experts."

During his professional career, Ulrich did most of his work in multiagency settings. It is not by chance, therefore, that CSH seems most at home in these environments. There is further evidence of this in the fact that all four of Midgley's (2000) case studies, using his version of boundary critique, were undertaken in multiagency settings as part of his "community operational research" practice. Why CSH is valuable in the multiagency context is, I think, evident from what has been said and the examples provided. It can help give clarity to the viewpoints of the different stakeholders, gain their informed consent and commitment to agreed changes, and take account of the wishes of those who are affected but may have limited opportunity otherwise to have their perspectives respected. There are question marks about whether CSH can be applied as easily in more hierarchical settings. That said, managers in all spheres of activity are increasingly being asked to justify the performance of their organizations on matters such as equality of treatment for people of different genders, races and sexual orientations, provision for the disabled, equal opportunities for people of all social classes, and protecting the environment. Perhaps if they incorporated CSH into their repertoires, they could be more certain that they were acting responsibly by avoiding discrimination of all kinds, and their organizations would benefit as a result.

We will achieve a fuller understanding of the strengths and weaknesses of CSH if we now undertake a second-order critique based upon social theory. Ulrich (1983), as was seen, makes clear his dissatisfaction with "systems science" for subordinating the systems approach to "instrumental reason." In this guise, it can only help us answer questions about *how to do things*. His own ambition is to employ systems thinking to provide a way forward in the face of the acute problem of practical reason – how to decide *what we ought to do*. In social theory terms, he sees the challenge as follows:

> From the 1970s it gradually became clear that if systems methodologies were to be able to deal systematically with issues related to people's differing world views, values, and interests, they would require some grounding in the 'interpretive' (hermeneutic) paradigm of the social sciences and the humanities, possibly also in additional traditions of thought such as critical social theory, language analysis, ethics, and philosophical pragmatism.
>
> *(Ulrich 2012a, p. 1233)*

Checkland's "soft systems methodology" (SSM) and Ulrich's CSH were responses to this challenge. Both adhere to a vision of social reality as constructed and reconstructed through the actions and interactions of social actors operating with multiple world-views, values, and interests. Both facilitate participative and methodologically well-structured explorations of multiple perspectives with a view to finding out what agreements about change are possible, at a particular point in time, while continuing to encourage a process of never ending learning. They can both be seen as systemic aids to "decision-making *based on* systematic problem structuring" (Ulrich 2012a, p. 1235). Nevertheless, Ulrich also sees important differences between SSM and CSH. Checkland recognized that hard systems thinking could not cope with ill-defined managerial and organizational problems and sought to develop a methodology that could. This required a shift from problem definition and optimization to problem exploration and learning. To this end, he found theoretical inspiration in the work of Vickers, Weber, Dilthey, Husserl, and Schutz. Checkland's purpose was well served by what can broadly be called interpretive social theory. For Ulrich (2012a, p. 1235), however, more is needed to fulfill his ambition of "dealing with the normative core of 'good' professional practice" than simply shifting from problem-solving to problem-structuring. The outcomes of any problem structuring exercise will themselves have value implications which need reflecting upon to ensure they are fully understood and do not exclude any relevant stakeholders from consideration. Otherwise, methodologies such as SSM risk embracing only "a managerialist notion of good practice." An approach is required that is capable of revealing and dealing critically with the "normative content and consequences" of professional interventions. In establishing CSH for this purpose, he found that he had to take his theoretical lead from Kant's critical philosophy, Habermas' discourse theory, and Churchman's philosophical pragmatism.

Having set out Ulrich's position, we can now relate CSH to the different sociological paradigms identified in Chapter 4. *In Critical Heuristics of Social Planning* (1983), Ulrich was concerned to distance CSH from any form of functionalist systems thinking. He saw the methodologies of hard systems thinking, system dynamics, and cybernetics as resting on mechanistic and organismic analogies, which were dangerous when applied to the social domain. This "systems science" was in thrall to instrumental reason and was predisposed to making false claims for the objectivity of its results. In building his case, it sometimes seemed as if Ulrich was presenting systems science as an enemy to be defeated by CSH. Soon, however, he began to communicate a more nuanced perspective, and it is clear that this was never his intention. Drawing upon Habermas' work on "types of action," he argues in favor of a three-level framework of rational systems practice (Ulrich 1988b). At the first level, called "operational systems management," systems methodologies that assist with the efficient management of scarce resources are seen as vital. At the second level, of "strategic systems management," systems approaches that can foster systemic steering capacities are thought to be indispensable in dealing with increased complexity and uncertainty. At both levels, however, the use of systems science can only be justified if it is embedded in a broader purposeful systems paradigm, which ensures that its limitations and normative implications are laid bare and debated. CSH is an essential critical instrument that must always be employed to ensure that systems science is applied in an appropriate manner. A third level, of "normative systems management," is presented as the unique domain of CSH, where it provides the means to carry out an investigation into the normative implications of a systems design, consider alternatives, and helps structure a debate about boundary judgments always ensuring that the views of the affected are considered.

The role of CSH with regard to systems thinking of a functionalist orientation is, therefore, to provide a critique of the normative implications that enter into the systems designs that it advocates. We should ask how useful such critique is. The answer, from the functionalist perspective, is not particularly useful. Functionalists seek to use methods akin to those employed in the natural sciences to uncover the laws which account for the behavior of systems. This is the case for a hard systems thinker working on the mathematics that can describe the best routing system for a company. It is true for the system dynamicist seeking to establish which factors to include in a model, and how the relationships between them operate, in order to understand how some real-world system works. They do not take well to being told that they need to take different value systems into account when reaching their conclusions. As an example, we can consider Beer's response to Ulrich's critique of his work in Chile. Ulrich

(1981b) criticized aspects of that project because it was based on an organismic rather than purposeful systems approach. Beer replied:

> When Dr. Ulrich was kind enough to send me a draft of his paper, I consumed a lot of audio tape in trying to convince him that in filtering our Chilean work through a Kantian epistemology he had badly misled himself as to what we actually did.
>
> *(2009, p. 113)*

He goes on to state that he has been engaged in a 30-year quest to discover how systems are viable. The outcome of this scientific investigation, the viable system model, is falsifiable in the Popperian sense and has not yet been falsified. As to the normative standards that are built into the model:

> Who or what has done this in the case of the class of viable systems? Clearly, they have set up the norms of viability themselves. The model is the outcome of trying to recognize what this norm actually is.
>
> *(Beer 2009, p. 115)*

As Beer was fond of stating, the law of requisite variety is to management what the law of gravity is to physics. They are not laws that you can quarrel with. In terms of Ulrich's "eternal triangle," for functionalists facts do not change just because boundaries shift and new values are taken into account.

Of course, just because functionalists do not regard Ulrich's form of critique as useful does not mean that others will not find it valuable. What we are forced to recognize, however, is that Ulrich's CSH offers its critique from a particular theoretical perspective based on an amalgam of Kant, Habermas' discourse theory, and Churchman's pragmatism. It will be a constant refrain, in what follows, that this is a limited type of critique. Critiques of social systems designs can be launched from the perspectives of all the social theories discussed in Chapter 4 – functionalist, interpretive, radical change, postmodern, etc. All are valuable. Ulrich's CSH has its place but cannot serve the purposes of critical systems practice as described in Chapter 21.

How does CSH relate to interpretive social theory? Ulrich sees much in common between his approach and other "problem structuring" methodologies that rest on interpretive foundations, such as Checkland's SSM. The reasons become clear when we remember that one of the most important uses of CSH is in a *dialogical* manner to improve mutual understanding between those who take different perspectives on a situation. Here the 12 boundary questions operate, like CATWOE in SSM (see Section 16.2.4), to bring clarity to the assumptions underlying different worldviews and to encourage richer and more meaningful debate. In Ulrich's words, they help promote "mutual recognition, tolerance, and cooperation," and suggest where agreement might be achieved over what changes to make. In comparison with other soft systems approaches, however, Flood and Jackson (1991a) accuse CSH of methodological immaturity. Although Ulrich (1988b) has called for it, there has been no consistent action research program that has driven improvement in CSH methodology and methods akin to the one associated with SSM over many decades. And despite the refinements commented on in Ulrich and Reynolds (2010), problems persist with using the approach.

Many commentators (e.g. Larsen 2011) have commented on the rigidity it imposes and difficulties with interpreting the meaning of the questions. Furthermore, it is hard to see the point of pursuing boundary critique in the "is" mode. The process of "actual mapping" gives the impression that there are "right" answers to the questions about the nature of the system of concern. If the point is simply to demonstrate that this is not the case then it can be better made by contrasting the different reference systems that exist, and exposing the assumptions of each, using the "ought" mode. There are other important points that Midgley (1997a) would like to see addressed. For example, there is the danger that, even in conditions where open debate is possible, CSH can introduce its own forms of coercion because those who are inarticulate, lack confidence or suffer from learning disabilities might be unable to engage effectively in rational argumentation.

CSH, however, cannot just be evaluated as a possible replacement for other soft systems methodologies. It must also be considered as a critical device that adds something important to them. Extending the thinking in an earlier paper (Ulrich 1988b) – and drawing upon some work of Ormerod – Ulrich (2012b) presents a framework in which successful professional intervention is shown to depend on *analytical competence, process competence*, and *context competence*. Analytical competence delivers "smart bits" based on quantitative analysis and modeling. Process competence relies on soft methodologies and methods to supply "helpful ways" of conducting an intervention. Context competence needs CSH to provide an awareness of "things that matter." Good professional practice, therefore, promotes action that is efficacious (it works on the basis of relevant knowledge); is purposeful (it is undertaken in the context of mutual understanding among the participants); and is legitimate (it adequately considers the context by engaging in boundary critique). Ulrich has reached the conclusion that CSH is best used in conjunction with other systems approaches to supply them with legitimacy (see also 2005, p. 5). What it adds is the "contextual sophistication" that can be provided by boundary critique. In order to ensure good professional practice, therefore, the outcomes of problem structuring interventions, just as the results of problem-solving exercises, must be subject to critical reflection and discourse. They will be selective in terms of the facts and values that have been taken into account and this will have consequences for stakeholders, including those affected but not involved. CSH, Ulrich argues, supplies the critique that can prevent methodologies like SSM unquestionably accepting "a managerialist notion of good practice" and provide a justification for their outputs. An important question again surrounds the significance of this type of critique. From the point of view of interpretive systems thinkers the answer is, I think, that it can be helpful but somewhat predictable. This becomes clear if we consider the example of the application of CSH to evaluate participatory planning projects in Botswana, as reported in Ulrich and Reynolds. CSH was chosen to check that the projects were inclusive, holistic and participatory, and to discover whose interests were promoted and whose marginalized (Ulrich and Reynolds 2010, p. 249). The researchers knew exactly what kind of critique they were going to get. Inevitably, the outcomes will also be predictable. In every case, soft systems thinkers will be told that they should try to ensure maximum participation in their projects and to look out for marginalized stakeholders. The detail should be of interest. But, eventually they will turn round and counter that they understand the basic issues involved and that their methodologies are designed to do the best they can in these respects in the less than perfect, often hierarchical contexts in which they have necessarily to operate. Critical systems thinkers of the Hull persuasion, employing a much wider range of theoretical positions from which to launch critique, will again insist that the problem with CSH is that it can supply only a limited type of critique based upon its own restrictive theoretical foundations. To take the same example, they would insist that the projects in Botswana could be usefully evaluated in many other ways – for example in terms of their efficiency, viability, and the structural biases in society that automatically favor the values and interests of some groups over those of others.

It is the relationship between Ulrich's work and the sociology of radical change that has caused most controversy. When CSH emerged it was hailed (Jackson 1985a) as the first "emancipatory" systems approach. It was represented as providing a methodology that sought to address many of the concerns of that paradigm – conflict, marginalization, the need for a fairer society. Since then, there have been numerous critiques of CSH claiming that it is less well suited than originally thought to doing this. Ulrich's own position is relatively clear. Social systems design, to be regarded as legitimate, must take into account the perspectives of the affected but not involved. If it does not do so, then it is proper to engage in "emancipatory boundary critique" to assist those who are excluded. The polemical employment of boundary judgments helps ordinary citizens to resist any claim to power made by planners and associated experts on the basis of the "objectivity" of their proposals. It reveals that their plans depend just as much on making particular boundary judgments as do the counter-proposals that can be brought forward by other stakeholders. This strengthens the position of the disadvantaged because it puts all the different perspectives about "what ought to be done" on an equal footing. It can secure, for all parties, a gain in "symmetry of critical competence" (Ulrich 2005, p. 14). It becomes difficult for more powerful stakeholders to ignore what is being said by those who oppose them. The "radical change" criticisms of CSH can now be assessed.

The first line of criticism paints Ulrich's CSH as just as utopian as Habermas' arguments for an "ideal speech situation." The critics ask: "Why should the involved bother to take account of the views and interests of those who are affected but not involved?" Ulrich observes that an equal position for reasonable dialogue can only be achieved if all parties, including the powerful, renounce the dogmatic or cynical employment of boundary judgments, but he has little to say about how this can be brought about. Willmott highlights the problem:

> Absent is any consideration of how such a renunciation is to be achieved, the conditions which promote it, or the steps which need to be taken to create such conditions.
>
> *(1989, p. 74)*

Midgley, in the course of a review of 11 publications between 1985 and 1996 that offer critiques of CSH, is able to add to this. In his opinion, the successful use of CSH depends on a situation already existing in which participative debate is possible:

> In short, CSH can be of use only when communication is possible, either face to face or via an arbitrator, and only then if all participants are able to handle involvement in rational argumentation.
>
> *(Midgley 1997a, p. 48)*

Since coercion is best defined as the closure of debate, CSH can hardly tackle problem situations in which coercion reigns.

Ulrich (2003) goes to some length to counter this criticism. In his view, the critics fix the argument by referring to a conceptual limiting case in which the possibilities for all discourse are closed down, and by demanding "emancipation here and now." Both these conditions make impossible demands of an emancipatory systems approach. If however we reconnect the notion of critical discourse with the wider system in which it takes place, then the emancipatory potential of CSH becomes much clearer. The wider system which Ulrich sees as coming to his aid is the public sphere:

> The public sphere is that realm of discursive chances which lies between the realms of the entirely private and the state's political administrative system. To the extent that it is accessible to ordinary people, it offers them an arena for unrestricted articulation and discussion of concerns that have been suppressed elsewhere.
>
> *(Ulrich 2003, p. 331)*

The public sphere is not a perfect context for unrestricted discourse. It will be characterized by any number of asymmetries of power, resource, and influence. But it offers a large space, between the two limiting cases of Habermas' "ideal speech situation" and the complete closure of debate, in which CSH can exercise its emancipatory potential. Further, because it is constituted by different discursive spheres, it offers many and varied opportunities for citizens to express their concerns. If they are blocked in one sphere, they can turn to another and so, even if it takes time, they have a realistic chance of getting their concerns voiced. Ulrich (2003, p. 331) offers the example of individual citizens bringing their worries about genetically modified foods into the public arena. Once they become subject to public discussion, such issues are not easily suppressed. CSH, of course, seeks to equip citizens with the critical competence which will improve their prospects of having their voices heard. CSH and civil society, therefore, go hand in hand. CSH requires a well-functioning public sphere and is, at the same time, a powerful tool for building an appropriate public sphere where its emancipatory potential can be realized – even if not "here and now."

Radical change critics will not be convinced that shifting the analysis up a level, to the wider system, helps Ulrich's case. They will just point to the significant inequalities in power, wealth, status, and influence that exist at the societal level and give rise to the problems encountered when CSH is used in more local contexts. In recent times, capitalism, left to run free in the "open society," seems more than ever to be concentrating wealth in the hands of the few. We saw Bourdieu arguing in Section 4.6 that, at the bottom of society, certain groups suffer from multiple disadvantages. If an individual has no access to monetary resources, it is likely that they will be similarly disadvantaged in relation to other forms of capital provided, for example, by social networks and education. Bourdieu sees the structured nature of this inequality as due to the continuing high degree of class-based domination in contemporary society. But similar arguments are made by other sociologists drawing attention to, for example, gender or race inequality. Three case studies allow me to illustrate the points at issue.

In the first, Kathy Luckett (2006) examines the use of CSH in a policy-making process to design an appropriate quality assurance (QA) system for higher education in South Africa. The system was to be operated by an external agency, the Higher Education

Quality Committee (HEQC). The research revealed that there were, in very general terms, three reference systems at play associated with three different types of institution:

- Historically English-language institutions with strong collegial and research cultures and a history of opposition to apartheid. They supported a facilitative, "hands-off" approach to QA
- Historically Afrikaans-language institutions with managerial cultures and a history of compliance with the apartheid state. They would, reluctantly, comply with a new "top-down" approach to QA
- Historically black institutions often with a poor reputation and a history of student protest and repression. They supported an accountability-driven approach to QA from the HEQC because they believed it could help promote the government's transformation agenda for higher education

Within each of these broad categories, different perspectives existed between university decision-makers, experts, academics, and students. Luckett summarizes the achievements of her project as follows:

> It is claimed that the use of CSH enabled the researcher to elicit rich qualitative data that captured different world-views, values and boundary judgments of social actors representing the different groups of stakeholders involved in or affected by the policy. The use of CSH contributed to a policy settlement that involved substantial improvement to the reformulated policies. However, it is argued that, in the context of a new democracy with a weak public sphere, CSH was unable to counter the unequal power relations between the government agency and the universities and the historical-political fractures of South African society that caused divisions within and across Ulrich's social roles.
>
> *(2006, p. 503)*

In short, the report compiled and sent to the HEQC contained valuable information, which contributed to better policy instruments for institutional audit and program accreditation. However, the intervention was not so successful in enabling rational dialogue between the affected and the involved. In particular, the perspective favoring a facilitative "hands-off" approach to QA gained little traction because HEQC was constrained by the Department of Education's commitment to a process that could help it carry through its political agenda of "transformation":

> However, it remained a weak intervention in terms of altering power relations in the policy field. In the end, the state agency determined the 'better argument' (its own) and claimed that the consultation process undertaken legitimated its (albeit improved) final versions of the policy instruments.
>
> *(Luckett 2006, p. 516)*

In Luckett's view the context, particularly the underdevelopment of the public sphere, constrained the effectiveness of CSH and emasculated its emancipatory potential in this project.

In another study in South Africa, Kriek et al. (2010) applied CSH to two mobile learning projects with the aim of discovering the emancipatory potential of the projects and the methodology. The two projects were conducted in very different socioeconomic contexts. In one, the ability of CSH to ensure participative debate was severely hampered by power structures and authority relations deriving from history, race, and class. The authors reiterate Luckett's point that a fully functioning public sphere is necessary if the emancipatory potential of CSH is to be realized. The third case study comes from the Philippines (Larsen 2011). The researchers sought to use "dialogical boundary critique" in a social learning project designed to address certain deep-seated dilemmas associated with coastal resource management. Close to 100 people were involved in a series of workshops, which used CSH as a framework for exploratory enquiry. In this instance, the boundary questioning proved completely ineffective when it came to encouraging debate about and exposing the existing system of patronage and the stakeholder coalitions that participants felt were determining what happened in the system. An intricate web of stakeholder alliances competed for financial assets, opportunity, and power. Illegal activity was rife, and tactics such as bribery, intimidation, and threat were common. As far as the participants were concerned, the discussions possible using CSH were detached from the reality. In Larsen's view, this created

> ... a 'dialogical vacuum' for stakeholders negotiating the actual resource use practices as the boundary judgements are at odds at the very level of determining what can be acceptably discussed in public .... This carves out a void where marginalized stakeholders are further oppressed as they cannot engage in dialogue from fear of suppression. In this vacuum the actual powers and patronage politics over resource use are regulated by informal negotiations which cannot be formally addressed.
>
> *(2011, p. 72)*

The three cases could be seen to support Ulrich's own thinking that CSH and a well-developed civil society go hand in hand. But, from a radical change perspective, they yield a more damning conclusion. It seems that if the public sphere is not operating as Ulrich envisions it, CSH can easily shift from an emancipatory to a functionalist instrument that buttresses the position of the powerful. The HEQC, for example, ended up with improved means of domination, from the point of view of some of the witnesses, and was able to close down the debate when it wanted.

We can now move onto a second major criticism of CSH from the point of view of radical change theory. Luckett ends her account with this insightful comment:

> The analysis of this case study suggests that CSH should be applied flexibly and creatively, taking into account the specificity and historicity of the policymaking context to which it is applied; recognising that its claims will always be contingent and fallible and that its gains will always be limited by unequal and nonrational power relations.
>
> *(2006, p. 517)*

The trouble is that CSH has no social theory, which can explain how "unequal and nonrational power relations" arise and what might be done about them. If Ulrich wanted his methodology to be genuinely emancipatory, he would have to develop an appropriate theory and then confront the forces in organizations and society that prevent rational argumentation and participative decision-making taking place. CSH is prevented from doing so, the critics argue (Jackson 1985a, 2000; Midgley 1997a) because of the type of critical thinking, emanating from Kant and Churchman, that it embraces. Kant and Churchman are "idealists" and provide Ulrich with a form of critique that is only able to reflect on the values and ideas that enter into social systems designs. Radical change theorists of a more "objectivist" persuasion (e.g. the later Marx and some theorists associated with the Frankfurt School, including Habermas in his early work) want to extend critique to the material conditions that give rise to particular values and beliefs. They argue that the social positions of stakeholders, in organizations and society, go a long way to explaining the ideas that they hold. A critical social theory would reveal how values and beliefs arise; how they are related to the political and economic aspects of the totality; and how power, deriving from the very structures of society, determines that certain ideologies dominate at particular times. Ulrich (2012a, p. 1240) is explicit that his approach requires no such "critical theory of society" or any other kind of social theory. Instead he is aiming at a "critical heuristics of social practice." But without such a theory, it is only possible to answer the 12 boundary questions in the "is" and "ought" mode from "commonsense" perspectives that may be severely flawed. In Mejia's (2001) terms, CSH provides a "content-less" form of critique. The answers to the 12 critically heuristic categories need "filling in," perhaps by the stakeholders in the design situation, perhaps by those who bring CSH to that situation. In the former case, this may simply be on the basis of the knowledge held tacitly and unquestioningly by those stakeholders who may not, for example, recognize that they are discriminating against women. In the latter case, it can lead to the imposition of the analyst's predetermined views in the name of critique. Radical change theorists, who know oppression when they see it, are not content to leave it to chance that the concerns of the disadvantaged groups identified by their social theories are addressed. McIntyre-Mills, as we saw, employs CSH on behalf of groups she knows to be disadvantaged. Stephens insists that CSH must be buttressed by eco-feminist theory to ensure that the previously neglected interests of women and the environment are brought to the fore.

Postmodernists criticize Habermas as the archetypal "critical modernist" aiming at human emancipation directed by universal consensus arrived at in the "ideal speech situation." Ulrich's rejection of his theoretical solution to the problem of practical reason, and acceptance that multiple interpretations will always exist, would therefore find favor with them. And so would his insistence that the systems designs of the involved be viewed in terms of their impact on the affected but not involved. On the other hand, there in not enough in Ulrich's work about the deceptiveness of language and the links between discourses and power relations in society for him to be classified as a postmodernist or poststructuralist thinker. Organizations are, for Ulrich, arenas displaying a multitude of worldviews but not, as Foucault would have it, inextricably linked with the play of dominations.

Luhmann's social theory points to a new meta-code of inclusion/exclusion which is becoming increasingly significant in society. Ulrich's boundary critique offers a possible means of exploring this. On the other hand, there is little sense in CSH of the powerful function systems in society which construct reality according to their own logics, taking account of individual perspectives only as irritants. In doing so, and in generating multiple exclusions, they severely restrict the operation of the fully functioning public sphere which CSH requires to realize its emancipatory potential.

It remains in this section to position CSH on the system of systems methodologies (SOSM). When it first appeared on the scene, Ulrich's work was heralded as filling a gap in the "periodic table" because it took seriously some of the concerns highlighted by the idea of "coercive" problem contexts and made suggestions as to how they could be addressed. It went beyond soft systems thinking in this respect and was heralded as the first "emancipatory" systems approach. CSH was unique in extending the scope of systems practice along the horizontal dimension of the grid by devoting explicit attention to the interests of those who might otherwise be excluded from debate about social systems designs. The values of the "affected but not involved," those who have to live the consequences of the designs, have to be taken into account before the plans brought forward by the involved can be deemed rational. The debate about what designs to pursue, and their likely consequences have to be made accessible to them and they must be equipped to participate without feeling cowed by the expertise or power of others. CSH could, therefore, be seen as seeking to emancipate all "citizens" by empowering them to take part in dialogue about the shape and direction of the purposeful systems in which they live. In recognition of its "achievements" it was classified as possessing simple-coercive assumptions. These would secure an important role for it in circumstances where the conditions for free and open debate already exist or where the polemical employment of boundary judgments, on behalf of the disadvantaged, was enough to make the powerful listen. CSH could not extend its range to complex-coercive contexts, it was argued, because it lacked a social theory capable of enabling it to identify and tackle power relationships that were structurally embedded in organizations and society.

Little, it seems, has annoyed Ulrich more about the way CSH has been interpreted than this "simple-coercive" positioning on the SOSM:

> The only way CSH could apparently be adjusted to the logic of the SOSM was by narrowing its notion of CST down to a *merely* 'emancipatory' purpose, as distinguished from the overall 'critical' purpose of the SOSM. This was achieved by associating it with a 'prison' metaphor that supposedly made it adequate for 'coercive' problem contexts only. CSH could thus be integrated into the SOSM, but at the expense of treating it as a self-contained methodology that seemingly was to be chosen (or not) as an *alternative* to soft and hard systems methodologies. As a result, its concern for the normative core of *all* practice moved out of sight.
>
> (Ulrich 2012a, pp. 1240–1241)

Ulrich's point is that the principle of boundary critique is essential for all forms of rational practice. As we have seen, it can be used by hard systems thinkers, as part of their reflective practice, to reveal the partiality of their proposed designs. Soft systems thinkers can employ it in the dialogical mode to unearth the boundary judgments entering into the debate they are facilitating and to test the selectivity of any actions that are agreed. In short, CSH supports reflective practice in all kinds of problem context regardless of the type of methodology used and the paradigm on which it is based. This is certainly true and it was misleading, in earlier accounts of CSH, to present its usefulness as confined to coercive contexts. But how does this actually effect the positioning of CSH on the SOSM? In my view it doesn't. The SOSM, aided by second-order analysis from a social theory point of view, is designed to bring clarity to the assumptions made by the different systems methodologies. This is not difficult with Ulrich because he puts the philosophical and theoretical foundations of his thinking upfront – the work of Kant, Habermas, and Churchman. These are very different from the largely functionalist assumptions made by hard systems thinkers, system dynamicists, and first-order cyberneticians. They also differ in certain respects from the interpretive social theory that, implicitly or explicitly, underpins soft systems thinking. They diverge considerably from those sociologists of a radical change persuasion who seek to provide an objectivist account of social reality. It is clear, therefore, that CSH can occupy only a small portion of a grid that must encompass all these different theoretical positions and their related methodologies. The fact that CSH offers its services from that position, as a critique of other methodologies, is neither here nor there. The same can be said of all the different methodologies and theories from their respective positions. Thus there can be critiques of systems practice from the hard, soft, and radical change perspectives as well as using CSH.

Another argument used to argue that CSH has a much broader range of application than the SOSM might suggest is that society is, at most times and places, much closer to the way CSH represents it than it is to what other methodologies assume. As a result, it is inevitable that CSH will be called upon more often and will work better. Ramage and Shipp put it like this:

> Ulrich has long argued that in fact most situations, in both the workplace and society at large, carry such inequalities of power that those form the norm rather than the exception.
>
> *(2009, p. 162)*

If you go on to insist that societies usually have a functioning public sphere, which offer multiple opportunities for ensuring that all voices are taken into account, there is the basis for an argument. Unfortunately, it is not one that would be endorsed by social scientists and systems thinkers who advocate alternative accounts of social reality. And, of course, the argument carries no weight when it comes to positioning CSH on the SOSM. The SOSM does not consider what the social world is actually like but is constructed taking into consideration the full range of perspectives systems practitioners, implicitly or explicitly drawing on social theory, use to understand it, based on an analysis of the methodologies they employ to try to improve it.

With these arguments out of the way, we can now get on with finding a positioning for CSH on the revised SOSM. Clearly, CSH does not see its role as managing complexity along the "systems dimension" of the SOSM, assisting with matters of system organization and structure, although it is able to offer a critique of the results produced by those who do. Along the horizontal dimension its interest is in pluralist contexts. Ulrich sees CSH working best in societies where there is a fully functioning public sphere in which pluralism thrives. In these contexts, used in the dialogical mode, it can illuminate different perspectives and help improve mutual understanding. It therefore shares much common ground with soft systems thinking and offers an alternative soft systems approach. Here, although not specifically relevant to the SOSM discussion, I would judge that CSH is more philosophically sophisticated but less methodologically well-formulated than SAST, Ackoff's interactive planning, and Checkland's SSM. The

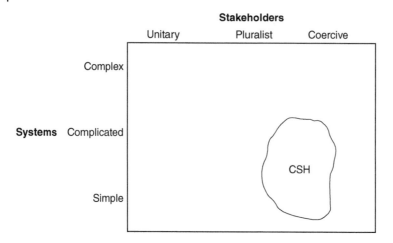

Figure 18.2 The positioning of critical systems heuristics on the SOSM.

question then arises of the extent to which we grant CSH access to the territory of coercive contexts. Midgley argues that coercion is usually characterized by closure of debate. Those with power simply refuse to talk to others, decide who participates, fix the agenda and, if necessary, ignore the outcomes (Midgley 2000, p. 208). CSH fails to recognize and respond to this and so is not suited to such contexts:

> Therefore CSH, which depends upon the possibility of communication (or arbitration) between stakeholder groups becomes redundant when coercion is experienced.
>
> *(Midgley 1997a, p. 37)*

Here my sympathies are with Ulrich. Midgley presents a conceptual limiting case, the extreme end of the coercive spectrum, in which CSH is bound to fail. There will be many situations, in somewhat more democratic settings, in which the polemical employment of boundary judgments can be brought to bear to expose the selectivity and partiality of the social systems designs favored by the powerful (supported by their experts) and to help the disadvantaged reinforce their arguments and gain a hearing. In my view, what Ulrich calls "emancipatory boundary critique" deserves its name and offers something new to the systems approach. That said, CSH clearly lacks a social theory to explain how inequalities of class, status, gender, etc., arise in the first place and are sustained. To many of a radical change persuasion, this is essential if a successful challenge is to be mounted to the established order. CSH therefore, according to the logic of the SOSM, makes only a limited incursion into coercive territory. Its position is shown in Figure 18.2.

18.5 Comments

There are two versions of the systems approach that lay claim to the title of "critical systems thinking" (CST). The first is Ulrich's CSH, as described in this chapter, and which he also refers to as CST. The second is the version largely, but not exclusively, developed at the Centre for Systems Studies at the University of Hull. This has been consistently called CST although, for a time, it was also identified with the multimethodology known as "total systems intervention" (TSI). It is fully described in Part IV. Of course, it is open to systems thinkers to call their theoretical and methodological formulations whatever they want. Nevertheless, I will argue here that the Hull version of CST, certainly in its later manifestations, goes beyond what Ulrich conceives of in CSH and offers a more rounded and powerful version of critique. For this reason, I believe it aids conceptual clarity to reserve the title of CST for the Hull work and to stick to calling Ulrich's endeavor by the name of CSH. To make this case I need to consider how Ulrich sees the relationship between his thinking and that emanating from Hull and then present my reasons for believing that the Hull work provides a broader and more useful basis for critique than CSH. It makes it easier for the reader if, to start with, I follow Ulrich and call the Hull approach TSI.

In a paper written in 2003, Ulrich seeks to portray TSI as simply "methodology choice":

> Based on Jackson and Key's 'system of systems methodologies' (SOSM), it sought to develop a 'meta-methodological' framework that should guide the choice and application of systems methodologies according to the kind of problem situation at hand.
>
> *(2003, p. 327)*

In his view, this is *shallow methodological complementarism*. It is a limited form of critique because it restricts the critical moment to the point at which a decision is made about which methodology to use. CSH is richer because it promotes reflective professional practice throughout the process of planning for a systems design and pays particular attention to the partiality of any recommendations that emerge. It can, moreover, take and maintain a critical stance in relation to interventions that employ any sort of approach:

> Consequently, we may develop and practice skills of boundary critique in conjunction with any kind of methodology, whether it is a 'hard', 'soft' or 'critical' systems methodology or any other kind of approach. Developing competence in boundary critique thus goes hand in hand with a methodological stance of 'deep' methodological complementarism.
>
> *(Ulrich and Reynolds 2010, p. 289)*

In two later, more conciliatory, papers seeking to demonstrate that both CSH and TSI have something to offer to OR, Ulrich (2012a,b) suggests that later versions of TSI have moved on and, in fact, come closer to CSH. They have done this by being more reflective about the context of application, dropping the claim to "meta-paradigmatic" status, and introducing a "reflection phase" in which the outcomes of an intervention are evaluated. This is a reasonable description of TSI's direction of travel. Nevertheless, Ulrich still suggests that while CSH exhibits a "deep complementarism" that makes it appropriate for undertaking the significant task of "dealing critically with the *normative content and consequences* of professional findings and conclusions," TSI's notion of critique remains restricted to dealing "critically with the *theoretical content and limitations* of professional methods and tools" (Ulrich 2012a, p. 1235). This, in my view, is a misrepresentation of later versions of TSI such as "critical systems practice" (see Jackson 2003). However, this matter must be left to Part IV, where critical systems practice is described. The reader can then decide.

For the moment I turn to the weaknesses of CSH viewed from a mature Hull perspective. My argument will consist of showing that, while CSH has moved closer to the kind of thinking that has emerged from Hull, it is still offers a limited form of critique compared to what the Hull approach has to offer. From this point, I refer to the Hull work as CST and Ulrich's as CSH.

Ulrich (2012b), as previously mentioned, draws on Ormerod's insight that good professional practice in OR requires *analytical competence, process competence,* and *context competence* that, respectively, deliver "smart bits" and "helpful ways" and make us aware of "things that matter." In the same paper, he deepens the argument by relating this framework to Habermas' three dimensions of reason. Theoretical-instrumental rationality seeks theoretical knowledge and efficacious action. Practical-normative rationality seeks mutual understanding about ends and values. Communicative rationality seeks to secure the conditions for rational debate and undistorted discourse. The three types of reason require three different forms of argumentation to justify them. Simply put, these might be supplied, respectively, by the practices of "systems science," "soft systems thinking" and CSH. This, as we shall see, is exactly where CST started. When in this "Hull mood," Ulrich evinces many of the concerns that CST has been trying to deal with ever since. I quote extensively from Ulrich to make the point:

> The three argumentation tasks thus emerge as constitutive competencies in dealing with the analytical (or quantitative), process-oriented (or facilitating) and contextual (or integrative) issues that professionals face in dealing with problem situations characterized by high and ever increasing complexity and diversity.
> *(Ulrich 2012b, p. 1320)*

> A view of good practice emerges that revolves around the idea of attending systematically to these three argumentation tasks and striking a balance between them.
> *(Ulrich 2012b, p. 1320)*

> Ormerod's trilogy ... avoids playing off any of the three core competencies against others. It treats the three archetypes of professional services and related skills as a matter of changing *emphasis* rather than choice. The specific tools used and the relative importance of 'technical' (or analytic), 'process' (facilitating) and 'political' (contextual skills) will vary, yet only *together* they constitute a well-developed competence profile.
> *(Ulrich 2012b, p. 1310)*

> *Good OR practice,* then, will basically consist in deploying the three core competencies according to the demands of the situation (flexibility) with a view to engaging with the decision-maker's problem as a whole (combination of

competencies and service doing justice to interdependencies) and to maintaining professional standards throughout (professional attitude and methodological discipline).

(Ulrich 2012b, p. 1312)

We can often simplify the job by temporarily 'bracketing' (suspending) two of the three argumentation tasks while examining the third (e.g. we take them for granted or keep the conditions in question stable); but in the end claims to practical rationality have to meet all three argumentation tasks or it will be difficult to uphold them against challenges.

(Ulrich 2012b, p. 1314)

To reiterate, CSH seems to have moved much closer to CST. The issues raised by these quotations have been extensively discussed in CST publications since at least 1991 (e.g. Jackson 1991a,b).

If CSH and CST have come closer, why then deny CSH the label CST? The answer is that CSH, despite apparently not wishing to (see Ulrich 2003, p. 339), focuses one-sidedly on a "single sphere of rationality" at the expense of the others. The critique offered by CSH is grounded on a pragmatization of Habermas' discursive approach derived principally from his readings of Kant, Habermas, of course, and Churchman. In other words, Ulrich privileges his own theoretical and methodological position and uses it to criticize the way methodologies based on alternative positions are used and the outcomes they yield. The limited scope of this form of critique is revealed if we look again at Ulrich's "eternal triangle" in which boundary judgments, facts, and values are seen to be so interdependent that a change in one element inevitably leads to changes in the others. Constant reflection on these interrelationships during debates involving different reference systems, including those of the affected but not involved, is taken to guarantee rationality. The justification for seeing these elements as so interdependent is stated to lie in the work of Kant and Habermas. Kant is said to uphold the idea that the theoretical-instrumental and the practical-ethical dimensions of rationality are inseparable (Ulrich 2003, p. 335). Habermas claims that in all communication, we *simultaneously* raise three validity claims – that what we say is "true" (factual and accurate), "right" (acceptable and legitimate), and "authentic" (genuine and undistorted) (Ulrich 2012b, p. 1317). Ulrich takes from these sources the idea that all three dimensions of reason have to be addressed in good professional practice and only together can ensure rationality:

The core of this interdependence is found through an analysis of the functions of speech that all argumentation involves. A view of good practice emerges that revolves around the idea of attending systematically to these three argumentation tasks and striking a balance between them.

(Ulrich 2012b, p. 1320)

But the proper way to do this is not to try to incorporate all three in a "triangle," which is incapable of accommodating them. Kant regarded reason as two dimensional and insisted that the dimensions needed different forms of justification. Habermas, as

Ulrich himself makes clear, regards the three validity claims raised in communication as having different argumentative requirements:

> The three claims stand for different ways in which we relate to the world: we refer to 'the' world of phenomena, to 'our' world of human relationships and to 'my' world of subjective experience. Each kind of reference requires its own type of evidence – empirical evidence, reference to mutual understanding, or consistency with the speaker's behavior – and its own form of rational argumentation – aiming at theoretical-instrumental, practical-normative and communicative rationality, respectively.
>
> *(Ulrich 2012b, p. 1317)*

So, I would suggest, even the conclusions of Kant and Habermas break open Ulrich's "eternal triangle." In Habermas' case, the claim to theoretical-instrumental knowledge must indeed be discursively redeemed but through rigorous debate within the scientific community, not open, participative discussion between different worldviews. He is not too far from Popper in this respect. In terms of the other forms of rationality, Habermas insists on the use of the critical standard of the ideal speech situation. Before a claim to "rightness" can be upheld, any factors that give rise to distorted communication must be identified and eliminated. At least in his early work, this meant making use of some explicit social theory that could uncover the sources of inequality and alienation in society that get in the way of undistorted communication. The actors in the social world, Habermas thought, are often in the same position as neurotic patients undergoing psychoanalysis – they suffer from false consciousness and do not truly comprehend their situation. It is incumbent on the critical theorist, therefore, to employ a social theory capable of explaining the ideologies preventing oppressed groups in society from recognizing their true interests:

> The theory serves primarily to enlighten those to whom it is addressed about the position they occupy in an antagonistic social system and about the interests of which they must become conscious in this situation as being objectively theirs.
>
> *(Habermas 1974, p. 32)*

Ulrich is only able to incorporate Habermas' thinking into his triangle by losing the most radical aspects of it, in particular the standard for critique provided by the notion of the ideal speech situation and the notion of a society free from domination that this entails.

If we move outside the realm of Ulrich's favored thinkers, the "eternal triangle" is shattered rather than just broken apart. According to functionalist theorists, of both the positivist or structuralist orientation, facts do not change just because values do. We rehearsed this argument in the earlier "critique of CSH" section with regard to hard systems thinking, system dynamics and, especially, Beer's cybernetics, and will not repeat it here. But, it is worth reminding ourselves that many radical change theorists also argue that society can take on "objective" forms, which we need to understand if we are going to change things for the better. Mingers' description of Bhaskar's "critical realism" will suffice:

> The realities of oppression and suppression still force themselves upon us in ways that go beyond mere moral debate. Critical realists would insist on the

ontological reality of social structure and their effects, and would insist on the necessity for an emancipatory social science to go beyond the everyday assumptions of existing moralities. This would seem to place critical realism above or beyond discourse ethics.

(2014, p. 213)

It is also worth noting, in respect of the triangle, that values do not necessarily adjust because new facts are taken into account. They are notoriously difficult to change as Churchman recognized when he stated that no data can ever fatally destroy a *Weltanschauung*. SAST, Ackoff's "interactive planning," and Checkland's SSM do not waste much time digging up new facts expecting them to change people's perceptions of the world. Rather, they challenge worldviews by exposing them to radically different perspectives.

Ulrich's CSH can, as it claims, offer a critique of any kind of systems methodology and its impact, but it can do so only from its own, limited theoretical position. Boundary critique amounts to an endorsement of the particular philosophy and theory that fits his "eternal triangle." Operating from such a narrow theoretical base, CSH can hardly operationalize any form of theoretical or methodological complementarism. CST, by contrast, employs a broader and more powerful form of critique by making use of the full range of theoretical positions and the various systems approaches based on them. It does not doubt that a social systems design needs to be judged on the basis of Habermas' three dimensions of reason but seeks to do so by treating them in an even-handed manner and, rather than emasculating them by forcing them into the "triangle," gives equal attention to the different modes of argumentation to which they give rise. CST launches a pluralist mode of critique using the different assumptions underlying the alternative sociological paradigms and from all points in the space specified by the SOSM. We have been carrying out, in this part of the book, just this kind of critical, second-order evaluation of 10 systems methodologies. A soft systems intervention, to take an example, can be evaluated not just on its own terms, and those of Ulrich, but from the position of the functionalist paradigm and from the point of view of whether it contributes anything to managing systemic complexity in SOSM terms. What Ulrich apparently fails to recognize is that all systems approaches carry with them explicit or implicit critical standards that can be used to interrogate how interventions using other systems methodologies are conducted and to evaluate their outcomes. Hard approaches can be used to make judgments about whether an intervention has increased the efficient performance of a social system; system dynamics about whether all the relevant factors impacting on a preferred outcome are taken into account; the viable system model about whether the intervention has contributed to long-term viability and sustainability. Soft systems thinking will evaluate an intervention on the basis of whether it has increased mutual understanding among the participants, and CSH will ask further questions as to whether the interests of the affected but not involved have been taken into account. Radical change theory will take a systems design to court if it fails to recognize and do something about the structural features of society that lead to and sustain inequality between different groups and classes. CST respects all of these measures, derived from the variety of systems approaches and their underpinning social theories, and makes use of them all, as we shall see in Part IV, to promote critical reflection on the context of the intervention, the methodological approaches adopted, and the results of any action that ensues. This critique "between paradigms" is central to modern CST and offers the "deep complementarism" that Ulrich's approach promises but can't deliver. It is a critically reflective practice that takes far more into account than CSH can do. CSH will always be employed as one aspect of this kind of critique but alongside critical standards emanating from alternative theoretical positions – positivist, structuralist, interpretive, radical change, etc. I am not sure why Ulrich fails to see the power of CST in this respect. But miss it he does. Perhaps, as he appears to suggest (Ulrich 2003, p. 339), he is too eager to redress an historical imbalance and "promote the weaker cognitive interests (or validity claims)."

The argument can be simplified. I was dean of Hull University Business School for around 12 years. During that time I was pleased if the rationality of any proposed action was challenged because we had failed to think through the boundary judgments on which it was based and forgotten its impact on certain stakeholders. On the other hand, I also found it useful to know whether the action would increase the efficiency of the school, avoid unexpected consequences of an undesirable sort, and contribute to its antifragility in the turbulent environment it faced home and abroad. I wanted to know, as well, that what was being done, and the reasons for it, were fully understood by all those involved. And it was important to consider whether, in taking the action (perhaps accepting certain types of students or teaching particular things) the school was reinforcing inequality in society or helping to improve everyone's future prospects. This type of CST approach can, I believe yield insights and guide action at all levels. For example, we can consider whether the attempts to raise the living standards of the poor in Venezuela, by emancipatory political action, have been beneficial to them and others in that country when set against the apparent losses in efficiency, the ability of the country to prosper in the wider world, the new and extreme divisions in worldviews created, and a possible loss in the ability to incrementally improve the conditions of the poor over the longer term using other strategies. Where should the balance be struck?

18.6 The Value of Critical Systems Heuristics to Managers

CSH's value to managers can be summarized in the following five points:

- It offers an "inclusive" systems approach that emphasizes the benefits of incorporating the values of all stakeholders in planning and decision-making
- CSH puts the concept of "boundary" at the center of systems thinking and makes it easy to see that drawing the boundary around a problem situation in different ways impacts massively on how it is seen and what is done
- It allows managers and others to question whose values are being respected and whose interests served by particular systems designs
- CSH forewarns managers by drawing to their attention the perspectives of different stakeholders, especially those affected by a design but not involved in it
- CSH empowers stakeholders by undermining the notion that expertise rules in planning and design and allows them to make the case that they should participate fully in discussions and decisions about purposes

18.7 Conclusion

The systems tradition has long possessed methodologies that can help with the efficient design of systems to achieve known goals. More recently, soft systems thinking has provided approaches capable, in pluralistic situations, of achieving a sufficient accommodation among stakeholders for some agreed course of action to follow. Until the publication of Ulrich's *Critical Heuristics of Social Planning*, however, there was no systems approach that provided a means for critically reflecting either upon the goals sought and the means used in hard systems thinking or on the nature of the accommodations achieved and the changes brought about through soft systems thinking. CSH also possesses a unique competence in bringing disadvantaged stakeholders into the debate and providing them with the opportunity to state their case with the same confidence and competence as powerful stakeholders and their self-proclaimed experts. It has, therefore, helped to fill a significant gap in systems thinking.

CSH and Team Syntegrity, the subject of the previous chapter, are the best known systems methodologies for addressing aspects of what we have called "coercive complexity." The discussions have suggested that, from the point of view of those who see society as coercive, and seek radical change, they do not go far enough. To Ulrich, in normal conditions when there is a functioning public sphere, CSH achieves all that one could ask of it. In his view

> … once the selectivity of claims has become transparent, democratically institutionalized processes of decision-making can work in a meaningful way.
>
> *(2005, p. 4)*

Midgley (1997a), however, argues that systems thinking could do more to help if it added to its repertoire means of supporting campaigning and nonviolent direct action. Both have a legitimate role to play in democratic societies. Checkland (2003) makes the point that, in coercive situations, no mere methodology – which is only some words on paper – can guarantee the overthrow of a tyrant. While it is true that systems thinking will not trouble tyrants, perhaps it could be more helpful to those who are seeking their overthrow. We will revisit these questions in Section 21.2.2.

Part IV

Critical Systems Thinking

This is the source of the trouble. Persons tend to think and feel exclusively in one mode or the other and in so doing tend to misunderstand and underestimate what the other mode is all about. But no one is willing to give up the truth as he sees it, and as far as I know, no one now living has any real reconciliation of these truths or modes. There is no point at which these visions of reality are unified. …. To reject that part of the Buddha that attends to the analysis of motorcycles is to miss the Buddha entirely

(Pirsig 1974, pp. 74–75)

Early approaches to applied systems thinking, such as operational research (OR), systems analysis, and systems engineering, were suitable for tackling well-defined problems of a technical nature, but were found to have limitations when faced with complex problem situations involving people with multiple perspectives and frequently at odds with one another. Systems thinkers, as we saw in Part III, responded by developing the Vanguard Method to deal with process complexity; system dynamics to manage structural complexity; socio-technical systems thinking and organizational cybernetics to handle organizational complexity; strategic assumption surfacing and testing, interactive planning and soft systems methodology to cope with people complexity; and team syntegrity and critical systems heuristics to empower stakeholders in problem situations when they might not always be listened to. In terms of the ideal-type grid that constitutes the system of systems methodologies (SOSM), there has been a corresponding increase in the number of problem contexts to which systems methodologies can be allocated. This is shown in rough outline in the Figure below. As stated, many times, the grid is not meant to suggest that the real-world is somehow divisible into easily identifiable problem contexts of the six types. It says nothing about the nature of "reality." The grid was constructed to reveal the assumptions made by the different systems methodologies about the nature of problem contexts. It reflects how they view social systems and the form of complexity that they prioritize. Further, it does not offer grounds for arguing that any methodology is more useful than any other. For example, a system dynamicist can continue to promote and justify using their favored methodology on the basis that the great majority of problem situations are "complicated-unitary." Finally, it should not be taken to imply that methodologies are fixed in just one position. The SOSM here shows the traditional assumptions about "systems" and "stakeholders" made by the

Critical Systems Thinking and the Management of Complexity, First Edition. Michael C. Jackson.
© 2019 John Wiley & Sons Ltd. Published 2019 by John Wiley & Sons Ltd.

major systems methodologies. A number of these methodologies have been further developed, changing some original assumption(s), and as a result have moved across the territory defined by the SOSM. For example, some adherents of SD and the viable system model (VSM) have sought to "soften" them up to deal with the "pluralism" that appeared to be hindering the successful implementation of results from using the original methodology. Whether the methodology in question then loses its traditional strengths, gaining little in return, has been subject to discussion in the relevant chapters of Part III.

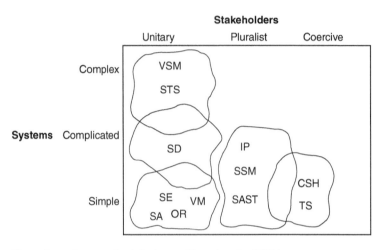

The major systems methodologies positioned on the SOSM.

Although, by the mid-1980s, systems thinking had significantly extended its scope and competence in this way, the wide variety of methodologies that had become available was confusing to decision-makers. This was especially the case because the newer approaches found themselves at war not only with traditional, hard systems thinking but also with each other. They were opposed on fundamental matters concerning the nature and purpose of systems thinking. Management science was deemed to be in a "Kuhnian crisis" (Dando and Bennett 1981). The various tendencies in systems thinking attracted different groups of adherents, put different emphases on the subject matter and key concepts of the field, and sometimes even harbored different interpretations of its purpose. In short, the different versions of systems thinking rested upon alternative philosophical and theoretical assumptions – they were based in different sociological paradigms.

It was apparent that something new had to happen for systems thinking to realize its potential as a guide for leaders and managers. It was Critical Systems Thinking (CST), using devices such as the SOSM, which provided the "something new." Given the complexity, turbulence, and diversity of most problem situations confronting decision-makers in the twenty-first century, it is hardly surprising that no one systems approach can supply the answer. CST sets out how the variety of systems methodologies, methods, and models that have been developed can be used in combination to promote more successful interventions in complex organizational and societal problem situations. The SOSM continues to provide its guidance about the strengths and weaknesses of the various methodologies and to suggest useful combinations, even for those areas of the grid of problem contexts where individual methodologies cannot reach. CST, by which I mean the combination of critical systems theory (always written out in full) and critical systems practice (CSP), has supplied the bigger picture that has allowed systems thinking to further mature and progress as a transdiscipline.

Part IV of the book provides the history of CST and sets out its aspirations in three chapters. Chapter 19 describes the origins of critical systems theory and demonstrates its importance for systems thinking and the management sciences more generally. It takes the story to the point where the philosophy and principles had developed sufficiently to allow thought to be given to how they could be applied in practice. Chapter 20 details three "multimethodologies," based on critical systems theory, which seek to show how the different systems approaches, methodologies, and methods can be employed in combination: Flood and Jackson's "total systems intervention," Midgley's "systemic intervention," and Mingers' "critical realist' version of multimethodology." Chapter 21, "Critical Systems Practice", brings the reader up to date in terms of the latest research. CSP takes its inspiration from the entire history of CST, criticisms leveled at it, some new theoretical ideas, and the lessons learnt from the earlier attempts to develop multimethodological practice. Three case studies are presented showing CSP in action.

19

Critical Systems Theory

This paper sets out to consider OR as a problem-solving methodology in relation to other systems-based problem-solving methodologies. A 'system of systems methodologies' is developed as the interrelationship between different methodologies is examined along with their relative efficacy in solving problems in various real-world problem contexts .... The analysis points to the need for a co-ordinated research programme designed to deepen understanding of different problem contexts and the type of problem-solving methodology appropriate to each

(Jackson and Keys 1984, p. 473)

19.1   Introduction

After completing my MA Management Systems at Lancaster, and a brief spell at Warwick studying for a PhD, I got my first academic job at the University of Hull, in 1979, in the Department of Operational Research (OR). The Department was looking to spread its wings into "business" education and had already hired an organizational behavior specialist, Patrick Maclagan. I guess my systems interests seemed to offer a bridge between their concerns and his. Once Paul Keys arrived a year later and informed me that I was supposed to do research in all the spare time I seemed to have, things began to take off. My 1978 Lancaster dissertation "Considerations on Method" was an analysis of applied systems thinking from the point of view of social science and argued for a critical systems approach. I now began to write and publish academic papers setting out the path to critical systems thinking (CST) as I saw it (Jackson 1982a,b, 1985b, 1987a,b; Jackson and Keys 1984). This, from a paper, called "Social Systems Science: A Critical Approach," sets the tone:

> In this final section of the paper an alternative approach to validating work in social systems science is considered. This is a 'critical' approach, derived from the work of Jurgen Habermas, which (it is claimed) builds on the strengths and overcomes the weaknesses of both the hard and soft versions.
>
> *(Jackson 1982b, p. 671)*

The MSc in OR offered by the Department at Hull already involved a project experience. In 1984, Paul Keys and I founded an MA Management Systems, with a five-month project element, to try out systems ideas. I well remember the shock on my OR colleagues' faces when I returned from Humberside Airport having set up one of our first studies. I had been sent to explore the possibility of a project on lengthening the runway. I came back with one looking at options for bird scaring. I explained what I had learned at Lancaster: that the starting point for a project should be what the client is actually interested in. The project provided an excellent soft systems methodology experience.

A new upsurge in energy followed the appointment of Bob Flood as Professor of Systems Science in 1989. In 1992, he and I founded the "Centre for Systems Studies," which became the home of CST. It hosted, for many years, the journal *Systems Practice* (later *Systemic Practice and Action Research*), which championed CST, and has been the base for *Systems Research and Behavioral Science*, the journal of the International Federation for Systems Research, for over 20 years. The Centre has been directed, since its formation, by myself, Bob Flood, Gerald Midgley, and Yasmin Merali. It has provided a base for the work of such as Amanda Gregory, John Oliga, Norma Romm, Keith Ellis, Wendy Gregory, Jennifer Wilby, Zhu Zhichang, Angela Espinosa, Giles Hindle, Mandy Brown, and Peter Dudley. There have, of course, also been significant contributions to the development of CST from outside the Hull circle, notably from John Mingers, Werner Ulrich, and Ramses Fuenmayor.

This chapter is divided into two main parts. It begins by providing an overview of the origins of critical systems theory. This takes us to around 1991 – the point at which its philosophy and principles were well enough developed to give rise to methodologies seeking to show how the ideas could be put into practice. These "multimethodologies" are the topic of Chapter 20. The second part of this chapter seeks to demonstrate the significance of critical systems theory for the management sciences generally. It has been my long-standing belief that CST has an importance beyond the transdiscipline of systems thinking. It can provide a theoretical basis and practical guidelines for other applied management disciplines, for example, information systems, quality management, and project management.

19.2 The Origins of Critical Systems Theory

The ideas that have inspired critical systems theory derive from two sources – social theory and the systems approach. Of particular importance, from the social science side, has been work that allows an overview to be taken of different ways of analyzing and intervening in organizations. For example, Burrell and Morgan's (1979) tome on sociological paradigms and organizational analysis and Morgan's (1986) take on various "images of organization" have supported critique of the assumptions different systems approaches make about social reality and social science. Critical social theory, from Marx through to Habermas and Foucault, has also had a significant role to play. From the systems approach, critical systems theory inherited the philosophy of holism, a powerful set of concepts and a range of methodologies and methods, based upon those concepts, for intervening in and seeking to improve problem situations. Critical systems theory seeks to draw upon the complementary strengths possessed by social theory and systems thinking. The social sciences are strong on theory, on thinking about the ontological and epistemological assumptions that go into attempts to gain knowledge and to use it in practice. They can provide systems thinking with a means for reflecting on its interventions and improving its methodologies and methods. From the other side, systems thinking can assist in the task of translating the findings of social theory into a practical form and encapsulating them in well-worked out methodologies for bringing about change. This is especially the case because problems in the real-world rarely correspond to traditional disciplinary boundaries and systems ideas encourage multidisciplinary and transdisciplinary practice.

Commentators usually explain the historical development of critical systems theory by referring to its "commitments." Schecter (1991) argues that it is defined by three commitments: critique of the various forms of systems thinking; the emancipation of human beings so they can realize their full potential; and pluralism, which is a commitment to develop a broad framework encompassing the achievements of all the different schools of systems thought. Flood and Jackson (1991b) recognize the philosophy of critical systems theory as embracing complementarism, making use of all types of systems approach as appropriate; sociological awareness of the pressures to choose particular methodologies and the social consequences of their use; and human well-being and emancipation. In the introduction to a set of readings on critical systems theory (Flood and Jackson 1991a), the same authors represent it as based upon three interrelated intentions – complementarism, emancipation, and critical reflection. In that same busy year, Jackson (1991a,b) argued that critical systems theory is built upon the five pillars of critical awareness, social awareness, complementarism at the methodological level, complementarism at the theoretical level, and dedication to human well-being and emancipation. Conducting a review, in 2000, Jackson concluded that three commitments dominated the lists. These are to "critical awareness," "pluralism," and "emancipation or improvement." These are considered in turn but bearing in mind Midgley's (1996) point that there are still no consensually agreed definitions and that the approach is best seen as a continuing debate around these three themes.

19.2.1 Critical Awareness

Critical awareness embraces "theoretical awareness" and "social awareness." Theoretical awareness grew out of the criticisms launched at proponents of particular systems methodologies by advocates of other approaches. A reasonable starting point for the discussion is the assault, in the 1970s, by soft systems thinkers (Checkland 1978, 1981; Ackoff 1979a; Churchman 1979a) on hard systems thinking (HST). Checkland, for example, argued that the assumptions made by the hard approach severely limit its domain of effective application. Making explicit reference to Burrell and Morgan's work on sociological paradigms, Checkland (1981) showed that HST is guided by functionalist assumptions. The world is seen as made up of systems that can be studied objectively and that have clearly identifiable purposes. These systems can be understood and modeled. Thus, HST can be employed to present decision-makers with the means to optimize the systems under their control. The problem, Checkland argues, is that very few real-world problem situations present themselves in terms of systems with unambiguous boundaries and with clearly defined goals and objectives. HST will prove ineffective in most contexts. Worse, its advocates may be tempted to distort problem situations so that they "fit" the demands of their methodology.

A more extensive critical approach was elaborated when Mingers (1980, 1984) and Jackson (1982a) stressed the importance of looking at the social theory underpinning all applied systems approaches and asked specific questions about the assumptions embedded in soft systems thinking and how these impacted upon its effectiveness. My own critique (Jackson 1982a) of the work of Churchman, Ackoff, and Checkland argued that the assumptions made by these authors about the nature of systems thinking and social systems constrained the ability of their methodologies to intervene, in the manner intended, in many problem situations. Soft systems thinking, too, had a limited domain of applicability. Using Burrell and Morgan's framework, I argued that it was based upon interpretive (subjective and regulative) assumptions. With Churchman, Ackoff, and Checkland, systems thinking took a subjectivist turn with the emphasis shifting from attempting to model systems "out there" in the world toward using systems models to capture possible perceptions of the world. In Checkland's SSM, for example, systems models of possible human activity systems are used to structure and enhance debate among stakeholders so that an accommodation about action to be taken can emerge. The recommendations emerging from a soft systems study are likely to be "managerialist" and regulative, it was argued, because no attempt is made to ensure that the conditions for a genuine debate involving all stakeholders are provided. The kind of open, participative discussion essential for the success of the soft systems approach, and the primary justification for its outcomes, is impossible to obtain in problem situations where there is fundamental conflict between interest groups that have access to unequal power resources. Soft systems thinking either has to walk away from these problem situations or it has to fly in the face of its own philosophical principles and acquiesce in proposed changes emerging from limited debates characterized by distorted communication.

It was apparent that all systems methodologies however useful they appeared, especially to their advocates, had their limitations and that these were closely related to the particular social theory that a methodology embraced, either explicitly or implicitly. From this starting point, I proceeded to offer critiques of a variety of different systems approaches – critical systems heuristics (CSH) (1985a), the viable system model (VSM) (1988a), and strategic assumption surfacing and testing (SAST) (1989). The process culminated in a book reviewing five strands of systems thinking – "organizations as systems," "hard," "cybernetic," "soft," and "emancipatory" – from the point of view of relevant social theory (Jackson 1991a). In that volume, the social theory employed to ground the critique took in Burrell and Morgan's work on sociological paradigms, Morgan's thinking on "images of organization," Habermas' theory of the three human interests and the "ideal speech situation," and the modernism versus postmodernism debate. The system of systems methodologies (SOSMs) was used to interpret the results in terms familiar to those working in the management sciences. The aim of the book was to provide a richer appreciation of the theoretical assumptions lying behind the different systems methodologies and heavily influencing their strengths and weaknesses.

As well as permitting the theoretical underpinnings of systems methodologies to be unearthed, the social sciences drew the attention of critics to the importance of the social context in which methodologies are used. This type of understanding was called "social awareness" (Jackson 1991a). Social awareness points to the societal and organizational pressures that lead to certain systems theories and methodologies being popular for guiding interventions at particular times. For example, hard and cybernetic approaches flourished in a Soviet Union dominated by the bureaucratic dictates of the one-party system. As Bahro (1978) remarked about the Soviet ideology as a whole, and would have done about the systems approaches that served it: they appear as "true" and "scientific" precisely to the extent that the compulsion functions effectively. With the change toward free-market capitalism and political pluralism, however, the circumstances that allowed this type of systems thinking to "succeed" changed and softer methodologies came to the fore. Rosenhead and Thunhurst (1982) offer an account of the development of OR in terms of the historical and material development of the capitalist mode of production. Foucault's arguments, about the relationship between power and knowledge at the microlevel, have added to our understanding of the importance of social context. In this vein, Flood and Gregory (1989) and Flood (1990) argue for a "genealogical" perspective on why certain systems theories and methodologies become popular and others fall out of favor. Localized power relations outside of discourse, they suggest, can affect the success and lead to the subjugation of particular systems approaches.

Social awareness should also make users of systems methodologies contemplate the consequences of use of the approaches they employ. For example, the choice of a hard approach usually implies that one goal or objective is being privileged at the expense of other possibilities. Li and Zhu take up the case of OR in China:

> OR is not merely a neutral tool for solving technical problems, but a world-building discourse that shapes society. The future of OR, especially Soft OR, in China will be determined by whether OR workers are willing and capable to act as institutional entrepreneurs promoting scientific and democratic decision-making that deepens the reform toward an open, just, and prosperous society.
>
> *(2014, p. 427)*

Similarly, the use of soft systems methodologies, which are dependent upon open and free debate to justify their results, might have detrimental social consequences for some stakeholders if the conditions for such debate are absent. The need for this form of social awareness was highlighted in my research (1982a, 1985b, 1988a,b, 1989) and provides the rationale for Ulrich's (1983) demand that the systems rationality of planners should always be exposed to the social rationality of the affected.

Critical awareness, incorporating theoretical and social awareness, became one of the central commitments of critical systems theory and remains so to this day. The main problem, of course, is that the apparent strengths and weaknesses of any particular systems approach will vary dramatically depending upon the paradigm from which it is observed and judged. The best way to proceed, it seems to me, is to undertake a rounded critique by taking into account the perspectives of all the major sociological paradigms. This is what was attempted in Part III.

19.2.2 Pluralism

The rise of critical systems theory is inseparable from the emergence of pluralism in systems thinking and practice. Pondering the variety of systems methodologies available, Jackson and Keys (1984) were motivated to explore the relationships between them with a view to better understanding their respective strengths and weaknesses and assisting practitioners to choose an appropriate methodology for an intervention. To this end, during 1983/1984, a research program was established at the University of Hull that used as its primary research tool the SOSM described in Section 8.3. This research program was successful enough to open up a new perspective on the development of systems thinking. Previously, it had seemed as if it was undergoing a "paradigm crisis" (Kuhn 1970) as HST encountered increasing anomalies and was challenged by other approaches. The SOSM, by contrast, showed that alternative systems approaches could be seen as complementary rather than as in competition. Each systems approach is useful for certain purposes, and in particular types of problem situation. The diversity of approaches, therefore, heralds not a crisis but increased competence in a variety of problem contexts. The SOSM offered a way forward from the prevailing systems thinking "in crisis" debate and encouraged mutual respect between proponents of different approaches who had previously seen themselves as being at war with one another. In doing so, it established pluralism as a central tenet of critical systems theory.

My own interest in pluralism continued with an attempt to demonstrate that it offered the best way forward for management science. A paper (Jackson 1987a) looked at the breakdown in confidence in traditional management science and the growth of the soft systems, organizational cybernetic and critical systems alternatives to this orthodoxy. It suggested that each of the alternatives had a significant contribution to make to the discipline and asked how the relationships between traditional management science and the new alternatives could best be theorized so that management science could make the most beneficial contribution to organizations and society. Borrowing a way of thinking and modifying some terms used in Reed's (1985) account of possible "redirections in organizational analysis," four developmental strategies for management science were put under the microscope – isolationism, imperialism, pragmatism, and pluralism.

Isolationists see their own approach to management science as being self-sufficient. They believe that there is nothing to learn from other perspectives, which they see as unhelpful for the task the discipline sets itself. Isolationists were identified as being strong in HST and organizational cybernetics. The isolationist strategy would lead to the different strands of systems thinking continuing to go their own ways, developing independently on the basis of their own presuppositions and with minimal contact between them. "Paradigm incommensurability" can be adduced in support of the isolationist strategy but, I argued, isolationism should be dismissed because it divides the discipline, forestalls the possibility of "reflective conversation" between the different strands, and discredits the profession in the eyes of clients who do not believe that one methodology can solve all their problems.

Imperialism represents a fundamental commitment to one theoretical position but a willingness to incorporate other strands of systems thinking if they seem to be useful because they add strength to the preferred position. Insights from other tendencies will be integrated into the edifice of the favored approach as long as they do not threaten its central tenets. Imperialists believe that they can explain the existence of alternative strands, and analyze the limited sphere of application of these alternatives, in terms of the approach to which they grant hegemony. Strong imperialist aspirations exist in soft systems thinking and emancipatory systems thinking. This strategy for the development of management science was dismissed because methodologies and methods developed in the service of one paradigm would be "denatured" if used under the auspices of another and so the full potential available to management science, if it capitalized on all the paradigms, would not be realized. It was argued, however, that the imperialist scenario might come to pass if extra-disciplinary, broader, societal influences favored one approach at the expense of the alternatives, squeezing the opportunities available to them.

The pragmatist strategy (the word was given its "everyday" not its philosophical meaning) is to develop management science by bringing together the best elements of the different strands on the criterion of what "works." Pragmatists do not worry about "artificial" theoretical distinctions. They concentrate on building up a "tool kit" of methods and techniques, drawn from the different strands of systems thinking, and are prepared to use them together in the course of problem-solving if the situation warrants it. The choice of techniques and the whole procedure is justified to the extent that it seems to bring results in practice. The attractiveness of the pragmatist option was recognized and its support among traditional management scientists and a few soft systems thinkers detailed. It was dismissed, however, because it could not support the development of management science as a discipline. Theory, which this pragmatist strategy eschews, is necessary if we are to understand why particular methods work and others do not, so that we can learn from experience, and so we can pass our knowledge on to future generations. Furthermore, pragmatism is dangerous in the social domain. It can lead to costly mistakes, which theoretical understanding might have helped to avoid, and it can lead to acquiescence in the use of the methods that appear to work only because they reinforce the position of the powerful and implementation is therefore ensured.

In contrast to the other three options available to management science, the pluralist strategy was seen as offering excellent opportunities for successful future development. Pluralism would seek to respect the different strengths of the various trends in systems thinking, encourage their theoretical development, and suggest ways in which they can be appropriately fitted to the variety of management problems that arise. In these circumstances, the diversity of theory and methods in systems thinking could be seen as having led to increased competence and effectiveness in a variety of different problem situations. Jackson and Keys' SOSM was identified as the most formal statement of this pluralist position. It was argued that a "meta-methodology" was needed ("total systems intervention," was still to come), which could guide theoretical endeavor and advise analysts, confronted with different problem-situations, which approach is more useful. Methodological pluralism was defended against the advocates of paradigm incommensurability on the grounds that the different strands of management science are necessary as supports for the anthropologically based cognitive interests of the human species, as identified by Habermas. Pluralism, it was stated

> ... offers the best hope of re-establishing management science as a cohesive discipline and profession – *and* on firmer foundations than those which supported the traditional version.
>
> *(Jackson 1987a, p. 464)*

Following the publication of the 1987 paper, debate about the possibility of pluralism in systems thinking began to concentrate at the theoretical level. This was because of the philosophical difficulties posed for the pluralist position by the arguments in favor of "paradigm incommensurability" derived from Kuhn (1970) and Burrell and Morgan (1979). It seemed inconceivable to proponents of paradigm incommensurability that different systems methodologies, based on what were irreconcilable theoretical assumptions, could ever be employed together in any kind of complementarist way. This would require standing "above" the paradigms. How could such a privileged position be attained?

It was clear enough in what direction critical systems theory was looking for answers. The preferred vehicle to support pluralism at the theoretical level, and therefore to give coherence to the SOSM, was Habermas' theory of human interests. Jackson (1985b, 1987a, 1988b) linked the technical interest to the concern shown by some systems methodologies for predicting and controlling the systems with which they deal, and the practical and emancipatory interests with the need to manage pluralism and engage with coercion. It followed that the two dimensions of the SOSM could be justified from Habermas's work and the different systems methodologies represented as serving, in a complementary way, different human species imperatives. Oliga argued that Habermas' interest-constitution theory is an important improvement over the interparadigmatic-incommensurability position of Burrell and Morgan, since

> ... whereas Burrell and Morgan merely explain the different paradigmatic categories, Habermas explains and reconciles the interest categories in terms of their being individually necessary (although insufficient) as human species, universal and invariant (ontological) forms of activity – namely labor, human interaction, and authority relations.
>
> *(Oliga 1988, p. 97)*

He then goes on to conduct his own survey of how well the technical, practical, and emancipatory interests are served by systems methodologies. Ulrich (1988b) used Habermas' taxonomy of types of action – instrumental, strategic, and communicative – to specify three complementary levels of systems practice, roughly parallel to the requirements of operational (or tactical), strategic, and normative planning. Different systems approaches are then allocated as appropriate to service operational, strategic, and normative systems management. Flood and Ulrich (1990) draw upon Habermas' work to reach similar conclusions.

By 1991, based on Habermas' thinking, it was possible for Flood and Jackson (1991b) to suggest that the concern about paradigm incommensurability could be resolved at the level of human interests. As a result, the SOSM could be rescued as a vehicle for promoting methodological pluralism. Complementarism at the theoretical level provided the basis and justification for complementarism at the methodological level. The SOSM could point to the strengths and weaknesses of different strands of systems thinking and put them to work in a way that respects and takes advantage of their own particular theoretical predispositions in the service of appropriate human interests.

19.2.3 Emancipation or Improvement

There has been a confusion about the relationship between critical systems theory and what I call "emancipatory systems thinking" that continues to exist, among some commentators, to this day. Thus, Ramage and Shipp (2009, p. 169) begin their account of critical systems theory by stating that it "emphasizes the importance of politics and power in organizations." While this is certainly true of both emancipatory and critical systems thinking, it is a defining feature of the former rather than the latter. Once critical systems theory had attached itself to Habermas' theory of three human interests, the technical, practical, and emancipatory, it began to see its "emancipatory commitment" (as some called it) in terms of a much broader dedication to human "improvement." Flood and Jackson (1991b) take this to mean bringing about those circumstances in which all individuals can achieve the maximum development of their potential. This requires raising their standard of living, their freedom to communicate, and their ability to influence the society and organizations in which they live. To put it bluntly, "emancipating" individuals from oppressive social relations is essential but is not the only important aspect of helping people realize their full potential. The link to Habermas is that "improvement" in this broad sense requires that each of his three "human interests" must be served by appropriate systems methodologies. Methodologies that serve the technical interest are essential to assist with material well-being by improving the productive potential and steering capacities of social systems. Methodologies that serve the practical interest are important to promote and expand mutual understanding among the individuals and groups participating in social systems. Methodologies serving the emancipatory interest are crucial to protect the domains of the technical and practical interests and to expose situations where the existence of distorted communication prevents the open and free discussion necessary for their successful operation. Habermas argued that all human beings have a technical, practical, and emancipatory interest in the functioning of organizations and society. So a systems perspective that can support all these various interests must necessarily have an important role to play in human well-being and emancipation. Critical systems thinkers make the point that this is exactly what their approach hopes to achieve. It wants to put hard, system dynamics, and cybernetic methodologies to work to support the technical interest; make available soft systems methodologies to assist with the practical interest; and it is developing emancipatory approaches to help with the emancipatory interest.

The confusion between emancipatory and critical systems thinking came about, I think, because many of those involved in the creation of critical systems theory, such as Jackson, Mingers, and Oliga, were also vociferous in demanding an extension of systems thinking to take on a radical change agenda. This was in part due to the political affiliations of the individuals involved. There was also, however, a good theoretical reason which had been revealed by the SOSM. As was stated in the original 1984 article, the "participants" or "stakeholders" dimension could, in "ideal-type" terms, be extended to embrace coercive contexts (an extension formally made by Jackson 1987b). At the time, Jackson and Keys did not know of any systems methodologies that acted as though problem contexts might be "coercive" – defined as situations where there is fundamental conflict between stakeholders and the only consensus that can be achieved arises from the exercise of power. From the critical point of view, this was obviously a weakness in the overall capability of systems thinking and made the construction of such approaches imperative. This led to explicit calls (e.g. Jackson 1982b, 1985b) for a "critical approach," which would take account of them. Thus, an interest in whether systems thinking could be used to help "emancipate" stakeholders in "coercive contexts" became dominant in the literature of critical systems theory. This obsession with a possible emancipatory systems approach was further fueled by the discovery of Ulrich's CSH. Although Ulrich's (1983) CSH represented an independently developed strand of systems thinking, deriving from Kantian idealism and Churchman's pragmatism, it seemed neatly to meet the requirements of coercive contexts in the SOSM. CSH was arguably capable, where soft systems thinking was not, of providing guidelines for action in certain kinds of coercive situation. It enabled systems designs or proposed designs to be carefully interrogated as to their partiality and set down criteria for genuine debates between stakeholders which had to include both those involved in systems designs and those affected but not involved.

To be clear, critical systems theory's commitment to "emancipation," as defined here, is about putting all the different systems approaches to work, according to their strengths and weaknesses, and the social conditions prevailing, in the service of a more general project of "improvement." Emancipatory systems thinking is narrower. Its role is to provide methodologies that, by challenging social arrangements that create inequality and discriminate against particular social groups, help to "emancipate" those who are disadvantaged and put them more in control of their own destiny. The domain of effective application of emancipatory systems thinking is coercive problem contexts. By 1991, it was possible to state the matter clearly:

> Critical systems thinking recognizes its overall emancipatory responsibility and seeks to fulfill this by adequately servicing, with appropriate systems methodologies, each of Habermas' human interests .... At the same time it perceives a special need, because of previous neglect, to nurture the development of emancipatory systems thinking. In theory, this means encouraging the use of specifically emancipatory systems methodologies suitable for coercive contexts.
>
> *(Jackson 1991a, p. 206)*

19.3   Critical Systems Theory and the Management Sciences

This part of the chapter looks at the contribution critical systems theory can make to the management sciences generally. As was stated in *Systems Methodology for the Management Sciences*:

> One of the aims of this book is to reconstruct systems thinking as a unified approach so that it can again occupy a position at the leading edge of research and practice in the management sciences.
>
> *(Jackson 1991a, p. 277)*

I believe that CST is that reconstruction and that it is ready to fulfill its potential in this respect. The capacity of CST to provide appropriate theoretical and practical guidance throughout the management sciences is, in my view, not surprising. All applied disciplines in this area have to cope with "systems" that exhibit massively increasing complexity and for which only multiparadigm thinking and multimethodological practice is fit for purpose. I will pursue my argument with brief reference to operational research, quality management, evaluation research, information systems, knowledge management (KM), project management, logistics, and health systems research. These are all fields to which I have contributed some thoughts on the matter in hand. I am conscious that others have shown the relevance of CST for improving the theory and practice of enterprise architecture, business process reengineering, organizational learning, change management, service systems science, and epidemiology. I must leave the interested reader to research that work by themselves.

Many of the early pioneers of OR became disillusioned with the way the discipline and profession developed. Churchman and Ackoff, for example, thought that the original intention to build a holistic, interdisciplinary, experimental science addressed to strategic problems in social systems had been betrayed as OR degenerated, in the 1960s, into little more than mathematical modeling. I have long argued for an "enhanced OR," drawing upon the various methodologies that have emerged under the banner of systems thinking. The rationale is that this would allow OR to live up to the ambitions of its founders. The case was first made in *Towards a System of Systems Methodologies* (Jackson and Keys 1984), which was the most cited paper published in the *Journal of the Operational Research Society* (JORS) between 1981 and 1999 (Ranyard 2001, p. 2). The "Community OR" initiative, launched in the United Kingdom in the mid 1980s, offered another opportunity. The initiative aimed to enrich OR methodology and extend it to a wider range of clients. I argued (Jackson 1987c) that the SOSM could be employed to suggest what new types of problem situation would be encountered and how OR could be enhanced, by approaches such as the VSM, SSM, and CSH, in order to deal with them. In a paper (Jackson 1993a), written for *JORS*, I continued the argument, suggesting that a social theory perspective (based on the work of Habermas, Burrell, and Morgan on "paradigms," and Morgan on "images") could provide a much enhanced appreciation of the relative strengths of OR and other system-based methodologies. In a 2006 paper, in the same journal, I argued that "Soft-OR" was only one way of adding to the positivist/quantitative approach of classical OR/MS. OR had an obligation to explore a wider range of theoretical options and to convert the possibilities offered by all of these into a form that managers and management scientists can use. Only in this way could it be reinvented to live up to the ambitions of its founders. A further paper in *JORS* (Jackson 2009) restated the case using Boulding's "hierarchy of complexity" (see Section 5.5) as a critical device to show why OR practitioners needed to move beyond "machine" thinking and add a variety of systems approaches to their intervention strategies. In my "Beale Lecture" to the OR Society (February 2019), I again argued for an "enhanced OR" capable of helping decision-makers meet the challenges posed by modern day complexity. It could achieve this by drawing on the theoretical and methodological advances made in "Soft OR" and "multimethodology" and upon developments in the related transdiscipline of systems thinking.

In the 1980s, total quality management (TQM) became a mantra among managers although many organizations experienced unexpected difficulties when trying to put the ideas into practice. In a paper, Flood and Jackson (1991d) subjected it to a critique using "metaphor analysis." At much greater length, in his book *Beyond TQM* (1993), Bob Flood argued that the problems encountered by TQM could only be overcome if it was reconceptualized in terms of a new management philosophy based on critical systems theory and implemented using the meta-methodology of "total systems intervention." The book was nominated for the Institute of Management Consultant's "Management Book of the Year 1993." Another influential introduction to "quality," by John Beckford (1998) uses the framework provided by critical systems theory to compare traditional approaches to quality to contemporary approaches based on systems thinking.

Two papers, written in 1988 (Clemson and Jackson 1988; Jackson and Medjedoub 1988) focus on the problem of designing appropriate evaluation systems for organizations with multiple goals and describe an intervention in Beverley Council for Voluntary Service. They recommend a contingency approach to evaluation based upon the SOSM. Different approaches to evaluation are seen to be relevant in different problem contexts. This work was taken forward by Amanda Gregory in a national project with the National Association of Councils for Voluntary Service (NACVS), sponsored by the Leverhulme Trust, on the evaluation of Councils for Voluntary Service. A first paper from the research (Gregory and Jackson 1992a) responded to the proliferation of different evaluation approaches. It formulated a simple classification of the available approaches, which matched them to the contexts in which they were most appropriate. A follow-up (Gregory and Jackson 1992a) presented an analysis of the theoretical underpinnings of the four main forms of evaluation and set out a system of evaluation methodologies again linked to the circumstances in which they might be used. A final report to the NACVS was published (Gregory and Jackson 1992c), and an update on the Beverley intervention was given by Gregory et al. (1994). In a later paper stemming from the project, Gregory (1996) identifies four paradigms in evaluation theory – goal-based, system-resource-based, multiactor based, and culture-based – and suggests that for much of their history they have remained in "isolation," warring with one another. Using the thinking behind the SOSM and "total systems intervention," however, it is possible to see how they might be used in a complementary way. Dismissive of attempts to "integrate" the different approaches, Gregory argues in favor of "multidimensional evaluation," where methodologies are used together in parallel. This protects the different contributions they can offer according to their distinctive theoretical underpinnings.

There has been considerable clamor for complementarism in methodology use in the information systems field. Wood-Harper et al. (1985), Avison and Wood-Harper (1990), Hirschheim (1985), Lyytinen and Klein (1985), Walsham (1991), and Walsham and Han (1991) have studied the options for combining methodologies in practice and pointed out the difficulties. The debates have reached a high level of sophistication and draw upon the work of social scientists such as Morgan, Habermas, and Giddens. As a result, we find information systems researchers grappling with the same problems as systems thinkers in trying to operationalize pluralism. I entered the fray (Jackson 1987d), on behalf of critical systems theory, with a paper setting out the opportunities for information management specialists if they looked beyond the traditional model of the organization as a machine. If information systems specialists widened their thinking and looked at the debates taking place in organization theory and systems thinking, it would enable them to explore alternative strategies and access specific guidance of direct relevance to their practical activities. A later paper (Jackson 1992b) discusses the "commitments" enshrined in critical systems theory and shows that each of them has, at one time or another, received attention in the literature on information systems. It argues that addressing them as a whole, guided by critical systems theory, can yield an integrated program for critical thinking in information systems research. A final paper, in this area (Jackson 1997b), suggests that critical systems theory provides a natural underpinning for work in information systems research and gives an example of its use in an information systems strategy project.

KM, which came to prominence in the 1990s, lent itself easily to critique from a critical systems perspective. Japanese authors saw the limitations of the Western approach to KM as deriving from its reliance on the model of the organization as an information processing machine. This opened the opportunity to argue that it was worth examining the theoretical assumptions underlying different versions of KM in order to understand them better and to recommend how it could be enhanced. I wrote a paper (Jackson 2005a), building on Morgan's work on "images of organization," which identified five "systems metaphors," which were influential in thinking about organizations. The article suggested how these had impacted upon different approaches to KM and derived various lessons that could help refine practice in the field. Another paper of the same year (Jackson 2005b) argued that the "theoretical awareness" and "methodological sophistication" of CST could be useful for analyzing the organizational models and well-known concepts employed in KM and for improving its chances of success:

> By clarifying the organisational models that KM currently embraces, by offering to it a range of alternative perspectives, and by pushing KM in the direction of the pluralist use of a variety of theories and methodologies, CST has much to contribute to the theoretical awareness of KM .... In terms of practice .... CST possesses a significant armoury of methodologies and methods that can help KM practitioners make their insights happen.
>
> *(Jackson 2005b, pp. 192 and 195)*

In the field of project management, there has been a growing awareness that the "rational," "positivist," "reductionist" approach that dominates is poorly equipped to cope with the challenges posed by complex projects. Cooke-Davies et al. characterize the current "paradigm" as resting upon

> ... a mechanistic world view deriving from Cartesian philosophy, a Newtonian understanding of the nature of reality and an Enlightenment epistemology whereby the nature of the world we live in will be ultimately comprehensible through empirical research.
>
> *(2007, p. 52)*

They suggest that the findings of complexity theory might provide a way forward. A book by Remington and Pollack (2007) identifies "technical," "structural," "temporal," and "directional" complexity as present in all complex projects and suggests "systemic pluralism" as a way to improve their management. In 2007, the International Centre for Complex Project Management (ICCPM) was established with the specific purpose of improving the delivery of complex projects by applying systems and complexity ideas to project management. Research on CST, and particularly the book *Systems Thinking: Creative Holism for Managers* (Jackson 2003), was a key input in the design of ICCPM's *Complex Project Manager Competency Standards* (see Version 4.1, 2012). The book was also central to the design of an Executive Masters Programme in "Complex Project Management," jointly developed by the ICCPM, Queensland University of Technology, and the Defence Materiel Organisation (DMO) of Australia, and launched in 2008. The DMO formally audited the programme and identified "a 5% cost/benefit realization if graduates from the program are placed on major DMO projects after completing the program." The program is now also taught, under the guidance of Stephane Tywoniak, at the Telfer School of Management, University of Ottawa, Canada. A CST-based course, "Systems Thinking and Complex Project Management" was the most successful of the early executive education programs run by the ICCPM.

In 1997, Bridget Mears-Young and I conducted a critical systems theory examination, using Burrell and Morgan's work on sociological paradigms, of both "traditional" and "nontraditional" logistics. It revealed them to be "functionalist" in nature and argued that the main problems faced by logistics derive from this. Logisticians were urged to "call in the revolutionaries" to help them achieve an "epistemological break" with functionalism. Only if they were willing to explore other paradigms would they be able to make progress. It has taken some time, but the paper is now starting to get noticed in the field.

Luis Sambo served two terms as the World Health Organization Regional Director for Africa and led the local response to the 2014 Ebola crisis in West Africa. His PhD thesis (Sambo 2009), under my supervision, inquires into the current state of health systems research and argues in favor of the "critical health systems thinking" which he had begun to implement. A series of papers developing this approach is in preparation.

As a final point, it is worth mentioning that systems theorists have offered recent developments in critical systems theory as the basis for a partnership with organization theorists. Organization theory is, in general, more theoretically sophisticated than systems thinking, so the partnership in this instance is one based on the practical contribution that systems thinking can make:

> From the systems point of view, organization theory, while masquerading as an applied discipline, actually gives very little assistance to those who wish to know how its conclusions can be put to work.
>
> *(Galliers et al. 1997, p. 273)*

Systems thinking, it is argued, with its emphasis on practical intervention, multi-methodology usage, and "boundary critique," can contribute significantly here. We are back to where we started in this chapter – looking for a mutually informing relationship between systems thinking and the social sciences. It has to be said that there has been little response from organization theorists. In my view, much to the detriment of their discipline.

19.4 Conclusion

The early years of CST were ones of considerable intellectual excitement, providing for the very rapid theoretical and practical development of the approach. In Schecter's opinion, by 1991 critical systems theory had already brought greater theoretical depth to discussions in the field of systems thinking and produced some strong original work in meta-theory; it had provided challenging critiques of earlier systems work; it had put issues of power and human emancipation on the agenda and put its commitment to emancipation into action; it had produced a framework for the complementary development of all the different systems approaches; and it had championed a commitment to careful, critical, self-reflective thinking. In that year, three books, inspired by critical systems theory, were published which readers can examine to check the validity of Schecter's flattering viewpoint. Flood and Jackson's *Critical Systems Thinking: Directed*

Readings (1991) was a collection of papers, accompanied by a commentary, that traced the origins and development of CST. Jackson's *Systems Methodology for the Management Sciences* (1991a) aimed to provide a comprehensive critique of the different systems approaches, drawing on the social sciences as the basis for the critique. Once critical systems theory had been formulated, as a philosophy and theory, it needed guidelines that would enable it to be applied in practice. These were provided in 1991 with the publication of Flood and Jackson's *Creative Problem Solving: Total Systems Intervention* (1991b). Total Systems Intervention (TSI) was heralded as a new approach to planning, designing, problem-solving, and evaluation based on critical systems theory. The next chapter continues the story by detailing TSI and two other "multimethodological" approaches which draw heavily upon critical systems theory.

Departing a little from the historical narrative, this chapter has also established that CST can contribute to the theoretical and methodological development of other sub-disciplines in the management sciences. CST is, I believe, up to the task, although its actual employment in this way remains in its infancy. There are successful academic and professional careers to be built on enhancing different branches of the management sciences using a critical systems theory perspective. Finally, it is worth mentioning that the role played by CST, outside of systems thinking per se, puts to bed the notion that it is just about "methodology choice." Critical systems theory can provide a critique of any intervention strategy and the results that it produces.

20

Critical Systems Thinking and Multimethodology

The fundamental tenet of multimethodology is the importance of combining together methods or sometimes methodologies in dealing with real-world situations, whether the purpose is pure research or a practical intervention .... The first argument in favour of multimethodology is that the real world is complex and multi-dimensional, while particular research or intervention methodologies focus only on specific aspects .... The second ... is that research and intervention is not a discrete event but a process that has phases (or, rather, different types of activities) that will predominate at different times .... A third argument ... is that it encourages triangulation of research and generates more interesting and stimulating results

(Mingers 2014, p. 184 et seq)

20.1   Introduction

Put as simply as possible, the idea of multimethodology (MM) is to use a combination of methodologies (possibly from different paradigms) and methods together in a single intervention. During the 1980s, as was noted in the previous chapter, interest in pluralism and MM escalated in a number of academic areas – for example, in organization theory (Reed 1985), systems thinking (Jackson and Keys 1984; Jackson 1987a), and information systems (IS) (Hirschheim 1985; Lyytinen and Klein 1985; Wood-Harper et al. 1985; Avison and Wood-Harper 1990; Walsham 1991; Walsham and Han 1991). Practitioners were also increasingly combining different methods in their work. Richard Ormerod (1992, 1994) has provided accounts of his own "pluralistic practice" as a consultant and, when he entered academia, he became a leading exponent of this approach in operational research (OR) (Ormerod 1995). The interest of a range of academics in MM, and the fact that multimethod use was becoming common in practice, spurred on its advocates. The year 1997 saw the publication of Mingers and Gill's (eds) *MultiMethodology – The Theory and Practice of Combining Management Science Methodologies*. This was a definitive collection of papers on MM addressed to an OR/management science (MS)/systems audience. It drew on theoretical work in critical systems thinking (CST) as well as on multimethod applications by consultants. The result was the emergence of a community of people interested in MM and multimethod

Critical Systems Thinking and the Management of Complexity, First Edition. Michael C. Jackson.
© 2019 John Wiley & Sons Ltd. Published 2019 by John Wiley & Sons Ltd.

theory and practice. By 2002, Munro and Mingers were able to report on a survey of OR/MS practitioners showing just how widespread the employment of multimethods, during the course of one intervention, had become in that field. For those working on real-world problems, pluralism seemed to be necessary and, judging from the responses from Munro and Mingers's sample, combining methods brought success. A more recent review of the influence of systems thinking in OR/MS (Mingers and White 2010) demonstrates that a considerable amount of work is continuing in the CST and MM traditions, with applications in areas such as IS planning, knowledge management, quality management, and designing a user support service. There have been many more recent applications of CST and MM, including in community development in Colombia (Henao and Franco 2016), community engagement in India (Chowdhury and Jangle 2018), and strategy formulation in Serbia (Zlatanovic 2017).

There are actually five main reasons given for why MM thinking and practice is desirable and has become popular. The first two are found in Mingers and Brocklesby (1997). The third exists in that same source and in Jackson (1997a). The fourth comes from Jackson (1997a). The fifth is due to Mingers (2014).

The first is that the world we confront and seek to intervene in is "multidimensional." Mingers and Brocklesby basing their thinking on Habermas's theory of "three worlds," *material*, *social*, and *personal*, argue that any real-world situation is a "complex interaction of substantively different elements." Mingers summarizes their position:

> There will be aspects that are relatively hard and observer-independent, particularly material and physical processes, that we can observe and model. There will be aspects that are socially constituted, dependent on particular cultures, social practices, languages, and power structures, that we must come to share and participate in. Finally, there will be aspects that are individual such as beliefs, values, fears, and emotions that we must try to express and understand.
>
> *(1997b, p. 11)*

Different paradigms will focus on different aspects of a problem situation, highlighting some features and hiding others. They will often offer incompatible accounts of its main characteristics. It therefore makes sense to work with multiple paradigms, looking through a variety of theoretical lenses and employing a range of methodologies, to try to capture as many facets of the problem situation as possible and intervene in a more comprehensive manner. The second argument is that research and intervention are processes that are likely to proceed through a number of phases. Each phase – from gaining an appreciation of a problem area to testing hypotheses or taking action – will involve different tasks and pose different problems for researchers and/or practitioners. It is good for them, therefore, to have a variety of methodologies and methods at their disposal. A third argument, evinced by Mingers and Brocklesby, and Jackson, is that MM chimed with the postmodernist argument against "grand narratives," which was fashionable in the late twentieth century. The spirit of the times was to oppose "totalizing" discourses that claim to know "the truth" about things. As Lyotard put it:

> Simplifying to the extreme, I define *postmodern* as incredulity toward metanarratives … Postmodern knowledge is not simply a tool of the authorities; it refines our

sensitivity to differences and reinforces our ability to tolerate the incommensurable. Its principle is not the expert's homology, but the inventor's paralogy.

(1984, pp. xxiv–xxv)

A willingness to entertain and promote multiple perspectives, methodologies, and methods is clearly in tune with this sensibility. The fourth argument rests on the lack of success of traditional methodologies and methods to deliver what was promised in practice. The failure of hard systems approaches when confronted with strategic problems in social systems became well known. Confidence was declining in the ability of IS, designed according to traditional structured design methods, to serve the purposes of their users and bring competitive advantage to organizations. As a result, academics and practitioners began to challenge the old ways and to advance various new approaches for developing and increasing the competence of their disciplines and professions. It then became a matter of how the new approaches fitted both with the old and with other new approaches. It became important to give thought to the possibility of MM and multimethod practice. The final argument, presented by Mingers, is that MM permits a form of "triangulation." A surveyor may use three different lines of sight to accurately fix the position of an object. If we get the same results from using different research approaches then we can have more confidence in the validity of our findings. If the conclusions differ, well, that itself is interesting and stimulating. Mingers makes the point that looking at a phenomenon from various perspectives provides a greater appreciation of its complexities, especially if what we see differs. The same logic holds with using different approaches to look at how an intervention is progressing and to review the outcomes it yields.

We are now in a position to look at what kind of MM approach is needed to deliver on these potential benefits and consider any difficulties in the way of realizing it. Critical systems thinkers agree on the starting point. The MM, or pluralist, approach opposes the "isolationist" position we examined in Chapter 19. Isolationists believe in the value of their own theory and methodology whatever context they are operating in and see little point in trying to communicate with those of a different persuasion. One important step toward pluralism is to accept that different theoretical and methodological approaches are legitimate and are able to address problem situations differently. It then becomes possible for decision-makers to think through what the appropriate way of proceeding is for the particular problem situation they are facing. This is an important step, but it is not big enough. The breakthrough to fully fledged pluralism comes when we recognize the need to use different approaches in combination to tackle one problem situation, because of its multidimensional characteristics. Also, it may be necessary to alter our choice of methods during each stage of an intervention. An obvious issue here, raised for example by Eden et al. (2009), is the extreme demands placed on decision-makers and management scientists who have to know about and combine approaches that may rest on different social theories. They are rightly concerned that multi-methodological approaches will come to grief faced with leaders and managers keen to reach a rapid agreement on what to do. This is a problem well understood in CST and we will return to it in Chapter 21.

From this point on there are differences between critical systems thinkers about the best way to pursue the MM approach. I will begin by setting out an "ideal-type" version of MM (Jackson 1999, 2001, 2003). In my view, the complexity, heterogeneity, and turbulence of many problem situations demand that pluralism must be exercised at three levels – methods, methodologies, and paradigms.

The first requirement for pluralism is that it is "multimethod." It must encourage flexibility in the use of the widest variety of methods, models, tools, and techniques in any intervention. This will enable practitioners to respond to the complexities of a problem situation and the exigencies it throws up during an intervention. It may require the "decomposition" of host methodologies, with which particular methods are usually associated, into their components and releasing these to be employed in the most useful way in the context in question. For example, the "rich picture" technique could be extracted from soft systems methodology (SSM) in order to provide initial structuring in a hard systems study. The notion of "partitioning" methodologies into their parts, and combining parts from different methodologies in an intervention, was first proposed by Midgley (1990, 1997b) as part of what he calls "creative methodology design." Mingers and Brocklesby (1996, pp. 503–506) and Mingers (2006, p. 237) discuss such "multimethod" practice at length and provide more cases of "decomposition." For example, a system dynamics (SD) model, which might usually be used to capture the behavior of a system, could be employed as a detailed cognitive map for the purpose of enhancing debate as part of a soft systems approach. The procedure is possible because methods are less restricted by theoretical assumptions. They involve specific activities designed to achieve particular outcomes – in the last example, mapping causal relationships. Owing little debt to particular theories, they can easily be transferred from their normal host methodology to serve alternative rationales. Not surprisingly, as Ormerod (1994, 1997a) observes, this "mixing and matching" of methods is common among management consultants. They will inevitably react to their client's demands and the changing problem situation by freely using the variety of methods, models, and techniques in a responsive fashion.

The dangers of an unrestricted multimethod approach are the same as those we identified with "pragmatism" in Chapter 19. Pragmatists are happy to draw upon the "tool kit" of methods developed in different strands of systems thinking and employ whichever ones appear to bring good results in the problem-situation with which they are engaged. They do not worry about methodological and theoretical distinctions. There are two good reasons to be concerned about this. The first should worry practitioners. It is that they may end up limiting the scope of their investigation by unreflectively using their tool kit of methods in the service of only one theoretical position. In doing so, they lose out on perhaps the key advantages of pluralism – that it allows problem situations to be viewed from the point of view of different paradigms and addressed using methodologies that reflect the strengths and concerns of the different paradigms. Thus, approaches that are solely multimethod represent a very impoverished form of pluralism. Unreflective multimethod work can isolate itself in one paradigm. The second concern is mainly for academics. Not all methods are equally "free-floating." Some serve particular methodologies better than others. A purely multimethod approach prevents us from doing research on the value and usefulness of the methods in terms of whatever broader purpose we might be trying to achieve; be that understanding how a social system functions, increasing mutual understanding, or serving some disadvantaged group. Only by consciously using methods under the control of a methodology that supports one paradigm can we discover how they perform for that methodology and go on to test their usefulness when incorporated in different methodologies. We then have a chance of finding out, for example, whether the Viable System Model (VSM), which seems to be a suitable model for designing effective organizational systems, is equally at home when employed as a "hermeneutic enabler." Ormerod (1994) had hoped that a stream of quality publications would ensue from his multimethod consultancy experiences. He is honest enough to admit that, although the experiences were strong, the publications were weak in the sense of lacking theoretical underpinning. Unless there is an explicit theoretical underpinning to their work, academics cannot learn much from multimethod practice and will have little to teach their students.

A genuinely pluralist approach must, therefore, be multimethodological as well as multimethod. The multidimensionality of problem situations requires that systems practitioners operate with different methodologies, based on alternative paradigms, in combination. Good methodologies translate the theory embedded in the paradigm they serve into practice and provide principles for the coherent use of different methods, models, and techniques to ensure that this occurs. Pluralism gains much of its power from insisting on the deployment of a variety of methodologies, based upon different paradigmatic assumptions, and using them in a way that gets the most out of each. We must seek to benefit from what they all have to offer. The link between methodologies and their embedded methods is, as we have just seen, a relatively weak one. This is not the case with methodologies and their governing paradigms. To ensure methodological diversity, it is essential to be strict in linking particular methodologies back to those paradigms. This requires a precise understanding of the theoretical underpinnings of the different methodologies. The "theoretical awareness" championed by critical systems theory is essential. If this is lacking, then it is impossible to guarantee that a MM intervention will occur. Methodologies owing allegiance to the same paradigm might be employed together in the mistaken belief that genuine pluralism was being practiced. Or alternative methodologies, originally designed to serve other paradigms, might be denatured to fit in with the favored theoretical position. This latter case is what we referred to in Section 19.2.2 as "imperialism." One paradigm and its associated methodology or methodologies is granted a superior status and other methodologies and methods are used to buttress that approach rather than to offer an alternative to it. Another important point, mainly for academics, is that unless we understand fully the relationship between methodologies and their underlying paradigms we cannot do the research that might allow us to improve our methodologies in the light of new theoretical insights. Nor can we test the different theoretical positions by seeking to apply them, using methodologies, to drive real-world interventions. Methodologies informed by theory are essential for ensuring a healthy link between theory and practice.

The third requirement for our "ideal-type" MM approach follows. It must be multiparadigm. We cannot be dismissive of the claims of any sociological paradigm. We have to bring to bear, in our pluralistic theory and practice, what we can learn from the functionalist, interpretive, radical change, and other theoretical positions. In order to ensure that the fullest appreciation of a problem situation is obtained, we have to view it through a wide variety of paradigmatic lenses. Being multiparadigm in our thinking will also encourage and enable us to address the issues raised by a problem situation in a truly multimethodological manner – with a range of methodologies reflecting the different paradigms. It will also ensure that we evaluate the results obtained from an intervention from the perspectives of the different paradigms. From critical systems theory, of course, we learn that in order to protect paradigm diversity we need to exhibit "social awareness." The influence of power, politics, and culture can limit the range of paradigms and methodologies that it seems feasible to employ and so reduce the potency of pluralism. This is particularly the case with the radical change paradigm and methodologies associated with it. Because systems practitioners often work for powerful clients, there will be a tendency to slip into employing methodologies that support the status quo. Paradigm diversity demands that MM practice be protected from this by requiring it to give proper attention to the development and use of alternative methodologies based on the radical change paradigm.

There is a major difficulty facing those who want to operate in a multiparadigm manner. This is the problem of "paradigm incommensurability." Mingers and Brocklesby put it well:

> The paradigm incommensurability thesis asserts that because paradigms differ in terms of the fundamental assumptions that they bring to organizational inquiry, agents must choose the rules under which they practice from among the various alternatives on offer. They must then commit themselves to a single paradigm, although sequential movement over time from one paradigm to another is permissible. The main reason why multi-paradigm research is proscribed is because of the supposed irreconcilable objectivist/subjectivist ontological and epistemological dichotomies that exist …. But … there are other related dichotomies such as structure versus agency, determinism versus voluntarism, and causation versus meaning. The opposing positions in each dichotomy represent alternative competing 'truths' about the world and, as such, they resist reconciliation or synthesis.
>
> *(1997, p. 496)*

The sort of MM approach needed has to acknowledge this problem of paradigm incommensurability and find ways of dealing with it. Different critical systems thinkers have brought alternative options to the table, for example, seeking to resolve the issue using Habermas's theory of human interests, creating a new CST paradigm that sustains methodological pluralism, and subsuming the dichotomies under another theoretical position that seems capable of containing them. We will consider these possibilities later. Multiparadigm working also means that systems practitioners will often have to wrestle with contradictory understandings of a problem situation and contradictory recommendations for change.

There is one final distinction to be made which will help our thinking about MM practice. This is between what Pollack (2009) calls the "parallel" and "serial" uses of different methodologies. It is necessary to insist that methodological pluralism is employed according to "parallel" principles and applied at all stages of an intervention – the analysis stage right through to the action and reflection stages. It may be tempting to adopt a "serial" approach and allocate different methodologies to the various phases of an intervention because they seem most suitable to a particular phase, for example front-ending a study with a soft systems approach to deal with multiple perceptions before moving onto a harder methodology for implementation. IS researchers have found it very difficult to resist bolting SSM onto the front of structured design methods (see Mingers 1992). But there is no theoretical justification for this procedure. To those of an objectivist persuasion, carrying out the analysis stage according to a subjectivist rationale does not make it any "richer." It simply confuses those involved about what is actually going on and what is required. It will delay taking action and make implementation more difficult down the line. To subjectivists, the issues of culture, politics, and power cannot simply be made to disappear at the beginning of a project, never to be seen again. They will remain as a crucial backdrop in any intervention and must be attended to continuously as it progresses. Critical systems practice, therefore, insists upon the observance of methodological pluralism at each and every stage of a project. Of course, adopting a parallel approach to MM makes life more complicated. This is another problem with which critical systems practice has to cope.

We have argued that an "ideal-type" form of MM, one that can bring the greatest benefit to systems thinkers in managing complexity, requires that we operate in a multiparadigm, multimethodological, and multimethod manner at each stage of an intervention. This allows us to consider and rule out certain "candidate" multimethodologies from further consideration.

The door to MM was opened by the system of systems methodologies (SOSM). It demonstrated that the variety of methodologies was a strong point not a weakness of systems thinking. Different methodologies, based upon alternative paradigms, were suitable for different ideal-type problem contexts. However, it did not offer a fully developed MM approach. There is some attention given to viewing problem situations through different paradigmatic lenses:

> The problem solver needs to stand back and examine problem contexts in the light of different 'Ws' [*weltanshauungen*]. Perhaps he can then decide which 'W' seems to capture the essence of the particular problem context he is faced with. This whole process needs formalising if it is to be carried out successfully. The problem solver needs to be aware of different paradigms in the social sciences, and he must be prepared to view the problem context through each of these paradigms.
>
> *(Jackson and Keys 1984, p. 484)*

Beyond that, however, the SOSM is correctly described, by Mingers and Brocklesby, as being a case of "methodology selection" (1997, p. 491). Its emphasis is on selecting a whole methodology as appropriate for a particular situation. The use of different methodologies in the same intervention is not considered. Its pluralism is restricted to different interventions. A further criticism is the lack of distinction the SOSM makes between "methodology" and "method." As a result, methods are not seen as separable from the methodology with which they are commonly associated. This prevents multimethod practice. If you chose Checkland's SSM, you inevitably have to take rich pictures, root definitions, conceptual models, etc., as well.

White and Taket (1997) argue for a strategy of "pragmatic pluralism" in OR. This essentially allies MM with postmodernism. It makes sense given that one of the reasons provided for the growth of interest in MM is that it chimed with the postmodernist opposition to all "grand narratives." Postmodernism challenges totalizing endeavors and commits itself to promoting "difference" in a world it claims cannot be represented by the old paradigms and in which it has become impossible to guide our actions on the basis of the old moralities. Their experience in conducting evaluations

> ... has been that we need to find ways of working in situations which have a high degree of variety and in which acceptance and respect for difference is important .... Such situations display a high degree of heterogeneity .... The pluralist strategy described is based on the acknowledgement and respect of difference, rather than its rationalization.
>
> *(Taket and White 1995, p. 517)*

This strategy demands "judicious mix and match" of parts of different OR/systems methodologies to fit the requirements of each particular situation. The great merit of "pragmatic pluralism" is that it allows practitioners flexibility of method use so that they can cleave closely to what is appropriate in the problem situation and to the twists and turns required in an intervention. It is certainly multimethod but it is not multi-methodological or multiparadigm and, as a result, its weaknesses far outweigh its strengths. The main problems have already been identified. With this sort of a-theoretical "pragmatism," there is no guarantee of a diversity of methodology and paradigm usage. All the methods employed may be used according to one implicit paradigm. We lose the benefits offered by other paradigms and the advantages gained by employing them in combination. Nor can we guarantee attention is given to the possible need for practice aiming to bring about radical change. White and Taket's (1997) injunction to do "what feels good" hardly provides sufficient safeguard. Under "pragmatism," unless it happens to feel good, pluralism loses its radical potential. Finally, if we use methods without reference to the methodology and paradigm they are reinforcing, it becomes impossible to research their effectiveness for this and other purposes. There is no significant difference between "pragmatic pluralism" and management consultancy which, Ormerod (1996) believes, could also adopt postmodernism as an underlying philosophical stance.

Drawing upon the Oriental tradition, Gu and Zhu (2000) have translated their experiences of using OR/MS in China into a MM known as *wuli-shili-renli* (WSR). Zhu (2000) explains:

> WSR suggests that it is useful to see real-world projects as conditioned by a differentiable whole, i.e. the interplay among *wuli* (relations within the world), *shili* (relations between the self and the world) and *renli* (relations between the self and others). In conducting operations research (OR) and management projects, we are ideally inquiring into three domains, i.e. to investigate and model objective existence, to consider and reflect on subjective ways of seeing and doing, and to manage the working of intersubjective human relations. We call this knowing *wuli*, sensing *shili*, and caring, *renli* ... WSR urges practitioners to bring all *wu*, *shi*, and *ren* elements and perspectives into a holistic consideration, and accordingly to search for appropriate methods to address various *lis* as well as their dynamic interactions.
>
> *(2000, p. 184)*

The similarity of the concerns evinced by WSR and UK variants of MM are clear, although the differences in philosophical and cultural background make direct comparisons difficult. In general, it is probably fair to say that WSR adheres to a sophisticated version of "pragmatism." The same can be said of Sushil's (2018) "flexible systems methodology." This is a "mixed-method and multi-method" research framework, which begins by investigating a research problem on multiple dimensions and then selects research methods to suit the characteristics of the problem. The *Global Journal of Flexible Systems Management* has carried many applications.

Another variant of MM, developed by Bosch and Nguyen (2015) and called "Think2Impact," has been attracting attention recently. It is designed to address complex problems and to help governments and organizations to operate and thrive in an increasingly complex environment. The approach engages participants in "Evolutionary Learning Laboratories" (ELLabs) involving seven stages:

- A workshop gathering stakeholders' mental models
- Sessions in which stakeholders learn to integrate their models into a systems structure
- Developing "causal loop diagrams" showing components and their interconnections and assigning roles and responsibilities to different stakeholder groups
- Identifying leverage points for systemic interventions (SIs) that will help stakeholders achieve their goals
- Developing an "Integrated Master Plan," with systemically defined interventions, using Bayesian Belief Network modeling
- Implementing the strategies that will have the greatest impact, according to the evaluation criteria, and monitoring the outcomes
- Reflecting on the outcomes and identifying unintended consequences

Think2Impact has been used as the methodology for a number of successful applications in Australia, Vietnam, and Africa (for example, Banson et al. 2015). It does however fall short of being a fully fledged MM, on the basis of our "ideal-type," for a number of reasons. First, it is a "serial" application of different systems approaches, allocating particular methodologies to different stages and, therefore, missing out on what other methodologies might have to offer at each stage. Second, it conducts its serial applications using the same methodologies in the same order. This restricts the flexibility of the approach to deal with a variety of circumstances, leads to the initial methodology conditioning how the others are used, and hampers the ability to respond to changes as the intervention proceeds. Finally, it makes a limited selection from the range of systems approaches available.

In critiquing these "candidate" multimethodologies, from the point of view of the "ideal-type," it has to be recognized that what is an ideal-type in theory turns into an impossible dream once we try to apply it in practice. What can actually be achieved in an intervention will depend on the particular context, the need to achieve some objectives quickly, the knowledge and skills of those involved, etc. For example, WSR applied in the Chinese context may well have much more success than some imported MM that more closely resembles the ideal-type. It comes from the Oriental tradition of thought and simply makes more sense in that situation. Similarly, we should not decry the usefulness of sometimes employing just one methodology, embodying a particular paradigm, to guide the use of a variety of methods, tools, and techniques. That may be just what the context demands. We do insist, however, that the decision to proceed in this way is taken carefully. What the "ideal-type" provides is a critical standard that allows the practitioner to be self-reflective about how and why they are falling short of the ideal in a particular intervention and the consequences that might have. As we shall see in what follows, and in Chapter 21, critical systems thinkers have, in fact, suggested forms of MM that both adhere closely to the ideal and are more practitioner-friendly than initial sight of the "ideal-type" might lead us to think is possible. We will now consider three variants on MM theory and practice which, in my opinion, move us a long way in that direction: they are theoretically justifiable and can also be employed in practice. In each case, some background is given, the MM is set out, an example of its use is described, and a critique is provided.

20.2 Total Systems Intervention

20.2.1 Background

Once CST had been formulated as a philosophy and theory, it needed guidelines that would enable it to be applied in practice. These were provided in 1991a with the publication of Flood and Jackson's *Creative Problem Solving: Total Systems Intervention* – the first critical systems MM. Total Systems Intervention (TSI) was presented as a new approach to planning, designing, problem-solving, and evaluation based on critical systems theory.

The early days of CST, as we witnessed in Chapter 19, saw it providing theoretically informed critiques of different systems methodologies; debating pluralism in theory and practice; exploring the nature of the "improvement" it wanted to bring about, including how to act in "coercive" contexts; and engaging closely with social theory, especially the work of Habermas. TSI signaled its embrace of critical systems theory by declaring its commitment to "sociological awareness," "complementarism," and "human well-being and emancipation." It sought to pragmatize but not to compromise the theory underpinning CST. To this end, TSI saw itself as a "meta-methodology" which could organize and employ, in an appropriate manner, other systems methodologies. It developed as an approach for putting all the different systems methodologies to work in a coherent way, according to their strengths and weaknesses, and the social conditions prevailing, in the service of a general project of improving complex organizational and societal systems. This general project embraced efficiency, effectiveness, and viability; sought to promote mutual understanding; and, at the same time, to give attention to issues of empowerment and emancipation.

Flood and Jackson (1991a) and Jackson (1991a) see seven principles as underpinning the meta-methodology of TSI:

- Problem situations are too complicated to understand from one perspective and the issues they throw up too complex to tackle with quick fixes
- Problem situations, and the concerns and issues they embody, should therefore be investigated from a variety of perspectives
- Once the major issues and problems have been highlighted, it is necessary to make a suitable choice of systems methodology or methodologies to guide intervention
- It is necessary to appreciate the relative strengths and weaknesses of different systems methodologies and to use this knowledge, together with an understanding of the main issues and concerns, to guide choice of appropriate methodologies
- Different perspectives and systems methodologies should be used in a complementary way to highlight and address different aspects of problem situations
- TSI sets out a systemic cycle of inquiry with interaction back and forth between its three phases
- Facilitators and participants are engaged at all stages of the TSI process

Creative Problem Solving: Total Systems Intervention provided descriptions of TSI in action and, in later books, Jackson (1991a, 2003) and Flood (1995) provide many more examples. TSI attracted attention all over the world with further applications in areas such as organizational design, organizational learning, evaluation, sustainability, social change, and IS design.

20.2.2 Multimethodology

TSI aims to put into practice the commitments adhered to by critical systems theory. Briefly, it regards problem situations as messes that cannot be understood from only one viewpoint. For this reason, it advocates analyzing them from a variety of perspectives. Once agreement is reached among the facilitators and participants about the major issues and problems they are confronting, an appropriate choice needs to be made of systems methodology, or set of systems methodologies, for managing the mess and tackling the issues that have been identified. This choice should be made in the full knowledge of the strengths and weaknesses of available systems approaches as revealed, for example, by the SOSM. When selecting methodologies, it is important that the idea of pluralism is kept in mind. Different methodologies can be used to address different aspects of problem situations and to ensure that the technical, practical, and emancipatory interests are all given proper consideration. Furthermore, the initial choice of

Table 20.1 The total systems intervention meta-methodology.

Creativity	
Task	To highlight significant concerns, issues, and problems
Tools	Creativity-enhancing devices, especially systems metaphors
Outcome	Dominant and dependent concerns, issues, and problems identified
Choice	
Task	To choose an appropriate systems intervention methodology or methodologies
Tools	Methods for revealing the strengths and weaknesses of different systems methodologies (e.g. the SOSM)
Outcome	Dominant and dependent methodologies chosen for use
Implementation	
Task	To arrive at and implement specific positive change proposals
Tools	Systems methodologies employed according to the logic of TSI
Outcome	Highly relevant and coordinated change that secures significant improvement in the problem situation

methodology or methodologies must be kept constantly under review and may need to change as the nature of the mess itself changes. In this way, TSI seeks to guide intervention in such a way that it continually addresses the major issues and problems faced in an organization or multiagency situation.

The TSI meta-methodology has three phases, which are labeled *creativity, choice,* and *implementation*. It is summarized in Table 20.1.

The task during the *creativity* phase is to highlight the major concerns and issues that surround the problem context that is being addressed. Creativity-enhancing devices are needed to assist decision-makers and other stakeholders during this phase. In theory a full "paradigm analysis," drawing on the range of available social theory, is required. In practice something simpler, but serving the same purpose, is a must. One possibility was to make use of Linstone's "multi-perspective" research. In 1984 Linstone, another acolyte of Churchman's, had released a book, *Multiple Perspectives for Decision Making*, showing how taking three different viewpoints could yield a rich appreciation of the nature of problem situations. The Traditional or technical (T) perspective, dependent on data and model-based analysis, is augmented by an Organizational (O) or societal perspective, and a Personal (P) or individual perspective. The T, O, and P perspectives act as filters through which systems are viewed and each yields insights that are not attainable with the others. Linstone argues that the different perspectives are most powerfully employed when they are clearly differentiated from one another in terms of the emphasis they bring to the analysis but are used together to interrogate the same complex problem. One should not expect consistency in the findings – two perspectives may reinforce one another but may equally cancel each other out. In the event, it was decided to build on Morgan's (1986) conception of "images of organization" as this offered a wider range of different perspectives as well as providing a closer link to social theory. As a result, TSI came to adopt "systems metaphors" as its favored method to encourage decision-makers and other stakeholders to think creatively about the problem situation they were facing. It asks them to view the problem situation through the lenses provided by a rich array of systems metaphors. Different metaphors focus attention on different aspects of the problem context. Some concentrate on structure, while others highlight human and political aspects. By insisting on the employment of a comprehensive set of Morgan's "images," TSI ensures that those using it gain a holistic appreciation of the problem situation and take on broad perspectives that draw their meaning from different paradigms.

The systems metaphors developed for use by TSI were those of the organization as a:

- Machine
- Organism
- Brain
- Culture
- Political System ("team", "coalition", and "prison" settings)

These were seen to capture, at a general level, the insights of almost all management and organization theory. The sorts of question it is useful to ask during metaphor analysis are:

- What metaphors throw light onto this problem situation?
- What are the main concerns, issues, and problems revealed by each metaphor?

- In the light of the metaphor analysis, what concerns, issues, and problems are currently crucial for improving the problem situation?

The key aspects of the problem situation revealed are subject to discussion and debate among the facilitators, decision-makers, and other stakeholders. If all the metaphors reveal serious problems then, obviously, the organization is in a crisis state! The outcome (what is expected to emerge) from the creativity phase is a range of significant issues and concerns. There may be other important but less immediately crucial problems that it is also sensible to record and pursue into the next phase. These "dominant" and "dependent" issues and problems then become the basis for designing an appropriate systems intervention approach.

The second phase is known as the *choice* phase. The task during this phase is to construct a suitable intervention strategy around a systems methodology or combination of systems methodologies. Choice will be guided by the characteristics of the problem situation, as discovered during the examination conducted in the creativity phase, and knowledge of the particular strengths and weaknesses of the different systems methodologies. A critical device is therefore needed that is capable of interrogating these methodologies to reveal what they do well and what they are less good at. Jackson (1991a) showed that metaphors, sociological paradigms, Habermas's three human interests, and positioning in the modernism versus postmodernism debate, could all be called upon for this purpose. Again, however, TSI needed something that was easier for practitioners to grasp but which would not compromise the principles of CST. The SOSM was, therefore, the obvious tool for TSI to employ in the choice phase. As we know, it unearths the assumptions underlying different systems approaches by asking what each assumes about the system(s) in which it hopes to intervene and about the relationship between the stakeholders associated with that system. Combining the information gained about the problem context during the creativity phase, and the knowledge provided by the SOSM about the strengths and weaknesses of different systems approaches, it is possible to move toward an informed choice of systems intervention strategy. For example, if the most urgent issues in the problem context have been revealed as technical ones that need fixing, and there is little dispute about these, then a hard systems methodology based on simple–unitary assumptions can be chosen at the beginning of the intervention with every hope of it bringing improvement. The most probable outcome of the choice phase is that there will be a dominant methodology chosen, which will be supported if necessary by dependent methodologies to help with secondary problem areas.

The third phase of TSI is *implementation*. The task is to employ the selected systems methodology or methodologies with a view to bringing about positive change. If, as is usual, one methodology has been deemed dominant, it will be the primary tool used to address the problem situation. TSI stipulates, however, the need always to be aware of the possibilities offered by other systems methodologies. For example, the key problems in an organization suffering from an inability to learn and adapt may be structural, as revealed by the organism and brain metaphors. But, the cultural metaphor might also appear illuminating albeit in a subordinate way given the immediate crisis. In these circumstances, organizational cybernetics might be chosen to guide the intervention but with a SSM taking on other issues in the background. Of course, as the problem situation changes, it will be necessary to reassess the state of the problem situation, by re-entering the creativity phase, and possibly selecting an alternative methodology as dominant. The idea of naming dominant and dependent methodologies, and switching between then as a study proceeds, is another device introduced by TSI to make it more usable in practice while adhering to the philosophy of CST. It is argued that the practical difficulties associated with multiparadigm working can be managed if an initial choice of dominant methodology is made, to run the intervention, with a dependent methodology (or methodologies), reflecting alternative paradigms, in the background. The relationship between dominant and dependent methodologies can then change as the intervention proceeds in order to maintain flexibility at the methodology level. This remains, for me, an extremely powerful idea because it allows the intervention to proceed in a theoretically informed way, making research possible, and protects paradigm diversity while being less confusing to participants. As long as we are explicit about our initially dominant methodology, and are ready to switch as necessary, proceeding initially with one methodology is not damaging. That initial choice does not exclude us from introducing alternative methodologies, based on different paradigms, as the situation demands.

For completeness, it should be remarked that there is some theoretical justification for the strategy of alternating dominant and dependent methodologies. The philosopher Althusser (see Jackson 2000, p. 45) conceived of the social totality as a system that, at various times, is dominated by one of its "instances": economic, political, theoretical, ideological, etc. In order to act to change the social totality, Althusser argued, you need to understand the relationships between these "instances," how each is developing and which one is dominant in a particular era of history. We are simply replacing the notion of alternating dominant "instances" with the idea of improving efficiency, viability, mutual understanding, fairness, etc., with each becoming the dominant concern in a problem situation for some period of time. Over the longer time period all will need to be addressed. It is just impossible, in practice, to do everything at once.

The immediate outcome of the implementation phase should be coordinated change brought about in those aspects of the problem situation currently most in need of improvement. The longer-term prognosis is of more work so that the process of improvement continues in the face of an increasingly turbulent environment. It is important to stress that TSI is a systemic and iterative process. As the problem situation changes in the eyes of the participants, a new intervention strategy will have to be devised. The only way to attend to these matters is to continually cycle around creativity, choice, and implementation, ready to change those methodologies that are dominant and dependent. TSI is a very dynamic meta-methodology:

> The essence of TSI is to encourage highly creative thinking about the nature of any problem situation before a decision is taken about the character of the main difficulties to be addressed. Once the decision has been taken, TSI will steer the manager or analyst towards the type of systems methodology most appropriate for dealing with the kind of difficulties identified as being most significant. As the intervention proceeds, using TSI, so the nature of the problem situation will be continually reviewed, as will the choice of appropriate systems methodology. In highly complex problem situations it is advisable to address at the same time different aspects revealed by taking different perspectives on it. This involves employing a number of systems methodologies in combination. In these

circumstances it is necessary to nominate one methodology as 'dominant' and others as 'supportive', although these relationships may change as the study progresses.

(Flood and Jackson 1991a, pp. xiii–xiv)

Since its formulation TSI has been taken in rather different directions by its two originators, Flood and Jackson. In a 1995 book, Flood suggests additions to the methods that can be used in each of the three phases of the meta-methodology and specifies three "modes" in which TSI can be used. These are the traditional "problem-solving" mode, the "critical review" mode, and the "critical reflection" mode. The critical review mode applies TSI to the assessment of candidate methodologies that might be incorporated in the meta-methodology. It is an elaboration of critical awareness. The critical reflection mode sees TSI used to evaluate its own interventions after the event in order to improve TSI itself. This is indeed essential for ensuring that TSI fulfills its obligation to pursue research as well as practice. In further books (Flood and Romm 1996; Flood 1999), Flood has sought to expand on the relationships between TSI and postmodernism, and TSI and chaos and complexity theory. My own more recent work has focused on strengthening TSI and dealing with its weaknesses. This led me through "creative holism" (Jackson 2003) to what I now call "critical systems practice." This development is covered in Chapter 21.

20.2.3 Case Study

This study was undertaken in Hull Council for Voluntary Service in 1997. The contribution of Mary Ashton, as principal researcher, was crucial to its success. A fuller account of the project can be found in Flood and Jackson (1991a, pp. 224–240).

Councils for Voluntary Service (CVS) act as umbrella organizations for a wide variety of other voluntary bodies they have in their membership. They are local development agencies that are non-profit-making and nongovernmental. There were about 200 CVS in England and Wales at the time. CVS aim to promote more and better voluntary action in their areas. Hull was a large CVS. It was founded in 1980 and had grown rapidly in size and influence. By the time of the study, it had more than 300 voluntary and community organizations under its umbrella and employed around 80 staff.

Hull CVS had a number of problems, but the one on which the project came to focus concerned certain difficulties faced by its Executive Committee. This committee was experiencing problems trying to oversee and control what was a rapidly expanding organization in a turbulent environment. It continued to operate as it did when the CVS was first founded, meeting every six weeks for a programmed two hours. The committee did not possess the flexibility to respond to the needs of the CVS Hull had become, and was widely perceived to be ineffective. The Executive Committee was subordinate to a Council in the CVS hierarchy. The Council consisted mainly of representatives of member organizations and met once a year at an Annual General Meeting (AGM). The Executive Committee was democratically elected by the Council at the AGM. Each member of the Executive was elected for a three-year term and one-third of the members were elected each year. They were, of course, unpaid. Two induction sessions were held for new Executive members. The Executive's job was to make policy by representing and refining the broad judgment of the Council and translating this into specific guidelines for action by the CVS. These relationships are shown in Figure 20.1.

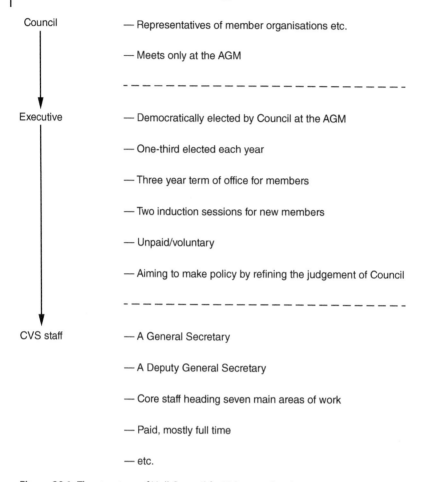

Council — Representatives of member organisations etc.

— Meets only at the AGM

Executive — Democratically elected by Council at the AGM

— One-third elected each year

— Three year term of office for members

— Two induction sessions for new members

— Unpaid/voluntary

— Aiming to make policy by refining the judgement of Council

CVS staff — A General Secretary

— A Deputy General Secretary

— Core staff heading seven main areas of work

— Paid, mostly full time

— etc.

Figure 20.1 The structure of Hull Council for Voluntary Service.

The Executive consisted of a Chair, Vice-Chair, Treasurer, 18 other representatives elected by Council, and 4 co-opted members (a City Councilor, a County Councilor, and two direct from Council). The General Secretary of the CVS took the minutes and, although he had no voting powers, he exercised a strong influence through his knowledge (as the senior full-time person on the executive) and power of recommendation. Other CVS staff could attend but, again, had no voting rights. The Executive was supported in its work by a Finance Committee. This had certain delegated powers, also met once every six weeks and provided reports and recommendations for the Executive to consider. It was serviced by the Deputy General Secretary of the CVS, and was made up of five Executive representatives and a voting staff representative. Other *ad hoc* groups were occasionally set up by the Executive to oversee and report on particular developments. This structure is shown in Figure 20.2.

Items found their way on to the agenda of Executive Committee meetings from a variety of places. They could emerge from the General Secretary, from the Finance Committee, the *ad hoc* groups, from core-staff members heading areas of work, or simply as Executive submissions (especially from the Chair and Executive members on

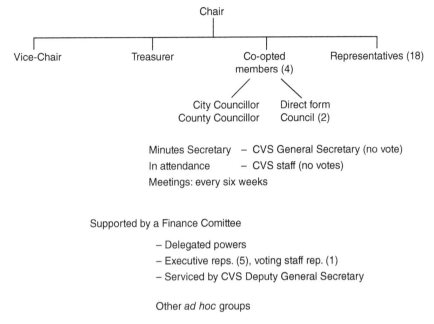

Figure 20.2 The structure of Hull CVS Executive Committee.

outside committees). There was some filtering of the items for discussion, usually by the General Secretary, but nevertheless the agenda was severely overloaded. Meetings had become lengthy and acrimonious. The Executive and staff both felt that the Committee was not providing the direction required. Two-thirds of the issues that came up for discussion concerned "management" rather than the policy matters on which the Executive was supposed to concentrate. After Executive meetings, minutes were circulated to all section heads for discussion. The General Secretary and Deputy General Secretary were responsible for monitoring the execution of policy as decided at the meetings.

The project began with the gathering of information about the Executive Committee, its structure and role, and the views held of it by both Executive Committee members themselves and CVS staff. This "creativity" stage of the intervention consisted of lengthy interviews carried out with eight key Executive members and nine staff. Each of these was followed up with a second interview, with a check made that the interviewer had fully grasped the subject's meaning. The "cognitive mapping" technique, developed by Eden et al. (1979) as part of their Soft-OR approach, was employed for this purpose. Questionnaires were sent to all other Executive members and to staff who had recently attended Executive meetings (35 questionnaires in all). Summarizing greatly, the views of Executive members divided into those with an "efficiency orientation" and those with a "suspicion orientation." The efficiency-oriented were frustrated by the inability of the Executive to get important business done. They felt that agendas were too lengthy and not prioritized. They wanted each item to come with a clear recommendation. Meetings became bogged down, they thought, because of the massive overload of work, the chronic shortage of time for discussion, and because people attended "cold" without digesting the information. Worse, there was poor committee discipline, with lots of sidetracking and standing orders not enforced. The efficiency-orientated felt that the Executive was not concentrating enough on policy. It was wasting time on management issues that the staff should take care of. Unfortunately, they thought, some members did not trust the key staff enough to "keep their noses out" of management issues. The suspicion-oriented were distrustful of key figures on the Executive and of the Executive's role. They wanted to promote more debate to find out what was going on. They felt that key officers had too much power and that agenda filtering was preventing important issues reaching the Executive. Executive meetings were too formal and hierarchical and this suppressed participation and creativity from ordinary members. The suspicion-oriented wanted to know more about what was actually going on in the CVS. Staff views could also be characterized as falling into two broad types. One group saw Executive shortcomings as having developed as a natural result of the rapid growth in activities of the CVS. They felt that the CVS had grown too rapidly and it was now time to apply the brakes. The Executive was overloaded and could not give enough consideration to staff views. Further, because of bad filtering, staff views did not get through to the Executive. For these reasons, it was not surprising if some Executive members felt out of touch. There was a need for more contact between Executive and staff. This could provide learning for the Executive and support for the staff. The second group was much more critical of the Executive and saw its shortcomings as self-inflicted. They thought that the Executive had not adapted to change and was, therefore, a stumbling block to getting things done. It was largely cut off from the reality of CVS work on the ground. More contact with staff was needed to "wake up" Executive members. From the cognitive maps, interviews, and other information gathered, it seemed that the following were very significant issues that had to be addressed:

- Providing more time for the Executive committee to deal with policy issues
- Increasing the professionalism of the committee's handling of management issues
- Improving the handling of committee business
- Generating mutual respect between committee members and staff
- Making the committee more aware of staff work
- Increasing committee contact with staff
- Increasing staff confidence in the committee

We now entered the "choice" phase of the study. Which systems methodologies, methods, models, etc., would allow Hull CVS to best manage the problems it faced? The SOSM provided the basis for this. It was easy to see that the CVS had become a "complex" organization, which was having difficulty trying to grow at the same time as adapt in a highly turbulent environment. This emphasis on complexity initially drew us toward Beer's VSM. The Executive of the CVS could be seen as the brain of an organism that did not possess enough requisite variety to control the system it was supposed to direct and manage. It had somehow to be equipped with the various functions of management exhibited by Beer's VSM. As we shall see, this proved a particularly useful metaphor. We were, however, "critically aware" of the weaknesses of the VSM. Presenting a report based upon a viable systems diagnosis might have provided some useful guidelines on how to deal with the structural problems of the Executive Committee, but it would have ignored other very important aspects of the problem context. For the CVS was nothing if not a pluralistic coalition of different groups, all with somewhat different interests and ways of perceiving the situation. A way had to be found to generate a consensus for change among the elements of this coalition. What was surely needed was a change in the culture of the organization, so that it was ready to accept change and particularly to rethink the way the Executive functioned. If these matters were not tackled then any "rationalistic" report might fail to achieve improvements because of the opposition it generated or because it failed to gain the commitment and enthusiasm of the most involved agents. There had, in fact, been previous internally generated suggestions for improving the performance of the Executive. These were sound enough and we were unlikely, as outsiders, to discover any magical solutions that had not occurred to those already living in the situation. The role of the project, therefore, had to be seen as generating a "culture" for change in the organization; and change that did not offend any of the groups in the coalition. At the same time, we felt that the eventual design had to meet cybernetic criteria of viability. The Executive had to become an effective "brain." This led to the choice of SSM as the dominant methodology with the VSM as a dependent model. SSM rests upon pluralist assumptions and articulates particularly well the concerns of the culture metaphor in its procedures. The VSM could be supportive in dealing with complexity because of the explicit understanding it brought of "brain-related" issues – learning, forward thinking, etc. Finally, it seemed as well to be aware of the significant political aspects to the problem context. These, at times, threatened to take the situation beyond the pluralistic toward the conflict and coercion end of the "stakeholders" dimension of the SOSM. There was conflict between some on the Executive and some staff, and on the Executive between those happy to leave responsibility for "running the show" to the staff and those who tended to suspect the motives of the key officers. The General Secretary's position as an important broker between Executive Committee and staff could clearly be threatened by any suggested redesign of the Executive. Particular attention had to be paid to the most influential individuals associated with the various interest groups in order to gain their support and trust. However, in this case, we thought, the political aspect could be handled informally within the bounds of SSM.

Embarking on the "intervention" phase, using SSM, the first task was to build a "rich picture" of the problem situation. In this case, it was deemed sufficient to work with the verbal rich picture assembled during the interviews and the cognitive mapping. The next step, therefore, was to consider all this information and to draw from it some insightful ways of looking at the work of the Executive Committee and the problems that it faced. Six "relevant systems" were proposed:

(a) *A policy processing system*: handling policy so that other bodies can execute or be guided by it
(b) *A need-seeking and idea-generating system*: helping the organization to seek and develop new initiatives
(c) *A representation system*: expanding the voluntary sector's voice on other crucial decision-making bodies
(d) *A monitoring and controlling system*: embracing the classical role of a management committee
(e) *An accountability system*: ensuring that the organization is seen to conduct its affairs competently in the eyes of those to whom it is accountable
(f) *A staff support system*: practically demonstrating support toward those appointed to manage the organization

During the course of the work on relevant systems, it occurred to the analysts that there were links between what was emerging and the roles any management or "meta-system" would have to perform in seeking to control a highly complex set of operations. In other words, it was worth thinking in "cybernetic" terms. It was decided to make this explicit and to see whether our relevant systems "covered the ground" in terms of the functions a meta-system has to fulfill. This was done by locating the function of each relevant system on Beer's VSM. Relevant system (a), *policy processing*, was clearly at System 5 level; relevant systems (b) and (c), *need-seeking* and *representation*, were at System 4 level. Relevant systems (d) and (e), *monitoring and controlling* and *accountability*, were System 3 and System 3* audit functions, respectively; while relevant system (f) *staff support*, operated at the System 3 and System 1 levels. This information is captured in Figure 20.3.

The fact that our choice of relevant systems had this additional cybernetic legitimacy gave us confidence. We were clearly thinking along the right lines if we wanted to provide the Executive with the "requisite variety" to manage the organization. Throughout the study we continued to employ this cybernetic rationality in a subordinate role, thinking about how the VSM functions could be carried out in Hull CVS. From the outside, indeed, it might look as though we could have taken a short cut round all the information-gathering, interviews, and questionnaires, and used Beer's VSM directly to pin-point cybernetic faults. This, though, would have been a mistake. We had to engender change in the culture of the organization to create a momentum for "redesign," and we had to hold together the various factions in the coalition, securing the support

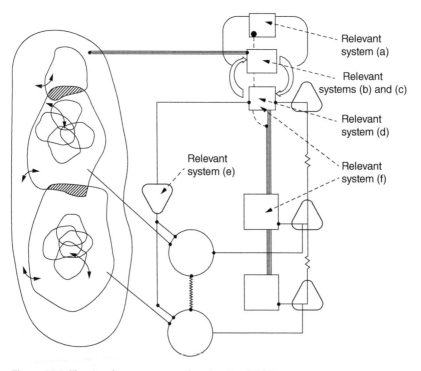

Figure 20.3 The six relevant systems related to Beer's VSM.

of each for the proposals. At the same time we had to sidestep and manage the political problems. Only constant working with the people in the organization, so that they were fully involved in generating proposals and came to own the suggested solutions, could address these issues. SSM had to remain dominant if anything was going to change in Hull CVS. At no time was the "brain" driven logic allowed to take over from the "culture" and "coalition" emphasis supplied by SSM.

A preliminary pass through SSM revealed that four of the six relevant systems were deemed most significant to the main area of concern – the ability of the Executive to control the organization. According to the requirements of SSM, these four were then built into "root definitions" and "conceptual models." By way of example, the root definition and CATWOE check for relevant system (a), and the conceptual model derived from this root definition, is supplied as Figure 20.4.

The implications of the four root definitions and conceptual models were then fully discussed with members of the Executive Committee and the staff. Using a set of guiding questions, a comparison was drawn up between the conceptual models and the "real-world situation" as expressed in the rich picture. Those interviewed were asked whether the activities in the conceptual models existed in the "real world" or not; if they did exist, "how are they done at present and are they done well?"; and if they did not exist or were done badly, "are they feasible and how might they be carried out effectively in the 'real-world' situation of Hull CVS?." From the comparisons, an agenda for further and wider debate was drawn up. By way of example, I can set out the points that emerged from conceptual model (a), *policy processing*:

- The need for more time for the Executive to consider major policy issues and to review implementation
- The need for a better way of consulting core-staff about basic matters of policy implementation
- The need for a substructure within which the Executive could improve its knowledge and understanding of the organization they make policy for
- Requirements for a method by which the Executive could more closely monitor the execution of policy and be aware of any necessary control action

On the basis of the discussions that took place around the agenda generated by all four conceptual models, we began to think about possible changes that would help to alleviate the difficulties. Most of the proposals came from suggestions made by various CVS officers in discussion with us. We would then bring these up in meetings with other personnel to gauge their reaction. This was particularly the case with specific suggestions for change, such as one controversial idea to establish a "management committee" to aid the Executive. We were acting as brokers between the interested parties, and moving forward only with ideas that seemed to attract general assent or, at least, failed to provoke severe disagreement. The political situation made it simply impossible to bring all the significant actors together, at one time and place, to hammer out an agreed set of proposals. The likelihood of such a meeting breaking up in disarray, and taking the proposals down with it, was too great. We also had our own "expert-driven" cybernetic agenda that contained a set of minimum specifications we felt any changes should meet in order to make the organization "viable." Top of this agenda was to see management issues handled lower down, thus reducing the "variety" flooding up to the Executive and exhausting its capacity to handle significant policy issues.

Root definition (a): A policy processing system

"A Hull CVS owned policy processing system which aims to represent the broad judgment of the Council and is thus able to create, develop and put into effect execution of CVS policy on its behalf; within the constraints of time and resources available to Executive Committee members and the organization."

C = CVS employees, the Council
A = The Executive Committee
T =

| Broad judgement of the Council | → | Created, developed and "put into effect" CVS policy |

W = The Executive Committee carries the ultimate responsibility for committing the CVS to a particular course of action
O = Hull CVS, that is, The Council (the Membership)
E = Time and resources available to the Executive Committee members and the organization

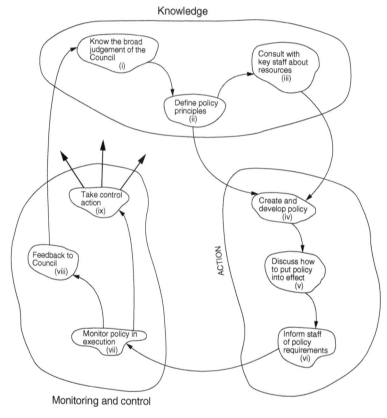

Figure 20.4 Conceptual model of the policy processing system (root definition a) for the CVS example.

Eventually, through a long drawn out and time-consuming process of going back and forth between important Executive members and staff, and constantly modifying the recommendations, we arrived at proposals we believed had general support and, to some degree, met the cybernetic criteria. A report was prepared with recommendations in three areas: structural, procedural, and attitudinal. These are set out in detail in Flood and Jackson (1991a). They included moving management and auditing tasks down to specialist committees, thus leaving time for policy discussion at the Executive Committee meetings; setting up Committee "support groups" for CVS staff in important areas of work; and providing induction and training sessions for new Executive Committee members. We were rewarded for this hard work when the recommendations were presented at an Executive meeting. The significant actors had come to own them as their own, and there was no opposition to the setting up of a subcommittee charged to oversee their implementation. The operating procedures of the CVS Executive Committee were substantially restructured as a result.

The Hull CVS intervention used an early form of TSI as the means of operationalizing CST. The "creativity" phase used interviewing and cognitive mapping as its main instruments. Metaphor analysis had not yet become a formal element in TSI. Looking back, however, it is clear that we were initially attracted to the "organism" and "brain" metaphors, but eventually allowed the "culture" and "politics" metaphors the primary influence. At the "choice" stage, governed by the SOSM, SSM was taken as the dominant methodology and this remained the case throughout. The VSM did however play a strong supporting role. We also had to reflect constantly on the complicated politics of the problem situation, although it did not become necessary to adopt a systems methodology specifically to address these issues. In general terms, I think it is fair to say that the Hull CVS case study demonstrates most of the facets of CST as it was understood at that time. It was conducted with constant critical reflection upon the relative strengths and weaknesses of the systems methodologies used. Different systems approaches were employed in a complementarist manner at both the theoretical and methodological levels. The project was emancipatory at least in the sense that it was assisting an organization committed to helping the most disadvantaged members of the community.

20.2.4 Critique

We begin by assessing TSI using the ideal-type of MM theory and practice and then move on to other criticisms of the approach.

TSI dismisses "isolationism." It is a multimethodolgy and multiparadigm approach. It insists on the value of the various systems methodologies that have been developed and seeks to put them to use in situations and for purposes for which they are most suited. Nor does it have any truck with "imperialism." Imperialists, it will be recalled from Chapter 19, are prepared to entertain the use of different methodologies and methods but only if they can be forced to fit their own favored paradigmatic assumptions. TSI suggests a way of managing very different methodologies in a manner that respects their foundations in alternative paradigms. This, I would argue, puts it a step ahead of other systems approaches in dealing with the complexity of the problem situations that decision-makers face today. At the same time, TSI rejects "pragmatism," insisting that the use of a variety of methodologies and methods must remain theoretically informed. Only then is it possible to guarantee paradigm and methodological diversity and ensure that learning can take place and be passed on to others.

As a MM approach, the breakthrough achieved by TSI, noted by Mingers and Brocklesby (1996, p. 492), is to advocate using methodologies adhering to different paradigms in the same intervention. This takes it beyond SOSM's "methodology selection" of the most appropriate methodology for a particular problem context. TSI is a meta-methodology that seeks to manage, in a coherent way, very different methodologies premised upon alternative theoretical assumptions to address the same problem situation. TSI suggests that it is more correct theoretically to use different methodologies alongside one another when dealing with highly complex problem situations. But, if this proves too difficult it has a way forward. The best way, in practice, to handle methodological pluralism is to clearly state that one methodology is being taken as "dominant," with others "dependent" for some period of time, being always willing to alter the relationship between dominant and dependent methodologies as the situation changes. One methodology, encapsulating the presuppositions of a particular paradigm, is granted "imperialistic status," but only temporarily. Its dominance is kept under continual review.

Another major strength of TSI, in line with the ideal-type, is its "parallel" use of different paradigmatic perspectives at the "creativity" stage as well as during its "choice" and "implementation" phases. Using "systems metaphors," and other devices as necessary, TSI is able to look at problem situations from a variety of very different viewpoints. This redeemed the pledge made by Jackson and Keys (1984, p. 484) to formalize the process whereby the problem solver stands back to examine problem contexts in the light of different "Ws." Finessing the dominant/dependent maneuver, TSI can then translate its findings into the later stages of its MM when different methodologies are employed in the best practical approximation possible to the parallel strategy advocated in our ideal-type of MM. It seems to me that using the different systems metaphors at the "creativity" stage answers, to some extent, Ulrich's (2003, 2012a,b) criticism that TSI only concerns itself with "methodology choice." In his view, TSI offers a "contingency framework for methodology choice and use" (Ulrich 2012a, p. 1243), which restricts its moment of critical reflection to the point at which a professional makes their choice of an appropriate methodology and associated methods:

> TSI ... consequently puts the critical focus on the *theoretical* underpinnings of alternative research paradigms rather than, as does critical systems heuristics (CSH), on the normative core of professional practice. Critical systems thinking, thus understood, promotes reflective practice with respect to these theoretical underpinnings; the central idea is to support a theoretically informed process of matching methodologies with problem contexts.
>
> *(Ulrich 2012a, p. 1239)*

In fact, TSI also requires reflective practice at the "creativity" stage where it brings different paradigmatic perspectives, in the shape of systems metaphors, to its analysis of problem situations. It can be argued that this reflection is deeper than that which CSH can offer because it is multiparadigm rather than being limited by a particular interpretation of critical thinking derived from Kant, Habermas, and Churchman. In his more recent papers (2012a,b), Ulrich does seem to acknowledge this reflective aspect of TSI's "creativity" stage. All that said, Ulrich makes one important criticism of TSI that is undoubtedly correct. In his view, critical reflection must extend to all stages of a professional engagement and TSI does not, in its original form, extend it to examining the results of its interventions. As Ulrich puts it:

> Critical methodology choice does not preclude a chosen methodology to produce, in each specific application, consequences in need of critical reflection. Whom do these consequences affect and in what way? Who bears the cost? Who gets the benefits? What ethical conflicts are involved? What is the basis of legitimation? What kind of improvement, if any, results? What other definitions of improvement would result if different contexts of application, and hence different facts and values were considered? And so on. From a critically-heuristic perspective it appears obvious that critique is more – much more – than informed methodology choice.
>
> *(2003, p. 336)*

It is embarrassing to have to admit that, in its original form, TSI's only means of addressing this issue would have been the clumsy one of re-entering the methodology and starting a new intervention. This matter was not fully addressed until Jackson (2003) incorporated a "reflection" phase into the MM as part of his development of TSI into "critical systems practice." Ulrich (2012a, p. 1242) recognized this reflection phase as bringing critical systems practice "a bit closer to CSH's focus on reflective practice." In my view, its multiparadigm orientation makes it a richer form of evaluation than CSH can muster. It also completes the job of ensuring that at least one MM can achieve the "parallel" use of different paradigmatic perspectives at all its stages – "creativity," "choice," "implementation," and "reflection."

If TSI concentrates at the meta-methodological level, to ensure that methodologies embodying different paradigmatic assumptions are used in combination, and at all stages of an intervention, this may also have led to one of its main weaknesses. It gives little explicit attention to "multimethod" practice – one of the key requirements of the ideal-type of MM. According to Mingers (1997c), although some case studies exhibit the linking of parts of different methodologies TSI, like the SOSM, talks as though it requires the use of "whole" methodologies. For this reason, Mingers and Brocklesby (1996, p. 492), reasonably enough, refer to TSI as an example of "whole methodology management." Midgley (2000, pp. 225–226) makes the same point. Once SSM is chosen as the dominant methodology, for example, it seems that you must only employ the particular methods and techniques set out in that approach. You can't suddenly start using techniques, perhaps causal loop diagramming or assumption surfacing, extracted from other methodologies. There is an unnecessary lack of flexibility here which had to be addressed. There is nothing theoretically wrong with using a selection of methods and techniques as long as they are consciously employed according to an explicit logic, interpretive in this case, and this allows a much greater responsiveness to the peculiarities of each problem situation in the beginning and as it evolves during an intervention.

Tsoukas (1992, 1993a,b) has provided a detailed critique of TSI that takes us beyond its capacity just to function adequately as a MM. Although most of his points have been rebutted (Green 1993; Jackson 1993b; Midgley 1993), and only a few will be repeated here, a criticism directed at its philosophy and theory deserves careful consideration.

As we saw, TSI grounds its pluralism, or "complementarism," on Habermas's early theory of human interests. This theory suggests that TSI can, on the basis of the three interests, stand "above the paradigms" and pick out appropriate methodologies according to the particular human interest to be served. Tsoukas, however, notes that different paradigms constitute different realities and, therefore, seek to provide answers to all three human interests:

> Different paradigms constitute different realities, and as such, they provide answers, either explicitly or implicitly, to *all* three human interests. Positivist problem-solving, for example, is not simply useful for achieving technical mastery over social processes. In attempting to do so, it also provides answers to the inextricably interwoven questions of interaction and power.
>
> *(Tsoukas 1993b, p. 314)*

Tsoukas is arguing that Habermas' theory cannot simply dissolve the arguments between the paradigms. Midgley (1989, 2000) agrees, arguing that Flood and Jackson are wrong to see it as something that can take us above and beyond interparadigm debate. Mingers (1997c, p. 412) adds that the theory of human interests was always an epistemological study of different forms of knowledge and "never claimed to overcome all problems of paradigm incommensurability."

I think this argument, criticizing TSI's use of Habermas's theory as a "meta-paradigm", is unanswerable. If evidence is needed, it comes in the form of the numerous criticisms that have been leveled at TSI from postmodernist and poststructuralist theoretical positions. These "paradigms," it seems, are very reluctant to be included within the Habermas/TSI repository of theories. Postmodernists are opposed to all "grand narratives" and regard Habermas' version of Enlightenment philosophy, embracing the progressive emancipation of individuals so they can take charge of their own destiny, as one of the most influential and dangerous of these. Taket and White (2000) view TSI as a "totalizing" endeavor that is in thrall to this narrative. Despite its proclamations to the contrary, they see it as an approach that seeks to tame pluralism and diversity rather than provoke them. The emphasis on rigor and formalized thinking in TSI sets up a tension, they believe, with the espoused purpose of employing a plurality of methodologies and methods. A deconstruction of the language of TSI reveals a contradiction between statements that imply closure and those which encourage openness to other approaches and ways of proceeding. Taket and White also worry that the emphasis on rationality and abstraction in TSI leads to the privileging of methods that are verbally based and that this can hinder the participation of some individuals and groups. Another problem with giving primacy to rationality is that feelings and emotions get ignored. For Tsoukas (1992), TSI's commitment to "human well-being and emancipation" is all very well, but where is the unitary social group waiting to be liberated? And how can we know what emancipation means for individuals unless we hold to some a-social, essentialist conception of human beings. Mingers (1997c) argues that Foucault's work on power/knowledge, from a poststructuralist position, undermines the modernist notion of rationality. Knowledge, indeed rationality itself, is always intertwined with power relationships. There is no standpoint, outside of history and particular social relations, where some pure form of knowledge can manifest itself. Mingers, while taking Foucault's conclusions seriously, believes that they do not "totally destroy the critical potential of knowledge" (1997c, p. 417). I take the same view of the whole postmodernist/poststructuralist critique of the Enlightenment project. I am with Habermas (1987) that the critique points to problems to be overcome in pursuit of that project rather than providing a cause to abandon it.

Another frequent criticism of TSI is that it says little about the "agent" carrying out the intervention using the meta-methodology. Taket and White (2000) believe there is insufficient discussion in the literature on the "roles" and "styles" that facilitators can adopt. It is certainly true that TSI, which demands multimethodological and multi-method competence, asks a great deal from would-be users and does not detail how the relevant competence can be obtained. Ormerod (1997b), by contrast, makes the "intervention competence" of the analyst central to his "transformation competence" approach to multimethod practice and reinforces the argument by setting out the development of his own intervention competence in the course of seven consultancies. Stacey and Mowles make the point that TSI represents its practitioners as "autonomous individuals subject to the causality of free choice" while, once a methodology is chosen, everyone becomes "subject to the formative causality of the system they have chosen" (2016, p. 215). They are right in the first half of this even if Stacey's continued failure to understand soft and critical systems thinking leads them astray thereafter. Midgley (1997a) argues that any new approach to pluralism in systems thinking must take into account the dynamic interaction that occurs between the subject who wishes to take action and the power-knowledge formations that form the identity of the subject. Mingers (1997c) also makes the agent the focus of his "critical pluralism." For him, any consideration of critical action has to focus upon "the actual, embodied, and embedded agent(s)." The fact that methodologies carry the critical tag, and prescribe emancipatory practice, cannot guarantee their critical employment – the commitment of the user, embedded in a particular social context, is crucial.

A final set of criticisms of TSI comes from those (Tsoukas 1993a,b; Mingers 1997c; Taket and White 2000) who question whether TSI lives up to its claims to promote "emancipation." It is accused of being strong on assertion but weak in terms of demonstrating any such commitment. Midgley (2000), developing this criticism, charges TSI with partial radicalism. It declares itself for human emancipation but ignores environmental concerns. For Midgley the two are inextricably linked. CST needs to have more to say about building sustainable relationships between people and the nonhuman environment.

In conclusion, we must return to our acceptance that TSI can no longer rely on Habermas' theory of knowledge-constitutive interests for its theoretical justification. Even Habermas, as Spaul (1997) recounts, believes the theory is no longer defensible. But if CST cannot be regarded as a "meta-paradigm," standing above other theoretical positions and allocating them to appropriate tasks, then what exactly is it? Midgley, having posed the question, answers that it is best seen as a paradigm in its own right. He objects to Flood and Jackson's (1991b) claim that CST is meta-paradigmatic:

> I would argue that this cannot be the case given that Flood and Jackson make assumptions about human knowledge drawn from Habermas' ... theory of knowledge-constitutive interests. These assumptions are alien to, and incommensurable with, assumptions made by the proponents of other systems paradigms. Far from being meta-paradigmatic, CST embodies its own unique

assumptions, meaning that its proponents are trying to establish the foundations for a *new* paradigm.

(1996, p. 15)

Mingers agrees and starts to outline what this new paradigm would be like. It would need to subsume within it other subparadigms:

This would allow the competing assumptions to be reconciled within some wider framework; methodologies to be partitioned and mixed; and the foundational assumptions of the new paradigm to be developed and explored.

(Mingers 1997c, p. 411)

In other words, we have to find a new paradigm that respects all that is good about CST and MM practice. If Habermas' thinking doesn't provide adequate guidance in this respect, then what else is available? In the rest of this chapter we consider two possible alternatives.

20.3 Systemic Intervention

20.3.1 Background

Gerald Midgley was well trained in TSI, with its various commitments, and the SOSM, but began to have doubts about its theoretical underpinnings and the direction in which these were taking CST. He explains his doubts in a 1996 paper named "What Is This Thing Called CST?" We have already met some of these but can summarize them here as being of three kinds. First is TSI's claim to be meta-paradigmatic. In Midgley's view, this cannot be sustained because TSI embodies its own unique assumptions with which other theorists do not agree. Second and related, of course, is its allegiance to Habermas' theory of knowledge-constitutive interests. This theory, in Midgley's opinion, rests on an untenable belief in the "march of progress" toward universal human emancipation and risks ignoring the need for sustainable improvement that takes the natural environment into account. Third, Midgley agrees with Ulrich that TSI has a restricted conception of the role of critique. It only seems to bring it into play when coercive contexts are identified. Critique is actually needed at all times, not least to identify cases of coercion. This is especially the case when undertaking paid work for clients when there are pressures to take organizational boundaries for granted and not to consider the impact of interventions on other stakeholders.

To resolve the difficulties he identifies with TSI, Midgley turns to the work of Churchman and Ulrich. Churchman noted that the boundaries drawn around a system determine what is taken into account, which stakeholders are considered, how improvement is defined, and what action is contemplated. Further, Churchman revolutionized systems thinking by arguing that the "boundaries" of systems are not fixed by the structure of reality but by the knowledge and values of observers. It follows that, in order to justify systems interventions and their results, we have to "sweep" in the perspectives of a wide variety of stakeholders, giving particular attention to those who tend to be excluded. Taking on the perspectives of the "enemies" of the systems approach is a good way of reflecting upon the inevitable lack of comprehensiveness of systems designs and reminding ourselves that we must constantly keep boundary judgments under review. Ulrich combined the insights he took from Churchman with the knowledge he derived from Habermas's work on discursive rationality, and translated them into a useable form in his CSH methodology. This offers a means of supporting rational professional practice that depends neither on the impossible "sweeping in" of all stakeholders nor the utopian vision of an ideal speech situation. CSH insists that appropriate boundary judgments can only be established through dialogue between those involved in designing systems and those affected by the design, and provides a means of detailing the nature of different boundary judgments and allowing them to be rigorously debated. Midgley has sought to draw upon and extend the work of Churchman and Ulrich on boundary judgments without losing those aspects of "methodological pluralism," central to TSI, which he still values and feels are legitimate. This leads him to propose a reconstitution of CST, with a MM called "Systemic Intervention" (SI) at its heart.

20.3.2 Multimethodology

SI is defined by Midgley (2014) as "purposeful action by an agent to create change in *relation to reflection upon boundaries*." The first phase of his MM focuses on reflection upon boundary judgments. The second involves bringing about change through the "creative design of methods." This phase is pluralistic and makes use of the variety of available methods found in the systems literature, and beyond, to create a flexible and responsive approach to intervention.

The first phase draws heavily on the work of Churchman and Ulrich. Ulrich insists that the critique provided by CSH has to be used alongside all methodologies and methods if systems interventions are to be rationally grounded:

> The conclusion is equally clear: boundary critique must not be subordinated to methodology choice but needs to precede it. *The separation of the two methodological concerns of critique and emancipation is not tenable on critical grounds.*
> *(Ulrich 2003, p. 337)*

Midgley contends that by following Ulrich, and putting boundary critique up-front in any intervention, he can overcome the problems identified with TSI. A new paradigm, he argues, can be built around the ideas of "boundary" and "boundary critique," which is particularly suited to CST and its ambitions. The concept of boundary is worthy of this role because it enables us to overcome subject/object dualism. This he sees as the most important enemy of systems thinking and one that has so far not been dealt with. The problem of subject/object dualism can be eliminated if we make "boundary judgments" the cornerstone of a process philosophy:

> If we regard the process of making boundary judgements as analytically prime, rather than a particular kind of knowledge generating system, then *subjects come to be defined in exactly the same way as objects – by a boundary judgement.*
> *(Midgley 2000, p. 79)*

In other words, a boundary judgment performs a "first-order distinction" and a "second-order distinction" at the same time. It defines the way the world appears looking outward and, looking inward, defines the "knowledge generating system" that produces that view of the world. In Midgley's opinion, being clear that SI is based on a new paradigm, and placing boundary critique up-front, brings important theoretical and practical benefits in its wake. Indeed, it allows the problems associated with TSI to be banished.

Theoretically, CST no longer has to make the unsustainable claim that it is multiparadigmatic. It can proudly declare that it has its own process paradigm based on boundary judgments. This paradigm will be subject to attack from nonbelievers, especially those who continue to adhere to various forms of dualism, and so CST is again subject to paradigm disputes. However its own paradigm, Midgley argues, is superior to others:

> Process philosophy ... allows us to talk about the real world, social construction and subjective understanding without contradicting ourselves ...[It]... is *preferable* to realism, idealism and social constructionism because it can offer an alternative grounding for most of the work from these other paradigms without the need to alter their most important contributions.
>
> *(Midgley 2000, pp. 90 and 92)*

It goes without saying therefore that this paradigm provides for the theoretical and methodological pluralism upon which CST puts much store. It can live and work with "understandings of systems and their boundaries as real-world entities, personal constructs *and/or* dialogical phenomena" (Midgley 2000, p. 150).

The new paradigm also brings practical advantages. Putting boundary critique up-front makes it apparent that what constitutes improvement depends on different perspectives governed by different boundary judgments. There is no need to refer to some ill-defined and impossible ideal of universal emancipation. It is clear that defining improvement requires the involvement of all those with a relevant perspective on the systems design and thus "sweeping in," as Churchman has it, a wide variety of stakeholder viewpoints. Issues of coercion can be identified at the beginning of a study by questioning the unreflective boundary judgments of decision-makers and ensuring that marginalized groups are identified and their concerns are brought to the fore. Examples, in projects conducted from the Centre for Systems Studies, include the children themselves in developing services for young people living on the streets (Boyd et al. 1999) and members of the actual client group in a diversion from custody project for mentally disordered offenders (Cohen and Midgley 1994). Further, the practitioner is pushed, as a matter of course, to consider matters outside of traditional organizational boundaries and to give due attention to the natural environment and the requirements of future generations. With boundary critique up-front, and shorn of the need to hide behind Habermas' work as a prop for its MM practice, CST is in a much better position to enact critique and its commitment to improvement, especially sustainable improvement.

In considering boundary critique, as the first phase of SI, we should note one significant addition that Midgley (1992) makes to Ulrich's understanding of the procedure. He argues that boundary judgments can become stabilized by social rituals that reinforce stereotypical attitudes. This can lead to fixed positions, which give rise to a narrow "primary" boundary and a wider "secondary" boundary. Between the two will be a "marginal" area. If the primary boundary is privileged, elements in the marginal area can be disparaged and become "profane." If the secondary boundary attracts attention, and is reinforced, then the marginal elements become the focus of attention and are made "sacred." He provides, as examples, the profane status often accorded to the unemployed under capitalism and the sacred status usually associated with the customer. The tendency to unreflectively accept stabilized boundary judgments, and in doing so to buttress the status quo, needs addressing at the beginning of SI and boundary critique can help us to do so. However, as Midgley concedes, stabilized boundary judgments may be very difficult to move if they are "held in place" because they express "wider struggles between competing discourses" (2000, p. 145). This leads him to consider whether a sociological theory is desirable to explain the origins of competing discourses, how some gain power, and what kind of action might be necessary, beyond boundary critique, to counter the social forces that lead to certain discourses assuming dominance. The sheer variety of social theories available gives him pause but he commits, eventually, to the development of a new social theory to complement boundary critique.

After "sweeping in" a wide variety of stakeholder viewpoints, examining them, and addressing issues of marginalization, there may be sufficient agreement among the stakeholders about appropriate boundary judgments for SI to move on to its second phase. Here, Midgley recommends proceeding by the "creative design of methods" (1997b). This is his alternative to using the SOSM for the purpose of methodology choice. The SOSM requires meta-paradigmatic thinking, which guides the choice of whole methodologies based on those paradigms. Because he does not believe meta-paradigmatic thinking is possible, Midgley (2000, p. 226) has changed his emphasis from creative methodology design to a flexible multimethod approach. This, in his view, still honors CST's commitment to pluralism. It involves, first of all, making judgments about the appropriate methods to use in the intervention. In negotiation with the various stakeholders, research questions are identified, each of which may require resolution using a different type of method. Successful choice of methods will depend on practitioners who are willing to entertain a range of options, often learned from others. Midgley (2014) sees SD, the VSM, interactive planning (IP), SSM, and CSH as being vital sources of methods for systemic action research. This demands skilled "methodological partitioning," of which Midgley (1990) was an early advocate. Rather than accepting "off-the-shelf" methodologies, and using the methods traditionally associated with them, we can "decompose" methodologies into their component parts. This provides a rich toolkit of methods that we can use flexibly to fit the circumstances and respond to the peculiarities of problem contexts. The actual choice of suitable methods will depend on the purposes being pursued, the theories, ideologies, and principles they are seen to embrace, and the practical results achieved with them in previous interventions. Inevitably, the research questions agreed upon will be systemically interrelated and, therefore, it is necessary to achieve a synthesis among the methods used to address them. Once appropriate methods have been chosen and organized into a coherent set, action can be undertaken that will bring about change that hopefully all, or almost all, of the stakeholders will see as beneficial.

Effort needs expending, throughout a project, to ensure the continued engagement of marginalized groups. If their involvement is threatened by the exercise of power, the researcher will need to consider whether there are other ways of ensuring their views are heard. For example, in the project dealing with homeless children, the young people were provided with space away from the professionals to develop their own thinking. If nothing can be done to support the disadvantaged, a researcher may have to contemplate withdrawing his or her services.

20.3.3   Case Study

The intervention described was conducted by Gerald Midgley, Isaac Munlo, and Mandy Brown of the Centre for Systems Studies, University of Hull, and was funded by the Joseph Rowntree Foundation (see Midgley et al. 1997, 1998). Its initial aim was to review how policy for the development and provision of housing services for older people was informed by data aggregated from assessments made of individual applicants. Overall, it was hoped that improvements to the process of information collection, handling, and use would lead to more general improvements in housing services for older people. The research was to cover housing itself, possible adaptations to properties or other assistance needed to enable older people to stay in their homes, and to involve all forms of provision, whether public, private, or by the voluntary sector. The researchers suggested a two-stage approach. In a first phase, a wide perspective would be taken on problems associated with the identification of need, the handling of information and its use for planning. In a second phase, various stakeholders would be engaged in designing actual improvements to the information provision aspects of the problem situation. The researchers were clear that, since this purposeful system was about designing housing services for older people, it was essential that the affected, the older people who were clients, should participate in the design.

 The first stage consisted of interviews with relevant stakeholders in two geographical areas. Rather than trying to determine at the beginning exactly which stakeholders to interview, the researchers used the approach of "rolling" out the boundaries of the people they should talk to. This required starting off with an obvious set of interviewees and inviting them, as part of the interview, to suggest who else might reasonably be questioned. Interviewees were asked which other stakeholders were relevant, perhaps because they held a different opinion on the problem situation, and in particular who else was involved in or affected by the interviewee's activities. This process continued until no new names came forward and produced some surprising additions to the original list. In the end, 131 interviews were conducted with clients, potential clients, carers, councilors, senior and middle managers, wardens, and assessment officers. Local government departments, health purchasers and providers, housing associations, voluntary organizations, private providers, building companies, users, and other stakeholders were all represented.

 After about 20 of these interviews, it was clear that an important choice had to be made which would impact on the whole of the rest of the study. The interviews revealed that the second phase of the intervention would miss many of the most significant problems that concerned stakeholders if it remained limited to looking only at issues of information provision. This was especially the case because the procedures for collecting information meant that needs that could not be met within current spending priorities were not even recorded. Older people were asking why there was such a mismatch between what they asked for in terms of housing services and what they got. Further, a number of the agencies involved were highlighting, as their primary concern, the difficulties they encountered in co-operating with other interested parties, i.e. in multiagency working. Neither of these issues could be addressed given the existing narrow specification of the second stage of the project.

The researchers saw this as an ethical issue about where the boundaries were drawn in this problem situation. If they followed the narrow specification they would only be able to increase efficiency and efficacy in the context of current resource distribution. This might lead to improvements in the provision of housing services for older people from the point of view of those who believed current levels of spending appropriate. It could hardly bring about improvements from the perspectives of older people or many of the managers or planners. Accepting the narrow boundaries of the study as originally prescribed could detract from rather than enhance improvements that satisfied all the stakeholders. The researchers called a meeting of the Advisory Group for the project at which the ethical consequences of different boundary judgments were considered. It was decided that the boundaries of the research should be widened to ensure that the larger problems now identified by some of the stakeholders could be embraced in the next stage of the study. The first phase ended with the production of two "problem maps" that showed the interrelationships between key issues, including issues of multiagency cooperation, and sought to capture the stated needs of older people.

A workshop was now suggested to discuss how the second phase would be conducted. During the preparations for this, however, it became clear that, because of certain sensitivities and the internal politics of the situation, the Housing and Social Services Departments wanted the workshop to be restricted to their own managers. The researchers were concerned that this would lead to the marginalization of the values and interests of users and other involved stakeholders and that they would be implicated in reinforcing this marginalization if they went ahead. Weighing the ethical pros and cons they decided to let the workshop proceed, but put in place certain safeguards to ensure that the boundaries remained wide. One of the researchers was given the role at the workshop of being advocate for the stakeholders who were not directly represented. And all of the researchers insisted throughout the workshop that the managers who were present try to put themselves in the position of other stakeholders and speak on their behalf. The decision to go ahead was justified and the tactics vindicated when it was clearly specified that the improvements sought in the second phase would be based on the desires of all stakeholders including older people and their carers.

The discussions at the workshop now turned to the methods that would be used to design improvements in the second phase. This again required care on the part of the researchers, lest their knowledge of the intricacies of systems methodologies and use of systems jargon marginalized the managers who would be responsible for the outcomes. The "systems discussion" was initially conducted by the researchers alone, with the permission and in the presence of the managers. The results were then explained to the managers and further general debate took place before actual decisions on methods were taken.

The second phase of the intervention had two parts. In the first, Ackoff's "idealized design" technique was used alongside Ulrich's "boundary questions" to discover what the stakeholders felt an ideal housing system for older people ought to be like. This was in response to the outcome of the first phase, which had shown the various issues to be so interrelated that only a "problem dissolving" approach could address them. It consisted of workshops held separately with three stakeholder groups: older people in receipt of housing services; carers and representatives of relevant community groups and voluntary organizations; and managers and front line professionals working for statutory agencies concerned with housing. The choice of these three groups of affected and involved stakeholders ensured the boundaries continued to be wide. Holding the workshops separately, so that each group had its own space, ensured that "professional discourses" did not dominate over the ordinary language of users. It avoided the risk of drawing the boundaries back in again. Employing Ackoff's idealized design technique (see Chapter 15), each group was asked to design an "ideal" housing system on the basis that the existing service system had disappeared the night before and they could construct its replacement as they wished, so long as it was technologically feasible, viable, and adaptable. The natural "boundary-busting" features of Ackoff's approach were enhanced by using Ulrich's boundary questions to lead the discussions. For example, it was asked "who should benefit from the provision of housing services to older people?," "who should be considered an expert?," etc. Ulrich's questions ensured that important boundary matters, which might otherwise have been taken for granted, were raised and debated. To ensure the maximum benefit was obtained from the questions, they were translated into plain English and phrased specifically to relate to housing for older people. The workshops produced three long lists of the desired properties of the housing service from the perspectives of the three stakeholder groups. The three lists defining ideal housing systems were so similar that it was possible for the researchers themselves to produce a first draft synthesis reflecting a single vision. This, together with the disagreements that did exist, was discussed and debated at a further workshop of managers from relevant agencies. At this workshop, the researchers acted as advocates for those groups not directly represented and asked the managers to be careful that the finalized list of desired properties reflected the concerns of users and carers.

In the second part of the second phase, the same managers and the researchers used Beer's VSM (see Chapter 13) to construct an organizational structure capable of delivering the ideal service while overcoming the existing problems of multiagency working. Beer's model was originally chosen because it was nonhierarchical in the sense of demanding that the primary focus should be on facilitating the work of those directly providing the service (the System 1 elements). During the workshop, the model was discussed in simple English to ensure that the expertise of the researchers did not dominate and some changes were made to it as a result. The managers were then asked to use it to provide an organizational design in which each of the five key functions described in the VSM was performed in accordance with the list of desired properties. At the end of the process, the final design was validated by systematically checking it against the lists of desired properties of the housing service system produced by the various stakeholders.

This case study offers an excellent example of a systems design, the provision of housing services for older people, in which a wide range of involved and affected stakeholders participated to ensure improvement in the system from multiple perspectives. Ulrich's boundary questions were employed to ensure critical reflection and the researchers also took advantage of the "polemical employment of boundary judgments" technique while acting as advocates for the users. The "creative design of methods" led to the choice of "idealized design" and the VSM as appropriate vehicles for taking the project forward. Midgley and his coresearchers added an acute awareness of the nature of the facilitation required to ensure that vulnerable stakeholder groups were not marginalized.

20.3.4 Critique

Midgley's (1992, 2000) argument in favor of MM is a familiar one derived from the SOSM and TSI. It states that we will never find one systems approach capable of tackling "as a whole" the interdependent nature of the complexity we face in the modern world. However, different systems methods have evolved to handle different aspects of complexity. If we use systems approaches with different strengths and weaknesses in informed combination, we will have a better chance of addressing different features of complexity as they come to prominence and impact what we are trying to achieve. Midgley does not believe however, in contrast to the TSI approach, that we can do this by operating at a meta-paradigm level and allocating different methodologies to suitable problem situations according to our purposes. Instead, he argues for a version of CST that is a paradigm in its own right while still being capable of housing pluralism. This is seen as resolving the difficulty of trying to combine methodologies based upon divergent philosophical and sociological assumptions while still protecting MM practice.

The new paradigm proposed for CST is based on a process philosophy that takes advantage of the capacity of the concept of "boundary judgments" to overcome subject/object dualism. The SI MM puts Ulrich's boundary critique up-front to ensure that existing boundary judgments are reviewed, marginalized groups are identified and included, and the intervention looks beyond traditional organizational boundaries. Once an accommodation is reached on the boundary questions, discussions can proceed on the, inevitably, interrelated issues that the stakeholders wish to address. The "creative design of methods" puts together a coherent set of methods, models, and techniques that have the best chance of tackling those issues and bringing about improvement from the perspectives of the different stakeholders.

SI is building up a substantial body of practice, with four extensive interventions reported in Midgley (2000). Most of the work is in the Community OR tradition of intervention to assist community development. This originated, formally at least, from Jonathan Rosenhead's initiative as President of the OR Society, in the mid-1980s, and was taken up enthusiastically in the Centre for Systems Studies and elsewhere. The four cases mentioned are all in multiagency settings.

The obvious weakness with "one paradigm multi-methodology" is that, unless the new paradigm is really capable of containing divergent theories and methodologies, the power of paradigm and methodological pluralism is constrained. We shall now consider SI in terms of our ideal-type of MM to assess exactly what loss of diversity it entails.

Midgley (e.g. 1989) has long argued that "partitioning" methodologies, and using the methods extracted from different methodologies together, can increase the flexibility and responsiveness of MM. This way of proceeding is possible because methods often carry little theoretical baggage and so can be used to deliver their benefits in the cause of a variety of paradigms. It is legitimate to consider all methods, models, and techniques, whichever methodology they were originally developed to serve, as candidates for use in an intervention. SI does this and so is strong as a multimethod approach. A downside, as noted by the ideal-type, is that following this strategy makes it difficult to gather knowledge about the usefulness of methods in support of different theoretical rationales. For example, exactly how well do "rich pictures" work as part of a functionalist approach?

The ideal-type warns of a significant danger that comes with emphasizing multimethod working at the expense of theoretically informed multimethodological and multiparadigm practice. If we are not careful we end up using methods in support of a single theoretical position that we have adopted without thinking about it. We lose sight of the need to draw upon the full range of paradigms, and their linked methodologies, to manage the complexity of multidimensional problem situations. This is the route to "pragmatism." Midgley does not want to fall into pragmatism. He employs SI as a means of theoretically steering interventions in a way that, he feels, protects methodological pluralism. But how successful is SI in this?

The dominant methodology employed in SI is Ulrich's CSH. We argued, in Chapter 18, that this was based on narrow theoretical presuppositions (drawn from Kant, Churchman, and a particular interpretation of Habermas) and provided for only a limited form of critique. Putting boundary critique up-front in an intervention must mean, therefore, that other methodologies are simply subsumed under the theoretical logic that drives Ulrich's approach. As Midgley (1989) seems happy to admit, this is a form of "imperialism" and must limit to some extent the capacity of those other methodologies to be used to full advantage. He does not seem unduly worried by this because he believes that the process philosophy underpinning SI is comprehensive enough to protect the essential contributions that different methodologies can make. There is some evidence for this. The examples he provides of SI in action do show aspects of cybernetic thinking (the VSM), IP, and SSM contributing to the success of his interventions.

Taking the argument to the level of paradigms, however, I feel that the evidence points to SI's theoretical imperialism inhibiting the multimethodological and multiparadigm approach that CST seeks to promote. SI, Midgley argues, constitutes its own process paradigm. He claims that this paradigm overcomes "dualism" because it recognizes that "first-order distinctions" and "second-order distinctions" are produced simultaneously when boundary judgments are made. We can, therefore, adopt realist, social constructionist, and subjective understandings of the world without contradicting ourselves. However, the population of those adhering to "dualism" remains large and the theoretical positions that cannot be incorporated in SI's paradigm are numerous. We can count among that population those of a functionalist persuasion – hard systems thinkers, most system dynamicists, many advocates of the VSM, etc. – who believe that they can provide real knowledge to decision-makers about how systems work and that this knowledge remains fixed however you view the system. Similarly, there are many radical change theorists, of an "objectivist" persuasion, who think they understand how society functions to privilege certain social groups and classes and disadvantage others. I personally find the external, material world something of an annoyance, and am constantly being bruised by it, and the social world is certainly resistant to many of the changes I would like to make. In fact, Midgley is sorely tempted by the "radical change" line of thought. As we saw, he points out to Ulrich that boundary judgments can become stabilized because they reflect struggles between competing discourses which are, themselves, heavily influenced by wider social forces and the power/knowledge nexus. This leads him to think that, some kind of political/sociological theory is necessary to complement boundary critique (2000, p. 148). In Larsen's (2011, p. 73) view this would indeed enable Midgley to think through the relationship between cognitive boundary issues and the "normative and observational realities of stakeholders." And it could provide greater insight into why different people think in such different ways, which "might explain their preferences and choices"

(Rajagopalan and Midgley 2015, p. 548). But we are not there yet. There are just too many theories. But is it really so difficult to be certain that women have and still are being discriminated against compared to men? And that current trends in capitalism are increasing inequality to levels never before seen? It is hard to see how, without theory development, these matters can be recognized and thought through, and how people will ever be inspired to take appropriate action to do something about them.

Midgley is further hamstrung because he follows Ulrich's assertion that boundary critique needs to precede methodology choice not be subordinated to it. As we know, even the concept of "system" means different things depending on the paradigm in which it is used. This is even more the case with "boundary" and "boundary critique." There are, therefore, various forms of boundary critique, all valuable, depending upon methodology choice. A SD practitioner would criticize a boundary, drawn to frame a causal model, which excluded elements that had an impact on real-world system behavior. Ulrich adopts a particular, subjective reading of boundary critique, which excludes all "realist" interpretations of the concept of a boundary and the specific critiques that would follow from these. CST must insist on taking seriously the variety of forms of boundary critique – including Ulrich's – that can be developed from different paradigmatic positions. The same argument holds, incidentally, against all attempts to overcome the philosophical and theoretical differences between sociological paradigms, and their linked methodologies, by appeal to particular systems concepts. A classic attempt is that of Cabrera et al. (2015) who use four underlying "rules" – distinctions, systems, relationships, and perspectives (DSRP) – to unify and organize the field of systems thinking. These concepts mean different things according to the paradigm in which they are located. CST just has to learn to live with and manage paradigm incommensurability.

If SI is formulated on a relatively restricted theoretical position, owing much to CSH, what impact does this have on what it can achieve in practice? We can attempt to answer this using another requirement for MM identified by the ideal type – parallel rather than serial use of systems methodologies. With boundary critique up-front it is not surprising that SI, at the beginning of an intervention, privileges a version of what we have called "emancipatory systems thinking." Above all else, it wants to ensure that those affected by a system, as well as those involved in its design, are included in a participative process of debate, which must precede decisions being taken. In most of the SI case studies, it is the supposed beneficiaries of the design who have to be brought into the discussions – young homeless people, the actual client group of mentally disordered offenders, and older people with housing needs. However laudable this is, it provides for a restricted examination of the problem situation. Matters such as efficiency, effectiveness, viability, the reinforcing of structural inequalities in the social domain, have to take a back seat. Although methodologies that express the concerns of other paradigms can surface later in an SI study they remain, as we have argued, denatured by the dominant theoretical rationale and there is no prospect of them assuming parallel importance. SI has relatively little to say about how the outcomes of its interventions should be evaluated. It seems willing to entertain the use of quantitative as well as qualitative measures for this purpose (Midgley 2000, p. 348) but this is not decreed or made systematic. A parallel use of methodologies would insist on evaluating the results of an intervention from the perspectives of the full range of paradigmatic positions. SI falls well short of offering a parallel version of MM.

Given that it cleaves so closely to Ulrich's approach, other criticisms of CSH might also be relevant to SI. In particular, it could be argued that, like CSH, it depends for its efficacy on there already being in existence the possibility of participative involvement and debate. To be fair, Midgley (1997a) has sought to answer this by arguing that where there is closure of debate we must widen our definition of systems practice to include direct political action and campaigning. In other words, we should take sides with those deemed to be marginalized and engage in direct action in order to create the conditions under which genuine debate becomes possible. The question does remain though: on the basis of what social theory do the campaigners justify their ability to recognize marginalized groups and privilege their interests? For completeness we must also mention Stacey and Mowles's (2016) critique of SI. Their most salient point concerns the infinite regress that occurs as more boundary critique is required to evaluate boundaries that have been settled upon on the basis of previous boundary critique. It is easy to think that this process would soon be terminated in the business context where SI has been little used and issues of efficiency and effectiveness tend to dominate.

20.4   Critical Realism and Multimethodology

20.4.1   Background

In an influential paper, *Multimethodology: Towards a Framework for Mixing Methodologies*, Mingers and Brocklesby (1997) set out their vision for MM. We have previously noted a number of points that they make but will expand on them here. This account also draws on the extensions made to the original ideas as reported in Mingers (1997b,c).

In essence, they recognize three main "varieties of multimethodology" – "methodology selection," "whole methodology management," and "multi-paradigm multimethodology." Methodology selection, associated with the SOSM, sees the systems practitioner valuing methodologies based on different paradigms and selecting an appropriate whole methodology to use in an intervention depending on the characteristics of the problem context. Whole methodology management retains the emphasis on using whole methodologies, expressing different paradigms, but wants to employ them together in the same intervention. TSI focused on how this can be managed. Mingers and Brocklesby's preference is for the multiparadigm multimethodology variety. This involves using parts of different methodologies, possibly from different paradigms, together on the same problem situation. Here the whole methodologies are "broken up" and the methods, models, and techniques usually associated with them are brought together in new combinations according to the requirements of the particular intervention. Others apparently in support of "multi-paradigm multimethodology" are many OR/MS practitioners and consultants, who want the freedom to use whatever methods seem appropriate at the time, and postmodernists who are quite at home "mutilating" methodologies to achieve a judicious "mix and match" of methods. The great merit of allowing methods, models, tools, and techniques to be detached from their usual methodologies, and employed flexibly, is that practitioners gain the maximum freedom to respond to the needs of the problem situation and to the twists and turns taken by the intervention.

To make multiparadigm multimethodology feasible, it is necessary to know what the most important elements of the different methodologies are useful for. To this end, Mingers and Brocklesby (1997) suggest a framework for MM design that links different methodologies and their parts to the important matters that will need to be addressed in any intervention and the stages that it will pass through. We have already noted how they categorize the multidimensional world in which MM is designed to operate. They follow Habermas' theory of "three worlds" and see all problem situations as possessing three interlinked elements – material, social, and personal. These will have to be confronted at all stages of an intervention. They then suggest, with the usual caveats, that the process of intervening in problem situations can be seen as involving the four activities of "appreciation," "analysis," "assessment," and "action." In essence, these refer to "what is happening," "why is it happening," "how could the situation be different," and "how could the difference be made real" (Mingers 2014). It then becomes possible to produce a grid based on the three elements and four activities. This contains 12 boxes, and different methodologies and methods can be assessed for their usefulness in addressing the issues raised in each box and allocated accordingly. For example, in the box defined as "appreciation of the social" we will be reviewing social practices and power relations for which aspects of SSM and CSH should be helpful. When we reach "assessment of the material" we will be reviewing alternative physical and structural arrangements and the VSM should prove useful. The "Soft-OR" approach of "strategic choice" is deemed a good option when we are considering "action relevant to the personal." This framework draws the attention of practitioners to the strengths and weaknesses of the different methods and helps them construct an appropriate multiparadigm multimethodology, which takes account of the multidimensional nature of a problem situation at each of the stages of an intervention. This "framework for mapping methods" is used by Mingers (2006, pp. 227–232) to provide a detailed analysis of the major OR and systems methodologies in which he uses shading to indicate where the strengths of particular methodologies and methods lie.

Attention is then given to whether their multiparadigm multimethodology approach is philosophically sustainable in the light of the argument that different paradigms are incommensurable. Mingers and Brocklesby believe that it is, but not on the basis of the theory of knowledge-constitutive interests proposed by Habermas and employed to provide theoretical support to TSI. This solution is ruled out "primarily because it does not make any new ontological statement that is capable of subsuming the incommensurable ontological assumptions of the original paradigms" (1997, p. 497). Instead they regard the "integrationist" social theories offered by Giddens and Bhaskar, and discussed in Section 4.6, as offering greater promise. This is not altogether surprising given that the integrationists believe they can subsume in their theories the objective–subjective dichotomy which is the greatest cause of dispute between sociological paradigms. Indeed, Walsham and Han (1991) had already suggested that Giddens' "structuration theory" might be a meta-theory capable of containing the multiple theories and methodologies found in the field of information theory. Mingers and Brocklesby turn to Giddens and Bhaskar because both these theorists:

> ... dispute the claim that we must choose between the competing realities offered by realist or nominalist thinking. These authors suggest that structure and meaning coexist in a dialectical relationship. Such a position contends the very basis

upon which paradigms have traditionally been described, and obviates the need to reconcile hard and soft methods by appealing to higher levels of reasoning.

(Mingers and Brocklesby 1997, p. 506)

Mingers initially agreed with Midgley that CST needed to develop its own paradigm. The role of integrationist social theory was to offer useful guidance to CST as it sought to establish itself as a paradigm capable of subsuming other paradigms within it. The new paradigm:

> ... would allow the competing assumptions to be reconciled within some wider framework; methodologies to be partitioned and mixed; and the foundational assumptions of the new paradigm to be developed and explored.
>
> *(Mingers 1997c, p. 411)*

In his later work (Mingers 2014), as will be seen, he came to regard Bhaskar's critical realism (CR) as adequate, on its own, to provide the philosophical and theoretical support required by multiparadigm MM.

Finally, Mingers argues that the design of any intervention requires thought to be given to the "Intervention System" (the agent or agents engaged with the problem situation), the "Problem Content System" (the real-world situation of concern), and the "Intellectual Resources System" (the available theories and methodologies). He is particularly concerned that the writings on MM have paid too little attention to the role of "the actual, embodied, and embedded agent(s)" who are involved in the intervention (1997c, p. 427). We should consider their relationships with two other systems – the problem situation and the intellectual resources at their disposal. It is essential to take into account their values and the impact they might have on the power/knowledge formations playing out in the problem situation.

20.4.2 Multimethodology

In his later work, Mingers firmly bases his MM on CR and, especially, on the writings of Bhaskar. As was described in Sections 2.4 and 4.6, CR gives primacy to ontology and posits that a real world of objects and structures exists, independent of our observations, that cause what happens and what doesn't happen. This is held to be true in the social as well as the natural domain. Although social facts and structures emerge as a result of social action and interaction, they are no less real and possess properties that are separate from individuals and are experienced as objective social forces affording the opportunity for some types of activity while constraining others. CR believes that it is the job of both the natural and social sciences to discover the underlying causal mechanisms that give rise to recurrent events and activities. In the social realm, CR is also said to provide a moral justification for emancipatory social science and practice.

Mingers (2014, p. 187) is convinced that CR "licenses" MM because it is pluralist in terms of both ontology and epistemology. In ontological terms, it accepts that a range of factors have a causal effect and that this includes social categories and structures, emerging from ideas, meanings, and interactions, as well as physical objects. Inevitably, therefore, we need different means of finding out about these causal factors – a variety of epistemologies. CR, accordingly, is willing to make use of positivism and interpretivism, in pursuit of its "realist" project, while being aware of their limitations. Positivism uses observation to reveal patterns of behavior on which it can base its predictions. While this gives it access to only a small subset of events – those that occur regularly and can be measured – it may give pointers to the underlying causal mechanisms that produce the patterns it studies. Interpretivism provides us with access to nonmeasurable aspects of the social such as meaning, social action, different points of view, social norms, and discourse. This is a good starting point for analysis because these things constitute the social realm. However, we must not commit the "epistemic fallacy" of believing that the social world is the same as our conceptions of it. It has emergent properties and takes on a life of its own. The causal force of the nondiscursive structures that are produced must then become the focal point of inquiry in their own right. Mingers puts it like this:

> Thus with CR we can avoid the limitations of both empiricism and constructionism whilst still making use of them as necessary. We begin with some events or occurrences that are puzzling or unwanted and then look for causal mechanisms that might be responsible for generating or maintaining them. This may well begin with data and/or discourse but will go beyond or beneath this to try and explain the phenomena not simply record it …
>
> *(2015, p. 322)*

Of course, if positivism/empiricism and interpretivism/constructionism are both useful epistemologies, then we can also make good use of all the quantitative and qualitative methodologies and methods that serve them.

There is a further, "emancipatory," dimension to Bhaskar's CR. It emerges from a complicated argument that Mingers (2014) describes well. Social science is regarded as inevitably evaluative because of its subject matter. For example, statements such as "X was murdered," essential in the explanation of complex social behavior, are value-laden. The social world, not surprisingly, is subject to a cacophony of interpretations based on individual perspectives and cultural traditions. Fortunately, beneath all this there are "moral truths" that are grounded in the characteristics of human nature. Through social science, it is argued, we can find out what these moral truths are. In a word, they relate to the need for "universalized freedom for all." The CR approach then allows us to recognize false interpretations of what is happening in the social world and uncover the causal structures that maintain the false beliefs that stand in the way of us achieving a society based on moral truths. We can also use CR methodology to identify what action is necessary to remove those negative causal mechanisms. According to Bhaskar, once we judge that something is immoral we are committed to taking action to change it. Further, this commitment can be universalized. We must act to remove similar ills and change whatever constraints imposed by society bring them about.

Drawing principally on the work of Wynn and Williams, Mingers (2014, p. 190) is able to outline the basics of a CR MM. It would have five stages with multimethod use and triangulation at each. These stages, according to my interpretation are:

- Identification of the situation to be explained (using observation, measurement, interviews, etc.)
- Analysis of the relevant material and social context and structure (all the systems, components, relationships, and processes that are active in the situation)

- Generation of hypotheses about potential generative mechanisms (using insight and imagination in a process of "abduction" from what was discovered in the first stage to likely explanations based on the outcomes of stage 2)
- Selection of the generative mechanism that offers the best and strongest explanatory power in relation to what is going on in the situation (making use of empirical corroboration via statistical analysis, prediction, etc., and also interpretive understanding of behavior)
- Determining and taking action to bring about the preferred outcomes

20.4.3 Case Study

Mingers (2014) illustrates what this would look like in practice by building on a case study from Malcolm Gladwell's book *The Tipping Point*.

The situation to be explained was a 300% increase in reported cases of sexually transmitted diseases (STDs) in Baltimore in the mid-1990s. Analysis of the context led to the identification of various component factors that generate cases of STDs (e.g. rate of exposure, rate of infection) and components that reduce them (e.g. medical treatment, speed of recovery). Three different eminent sources suggested three plausible generative mechanisms for the increase in STDs:

- Increased use of crack cocaine (increases risky sexual behavior; brings more people into poorer areas to buy the drug, increasing their exposure to STDs)
- The cutback in medical services at STD clinics – staff reduced from 20 to 10, patient visits from 36 000 to 21 000 – due to cuts in the city's budget (reduces number being treated and speed of treatment resulting in carriers infecting more people)
- The City's slum clearance program which, during the period in question, led to inhabitants of poorer areas, more likely to suffer from STDs, being spread across the city (increases the numbers of susceptible people who can come into contact with STDs)

Mingers suggests that a CR approach would have gone on to conduct further research such as mapping out movements, looking for patterns of sexual behavior, interviewing those involved. This might have allowed the relevant system to be simulated and the strengths of the different causal elements determined on the basis of what they would have to be to produce the observed behavior. An informed intervention strategy could then have been put in place. This stage was not carried out but a multidimensional change program, involving education and awareness training and more medical staff, did bring the situation back under control.

20.4.4 Critique

Mingers and Brocklesby (1997) define their favored multiparadigm MM approach as using parts of different methodologies, possibly from different paradigms, together in the same intervention according to the requirements of the problem situation. Whole methodologies, such as SSM, can be decomposed to add to the toolkit of suitable systems methods. They also provide a framework pointing to the strengths and weaknesses of the different methods according to the multidimensional nature of problem situations and the various stages that interventions pass through. Henao and Franco (2016) report that they have found this framework useful in a MM intervention supporting a management team in their planning in a community development context.

Castellini and Paucar-Caceres (2019) find the matrix a good starting point for a MM intervention in a textile SME. It should be noted, however, that if the framework is meant to map methodologies as well as methods then it falls foul of a criticism similar to one that Tsoukas levels at TSI. You can't simply allocate different methodologies, representing different paradigms, to different aspects of a problem situation (material, social, personal) because each paradigm thinks it can deal with *all* of those aspects.

In terms of our ideal-type of MM, the Mingers and Brocklesby approach has a powerful multimethod element to it. But if this is all they are aiming at then the label multiparadigm MM is an obvious misnomer. As they themselves argue, methods are easily detached from particular methodologies and their underlying paradigms. Therefore, just using different methods cannot ensure that multiple methodologies and paradigms are in play. Instead, as we know only too well, there is a risk of a descent into "pragmatism" with multiple methods used unreflectively in the service of one theoretical rationale. And it provides no prospect of us learning about the relative efficacy of the methods in serving different paradigms. If MM practice is to realize its full potential, coping with the multidimensional nature of the world and yielding a rich array of research outcomes, it needs to be genuinely multimethodological as well as multimethod in character. Because of the close link that exists between methodologies and paradigms, this should also ensure that different paradigms, espousing radically different worldviews, are in the arena.

The presumption must be that Mingers and Brocklesby do want more than just a multimethod approach and that the confusion arises because they sometimes do not make a sufficient distinction between "methods" and "methodologies," more particularly, that when they use the word methods, they frequently mean methodologies. The evidence comes from a number of sources: in the name they give their favored multi-methodological approach; in the presence of whole methodologies, alongside methods, in their framework; in Mingers (1997b, p. 7) calling for a "strong pluralism" that employs "a blend of methodologies from different paradigms"; and in Mingers's (1997c, p. 418) concerns about Taket and White's "pragmatic pluralism" because of its denial of the value of theory and support for just "doing what feels good."

Mingers's intentions become clearer when he embraces CR as the philosophy underpinning MM practice. He is happy to endorse the ideal-type of MM in most respects but differs on one fundamental issue:

> The main point of disagreement, however …[is]… that we have to accept the validity and in some sense equality of currently existing paradigms rather than try to go beyond them. I see no justification for accepting the validity of the paradigms as currently formulated since each has been legitimately critiqued by the others. Indeed, Jackson's book (2000) does a reasonable job of pointing out the limitations of each of his four paradigms.
>
> *(Mingers 2006, p. 209)*

As we have seen, Mingers regards CR as a superior paradigm. Moreover, it can provide the space for the full range of methodologies and methods, drawing upon theories as diverse as positivism and interpretivism, to operate and show their strengths:

> CR works as an underpinning philosophy that allows us to accept the strengths of both empiricism and constructionism without having to accept their limitations in terms of real-world problem solving. It asserts the existence of a causally

efficacious real world while accepting the inevitably limited nature of our access to that world. It thereby directs our attention to go beyond simply describing what we experience, whether in terms of empirical observations or personal values and beliefs, towards trying to explain the causality behind these experiences and thus towards being able to change them.

(Mingers 2015, p. 328)

The problem with this is that paradigms other than CR are never fully respected and are always used with a view to them ultimately contributing to the purposes of CR. So, positivist methodologies have a role to play but only in discovering patterns that might point to the presence of underlying causal structures. Interpretive methodologies can reveal beliefs about the social world but these may well prove to be misguided when set against the understanding CR can give us of the underlying causal structures that really determine the nature of social reality.

The extent to which CR can limit other methodologies from operating on their own terms is clear from the case study Mingers offers to demonstrate how MM might operate in practice. The Baltimore project on STDs is supposed to illustrate how a multi-methodological approach might work but actually looks more like an example of SD in action. In his description of the case, Mingers uses causal loop diagrams to bring clarity to the relationships between influential components in the problem situation, talks of balancing feedback loops, and suggests that a simulation would help determine the strength of the causal components. It seems that the CR version of MM is easily reduced to SD with other methodologies hardly entering the frame. This might interest SD thinkers such as Lane who, as we saw in Chapter 11, argues that integrationist social theory provides a natural theoretical home for SD. But the correspondences between SD and CR will disturb those who hope that the latter can provide a basis for MM practice. Other methodologies have to be radically denatured to fit into the CR paradigm.

The problem is that CR, far from being a paradigm that is welcoming of other theoretical traditions, actually occupies a quite restricted space in social theory. Its assumptions are fundamentally different to those of positivism, interpretivism, "subjectivist" radical change theory, and postmodernism. None of these other paradigms commit themselves to the existence of underlying generative mechanisms which can explain the events that are experienced. If CR is one paradigm of thought, with its central tenets opposed by a number of others, it is impossible to see how it can support a multiparadigm MM approach. Mingers has invented a form of CR imperialism to set alongside Checkland's "interpretive imperialism" and Midgley's "process philosophy imperialism." And it is not a very benevolent form of imperialism.

The point about the narrowness of Bhaskar's theoretical position can be made by considering what thinkers with different philosophies would say about his work. Here I take the example of his vision of the ideal society. To do so I draw on Mingers's (2014) account but must be clear that, in this instance, he is offering a review of Bhaskar's thinking not necessarily an endorsement of it. Bhaskar has expressed his views on the ideal society we should be seeking to bring about:

Bhaskar uses the Greek term *eudaimonia* to describe a happy and flourishing society in which everyone is free from unnecessary constraints on their freedom. This recognizes that people should be free to be different to the extent that this does not restrict the freedom of others. The freedom of each is a necessary condition for the freedom of others.

(Mingers 2014, p. 210)

To reach this utopian state CR must launch an "explanatory critique" of the false beliefs and oppressive power relations that exist in the social world by uncovering the causal mechanisms that generate them. It can then suggest actions to change those causal mechanisms and bring society closer to the ideal. Once wrongs are brought to people's attention by CR they automatically have a moral responsibility to remove the structural constraints that generate them. Well, there are so many opposing arguments that it is difficult to know where to start. Complexity thinkers, and the advocates of MM who essentially share their conception of the world, would argue that there are so many factors involved in the creation of social reality, they are so closely intertwined, and the outcomes they produce are so unpredictable that the identification of particular causal mechanisms is impossible, and we cannot possibly know the outcome of trying to change things. We could end up with something much worse than we have now. Interpretive theorists deny that the social world has objective characteristics that can be made the subject of scientific inquiry. In these circumstances, how can we be sure that particular beliefs are wrong? There are many philosophers, dating back to Hume, who insist that it is impossible to derive an "ought" from an "is." They would argue that no account of how society "is" can commit us to trying to change it. Functionalists will wonder just how well the ideal society will work if their knowledge about how to operate efficiently, and design effective organizational structures that can respond to change, is ignored. Postmodernists will be amazed at the prospect of a utopia where dispute and conflict disappear because we all suddenly agree on a universal set of norms.

This point, that even MM approaches can be in Churchman's words "terribly restricted," can be reinforced if we draw upon Mingers's comparison of Habermas' and Bhaskar's views on ethics to look at the very different approaches taken by Midgley's SI and Bhaskar's CR to moral questions about what is good for all. SI, basing itself on the thinking of Habermas and Ulrich, requires moral truths to emerge from free argumentation and debate and to meet with the approval of all those affected through their participation in something akin to an "ideal speech situation." Valid norms are constructed by people reaching agreement through a process of discourse (Mingers 2014, p. 213). CR, by contrast, holds to a position of "moral realism" in which moral truths exist independently of the subjective views of individuals and traditions. Social science can look from the outside and determine which moral beliefs are right and which are wrong. Two supposedly MM approaches differ profoundly on one of the key issues dividing the paradigms. The restrictive viewpoint of each is revealed by considering the other.

One more point can be made in relation to the ideal-type of MM. It requires methodologies and methods to be used in a parallel not serial manner. Mingers and Brocklesby's "framework for multi-methodology design," with its emphasis on the need to address the multidimensionality of problem situations at each stage of an intervention, suggests that parallel usage is recommended. However, the framework also maps methodologies and methods according to their appropriateness for different stages of an intervention. This suggests a serial usage which, it was argued, detracts from successful MM practice.

In taking leave of the CR version of MM, it is worth drawing attention to two further matters. The first is Mingers's identification of debates in moral philosophy as significant for CST. He discusses the three main strands of ethical theory – utilitarianism, Kant's deontological approach, and virtue ethics – and uses them to discuss, compare, and contrast the ethical implications of the writings of Habermas and Bhaskar. CST can only benefit from a greater awareness of moral philosophy and more work needs doing in this direction. It is also possible that ethical theory might benefit from CST. The second is the attention Mingers plays to the role of the agent(s) in any multimethodological intervention. This is important and will need to be discussed further in the next chapter.

20.5 Conclusion

Bowers, an ex-Hull PhD student, provides us with three fitting conclusions to this chapter. Talking of any attempt to adopt a meta-paradigmatic position, in the manner of TSI, he writes:

> There is no neutral or extra paradigmatic platform on which to stand for any valid comparison, translation or arbitration between [paradigms] – any such point of view would necessarily itself be paradigmatic.
>
> *(2011, p. 539)*

His second message addresses the impossibility of one-paradigm pluralism. It is simply not pluralism and cannot support the sort of MM practice that the ideal-type sees as essential. In a word, any paradigm that from "above" seeks to include multiple methodologies "necessarily corrupts our view of them" (2011, p. 540). In that sense, the version of Habermas' thinking employed by TSI, for all its faults, is less corrupting than the two other possible theoretical supports for MM discussed in this chapter. His third point issues a challenge for the final chapter:

> There is currently no widely accepted and proper unifying theory that completely supports, describes, explains, guides and directs us to operate in a world with multiple paradigms.
>
> *(2011, p. 540)*

21

Critical Systems Practice

The short ice age inside myself is over. Everything is back to normal again. All truths continue to be provisional. Searching for an overview can begin once again

(Mankell 2016, p. 47)

21.1 Prologue

In 1999, I received an offer from the University of Hull to establish a new business school. I had left the University in 1994 largely because of a lack of progress in this regard. During those five years I was at the University of Humberside (also in Hull) which became the University of Lincolnshire and Humberside. I learned to be a dean, established a Centre for Systems Research (with such luminaries as Raul Espejo and Gerard de Zeeuw) and, at the Lincoln campus, appointed Rebecca Herron to run the national Community Operational Research (OR) Unit, which moved from Northern College. But the new university was intent on shifting everything to Lincoln. The offer from Hull was one that I could not refuse. I had no wish to move out of Hull – why would you? I was Dean of Hull University Business School (HUBS) for nearly 12 years.

The plan for HUBS was a good one. On the ground, however, things were desperate. Two departments, "Management Systems and Sciences" and "Accounting and Finance," had agreed to join the new school but had not, historically, got on well together. The Department of Economics was reluctant to join and it took two year's work to get them in. Further along the line, the University took on the responsibility for a College of Higher Education in Scarborough and a small management group from that institution had to be integrated into HUBS. Taking all these units into account, the base staffing level was around 45, student numbers about 600 (not counting validated programmes), and turnover about £5 million per annum. At Hull there were small undergraduate programmes in Management Systems, Accounting, and Economics. There was virtually nothing on-campus at postgraduate level. Most of the income came from MBAs operated overseas through a variety of agents, in many different countries. Some perverse incentives for those responsible meant that there were over 30 different programmes, sometimes 4 or 5 in the same country. The MBAs in Hong Kong were featured, not in a favorable light, on the BBC's *Dispatches* programme. There were further MBAs, validated by Hull, at private colleges and local higher education colleges throughout the United Kingdom. The MBAs had become the main business. There was no interest in

Critical Systems Thinking and the Management of Complexity, First Edition. Michael C. Jackson.
© 2019 John Wiley & Sons Ltd. Published 2019 by John Wiley & Sons Ltd.

the Hull departments in "teaching quality," which was about to become a major issue in UK higher education. Research performance was poor and was being investigated. There were virtually no links with local, national, or international businesses. In 1999, when HUBS was formed, it was part of a Faculty of Social Sciences and had no separate space of its own. It was a "mess."

By 2011 everything had changed. HUBS was a faculty in its own right and had moved into new and refurbished buildings which were frequently described as "world-class." An academic leadership team, including the dean, deputy dean, and directors of teaching and learning, research, and external business, had oversight over the work of six subject groups (one in Scarborough) covering all the major business and management disciplines. HUBS had its own support units, under an administrative director, responsible for general administration, marketing, finance, HR, recruitment, postgraduate admissions, alumni, business links, and IT. There were approximately 120 academic staff and a further 45 administrative staff. Student numbers had grown to over 3500, including 900 at the postgraduate level, and turnover was at £28 million per annum. HUBS had over 60% of the overseas students in the University. There were thriving full- and part-time undergraduate programmes, offering the possibility of a year in industry or abroad, a range of very popular MSc programmes, and successful full- and part-time, on-campus MBA programmes. The overseas MBAs had been rationalized to just four with two agents for the Middle East and Far East. There were a few validations, all at undergraduate level, at local colleges. Efficiency was emphasized and the University received an annual contribution of around 45% of HUBS' income to cover overheads and make investments. Detailed attention was given to the programmes and to teaching quality. An early quality assurance assessment of the school gave a score of 23 out of 24. The MBAs, full- and part-time, home and overseas were AMBA accredited. Research was encouraged and reorganized. The Centre for Systems Studies did well and a £9 million grant was obtained from Yorkshire Forward to establish a Logistics Institute. In the 2008 Research Assessment Exercise, HUBS came 16th in the United Kingdom for "research power" (number of staff entered multiplied by the average quality score given to each piece of work submitted). There was a large population of PhD students tied into the research activity. HUBS had grown its corporate connections with the help of a strong Advisory Board. There was significant executive, in-house and consultancy activity with local and national organizations. The success of HUBS led to EQUIS accreditation (from the major European accreditation body for business schools) and AACSB accreditation (from the US-based business school accrediting association). HUBS was the first business school in Yorkshire, and only the 13th in the United Kingdom, to gain "triple-crown" (AMBA, EQUIS, and AACSB) accreditation, putting it in the top 1% of business schools in the world.

As Dean during this period, I had some part in the sea change in performance and was the person who received credit by being elected a Companion of the Association of Business Schools and awarded an OBE for "services to higher education and business" in the Queen's New Year's Honours List of 2011. Five senior staff of HUBS went on to be business school deans at other universities. My immediate boss, in the early days of the growth of HUBS, Deputy Vice-chancellor Howell Lloyd, says:

> In 1999 the University of Hull set about consolidating its business-related programmes. Twelve years later, under Mike Jackson's leadership, those plans had

resulted, beyond expectations, in a triple-crown accredited business school with some 3,500 students. The HUBS story is a remarkable one and in this chapter he describes some of the ways in which success was achieved.

(2018)

The Vice-chancellor of the University, between 1999 and 2009, David Drewry, states:

One of the most impressive developments in the University, during my time as VC, was the rapid growth and remarkable success of the business school. Working hand-in-hand with the University, and using systems thinking, HUBS formulated a vision of promoting 'responsible leadership for a complex world' and gained national and international recognition for what it achieved.

(2018)

A number of things made the transformation possible:

- It was a period in which business schools generally were booming and business subjects were popular with both home and overseas students
- There was no business school at the University of Hull in 1999 and this made it possible to build one from scratch in a way that took advantage of the favorable environment
- The City of Hull was crying out for a business school and there was strong local support
- The leadership of the University was strong and farsighted enough to encourage faculties to take responsibility for their own futures and not see their success as a threat to central control
- HUBS was blessed with a number of extremely competent, imaginative, and hard-working staff who generated and led their own initiatives on behalf of the school
- A significant proportion of the staff, as well as Pauline (my wife) and myself, were from Hull or the immediate vicinity. We were not going anywhere and it was obvious to stakeholders that we were not putting effort in just to advance our careers and move elsewhere

In a case study, later in this chapter, I hope to demonstrate that Critical Systems Practice (CSP) also played a part. HUBS became dedicated to the development of "responsible leadership for a complex world." It had a clear mission to equip decision-makers to make a sustainable difference in a fast-changing and interconnected environment. To do so HUBS sought to maximize the connections between excellent research, learning and teaching across disciplines, and corporate engagement, working with partners who shared our aspirations and values. Occasionally, as will be seen, systems thinking, with its methodologies and methods, was used.

21.2 Description of Critical Systems Practice

21.2.1 Historical Development

We can take up the story of developing a suitable multimethodology for CSP from where TSI was left in the previous chapter. Flood and Jackson (1991a, pp. 241–244) finish *Creative Problem Solving: Total Systems Intervention* with an auto-critique of TSI. That critique mentions the need to pay more attention to nonhuman beings and environmental issues and to take into account feminist and non-Western ideas. It warns against the "positivist" reading of the system of systems methodologies (SOSM) as an attempt to represent real-world problem contexts and it discusses the dilemma facing consultants who try to use TSI, with its commitment to human well-being and emancipation, with powerful clients. Soon afterwards, Flood et al. (1992) followed up by outlining a research program which they felt would address other outstanding issues. Four areas were identified as needing attention. The first required researchers to look for other creativity enhancing devices that could complement or replace metaphor analysis during the creativity phase. The second asked whether the SOSM should remain the primary vehicle for the choice phase. The third asked, in effect, for a set of "generic" methodologies, clearly related to paradigms, to replace the specific methodologies (SD, SSM, CSH, etc.) that TSI had inherited. The fourth required more attention be given to the process of using TSI. Since that time, as the previous chapter revealed, there has been a considerable amount of external criticism of TSI; alternative multimethodological approaches, such as those of Midgley and Mingers, have blossomed; and new theoretical and methodological developments have come to the fore, which any critical systems approach must take into account.

In my own work, with which this chapter is concerned, I have sought to address the issues mentioned in the auto-critique and research agenda, to respond to external criticism of TSI, to take into account new theoretical and methodological ideas, and to reflect what has been learned from CSP practice. *Systems Approaches to Management* (Jackson 2000) provided a much fuller account of critical systems theory than I had attempted previously and a thorough reworking of TSI, which I renamed CSP. There was an attempt to address telling aspects of the postmodernist critique and, in response to Tsoukas' (1992, 1993a) criticism, the metaparadigm approach based on Habermas' work was dropped and replaced by "critique between the paradigms." The "ideal type" of multimethodology was framed, including recognition of the need for multimethod pluralism. "Generic" functionalist, interpretive, emancipatory, and postmodern methodologies were formulated, constitutive rules for the use of CSP set out, and Mode 1 and Mode 2 ways of employing CSP discussed. In *Systems Thinking: Creative Holism for Managers* (Jackson 2003), these developments were consolidated and explained in a more practitioner-friendly manner. In addition, there were further changes to the critical systems thinking (CST) "commitments," a "reflection" stage was formally added to CSP, and the notion of "critique between the paradigms" extended to all stages of the multimethodology, including to the results of an intervention. Other papers have tried to extend "critical awareness" (Jackson 2006, 2009; Jackson et al. 2008), establish the ideal type of multimethodology (Jackson 1999), and promote and defend CST as an approach (Jackson 2001, 2010).

In this chapter, I am not going to provide a detailed account of the emergence of CSP and the developments that have occurred since. That would be of historical interest only. Instead, I want to present CSP in its most up-to-date form. In setting out its contemporary "philosophy and theory," "multimethodology," "methodologies," and "methods," I will, however, note where new theoretical and methodological developments have had an impact, where changes have been made in response to criticisms, and where CSP practice has required modifications in the approach. Three case studies of

CSP in action are provided. There is also a "critique" section in which I detail some remaining concerns and how CSP might respond to them. The chapter follows the pattern of the earlier chapters on individual methodologies, in Part III, with "comments" and "value to managers" sections. What I have to say will not deal with every issue and convince everyone. Readers will have to make their own minds up on the extent to which I have succeeded in addressing the concerns that have been raised. I do hope to show that CSP continues to learn from theory, practice, and criticism and has become more theoretically coherent and practically useful as a result.

21.2.2 Philosophy and Theory

To make the new developments clear it is best to continue to describe the philosophy and theory of CST in terms of *critical awareness*, *pluralism*, and *improvement*.

Critical awareness has traditionally included "theoretical awareness" and "social awareness." This book has, hopefully, continued the journey toward "theoretical awareness." It has sought to uncover the roots of systems thinking in philosophy, the physical, life and social sciences, and to explain its foundations in general systems theory, cybernetics, and complexity theory. Part III engaged in a second-order analysis of applied systems thinking, looking in depth at 10 systems methodologies from the point of view of a broader appreciation of social theory and a refined version of the SOSM. The latter was used in its appropriate role to translate the findings of the social sciences into a more digestible form for systems practitioners and other management scientists. The crucial importance of second-order thinking first came to the fore in cybernetics (see Section 6.4). However, cybernetics was limited in what it could achieve in this respect because it largely ignored the social sciences. CST has always engaged in "second-order" critique. Using social theory and the SOSM to reveal the underlying assumptions made by different strands of systems thinking was one of its main points. It makes it possible to engage in informed pluralism and to promote comprehensive "improvement." Recent calls for second-order systems science are, therefore, welcome as long as they recognize and build on what has already been achieved by CST.

As we saw in Section 4.7, Luhmann's sociology provides strong theoretical support for theoretical awareness based on second-order analysis. In his view, social theory must give up its quest for ontological certainty and become the study of how first-order observers observe. Such "second-order" observation represents a shift from ontology to epistemology. Instead of trying to uphold claims about the nature of social reality, sociologists should concentrate on how different social theorists and theories construct societal issues and problems from the distinctions they employ. Using second-order observation, we are able to understand how the first-order theory we are studying observes, and what it sees and possibly does not see. This requires making the fundamental "distinctions" made by theories as clear as possible. Providing an example of what is needed, Luhmann states:

> These Hegelians are so familiar with the modes of argumentation and the forms of development of his theory that they know them by heart. They can speak like Hegel. But they have no language left but the dialectical one. The countermeasure I have in mind is to make the theory decisions as transparent as possible. To do this, it is necessary to single out the following questions at every juncture.

What are the different options? What is connected to the decision in favor of *this* concept as opposed to another one? Where is there an exit point? Where does one have the freedom to choose something else in order to see what else needs to be changed if one revises a certain decision?

(Luhmann 2013, p. 254)

Luhmann goes on to argue that, because social theories can have an impact on society, we must always take responsibility for our theoretical decisions. Second-order analysis can help us to understand the consequences of adopting certain theories and the concepts associated with them. We must try to observe what really happens "when organizations, families, or therapies" are planned on the basis of a particular terminology (Luhmann 2013, p. 131). We have here a restatement of one aspect of CST's commitment to "social awareness." Contrary to what Zhu (2011, p. 788) says, CST has never regarded systems methodologies as "innocent." It has always seen them as "transformative" – as bringing consequences in their wake – and has insisted that these be considered when choosing methodologies in a particular intervention. The other aspect of "social awareness," highlighted in CST, is the need to reflect on the societal and organizational "climate" within which we intend to use systems methodologies. In order to protect the pluralism CST demands, we have to be extremely watchful. Cultural and cognitive, as well as philosophical constraints can delimit the range of methodologies it is possible to use and so reduce the potency of pluralism (Mingers and Brocklesby 1997; Brocklesby 1997). We have to be particularly careful that pluralism maintains a radical edge to it. Because management scientists often work in a paid capacity, for powerful clients, there will be a tendency to employ methodologies that support the status quo. This will impact on which methodologies are favored and the results they can generate. Flood (1990; Flood and Romm 1996), taking his lead from Foucault's work, has noted that CST's initial thoughts on these matters must be extended to include consideration of the effects that power at the micro-level can have on the reception and use of knowledge.

Despite Flood and Jackson's auto-critique of TSI, and Midgley's (2000) promptings, it is arguable that CST still fails to shout loud enough about the natural world and environmental issues. In a recent paper, Stephens et al. (2019) specifically address the need to provide "ecological justice for nature" in CST. Historically, they say, CST has taken an anthropocentric perspective. This is, in part, due to the fact that it has sought to ground its critique on Habermas' notion of "communicative competence" based on the "ideal speech situation." Non-human entities, such as animals, plants, and the natural environment, are marginalized by such a theory. Stephens et al. argue that Ulrich's critical systems heuristics (CSH), together with aspects of ecofeminist theory, should be used to question anthropomorphic assumptions when they arise and to suggest how nonhuman entities might be involved in decision processes that affect them. They draw attention to the increasing amount of research demonstrating that animals, plants, and ecosystems have sentience, agency and, even, consciousness and feelings. This, together with the havoc we are causing by ignoring the voice of the environment, makes it imperative that we find ways of accessing the communicative capability of nature. There is ample evidence that plants take notice of us. Is it possible, they ask, that we "can better learn how to know plants?." Even if we cannot yet directly involve them in decision-making, we should ensure they are represented by others who can speak on their behalf.

An important step, Stephens et al. suggest, in ensuring that nature is included in a just and ethical manner in our systems designs, is to change CST's adherence to "social awareness" and "human emancipation" to commitments to "socio-ecological awareness" and "emancipation" in a wider sense. It is necessary to respond, and I think it is sensible to do so by following their advice and adding "ecological awareness" as another aspect of critical awareness. The commitment to "human emancipation" in CST has long given way to a wider vision of "improvement." There is no need to change this name as long as we ensure that, in talking about improvement, we take into account the natural world and people's relationships with it. Incidentally, Stephens et al. suggest that a similar issue of inclusion may arise in the future with artificial intelligence.

I have on occasion thought to collapse social awareness, and provisional thoughts about ecological awareness, into "theoretical awareness," leaving that as the sole element of critical awareness. This would be possible if the many social theories that CST makes use of in its theoretical critique ensured, between them, that sufficient attention was drawn to the issues raised by "social awareness" and "ecological awareness" – if they paid heed to the relationships between social science, the society in which it is developed and used, and the natural environment. Social science does not appear to do this yet on a consistent basis. Until it does, it is as well to keep social awareness and ecological awareness as back-up forms of critique, within critical awareness, to ensure these matters do receive the necessary scrutiny.

The second commitment of CST is to *pluralism*. The ideal type of "multimethodology," developed in the previous chapter, insists on multimethod, multimethodology, and multiparadigm practice at every stage of an intervention.

CSP accepts the argument (Midgley 1989; Mingers and Brocklesby 1997; Taket and White 1995) that there is considerable merit in allowing systems methodologies to be decomposed and using parts of different methodologies, together with any other useful tools and techniques, in new combinations according to what is required in an intervention. This multimethod way of working increases the flexibility and responsiveness of practitioners to address the detail of rapidly changing problem situations. The strength becomes a weakness, however, if multimethod practice is conducted without reference to the methodologies and paradigms behind their use. We cannot learn about the effectiveness of particular methods in supporting particular theoretical rationales. Even more importantly, there will be an almost inevitable relapse into "pragmatism." In the absence of methodological guidance, some implicit paradigm of analysis will come to dominate the intervention by default and the benefits of exploring what outcomes might be achieved using methodologies representing alternative paradigms will be lost.

For these reasons, a second requirement of the ideal type is that methodologies owing allegiance to different paradigms should be employed in the same intervention unless an alternative way of proceeding can be justified. It is the complexity, heterogeneity, and turbulence of problem situations that suggest systems practitioners need a pluralism that encourages the use, together, of different methodologies based upon alternative paradigms. We should seek to benefit from what each paradigm has to offer. Multimethodology can provide its greatest benefits only in the context of methodology and paradigm diversity. This is not to dismiss the usefulness of sometimes employing just one methodology, embodying a particular paradigm, to guide the use of a variety of methods, tools, and techniques. It is just to insist that such an approach needs to be followed self-consciously and to permit changes of paradigmatic orientation. If it occurs without due consideration, it deprives multimethodology practice of the vitality it gains from being able to deploy a variety of methodologies, based upon different paradigmatic assumptions, to their true potential. Multimethodology working demands, in turn, a precise understanding of the theoretical underpinnings of the different methodologies. If such theoretical understanding is neglected then proper paradigm diversity cannot be guaranteed. At best, the outcome is likely to be "imperialism." Other methodologies will be forced to fit into some favored approach and will lose their unique strengths. No "genuine" pluralism will be achieved.

The third requirement of the ideal type of multimethodology follows. To cope with multidimensional complexity, systems practitioners must learn to appreciate a wide variety of sociological paradigms and what each has to offer. They need to work with maximum paradigm diversity and understand the different ways each paradigm views the world. They should understand that paradigm incommensurability is a good thing because each paradigm shows them different things, suggests different issues to address, and presents alternative ways of intervening in a problem situation. Archer's argument (Section 4.6), that structure and agency are interdependent features of social reality but they must be studied separately, provides further theoretical backing for our multiparadigm approach. In her view, we can only give due attention to the influence of both if we break them up for analytical purposes. However, intertwined structure and agency may be in "reality," we must adhere to "analytic dualism" when studying their relationship or our analysis will inevitably collapse into either functionalism or interpretive social theory.

The theoretical issues addressed by Archer are tough enough. But the problems are magnified for the systems practitioner who wants to work with different paradigms in the course of an intervention designed to bring about improvement in a problem situation. How is it possible to manage paradigm diversity to maximum benefit in a multimethodological project?

Various suggestions have been made by advocates of CST and rejected. The reader will recall that one of the main criticisms of TSI was that it uncritically adhered to Habermas' early theory of human interests. This theory seemed to allow it to operate at a meta-level to the paradigms, allocating appropriate methodologies to different aspects of a problem situation as appropriate. The difficulty, as Tsoukas puts it, is that:

> Reality-shaping paradigms ... are not *a la carte* menus; you don't just pick whatever suits you at any time.
>
> *(1993a, p. 315)*

Or, in Luhmann's words:

> The observer does not exist somewhere high above reality. He does not hover above things and does not look down from above in order to observe what is going on. Nor is he a subject... outside the world of objects. Instead, he is in the middle of it all ...
>
> *(2013, p. 101)*

Recognizing the flaw in basing multimethodological practice on Habermas' theory of human interests, critical systems thinkers have been tempted, as we witnessed in the previous chapter, to declare CST a new paradigm or to find it a place in a particularly accommodating existing paradigm. The first strategy, endorsed by Midgley (2000) declares that a process philosophy, with boundary judgments at its nondualist core, can make of CST a paradigm in its own right. Boundary critique, sweeping in the viewpoints of many stakeholders, is put up front in any study. Other methodologies and methods are then subsumed into this philosophy and employed in creative combination for particular aspects of an application. Mingers, championing the second strategy, argues that:

> What is required is an underpinning philosophical framework that can encompass the different paradigms, and guidance on appropriate ways to mix different research methods.
>
> *(1997a, p. 14)*

As we saw, he came to see critical realism as an existing paradigm that could contain and get the best from pluralism. For Midgley and Mingers, therefore, CST cannot be metaparadigmatic. It must build a paradigm of its own or use a suitable existing one to serve its purpose. The strength of these options is that they resolve the difficulty of having to combine methodologies based upon divergent philosophical and sociological assumptions. A new paradigm is proposed that can house all or, at least, most of the others. The obvious weakness is that, unless we accept that the new paradigm is capable of containing a divergent range of methodologies, the power of paradigm diversity is constrained. As we saw, there are many who can make convincing arguments, from alternative paradigms, against Midgley's and Mingers' versions of multimethodology. What they propose is more restrictive than the TSI approach – not a way forward but a way backwards. If it is to protect paradigm diversity, CST cannot sell itself to any one paradigm. One-paradigm pluralism is simply not pluralism. CST must continue to encourage and struggle with paradigm incommensurability.

A more promising way forward is offered by the "discordant pluralism" of Wendy Gregory. Gregory (1992, 1996a,b) argues for discordant pluralism against the "complementarism" she sees as dominating TSI:

> To an extent, the main difference between these two pluralist positions is captured in their titles: the complementarist wishes to use theoretical approaches in complementary ways, whilst the *discordant pluralist* would allow discordant theoretical approaches to *both challenge and supplement* one another.
>
> *(Gregory 1992, p. 621)*

Discordant pluralism, therefore, suggests that the differences between paradigms should be emphasized rather than "rationalized away." This is useful and points the way forward to the kind of pluralism that can deliver the greatest benefits for systems thinking. CST requires a multimethodology that accepts and protects paradigm diversity and handles the relationships between the divergent paradigms. It should accept that paradigms are based on incompatible philosophical assumptions and that they cannot be integrated without something being lost. It needs to manage the paradigms, not by aspiring to metaparadigmatic status or creating a separate paradigm, but by using them critically. Paradigms have to confront one another. Critique is managed *between* the paradigms. No paradigm is allowed to escape unquestioned because it is continually challenged by the alternative rationales offered by other paradigms. CST no longer aspires to metaparadigmatic status or, indeed, to its own paradigm. Its job instead is to protect paradigm diversity and encourage critique between the paradigms.

A similar line of thinking has been advanced by Morgan (1983). He rejects the possibility of finding some independent point of reference from which the claims of different research strategies can be evaluated. Instead, researchers should engage in critical reflection on their own strategy and the course of action it suggests. This is best done by means of a "reflective conversation" in which its taken-for-granted assumptions are set down and compared with those of competing perspectives. This should reveal the nature, strengths, and limitations of the favored strategy. Employing reflective conversation aids us in appreciating other approaches and helps us refine our own, while becoming more modest about what it can achieve. I explicitly adopted the method of "reflective conversation" to conduct a critical evaluation of Beer's Viable System Model (VSM) (Jackson 1988a) and, convinced by Gregory's arguments, soon after began to develop CSP as a process of encouraging paradigm diversity and managing debate between the paradigms. Luhmann describes the difference between this second-order thinking and the classical approach to "collecting" knowledge precisely and, in doing so, makes clear the need for "reflective conversation" and "discordant pluralism":

> When handling a distinction, you always have a blind spot or something invisible behind your back. You cannot observe yourself as the one who handles the distinction. Rather you must make yourself invisible if you want to observe .... In all observing, therefore, something invisible is produced at the same time. The observer must make himself invisible as the element of the distinction between the observer and the observed. For this reason, there are only shifts between that which one sees and that which one does not see, but there is no comprehensive enlightening or scientific elucidation of the world as a totality of things or forms or essences that could be worked through piece by piece, not even if the task is seen as infinite.
>
> *(Luhmann 2013, pp. 104–105)*

The ideal type of multimethodology insists that methodologies adhering to different paradigms should be used at all stages in the same intervention unless there are good reasons for temporarily adopting an imperialist stance. The notion of "critique between the paradigms" helps to clarify how this might be done. It can be illustrated if we jump ahead a little and consider the four phases of CSP, described more fully in the next section. A practitioner, in pursuing "creativity," should consider a problem situation from the perspectives of the different paradigms and compare the results. Engaging in "choice," they will be looking at what they hope to achieve in terms of the strengths and weaknesses of different systems methodologies as seen from the paradigms they represent *and* alternative paradigms. Similarly, the way "implementation" proceeds will be continually critiqued through the lenses offered by different paradigms. And the results obtained from the intervention will be evaluated according to the concerns evinced by alternative paradigms. A successful project will have to score well on improving efficiency, efficacy, effectiveness, viability, mutual understanding, empowerment, emancipation, etc.

TSI treated Habermas' theory of human interests as a metaparadigmatic standpoint from which the various sociological paradigms can be understood and the methodologies based on them suitably employed. Tsoukas argued that it is impossible to dissolve the arguments between paradigms in this way because they constitute different realities and each believes that it can provide answers to all three human interests. Adopting a "critique between paradigms" approach allows CST to sidestep Tsoukas's censure. The different paradigms are allowed full rein, unrestricted by any metaparadigm or, indeed, by some "super-paradigm" that is deemed capable of containing all the others. At a stroke, however, CST loses whatever theoretical justification it once obtained from Habermas' work, or from process philosophy, or from critical realism. A new and serious problem arises: How can CST avoid relativism?

Barton (1999) and Zhu (2011, 2012) have suggested that pragmatism can provide appropriate philosophical foundations for, respectively, systems thinking and "innovative mixing-methodology practice." Noting that Zhu sees this as a better alternative, rather than a support, to paradigm theorizing, I still agree with both of them that this is the place to look if we are to find a justification for CSP. In talking about pragmatism here I am referring to the theoretically developed form deriving from Pierce, James, and Dewey, and which Zhu also relates to Confucius, rather than the everyday, shorthand version of "if it works, use it" that I have used to contrast with isolationism, imperialism, and pluralism in earlier debates, or the postmodern "pragmatic pluralism" of Taket and White. My discussion is based on the brief description of philosophical pragmatism presented in Chapter 1, and on the articles by Barton and Zhu.

Pragmatism rejects the "spectator theory of knowledge." It is humans who impose a structure on the complexity of the world. Because we do not have direct access to the external world we cannot judge our theories in terms of whether they correspond to it. Rather, we must seek justification for our beliefs and actions in terms of their practical effectiveness. Further, because reality is not static, and our beliefs are important in constructing the world, we need to look for and employ concepts that are effective in helping us to achieve our goals and in bringing long-term benefits. Pragmatism, therefore, is not atheoretical. It has to know what theories it is using to understand and act upon the world, in order that it can decide which of them enable objectives to be achieved and which don't. Indeed, it needs what Zhu calls "ontological flexibility" so that it has a variety of theoretical positions at its disposal, which it can test in terms of the benefits to which they give rise:

> In some circumstances and for certain tasks at hand, ontology of process, of events and happenings, is telling; in other circumstances and for other tasks, ontology of things, of contents and substances, makes more sense .... Pragmatist ontology reaches out to all human constructs .... The result is enlarged human competence for coping with life problems resourcefully, robustly .... Such a pragmatic imagination respects all kinds of ontology but accepts obligation to no one. It puts into use diverse ontologies in the face of changing circumstances. It examines and refines them in the light of practical consequences.
>
> *(Zhu 2012, pp. 3–4)*

Ontologies cannot be judged on whether they capture the "essence" of reality. Rather, their justification resides in their ability to serve our purposes in the particular circumstances in which we find ourselves. In the complex world we face, when we cannot be certain what will unfold and how our own actions will contribute to what happens, we should embrace a plurality of perspectives, methodologies, and methods and see which work for us. As Barton makes clear, Pierce wanted to rest pragmatic inquiry on an "experimentalism" not so far divorced from the scientific method as later conceived of by Popper. Plausible hypotheses should be arrived at through a process of abduction and tested for their predictive success. The relevant "community of inquiry," in this case the "scientific community," decides whether they pass the tests set for them. If they are judged to do so then they contribute to the extension of knowledge. James and Dewey, concerned with dealing with human problems, want a larger community of those impacted to confirm that the problems have been overcome before they are willing to declare the knowledge used as legitimate. I have little difficulty in endorsing Zhu's vision of a "mixing-methodology practice" based on promoting ontological flexibility and methodology-in-use:

> Incorporating ontological flexibility, it allows OR workers to enact multiple realities, craft ontology-in-use, weave available methods with situated particulars, justify methodologies based on practical consequences, so as to get jobs done and enhance competences.
>
> *(Zhu 2011, p. 784)*

CSP, as we shall see, offers the guidance necessary to make "ontological flexibility" possible and to translate it into practice at every stage of an intervention. It offers appropriate means of judging whether theories and methodologies can be justified – whether that is because they are falsifiable, lead to an increase in mutual understanding, or ensure the empowerment of the oppressed. In return it is more than happy to see "interminable metaphysical disputes" resolved on pragmatic criteria.

In closing this discussion of pluralism as a commitment of CSP, it is worth drawing attention to two other ways of promoting diversity in an intervention that are brought to the fore by Taket and White (2000) using their postmodern lens. Many interventions can be improved if attention is given to pluralism "in the modes of representation employed" (verbal, visual, and role-playing) and "in the facilitation process" (flexibility, forthrightness, focus, and fairness).

The third commitment of CST is to *improvement*. CST has, since its inception, often made grand statements about being dedicated to "emancipation" and, despite the hyperbole, it is reasonable to suggest that putting fairness and empowerment on the agenda of systems thinkers has been one of its main achievements. As we saw in Chapter 19, however, this led to some initial confusion between critical and emancipatory systems thinking. Eventually, it was made clear that "emancipation" was only one of the three human interests that, following Habermas, CST needed to reinforce. CST in 1991, therefore, still embraced emancipatory practice but as part of a broader commitment that also saw it lend its support to the "technical" interest in prediction and control and the "practical" interest in promoting mutual understanding. It was argued that success in pursuing all three interests was essential to realizing those circumstances in which individuals could fulfill their full potential (Flood and Jackson 1991a).

Following the attack on the "grand narrative" of universal liberation, conducted by postmodernists such as Lyotard, critical systems thinkers became even more circumspect in the language they used to address this topic. It was accepted that Habermas' position, based on achieving universal consensus in the context of the "ideal speech situation," was untenable. It became more normal to talk in terms of achieving "local improvement." Brocklesby and Cummings (1996) introduced the idea of a CST built on an alternative philosophical tradition. This also began with Kant but took a postmodern turn, via Nietzsche and Heidegger, to its culmination in the work of Foucault. The dominant theme here is *self*-emancipation rather than, as with Habermas, universal human emancipation.

Nowadays, critical systems thinkers tend to talk in terms of "improvement" rather than emancipation. In the CSP multimethodology, as we shall see, the aim is to conduct interventions which lead to "improvement," over time, from the points of view of all the different paradigms. The improvement should embrace efficiency, efficacy and effectiveness, viability and sustainability, mutual understanding, empowerment, and emancipation. However, there still remains the need for a powerful strand of systems thinking designed to highlight and do something about emancipatory concerns. First, to ensure that interventions primarily meant to address matters such as efficiency and viability are critiqued from the viewpoints of all affected stakeholders. Whatever benefits might be forthcoming from a functionalist worldview, a project that threatens to worsen the position of the disadvantaged, from the emancipatory perspective, may need to be modified or abandoned. Second, to guarantee that the needs of disadvantaged stakeholders are always considered up-front in an intervention and that their voices are given due weight in decision-making. Improvement in the position of these stakeholders, from an emancipatory perspective, should count just as much as any other benefits deemed to flow from an intervention. Finally, because there will be circumstances where some form of emancipation is the main purpose of the intervention. Emancipatory systems thinking will then be called upon to guide the study and emancipatory methodologies will assume dominance.

In *Systems Approaches to Management* (Jackson 2000), I considered two main types of emancipatory systems practice, labeled "emancipation through discursive rationality" and "emancipation as liberation." The first type derived originally from Marx's early writings but was led, by Habermas, along a road that took it to a new critical standard – communication free from domination. As was outlined, in Section 4.4, for him emancipation became associated with the "ideal speech situation" in which citizens determine their true interests free from "distorted communication." In the ideal speech situation, it is the better argument that prevails and not the ideology of the powerful. Craib helpfully summarizes Habermas' position:

> First, rationality ... is there in our language itself .... Second, there is an implicit ethics which Habermas attempts to draw out – a universal ethics .... It is often referred to as a procedural ethic which directs not to the content of a norm but to the way it is arrived at. It is arrived at through free rational discussion .... Third ... is the implication of a radically democratic society, in which each has access to the tools of reason, the opportunity to contribute to the argument, to be heard and to be included in the final decision.
>
> *(1992, pp. 234–235)*

CST is lucky to have available the work of two systems thinkers which, from a practical point of view, enhances that of Habermas. Beer's "team syntegrity" (TS) owes everything to organizational cybernetics and nothing to Habermas. Nevertheless, it provides a set of procedures designed to promote open, nonhierarchical, participative decision-making around issues wherever they arise in organizations or, indeed, in multiorganizational settings. Ulrich's CSH is heavily influenced by Habermas' thought, but tries to take it in a more practical direction. Boundary judgments, he argues, provide us with an access point to the normative implications of systems designs and boundary critique provides a method which all stakeholders, including the affected but not involved, can use to reveal the partiality of all boundary judgments.

In examining the criticisms of both TS and CSH, however, we came up against a significant objection. Both presuppose the very thing they seek to bring about. They assume a society in which decisions in organizations and society are actually taken on the basis of discursive rationality. They tend to take it for granted that the conditions necessary for communicative competence are close to being realized, that there is something like a well-functioning public sphere, and that all citizens are equipped to take part equally in participative debate. To critics this is an unrealistic picture of the nature of current social systems. In societies stratified on the basis of class, race, gender, etc., and in organizations run for the benefit of shareholders and senior managers, why should the powerful bother to take account of the views and interests of the disadvantaged? Even if those discriminated against are allowed to participate in a debate, can we expect that they will be competent enough to represent their own true interests and that they will be listened to? Isn't it more likely that they will lack the resources to participate as equals and will suffer from "false consciousness," allowing the ideologies of the powerful to dominate decision-making and what happens as a result? Radical change critics maintain that genuinely emancipatory practice needs a social theory that can explain sources of inequality and discrimination in society and can indicate how best to confront the forces that prevent rational argumentation and participative decision-making from taking place. In the case of CSH, for example, it was argued that it may allow us to reflect upon the ideas that enter into any social systems design but it does not help us to understand the material conditions in society that give rise to those ideas and that lead to certain ideas holding sway.

If the construction of explicit social theories is seen by some radical change theorists – those of a more objectivist persuasion – as essential to emancipatory practice, then CST should seek to reflect this in the methodologies it makes available. In fact, too little attention has been given to developing this aspect of CST. This is indicated by the empty space on the right-hand side, especially the upper right-hand side, of the SOSM figure at the beginning of Part IV. More must be done if CST is to provide a fully rounded picture of what "improvement" entails. The second type of emancipatory practice discussed in *Systems Approaches to Management*, "emancipation as liberation," helps to address this issue because it seeks explicit theories from which a critique of the existing social system can be launched and recommendations made for alternative, improved social arrangements. In particular, it identifies particular groups or interests that need liberating from oppressive social structures and relationships and suggests how their liberation can be brought about. Its inspiration is the later work of Marx. I will call this second approach "liberating systems theory" (LST) in what follows, with a nod to Flood (1990).

In the three volumes of *Capital*, Marx (1961) provides a "scientific" explanation of how the capitalist mode of production leads to class struggle between the capitalist class, who own the means of production, and the exploited working class from whose labor profit is extracted. This will eventually result in the overthrow of capitalism by the workers in a revolution that will usher in a communist society. More recently, sociologists have identified other groups who are discriminated against in society and organizations, on the grounds of gender, race, sexual orientation, disability, etc. This form of critique has also been extended to our treatment of the natural world. Capra (1996), for example, attacks our current approach to "social ecology" and the reductionist world view that sustains it. He then provides a new theory of living systems that will allow us to "reconnect" with the web of life and build sustainable communities which meet our current needs without diminishing the opportunities available to future generations. This critique allows him to propose new social arrangements that promote "maximum sustainability" by learning from the principles that he sees as underpinning the pattern and structure of the ecological system.

Anne Stephens (2013) champions LST from an explicitly critical systems perspective. Specifically, she seeks to demonstrate how cultural ecofeminism and the critical systems approach can benefit from each other. From Churchman and Ulrich, ecofeminism can learn about the importance of boundary judgments and how they can be critiqued. CST, especially the work of Midgley, can contribute insights into how to use different methodologies and methods in combination in a participative manner. In the other direction, Stephens notes that little attention has been given to gender specific concerns and ecological issues in CST. Cultural ecofeminism offers an holistic approach that seeks to take account of multiple linked oppressions, for example, of women and the environment. These dual oppressions are seen to flow from the same patriarchal perspective based upon positivism and dualism. Cultural ecofeminism can, therefore, help inform CST about aspects of human and nonhuman exploitations and point to the sources that produce the oppression. Drawing upon both traditions of thought, Stephens is able to suggest five "feminist-systems thinking principles" to guide participatory action research:

- Be gender sensitive
- Value the voices from the margins
- Center nature
- Select appropriate methods/methodologies
- Bring about social change

CST should absorb social and ecological theories that demonstrate the need for LST on behalf of oppressed groups.

But how to justify such theories and use them in the context of CST's broader project of "improvement"? One solution is to present the appropriate theory to the oppressed group and see whether it enables them to see through false consciousness, recognize their true interests in an antagonistic social system, and do something to change that system. Freire (1970) advocates this approach. A Marxist reading of the situation is introduced to an oppressed group through "critical pedagogy." This is a form of democratic dialogue conducted by those who understand the causes of oppression with those actually oppressed. The latter should become aware, during this process, of the reality of the structures of domination from which they suffer, feel the need to liberate themselves and start to become politically engaged to that end. Freire has been criticized for downplaying other forms of oppression, based on such as race and gender. An explicitly systems-based approach, "interpretive systemology" (Fuenmayor 1991; Fuenmayor and Lopez-Garay 1991), has similarities with Freire's way of proceeding. Interpretive systemologists start with how particular individuals and groups make sense of their world before gradually "unfolding the scene," introducing different interpretations of the context. Debate is encouraged about the different interpretations and this leads to "a state of enriched consciousness" among those involved. In practice, despite the phenomenological foundations of interpretive systemology, it is assumed that a Marxist perspective will become accepted as the one that best explains "the facts" and action will happen on that basis (Jackson 2000, pp. 298–299). This kind of approach to justifying LST can be extended to environmental concerns using theory to convince more and more people to become representatives of the interests of the natural world. Another possible way to justify LST is to follow the route charted by Bhaskar with his critical realism. Bhaskar, in many ways a disciple of the later "scientific" version of Marxism, believes that social science can reveal how certain structural mechanisms give rise to false interpretations of social reality which get in the way of us building a society based on moral truths in which "universalized freedom for all" is achieved. Once this becomes clear, we are morally committed to take action to remove the negative causal mechanisms.

These reflections on the LST approach allow us to complete the SOSM figure at the beginning of Part IV as in Figure 21.1. This positioning is simply trying to capture the assumptions LST makes in terms of the grid. It has a focus on forms of "coercion" that derive and are intertwined with complex structural mechanisms in society that give rise to and sustain oppression. Further, the thinking extends to relationships between social systems and the natural environment, their interdependence, and the tendency of the former to ignore the requirements of the latter.

The positioning says nothing, of course, about particular "real-world" problem contexts or about how frequently they occur. And if a decision is made to adopt the LST approach, then it must be subject to the same "critique between paradigms" to which it

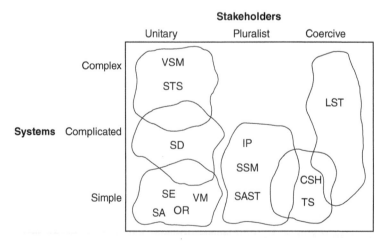

Figure 21.1 The positioning of liberating systems theory on the SOSM.

is itself designed to contribute. For example, functionalists will be far from impressed with Freire and Macedo's (1987) claim that an adult literacy program in Guinea-Bissau was a success because it raised people's awareness of their situation even though it failed in terms of the numbers that were taught to read and write. To interpretive social theorists and supporters of "emancipation through discursive rationality," LST appears élitist in character. The critique it offers must rest upon a social theory which, it is claimed, offers a superior perspective on the social world to that possessed by those actually living the social reality. This theory purports to provide genuine knowledge which can be used to unmask ideologies and liberate the oppressed from "false consciousness." The critics want to know how such theorists are able to step outside their own social situation and gain a privileged viewpoint. As Munro (1997) puts it, every therapist must also seek therapy and every judge can also be judged. Furthermore, from the perspective of Foucault, any claim to knowledge is also a claim to power over others:

> Foucault argues that knowledge is a power over others, the power to define others. Knowledge ceases to be liberation and becomes enslavement .... A discourse embodies knowledge ... and therefore embodies power.
>
> *(Craib 1992, p. 186)*

LST may be inviting the oppressed simply to enter into new relations of domination. To Luhmann, the emphasis on stratification in much of LST is redolent of Old European thought. It fails to recognize the decisive shift in society toward functional differentiation.

To end this discussion of "improvement" as a commitment of CST, it is worth noting that the literature of Community OR (e.g. Carter et al. 1987; Ritchie et al. 1994; Midgley and Ochoa-Arias 2004; Johnson et al. 2018) adds much of value to the debate, especially in relation to emancipatory practice.

21.2.3 Multimethodology

TSI went a long way toward providing a suitable multimethodology for CST. In considering CSP as the revised multimethodology, it is worth noting that much of the terminology remains the same and that there is some basic continuity. The philosophy embraced can still be described in terms of three commitments – "critical awareness," "pluralism," and "improvement" – although significantly modified as in the discussion above. The principles have largely stood the test of time (the reader is referred back to Chapter 20 for these). Intervention still has its "creativity," "choice," and "implementation" phases, although a fourth "reflection" phase has now been added. To orientate the reader for the following discussion, CSP is presented in outline in Table 21.1.

The four phases will now be described in detail before some general points are made about CSP. It should be remembered that the aim, during each phase, is to remain true to the philosophy and theory of CST while producing a set of multimethodological guidelines that are useable in practice.

The task in the *creativity* phase is to highlight the major issues in the problem situation that need addressing. Any creativity-enhancing devices (e.g. brainstorming, the "nominal group technique," "rich pictures") can be employed but they must be used alongside various "systems perspectives" (SPs). I am calling them SPs because together

Table 21.1 The Critical Systems Practice multimethodology.

Creativity	
Task	To highlight significant issues in the problem situation
Tools	Systems perspectives, and other creativity-enhancing devices, employed to ensure that the viewpoints of different paradigms receive proper attention
Desired outcome	Identification of primary and secondary issues that need to be addressed
Choice	
Task	To choose appropriate systems methodologies (perhaps in generic form), reflecting different paradigms, and a variety of suitable methods, models, tools, and techniques to use in the problem situation
Tools	Critical awareness of the strengths and weaknesses of systems methodologies, methods, etc., obtained from social theory and the SOSM; and also from previous experience
Desired outcome	Primary and secondary systems methodologies and appropriate methods, etc. chosen for use
Implementation	
Task	To take action that addresses the issues identified during the "creativity" phase
Tools	Appropriate systems methodologies, suitable methods, etc., employed according to the logic of CSP
Desired outcome	Changes which secure improvement in the problem situation from the points of view of the different paradigms
Reflection	
Task	• To evaluate the changes made according to the concerns of the different paradigms • To obtain learning about the CSP multimethodology, the systems methodologies, the methods, etc. used
Tools	• Suitable evaluation methods derived from the different paradigms • Clear understanding of the current state of knowledge about CSP, etc., and reflection on what happened in the intervention
Desired outcome	• Start using CSP again to address whatever issues are now significant • Research findings that feedback into improving CSP

they provide a systemic appreciation of the problem situation. I have dropped the TSI terminology of "systems metaphors" because some of the SPs do not take the form of metaphors. Using the set of SPs ensures that a problem situation is looked at from the point of view of the variety of sociological perspectives. We cannot expect participants to possess sufficient knowledge of social theory to carry out a full-blown paradigm analysis. SPs are easier to understand and use. They are, nevertheless, closely related to paradigms so that, if a comprehensive set of appropriate SPs is chosen, little is lost in terms of a pluralistic appreciation of the problem situation. Another reason for insisting on the use of a particular set of SPs is that they have proven their worth. They are drawn from the history of social theory and, as Morgan's (1986) work on "images of organization" suggests, they have all, at one time or another, been found useful in thinking about leadership and management. Problems have tended to occur when one has dominated people's thinking at the expense of all the others as, for example, with "scientific management." As long as we are aware of what each SP highlights and hides, as we will be if we use a range, then we should seek to take advantage of the stock of knowledge that they encapsulate.

I want to reinforce this point with a short digression on a much neglected contribution by Pepper (1942) to systems thinking. Pepper describes how the "root metaphor method" has, through "the traditional analogical method of generating world theories," led to a number of "world hypotheses" constituting the "principal metaphysical systems" which attempt to make sense of and explain the "world experience." This method is defined by Pepper in the following terms:

> A man desiring to understand the world looks about for a clue to its comprehension. He pitches upon some area of common-sense fact and tries to see if he cannot understand other areas in terms of this one. This original area becomes then his basic analogy or root metaphor. He describes as best he can the characteristics of this area, or, if you will, discriminates its structure. A list of its structural characteristics becomes his basic concepts of explanation and description. We call them a set of categories. In terms of these categories he proceeds to study all other areas as fact .... He undertakes to interpret all facts in terms of these categories.
>
> *(1942, pp. 91–92)*

Pepper identifies six world hypotheses: "mysticism," "dogmatism," "formism," "mechanism," "organicism," and "contextualism"; although only the final four have proved capable of generating adequate world theories. Formism (or "realism" or "Platonic idealism") has "similarity" as its root metaphor. All specific objects of experience are seen as copies of ideal forms. Mechanism (or "naturalism" or "materialism") has "the machine" as its root metaphor. This world hypothesis sees reality as totally mechanistic, as operating under physical laws and thus being completely determined. Organicism (or "absolute or objective idealism") has "organism" and "integration" as its root metaphors. It notes the steps involved in the organic process and the principal features of the organic structure that result. Contextualism (or "pragmatism") presents the world as a complex characterized by change and novelty, order and disorder. The contextualist is concerned with "an act in its context." Acts are

> ... composed of interconnected activities with continuously changing patterns .... They are literally the incidents of life. The contextualist finds that everything in the world consists of such incidents.
>
> *(Pepper 1942, p. 233)*

Within such a complex state of flux, it is difficult to attain meaning and we have to select "contexts" that organize and attribute meaning to the world. Successful contexts have "quality" and "texture."

In Pepper's work, we have confirmation of the value that philosophers, scientists, and others have found in certain "root metaphors" when they seek to bring order to the world. This cannot be easily dismissed. It is also interesting that Emery, in his introduction to an important early collection of papers on systems theory, sees Pepper's "root metaphors" as "clearly operating" in the work of different systems thinkers. This, he thinks, accounts for much "of the mutual incomprehension that exists among them" (1969, p. 15). He goes on to declare "contextualism" as the root metaphor that best expresses the bias employed in selecting the papers for his collection. When all is said and done, if there is no prospect of "neutral" observation we may as well be up-front about the lenses we are using at the beginning of an intervention.

Building on this thinking, and recognizing the need to pay attention to "social" and "ecological awareness," the SPs that I feel are most useful for viewing problem situations are:

- The machine perspective
- The organism perspective
- The cultural perspective
- The political perspective
- The "coercive system" perspective
- The environmental perspective
- The interrelationships perspective

For illustrative purposes, we will consider what these various SPs pay attention to when looking at organizations. This thinking will have to be adapted in other contexts such as multiorganizational settings.

The machine perspective views organizations as rational instruments designed to be effective in achieving the goals of their controllers. They consist of parts that must be organized in a hierarchy to ensure that those goals are achieved. Processes delivering the outputs of the organization must be as efficient as possible. Control and coordination is exercised by the hierarchy and through rules. Information is processed up and down the hierarchy. The organization is judged on whether it demonstrates efficiency and efficacy.

The organism perspective portrays organizations as complex systems seeking to survive. They have needs that must be met by their relatively autonomous subsystems. They are open-systems, dependent on their environments, and must find a suitable environmental niche and respond rapidly to changes that occur. The organism has a brain, or managerial subsystem, which is responsible for decision-making and learning and, in particular, for integrating the subsystems and securing favorable interchanges with the environment. The organization is evaluated on whether the parts are functioning well to meet the requirements of the whole, on whether they are coordinated, and whether viability can be secured in the face of a turbulent environment.

The cultural perspective refers to familiar and persistent ways of seeing and acting in organizations. It views human beings as the most important element contributing to their success. Humans attach meaning to the situations they find themselves in and can attribute different purposes to the organizations they are members of. Organizations are seen as processes in which different perceptions of reality are continuously negotiated and renegotiated. Leaders must pay attention to different perspectives and values and motivate people toward a shared purpose. The success of the organization will depend on its leaders creating a culture that is sufficiently robust but also inspires and does not drive out original thinking and innovation.

The political perspective sees organizations as loose coalitions of different individuals and groups with diverse interests. It focuses on the everyday politics of organizational life, conflict, how different interests are reconciled, and how power is obtained and used. To the many whose working lives are blighted by squabbles between departments and by power struggles among senior managers, it is an extremely realistic perspective. Organizational politics can lead to an organization losing its capacity to function effectively and respond to its environment. An organization, viewed in this way, will be evaluated in terms of the means it has for reconciling different interests and the extent to which these enable it to keep conflict under control.

The "coercive system" perspective views organizations as instruments of domination used by some to benefit themselves at the expense of others. Organizations are entities made up of different groups (class, status, gender, etc.) whose interests are almost impossible to bridge given the present structure of society. Relationships are exploitative and the organization only holds together because of the power of some group(s) to control the activities of others. Coercion may extend outside the boundaries of the organization, for example, along its supply chain. Those who are exploited may suffer from false consciousness which traps them in an alienating form of life or simply have little option but to remain part of the system to survive. The paradigm with which this perspective is linked insists that improvement can only come with the empowerment and emancipation of oppressed individuals and groups.

The environmental perspective is an extension of the organism perspective. The long-term future of organizations and their environments are seen as interdependent. Organizations ultimately depend upon the natural environment from which they draw their resources and which they impact with their products and waste. Their sustainability depends on them nurturing their environments and they are judged on the way their behavior takes into account the needs of the natural world and of future generations.

The interrelationships perspective is explicitly systems-based and is included because, from the systems viewpoint, the matters identified by the other perspectives are often highly interrelated. It reminds us of the need to look out for important linkages so that we can identify the most powerful leverage points for bringing about improvement and also anticipate the possible unintended consequences of intervention in just one area.

It will be clear that the machine and organism perspectives relate to the functionalist paradigm; the culture and political system perspectives to the interpretive paradigm; and the "coercive system" perspective to the radical change paradigm. The presence of the environmental perspective reflects CST's new "ecological awareness" and the failure of social theory to sufficiently engage with environmental concerns. The interrelationships perspective carries the systems message of connectivity and reminds us that things never go quite as planned or "settle down." A CSP intervention will, hopefully, improve a problem situation but it will inevitably bring new issues to the fore which will need addressing if further improvement is to be achieved. Perspectives representing other paradigms can, of course, be added if it seems useful. I once (Jackson 2003) made a case for including the "carnival" perspective to view organizations from a postmodern position. Carnivals encourage creativity and diversity and are places where people have fun. There is much about life in organizations that comes to the fore if we pay attention to playfulness, sex, irony, etc. The emphasis on seven SPs is meant to indicate the minimum number and range necessary to ensure that focused and productive creativity happens; it is not meant to close it down.

The reader may recall one of Churchman's principles of systems thinking: *The systems approach begins when first you see the world through the eyes of another.* He explains this further:

> Another way to say the same thing is that the systems approach begins with philosophy, because philosophy is the opportunity to see the world through the eyes of a Plato, a Leibniz, or a Kant. The reading of philosophy is not an abstract study; the serious student takes the burden of becoming convinced that each important philosophical position is right, absolutely right. He relives the intellectual vitality of the past. He feels to the utmost that the real-world is the modeled world; that the real-world is the experienced world; that the real-world is dialectical; and so on. He does all this without losing his own individuality.
>
> *(1979b, p. 232)*

It is of more direct usefulness to decision-makers to view problem situations through the lenses provided by the different SPs but, otherwise, Churchman's principle holds. The key to making the best use of the SPs is to learn to be comfortable inhabiting each of them and to look out at the world convinced that what each is telling you is the truth. Each SP will highlight issues that the others cannot "see" or, at least, provide a different explanation for the same issue.

The first phase of CSP draws to a close when the issues brought to the fore by the SPs are classified as "primary" and "secondary." This kind of prioritization usually begins to happen without direction from the facilitator. The primary issues are, obviously, those that need most urgent attention – perhaps because they seem to get to the "heart" of the matter. They will guide the initial choice of methodologies. The secondary issues are ruled out for immediate action but, CSP insists, they must be kept in mind and may assume greater importance, later in the intervention, when a fresh analysis using the SPs brings them into the limelight. The creativity phase takes the broadest possible look at the problem situation and gradually focuses down on those issues deemed most crucial at that point in its evolution.

The second, *choice*, phase is concerned with choosing the most suitable systems methodologies, and the best methods, models, tools, and techniques, to address the issues highlighted by the creativity stage. The choice of methodologies should ideally be done using all facets of "critical awareness" to identify their particular strengths and weaknesses in relation to the issues that are faced in the problem situation. In a real project, it is reasonable to rely on the SOSM's translation of philosophy and theory into practical guidelines, to make an initial choice. This will be aided by our knowledge of which SPs brought the primary issues forward. The machine perspective is naturally associated with simple-unitary assumptions. If the "primary issues" emerge from its use, the SOSM will point to methodologies which assume that problem contexts are simple-unitary. Hard systems thinking (HST) and Vanguard Method (VM) pay attention to the complexity that arises from technical and process issues in simple-unitary problem contexts and are well equipped to manage such matters. They would be a good place to start. To take another example, if the "primary issues" are revealed by the organism perspective, and focus on matters such as co-ordination between subsystems and antifragility, then we will be led by the SOSM toward socio-technical systems thinking (STS) or the VSM for initial guidance on how to proceed. The SOSM would regard them as suitable because they privilege the complexity arising in complex-unitary contexts. A similar logic leads us from issues raised by the cultural and political perspectives to soft systems methodologies; from the coercive SP to TS, CSH, and LST; from the environmental perspective to variants of LST; and from the inter-relationships viewpoint to system dynamics (SD). It is not an exact science and, of course, the knowledge and experience of the systems practitioner will have an important part to play.

During the choice phase, consideration must be given to using methodologies that owe allegiance to different paradigms. The complexity, heterogeneity, and turbulence of problem situations make it essential that we take advantage of what each paradigm has to offer. Pluralism can provide its greatest benefits only in the context of paradigm diversity. Nevertheless, there will circumstances when the "primary issues" are heavily concentrated in just one area, where it is appropriate to employ just one methodology to guide the use of a variety of methods, tools, and techniques. Such a decision should be taken self-consciously so that it permits change of methodological and paradigmatic orientation as the intervention proceeds. In practice, it is almost always necessary to name one methodology as "primary" and others as "secondary," guided by the classification of primary and secondary issues, in order to provide a structure to the intervention which participants can understand. Again, this ordering will need to be constantly reviewed.

With regard to the choice of methods, models, tools, and techniques, CSP encourages the maximum flexibility possible. It is happy to see systems methodologies "partitioned" or "decomposed" to give the practitioner access to a full range of methods to use in combination in support of the systems methodologies that are to be employed in the intervention. Methods that originate from outside systems thinking, perhaps from Soft-OR, should also be considered. Of course, the practitioner should be guided by what they have learned in previous interventions about which methods are best for supporting which methodologies. Some may carry theoretical baggage that makes them inappropriate for providing helpful service to certain paradigms. Williams and Hummelbrunner (2010) provide a treasure-trove of 19 systems- and non-systems-based methods, models, tools, and techniques, which the critical systems practitioner can choose from in constructing appropriate systemic responses to complex problem situations.

The *implementation* phase uses the chosen primary and secondary systems methodologies, and the preferred methods, etc., according to the logic of CSP, to take action that addresses the issues identified during the creativity phase. The phrase "according to the logic of CSP" refers to the practical difficulty, that we now hit head-on, of using different methodologies in parallel. In TSI, there exists the notion, which can be traced to Althusser, that the problems associated with multimethodology practice can be managed if an initial choice of dominant methodology is made, to run the intervention, with dependent methodologies, reflecting alternative paradigms, in the background. The relationship between dominant and dependent methodologies can be kept under review and changed as the intervention proceeds. In CSP, the phraseology changes from "dominant" and "dependent" to "primary" and "secondary" – the earlier terminology hinting at a form of "imperialism" – but otherwise this remains, for me, a powerful idea. It allows the intervention to proceed in a theoretically informed way (making research possible), and with less confusion to the participants, while as far as feasible protecting methodology and paradigm diversity. As long as we are explicit about our initial primary methodology, and are ready to switch, then the initial choice is not dangerously limiting and will not exclude us from introducing alternative methodologies, based on different paradigms, as required.

There is still the issue of what would lead us to switch methodology once an intervention has begun. Let us say we begin an intervention with a soft systems methodology (SSM) as primary. It is possible that an occasion will arise when a model introduced to enhance mutual understanding will appear to "capture" so well the logic of the problem situation that a shift to a functionalist methodology will seem justifiable. The model will then be taken as a representation of reality and a shift made which establishes a functionalist orientation as primary. Or, to provide another example, there might be an occasion when, upon revisiting creativity, the ethical standpoint of some stakeholders is so offended that the shift to an emancipatory rationale seems desirable. Enough horrors occur in organizations and societies to give anyone pause. The emancipatory option must remain high on the agenda. As Churchman (1970) argued, the professional systems scientist should consider whether it is desirable to help certain organizations to commit suicide. Implementation should end with coordinated improvements made to the problem situation that address the initial primary and secondary issues.

CSP is intended to be used in the action research mode. The *reflection* phase should, therefore, contribute to research as well as to evaluating the results of the intervention. Any use of the CSP multimethodology is, in principle, capable of yielding research findings about: how to manage the relationship between different paradigms; the methodologies and how to use them; the methods, models, tools, and techniques employed; and about the real-world problem situation investigated. It was to ensure that appropriate attention was given to research that the reflection phase was formally added (Jackson 2003). To make it work, the action researcher needs to be clear upfront about the present state of knowledge about the multimethodology, methodologies, etc., and think about any specific learning opportunities provided by the particular intervention. The other aspect of reflection, although hinted at in earlier work (Jackson 2000, 2003), has only recently been fully developed. It continues the theme of "pluralism at all stages" by requiring the results of a systems intervention to be evaluated according to the concerns of the different methodologies. It seeks to judge how successful it has been in bringing about improvement by taking into account what each paradigm rates as most significant. A functionalist logic will be brought into play to judge the intervention in terms of efficiency, efficacy, and the viability of the systems design. The priorities of the interpretive paradigm will consider whether there is improved effectiveness in the sense that longer term purposes are explored and desirable changes to achieve them are agreed and implemented. The radical change paradigm will evaluate whether the intervention has promoted empowerment and emancipation. Ecological awareness will be brought to bear to consider the impact of the changes on the environment and future generations. Postmodernism, if we wish, can ask whether diversity has been encouraged and whether it has all been good fun. We will not succeed on meeting all these criteria in any one intervention, but it is necessary to pay close attention to all of them over any longer period of involvement. A highly successful engagement would hope to demonstrate progress on all these fronts. This type of reflection will also suggest where future efforts should be concentrated and help at the start of the next CSP intervention.

There are three further recommendations that can enhance the way we use CSP. First, it needs to be employed flexibly. It can now be adapted to different problem situations both in terms of the methodologies it employs and the methods, models, and techniques it makes use of. Critical systems practitioners should also give thought in each intervention about who is to be involved and during which phases. The creativity phase naturally works best if there is wide participation and if a wide variety of viewpoints are represented. Later in the intervention, the particular methodologies employed will dictate the degree of participation that is desirable. Some methodologies have stages that must be expert-driven. Second, use of CSP should be iterative. Participants must be willing to cycle continuously around the four phases. As an intervention progresses, the issues that initially seem crucial may fade into the background and new ones emerge. This can be catered for by continually reentering the multimethodology, reconsidering what issues should be prioritized, working with different methodologies assuming the primary and secondary roles, and entertaining the possibility of using new methods. Whichever methodology is initially chosen as "primary" should, as implementation proceeds, be critiqued in terms of what is happening through the lenses offered by alternative paradigms. Switches in primary methodology should be common. Finally, Checkland's distinction between Mode 1 and Mode 2 uses of SSM (see Section 16.2.3) can, with benefit, be transferred to CSP. An academic, imbued with CSP and in a position to set up a study, is likely to start at the multimethodological level, choose primary and secondary methodologies and operate with a range of methods and models appropriate to the methodology dominant at a particular time. The academic, according to his or her inclinations, can then research the theoretical assumptions of the paradigms, the multimethodology, the robustness of the methodological rules, and the usefulness of certain tools and techniques for serving particular purposes. This would be a formal Mode 1 use of CSP, where the multimethodology guides the intervention. A CSP-aware practitioner will, on the other hand, be more likely to use CSP in a Mode 2 manner. The intervention will be dominated by the concerns and pressures of the immediate problem situation. The participants will employ whatever methods, tools, and techniques happen to come readily to hand. However, the multimethodology will be used, during the course of the intervention, to help those involved reflect on what is happening and open up new possibilities by moving between methodologies and paradigms. It will also be employed, periodically, to evaluate how the involvement is going from a wide variety of viewpoints. Most applications are likely to be somewhere between the extremes of Mode 1 and Mode 2.

It is now possible, drawing upon the framework of "constitutive rules" for methodologies formulated by Checkland and Scholes (1990) and Checkland (1999), to define how a CSP intervention should be conducted. This is set out in Table 21.2.

21.2.4 Methodologies

In order for CSP to ensure paradigm diversity in all its phases it needs methodologies which operationalize the assumptions of the different paradigms. For this purpose TSI used existing systems methodologies, such as HST, SSM, and CSH. This is likely to remain the favored option of practitioners. There are, however, advantages in specifying and using "generic methodologies" as the main methodological component of CSP. One advantage is that the theoretical link back to paradigms is more explicit and this

Table 21.2 Constitutive rules for the Critical Systems Practice multimethodology.

1.	The CSP multimethodology is a structured way of thinking which understands and respects the uniqueness of a wide variety of sociological paradigms and draws upon them to improve real-world problem situations
2.	CSP makes use of creativity-enhancing devices to discover and classify the main issues posed by a problem situation, ensuring minimally that it is examined through the lenses of the seven systems perspectives
3.	CSP uses systems methodologies, sometimes in generic form, which can be clearly related back to the different paradigms, as the basis for its intervention strategy – often employing the tactic of naming one methodology as primary and others as secondary, with the possibility of this relationship changing during the course of the intervention
4.	The methodologies used in CSP will themselves employ, in combination, methods, models, tools, and techniques drawn from a variety of different sources, including "decomposed" methodologies
5.	The choice of systems methodologies, and of the methods, models, tools, and techniques used in a particular intervention, will rest upon an appreciation of their different strengths and weaknesses as revealed by "critical awareness," learning from previous action research, and on the experience of the participants
6.	Since CSP, and the systems methodologies it employs, can be used in different ways in different situations, each use should exhibit conscious thought about how to adapt it to the particular circumstances
7.	Each use of CSP should yield research findings based on an understanding of relevant current knowledge as well as provide an evaluation of the outcomes derived from the concerns of the different methodologies and paradigms

allows us unambiguously to translate into practice, and test in real-world interventions, the hypotheses of particular paradigms. CSP has, since its inception (Jackson 2000, 2003), been making moves to free itself from the specific methodologies it inherited – many of which were underspecified as to their theoretical assumptions – and to establish such generic systems methodologies. In *Systems Approaches to Management* (2000), I set out "constitutive rules" for generic functionalist, interpretive, emancipatory, and postmodern systems methodologies. Slightly revised versions of the main elements of the first three of these are shown in Tables 21.3–21.5.

The generic methodologies have been derived from the clear dictates of the paradigms to which they correspond. They also take into account existing systems methodologies which, either explicitly or implicitly, follow the orientation provided by the relevant paradigm. For example, the "generic interpretive systems methodology" draws from the philosophy and theory of interpretive social theory as well as from strategic assumption surfacing and testing (SAST), IP, and SSM.

Further work is necessary on whether other paradigms can usefully be translated into generic systems methodologies; to evaluate the success of the methodologies in transferring the propositions of the different paradigms into practice; and to assess whether learning from practice is leading to adjustments to the paradigms.

Another advantage of generic systems methodologies is that they ease the incorporation of methods, models, and techniques taken from their usual host methodologies into other methodologies serving different paradigms. It can be difficult to "squeeze" potentially useful methods into existing methodologies alongside those they already house. They can now be employed more freely, as required by the problem situation, as

Table 21.3 Main elements of a generic functionalist systems methodology.

The claim to have used a systems methodology according to the functionalist paradigm must be justified according to the following guidelines:

a. an assumption is made that the real-world is systemic
b. analysis of the problem situation is conducted in systems terms
c. models aiming to capture the nature of the situation are constructed, enabling us to gain knowledge of the real-world
d. models are used to learn how best to improve the real-world and for the purposes of design
e. quantitative analysis can be useful since systems obey laws
f. the process of intervention is aimed at improving goal seeking and resilience
g. intervention is best conducted on the basis of expert knowledge
h. outcomes are tested primarily in terms of their efficiency, efficacy, and viability

Table 21.4 Main elements of a generic interpretive systems methodology.

The claim to have used a systems methodology according to the interpretive paradigm must be justified according to the following guidelines:

a. there is no assumption that the real-world is systemic
b. analysis of the problem situation is designed to be creative and is not conducted in systems terms
c. models are constructed that represent some possible "purposeful systems"
d. models are used to interrogate perceptions of the real-world and to structure debate about changes that are feasible and desirable
e. quantitative analysis is unlikely to be useful except in a subordinate role
f. the process of intervention is systemic, never-ending, and is aimed at generating learning and alleviating unease about the problem situation
g. the intervention is best conducted on the basis of stakeholder participation
h. outcomes are evaluated primarily in terms of effectiveness – on whether longer term purposes are explored and desirable changes to achieve them are agreed and implemented

Table 21.5 Main elements of a generic emancipatory systems methodology.

The claim to have used a systems methodology according to the emancipatory paradigm must be justified according to the following guidelines:

a. an assumption is made that the real-world has become systemic in a manner alienating to individuals and/or oppressive to particular social groups
b. analysis of the problem situation is designed to discover who is disadvantaged by current systemic arrangements
c. models are constructed that reveal the sources of alienation and oppression
d. models are used to "enlighten" the alienated and oppressed about their situation and to suggest how they should act to improve it
e. quantitative analysis may be useful especially to capture particular biases in existing systemic arrangements
f. the process of intervention is systemic and is aimed at improving the problem situation for the alienated and/or oppressed
g. the intervention is conducted in such a way that the alienated and/or oppressed begin to take responsibility for their own future
h. outcomes designed to improve the position of the alienated and/or oppressed are evaluated primarily in terms of empowerment and emancipation

long as this is consciously in the service of one or other of the generic methodologies. This facilitates multimethod practice. So, for example, the VSM, originally designed as a functionalist device, can be seamlessly fitted into the interpretive or emancipatory generic methodologies. In *"The Critical Kernel in Modern Systems Thinking"* (Jackson 1990), I suggested that both the VSM and SSM contained a "critical kernel" which could be extracted to support emancipatory systems practice. The VSM could be used to critique hierarchy and the misuse of power because it requires a democratic milieu for its full and satisfactory operation. The debate stages of SSM demand communicative competence as the philosophical foundation for the process they orchestrate. SSM could, therefore, critique institutional arrangements that lead to distorted communication. Flood and Romm have taken this further and indicated that, whether it possesses a critical kernel or not, any systems method, model, tool, or technique, can be employed for an emancipatory purpose:

> The given and immediate purpose of any method can be dominated by the given and immediate purpose of some other method so that, for example, with astute and careful handling a cybernetic or soft systems method can be employed to tackle emancipatory issues in a way which undercuts and redirects its original theoretical underpinning.
>
> *(Flood and Romm 1995, p. 378)*

They provide examples of the VSM being employed in this way to deal with corruption and coercion; and of Ackoff's interactive planning being used to establish fairer social relationships. They go on to argue that the particular use of systems methods of all kinds to support emancipatory ends is actually a special case of a broader strategy. Systems methods and models can be employed "obliquely," as they call it, in the service of any paradigm foreign to the one with which they were originally associated. Of course methods will have to prove their worth, particularly, when they find themselves serving purposes very far from those originally intended. The existence of generic systems methodologies facilitates research into this matter.

21.2.5   Methods

One of the most useful devices to employ alongside the SPs, in the creativity stage of CSP, is "rich picturing." Rich pictures (RPs) originated as part of Checkland's SSM, and an explanation of their nature and purpose, together with examples, has been provided in Section 16.2.4. It is possible to combine RPs and SPs in a number of fruitful ways. The participants can simply be asked to produce RPs without any reference to the SPs. The RPs can then be examined to see which SPs they pay most attention to and the participants asked to draw enhanced versions taking into account all the SPs. Another procedure, which I favor, is to get the participants to produce RPs which seek to capture how well the system of concern is performing as a machine and organism, and what cultural and political factors are prominent in the problem situation. Of course, in this case, these SPs have to be explained in advance. This way of working helps participants to achieve focused RPs more quickly and convinces them early of the value of CSP. The facilitator can then introduce the participants to the more "difficult" SPs – coercive system, environmental and interrelationships – and ask them to enrich their existing RPs, or draw new ones, taking these new viewpoints into account. This helps participants to gradually become adept at looking at the problem situation through different lenses. There are variations on these approaches depending on whether participants are asked to draw one RP, or multiple ones each from a different perspective.

I have nothing to add to what I have already said about the choice and implementation phases of CSP except to point to an analogy that I have used elsewhere (Jackson 2000, 2003) and which others, I believe, have found useful. The analogy compares the critical systems practitioner to a "holistic doctor." Confronted by a patient with pains in their stomach, the doctor might initially consider standard explanations, such as overindulgence, period pains, or irritable bowel syndrome. If the patient fails to respond to the treatment prescribed, and returns to the surgery, you would expect the doctor to entertain the possibility of a more deep-seated and dangerous malady. The patient might be sent for X-ray, body scan, or other tests designed to search for more fundamental problems. If nothing was found, a thoughtful conversation with the patient might suggest that the pains were a symptom of anxiety and/or depression. Various forms of counseling or psychological support could be offered. Or, perhaps, knowledge of the patient's domestic circumstances, and sight of their bruises, could lead the doctor to conclude that the patient was suffering at the hands of a violent partner and to have a discussion about involving the police. Finally, perhaps the patient just needs another interest – such as painting or golf – to take their mind off worries at work. We would expect a "holistic doctor" to be open to all these possibilities, to have appropriate responses and "treatments" available, and to use them flexibly and in combination as necessary. To my mind, the critical systems practitioner, when probing with functionalist (positivist and structuralist), interpretive, emancipatory, and postmodern perspectives, is similarly taking a systemic approach to organizational and societal problems. It has to be said that Zhu regards this analogy as culturally specific and it is indeed limited in this and other respects:

> ... 'the holistic doctor' is now promoted as a root metaphor to prescribe how problem solvers should use methodologies. But where are path-dependence and social embeddedness? Is the doctor living in Heaven, without linkages with communities, histories and power relations? Is the doctor universal, or is it a culturally projected identification of the solitary individual as the decision maker ... and characterization of human action as deliberate choices ...
>
> *(Zhu 2010, p. 219)*

Ray Ison (2017) has put forward an "isophor" of the systems practitioner as "juggler" which takes more account of the type of issue that Zhu raises. He calls it an isophor, rather than a metaphor, because it is the experience of juggling, not the abstract concept of it, that takes a person close to feeling what it is like to do systems practice. In Ison's isophor for effective systems practice four balls are juggled:

> The B-ball symbolizes the attributes of Being a practitioner with a particular tradition of understanding. The E-ball symbolizes the characteristics ascribed to the "real-world" situation that the juggler is Engaging with. The C-ball symbolizes the act of *Contextualizing* a particular approach to a new situation. The M-ball is about how the practitioner is Managing their involvement with the situation.
>
> *(Ison 2017, p. 60)*

This isophor pays more attention to the systems practitioner's social and cultural history, skills, and actual engagement with the situation, than does the holistic doctor analogy.

The reflection phase of CSP, when it addresses whether improvements to the problem situation have been made, acknowledges the concerns of a wide range of methodologies and paradigms. A reasonable starting point is to evaluate the success of an intervention on whether it has increased:

- *Efficiency*: Minimum use of resources to obtain the required outcomes; a concern of the functionalist paradigm and promoted by HST
- *Efficacy*: The means work to achieve the desired outcomes; a concern of the functionalist paradigm and promoted by VM
- *Awareness*: Of interrelationships in the problem situation and the possible unintended consequences of action; a concern of the functionalist paradigm and promoted by SD
- *Anti-fragility*: The system can develop in relation to its turbulent environment and increase its capabilities while responding to change; a concern of the functionalist paradigm and promoted by STS and the VSM
- *Effectiveness*: The exploration of longer term purposes and agreement on desirable changes to achieve them; a concern of the interpretive paradigm and promoted by soft systems methodologies such as SAST, IP, and SSM
- *Empowerment*: All stakeholders are represented in decision-making; a concern of the radical change paradigm and promoted by TS and CSH
- *Emancipation*: The position of disadvantaged stakeholders is enhanced; a concern of the radical change paradigm and promoted by LST
- *Sustainability*: The environment and future generations are taken into account; a concern of ecological awareness

It occurs to me that, before a CSP project begins, it is worth constructing a bar chart on which the current state of the system is portrayed. The projected gains and losses of the intervention in each area can then be estimated. That should encourage some deep, critical thinking about whether the project is really worthwhile. If the decision is to go ahead, the bar chart would be up-dated on a regular basis with estimates of the how things were going in terms of the various indicators. The extent and nature of the trade-offs would be a matter for serious discussion. At the end of the intervention, an assessment would be made of how the system is then performing in terms of the indicators. Over a long-term involvement, CSP would want to see gains on all the measures. The final chart can also suggest what needs attending to next. I do always carry this range of indicators in my head during an intervention, and have used them to carry out evaluations, but have not yet used bar charts in the formal way I am now suggesting.

I am aware that there are some superficial similarities between what is being proposed and other "rounded" frameworks for improving organizational performance such as "the balanced scorecard" and the European Foundation for Quality Management (EQFM) Excellence Model. These approaches also require the user to look at different aspects of performance in order to get a "complete" view. In practice, however, and in the absence of any understanding of the origins of different perspectives, it is the machine-like view of organizations that continues to dominate. They look at a broader range of issues but from the same machine viewpoint. Further, and related, they fail to make the link to the range of well-tested systems methodologies that would allow them to pursue improvement in a manner appropriate to the different aspects of organizations they identify. Finally, it is assumed that improvement in any one of the areas identified will lead to overall improvement. Despite claims to be "holistic" they ignore the fundamental systems message that the interrelationships between the parts are more important than the parts themselves – improvement in one aspect can lead to suboptimization overall.

21.3 Critical Systems Practice in Action

Three examples of CSP in action will be considered. Each intervention has had a significant impact upon the development of the multimethodology and shows how it came to take on its current form. They took place in North Yorkshire Police (NYP); in Kingston Turbines PLC, an engineering company; and in HUBS. The range of settings is another factor that influenced this choice of examples. In each case there is a brief introduction to the organization, discussion of the problem situation and what was done, and reflection on CSP and what was achieved and learned in terms of its "constitutive rules" (see Figure 21.1).

21.3.1 North Yorkshire Police

The NYP project used CSP to guide information systems development. It was carried out by a Chief Inspector, Steve Green, under my supervision. The original reporting on the project is available to interested readers (Green 1991) and the case has also been written up as an account of the use of TSI, as the multimethodology was still called at that time (Green 1992). The fullest version, apart from the original reporting, is in Jackson (1997b) and I shall follow that version closely in what follows.

NYP is, in terms of geographical area, the largest English, single-county force and extends over approximately 3200 square miles. It covers a largely rural territory and includes the beautiful area of the North Yorkshire Moors shown in the popular television series *Heartbeat* – a series which captures, apparently, what policing was like there in the 1960s. Also in its boundaries are urban settlements such as York and Scarborough. In 1991, NYP employed some 1400 police officers, supported by 500 civilian staff, and had a budget of more than £55 million for the year. In 1991, it had to cope with almost 170 000 separate incidents of crime, the predominant offenses being burglary and thefts of and from motor vehicles. At the time of the project, the NYP was undergoing a major reorganization of its structure. Its headquarters were in Northallerton and dealt with matters of policy, finance, personnel, complaints against the police, and research and development. Beyond headquarters, it had traditionally been divided into four divisions, centered on York, Harrogate, Richmond, and Scarborough, and these divisions then further divided into a total of 10 subdivisions. The reorganization was to see the disappearance of one level in this hierarchy and the amalgamation of a number of subdivisions, leaving only seven territorial divisions of the force (based at York, Selby, Harrogate, Skipton, Richmond, Malton, and Scarborough) reporting to headquarters. These changes were explicitly designed, by the Chief Constable, as a step toward forcing decision-making down the hierarchy of the organization. Managers of the various territorial divisions were to be given considerable local autonomy. The study of communications within NYP that we were undertaking was seen as a part of the overall change program. It was important to ensure that the information that flowed around NYP supported the new structure, aims, and purposes, and not the old. Apart from that most important aspect of the brief, our remit was to produce a workable strategy for communications within NYP which would rationalize information flows, and improve the quality of policy making and dissemination through the provision of accurate and timely information.

In relating what happened in NYP, I shall emphasize those aspects of the project where CSP was decisive in determining what occurred. Perhaps because it was in the very early days of trying to use the approach, and because we were very consciously developing and using strategies which were later to become second nature, the project clearly illustrates a number of the significant benefits that derive from employing CSP. Overall the project was successful but there were some hiccups along the way. Interestingly, at the point where the intervention nearly came to grief, the blame could be laid at our door for *not being critical enough*.

The project began with interviews of a cross-section of individuals at all levels in the organization – civilian support staff as well as police officers, headquarters staff as well as staff from one selected territorial division. This was the start of the "creativity" phase of the intervention. The interviews were fairly general and wide-ranging, but did make some use of the "cognitive mapping" technique. The impression gained from the interviews was of very general dissatisfaction with the existing communication and information flows. All interviewees recognized that the impending reorganization required new and improved information systems. The higher ranks in the organization were primarily concerned that the spirit of their policy initiatives seemed to get lost in the existing communication system and, therefore, implementation on the ground was never as intended. They could, to some extent, communicate the detail of what they wanted but not why they wanted it. The Chief Constable commented that:

> It is apparent from a number of changes I have made that, frequently, the letter of an instruction is complied with but, clearly, the philosophy has been lost.
> *(Quoted in Green 1991, p. 42)*

He provided the example of an attempt to control vehicle expenditure by imposing a cut in the overall mileage traveled. This was implemented in many subdivisions by allocating a target mileage to each vehicle on a "per shift," weekly or monthly basis. As a result, one officer was instructed to use a Land Rover, which still had "mileage to spare," rather than a Ford Fiesta which had exceeded its target miles. The senior ranks were particularly critical of middle managers whom they saw as bureaucratic and unwilling to make decisions. Middle managers themselves complained about the confusing nature of the various media of communication employed since these seemed to mix up important policy matters with minor administrative details. They also saw the Executive as being secretive and excluding them from decision-making, and of failing to provide them with instructions in a timely and accurate manner. The lower ranks were frustrated by their inability to get what they saw as important information passed upwards and acted upon. Organizational communication in NYP was compared to a game of "snakes and ladders" – information would get so far up the hierarchy but, before it reached its destination, it would hit a "snake" and tumble down again. Civilian employees saw problems of communication as closely intertwined with their perceived status as second-class citizens in the organization compared to police officers. A final point worthy of note was the feeling among staff representatives that the consultative processes of the force were not useful or meaningful. The staff associations representing the police officers and the trade union (NALGO), representing the civilian support staff, seemed only to be called in once decisions had been taken. They were used to disseminate information about decisions rather than as bodies to be consulted about the views of staff before decisions were taken.

TSI, the form that CSP then took, calls for a problem situation to be analyzed using systems metaphors and this method was used during the interviews. The consensus was that NYP operated like a bureaucratic machine. There was a functional division of labor at headquarters and in the divisions, a prevalence of charts showing the organizational hierarchy, detailed job descriptions, a formal discipline code, a recognizable "officer class" and, of course, uniforms and badges of rank. Many felt that this form of organization was inappropriate in a situation which required police officers to act flexibly to cope with an increasingly unpredictable and turbulent environment. It seemed, indeed, that the main source of the organization's problems, including communication problems, lay in its adherence to the machine model and machine thinking. As Green argues:

> Such perceived problems as middle management's adherence to bureaucratic methods and refusal to be more responsive and decisive, the organization's refusal to treat its members as individual human beings, the gulf which existed between the territorial divisions of the force and its headquarters, and the compartmentalization of specialist departments could be explained in terms of the shortcomings of the machine metaphor.
>
> *(1992, p. 585)*

Use of the organism metaphor led to reflection on how little attention NYP gave to its environment and to communication with the public it sought to serve. The brain metaphor revealed that the organization's current information flows were ill-suited to the promotion of local autonomy. It was clear that significant changes would be necessary if decision-making was to be delegated to lower levels, if information was to be conveyed in a manner suitable for learning, and if the organization as a whole was to become responsive in the face of its environment. Cultural analysis tended to support the findings of the machine metaphor, especially highlighting the bureaucratic thinking of middle managers. It was apparent that changing the culture of the organization, to ensure proper use of information flows based on a more decentralized structure, would be no easy task. The political and coercive system metaphors revealed no open conflict in NYP about the aims of the organization but drew our attention again to the issue of proper involvement of staff through the consultation process. What CST was revealing at this stage was that the kind of improvements sought could not be brought about by designing more efficient and effective information systems on the basis of the machine model. Making NYP a better machine would lead to things getting worse, not better. We were able, using metaphor analysis, to draw in findings from the social sciences to reconsider the nature of the organization for which we were designing the information systems. Instead of making NYP a better machine, we needed to rethink it as an "organism with a brain" and put in place information systems that supported local autonomy and decision-making, learning, responsiveness, and the ability to adapt. Without CST, I would suggest, it would have been easy to fall into the trap of employing an information systems development approach, premised on the machine model, which could not address the real issues; would in fact have made things worse.

We were now on to the "choice" phase. Choice is about selecting a systems methodology, or methodologies, best able to deal with the types of issues revealed in the creativity stage. We were aware of the difficulty of changing the organization's culture and aware of the political issues surrounding consultation, but we became obsessed with the notion that rapid progress to improving things could be made if we designed the communication and information systems to support a vision of NYP as an "organism with a brain." This was especially the case since this approach seemed to have senior management support. The SOSM told us that Beer's VSM was exactly what we needed for designing information systems on the basis of the "organism" and "brain" metaphors.

We proceeded apace to the "implementation" phase and to the application of the VSM as our "dominant methodology." Figure 21.2 shows NYP pictured as a viable system. The analysis we conducted and the recommendations that we were led to make then became very much standard VSM fare. Each of the seven new territorial divisions was to be given autonomy, developing its own statement of purpose and its own environmental scanning and planning capabilities. In VSM terms, each had to become a viable system in its own right. Figure 21.3 shows the Selby Division as a viable system. At the time, coordination was achieved by requiring strict adherence to commands from headquarters. This was no longer appropriate if the divisions were to develop their own identities. The nature of "force orders" was therefore clarified by separating out orders requiring strict adherence, policy guidelines (which the divisions could interpret according to their own local circumstances), and coordination matters. The coordination function should, it was suggested, cease to be controlled by the center. If it could be set up and maintained by the divisions, it would truly be seen as a service to them and not an authoritarian element of command. Removing direct control from headquarters of so much of the divisions' activities might be seen, by senior management, as a recipe for anarchy unless they could at least be sure of control over the outcomes. To this end, attention was given to the establishment of "monitoring channels" which let senior management know how the divisions were doing in terms of some key performance indicators. Senior management should henceforth exercise control on the basis of goals achieved but would not, generally, interfere in specifying the means used to achieve the goals. Information on performance, according to the indicators, should also be freely available to the divisions themselves so that they could adjust their own behavior. Finally, it was recommended that policy making at the top level of NYP could best be supported by the creation of a development function, which might be formed by merging the current "operational conference," which dealt with internal matters, with the "information technology conference," which showed some rudimentary interest in monitoring external affairs. This was necessary if the organization was ever to be appropriately responsive to the changing environment it faced.

The recommendations were written up in a discussion document, *Divisional Autonomy – The Viable System Perspective*, which was circulated to force senior

Figure 21.2 North Yorkshire
Police as a viable system.
Source: From Green (1991).

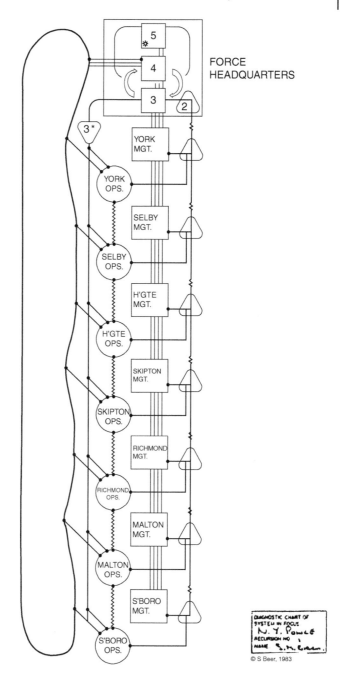

management prior to presentation to members of the Steering Committee overseeing
the reorganization, and which consisted of the Chief Constable, the Assistant Chief
Constable Operations, the Chairman of the Police Authority, representatives of the
Staff Associations, and members of the Implementation Team. We were unperturbed at
a trickle of feedback before the meeting suggesting that the document had not been well

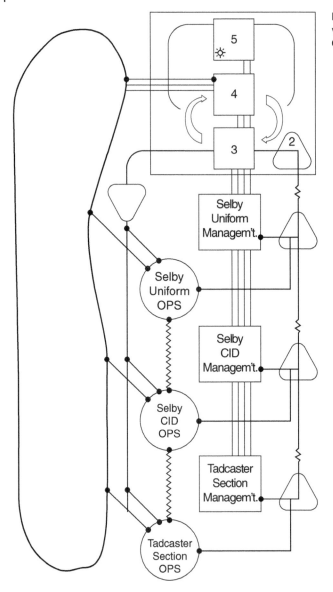

Figure 21.3 Selby division as a viable system. *Source:* From Green (1991).

received. We felt that, logically speaking, the VSM provided exactly the information flows NYP needed – giving the autonomy to the divisions and the information to make their own decisions, ensuring coordination, and providing senior management with the means to ensure proper internal control and to see that NYP was adaptable to external developments. It was disturbing, however, when Steve Green was asked by one middle manager, on the morning of the meeting, whether he had "gone mad." The project was lucky to survive this meeting. It was clear that there was no general understanding of how local autonomy could work in an organization like NYP, how coordination could be maintained, or why the headquarters meeting structure should be rearranged to ensure responsive policy making as we envisaged it. There was no doubt also that the distrust that had grown up, because of the history of lack of consultation over important decisions, continued to poison the atmosphere. It was probably only the fact that the Chief Constable had not read the report before the meeting, and was impressed by some of the points made in discussion, that saved the day.

As critical systems thinkers, we knew about the strengths and weaknesses of the VSM, but we had not been *critical enough* of how its weaknesses would detract from the success of this project. In particular, the VSM did not keep us alert as to whether the changes recommended would be culturally and politically acceptable. The intervention had taken too much of an expert-driven form with insufficient participation from those whose minds had to be won over if change was to be accepted. Furthermore, it had not provided us with the means to address the consultation issue. It was just as well we were forced to think again. Even if we had been allowed to design the recommended information systems, they would have been sabotaged. Re-entering the "choice" phase, we felt we now needed a methodology that would bring about a sufficient cultural shift in NYP to make feasible the kinds of changes that seemed necessary. We were also determined that it be able to address the consultation issue. To these ends, SSM was selected. This methodology was followed through in a fairly conventional, Mode 1, form. Rich pictures of the problem situation and of important aspects of the problem situation were drawn (see Green 1991). Four relevant systems were eventually chosen for consideration:

- A system to develop a concept of local autonomy appropriate for implementation in NYP
- A system to provide for the coordinated implementation of policy
- A system to make policy in a manner which balances the demands of the present with the needs of the future
- A system to provide a consultative style of decision-making

Three of these were directed at issues that had troubled participants in the meeting on the recommendations derived from the VSM and the fourth at the consultation issue. Root definitions and conceptual models were constructed. The conceptual models "to provide for the coordinated implementation of policy" and "to provide a consultative style of decision-making" are included here, as Figures 21.4 and 21.5, for the interest of readers.

Following the philosophy of SSM, the new "dominant methodology," all stages were conducted in as participative a way as possible. The original interviewees (with one or two notable additions) were revisited, the various issues were discussed on the basis of the root definitions and conceptual models, and the models amended either in the presence of the interviewees or, later, in the light of the discussions. Considerable debate was generated among those involved as they began to learn their way to their own understanding of matters such as what autonomy might mean in NYP. Finally a consultancy report was prepared containing conclusions expressed in simple, real-world, terms. It has to be said that the recommendations in this report were little different from those in the previous one, but the reception of the report was completely different. If anything the response now was "yes, very good, this is obvious, what have you been spending your time doing?" This is a response that is disturbing to inexperienced users of SSM but it is actually just about the highest level of praise an SSM practitioner can

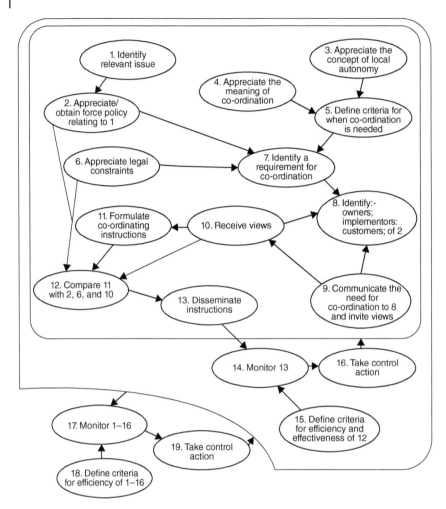

Figure 21.4 Conceptual model to provide for coordinated implementation of policy. *Source:* From Green (1991).

receive. When something is "obvious" it has become part of the culture of the organization, people act accordingly and things get implemented.

The intervention was successful and brought about considerable change in NYP. Green (1992) mentions much greater autonomy to the divisions; the formation of a force "Strategy Group" which identified corporate goals but did not dictate how they should be achieved; and a better appreciation of what constitutes meaningful consultation and where it fits in decision-making. We continued to work with NYP on a number of CST engagements that stemmed from this original project. One, conducted by Keith Ellis and Andrew Humphreys (a divisional commander), was aimed at creating "a top-level corporate strategic planning process" and is reported on in Ellis (2002) and Jackson (2003, pp. 289–295). Another, directed by Bob Flood and Philip Green, was aimed at introducing Local Area Policing (LAP) into the York Division of NYP and is reported on in Flood and Green (1996).

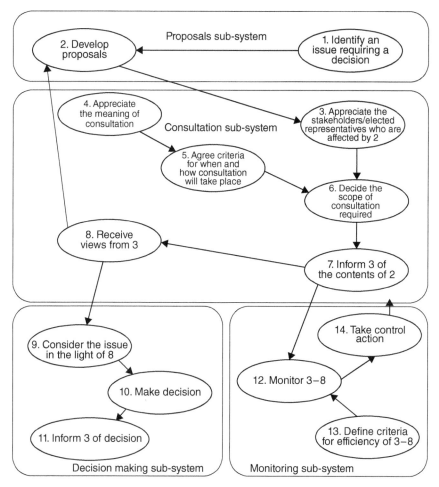

Figure 21.5 Conceptual model to provide a consultative style of decision-making. *Source:* From Green (1991).

I would argue that this was a project that made full use of CSP, as far as it had developed at that time, and would not have been as successful if it had not done so. A more detailed examination of this contention can be provided by examining it against the "Constitutive rules for the CSP multimethodology" set out in Table 21.2.

1. Functionalist, interpretive, and emancipatory theoretical positions were considered and adopted during the study.
2. The creativity phase was conducted according to CSP as it existed in 1991. Interviews and cognitive mapping were employed to surface significant issues but the main purpose of this was to provide material for a full-blown metaphor analysis. The metaphors were regarded as the vehicle whereby the different theoretical rationalities could be brought to bear to analyze the problem situation.
3. The VSM and SSM were adequate in ensuring that the functionalist and interpretive rationales were both employed in the intervention. No emancipatory methodology

was used but an "emancipatory gaze" was cast over proceedings. Generic systems methodologies were not available at the time. The initial choice of organizational cybernetics as the "dominant" methodology, reflected in the VSM, was perhaps unfortunate. However, the flexibility of the multimethodology allowed us to change tack and elevate SSM to the position of dominance at a later stage. The project should certainly not be read just as a case study on the need for effective participant involvement in change processes, as was secured using SSM. The coherence of the recommendations produced on the basis of the VSM diagnosis and design contributed equally to overall success. That said, we were perhaps lucky that the VSM and SSM converged on similar conclusions and we were spared from having to choose between contradictory recommendations derived from alternative rationalities.

4. The systems methods, models, tools, and techniques employed were limited to those usually associated with organizational cybernetics and SSM. The only exception was the use of cognitive mapping in the creativity phase. This is not an example of advanced multimethod practice.

5. The project demonstrated critical awareness of the strengths and weaknesses of the methodologies and methods and what they can most appropriately be employed to do. However, we were, initially, carried away with what the VSM seemed able to offer. We should have recognized that it was a mistake to try to push through the conclusions derived from the VSM even if these did seem to command senior management support.

6. Both the multimethodology and the methodologies used were adapted to the particular circumstances of the project – in the case of organizational cybernetics not sufficiently in the first instance. In terms of Checkland's distinction between whether an intervention is methodology-driven (Mode 1) or situation-driven (Mode 2), this project was predominantly Mode 1. There were, however, occasions when the demands of the situation took over and reference to what the multimethodology required was suspended until the emergency was dealt with. Once the immediate crisis had been overcome, it would then be employed in a Mode 2 form to help work out what had gone wrong and what new direction was needed. The prime example was the rethinking of which methodology to take as "dominant" after the failure of the meeting at which the VSM recommendations were presented. Our account of the intervention does not emphasize it, but there was thought given to how the methods used were best adapted to the circumstances. For example, various kinds of visual device, one using the "scales of justice" to think through the balance between control and autonomy, were used alongside rich pictures during the SSM process.

7. Improvement was primarily evaluated in terms of efficacy, effectiveness, and antifragility. However, empowerment and emancipation were also considered. If nobody was "liberated" according to the usual use of that term, then at least attention was given to improving the involvement of ordinary staff in decision-making through new consultative procedures which had the support of the staff associations and trade union. The intervention generated a considerable number of research findings. This was one of the first significant projects to formally use a CST multimethodology and Green (1992) was able to reflect on and help develop this approach. Sections in the original reporting (Green 1991) were devoted to what had been learned about the VSM and SSM. Metaphor analysis, as part of the creativity phase, was one of the

methods about which most was discovered. It began to be seen as important not as a replacement for other creativity enhancing techniques, but as an assurance that the ground had been covered in exploring the problem situation. For example, looking at the problem situation through the "coercive system" metaphor ensured that the emancipatory perspective was not forgotten. A good deal was learned about the problem situation in NYP. This assisted Green in his everyday activities as a manager and facilitated success in the later projects.

21.3.2 Kingston Gas Turbines

The project took place in Kingston Gas Turbines (KGT) over a three-year period, between 1997 and 2000. KGT has been in the business of manufacturing gas turbines since 1946 and, at the time of the study, employed around 2000 personnel. KGT's existing structure was essentially traditional, with design, proposals, sales, and production departments supported by functional units such as finance, quality, human resources, information technology, plant maintenance, and contracts. However, a recent restructuring had seen moves in the direction of a "matrix" with staff distributed in multidisciplinary teams across the company. In 1997, KGT embarked on an ambitious "double the business" strategy. As part of this, it required all staff to participate in teams which would seek to analyze and solve the company's problems. The intervention was carried out by Alvaro Carrizosa, first as an MSc student and then as a change agent/researcher, funded by the company and, at the same time, undertaking a PhD. I was his academic supervisor. It has been described before – in detail in Carrizosa (2002) and in outline in Jackson (2000, pp. 416–421). The aim here is not to replicate these reports, but to use the intervention to illustrate the main elements of CSP as set out earlier in this chapter. The three-year intervention can be seen as evolving through five "projects," labeled here Projects 1–5.

Project 1 was undertaken for the Proposals Department, which was responsible for working up formal tenders to submit to customers for jobs. This "proposals" project was concerned with the consequences of the "double the business" strategy for the way the department managed its internal and external relations. A brief consideration of HST and organizational cybernetics as dominant methodologies ended because of the degree of pluralism in the problem situation. The SOSM pointed to a soft, or interpretive, approach as an appropriate response, and this remained dominant during the rest of Project 1. The various issues and problems facing the Proposals Department were unearthed through interviews and informal conversations and captured in a rich picture. Discussion of the rich picture yielded various themes that demanded further consideration: communication themes, structure themes, uncertainty themes, efficiency themes, and roles themes. Metaphor analysis was used to engender creative thinking about possible futures that would dissolve the problems as they currently existed. Participants were encouraged to develop their own original metaphors, but the TSI "set" was also employed to ensure that the perspectives of the different paradigms were all taken into account. In general terms, the participants favored the "organism" metaphor as a way of viewing what the future should be like. It seemed necessary to become more customer- and market-orientated in order to survive and prosper in what was becoming an extremely turbulent environment. The project, at this point, inevitably started to involve the Sales Department, and five "relevant" systems were outlined that described structures for sales that would enable it to react more flexibly to the environment. The VSM was used to explore one of these alternatives. The outcome of considering possible "feasible and desirable" changes was recommendations for changing the structure of the Sales Department to give it more of a project management orientation.

The success of this project led to Alvaro being hired by KGT to help with Project 2, which was about organization structure. This was called the PIT project because it was undertaken by a group of middle managers known as the Process Implementation Team (PIT Group). It arose as a result of dissatisfaction with the earlier restructuring which had led to the establishment of the multidisciplinary teams spread across the company. There was a feeling that the technical knowledge, on which the engineers prided themselves, was becoming diluted because they were no longer in such close contact with one another. Moreover, the capacity to transmit information and learning from one part of the organization to another also seemed to be reduced. Project 2 was governed throughout, we can see now, by functionalism. The main impact of systems thinking was to significantly expand the boundaries of what was being looked at. So, what began as a search for solutions in terms of the layout of offices and departments turned into a review of communication systems generally, then of organizational structures, and then of the vision and strategy needed to guide structural considerations. Receiving little help from top management on vision and strategy, the PIT Group seized the opportunity to focus on five business processes that had a direct link to the market. A new organizational structure was proposed built around these business processes. The new arrangements were presented to top management and, following some minor changes, were approved for immediate implementation without further consultation.

Project 3 – "the thinking space" – was born of the realization by those middle managers charged with implementation that not all members of the organization understood, let alone agreed with, the changes proposed by the PIT Group and now adopted by senior management. They determined to set up a forum for discussion about how the reorganization should actually be implemented. They hoped this would help ameliorate any negative consequences that might follow from the autocratic way in which plans for the new structure were adopted and imposed. The focus became what characteristics a thinking space should have in order to permit open discussion of how implementation might take place. It was felt that all participants should be able to disseminate their views and reflections on the evolving situation. This sharing of multiple perspectives would help promote and enrich communication, reflection and learning, and eventually encourage co-operation among those involved. Alvaro's role was to constantly bring new perspectives to bear on the issues being discussed. Generic systems methodologies were employed, during Project 3, with the interpretive and emancipatory in the dominant positions and postmodernism in a subsidiary role. This can be seen as a reaction to the overwhelming functionalist rationale that dominated Project 2. The aim became to increase the "collective competence" of those directly contributing to the thinking space, and even among those who were not contributing directly but were nevertheless interacting with those involved. To assist in achieving this, rich pictures, root definitions, conceptual models, the VSM, system dynamics models and metaphors were all introduced and used. Carrizosa (2000) lists the "properties and characteristics" of the thinking space as they were eventually co-defined by the researcher and the participants:

- An action language, focusing on actors and activities in everyday work
- Structured conversations that helped the actors address the most relevant issues
- Actors engaging in equal participation and able to freely express their viewpoints
- A systems approach that helps actors define what is important to them
- An activity, a way of doing and acting, not another company program
- The researcher as actor
- A dynamic process

The thinking space, enacted on this basis, played an important role in encouraging participation and learning during the various projects undertaken as the new organizational structure was implemented. Its success led to the idea that it should become a permanent part of continuous learning in the company beyond the implementation activity. Projects 4 and 5 stemmed from this ambition.

Project 4 – "the book" – consisted of writing a book about the experiences of participants in implementing the new organizational structure. The book was produced in an interactive manner with different actors contributing their thoughts on change processes and how they could be brought about effectively. The multiple perspectives available from the participants were further enriched by discussing Senge's ideas on organizational learning (see Section 11.2.4) and relevant aspects of complexity theory derived from the work of Stacey (see Section 7.5). As well as allowing self-reflection, and exchange and enhancement of viewpoints, the book permitted issues of power relations and constraints on action to be addressed. It enabled the participants to structure and share their thoughts on Project 3 and so consolidate their learning. The new "organization theory" that was then held in common, and objectified in the book, could become the basis for new purposeful action. The book project continued the pursuit of the interpretive and emancipatory rationales established in Project 3 (with the postmodern "dependent") through the very novel device of co-writing a book. It remained open as to how the experience of co-writing and the actual content of the book would be taken forward in later problem-resolving exercises.

In fact, Project 5, addressing the problem of creating "an integrated business approach" in KGT, drew upon both the book and the knowledge gained of what was required for a thinking space. Known as "the walls workshops," Project 5 allowed participants to engage in completely open communication about the complex issues facing KGT and how to tackle them. Everyone was encouraged to contribute and, once discussions on a particular issue had reached a certain level, to represent the results on walls accessible to all actors. The representations were in the form of systems diagrams and various other visual artifacts. They expressed possible options and actions to be taken, new perceptions and interpretations of problems, possible causes and effects, suggestions for local and more global improvements, etc. The representations could be continuously modified and updated. The discussions taking place were therefore available for scrutiny, validation or revision, and feedback. By the end of the project this approach had become readily accepted in KGT. The walls workshops project clearly had a strong postmodern element to it, as well as continuing the emancipatory and interpretive themes established in Projects 3 and 4.

The KGT intervention provided an exhilarating experience. The period 1997–2000 was a time of great turbulence in KGT, with new ownership structures, changes in organization design and major initiatives of all kinds. At times it seemed that the only thing that did remain constant was the existence of our critical systems study, exploring purposeful change in KGT. The change agent/researcher and his supervisor emerged even more convinced of the need to employ, as CSP advocates, a pluralism of perspectives, theoretical positions, and methodologies, and to use methods and models according to the needs of the particular moment. The intervention illustrates most aspects of good CSP as we can see if we examine it in terms of the "constitutive rules" (see Table 21.2).

1. Aspects of functionalist, interpretive, emancipatory, and postmodern thinking were all employed, as we have seen. The project occurred at the height of CSP's entanglement with postmodernism. Postmodernism asks us to encourage diversity, to challenge power relations and to seek improvement on the basis of local knowledge. Much of what happened in Projects 3–5, "The Thinking Space," "The Book," and "The Walls Workshops" becomes intelligible from a postmodern perspective. I have now dropped postmodernism as an essential element in CSP but continue to see it as offering some useful adjuncts to methodologies based on other paradigms.

2. Various creativity enhancing devices were used to try to understand what was going on in the problem situation. For example, a form of metaphor analysis that went beyond the original TSI set was employed. Nevertheless the formal TSI metaphors were also brought to bear to ensure that the problem situation was examined from the point of view of multiple paradigms.

3. There was a strict adherence to methodological pluralism. Most of the projects had clearly defined "creativity," "choice," and "implementation" phases (though not always in that order!) and pluralism was used in each of these. Generic systems methodologies began to be employed in the intervention and were used to reflect on what happened. There were numerous shifts of "dominant" and "dependent" methodologies during the project. It began in a functionalist manner and this rationality reasserted itself during Project 2, which led to the imposition of the new organizational structure. Project 1, once it got into its stride, was governed by the interpretive approach and this was also significant during "The Thinking Space" intervention. Emancipatory concerns achieved, it seems to me, equal status to those of the interpretive approach in "The Thinking Space" project and were also to the fore in "The Walls Workshops." Postmodernism was prominent, as we have noted, in Projects 3–5.

4. At the level of methods, the freedom to mix and match was fully exploited. There was an instance (Project 1) of the VSM used in support of a dominant interpretive methodology. In Project 3, system dynamics models and the VSM were employed alongside rich pictures, root definitions, conceptual models, etc.

5. The change agent/researcher brought with him some knowledge of the strengths and weaknesses of different systems approaches and this was revisited, by himself and others, as experience was gained of their use in the particular circumstances of KGT. Thus, the project demonstrated "critical awareness" of the systems methodologies and methods used.

6. CSP showed itself to be responsive to the changing circumstances in KGT, adapting its dominant methodologies as necessary and using a vast array of methods, models,

tools, and techniques as they became suitable. Its natural flexibility was added to, in postmodern terms, by pluralism in the use of different "modes of representation" and "facilitation." The KGT intervention began as a Mode 1 exercise but soon became more Mode 2. By the time of Project 3, the situation was very definitely determining what aspects of systems thinking could be usefully brought to the intervention. The change agent/researcher had to continuously learn more systems thinking in order to keep up and reflect on what was happening in the problem situation. At a further step removed from the action, I was trying, in a Mode 2 manner, to make sense of what was going on with reference to CST.

7. Finally, CSP proved itself as a way of doing action research. Improvement to the problem situation was, of course, a continuing theme and was evaluated, in the different projects, according to the concerns of a wide variety of paradigms and methodologies. As well as bringing about improvement in KGT, much was learned about CST and CSP. Carrizosa (2002) details learning in respect of "pluralism," "improvement," and the "role of the agent." Even more significant, perhaps, are his conjectures about the particular situations in which CSP works best. He calls these circumstances "platforms" and sets out two elements that must be present for such platforms to come into existence. First, facilitator(s) and participants must be willing to engage in pursuing "collective competence" – part of which will mean them becoming competent in multimethodology practice. Second, the involved actors need to see what they are doing as a continuous mutual research endeavor.

21.3.3 Hull University Business School

The Prologue to this chapter provided a brief introduction to HUBS as the problem situation and described the results of the intervention. The emphasis here is on how CSP played a part in helping achieve those successful outcomes. It would not be useful to provide an historical narrative of events. First, because the intervention lasted nearly 12 years, it would take another book to do so. Second, because the use of CSP was episodic and followed the Mode 2 style. Although it was always there in the background, it only broke the surface, and became explicit, if the situation seemed to demand it and the circumstances were propitious for that to happen. The solution I have adopted is to consider how aspects of CSP were used, both implicitly and explicitly, to address the six sources of complexity identified in the book as being the focus of attention of the different systems methodologies – technical, process, structural, organizational, people, and coercive complexity.

The efficiency of HUBS, over the last 10 years of my time as dean, meant that it was able to deliver a significant surplus to the University to cover overheads and for further investment. Two initiatives, in my view, contributed significantly to improving the performance of HUBS as a "machine" to deliver this surplus. The first was that the programs were organized for maximum efficiency. At the undergraduate level, a common first year was introduced across all the degrees and there was also sharing of modules in the second and third years. To take another example, a suite of Masters of Science degrees was established, each degree in an area of strong market demand, and they shared a minimum of four common modules. The second initiative was the "workload model." The model allocated time for duties, such as teaching, research, and administration, according to a precise formula. A simple linear programming model, from the OR stable, was employed in its construction. Each member of staff had a set number of hours (around 1600) to fill and would negotiate their workload with their head of department before the beginning of each academic year. In terms of efficiency, the model had the benefit of ensuring that everyone fulfilled at least their minimum workload requirements and did so in a way that was beneficial to HUBS' ambitions. For example, if there was no evidence of research output from a member of staff then there were no more hours allocated to that person for research. The managing of "technical complexity," through these two initiatives, produced an annual average return to the University, over those 10 years, of 45% of turnover.

There was, of course, opposition to the initiatives from a few staff. The workload model will be taken an example of how it was managed. There was no official model in the University at the time and so the HUBS model had, each year, to be agreed according to the democratic procedures of the business school. An extensive period of consultation was always involved and, as a result, I cannot remember a case when the model did not easily gain the consent of the School Board. The arguments for the model in terms of efficiency were overwhelming but they also had to be convincing from other perspectives. They were, and it was easy to make the case. From the cultural perspective, the model did much to increase mutual understanding. The outcomes of the workload planning process were made available to all and this transparency meant every individual could see what HUBS, as a whole, was doing with its staff resources and how each individual was contributing. This also led, looking at it from another perspective, to a feeling that the model encouraged "fairness." Everybody could be seen to be doing the same amount of work even if it was distributed differently for each individual. If no formal model existed, the workload would still have to be allocated. Left to happen by default the process was open to all sorts of accusations of bias, favoritism, and unfairness. The explicit HUBS model, and the way it was managed, removed the force of any such arguments. For individuals, the model created a greater sense of "freedom." They were able, within the constraints of what HUBS required, to direct their efforts into activities they were good at and enjoyed, for example, postgraduate rather than undergraduate teaching. Further, the workload model set a notional limit to what they needed to do to fulfill their obligations to HUBS and the University. Beyond that they were free to make a case that they were entitled to additional earnings from teaching on overseas programs, from executive education, and from consultancy. This had, for some time, been a controversial issue for the University. The existence of the workload model, which contributed significantly to the 45% return, made it easy to argue that contractual duties were being more than fulfilled and that extra payment was necessary for overtime that, after all, served the University's own interests since it brought in additional revenue. From the "organism" perspective, the workload model contributed to HUBS' ability to develop and respond to shifts in its environment. It made it easy to demonstrate a shortfall in HUBS' capacity to deliver its ambitions with existing resources and so provided the basis for negotiations with the University about the need for additional staff in growth areas. As priorities changed because of external demands, for example, as "research excellence framework" (REF) deadlines approached, the model made it possible to shift staff time appropriately. I would not want to claim that the model did not have problems. A few used it to "work to rule" but only the same people who would have done even less without the peer pressure to which the model gave rise. Over time, it became bureaucratic and, by the time I stood down as dean, was in need of drastic simplification. Nevertheless, for many years it served HUBS well.

HUBS managed a few of the "processes" that were essential to its success and depended on the University for providing others. The process of admitting postgraduate students was controlled in HUBS and was significant because the majority were from overseas and paid high fees. Using the VM, explicitly, to check on this process revealed that it involved only a few value steps from the point of view of the customer – receive and record application, check credentials, make an offer and include with that all the information the student might want to know. The turnaround time from receipt of the application to the offer going out became minimal and contributed to the success of the business school in attracting postgraduate overseas students – more than 70% of the total in the University. Other processes managed within HUBS came within the remit of an extremely competent school administrator and it was pleasing to hear reports of, for example, staff receiving the correct expenses and additional payments on time. Many key University processes, such as undergraduate admissions, also worked well. Frustration with others occasionally boiled over, however, and conceptual models of "a system to prevent HUBS making international links" and "a system to prevent HUBS recruiting good staff" were built and compared to what the international office and HR were doing. Taking a political perspective determined, to some extent, whether and how they were used.

A familiarity with the basic tenets of system dynamics in the school made it easy to talk in terms of positive and negative feedback loops when examining "structural complexity." HUBS, during my period as dean, operated in a positive growth cycle. Success in attracting students, and in research and outreach, brought in the money for recruitment of additional staff to strengthen existing activity and pursue growth in other areas. This was relatively easy to conduct and led, for example, to the establishment of new marketing, and organizational behavior and human resource management staff groupings. At the same time, a close eye was kept on the possible unintended consequences of initiatives that might set in train negative influences. For example, it became apparent that the REF imperative to recruit staff capable of producing highly rated research could also lead to a decline in teaching quality. This would be reflected in poor results in the National Student Survey, lead to a fall in league table position, a decrease in applications and student numbers, a fall in income and, eventually, a fall in the ability to sustain high-level research. Another balancing loop had to be introduced to reinforce good quality teaching. It will be apparent that the CSP injunction, to look at how initiatives are progressing from a variety of perspectives, also encourages the search for crucial interdependencies. For example, the workload model not only brought greater efficiency, but it also contributed to "fairness."

Organizational cybernetics and its primary model, the VSM, were constantly to the fore as HUBS sought to come to terms with "organizational complexity." In 1999, the business school was a "green field site" and the VSM could be employed in the design mode. An important first step was to define the elements of System 1 of HUBS as undergraduate teaching, postgraduate teaching, research, and "reach-out" (links with local colleges, executive education, consultancy activity, etc.). This is represented in Figure 21.6. Emphasizing these activities helped to focus minds on the business school as an entity rather than the departments from which it was constructed. This was reinforced by appointing directors for each of the areas and distributing resources, through

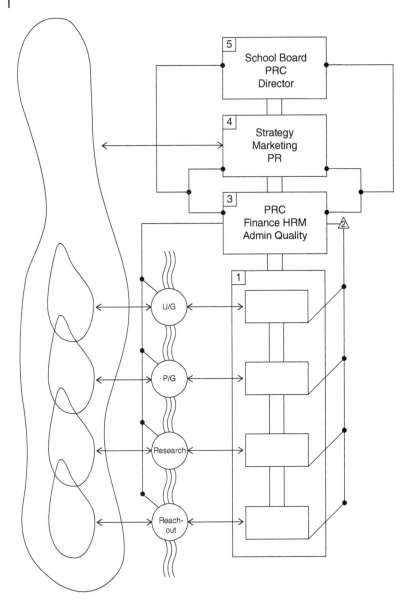

Figure 21.6 Hull University Business School as a viable system.

the directors, directly to these activities rather than to the departments. The departments were renamed "subject groups" to further reinforce the message. The subject groups only appeared at recursion level two of the VSM, responsible for supplying resources to the main activities while taking responsibility for keeping up-to-date with their subject areas, staff development, and workload planning. This worked extremely well for teaching and reach-out. For example, the degree programs became suitably multidisciplinary. Even students studying economics and accounting got a rounded management and business education to complement their specialist skills. It worked less well for research. Here the subject groups reasserted themselves and, apart from the Centre for Systems Studies and later the Logistics Institute, no trans- or multidisciplinary research groups took hold. Toward the end of my time as dean, I tried again by abolishing all the existing research groups and starting things from scratch with significant encouragement to new multidisciplinary centers. There were no REF deadlines looming and this attempt was inspired by complexity theory. Chaos could reign and, hopefully, new groups would self-organize and grow strong. It might have worked if I had stayed long enough to direct resources accordingly but, as it turned out, I was soon to stand down and, yet again, old disciplinary allegiances came to the fore.

Another advantage provided by the VSM, when used in the design mode, was that it helped HUBS to establish itself as the business school it wanted to be rather than the business school it appeared to be in its first year. We wanted to be strong in teaching at all levels, excellent at research, and to have good links with both local and national businesses. In 1999, there was some undergraduate teaching, a raft of overseas MBAs, the Centre for Systems Studies, and little else. It was, therefore, decided that the structures of the business school should reflect our ambitions, rather than the reality, and that appropriate management resource would be applied to those areas corresponding to the ambitions even if they were currently realized only in the most rudimentary form. This worked well during an initial period of very rapid growth. Of course, we did not know exactly what kind of business school we wanted to be in 12 years time or what sort of business school the environment would permit, and so it was impossible to plan in detail. Fortunately, the VSM proved flexible enough to offer continuing guidance on what changes were needed in primary units, levels of recursion, etc., as HUBS, and its environment, changed and it continued its evolution. The VSM provided a way of thinking that encouraged antifragility as well as adaptation. It helped us to co-evolve with our environment and gradually discover what sort of business school was possible in Hull. The VSM was "revealing" rather than "enframing." We learned our way to becoming a successful business school.

Thinking with the VSM in mind helped in other ways as well. It naturally focuses attention on the primary activities that the organization is directing to the market. This ensured that HUBS gave sufficient managerial attention, and the necessary support, to those aspects of its work that generated income. Students at the under- and postgraduate levels brought in the most money. Excellent research and outreach bring in some but, more importantly, contribute to improving reputation and gaining the external accreditations that lead to more applications from both home and overseas students. In the long term they make a very significant, if less direct, difference to income. This is another example of understanding and following sets of linkages. The VSM assisted in discussions and negotiations between HUBS and the University. A recurrent issue that arises in Universities with strong business schools is how to maintain some control over them, ensuring they operate for the benefit of the whole institution, while allowing them the freedom to flourish. At Hull, the VSM was used explicitly, during a senior management away day, to explain how the University could maintain overall cohesion while granting its parts significant autonomy. Figure 21.7 is the diagram that was used to start discussion on the relationship between the University of Hull and its faculties. It had an impact on thinking in the VC's office. One aspect of the tension that often exists between universities and their business schools concerns the extent to which support services should be managed at the local level. Directors of

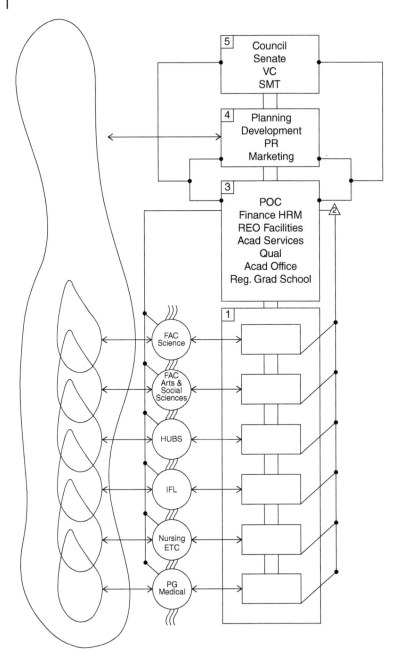

Figure 21.7 The University of Hull as a viable system.

support services usually make the case that centralization is good for the professional development of their staff. On the other hand, business schools face very different environments from other parts of universities and require specialist support staff with an awareness of the peculiarities of their markets. This is true when it comes to marketing, recruiting overseas students, engaging and winning contracts with external clients, career and professional development advice for students (especially MBAs), securing the best staff in a competitive market, and nurturing alumni. The argument is not easy to win for business schools in the face of opposition from other faculties and the directors of central functions. The VSM helped the HUBS leadership team frame the argument in terms of the business school's requirements as a viable system although, in this case, translated into the everyday language of why the business school needed devolved functions for its own success and so that it could contribute more to the finances and reputation of the University. HUBS won the argument and benefitted from its own marketing, finance, HR, recruitment, alumni, and business engagement offices. It must also be said that the VSM should make managers of decentralized units acutely aware of their responsibilities to the whole. I'm not sure how good I was, personally, at conveying this awareness.

STS complements the VSM by stressing the importance of leadership at all levels, autonomous work groups, and senior managers acting in a "boundary spanning" role to provide the stability within which other managers can operate to improve performance. All these matters were constantly on the agenda in HUBS.

Turning to "people complexity," HUBS arranged frequent meetings to discuss its overall mission, vision, and objectives and to develop strategic and operational plans. As a result, there was "mutual understanding" of what it was seeking to achieve and the means it was using to do so. The strap line "responsible leadership for a complex world" emerged from a school open day together with lots of ideas about the importance of "connectedness" and the benefits that this could bring to our various activities. There were annual "teaching conferences" and other, more social, events designed to bring the school together. When HUBS moved into its custom-designed premises, a large and attractive space had to be retained as part of a protection order. It was an easy decision to turn this into a HUBS coffee expanse. Whereas it might be worrying, in other faculties, that staff were spending too long having coffee, in HUBS it was a concern if they weren't spending enough. It was a forum for discussion and debate and senior managers could always be located there at particular times of the day. It probably contributed more to encouraging an open, constructive, and challenging culture in HUBS than anything else.

Preparing students to be responsible leaders in a complex world, and undertaking research and reach-out that contributed to responsible leadership, was at the pinnacle of HUBS' "idealized design," but it needed translating into something much more concrete if it was to provide guidance on what exactly to do and gain staff commitment to change. In practice, it became the basis for plans which informed well-received bids for accreditation to EQUIS, AMBA, and AACSB. The demands of these bodies that HUBS meet their detailed requirements helped, in return, to mold the plans into specifications for action at a level of detail to which staff could respond. There were weekly, agenda-free meetings of senior managers to review progress. The achievement of each of the accreditations in turn was a landmark which people felt they had contributed to and which they could celebrate together. The accreditations signified external recognition of everyone's efforts and the progress that was being made as a result.

Actual soft systems methodologies were used implicitly and, from time to time, explicitly to "manage" people complexity. Aspects of Checkland's SSM were used explicitly to structure debate around issues that were causing unease in the business school. Figures 21.8 and 21.9 are "rich pictures," which were drawn to start discussion,

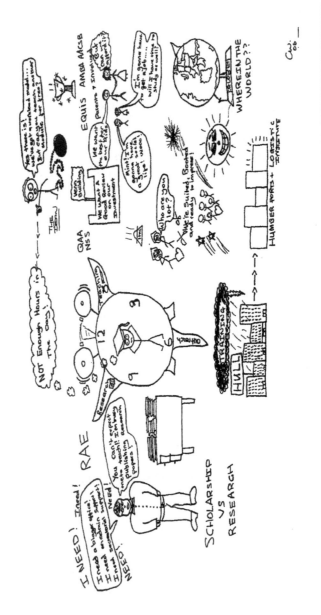

Figure 21.8 A rich picture outlining the multiple pressures on staff in HUBS. *Source:* Thanks to Amanda Gregory.

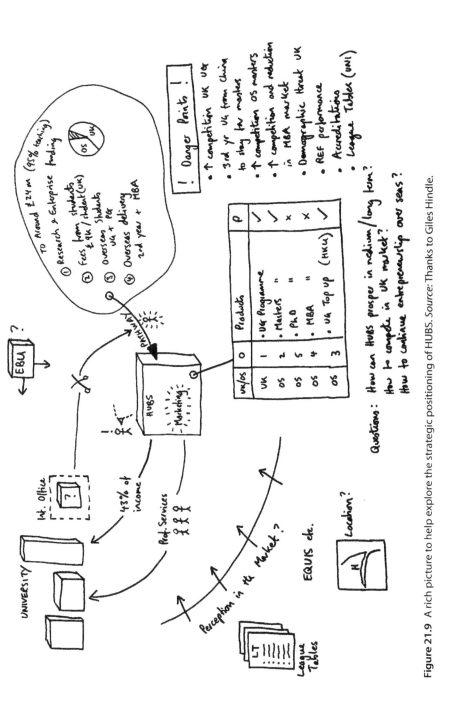

Figure 21.9 A rich picture to help explore the strategic positioning of HUBS. *Source:* Thanks to Giles Hindle.

respectively, about how to alleviate the "multiple pressures on staff in HUBS" and "to help explore the strategic positioning of HUBS." More usually, SSM was implicitly integrated into the everyday management practice of reviewing the current state of the "mess," developing possible ways forward, informally discussing what to do, and agreeing to take action to implement changes deemed "feasible and desirable."

Business schools have multiple stakeholders whose views they need to take into account if they are going to be successful – academic staff of different grades, administrative and support staff, students at all levels, the University, research funders, the local business community, other clients, accrediting bodies, etc. Although no formal SAST exercises were carried out, we were aware of the importance of reviewing the assumptions we were making about the relevant stakeholders in taking strategic decisions. For example, we knew that a decision to withdraw from the validation of other institution's MBAs would please the MBA accrediting body, AMBA, but worry the University because it would lose income. Reviewing the assumptions made about stakeholders assisted in taking and justifying complex decisions. It helped HUBS to understand, on the basis of CSH, that the different stakeholders were using different "boundary judgments" when considering what they wanted from the business school and assessing how it performed. We were keen to understand the relevance of these boundary judgments to future success and sought to learn about them using mechanisms such as staff–student meetings, regular meetings of an Advisory Board of local stakeholders, and regular contact with the accreditation bodies. In discussions with the University there were occasions when the boundary judgments of other stakeholders were employed in a "polemical" manner to support HUBS' position. For example, the perspective provided by EQUIS could be relied upon to support arguments for greater autonomy. The membership of the Advisory Board contained many influential figures whose views the University respected and would take account of.

HUBS' attempts to manage "coercive complexity" can be considered in terms of its embrace of "discursive rationality" and its modest adoption of certain of the themes of LST. HUBS encouraged an ethos of treating with equal respect the contributions made by different categories and grades of staff. Arrogant professors were put in their place. No academic – administrative divide was allowed to develop. Porters wanted to work in HUBS. The business school was an early signatory to the Principles for Responsible Management Education (PRME), maintained an active membership, and incorporated a commitment to global social responsibility in its teaching and research. For example, it was one of the first business schools to introduce undergraduate modules on "Business Ethics" and on sustainable business: "Sustainable Business: Principles and Practice of Green Management." HUBS collected data on female representation at senior levels in the school and performed well in this respect. The numbers of LGBT staff grew in a greater proportion than overall staff numbers and, as far as I am aware, there was no discrimination. There was one "industrial tribunal" case involving HUBS while I was dean. This involved an alleged failure by the University to deal with racist behavior by a couple of students. The case was dismissed and the HUBS investigation into the matter commended. The HUBS leadership group insisted that its coffee outlet be "fair trade." A series of programmes encouraging "women into management" were conducted. HUBS is located in a region of economic and social deprivation and responded to this in its actions. Links were made with local schools and specific information was provided to local students encouraging them to apply. Appropriate concessions were made. The growth of the business school provided direct employment opportunities and, through the numbers of additional students attracted, benefitted the city. The successful bid to Yorkshire Forward for £9 million to establish a Logistics Institute was conceived as the best possible way HUBS could support the regeneration of the regional economy. "Community OR" was encouraged as a means of making the management science and systems skills of staff available to the disadvantaged on a pro bono basis. In October 2018, in recognition of its commitment to developing "responsible leadership," the Chartered Association of Business Schools chose HUBS as a case study for a documentary seeking to demonstrate that some business schools actually do good in the world.

The intervention illustrates good CSP, as we can see if we examine it in terms of the "constitutive rules" (see Table 21.2).

1. The perspectives of the functionalist, interpretive, and radical change sociological paradigms, and aspects of postmodernism, were brought to bear to provide an understanding of the problem situation and to guide the intervention. It makes a difference. Anyone who does not believe this should read Martin Parker's account of two years as an employee in a business school run along strictly "functionalist" lines. For example:

    > I arrived at EBS just after the appointment of a new Dean who was committed to a top-down change project which was to take the school in an even more hierarchical direction …. This involved telling a particular story about the past, and positioning anyone who defended it as conservative, fearful or disruptive. It also involved tight control over the means of communication, including shutting down the means by which collective disenchantment could be voiced and ensuring that there was a small senior management team of well-paid academics (most of them appointed by the Dean himself) who would not question the direction that the school was taking.
    >
    > *(Parker 2014, p. 282)*

    Mingers (2015), too, argues that acknowledging the contributions of critical realism and systems thinking can change the way business schools go about their work.
2. The seven "SPs" were all used, along with other creativity-enhancing devices, to review issues in HUBS as they arose.
3. A variety of systems methodologies, relating back to the different paradigms, was used throughout the intervention, e.g. HST, VM, VSM, SSM, CSH, as well as the "generic systems methodologies". Different methodologies assumed the primary role at different times.
4. Methods were extracted from methodologies and used as appropriate to the situation. An example is the use of rich pictures simply to encourage creative thinking and not as part of SSM.
5. Academic and professional papers and reports commenting on the capabilities of different systems methodologies and methods were taken into account in deciding which to use.

6. Methodologies were employed in a Mode 1 or Mode 2 manner, as appropriate, and other adaptations made as necessary.

7. Learning about the strengths and weaknesses of the different systems methodologies took place and informed later stages of the intervention. The initiatives taken to change HUBS were constantly evaluated according to the concerns of the different paradigms and forms of "critical awareness": What impact might they have or were they having on improving efficiency, efficacy, effectiveness, viability, antifragility, mutual understanding, commitment to purposes, empowerment, emancipation, sustainability? Could any possible unintended consequences be predicted or discerned?

21.4   Critique of Critical Systems Practice

CSP has sought to base itself on defensible theoretical positions, and to remain true to the spirit of an "ideal type" of multimethodology, while pragmatizing both to make the ideas useable for practitioners. The extent to which it has been successful is for others to judge. Nevertheless, it is worth quoting Edwards who has examined the field of "metatheorizing" and argues that good metatheories will

> … i) often emerge out of a process of theory building that integrates, synthesizes or constructively analyses other theories and methods, ii) are themselves used to systematically review and critique other theories and iii) include new conceptual lenses that explore new ways of understanding.
>
> *(2014, p. 725)*

CST, he says, "exemplifies all three of these aspects." Ultimately, its success or failure will depend upon its ability to bring about "improvement" and to continually learn better ways of doing this. The early indications are good.

If CSP is now well-formulated and useful, an important reason is that it has learned from its critics. It has abandoned, in the face of criticism from Tsoukas, Gregory, Midgley, and Mingers and Brocklesby, any claim to metaparadigmatic status and seeks to manage relationships between the paradigms in a way that values and takes advantage of their differences. CSP has responded to Ulrich's argument that it is all about "methodology choice." The current version more clearly articulates a process of intervention in which a directed and wide-ranging form of critique is evinced at every stage – creativity, choice, implementation, and reflection. The reader is encouraged to take up the invitation, offered in Section 7.6, to revisit Stacey's critique and satisfy themselves with the responses provided there in the light of the full account of CSP that they can now draw upon. As encouraged by Midgley, and Mingers and Brocklesby, CSP recognizes the value of decomposing existing methodologies and embracing flexible multimethod practice. Stephens et al. have forced its hand into eventually accepting "ecological awareness" as a crucial aspect of "critical awareness."

Critics will, hopefully, continue to draw attention to issues that CSP has not fully addressed. We already know a few of these. Two, remaining from Mingers and Brocklesby (1997), Mingers (1997a), and Brocklesby (1997), concern the "cultural" and "cognitive" feasibility of multimethodology. The former relates to the current cultural constitution of the management science community:

> Management science contains a large number of ... highly fragmented subcultural communities, and many of these converge around methodologies or techniques which embody particular sets of values and beliefs.
>
> *(Mingers and Brocklesby 1997, p. 498)*

This difficulty is exacerbated because individuals find it hard to simply "switch" between paradigms. They undergo "lengthy socialization and acculturation processes" into particular ways of thinking and invest time and careers in becoming competent in chosen methodologies and methods. Mingers and Brocklesby conclude that individuals do have some freedom of choice in moving between paradigms but it "is by no means a simple matter" (1997, p. 499). Stacey, as noted in Section 7.6, makes a similar argument in more florid prose:

> People do not simply alter perspectives as if they did not matter – they kill each other for them because they are aspects of collective identity. Despite the concern with the social, with political action, power and freedom, the systemic way of looking at these does not accommodate their ordinary conflictual nature and it retains the primacy of the individual.
>
> *(Stacey and Mowles 2016, p. 216)*

If conflicts start because individuals fail to understand alternative perspectives then, it seems to me, that it is even more urgent to develop competence in inhabiting different worldviews, as required by CSP. "Cognitive" feasibility refers more to the difficulties faced by individuals in switching between paradigms because of their personality type, and data processing and research preferences. Brocklesby (1997) considers the issue in depth. His conclusion, based on an analysis using the work of Maturana and Varela, is that it is unlikely but by no means impossible for individuals to become multimethodology literate. Richard Bawden insists that, to produce "systemic individuals," a "critical learning system" is needed that demands:

> ... (a) we accept that we each 'use' particular paradigms to make sense of the world around us; (b) we are able to recognize the nature of these preferred paradigms; (c) that we can recognize and embrace other paradigms in addition to our preferred ones; and (d) that we are especially able to embrace a *systemic* paradigm so that we can make sense of (and make use of) the learning systems metaphor ...
>
> *(Bawden 1995, p. 27)*

Certainly, to overcome the cultural and cognitive constraints to multimethodology, it is necessary to put in place educational and training programs that embrace pluralism and the challenges of critical systems theory and practice. The situation is far from hopeless. Pollack describes a project in which "frequent swaps were made between the hard and soft paradigms by a single practitioner with little apparent difficulty" (2009, p. 162). Eden et al. (2009) mention a member of their research team who came to see that the social construction of reality is not "laughable" anymore. Bowers presents the argument that moving between paradigms requires some learning and practice but that "the mind is already accustomed to the relative nature of being and knowledge and is well versed in making these types of shift" (2011, p. 550). Midgley suggests a "model of learning" that can gradually lead practitioners to develop their understanding of different methodologies and methods. The emphasis is on

> ... learning over time, starting from the knowledge base the intervener has at the point at which s/he realizes the value of mixing methods. If this knowledge base consists of no more than one or two ideas from a single paradigm, then *that's a start* – s/he can reach out and begin learning from there.
>
> *(Midgley 2000, pp. 266–267)*

He has recently suggested that Cabrera et al.'s (2015) four key concepts of systems thinking – distinctions, systems, relationships, and perspectives – can make a good starting point. Interpreting an individual's experiences in terms of one or more relevant concepts can lead on to a deeper discussion of the systems methodologies and methods that are most closely aligned to the concept(s).

An interesting analogy for systems practitioners trying to find their way in a multidimensional world, which requires them to adopt multiple perspectives, is suggested by Valentinov et al. (2018) in their paper exploring the relationship between Luhmann's social systems theory (see Section 4.7) and stakeholder theory. In the position of the systems practitioner is the organization which is defined, according to Luhmann's theory, by complexity reduction and operational closure. Organizations often relate directly to the main function systems of society (research laboratories to science, schools to education, etc.) but they are also potentially multifunctional. Universities, for example, refer to the function systems of the economy, education, and science. Further, organizations are dependent on their environments for their survival. Although they must be selective in their relationships with their environments, their sustainability actually depends upon them safeguarding "metabolic dependencies" with their social and natural milieus. Organizations are therefore in a dilemma. They are necessarily operationally closed but have to navigate between different function systems and maintain appropriate relationships with a rapidly changing environment. Valentinov et al. suggest that organizations can best counteract the sustainability risks posed by operational closure if they pay attention to the interests and needs of their multiple stakeholders:

> If a firm is seen as a social system metabolically dependent on the environment constituted by its stakeholders, then the jointness of stakeholder interests reflect the fundamental and highly intuitive fact of system – environment interdependence. At a philosophical level, catering to the interests of stakeholders is tantamount to establishing an adequate metabolic provisioning of the firm as a social system.
>
> *(Valentinov et al. 2018, p. 12)*

Organizations, just as systems practitioners, can benefit from taking on board multiple-perspectives. This is difficult for organizations because of operational closure, just as it is difficult for systems practitioners due to "cognitive closure" (see Brocklesby 1997).

Returning to criticisms of CSP, Zhu (2011) has launched a scathing attack on what he calls its "paradigm mentality." He is all in favor of the use of multiple methodologies and methods but feels that it is being held back by paradigm-based theorizing. Instead he proposes a pragmatist alternative. I have already made clear that CSP needs to base itself on the sort of philosophical pragmatism briefly outlined in Section 1.4. I have also expressed agreement with Zhu's call for "ontological flexibility" and argued that this is exactly what CSP provides (see Section 21.2.2). Here, I answer to three further criticisms of his that, I feel, have some merit.

The first states that CST has become over-preoccupied with the wholesale philosophical legitimation of its pronouncements – something few practitioners care about. This has diminished its practical relevance:

> Paradigm-based theorizing is not working. It fails to make a practical difference. This is certified as OR workers manage to mix methodologies satisfactorily in intervention without theorists sorting out the paradigm incommensurability mess.
>
> *(Zhu 2011, p. 795)*

Apparently, success in a specific social context is justification enough. This criticism is similar to one raised by Ormerod (1997b). He argues against an emphasis being placed on the philosophical underpinnings of methodologies. His observation of consultants is that they are quite happy to mix and match methodologies and methods whatever their theoretical origins. Taking his lead from practice, Ormerod prefers to base any mechanism for choosing between methodologies and methods on their "transformational potential." In simple terms this means their ability to secure desirable outcomes in a particular context. For him:

> The combination of methods needs to work in the practical sense that the right people (and other resources) need to be involved in a process that results in the desired outcome.
>
> *(Ormerod 1997b, p. 430)*

While these arguments may be correct in terms of the short-term demands on consultants, and worthy in that they may lead to immediate benefits for practitioners, they seem to miss the point with regard to the bigger picture. To put it bluntly, there are too many vague terms used here – "satisfactorily" from Zhu, the "right people" and "desired outcome" from Ormerod. Worrying about what these should mean may well require an engagement with theory but the implications are just as "practical," and of a higher order than the concerns prioritized by Zhu and Ormerod. Further, for researchers and more reflective practitioners interested in finding out why different approaches work, and passing this understanding on to others, a more theoretical stance is a necessity. Theoretically informed methodologies are essential for ensuring a healthy link between theory and practice in systems thinking. Ormerod and Zhu are endorsing a simplistic, everyday form of pragmatism – do the means seem to work – rather than embracing the fully fledged, philosophical interpretation to which CSP looks.

I must also point out that, while critical systems thinkers have indeed demonstrated a significant concern with philosophy and theory, this has usually been driven by the purposes of those who want to use systems approaches in practice. CST is interested in theoretical arguments that make a difference. To take just one example, the third chapter of one theoretical tome is titled "relevant social theory" and clearly states the reason for looking at social theory:

> Not all the fine theoretical distinctions made by social scientists do make a difference, but some are of considerable importance and must be regarded as crucial for systems thinking.
>
> *(Jackson 2000, p. 21)*

The paradigms explicitly and implicitly endorsed by the different systems methodologies do make a significant difference to how interventions are conducted as, I hope, Part III of this book has demonstrated. That said, I do have some sympathy with those who find CST's interminable theoretical arguments irritating. It is time to get on with it.

The second of Zhu's criticisms relates to CSP's adherence to a small set of paradigms. In Jackson (2003), for example, the perspectives of the functionalist, interpretive, emancipatory, and postmodern paradigms are given prominence. Zhu asks:

> Why these four paradigms, why not three or five or a different four? .... [In CST] .... Pluralism, the unquestioned good, is all about the relationships of the neatly ordered socio-philosophical paradigms within a known 'small set'.
>
> *(2011, pp. 786–787)*

This echoes the criticism of Taket and White, from a postmodernist perspective, that TSI can be seen as a "totalizing endeavor" (see Section 20.2.4). The answer is that there do not have to be just four or any other limited set of paradigms. It is open to the systems practitioner to specify and introduce alternative perspectives and to test them to see if they add value. As Ormerod points out, in suggesting that pragmatism could provide an "overarching philosophy" for "Jackson's critical theory," that philosophy is flexible enough to accommodate other positions (2006, p. 908). However, we do need to be clear about the nature and assumptions made by any paradigms that are introduced. This is so that we can do research on what they can actually achieve. It is also to protect the "ontological flexibility" that Zhu values highly. We need to be certain that we are, indeed, operating with a range of paradigms supporting different ontological positions. It is also the case, and one made more fully in Section 21.2.3, that the "root metaphors" that underpin the paradigms and SPs used in CSP have proven themselves over time. They have been found useful by the philosophers, natural, life, and social scientists employing them. It is strange that Zhu, advocating a "pragmatist alternative," would want to abandon theoretical positions that have been shown to bring benefits when used to guide practice.

Zhu's third objection to the paradigm mentality is that:

> It means that current theorizing is focused on finding the globally right 'grids', 'typologies' and 'underpinnings' as the general theoretical solutions, from existing socio-philosophical 'paradigms', from an observer's perspective, and then 'translating' these theoretical solutions into 'a form that managers can use',

instead of the other way around – tracing and accounting for OR workers' situated experiences .... The local variety of mixing-methodology practice is so rich and dynamic that any systematic philosophical – cum – paradigmatic interpretation will turn out to be distorting, de-meaning and constraining .... In a pragmatist orientation, the primary role of researchers is to follow the various actors through their everyday enrolments in order to appreciate the many ontologies, methodologies, methods and tools in use, which may, but do not need to, fit with known categories.

(Zhu 2011, pp. 793–794)

To me this is a strong argument for the Mode 2 use of CSP that has already been endorsed and shown in action in the KGT and HUBS interventions described in Section 21.3. Beyond that I would want to argue that while it is impossible, in any theoretical framework, to capture the complexities and intricacies of the problem situations we face, we are meaning-bestowing animals. We are driven to do so and so we might as well make explicit the theoretical frameworks we employ. Again, that might take us somewhere toward finding out which are useful and which are not.

In short, I value Zhu's criticisms. They stem from a "pragmatist alternative" that wants to incorporate "ontological flexibility" and seeks to be "action oriented, multiplicity embracing, ethically concerned, and politically sensitive" (2011, p. 784). A more sympathetic reading of what CST is about might lead him to the conclusion that it already provides this "alternative."

21.5 Comments

I have taught CST to postgraduate students in universities and to executives, on short courses, for many years and all over the world. I just want to comment on what a satisfying experience that is. I always get pleasure from the insight they gain from viewing case studies from the different "SPs" or "systems metaphors" (as they were once called). The value of CST immediately becomes apparent as each perspective allows them to see different things and to supply different explanations for the same phenomenon. On the model of the Lancaster MA, when I was there, I have designed role playing exercises where groups of "consultants" use CSP to analyze a case and report back to the "senior managers" of a company. The students have to present the results of the "creativity" phase and justify their "choice" of methodologies to "managers" who do not know systems thinking. Playing the game over a week allows the students to have an impact on the "worldviews" of the managers and so the value of soft systems approaches becomes apparent. As the problem situation changes, a different choice of methodologies may become necessary. I have played the "Ambleside" case study, beloved of many intakes to the Hull MBA, with "Alan Green," the Managing Director, as a traditional male chauvinist. This provides some resemblance to a coercive context. If they are unable to shift his sexist attitudes, groups have to consider whether to continue with the "consultancy." I have had individuals storm out of the room. At the end, the "consultants" hand in their report to the company. The best reports will reflect everything the students have learned during the week.

21.6 The Value of Critical Systems Practice To Managers

The following five points are worth emphasizing:

- Proponents of CST have always warned managers not to be fooled by those who peddle fads and quick fixes. The problem situations they face are too complex and diverse to be handled by anything other than the systemic and creative form of intervention that CSP seeks to guide and structure
- CSP advocates and enables the maximum creativity when the problem situation is being analyzed
- Opposing the "one best way in all circumstances" mentality, CSP helps managers to evaluate the usefulness to them, in their situation, of different management solutions and, particularly, different systems methodologies and methods, and to use them in combination if necessary
- CSP points to the need to evaluate management action, in the long-term, using a variety of measures – efficiency, efficacy, the exploration of purposes and growth of mutual understanding, viability, antifragility, empowerment, emancipation, sustainability, and possible unintended consequences
- Because it is based on the "action research" philosophy, CSP provides a learning system that managers can tap into in order to reflect on what has been achieved and how to further improve their own practice

21.7 Conclusion

I will provide a proper conclusion to the book. For the moment though, I want to return to the end of Chapter 8. There we were at the beginning of a journey. We were meditating

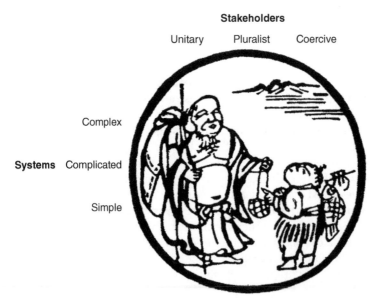

Figure 21.10 The positioning of Critical Systems Practice on the SOSM.

on the complexity of High Force viewed through the eyes of a Chinese "silent traveler" in the Yorkshire Dales. It is now the end of the long journey. We must come back into the world. The tenth stage of "spiritual oxherding," in the "Ten Oxherding Pictures," is called "entering the city with bliss-bestowing hands." The words that accompany it are as follows:

> Barechested and barefooted, he comes out into the market-place;
> Daubed with mud and ashes, how broadly he smiles!
> There is no need for the miraculous power of the gods,
> For he touches, and lo! The dead trees come into full bloom.
> (Suzuki 1973, p. 375)

As Luhmann reminded us, in Section 4.7, any second-order analysis should impart to us how little we know – not least because a third-order analysis of a second-order analysis is always possible. Systems thinking cannot work miracles but it can encourage us to continue on a journey to learn more and intervene, with modesty, to try to improve the state of the world. Where is CSP positioned on the SOSM? Please see Figure 21.10.

References

Achterbergh, J. and Vriens, D. (2011). Organizational cybernetics: is Beer's VSM sufficient for organizational regulation. *Journal of Organisational Transformation and Social Change* 8: 19–33.

Ackermann, F., Bawden, R., Bosch, O. et al. (2009). The case for soft O.R. *OR/MS Today* 36: 20–21.

Ackoff, R.L. (1967). Management misinformation systems. *Management Science* 14: B147–B156.

Ackoff, R.L. (1968). Toward an idealized university. *Management Science* 15: B121–B131.

Ackoff, R.L. (1970a). *A Concept of Corporate Planning*. New York: Wiley.

Ackoff, R.L. (1970b). A black Ghetto's research on a university. *Operations Research* 18: 761–771.

Ackoff, R.L. (1974a). The systems revolution. *Long Range Planning* 7: 2–20.

Ackoff, R.L. (1974b). *Redesigning the Future*. New York: Wiley.

Ackoff, R.L. (1975). A reply to the comments of Chesterton, Goodsman, Rosenhead and Thunhurst. *Operational Research Quarterly* 26: 96–99.

Ackoff, R.L. (1977). Optimization + objectivity = opt out. *European Journal of Operational Research* 1: 1–7.

Ackoff, R.L. (1978). *The Art of Problem Solving: Accompanied by Ackoff's Fables*. New York: Wiley.

Ackoff, R.L. (1979a). The future of operational research is past. *Journal of the Operational Research Society* 30: 93–104.

Ackoff, R.L. (1979b). Resurrecting the future of operational research. *Journal of the Operational Research Society* 30: 189–199.

Ackoff, R.L. (1981). *Creating the Corporate Future*. New York: Wiley.

Ackoff, R.L. (1982). On the hard headedness and soft heartedness of M.C. Jackson. *Journal of Applied Systems Analysis* 9: 31–33.

Ackoff, R.L. (1988). C. West Churchman. *Systems Practice* 1: 351–355.

Ackoff, R.L. (1999a). *Ackoff's Best: His Classic Writings on Management*. New York: Wiley.

Ackoff, R.L. (1999b). *Re-Creating the Corporation: A Design of Organizations for the 21st Century*. New York: Oxford University Press.

Ackoff, R.L. (2003). Personal communication.

Critical Systems Thinking and the Management of Complexity, First Edition. Michael C. Jackson.
© 2019 John Wiley & Sons Ltd. Published 2019 by John Wiley & Sons Ltd.

Ackoff, R.L. (2005). What constitutes leadership and why it can't be taught. In: *Handbook of Business Strategy*, 193–195. Bradford: Emerald Group.

Ackoff, R.L. (2008). *Systems Thinking for Curious Managers*. Axminster: Triarchy Press.

Ackoff, R.L. (2010). *Memories*. Axminster: Triarchy Press.

Ackoff, R.L. (2018). *Bell Lab*. Lecture. https://vimeo.com/148192220?ref=em-v-share (accessed 9 December 2018).

Ackoff, R.L. and Addison, H.J. (2007). *Management f-LAWS: How Organizations Really Work*. Axminster: Triarchy Press.

Ackoff, R.L. and Deane, W.B. (1984). The revitalization of Alcoa's tennessee operations. *National Productivity Review* 3 (3): 239–245.

Ackoff, R.L. and Emery, F.E. (1972). *On Purposeful Systems*. London: Tavistock.

Ackoff, R.L. and Gharajedaghi, J. (1996). Reflections on systems and their models. *Systems Research and Behavioral Science* 13: 13–22.

Ackoff, R.L. and Rovin, S. (2003). *Redesigning Society*. Stanford, CA: Stanford University Press.

Ackoff, R.L. and Rovin, S. (2005). *Beating the System: Using Creativity to Outsmart Bureaucracies*. San Francisco, CA: Berrett-Koehler.

Ackoff, R.L., Magidson, J., and Addison, H.J. (2006). *Idealized Design: Creating an Organization's Future*. Philadelphia, PA: Wharton School.

Adams, J. (1973). *Chile: Everything Under Control*, 4. Science For People et seq.

Agrell, P.S. and Leonarz, B. (2006). Churchman's contributions to the advancement of management science. In: *C West Churchman and Related Works Series: Wisdom, Knowledge and Management* (ed. J.P. van Gigch), 27–32. New York: Springer.

Algraini, S. and McIntyre-Mills, J. (2018). Human development in Saudi education: a critical systemic approach. *Systemic Practice and Action Research* 31: 121–157.

Althusser, L. and Balibar, E. (1970). *Reading Capital*. London: New Left Books.

Archer, M. (2017). Morphogenesis versus structuration: on combining structure and action. In: *Structure, Culture and Agency: Selected Papers of Margaret Archer*, Kindle Edition (ed. T. Brock et al.), 102–127. London: Routledge.

Aristotle (n.d.). 350 BCE. Metaphysics. Internet Classics Archive.

Armson, R. (2011). *Growing Wings on the Way: Systems Thinking for Messy Situations*, Kindle Edition. Axminster: Triarchy Press.

Aron, R. (1967). *Main Currents in Sociological Thought*, vol. 2. London: Basic Books.

Ashby, W.R. (1952). *Design for a Brain*. London: Chapman & Hall.

Ashby, W.R. (1956). *An Introduction to Cybernetics*. London: Methuen.

Avison, D. and Wood-Harper, A. (1990). *Multiview: An Exploration into Information Systems Development*. Oxford: Blackwell.

Baburoglu, O.N. (1992). Tracking the development of the Emery-Trist Systems Paradigm (ETSP). *Systems Practice* 5: 263–290.

Bahro, R. (1978). *The Alternative in Eastern Europe*. London: New Left Books.

Bakewell, S. (2017). *At the Existentialist Cafe: Freedom, Being and Apricot Cocktails*. London: Vintage.

Banson, K.E., Nguyen, N.C., Bosch, J.H., and Nguyen, T.V. (2015). A systems thinking approach to address the complexity of agribusiness for sustainable development in Africa: a case study in Ghana. *Systems Research and Behavioral Science* 32: 672–688.

Barabba, V.P. (2004). *Surviving Transformation: Lessons from GM's Surprising Turnaround*. Oxford: Oxford University Press.

Barabba, V.P. and Mitroff, I.I. (2014). *Business Strategies for a Messy World: Tools for Systemic Problem-Solving*. New York: Palgrave Macmillan.

Barnard, C. (1938). *The Functions of the Executive*. Cambridge: Harvard University Press.

Barthes, R. (1973). *Mythologies*. St. Albans: Granada Publishing.

Barton, J. (1999). *Pragmatism, Systems Thinking and System Dynamics*. Melbourne, unpublished manuscript.

Barton, J. and Selsky, J.W. (2000). Toward an Emery model of management: implications and prospects of Emery open systems theory. *Systemic Practice and Action Research* 13: 705–720.

Bateson, G. (1973). *Steps to an Ecology of Mind*. London: Paladin Books.

Bawden, R. (1995). *Systemic Development: A Learning Approach to Change*. University of Western Sydney.

Beckett, C. (2012). *Dark Eden*. London: Corvus.

Beckford, J. (1998). *Quality: A Critical Introduction*. London: Routledge.

Beer, S. (1966). *Decision and Control: The Meaning of Operational Research and Management Cybernetics*. Chichester: Wiley.

Beer, S. (1967). *Cybernetics and Management*, 2e. London: English Universities Press.

Beer, S. (1972). *Brain of the Firm*. London: Allen Lane.

Beer, S. (1974). *Designing Freedom*. Toronto: Canadian Broadcasting Corporation Publications.

Beer, S. (1975). *Platform for Change*. Chichester: Wiley.

Beer, S. (1979). *The Heart of Enterprise*. Chichester: Wiley.

Beer, S. (1981). *Brain of the Firm*, 2e. Chichester: Wiley.

Beer, S. (1985). *Diagnosing the System for Organizations*. Chichester: Wiley.

Beer, S. (1989a). The viable system model: its provenance, development, methodology and pathology. In: *The Viable System Model: Interpretations and Applications of Stafford Beer's VSM* (ed. R. Espejo and R.J. Harnden), 11–37. Chichester: Wiley.

Beer, S. (1989b). I am the Emperor – and I want dumplings. *Systems Practice* 2: 267–285.

Beer, S. (1989c). *Chronicles of Wizard Prang*, Wizardprang. Cwarel Isaf Institute. itsy.co.uk.

Beer, S. (1990). On suicidal rabbits: a relativity of systems. *Systems Practice* 3: 115–124.

Beer, S. (1994). *Beyond Dispute: The Invention of Team Syntegrity*. Chichester: Wiley.

Beer, S. (2009). *Think Before You Think: Social Complexity and Knowledge of Knowing*. Charlbury: Wavestone.

Begun, J.W. (1994). Chaos and complexity: frontiers of organisation science. *Journal of Management Inquiry* 3: 329–335.

Bell, I. (2013). *Once Upon a Time: The Lives of Bob Dylan*, Kindle Edition. London: Mainstream Publications.

Bell, S. and Morse, S. (2013). How people use rich pictures to help them think and act. *Systemic Practice and Action Research* 26: 331–348.

Bennett, S. (2002). Otto Mayr: contributions to the history of feedback control. *IEEE Control Systems Magazine* 29–33.

Bennett, K. (2016). *An Inside Job: The Vanguard Method in Financial Services*, 1e. The Vanguard Periodical.

Berg, T. (2015). Rich picture: the role of the facilitator. *Systemic Practice and Action Research* 28: 67–77.

Berger, P.L. and Luckmann, T. (1971). *The Social Construction of Reality*. Harmondsworth: Penguin.

von Bertalanffy, L. (1969). The theory of open systems in physics and biology. In: *Systems Thinking* (ed. F.E. Emery), 70–85. Harmondsworth: Penguin.

von Bertalanffy, L. (1971). *General System Theory*. Harmondsworth: Penguin.

Bettencourt, L.M.A. (2013). The kind of problem a city is. New Mexico: Santa Fe Institute, Working Paper.

Bevan, R.G. (1980). Social limits to planning. *Journal of the Operational Research Society* 31: 867–874.

Beyes, T.P. (2005). Observing observers: Von Foerster, Luhmann, and management thinking. *Kybernetes* 34: 448–459.

Blackler, F.H.M. and Brown, C.A. (1980). *Whatever Happened to Shell's New Philosophy of Management*. London: Saxon House.

Blake, W. (1815). Jerusalem.

Blohm, H., Beer, S., and Suzuki, D. (1986). *Pebbles to Computers: The Thread*. Toronto: Oxford University Press.

Boguslaw, R. (1981). *The New Utopians*. New York: Irvington.

Bolweg, J.F. (1976). *Job Design and Industrial Democracy*. Leiden: Nijhoff.

Borch, C. (2011). *Niklas Luhmann*, Kindle Edition. London: Routledge.

Bosch, O. and Nguyen, N. (2015). *Systems Thinking for EVERYONE: The Journey from Theory to Making an Impact*. Canberra: Think2Impact.

Boulding, K.E. (1961). *The Image*. Ann Arbor, MI: Michigan.

Boulding, K.E. (1968). General systems theory – the skeleton of science. In: *Modern Systems Research for the Behavioral Scientist* (ed. W. Buckley), 3–10. Chicago: Aldine.

Boulton, J., Allen, P., and Bowman, C. (2015). *Embracing Complexity: Strategic Perspectives for an Age of Turbulence*. Oxford: Oxford University Press.

Bowers, T. (2011). Towards a framework for multiparadigm multimethodologies. *Systems Research and Behavioral Science* 28: 537–552.

Boyd, A., Brown, M., and Midgley, G. (1999). *Home and Away: Developing Services with Young People Missing from Home or Care*. Hull: Centre for Systems Studies, University of Hull.

Brant, S.G. (2009). Russell Ackoff, 'Einstein of Problem Solving', Has Died. *Huffington Post* (1 November).

Britton, G.A. and McCallion, H. (1994). An overview of the Singer/Churchman/Ackoff school of thought. *Systems Practice* 7: 487–521.

Brocklesby, J. (1997). Becoming multimethodology literate: an assessment of the cognitive difficulties of working across paradigms. In: *MultiMethodology* (ed. J. Mingers and A. Gill), 189–216. Chichester: Wiley.

Brocklesby, J. (2007). The theoretical underpinnings of soft systems methodology – comparing the work of Geoffrey Vickers and Humberto Maturana. *Systems Research and Behavioral Science* 24: 157–168.

Brocklesby, J. (2012). Using the viable systems model to examine multi-agency arrangements for combatting transnational organised crime. *Journal of the Operational Research Society* 63: 418–430.

Brocklesby, J. and Cummings, S. (1996). Foucault plays Habermas: an alternative philosophical underpinning for critical systems thinking. *Journal of the Operational Research Society* 47: 741–754.

Brocklesby, J. and Mingers, J. (2005). The use of the concept autopoiesis in the theory of viable systems. *Systems Research and Behavioral Science* 22: 3–9.

Brooks, R. (2016). Beleaf it or not, trees chat to family and friends on the wood-wide web. *The Sunday Times* (11 September), p. 14.

Brooks, M. (2017). *The Quantum Astrologer's Handbook*. London: Scribe.

Brown, R.K. (1967). Research and consultancy in industrial enterprises: a review of the contribution of the Tavistock Institute of Human Relations to the development of industrial sociology. *Sociology* 1: 33–60.

Bryer, R.A. (1979). The status of the systems approach. *Omega* 7: 219–231.

Bryer, R.A. (1980). Some comments on Churchman and Ulrich's 'Reply' to 'The status of the systems approach. *Omega* 8: 280.

Buckle Henning, P. and Thomas, J. (2006). *A Boundary Critique of Gender in the Project Management Body of Knowledge*. International Society for the Systems Sciences.

Buckley, W. (1967). *Sociology and Modern Systems Theory*. Englewood Cliffs, NJ: Prentice-Hall.

Buckley, W. (1976). Society as a complex adaptive system. In: *Systems Behaviour*, 2e (ed. J. Beishon and G. Peters), 178–201. London: Harper & Row.

Burnes, B. (2005). Complexity theories and organizational change. *International Journal of Management Reviews* 7: 73–90.

Burns, T. and Stalker, G.M. (1961). *The Management of Innovation*. London: Tavistock.

Burrell, G. (1983). 'Systems thinking, systems practice': a review. *Journal of Applied Systems Analysis* 10: 121–125.

Burrell, G. and Morgan, G. (1979). *Sociological Paradigms and Organizational Analysis*. London: Heinemann.

Byrne, D. and Callaghan, G. (2014). *Complexity Theory and the Social Sciences: The State of the Art*, Kindle Edition. Oxford: Routledge.

Cabana, S., Emery, F., and Emery, M. (1997). The search for effective strategic planning is over. *The Journal for Quality and Participation* 20: 381–393.

Cabrera, D., Cabrera, L., and Powers, E. (2015). A unifying theory of systems thinking with psychosocial applications. *Systems Research and Behavioral Science* 32: 534–543.

Cannon, W.B. (1939). *The Wisdom of the Body*. London: Kegan Paul.

Capra, F. (1975). *The Tao of Physics*. Boston, MA: Shambhala.

Capra, F. (1996). *The Web of Life: A New Synthesis of Mind and Matter*. London: Harper Collins.

Capra, F. and Luisi, P.L. (2014). *The Systems View of Life: A Unifying Vision*. Cambridge: Cambridge University Press.

Carrizosa, A. (2000). *Enacting Thinking Spaces Towards Purposeful Actions: An Action Research Project*. Lincoln: University of Lincoln.

Carrizosa, A. (2002). Platforms for critical systems practice. PhD thesis. Lincoln: University of Lincoln.

Carrizosa, A. and Ortegon, M. (1998). *Using Systems Metaphors to Interpret the Edge of Chaos*. Beijing: International Society for the Systems Sciences.

Carter, P., Jackson, M.C., Jackson, N., and Keys, P. (1987). Community OR at Hull University. *Dragon* 2: Special issue.

Carvajal, R. (1983). The impact of a social systems scientist on a country. *Omega* 11: 559–565.

Castaneda, C. (1974). *Tales of Power*. New York: Simon & Schuster.

Castellini, M.A. and Paucar-Caceres, A. (2018). A conceptual framework for integrating methodologies in management: partial results of a systemic intervention in a textile SME in Argentina. *Systems Research and Behavioral Science*, 36.1, forthcoming.

Caulkin, S. (2007). Cool Judgment on the f-Laws of the business jungle. *The Observer* (11 February), p. 8.

Caulkin, S. (2010). *Managing for Better: Reflections on the Vanguard Leaders Summit, Milton Keynes, 25 November, 2010*. Buckingham: Vanguard Publications.

Cavaleri, S. and Obloj, K. (1993). *Management Systems: A Global Perspective*. Belmont, CA: Wadsworth.

Caws, P. (2015). General systems theory: its past and potential. *Systems Research and Behavioral Science* 32: 514–521.

Chalmers, A.F. (1982). *What Is this Thing Called science?* 2e. Milton Keynes: Open University Press.

Chapman, R. (2010). *Syd Barrett: A Very Irregular Head*, Kindle Edition. London: Faber & Faber.

Checkland, P.B. (1976). Towards a systems-based methodology for real-world problem-solving. In: *Systems Behaviour* (ed. J. Beishon and G. Peters), 51–77. London: Harper & Row.

Checkland, P.B. (1978). The origins and nature of 'Hard' systems thinking. *Journal of Applied Systems Analysis* 5: 99–110.

Checkland, P.B. (1980). Are organizations machines? *Futures* 12: 421–424.

Checkland, P.B. (1981). *Systems Thinking, Systems Practice*. Chichester: Wiley.

Checkland, P.B. (1983). OR and the systems movement: mappings and conflicts. *Journal of the Operational Research Society* 34: 661–675.

Checkland, P.B. (1985). From optimizing to learning: a development of systems thinking for the 1990s. *Journal of the Operational Research Society* 36: 757–767.

Checkland, P.B. (1988). Churchman's 'Anatomy of system Teleology' revisited. *Systems Practice* 1: 377–384.

Checkland, P.B. (1999). *Systems Thinking, Systems Practice, Including a 30-Year Retrospective*. Chichester: Wiley.

Checkland, P.B. (2000). New maps of knowledge, some animadversions (friendly) on: science (reductionist), social science (hermeneutic), research (unmanageable) and universities (unmanaged). *Systems Research and Bevavioral Science* 17: S59–S75.

Checkland, P.B. (2003). Personal communication.

Checkland, P.B. (2011). Autobiographical retrospectives. *International Journal of General Systems* 40: 487–512.

Checkland, P.B. (2012). Four conditions for serious systems thinking and action. *Systems Research and Behavioral Science* 29: 465–469.

Checkland, P.B. and Holwell, S. (1998). *Information, Systems and Information Systems*. Chichester: Wiley.

Checkland, P.B. and Holwell, S. (2004). 'Classic' OR and 'Soft' OR – an asymmetric complementarity. In: *Systems Modelling: Theory and Practice* (ed. M. Pidd), 44–60. Chichester: Wiley.

Checkland, P.B. and Poulter, J. (2006). *Learning for Action: A Short Definitive Account of Soft Systems Methodology and Its Use for Practitioners, Teachers and Students*. Chichester: Wiley.

Checkland, P.B. and Poulter, J. (2010). Soft systems methodology. In: *Systems Approaches to Managing Change: A Practical Guide* (ed. M. Reynolds and S. Holwell), 191–242. London: Springer.

Checkland, P.B. and Scholes, J. (1990). *Soft Systems Methodology in Action*. Chichester: Wiley.

Checkland, P.B. and Winter, M. (2006). Process and content: two ways of using SSM. *Journal of the Operational Research Society* 57: 1435–1441.

Cherns, A. (1976). The principles of socio-technical design. *Human Relations* 29: 783–792.

Cherns, A. (1987). Principles of socio-technical design revisited. *Human Relations* 40: 153–161.

Chesterton, K., Goodsman, R., Rosenhead, J., and Thunhurst, C. (1975). A comment on Ackoff's 'The social responsibility of OR'. *Operational Research Quarterly* 26: 91–95.

Chowdhury, R. (2015). Using interactive planning to create a child protection framework in an NGO setting. *Systems Practice and Action Research* 32: 547–574.

Chowdhury, R. and Jangle, N. (2018). Critical systems thinking towards enhancing community engagement in micro-insurance. *Global Journal of Flexible Systems Management* 19: 209–224.

Chowdhury, R. and Nobbs, A. (2008). Critical insights into NHS information systems deployment. In: *Management Practices in High-Tech Environments*, Chapter 14 (ed. D. Jemielniak and J. Kociatkiewicz). Hershey: Idea Group Inc.

Churchman, C.W. (1968). *Challenge to Reason*. New York: McGraw-Hill.

Churchman, C.W. (1970). Operations research as a profession. *Management Science* 17: B37–B53.

Churchman, C.W. (1971). *The Design of Inquiring Systems*. New York: Basic Books.

Churchman, C.W. (1974). Perspectives of the systems approach. *Interfaces* 4: 6–11.

Churchman, C.W. (1979a). Paradise regained: a hope for the future of systems design education. In: *Education in Systems Science* (ed. B.A. Bayraktar et al.), 17–22. London: Taylor & Francis.

Churchman, C.W. (1979b). *The Systems Approach*, 2e. New York: Dell Publishing.

Churchman, C.W. (1979c). *The Systems Approach and Its Enemies*. New York: Basic Books.

Churchman, C.W. (1982). *Thought and Wisdom*. Seaside, CA: Intersystems Publications.

Churchman, C.W. and Schainblatt, A.H. (1965). The researcher and the manager: a dialectic of implementation. *Management Science* 11: B69–B87.

Churchman, C.W., Ackoff, R., and Arnoff, E.L. (1957). *Introduction to Operations Research*. New York: Wiley.

Cilliers, P. (1998). *Complexity and Postmodernism: Understanding Complex Systems*, Kindle Edition. London: Routledge.

Clarke, S. and Lehaney, B. (1999). Organisational intervention and the problems of coercion. *Systemist* 27: 40–52.

Clegg, B. (2015). Fashioning a model for Zara. *Impact*, Autumn 7–12.

Clemson, B. (1984). *Cybernetics: A New Management Tool*. Tunbridge Wells: Abacus.

Clemson, M. and Jackson, M.C. (1988). Evaluating organizations with multiple goals. *OR Insight* 1: 2–5.

Coates, C. (2013). Communes Britannica. *Walden* (19 October).

Cohen, C. and Midgley, G. (1994). *The North Humberside Diversion from Custody Project for Mentally Disordered Offenders*. Hull: Centre for Systems Studies, University of Hull.

Coleman, J. (2006). *The Tavistock Institute of Human Relations: Shaping the Moral, Spiritual, Cultural, Political and Economic Decline of the USA*. New York.

Conant, R.C. and Ashby, W.R. (1970). Every good regulator of a system must be a model of that system. *International Journal of Systems Science* 1: 89–97.

Connell, N.A.D. (2001). Evaluating soft OR: some reflections on an apparently 'Unsuccessful' implementation using a Soft Systems Methodology (SSM) based approach. *Journal of the Operational Research Society* 52: 150–160.

Cooke-Davies, T., Cicmil, S., Crawford, L., and Richardson, K. (2007). We're not in Kansas anymore, Toto: mapping the strange landscape of complexity theory, and its relationship to project management. *Project Management Journal* 14: 50–61.

Cooley, C. (1909). *Social Organization: A Study of the Larger Mind*. New York: C Scribner's.

Cooper, R. and Burrell, G. (1988). Modernism, postmodernism and organisational analysis: an introduction. *Organisation Studies* 9: 91–112.

Cowen, R. (2015). *Common Ground*, Kindle Edition. New York: Hutchinson.

Craib, I. (1992). *Modern Social Theory: From Parsons to Habermas*. Hemel Hempstead: Harvester-Wheatsheaf.

Culler, J. (1976). *Saussure*. Glasgow: William Collins.

Cummings, N. (2011). Operational research - the beginning: how scientists evolved the techniques which became OR. *Inside OR*, July, pp. 22–25.

Curtis, A. (2011). Why ecosystems are not the whole truth. *The Observer* (29th May).

Dando, M.R. and Bennett, P.G. (1981). A Kuhnian crisis in management science? *Journal of the Operational Research Society* 32: 91–103.

Dawe, A. (1970). The two sociologies. *The British Journal of Sociology* 21: 207–218.

Dawkins, R. (1976). *The Selfish Gene*. Oxford: Oxford University Press.

Deming, W.E. (1982). *Out of Crisis*. Cambridge: Cambridge University Press.

Deming, W.E. (1994). *The New Economics: For Industry, Government, Education*. Cambridge: MIT Press.

Dent, E.B. and Umpleby, S.A. (1998). Underlying assumptions of several traditions in systems theory and cybernetics. In: *Cybernetics and Systems '98*, 513–518. Vienna: Austrian Society for Cybernetic Studies.

Descartes, R. (1897). *Oeuvres*. Paris: Adam & Tannery.

Descartes, R. (1968). *Discourse on Method and the Meditations*. Harmondsworth: Penguin.

Donaires, S. (2006). A critical heuristic approach to the establishment of a software development process. *Systemic Practice and Action Research* 19: 415–428.

Douglas, J. (2015). *Social Meanings of Suicide*. Princeton, NJ: Princeton University Press.

Drack, M. (2009). Ludwig von Bertalanffy's early systems approach. *Systems Research and Behavioral Science* 26: 563–572.

Drack, M. and Pouvreau, D. (2015). On the history of Ludwig von Bertalanffy's 'General Systemology', and on its relationship to cybernetics – part III: convergences and divergences. *International Journal of General Systems* 44: 523–571.

Drewry, D. (2018). Personal communication.

Durkheim, E. (1933). *The Division of Labour*. New York: Free Press.

Durkheim, E. (1938). *The Rules of Sociological Method*. New York: Free Press.

Du Sautoy, M. (2016). *What We Cannot Know: Explorations at the Edge of Knowledge*, Kindle Edition. London: Fourth Estate.

Dylan, B. (2001). Mississippi, Love and Theft Album. [Sound Recording].

Eaton, G. (2018). Project Cybersyn: The Afterlife of Chile's Socialist Internet. *The New Statesman* (22 August).

Eden, C. and Ackermann, F. (2018). Theory into practice, practice to theory: action research in method development. *European Journal of Operational Research* 271: 1145–1155.

Eden, C., Jones, S., and Sims, D. (1979). *Thinking in Organizations*. London: Macmillan.

Eden, C., Ackermann, F., Bryson, J.M. et al. (2009). Integrating modes of policy analysis and strategic management practice: requisite elements and dilemmas. *Journal of the Operational Research Society* 60: 2–13.

Edwards, M.G. (2014). Misunderstanding metatheorizing. *Systems Research and Behavioral Science* 31: 720–744.

Eliot, T.S. (1959). *East Coker: Four Quartets*. London: Faber & Faber.

Ellis, R.K. (2002). Toward a systemic theory of organisational change. PhD thesis. London: City University.

Emery, F.E. (ed.) (1969). *Systems Thinking*. Harmondsworth: Penguin.

Emery, F.E. (1981a). The emergence of ideal-seeking systems. In: *Systems Thinking*, vol. 2 (ed. F.E. Emery), 431–458. Harmondsworth: Penguin.

Emery, F.E. (1981b). Searching for common ground. In: *Systems Thinking*, vol. 2 (ed. F.E. Emery), 459–468. Harmondsworth: Penguin.

Emery, M. (2000). The current version of Emery's open systems theory. *Systemic Practice and Action Research* 13: 623–643.

Emery, M. (2010). Refutation of Kira & van Eijnatten's critique of the Emery's open systems theory. *Systems Research and Behavioral Science* 27: 697–712.

Emery, M. (2011). Fiddling while the planet burns: the scientific validity of chaordic systems thinking. *Systems Research and Behavioral Science* 28: 401–417.

Emery, F.E. and Emery, M. (1974). *Participative Design: Work and Community Life*. Canberra: Centre for Continuing Education, Australian National University.

Emery, M. and Purser, R.E. (1996). *The Search Conference: A Powerful Method for Planning Organizational Change and Community Action*. San Francisco, CA: Jossey-Bass.

Emery, F.E. and Thorsrud, E. (1969). *Form and Content in Industrial Democracy*. London: Tavistock.

Emery, F.E. and Thorsrud, E. (1976). *Democracy at Work*. Leiden: Nijhoff.

Emery, F.E. and Trist, E.L. (1969b). Socio-technical systems. In: *Systems Thinking* (ed. F.E. Emery), 281–296. Harmondsworth: Penguin.

Emery, F.E. and Trist, E.L. (1969a). The causal texture of organizational environments. In: *Systems Thinking* (ed. F.E. Emery), 21–32. Harmondsworth: Penguin.

Eno, B. (2009). Foreward. In: *Think Before You Think: Social Complexity and Knowledge of Knowing*, 7–12. Charlbury: Wavestone.

Espejo, R. (2008). Observing organizations: the use of identity and structural archetypes. *International Journal of Applied Systemic Studies* 2 (1/2): 6.

Espejo, R. (2017). Cybernetic argument for democratic governance: cybersyn and cyberfolk. In: *Cybernetics: State of the Art*, 34–57. Berlin: Universitatsverlag der TU Berlin.

Espejo, R. and Harnden, R.J. (1989a). The VSM: an ongoing conversation. In: *The Viable System Model: Interpretations and Applications of Beer's VSM* (ed. R. Espejo and R.J. Harnden), 441–460. Chichester: Wiley.

Espejo, R. and Harnden, R.J. (eds.) (1989b). *The Viable System Model: Interpretations and Applications of Stafford Beer's VSM*. Chichester: Wiley.

Espejo, R. and Reyes, A. (2011). *Organizational Systems: Managing Complexity with the Viable System Model*. Berlin: Springer-Verlag.

Espejo, R. and Schwaninger, M. (eds.) (1993). *Organizational Fitness: Corporate Effectiveness Through Management Cybernetics*. Frankfurt: Campus Verlag.

Espejo, R., Schuhmann, W., Schwaninger, M., and Bilello, U. (1996). *Organizational Transformation and Learning*. Chichester: Wiley.

Espinosa, A. (2002). Proyecto Consolidacion del Sistema de Informacion Ambiental Colombiano (SIAC). Bogota: United Nations Development Program, Working Paper.

Espinosa, A. (2003). *Team Syntegrity as a Tool to Promote Democratic Agreements: An Example from the National Environmental Sector in Colombia*. Crete: International Society for the Systems Sciences.

Espinosa, A. and Duque, C. (2018). Complexity management and multi-scale governance: a case study in an Amazonian indigenous association. *European Journal of Operational Research* 268: 1006–1020.

Espinosa, A. and Harnden, R. (2007). Team syntegrity and democratic group decision making: theory and practice. *Journal of the Operational Research Society* 58: 1056–1064.

Espinosa, A. and Walker, J. (2011). *A Complexity Approach to Sustainability: Theory and Application*. London: World Scientific.

Espinosa, A. and Walker, J. (2017). *A Complexity Approach to Sustainability: Theory and Application*, 2e. London: World Scientific.

Espinosa, A., Harnden, R., and Walker, J. (2008). A complexity approach to sustainability – stafford beer revisited. *European Journal of Operational Research* 187: 636–651.

Espinosa, A., Reficco, E., Martinez, A., and Guzman, D. (2015). A methodology for supporting strategy implementation based on the VSM: a case study in a Latin-American multi-national. *European Journal of Operational Research* 240: 202–212.

Estulin, D. (2015). *The Tavistock Institute: Social Engineering the Masses*. New York: Trine Day.

Farka Touré, A. (1994). *Talking Timbuktu* (with Ry Cooder). CD insert.

Fischlin, D., Heble, A., and Lipsitz, G. (2013). *The Fierce Urgency of Now: Improvisation, Rights, and the Ethics of Co-Creation*, Kindle Edition. Durham: Duke University Press.

Flood, R.L. (1990). *Liberating Systems Theory*. New York: Plenum.

Flood, R.L. (1993). *Beyond TQM*. Chichester: Wiley.

Flood, R.L. (1995). *Solving Problem Solving*. Chichester: Wiley.

Flood, R.L. (1999). *Rethinking the Fifth Discipline: Learning Within the Unknowable*. London: Routledge.

Flood, R.L. and Carson, E.R. (1988). *Dealing with Complexity: An Introduction to the Theory and Application of Systems Science*. New York: Plenum.

Flood, R.L. and Green, P. (1996). TSI in North Yorkshire police. In: *Critical Systems Thinking: Current Research and Practice* (ed. R.L. Flood and N.R. Romm), 217–234. New York: Plenum.

Flood, R.L. and Gregory, W.J. (1989). Systems: past, present, future. In: *Systems Prospects* (ed. R.L. Flood, M.C. Jackson and P. Keys), 55–60. New York: Plenum.

Flood, R.L. and Jackson, M.C. (eds.) (1991d). *Critical Systems Thinking: Directed Readings*. Chichester: Wiley.

Flood, R.L. and Jackson, M.C. (1991a). *Creative Problem Solving: Total Systems Intervention*. Chichester: Wiley.

Flood, R.L. and Jackson, M.C. (1991b). Critical systems heuristics: application of an emancipatory approach for police strategy toward the carrying of offensive weapons. *Systems Practice* 4: 283–302.

Flood, R.L. and Jackson, M.C. (1991c). Total systems intervention: a practical face to critical systems thinking. *Systems Practice* 4: 197–213.

Flood, R.L. and Romm, N.R. (1995). Enhancing the process of choice in TSI, and improving the chances of tackling coercion. *Systems Practice* 8: 377–408.

Flood, R.L. and Romm, R.A. (1996). *Diversity Management: Triple Loop Learning*. Chichester: Wiley.

Flood, R.L. and Ulrich, W. (1990). Testament to conversations on critical systems thinking between two systems practitioners. *Systems Practice* 3: 7–29.

Flood, R.L., Jackson, M.C., and Schecter, D. (1992). Total systems intervention: a research program. *Systems Practice* 5: 79–83.

Forrester, J.W. (1958). Industrial dynamics: a major breakthrough for decision makers. *Harvard Business Review* 37–48.

Forrester, J.W. (1961). *Industrial Dynamics*. Portland: Productivity Press.

Forrester, J.W. (1968). *Principles of Systems*. Portland: Productivity Press.

Forrester, J.W. (1969). *Urban Dynamics*. Portland: Productivity Press.

Forrester, J.W. (1971a). *World Dynamics*. Portland: Productivity Press.

Forrester, J.W. (1971b). Counterintuitive behavior of social systems. *Technology Review* 73: 52–68.

Fortuin, L., van Beek, P., and van Wassenhove, L. (eds.) (1996). *OR at Work*. London: Taylor & Francis.

Foucault, M. (1973). *The Order of Things: An Archaeology of the Human Sciences*. New York: Random House.

Foucault, M. (ed.) (1982). *I Pierre Riviere, Having Slaughtered My Mother, My Sister, and My Brother: A Case of Parricide in the 19th Century*. Lincoln: University of Nebraska Press.

Fox, W.M. (1995). Sociotechnical system principles and guidelines: past and present. *Journal of Applied Behavioral Science* 31: 91–105.

Freire, P. (1970). *The Pedagogy of the Oppressed*. New York: Seabury.

Freire, P. and Macedo, C. (1987). *Literacy: Reading the Word and the World*. South Hadley, MA: Bergin & Harvey.

Friend, J. (2012). The IOR legacy. *Inside OR*, April, pp. 20–21.

Friend, J. (2017). A friendly chat. *Inside OR*, February, pp. 20–21.

Friend, J. and Hickling, A. (2004). *Planning Under Pressure: The Strategic Choice Approach*, 3e. London: Routledge.

Friend, J.K., Norris, M.E., and Stringer, J. (1988). The institute for operational research: an initiative to extend the scope of OR. *Journal of the Operational Research Society* 39: 705–713.

Frisby, D. (ed.) (1976). *The Positivist Dispute in German Sociology*. London: Heinemann.

Fuenmayor, R. (1991). Truth and openness: an epistemology for interpretive systemology. *Systems Practice* 4: 473–490.

Fuenmayor, R. and Lopez-Garay, H. (1991). The scene for interpretive systemology. *Systems Practice* 4: 401–418.

Galliers, R., Mingers, J., and Jackson, M.C. (1997). Organization theory and systems thinking: the benefits of partnership. *Organization* 4: 269–278.

Garfinkel, H. (1984). *Studies in Ethnomethodology*. Cambridge: Polity Press.

Gartz, P.E. (1997). Commercial systems development in a changed world. *IEEE Transactions on Aerospace and Electronic Systems* 632–636.

Gaskell, C. (1997). The management of change in prisons. PhD thesis. Hull: University of Lincolnshire and Humberside.

Gaskell, C., Jackson, M.C., and Daly, R. (1996). The management of change in Hull prison: a new topic, a new research approach. *Prison Service Journal (Supplement)* 22–28.

Georgiou, I. (1999). Groundwork of a Sartrean input toward informing some concerns of critical systems thinking. *Systemic Practice and Action Research* 12: 585–606.

Georgiou, I. (2013). Open Letter to the Harvard Business Review. https://sites.google.com/site/iongeorgiou/open-letter-to-the-harvard-business-review (accessed 18 December 2018).

Gharajedaghi, J. (2011). *Systems Thinking: Managing Chaos and Complexity*, 3e. Burlington, MA: Elsevier.

Gibbons, M., Limoges, C., Nowotny, H. et al. (1994). *The New Production of Knowledge: The Dynamics of Science and Research in Contemporary Societies*. London: Sage.

Giddens, A. (1976). *New Rules of Sociological Method*. London: Hutchinson.

Glanville, R. (2003a). Second-order cybernetics, CEPA eprint 2326. https://www.univie.ac.at/ (accessed 18 December 2018).

Glanville, R. (2003b). Obituary: Heinz von Foerster. *Systems Research and Behavioral Science* 20: 85–89.

von Glasersfeld, E. (1984). An introduction to radical constructivism. In: *The Invented Reality* (ed. P. Watzlawick), 17–40. New York: Norton.

von Glasersfeld, E. (1990). Distinguishing the observer: an attempt at interpreting Maturana. *Methodologia* 8: 1–10.

Gleick, J. (1987). *Chaos: The Making of a New Science*. London: Abacus.

Golding, W. (1955). *The Inheritors*. London: Faber & Faber.

Goodwin, B. (1994). *How the Leopard Changed Its Spots: Evolution of Complexity*. NJ: Princeton University Press.

Gorelik, G. (1975a). Reemergence of Bogdanov's Tektology in soviet studies of organization. *Academy of Management Journal* 18: 345–357.

Gorelik, G. (1975b). Principle ideas of Bogdanov's 'Tektology': the Universal Science of Organization. *General Systems* 20: 3–13.

Gorelik, G. (1984). *Essays in Tektology* (translated from Bogdanov), 2e. Seaside, CA: Intersystems.

Government Office for Science (2018). *Computational Modelling: Technological Futures*. London: Council for Science and Technology.

Green, S.M. (1991). Total systems intervention: organisational communication in North Yorkshire Police. MA dissertation. Hull: University of Hull.

Green, S.M. (1992). Total systems intervention: organisational communication in North Yorkshire police. *Systems Practice* 5: 585–599.

Green, S.M. (1993). Total systems intervention: a trial by jury. *Systems Practice* 6: 295–299.

Gregory, W.J. (1992). Critical systems thinking and pluralism: a new constellation. PhD thesis. London: City University.

Gregory, A.J. (1996). The road to integration: reflections on the development of organizational evaluation theory and practice. *Omega* 24: 295–307.

Gregory, W.J. (1996). Discordant pluralism: a new strategy for critical systems thinking. *Systems Practice* 9: 605–625.

Gregory, A.J. and Jackson, M.C. (1992a). Evaluating organizations: a systems and contingency approach. *Systems Practice* 5: 37–60.

Gregory, A.J. and Jackson, M.C. (1992b). Evaluation methodologies: a system for use. *Journal of the Operational Research Society* 43: 19–28.

Gregory, A.J. and Jackson, M.C. (1992c). *NACVS Evaluation Project: Final Project Report*. Hull: Centre for Systems Studies, University of Hull.

Gregory, A.J., Jackson, M.C., and Clemson, M. (1994). Evaluation for Beverley CVS. In: *Community Works*, 193–199. Sheffield: Pavic Publications.

Gu, J. and Zhu, Z. (2000). Knowing Wuli, sensing Shili, caring for Renli: methodology of the WSR approach. *Systemic Practice and Action Research* 13: 11–20.

Guilfoyle, S. (2013). *Intelligent Policing: How Systems Thinking Methods Eclipse Conventional Management Practice*. Axminster: Triarchy Press.

Gyllenhammer, P. (1977). *People at Work*. Reading, MA: Addison-Wesley.

Habermas, J. (1970). Knowledge and interest. In: *Sociological Theory and Philosophical Analysis* (ed. D. Emmett and A. MacIntyre), 36–54. London: Macmillan.

Habermas, J. (1974). *Theory and Practice*. London: Heinemann.

Habermas, J. (1975). *Legitimation Crisis*. Boston, MA: Beacon Books.

Habermas, J. (1987). *The Philosophical Discourse of Modernity*. Oxford: Polity Press.

Habermas, J. (2011). A Philosopher's mission to save the EU. *Spiegal Online* (25 November).

Habermas, J. (2016). Core Europe to the rescue: a conversation with Jurgen Habermas about Brexit and the EU crisis. *Social Europe* (12 July).

Hafner, K. (2016). Jay W Forrester Dies at 98; a pioneer in computer models. *The New York Times (Technology)* (17 November), pp. 1–9.

Haftor, D.A. (2011). An evaluation of R. L. Ackoff's interactive planning: a case-based approach. *Systemic Practice and Action Research* 24: 355–377.

Hamel, G. (2007). *The Future of Management*. Boston, MA: Harvard Business School Press.

Hamel, G. and LaBarre, P. (2010). Dispatches from the front lines of management innovation. *McKinsey Quarterly* (November), pp. 1–7.

Hamilton, P. (1983). *Talcott Parsons*. Chichester: Ellis Horwood.

Hammond, D. (2003). *The Science of Synthesis: Exploring the Social Implications of General Systems Theory*. Boulder, CO: University Press of Colorado.

Hanafizadeh, P. and Mehrabioun, M. (2018). Application of SSM in tackling problematical situations from Academicians' viewpoints. *Systemic Practice and Action Research* 31: 179–220.

Haralambos, M. and Holborn, M. (1995). *Sociology: Themes and Perspectives*. London: Collins Educational.

Harnden, R.J. (1989). Outside and then: an interpretive approach to the VSM. In: *The Viable System Model: Interpretations and Applications of Beer's VSM* (ed. R. Espejo and R.J. Harnden), 383–404. Chichester: Wiley.

Harnden, R.J. (1990). The languaging of models. *Systems Practice* 3: 289–302.

Harnden, R.J. and Leonard, A. (eds.) (1994). *How Many Grapes Went into the Wine: Stafford Beer on the Art and Science of Holistic Management*. Chichester: Wiley.

Harwood, S.A. (2012). The methodology of change and the Viplan methodology in practice. *Journal of the Operational Research Society* 63: 748–761.

Heidegger, M. (1978). The question concerning technology. In: *Basic Writings*, 283–317. London: Routledge & Kegan Paul.

Heine, H. (1834, 2007). *On the History of Religion and Philosophy in Germany*. Cambridge: Cambridge University Press.

Hemingway, E. (1940). *For Whom the Bell Tolls*, Chapter 43. New York: Scribner.

Henao, F. and Franco, L.A. (2016). Unpacking multimethodology: impacts of a community development intervention. *European Journal of Operational Research* 253: 681–696.

Henderson, L.J. (1970). Sociology – 23 lectures – 1938–42. In: *L. J. Henderson on the Social System: Selected Writings*, 57–148. Chicago: University of Chicago Press.

Hill, P. (1971). *Towards a New Philosophy of Management*. Epping: Gower Press.

Hindle, G.A. (2011). Teaching soft systems methodology and a blueprint for a module. *Informs Transactions in Education* 12: 31–40.

Hindle, G.A. and Jackson, M.C. (1997). SSM within Humberside TEC: the isssue of sustainability. Lincoln: Lincoln School of Management, Working paper 13.

Hindle, G.A. and Vidgen, R. (2018). Developing a business analytics methodology: a case study in the Foodbank sector. *European Journal of Operational Research* 268: 836–851.

Hipel, K.W., Fang, L., and Heng, M. (2010). Systems of systems approach to policy development for global food security. *Journal of Systems Science and Systems Engineering* 19: 1–21.

Hirschheim, R. (1985). Information systems epistemology: an historical perspective. In: *Research Methods in Information Systems* (ed. E. Mumford et al.), 13–35. Amsterdam: North-Holland.

Ho, J.K. and Jackson, M.C. (1987). Building a 'Rich Picture' and assessing a 'Quality Management' program at Thornton printing company. *Cybernetics and Systems: An International Journal* 18: 381–405.

Hoare, P. (2017). Other minds: the octopus and the evolution of intelligent life by Peter Godfrey-Smith. *The Guardian* (18 March), p. 7.

Holland, J.H. (2014). *Complexity: A Very Short Introduction*. Oxford: Oxford University Press.

Holmberg, S.C. (1997). Team syntegrity assessment. *Systems Practice* 10: 241–254.

Honderich, T. (ed.) (1995). *The Oxford Companion to Philosophy*. Oxford: Oxford University Press.

Hoos, I.R. (1974). *Systems Analysis in Public Policy: A Critique*. Berkeley, CA: University of California Press.

Hordijk, L. (2007). What is systems analysis?. *Options Magazine* (Winter), pp. 1–2.

Horkheimer, M. (1976). Traditional and critical theory. In: *Critical Sociology* (ed. P. Connerton), 206–224. Harmondsworth: Penguin.

Hoverstadt, P. (2008). *The Fractal Organization: Creating Sustainable Organizations with the Viable System Model*. Chichester: Wiley.

Hoverstadt, P. (2010). The viable system model. In: *Systems Approaches to Managing Change: A Practical Guide* (ed. M. Reynolds and S. Holwell), 87–133. London: Springer.

Hoverstadt, P. (2011). Defining identity by structural coupling in VSM practice. *Systemist* 33: 5–19.

Hoverstadt, P. and Loh, L. (2017). *Patterns of Strategy*. London: Routledge.

Huhn, M. (2012). Cybernetic management paradigms. In: *Systemic Management for Intelligent Organizations* (ed. S.N. Grosser and R. Zeier), 3–19. Heidelberg: Springer-Verlag.

Husserl, E. (1970). *The Crisis of European Sciences and Transcendental Phenomenology*. Evanston, IL: Northwestern University Press.

Huxley, A. (1959). *The Doors of Perception*. Harmondsworth: Penguin.

IBM (2010). *Capitalizing on Complexity: Insights from the Global Chief Executive Officer Study*. USA: IBM.

ICCPM (2012). *Complex Project Manager Competency Standards*, Version 4.1. Australia: ICCPM.

INCOSE (2017). www.incose.org (accessed 22 October 2017).

Inglis, D. and Thorpe, C. (2012). *An Invitation to Social Theory*, Kindle Edition. Cambridge: Polity Press.

Ison, R. (2017). *Systems Practice: How to Act*, 2e. London: Springer.

Jackson, M.C. (1982a). The nature of soft systems thinking: the work of Churchman, Ackoff and Checkland. *Journal of Applied Systems Analysis* 9: 17–28.

Jackson, M.C. (1982b). Verifying social systems theory in practice: a critique. In: Proceedings of the SGSR, 668–673. Louisville, KY: Society for General Systems Research.

Jackson, M.C. (1983). The nature of soft systems thinking: comments on the three replies. *Journal of Applied Systems Analysis* 10: 109–113.

Jackson, M.C. (1985a). The itinerary of a critical approach: review of Ulrich's 'Critical heuristics of social planning'. *Journal of the Operational Research Society* 36: 878–881.

Jackson, M.C. (1985b). Social systems theory and practice: the need for a critical approach. *International Journal of General Systems* 10: 135–151.

Jackson, M.C. (1987a). Present positions and future prospects in management science. *Omega* 15: 455–466.

Jackson, M.C. (1987b). New directions in management science. In: *New Directions in Management Science* (ed. M.C. Jackson and P. Keys), 133–164. Aldershot: Gower.

Jackson, M.C. (1987c). Community operational research: purposes, theory and practice. *Dragon* 2: 47–73.

Jackson, M.C. (1987d). Systems strategies for information management in organizations which are not machines. *International Journal of Information Management* 7: 187–195.

Jackson, M.C. (1988a). An appreciation of Stafford Beer's 'Viable System' viewpoint on management practice. *Journal of Management Studies* 25: 557–573.

Jackson, M.C. (1988b). Systems methods for organizational analysis and design. *Systems Research* 5: 201–210.

Jackson, M.C. (1989). Assumptional analysis: an elucidation and appraisal for systems practitioners. *Systems Practice* 2: 11–28.

Jackson, M.C. (1990). The critical kernel in modern systems thinking. *Systems Practice* 3: 357–364.

Jackson, M.C. (1991a). *Systems Methodology for the Management Sciences*. London: Plenum Press.

Jackson, M.C. (1991b). The origins and nature of critical systems thinking. *Systems Practice* 4: 131–149.

Jackson, M.C. (1992a). The soul of the viable system model. *Systems Practice* 5: 561–564.

Jackson, M.C. (1992b). An integrated programme for critical thinking in information systems research. *Journal of Information Systems* 2: 83–95.

Jackson, M.C. (1993a). Social theory and operational research practice. *Journal of the Operational Research Society* 44: 563–577.

Jackson, M.C. (1993b). Don't bite my finger: Haridimos Tsoukas' critical evaluation of Total systems intervention. *Systems Practice* 6: 289–294.

Jackson, M.C. (1995). Beyond the fads: systems thinking for managers. *Systems Research* 12: 25–42.

Jackson, M.C. (1997a). Pluralism in systems thinking and practice. In: *MultiMethodology* (ed. J. Mingers and A. Gill), 347–378. Chichester: Wiley.

Jackson, M.C. (1997b). Critical systems thinking and information systems research. In: *Information Systems: An Emerging Discipline* (ed. J. Mingers and F. Stowell), 201–238. Maidenhead: McGraw-Hill.

Jackson, M.C. (1999). Towards coherent pluralism in management science. *Journal of the Operational Research Society* 50: 12–22.

Jackson, M.C. (2000). *Systems Approaches to Management.* New York: Kluwer/Plenum.

Jackson, M.C. (2001). Critical systems thinking and practice. *European Journal of Operational Research* 128: 233–244.

Jackson, M.C. (2003). *Systems Thinking: Creative Holism for Managers.* Chichester: Wiley.

Jackson, M.C. (2005a). Systems metaphors and knowledge management practice. *International Journal of Knowledge and Systems Sciences* 2: 19–24.

Jackson, M.C. (2005b). Reflections on knowledge management from a critical systems perspective. *Knowledge Management Research and Practice* 3: 187–196.

Jackson, M.C. (2006). Beyond problem structuring methods: reinventing the future of OR/MS. *Journal of the Operational Research Society* 57: 868–878.

Jackson, M.C. (2009). Fifty years of systems thinking for management. *Journal of the Operational Research Society* 60: S24–S32.

Jackson, M.C. (2010). Reflections on the development and contribution of critical systems thinking and practice. *Systems Research and Behavioral Science* 27: 133–139.

Jackson, N.V. and Carter, P. (1984). The attenuating function of myth in human understanding. *Human Relations* 37: 515–533.

Jackson, M.C. and Keys, P. (1984). Towards a system of systems methodologies. *Journal of the Operational Research Society* 35: 473–486.

Jackson, M.C. and Medjedoub, S. (1988). Designing evaluation systems: theoretical groundings and a practical intervention. In: *Cybernetics and Systems '88* (ed. R. Trappl), 165–171. Dordrecht: Kluwer.

Jackson, T. and Webster, R. (2016). *Limits Revisited: A Review of the Limits to Growth Debate.* London: All-Party Parliamentary Group on Limits to Growth.

Jackson, M.C., Keys, P., and Cropper, S. (eds.) (1989). *Operational Research and the Social Sciences.* New York: Plenum.

Jackson, M.C., Johnston, N., and Seddon, J. (2008). Evaluating systems thinking in housing. *Journal of the Operational Research Society* 59: 186–197.

Jenkins, G.M. (1972). The systems approach. In: *Systems Behaviour* (ed. J. Beishon and G. Peters), 78–104. London: Harper and Row.

Jimenez, J. (ed.) (2006). *Participation and Development: The Mexico of the Future.* Mexico City: Universidad Nacional Autonoma de Mexico.

Johnson, N. (2009). *Simply Complexity: A Clear Guide to Complexity Theory.* Oxford: Oneworld.

Johnson, J.L. and Burton, B.K. (1994). Chaos and complexity theory for management: caveat emptor. *Journal of Management Enquiry* 3: 320–328.

Johnson, M.P., Midgley, G., Wright, J., and Chichirau, G. (2018). Community operational research. *European Journal of Operational Research* 268 (3): 761–770.

Jones, O. (2014). *The Establishment.* London: Allen Lane.

Jopling, J. (2012). A complexity approach to sustainability - theory and application: review. *Feasta* (6 August).

Kahneman, D. (2011). *Thinking, Fast and Slow.* London: Penguin.

Kalanithi, P. (2016). *When Breath Becomes Air,* Kindle Edition. London: The Bodley Head.

Kalawsky, R.S. (2013). The next generation of grand challenges for systems engineering. *Procedia Computer Science* 834–843.

Kast, F.E. and Rosenzweig, J.E. (1981). *Organization and Management: A Systems and Contingency Approach,* 3e. New York: McGraw-Hill.

Katz, D. and Kahn, R.L. (1978). *The Social Psychology of Organizations*, 2e. New York: Wiley.

Kauffman, S. (1995). *At Home in the Universe*. New York: Oxford University Press.

Kemp, J. (1968). *The Philosophy of Kant*. Oxford: Oxford University Press.

Keys, P. (1991). *Operational Research and Systems: The Systemic Nature of Operational Research*. New York: Plenum.

Keys, P. (1995). OR as technology: some issues and implications. In: *Understanding the Process of Operational Research: Collected Readings* (ed. P. Keys), 323–334. Chichester: Wiley.

Kilmann, R.H. (1983). A dialectical approach to formulating and testing social science theories: assumptional analysis. *Human Relations* 36: 1–22.

Kingsnorth, P. (2015). *The Wake*. London: Unbound.

Kingsnorth, P. (2016). The call of the wild. *The Guardian* (23 July), pp. 15–16.

Kingsnorth, P. (2017). *Beast*. London: Faber & Faber.

Kira, M. and van Eijnatten, F.M. (2008). Socially sustainable work organizations: a Chaordic systems approach. *Systems Research and Behavioral Science* 25: 743–756.

Kira, M. and van Eijnatten, F.M. (2010). Socially sustainable work organizations and systems thinking. *Systems Research and Behvioral Science* 27: 713–721.

Kirby, M.W. (2003). *Operational Research in War and Peace: The British Experience from the 1930s to 1970*. London: Imperial College Press and the Operational Research Society.

Kirby, M.W. (2007). Paradigm change in operations research: thirty years of debate. *Operations Research* 45: 1–13.

Kirby, M.W. and Rosenhead, J. (2005). IFORS' operational research hall of fame: Russell L. Ackoff. *International Transactions in Operational Research* 12: 129–134.

Klir, G.J. (2001). *Facets of Systems Science*, 2e. New York: Kluwer/Plenum.

Kriek, L., Lotriet, H., and Machdel, M. (2010). *An Assessment of the Application of Critical Systems Heuristics in the Discovery of the Emancipatory Value of South African Mobile Learning Projects*. Oxford: United Kingdom Systems Society.

Kuhn, T.S. (1970). *The Structure of Scientific Revolutions*, 2e. Chicago: University of Chicago Press.

Kuhn, L. (2009). *Adventures in Complexity: For Organisations Near the Edge of Chaos*. Axminster: Triarchy Press.

Kurtz, C.F. and Snowden, D.J. (2003). The new dynamics of strategy: sense-making in a complex and complicated world. *IBM Systems Journal* 42: 462–483.

Laloux, F. (2014). *Reinventing Organizations: A Guide to Creating Organizations Inspired by the Next Stage in Human Consciousness*. Brussels: Nelson Parker.

Lane, D.C. (1999). Social theory and system dynamics practice. *European Journal of Operational Research* 113: 501–527.

Lane, D.C. (2000). Should system dynamics be described as a 'Hard' or 'Deterministic' systems approach. *Systems Research and Behavioral Science* 17: 3–22.

Lane, D.C. (2001a). Rerum Cognoscere Causas: part 1 – how do the ideas of system dynamics relate to traditional social theories and the voluntarism/determinism debate? *System Dynamics Review* 17: 97–118.

Lane, D.C. (2001b). Rerum Cognoscere Causas: part 11 – opportunities generated by the agency/structure debate and suggestions for clarifying the social theoretic position of system dynamics. *System Dynamics Review* 17: 293–309.

Lane, D.C. (2006). IFORS' operational research hall of fame: Jay Wright Forrester. *International Transactions in Operational Research* 13: 483–492.

Lane, D.C. (2008). The emergence and use of diagramming in system dynamics: a critical account. *Systems Research and Behavioral Science* 25: 3–23.

Lane, D.C. (2016). 'Till the muddle in my mind have cleared Awa': can we help shape policy using systems modelling? *Systems Research and Behavioral Science* 33: 633–650.

Lane, D.C. and Husemann, E. (2008). Steering without Circe: attending to reinforcing loops in social systems. *System Dynamics Review* 24: 37–61.

Lane, D.C. and Oliva, R. (1998). The greater whole: towards a synthesis of system dynamics and soft systems methodology. *European Journal of Operational Research* 107: 214–235.

Lane, D.C. and Sterman, J.D. (2018). A model simulator: the lives of Jay W Forrester. *Journal of Simulation* 12: 90–97.

Lane, D.C., Munro, E., and Husemann, E. (2016). Blending systems thinking approaches for organisational analysis: reviewing child protection in England. *European Journal of Operational Research* 251: 613–623.

Larsen, R.L. (2011). Critical systems thinking for the facilitation of conservation planning in Philippine coastal management. *Systems Research and Behavioral Science* 28: 63–76.

Lawrence, J.R. (ed.) (1966). *Operational Research and the Social Sciences*. London: Tavistock.

Ledington, P. and Donaldson, J. (1997). Soft OR and management practice: a study of the adoption and use of soft systems methodology. *Journal of the Operational Research Society* 48: 229–240.

Leeman, J.E. (2002). Applying interactive planning at DuPont. *Systems Practice and Action Research* 15: 85–109.

Lem, S. (1975). *The Cyberiad: Fables for the Cybernetic Age*. London: Secker & Warburg.

Leonard, A. (1996). Team syntegrity: a new methodology for group work. *European Management Journal* 14: 407–413.

Leonard, A. (2008). Integrating sustainability practices using the viable system model. *Systems Research and Behavioral Science* 25: 643–654.

Leonard, A. (2009). The viable system model and its application to complex organizations. *Systemic Practice and Action Research* 22: 223–233.

Lesmoir-Gordon, N., Rood, W., and Edney, R. (2009). *Introducing Fractals: A Graphic Guide*. London: Icon Books.

Levi-Strauss, C. (1968). *Structural Anthropology*. Penguin: Harmondsworth.

Lewens, T. (2015). *The Meaning of Science*, Kindle Edition. Louisiana: Pelican.

Lewin, K. (1967). Feedback problems of social diagnosis and action. In: *Modern Systems Research for the Behavioral Sciences* (ed. W. Buckley), 441–444. Chicago: Aldine.

Li, Y. and Zhu, Z. (2014). Soft OR in China: a critical report. *European Journal of Operational Research* 239: 427–434.

Lilienfeld, R. (1978). *The Rise of Systems Theory: An Ideological Analysis*. New York: Wiley.

Linstone, H.A. (1984). *Multiple Perspectives for Decision-making: Bridging the Gap Between Analysis and Action*. New York: North-Holland.

Lloyd, H. (2018). Personal communication.

Luckett, K. (2006). An assessment of the application of 'Critical Systems Heuristics' to a policy development process. *Systems Practice and Action Research* 19: 503–521.

Luhmann, N. (1989). *Ecological Communication*. Cambridge: Polity Press.

Luhmann, N. (2006a). The society of society. In: *Luhmann Explained: From Souls to Systems*, Kindle Edition (ed. H.-G. Moeller), 238–259. Chicago and La Salle: Open Court.

Luhmann, N. (2006b). Beyond barbarism. In: *Luhmann Explained: From Souls to Systems*, Kindle Edition (ed. H.-G. Moeller), 260–271. Chicago and La Salle: Open Court.

Luhmann, N. (2013). *Introduction to Systems Theory*. Cambridge: Polity Press.

Lukes, S. (1974). *Power: A Radical View*. London: Macmillan.

Lumbo, D.A. (2007). Applications of interactive planning methodology. MSc dissertation. Philadelphia, PA: University of Pennsylvania.

Luoma, J., Hamalainen, R., and Saarinen, E. (2011). Acting with systems intelligence: integrating complex responsive processes with the systems perspective. *Journal of the Operational Research Society* 62: 3–11.

Lyotard, J.-F. (1984). *The Postmodern Condition: A Reort on Knowledge*. Manchester: Manchester University Press.

Lyytinen, K.J. and Klein, H.K. (1985). The poverty of scientism in information systems. In: *Research Methods in Information Systems* (ed. E. Mumford et al.), 131–161. Amsterdam: North-Holland.

Maani, K. (2017). *Multi-Stakeholder Decision Making for Complex Problems*. Singapore: World Scientific.

Maani, K.E. and Cavana, R.Y. (2000). *Systems Thinking and Modelling*. New Zealand: Pearson Education.

MacDonald, I. (2008). *Revolution in the Head: The Beatles' Records and the Sixties*. London: Vintage.

Magnus, E., Knudtsen, M.S., Wist, G. et al. (2016). The search conference as a method in planning community health promotion actions. *Journal of Public Health Research* 5 (2): 621.

Malik Management (2018). Solutions in the public sector. https://malik-management.com/malik-solutions/solutions-in-the-public-sector (accessed 17 August 2018).

Mankell, H. (2016). *Quicksand: What It Means to Be a Human Being*, Kindle Edition. London: Harvill Secker.

Marsh, H. (2014). *Do No Harm: Stories of Life, Death and Brain Surgery*. London: Weidenfeld & Nicolson.

Martin-Cruz, N., Martin-Perez, V., Perez-Rios, J.M., and Velasco-Jimenez, I. (2014). Team syntegrity as a tool for efficient teamwork: an experimental evaluation in a business simulation. *Systems Research and Behavioral Science* 31: 215–226.

Maruyama, M. (1968). The second cybernetics: deviation-amplifying mutual causal processes. In: *Modern Systems Research for the Behavioral Scientist* (ed. W. Buckley), 304–313. Chicago: Aldine.

Marx, K. (1961). *Capital*. Moscow: Foreign Languages Publishing House.

Marx, K. (1973). *Surveys from Exile: The Eighteenth Brumaire of Louis Bonaparte*. Harmondsworth: Penguin.

Marx, K. (1975a). Economic and philosophical manuscripts of 1844. In: *Marx: Early Writings*, 279–400. Harmondsworth: Penguin.

Marx, K. (1975b). Theses on Feuerbach. In: *Marx Early Writings*, 423. Harmondsworth: Penguin.

Marx, K. and Engels, F. (1967). *The Communist Manifesto*. Harmondsworth: Penguin.

Mason, R.O. (1969). A dialectical approach to strategic planning. *Management Science* 15: B403–B414.

Mason, R.O. and Mitroff, I.I. (1981). *Challenging Strategic Planning Assumptions: Theory, Cases and Techniques*. Chichester: Wiley.

Mason, R.O. and Mitroff, I.I. (2014). Charles West Churchman - philosopher of management. *Journal of Management Inquiry* 23: 37–46.

Mathews, S. (2005). The Fun Palace: Cedric Price's experiment in architecture and technology. *Technoetic Arts: A Journal of Speculative Research* 3: 73–91.

Maturana, H.R. (1975). The organization of the living: a theory of the living organization. *International Journal of Man-Machine Studies* 7: 313–332.

Maturana, H.R. (1987). The biological foundations of self-consciousness and the physical domain of existence. In: *Physics of Cognitive Processes: Amalfi 1986* (ed. E.R. Caianiello), 324–380. Singapore: World Scientific.

Maturana, H.R. and Varela, F.J. (1980). *Autopoiesis and Cognition: The Realization of the Living*. Dordrecht: Reidel.

Maturana, H.R. and Varela, F.J. (1992). *The Tree of Knowledge: The Biological Roots of Human Understanding*. Boston, MA: Shambhala.

Maxwell, J.C. (1868). On Governors. *Proceedings of the Royal Society of London* 16 (100): 270–283.

May, R.M. (1974). Biological populations with nonoverlapping generations: stable cycles and chaos. *Science* 186: 645–647.

Mayr, O. (1975). *The Origins of Feedback Control*. Cambridge: MIT Press.

McCann, B. (2012). Wastewater reuse brings life back to Spain's Segura river. *Water* 21: 28–33.

McCarthy, T.A. (1973). A theory of communicative competence. *Philosophy of the Social Sciences* 135–156.

McIntyre-Mills, J. (2003). *Critical Systemic Praxis for Social and Environmental Justice: Participatory Policy Design and Governance for a Global Age*, Kindle Edition. New York: Kluwer/Plenum.

McIntyre-Mills, J. (ed.) (2006a). *C West Churchman and Related Works Series: Rescuing the Enlightenment from Itself*. New York: Springer.

McIntyre-Mills, J. (ed.) (2006b). *C West Churchman and Related Works Series: Systemic Governance and Accountability, Working and Re-Working the Conceptual and Spatial Boundaries*. New York: Springer.

McLaren, P. and Leonard, P. (1993). *Paolo Freire: A Critical Encounter*. London: Routledge.

Meadows, D.H. (2008). *Thinking in Systems: A Primer*. Vermont: Chelsea Green.

Meadows, D.H., Meadows, D.L., Randers, J., and Behrens, W.W. III (1972). *The Limits to Growth*. New York: Universe Books.

Meadows, D.H., Randers, J., and Meadows, D.L. (2004). *The Limits to Growth: The 30-Year Update*. London: Routledge.

Mears-Young, B. and Jackson, M.C. (1997). Integrated logistics – call in the revolutionaries! *Omega* 25: 605–618.

Medina, E. (2014). *Cybernetic Revolutionaries: Technology and Politics in Allende's Chile*. Cambridge: MIT Press.

Mejia, A. (2001). *The Problem of Knowledge Imposition: Paulo Freire and Critical Systems Thinking, Research Memorandum, 29*. Hull: Business School, University of Hull.

Mejia, A.D. and Espinosa, A. (2007). Team syntegrity as a learning tool: some considerations about its capacity to promote critical learning. *Systems Research and Behavioral Science* 24: 27–35.

Mensch, J. (2013). *Kant's Organicism: Epigenesis and the Development of Critical Philosophy*, Kindle Editon. Chicago: University of Chicago Press.

Metcalfe, S. (2017). The big idea that defines our era. *The Guardian* (19th August), pp. 29–31.

Middleton, P. (ed.) (2010). *Delivering Public Services That Work*, vol. 1. Axminster: Triarchy Press.

Midgley, G. (1989). Critical systems and the problem of pluralism. *Cybernetics and Systems* 20: 219–231.

Midgley, G. (1990). Creative methodology design. *Systemist* 12: 108–113.

Midgley, G. (1992). The sacred and profane in critical systems theory. *Systems Practice* 5: 5–16.

Midgley, G. (1993). A reply to Haridimos Tsoukas, the radical critic of radical critique. *Systems Practice* 6: 301–309.

Midgley, G. (1996). What is this thing called CST? In: *Critical Systems Thinking: Current Research and Practice* (ed. R.L. Flood and N.R. Romm), 11–24. New York: Plenum.

Midgley, G. (1997a). Dealing with coercion: critical systems heuristics and beyond. *Systems Practice* 10: 37–57.

Midgley, G. (1997b). Mixing methods: developing systemic intervention. In: *MultiMethodology* (ed. J. Mingers and A. Gill), 291–332. Chichester: Wiley.

Midgley, G. (2000). *Systemic Intervention: Philosophy, Methodology, and Practice*. New York: Kluwer/Plenum.

Midgley, G. (ed.) (2002). *Systems Thinking*, vol. 1–4. London: Sage.

Midgley, G. (2014). *Systemic Intervention, Research Memorandum 95*. Hull: Centre for Systems Studies, Hull University Business School.

Midgley, G. and Ochoa-Arias, A. (eds.) (2004). *Community Operational Research: OR and Systems Thinking for Community Development*. New York: Kluwer/Plenum.

Midgley, G. and Pinzon, L.A. (2011). Boundary critique and its implications for conflict prevention. *Journal of the Operational Research Society* 62: 1543–1554.

Midgley, G. and Richardson, K. (2007). Systems thinking for community involvement in policy analysis. *Emergence: Complexity and Organization* 9: 167–183.

Midgley, G., Munlo, I., and Brown, M. (1997). *Sharing Power*. Bristol: Policy Press.

Midgley, G., Munlo, I., and Brown, M. (1998). The theory and practice of boundary critique: developing housing services for older people. *Journal of the Operational Research Society* 49: 467–478.

Miller, J.G. (1978). *Living Systems*. New York: McGraw-Hill.

Miller, P. (2011). *The Smart Swarm: How to Work Efficiently, Communicate Effectively, and Make Better Decisions Using the Secrets of Flocks, Schools, and Colonies*. New York: Avery.

Miller, J.G. and Miller, J.L. (1990). Introduction: the nature of living systems. *Behavioral Science* 35: 157–163.

Miller, J.G. and Miller, J.L. (1995). Applications of living systems theory. *Systemic Practice and Action Research* 8: 19–45.

Miller, J.H. and Page, S.E. (2007). *Complex Adaptive Systems: An Introduction to Computational Models of Social Life*. Princeton, NJ: Princeton University Press.

Mingers, J. (1980). Towards an appropriate social theory for applied systems thinking: critical theory and soft systems methodology. *Journal of Applied Systems Analysis* 7: 41–49.

Mingers, J. (1984). Subjectivism and soft systems methodology – a critique. *Journal of Applied Systems Analysis* 11: 85–104.

Mingers, J. (1992). SSM and information systems: an overview. *Systemist* 14: 82–88.

Mingers, J. (1995). *Self-Producing Systems: Implications and Applications of Autopoiesis*. New York: Plenum.

Mingers, J. (1997a). Systems typologies in the light of autopoiesis: a reconceptualization of boulding's hierarchy and a typology of self-referential systems. *Systems Research and Behavioral Science* 14: 303–313.

Mingers, J. (1997b). Multi-paradigm multimethodology. In: *MultiMethodology* (ed. J. Mingers and A. Gill) 1–17. Chichester: Wiley.

Mingers, J. (1997c). Towards critical pluralism. In: *MultiMethodology*, 407–440. Chichester: Wiley.

Mingers, J. (2000). An idea ahead of its time: the history and development of soft systems methodology. *Systemic Practice and Action Research* 13: 733–755.

Mingers, J. (2006). *Realizing Systems Thinking: Knowledge and Action in Management Science*. New York: Springer.

Mingers, J. (2011). Soft O.R. comes of age – but not everywhere. *Omega* 39: 729–741.

Mingers, J. (2014). *Systems Thinking, Critical Realism and Philosophy*, Kindle Edition. London: Routledge.

Mingers, J. (2015). Helping business schools engage with real problems: the contribution of critical realism and systems thinking. *European Journal of Operational Research* 242: 316–331.

Mingers, J. and Brocklesby, J. (1996). Multimethodology: towards a framework for critical pluralism. *Systemist* 18: 101–132.

Mingers, J. and Brocklesby, J. (1997). Multimethodology: towards a framework for mixing methodologies. *Omega* 25: 489–509.

Mingers, J. and Gill, A. (eds.) (1997). *MultiMethodology: The Theory and Practice of Combining Management Science Methodologies*. Chichester: Wiley.

Mingers, J. and Rosenhead, J. (2004). Problem structuring methods in action. *European Journal of Operational Research* 152: 530–554.

Mingers, J. and Taylor, S. (1992). The use of soft systems methodology in practice. *Journal of the Operational Research Society* 43: 321–332.

Mingers, J. and White, L. (2010). A review of the recent contribution of systems thinking to operational research and management science. *European Journal of Operational Research* 207: 1147–1161.

Miser, H.J. (ed.) (1995). *Handbook of Systems Analysis: Cases*. New York: Wiley.

Miser, H.J. and Quade, E.S. (eds.) (1985). *Handbook of Systems Analysis: Overview of Uses, Procedures, Applications, and Practice*. New York: North-Holland.

Miser, H.J. and Quade, E.S. (eds.) (1988). *Handbook of Systems Analysis: Craft Issues and Procedural Choices*. New York: Wiley.

Mitchell, M. (2009). *Complexity: A Guided Tour*. Oxford: Oxford University Press.

Mitroff, I.I. and Linstone, H.A. (1993). *Unbounded Mind: Breaking the Chains of Traditional Business Thinking*. Oxford: Oxford University Press.

Mitroff, I.I. and Mason, R.O. (1987). Dialectical pragmatism: a progress report on an interdisciplinary program of research on dialectical inquiring systems. In: *Decision Making About Decision Making: Metamodels and Metasystems* (ed. J.P. van Gigch), 137–147. Cambridge: Abacus Press.

Mitroff, I.I., Barabba, C.P., and Kilmann, R.H. (1977). The application of behavioral and philosophical techniques to strategic planning: a case study in a Large Federal Agency. *Management Science* 24: 44–58.

Mitroff, I.I., Emshoff, J.R., and Kilmann, R.H. (1979). Assumption analysis: a methodology for strategic problem-solving. *Management Science* 25: 583–593.

Moeller, H.-G. (2006). *Luhmann Explained: From Souls to Systems*. Chicago: Open Court.

Moeller, H.-G. (2012). *The Radical Luhmann*, Kindle Edition. New York: Columbia University Press.

Molloy, K.J. and Best, D.P. (1980). *The Checkland Methodology Considered as a Theory Building Methodology*, 17. Washington, DC: Hemisphere et seq.

Monbiot, G. (2014). Drowning in money: the pig-headed policies that make flooding inevitable. *The Guardian* (14 January), p. 30.

Morecroft, J. (2010). System dynamics. In: *Systems Approaches to Managing Change: A Practical Guide* (ed. M. Reynolds and S. Holwell), 25–84. London: Springer.

Morecroft, J.W. and Sterman, J.D. (eds.) (1994). *Modelling for Learning Organizations*. Portland: Productivity Press.

Morgan, G. (ed.) (1983). *Beyond Method*. Beverly Hills, CA: Sage.

Morgan, G. (1986). *Images of Organization*. Beverly Hills, CA: Sage.

Morgan, G. (1997). *Images of Organization*, 2e. Beverly Hills, CA: Sage.

Morieux, Y. and Tollman, P. (2014). *Six Simple Rules: How to Manage Complexity Without Getting Complicated*. Boston, MA: Harvard Business Review Press.

Morozov, E. (2014). The planning machine: project cybersyn and the origins of the big data nation. *The New Yorker* (13 October).

Morris, J. (1983). The brain, the heart and the big toe. *Creativity and Innovation Network* 9: 25–30.

Morrison, J.M. (2009). Russell L. Ackoff, A Scholar who cared. *Philadelphia Daily News* (4 November).

Müller, K.H. and Riegler, A. (2016). Mapping the varieties of second-order cybernetics. *Constructivist Foundations* 11: 443–454.

Mumford, E. (1983). *Designing Participatively: Participative Approach to Computer Systems Design*. Manchester: Manchester University Press.

Mumford, E. (2003). *Redesigning Human Systems*. Hershey, PA: Information Science Publishers.

Mumford, E. (2006). The story of socio-technical design: reflections on its success, failures and potential. *Information Systems Journal* 16: 317–342.

Munro, I. (1997). An exploration of three emancipatory themes within OR and systems thinking. *Journal of the Operational Research Society* 48: 576–584.

Munro, E. (2010). *The Munro Review of Child Protection Part 1: A Systems Analysis*. London: TSO.

Munro, E. (2011). *The Munro Review of Child Protection: Final Report - A Child-centred System*. London: TSO.

Munro, I. and Mingers, J. (2002). The use of multimethodology in practice – results from a survey of practitioners. *Journal of the Operational Research Society* 53: 369–378.

Naughton, J. (1977). *The Checkland Methodology: A Reader's Guide*, 2e. Milton Keynes: Open University Systems Group.

Nelson, H.G. (2003). The Legacy of C. West Churchman: a framework for social systems assessments. *Systems Research and Bevavioral Science* 20: 463–473.

Netland, T.H., Knutstad, G., Ravn, J.E. et al. (2009). STS is Dead - Live STS! : Emphasising the Need for a Modern Sociotechnical System Approach on High-Tech Production Systems. Orlando: Abstract number 011–0226.

NHC (2006). *A Systematic Approach to Service Improvement - An Update: Evaluating the Sustainability of Systems Thinking in Housing*. Sunderland: NHC.

Nobel Prize Committee (1977). Press Release on the 1977 Nobel Prize.

Noble, D. (2007). UK must get with the systems to stay in front. *Times Higher* (2 February).

Noble, D. (2008). The man who mapped the heart. *Guardian* (16 December), p. 9.

O'Donovan, B. (2014). Editorial for special issue of SPAR: the Vanguard method in a systems thinking context. *Systemic Practice and Action Research* 27: 1–20.

O'Donovan, B. and Zokaei, K. (2011). Improving performance throughout a housing supply chain: portsmouth City Council's systems thinking transformation. In: *Systems Thinking: From Heresy to Practice* (ed. K. Zokaei et al.), 84–107. London: Palgrave Macmillan.

ODPM (2005). *A Systematic Approach to Service Improvement: Evaluating Systems Thinking in Housing*. London: ODPM Publications.

OECD (2017). *Systems Approaches to Public Sector Challenges: Working with Change*. Paris: OECD Publishing.

Ohno, T. (1988). *The Toyota Production System*. New York: Productivity Press.

Oliga, J.C. (1988). Methodological foundations of systems methodologies. *Systems Practice* 11: 87–112.

Open University (2016). *Systems Engineering: Challenging Complexity*, OU Course T837_1. Milton Keynes: Open University Press.

Ormerod, R. (1992). Combining hard and soft systems practice. *Systemist* 14: 160–165.

Ormerod, R. (1994). Combining management consultancy and research. *Systemist* 16: 41–53.

Ormerod, R. (1995). Putting soft OR methods to work: information systems strategy development at Sainsbury's. *Journal of the Operational Research Society* 46: 277–293.

Ormerod, R. (1996). Combining management consultancy and research. *Omega* 24: 1–12.

Ormerod, R. (1997a). Mixing methods in practice: a transformation-competence approach. In: *MultiMethodology* (ed. J. Mingers and A. Gill), 29–58. Chichester: Wiley.

Ormerod, R. (1997b). The design of organisational intervention: choosing the approach. *Omega* 25: 415–435.

Ormerod, R. (2006). The history and ideas of pragmatism. *Journal of the Operational Research Society* 57: 892–909.

Ormerod, R. (2011). The relationship between operational research and systems thinking. *Journal of the Operational Research Society* 62: 242–245.

Ortegón-Monroy, M.C. (2003). Chaos and complexity theory in management: an exploration from a critical systems thinking perspective. *Systems Research and Behavioral Science* 20: 387–400.

Owen, J. and Inman, R. (2017). Operational research at general motors. *Impact*, Spring 26–31.

Parker, M. (2014). University, Ltd: changing a business school. *Organization* 21: 281–292.

Parker, L. (2016). Natural selection. *The Guardian* (25 October), pp. 29–31.

Parsons, T. (1956). Suggestions for a sociological approach to the study of organizations-1. *Administrative Science Quarterly* 1: 63–85.

Parsons, T. (1960). *Structure and Process in Modern Society*. New York: Free Press.

Parsons, T. and Smelser, N.L. (1956). *Economy and Society*. London: Routledge & Kegan Paul.

Pasmore, W., Francis, C., Haldeman, J., and Shani, A. (1982). Socio-technical systems: a North American reflection on empirical studies of the seventies. *Human Relations* 35: 1179–1204.

Passmore, J. (1970). *A Hundred Years of Philosophy*. Harmondsworth: Penguin.

Paucar-Caceres, A. (2011). The development of management sciences/operational research discourses: surveying the trends in the US and the UK. *Journal of the Operational Research Society* 62: 1452–1470.

Pauli, G.S. (2010). *The Blue Economy: 10 Years, 100 Innovations, 100 Million Jobs*. Brookline, MA: Paradigm Publications.

Pearson, A. (1994). You drive for show but you putt for dough: a facilitator's perspective. In: *Beyond Dispute: The Invention of Team Syntegrity* (ed. S. Beer), 313–322. Chichester: Wiley.

Pell, C. (ed.) (2012). *Delivering Public Services that Work*, vol. 2. Axminster: Triarchy Press.

Pepper, S.C. (1942). *World Hypotheses: A Study in Evidence*. Berkeley, CA: University of California.

Pettit, P. (1977). *The Concept of Structuralism: A Critical Guide*. Oakland, CA: University of California Press.

Pfiffner, M. (2001). *Team Syntegrity: Using Cybernetics for Opinion Forming in Organizations*. St. Gallen: Malik Management.

Pfiffner, M. (2004). *From Workshop to Syntegration: The Genetic Code of Effective Communication*. St. Gallen: Malik Management.

Phillips, D.C. (1976). *Holistic Thought in Social Science*. Stanford, CA: Stanford University Press.

Piaget, J. (1973). *Main Trends in Interdisciplinary Research*. London: George Allen & Unwin.

Pickering, A. (2009). Beyond design: cybernetics, biological computers and hylozoism. *Synthese* 168: 469–491.

Pickering, A. (2010). *The Cybernetic Brain: Sketches of Another Future*, Kindle Edition. Chicago: University of Chicago Press.

Pirsig, R.M. (1974). *Zen and the Art of Motorcycle Maintenance*. London: Bodley Head.

Plato (1999). *The Essential Plato*. The Softback Preview.

Poerksen, B. (2003). 'At each and every moment, I can decide who I am': Heinz von Foerster on the observer, dialogic life, and a constructivist philosophy of distinctions. *Cybernetics and Human Knowing* 10: 9–26.

Pollack, J. (2009). Multimethodology in series and parallel: strategic planning using hard and soft OR. *Journal of the Operational Research Society* 60: 156–167.

Pourdehnad, J. and Hebb, A. (2002). Redesigning the academy of vocal arts (AVA). *Systems Research and Behavioral Science* 19: 331–338.

Pouvreau, D. (2014). The hermeneutical system of general systemology: bertalanffian and other early contributions to its foundations and development. In: *Traditions of Systems Theory: Major Figures and Contemporary Developments* (ed. D. Arnold), 81–136. New York: Routledge.

Prigogine, I. (1976). Order through fluctuation: self-organization and social system. In: *Evolution and Consciousness: Human Systems in Transition* (ed. C.H. Waddington and E. Jantsch), 93–130. Reading, MA: Addison-Wesley.

Prigogine, I. (1997). *The End of Certainty: Time, Chaos and the New Laws of Nature*. New York: The Free Press.

Prigogine, I. and Stengers, I. (1984). *Order Out of Chaos: Man's New Dialogue with Nature*. New York: Bantam Books.

Probst, G. and Bassi, A.M. (2014). *Tackling Complexity: A Systemic Approach for Decision-Makers*. Sheffield: Greenleaf.

Pruyt, E., Auping, W.L., and Kwakkel, J.H. (2015). Ebola in West Africa: model-based exploration of social psychological effects and interventions. *Systems Research and Behavioral Science* 32: 2–14.

Quade, E.S. and Miser, H.J. (1980). The context, nature and use of systems analysis. Laxenburg, Austria: IIASA, Working Paper WP-80-058.

Rajagopalan, R. and Midgley, G. (2015). Knowing differently in systemic intervention. *Systems Research and Behavioral Science* 32: 546–561.

Ramage, M. and Shipp, K. (2009). *Systems Thinkers*. London: Springer.

Ranyard, J.C. (2001). Editorial. *Journal of the Operational Research Society* 52: 1–3.

Ranyard, J.C., Fildes, R., and Hu, T. (2015). Reassessing the scope of O.R. practice: the influences of problem structuring methods and the analytics movement. *European Journal of Operational Research* 245: 1–13.

Rapley, C. (2016). The Anthropocene epoch: scientists declare dawn of human-influenced age. *The Guardian* (29 August).

Rapoport, A. (1968). Foreward. In: *Modern Systems Research for the Behavioral Scientist* (ed. W. Buckley), xiii–xxii. Chicago: Aldine.

Rapoport, R.N. (1970). Three dilemmas in action research. *Human Relations* 23: 499–513.

Reed, M. (1985). *Redirections in Organizational Analysis*. London: Tavistock.

Rehm, R. and Cebula, N. (1996). The Search Conference Method for Participative Planning (Adapted from 'The Search Conference: State of the Art' by Merrelyn Emery). *Elements UK - Library of Articles*, pp. 1–12.

Remington, K. and Pollack, J. (2007). *Tools for Complex Projects*. Aldershot: Gower.

Restrepo, M.J., Lelea, M.A., and Kaufmann, B. (2016). Second-order cybernetic analysis to reconstruct farmers' rationale when regulating milk production. *Systemic Practice and Action Research* 29: 449–468.

ReThink Health (2018). www.rethinkhealth.org (accessed 20 March 2018).

Reynolds, M. and Holwell, S. (eds.) (2010). *Systems Approaches to Managing Change: A Practical Guide*. New York: Springer.

Rice, A.K. (1958). *Productivity and Social Organization*. London: Tavistock Publications.

Rice, A.K. (1963). *The Enterprize and Its Environment*. London: Tavistock Publications.

Richardson, G. (1991). *Feedback Thought in Social Science and Systems Theory*. Philadelphia, PA: University of Pennsylvania Press.

Riswanda, Corcoran-Nantes, Y., and McIntyre-Mills, J. (2016). Re-framing prostitution in Indonesia: a critical systemic approach. *Systemic Practice and Action Research* 29: 517–539.

Riswanda, McIntyre-Mills, J., and Corcoran-Nantes, Y. (2017). Prostitution and human rights in Indonesia: a critical systemic review of policy discourses and scenarios. *Systemic Practice and Action Research* 30: 213–237.

Ritchie, C., Taket, A., and Bryant, J. (eds.) (1994). *Community Works*. Sheffield: Pavic Publications.

Rittel, H.W.J. and Webber, M.M. (1981). Dilemmas in a general theory of planning. In: *Systems Thinking*, vol. 2 (ed. F.E. Emery), 81–102. Harmondsworth: Penguin.

Roberts, A. (2015). *The Thing Itself*, Kindle Edition. London: Gollancz.

Robertson, B.J. (2016). *Holacracy: The Revolutionary Management System that Abolishes Hierarchy*. Harmondsworth: Penguin.

Robertson, M.M., Hettinger, L.J., Waterson, P.E. et al. (2015). Sociotechnical approaches to workplace safety: research needs and opportunities. *Ergonomics* 58: 650–658.

Rodenas, M.A. and Albacete, M. (2014). The river Segura: reclaimed water, recovered River. *Journal of Water Reuse and Desalination* 4: 50–57.

Roethlisberger, F.J. and Dickson, W.J. (1939). *Management and the Worker*. Cambridge: Harvard University Press.

Rose, S. (2017). On growth and form by D'Arcy Wentworth Thompson. *The Guardian Review* (22 July), p. 8.

Rosenblueth, A., Wiener, N., and Bigelow, J. (1968). Behavior, purpose and teleology. In: *Modern Systems Research for the Behavioral Scientist* (ed. W. Buckley), 221–225. Chicago: Aldine.

Rosenhead, J. (1976). Some further comments on 'The Social Responsibility of OR'. *Operational Research Quarterly* 17: 266–272.

Rosenhead, J. (1981). Operational research in urban planning. *Omega* 9: 345–364.

Rosenhead, J. (1984). Debating systems methodology: conflicting ideas about conflict and ideas. *Journal of Applied Systems Analysis* 11: 79–84.

Rosenhead, J. (1989a). Introduction: old and new paradigms of analysis. In: *Rational Analysis for a Problematic World* (ed. J. Rosenhead), 1–20. Chichester: Wiley.

Rosenhead, J. (ed.) (1989b). *Rational Analysis for a Problematic World*. Chichester: Wiley.

Rosenhead, J. (1998). Complexity theory and management practice. London: London School of Economics, Working Paper, 98.25.

Rosenhead, J. (2006). IFORS' operational research hall of fame: Stafford beer. *International Transactions in Operational Research* 13: 577–581.

Rosenhead, J. (2009). Reflections on fifty years of operational research. *Journal of the Operational Research Society* 60: S5–S15.

Rosenhead, J. and Mingers, J. (2001a). A new paradigm of analysis. In: *Rational Analysis for a Problematic World Revisited* (ed. J. Rosenhead and J. Mingers), 1–19. Chichester: Wiley.

Rosenhead, J. and Mingers, J. (eds.) (2001b). *Rational Analysis for a Problematic World Revisited*. Chichester: Wiley.

Rosenhead, J. and Thunhurst, C. (1982). A materialist analysis of operational research. *Journal of the Operational Research Society* 33: 111–122.

Rousseau, D. (2017a). Systems research and the quest for scientific systems principles. *Systems* 5: 25–33.

Rousseau, D. (2017b). Three general systems principles and their derivation: insights from the philosophy of science applied to systems concepts. In: *Disciplinary Convergence: Implications for Systems Engineering Research*, (ed. A.M. Madni et al.), Los Angeles, CA: Springer.

Rousseau, D., Wilby, J., Billingham, J., and Blachfellner, S. (2016). Manifesto for general systems transdisciplinarity. *Systema* 4: 4–14.

Rousseau, D., Wilby, J., Billingham, J., and Blachfellner, S. (2018). *General Systemology: Transdisciplinarity for Discovery, Insight and Innovation*. Singapore: Springer.

Rovelli, C. (2015). *Seven Brief Lessons on Physics*, Kindle Edition. Harmondsworth: Penguin.

Royston, G. (2013). Operational research for the real world: big questions from a small island. *Journal of the Operational Research Society* 64: 793–804.

Ruegg-Sturm, J. (2005). *The New St. Gallen Management Model: Basic Categories of an Approach to Integrated Management*. Basingstoke: Palgrave Macmillan.

Sambo, L. (2009). Health systems thinking: the need for a more critical approach. PhD thesis. Hull: University of Hull.

Santa Fe Institute (2017). Our mission. www.santafe.edu (accessed 6 June 2018).

Sardar, Z. and Abrams, I. (2008). *Introducing Chaos: A Graphic Guide*. London: Icon Books.

Sartre, J.-P. (1975). Sartre at seventy: an interview. *The New York Review* (7 August).

Schecter, D. (1991). Critical systems thinking in the 1980s: a connective summary. In: *Critical Systems Thinking: Directed Readings* (ed. R.L. Flood and M.C. Jackson), 213–226. Chichester: Wiley.

Schoderbeck, P.P., Schoderbeck, C.G., and Kefalas, A.G. (1985). *Management Systems: Conceptual Considerations*, 3e. Dallas, TX: Business Publications.

Schwaninger, M. (1997). Self organization and self reference in the cognition of organizations. In: *Interdisciplinary Approaches to a New Understanding of Cognition and Consciousness*, (ed. V. Braitenberdg et al.), 70–80. Augsburg: Unipress.

Schwaninger, M. (2001). System theory and cybernetics: a solid basis for transdisciplinarity in management education and research. *Kybernetes* 30: 1209–1222.

Schwaninger, M. (ed.) (2006). *Intelligent Organizations: Powerful Models for Systemic Management*. New York: Springer.

Schwaninger, M. (2009). System dynamics in the evolution of the systems approach. In: *Encyclopedia of Complexity and Systems Science*, 8974–8980. New York: Springer.

Schwaninger, M. (2018). Governance for intelligent organizations: a cybernetic contribution. *Kybernetes* https://doi.org/10.1108/K-01-2018-0019.

Schwaninger, M. and Scheef, C. (2016). A test of the viable system model: theoretical claim vs empirical evidence. *Cybernetics and Systems: An International Journal* 47: 544–569.

Scott, W.R. (1987). *Organizations: Rational, Natural, and Open Systems*, 2e. Englewood Cliffs, NJ: Prentice-Hall.

Seddon, J. (2005). *Freedom from Command and Control: A Better Way to Make the Work Work*, 2e. Buckingham: Vanguard Press.

Seddon, J. (2008). *Systems Thinking in the Public Sector*. Axminster: Triarchy Press.

Seddon, J. (2014). *The Whitehall Effect*. Axminster: Triarchy Press.

Seddon, J. (2017a). Universal credit. *Vanguard News*. Autumn.

Seddon, J. (2017b). No beds today. *Vanguard News* (6 March).

Seddon, J., O'Donovan, B., and Zokaei, K. (2011). Rethinking lean service. *Service Design and Delivery* 41–60.

Seligman, I. (2016). *Lines of Thought: Drawing from Michelangelo to Now*. London: Thames & Hudson, in collaboration with the British Museum.

Selznick, P. (1948). Foundations of the theory of organization. *American Sociological Review* 13: 25–35.

Senge, P. (1990). *The Fifth Discipline: The Art and Practice of the Learning Organization*. London: Random House.

Senge, P. (ed.) (1994). *The Fifth Discipline Fieldbook*. London: Century.

Senge, P. and Sterman, J.D. (1994). Systems thinking and organizational learning: acting locally and thinking globally in the organization of the future. In: *Modeling for Learning Organizations* (ed. J.D.W. Morecroft), 195–216. Portland: Productivity Press.

Shannon, C.E. and Weaver, W. (1949). *The Mathematical Theory of Communication*. Urbana: University of Illinois Press.

Sica, A. (1981). Review of 'The systems approach and its enemies'. *American Journal of Sociology* 87: 208–211.

Silverman, D. (1970). *The Theory of Organisations*. London: Heinemann.

Skinner, Q. (ed.) (1985). *The Return of Grand Theory in the Human Sciences*. Cambridge: Cambridge University Press.

Smagt, T.v.d. (2006). Causation and constitution in system dynamics: modelling a socially constituted world. *Systems Research and Behavioral Science* 23: 513–524.

Smith, J.M. (1984). Rottenness is all. *London Review of Books* (3 May), pp. 1–6.

Smith, P.A.C. and Pourdehnad, J. (2018). *Organizational Leadership for the Fourth Industrial Revolution: Emerging Research and Opportunities*. Hershey: IGI Global.

Snowden, D.J. and Boone, M.E. (2007). A Leader's framework for decision making. *Harvard Business Review* 69–76.

Spanish Society of Civil Engineers (2011). *Integrated Urban Water Reclamation and Reuse System in the Murcia Region, Spain: 'Acueducto de Segovia de Obra Civil y Medio Ambiente' Prize*. Murcia: Region de Murcia, Consejeria de Agricultra y Agua.

Spaul, M. (1997). Multimethodology and critical theory: an intersection of interests? In: *MultiMethodology* (ed. J. Mingers and A. Gill), 323–346. Chichester: Wiley.

Spencer, H. (1969). *Principles of Sociology*. London: Macmillan.

SPRU (1973). *Models of Doom: A Critique of the Limits to Growth*. London: Universe Books.

Stacey, R.D. (1992). *Managing Chaos*. London: Sage.

Stacey, R.D. (1996). *Complexity and Creativity in Organizations*. San Francisco, CA: Berret-Kohler.

Stacey, R.D. (2003). *Strategic Management and Organisational Dynamics*, 4e. Harlow: Pearson Education.

Stacey, R.D. and Mowles, C. (2016). *Strategic Management and Organisational Dynamics: The Challenge of Complexity to Ways of Thinking about Organisations*, 7e. Harlow: Pearson.

Stacey, R.D., Griffin, D., and Shaw, P. (2000). *Complexity and Management: Fad or Radical Challenge to Systems Thinking*. London: Routledge.

Stahl, B.C. (2007). ETHICS, morality and critique: an essay on Enid Mumford's socio-technical approach. *Journal of the Association for Information Systems* 8: 479–490.

Stephens, A. (2013). *Ecofeminism and Systems Thinking*, Kindle Edition. London: Routledge.

Stephens, A., Taket, A., and Gagliano, M. (2018). Ecological justice for nature in CST. *Systems Research and Behavioral Science*, 36.1, forthcoming.

Sterman, J.D. (2000). *Business Dynamics: Systems Thinking and Modeling for a Complex World*. New York: Irwin/McGraw-Hill.

Sterman, J.D. (2003). Learning in and about complex systems. In: *Systems Thinking*, vol. 3 (ed. G. Midgley), 330–364. London: Sage.

Stern, S. (2009). Fond farewell to a brilliant thinker. *Financial Times* (9 November).

Stowell, F.A. (2016). Soft not vague. On Peter B. Checkland, systems thinking, systems practice a 30 year retrospective. In: *Schlusselwerke der Systemtheorie* (ed. D. Baecker), 375–402. New York: Springer.

Stowell, F.A. and Welch, C. (2012). *The Manager's Guide to Systems Practice: Making Sense of Complex Problems*. Chichester: Wiley.

Strogatz, S. (2004). *Sync: The Emerging Science of Spontaneous Order*. London: Penguin Books.

Stroh, D.P. (2015). *Systems Thinking for Social Change*. Vermont: Chelsea Green Publishing.

Sunim, H. (2017). *The Things You Can See Only When You Slow Down: How to be Calm in a Busy World*, Kindle Edition. London: Penguin.

Sushil (2018). Flexible systems methodology: a mixed-method/multi-method research approach. *Global Journal of Flexible Systems Management* 19: 109–110.

Suzuki, D.T. (1973). *Essays in Zen Buddhism*. London: Rider.

Taket, A. and White, L. (1995). Working with heterogeneity: a pluralist strategy for evaluation. In: *Critical Issues in Systems Theory and Practice* (ed. K. Ellis et al.), 517–522. New York: Plenum.

Taket, A.R. and White, L.A. (2000). *Partnership and Participation: Decision-Making in a Multiagency Setting*. Chichester: Wiley.

Taleb, N.N. (2007). *The Black Swan: The Impact of the Highly Improbable*. New York: Random House.

Taleb, N.N. (2013). *Antifragile: How to Live in a World We Don't Understand*, Kindle Edition. London: Allen Lane.

Ter Meulen, B.C., Tavy, D., and Jacobs, B.C. (2009). From stroboscope to dream machine: a history of flicker-induced hallucinations. *European Neurology* 61: 316–320.

Thaler, R.H. (2015). *Misbehaving: The Making of Behavioural Economics*. Canada: Allen Lane.

Thomas, A. and Lockett, M. (1979). Marxism and systems research: values in practical action. In: *Proceedings of the SGSR*, 284–293. Louisville: Society for General Systems Research.

Tranfield, D. and Starkey, K. (1998). The nature, social organisation and promotion of management research: towards policy. *British Journal of Management* 9: 341–355.

Trist, E.L. (1981). *The Evolution of Socio-Technical Systems: A Conceptual Framework and an Action Research Program*. Ontario: Ontario Ministry of Labour.

Trist, E.L. and Bamforth, K.W. (1951). Some social and psychological consequences of the Longwall method of coal-getting. *Human Relations* 4: 3–38.

Trist, E.L., Higgin, G.W., Murray, H., and Pollock, A.B. (1963). *Organizational Choice: Capabilities of Groups at the Coal-Face under Changing Technologies*. London: Tavistock Publications.

Troncale, L. (2003). The future of general systems research. In: *Systems Thinking*, vol. 1 (ed. G. Midgley), 231–296. London: Sage.

Troncale, L. (2006). Towards a science of systems. *Systems Research and Behavioral Science* 23: 301–321.

Truss, J., Cullen, C., and Leonard, A. (2000). *The Coherent Architecture of Team Syntegrity: From Small to Mega Forms*. Toronto: Team Syntegrity Inc.

Tsoukas, H. (1992). Panoptic reason and the search for totality: a critical assessment of the critical systems perspective. *Human Relations* 45: 637–657.

Tsoukas, H. (1993a). The road to emancipation is through organizational development: a critical evaluation of total systems intervention. *Systems Practice* 6: 53–70.

Tsoukas, H. (1993b). 'By their fruits Ye shall know them': a reply to Jackson, green and Midgley. *Systems Practice* 6: 311–317.

Tsoukas, H. and Hatch, M.J. (2001). Complex thinking, complex practice: the case for a narrative approach to organizational complexity. *Human Relations* 54: 979–1013.

Turke, R.-E. (2008). *Governance: Systemic Foundation and Framework*. Heidelberg: Physica-Verlag.

Ulrich, W. (1981a). On blaming the messenger for the bad news: reply to Bryer's 'Comments'. *Omega* 9: 200–202.

Ulrich, W. (1981b). A critique of pure cybernetic reason: the Chilean experience with cybernetics. *Journal of Applied Systems Analysis* 8: 33–59.

Ulrich, W. (1983). *Critical Heuristics of Social Planning.* Bern: Haupt.

Ulrich, W. (1988a). Churchman's 'Process of Unfolding' – its significance for policy analysis and evaluation. *Systems Practice* 1: 415–428.

Ulrich, W. (1988b). Systems thinking, systems practice, and practical philosophy: a program of research. *Systems Practice* 1: 137–163.

Ulrich, W. (1998). Systems thinking as if people mattered: critical systems thinking for citizens and managers. Lincoln: University of Lincoln, Working Paper, 23.

Ulrich, W. (2003). Beyond methodology choice: critical systems thinking as critically systemic discourse. *Journal of the Operational Research Society* 54: 325–342.

Ulrich, W. (2005). A brief introduction to Critical Systems Heuristics (CSH). http://www. ecosensus.info/about/index.html (accessed 28 June 2018).

Ulrich, W. (2007). Philosophy for professionals: towards critical pragmatism. *Journal of the Operational Research Society* 58: 1109–1117.

Ulrich, W. (2012a). Operational research and critical systems thinking – an integrated perspective, part 1: OR as applied systems thinking. *Journal of the Operational Research Society* 63: 1228–1247.

Ulrich, W. (2012b). Operational research and critical systems thinking – an integrated perspective, part 2: OR as argumentative practice. *Journal of the Operational Research Society* 63: 1307–1322.

Ulrich, W. (2018). http://wulrich.com (accessed 29 May 2018).

Ulrich, W. and Reynolds, M. (2010). Critical systems heuristics. In: *Systems Approaches to Managing Change: A Practical Guide* (ed. M. Reynolds and S. Holwell), 243–292. London: Springer.

Umpleby, S.A. (2016). Second-order cybernetics as a fundamental revolution in science. *Constructivist Foundations* 11: 455–465.

Valentinov, V., Roth, S., and Will, M.G. (2018). Stakeholder theory: a Luhmannian perspective. *Administration & Society* 52: 1–24.

Van de Water, H., Schinkel, M., and Rozier, R. (2007). Fields of application of SSM: a categorization of publications. *Journal of the Operational Research Society* 58: 271–287.

Van Gigch, J.P. (ed.) (2006). *C West Churchman and Related Works Series: Wisdom, Knowledge and Management.* New York: Springer.

Varela, F. (1979). *Principles of Biological Autonomy.* New York: Elsevier-North Holland.

Varela, F.J., Maturana, H.R., and Uribe, R. (1974). Autopoiesis: the organisation of living systems. *Biosystems* 5: 187–196.

Vennix, J.A.M. (1996). *Group Model Building: Facilitating Team Learning Using System Dynamics.* Chichester: Wiley.

Venter, C. and Goede, R. (2017). The use of critical systems heuristics to surface and reconcile users' conflicting visions for a business intelligence system. *Systemic Practice and Action Research* 30: 407–432.

Vickers, G. (1965). *The Art of Judgement.* London: Chapman & Hall.

Vickers, G. (1970). *Value Systems and Social Process.* Harmondsworth: Pelican Books.

Vickers, G. (1972). *Freedom in a Rocking Boat.* Harmondsworth: Pelican Books.

Vickers, G. (1983). *Human Systems Are Different.* London: Harper and Row.

Vriens, D. and Achterbergh, J. (2006). The social dimension of system dynamics-based modelling. *Systems Research and Behavioral Science* 23: 553–563.

Walby, S. (2007). Complexity theory, systems theory, and multiple intersecting social inequalities. *Philosophy of the Social Sciences* 22: 449–470.

Walker, G. (2015). Come back sociotechnical systems theory, all is forgiven. *Civil Engineering and Environmental Systems* 32: 170–179.

Walker, M. (2017). The search for viability: a practitioner's view of how the viable systems model is helping transform english local government (and why it has passed unrecognised). *Systems Research and Behavioral Science* 34: 313–334.

Walker, J. (2018). The viable systems model: a guide for co-operatives and federations. www.scio.org.uk (accessed 3 February 2018).

Wallerstein, I. (2004). *World-Systems Analysis: An Introduction*. Durham: Duke University Press.

Walsham, G. (1991). Organizational metaphors and information systems research. *European Journal of Information Systems Research* 1: 83–94.

Walsham, G. and Han, C.-K. (1991). Structuration theory and information systems research. *Journal of Applied Systems Analysis* 18: 77–85.

Walter, W.G. (1953). *The Living Brain*. London: Norton.

Warfield, J.N. (2002). *Understanding Complexity: Thought and Behavior*. Washington, DC: AJAR Publishing.

Warfield, J.N. (2003). Systems movement: autobiographical perspectives. *International Journal of General Systems* 32: 525–563.

Weaver, W. (2003). Science and complexity. In: *Systems Thinking* (ed. F.E. Emery), 377–385. London: Sage.

Weber, M. (1948). Class, status and party. In: *From Max Weber* (ed. H.H. Gerth and C. Wright Mills), 180–195. London: Routledge & Kegan Paul.

Weber, M. (1949). *The Methodology of the Social Sciences*. New York: The Free Press.

Weber, M. (1964). *The Theory of Social and Economic Organization*. New York: Free Press.

Weinberg, G.M. (2011). *An Introduction to General Systems Thinking*. USA: Weinberg & Weinberg.

Werner, L.C. (ed.) (2017). *Cybernetics: State of the Art*. Berlin: TU Press.

West, G. (2017). *Scale: The Universal Laws of Life and Death in Organisms, Cities and Companies*, Kindle Edition. London: Weidenfeld & Nicolson.

Wheatley, M.J. (1992). *Leadership and the New Science: Learning About Organization from an Orderly Universe*. San Francisco, CA: Berrett-Koehler.

Wheeler, J. (2015). General relativity: the most beautiful theory. *The Economist* (28 November).

White, L.A. (1994). Let's syntegrate. *Operational Research Insight* 3: 13–18.

White, M. (1997). *Isaac Newton: The Last Sorcerer*. London: Fourth Estate.

White, L.A. (1998). Tinker, tailor, soldier, sailor: a syntegrity to explore London's diverse interests. *Operational Research Insight* 7: 12–16.

White, L.A. and Taket, A. (1997). Critiquing multimethodology as metamethodology: working towards pragmatic pluralism. In: *MultiMethodology* (ed. J. Mingers and A. Gill) 379–405. Chichester: Wiley.

Whittaker, D. (2003). *Stafford Beer: A Personal Memoir*. Charlbury: Wavestone.

Wiener, N. (1948). *Cybernetics*. New York: Wiley.

Wiener, N. (1950). *The Human Use of Human Beings*. London: Eyre & Spottiswoode.

Wijnhoven, F. (2009). *Information Management: An Informing Approach*. London: Routledge.

Williams, B. and Hummelbrunner, R. (2010). *Systems Concepts in Action: A Practitioner's Toolkit*. Stanford, CA: Stanford University Press.

Willmott, H. (1989). OR as a problem situation: from soft systems methodology to critical science. In: *Operational Research and the Social Sciences* (ed. M.C. Jackson, P. Keys and S. Cropper), 65–78. New York: Plenum.

Wilson, B. (1990). *Systems: Concepts, Methodologies and Applications*, 2e. Chichester: Wiley.

Wilson, B. (2001). *Soft Systems Methodology: Conceptual Model Building and Its Contribution*. Chichester: Wiley.

Wilson, B. and Van Haperen, K. (2015). *Soft Systems Thinking, Methodology and the Management of Change*. London: Palgrave.

Wolfe, N. (2011). *The Living Organization: Transforming Business to Create Extraordinary Results*. USA: Quantum Leaders Publishing.

Wolstenholme, E.F. (1990). *Systems Enquiry: A System Dynamics Approach*. Chichester: Wiley.

Wolstenholme, E.F. (2003). Towards the definition and use of a core set of archetypal structures in system dynamics. *System Dynamics Review* 19: 7–26.

Wood-Harper, A.T., Antill, L., and Avison, D.E. (1985). *Information Systems Definition: The Multiview Approach*. Oxford: Blackwell.

Wordsworth, W. (1814). The Excursion. https://archive.org (accesssed 24/2018).

Wordsworth, W. (1850). The Prelude. https://bartleby.com (accessed 26 October 2017).

Wulf, A. (2015). *The Invention of Nature: The Adventures of Alexander von Humboldt*. London: John Murray.

Yearworth, M. and White, L. (2014). The non-codified use of problem structuring methods and the need for a generic constitutive definition. *European Journal of Operational Research* 237: 932–945.

Yee, C. (1941). *The Silent Traveller in the Yorkshire Dales*. London: Methuen.

Zhu, Z. (2000). WSR: a systems approach for information systems development. *Systems Research and Behavioral Science* 17: 183–203.

Zhu, Z. (2007). Complexity science, systems thinking and pragmatic sensibility. *Systems Research and Behavioral Science* 24: 445–464.

Zhu, Z. (2010). Theorizing systems methodologies across cultures. *Systems Research and Behavioral Science* 27: 208–223.

Zhu, Z. (2011). After paradigm: why mixing-methodology theorising fails and how to make it work again. *Journal of the Operational Research Society* 62: 784–798.

Zhu, Z. (2012). Essai: process thinking without process ontology. University of Hull: Unpublished paper.

Zlatanovic, D. (2016). Combining the methodologies of strategic assumptions surfacing and testing and organizational cybernetics in managing problem situations in enterprises. *Economic Horizons* 19: 17–33.

Zlatanovic, D. (2017). A multi-methodological approach to complex problem solving: the case of a Serbian Enterprise. *Systems* 5 (2): 40. https://doi.org/10.3390/systems5020040.

Zokaei, K., Seddon, J., and O'Donovan, B. (eds.) (2011). *Systems Thinking: From Heresy to Practice*. Basingstoke: Palgrave Macmillan.

Conclusion

> The only wisdom we can hope to acquire
> Is the wisdom of humility: humility is endless.
> The houses are all gone under the sea.
> The dancers are all gone under the hill.
> (Eliot 1959, lines 97–100)

Almost all people want to leave the world a better place than they found it. And we have a few years on this earth when we can try to make a difference. The argument of this book is that critical systems thinking can help us to do so.

The overwhelming issue facing governments and organizations, in the twenty-first century, is complexity. This arises from increased size, interconnectivity and diversity, and from the rapid rate of change. It also stems from the increased differentiation of society, leading to divergent stakeholder perspectives, and the loss of faith in the "grand narratives" of religion, progress, economic growth, free-market capitalism, and communism. There is a general recognition that decision-makers need more than commonsense to meet the challenges posed by the growth in complexity and, also, that traditional analytic tools don't help much. Increasingly, international bodies, governments, businesses, public sector entities, and other organizations are looking to systems thinking for answers. Angel Gurria, the OECD Secretary General, declared, in March 2018, that "unless we adopt a systems approach, unless we employ systems thinking, we will fail to understand the world we are living in." The UN System Leadership Framework classes systems thinking as one of the four behaviors that international leaders need to adopt in their leadership practice. This book has sought to provide a comprehensive account of what systems thinking has to offer.

We began by looking at the emergence of systems thinking in philosophy and then how it had impacted the physical sciences, the life sciences, and the social sciences. It is apparent that the traditional disciplines have, in almost all cases, turned to systems thinking to help them come to terms with the complexity of their subject matter. Systems thinking is proving its worth across the spectrum of intellectual endeavor. The sense that all the disciplines face similar issues when dealing with complex systems has led to ambitious attempts to establish systems science as a new transdiscipline relevant to all the single disciplines and able to help them communicate and advance. The three main variants of systems science are general systems theory, cybernetics, and

Critical Systems Thinking and the Management of Complexity, First Edition. Michael C. Jackson.
© 2019 John Wiley & Sons Ltd. Published 2019 by John Wiley & Sons Ltd.

complexity theory. One finding of the systems sciences is that there are some useful analogies, some would say laws, that transfer across the physical, biological, and social domains. Another is that there are "emergent properties" that arise at different system levels and which make, for example, human and social systems different. The conclusion, from the point of view of this book, is that managers need to know about the common features systems possess and also the ways in which the social systems they manage differ from other types.

The book then started to explore systems practice – how systems ideas can be used by managers. By managers, I mean policy-makers, leaders, executives, and decision-makers of all kinds. If I had to be precise about that, I would combine Ackoff's (2005) definition of a manager with his definition of a leader, and say that it is a person who "directs others in pursuit of ends, and by the use of means, that s/he or they select or approve." It soon became clear that various ways of making use of systems ideas in management have been developed. They all use the systems concepts that originated in the disciplines and in the systems sciences, but they do so in different ways. They also take different positions on whether it is the similarities between systems that are of paramount importance or the emergent properties at different levels. This is not surprising because it is impossible for any one systems approach to make sense of the complexity of problem situations in the modern world. Each systems methodology ends up emphasizing certain aspects of complexity while largely ignoring others. To paraphrase Snowden and Boone (2007), "wise executives tailor their approach to fit the type of complexity exhibited by the circumstances they face."

To make sense of the diversity of systems approaches designed to help managers, it is necessary to carry out some sort of critical, or "second-order" analysis to reveal which aspects of complexity each focuses upon. Ten well-known systems methodologies were chosen for examination. These are all theoretically informed, well-formulated approaches that have been developed and tested in practice over many years. This distinguishes them from the many popular "management fads and panaceas" (Jackson 1995). The devices chosen to carry out the investigation were philosophy, the social sciences, and the "System of Systems Methodologies" (SOSM). Philosophy allows us to take a look at the assumptions made by the methodologies about ontology and epistemology. The social sciences help us to grasp how they use these assumptions to understand social systems and try to change them by employing suitable methodologies. The SOSM translates these findings into a language appropriate for managers by asking what each methodology assumes about the complexity of systems and the relationship between stakeholder perspectives – the two primary sources of complexity. Using these devices, the strengths and also the partiality of each methodology was revealed. In very broad terms, we categorized the 10 methodologies according to the types of complexity they highlight. Six types of complexity featured – technical, process, structural, organizational, people, and coercive. Each is addressed by one or more of the systems methodologies.

Critical systems thinking was then introduced. Critical systems thinking recognizes that it is impossible to understand the "whole system." Instead, it seeks to take advantage of the "critical awareness" gained about the strengths and limitations of the different methodologies to use them in informed combinations to bring about improvement over time. To do so, it employs a multimethodology called "critical systems practice." Critical systems practice makes it possible to employ a variety of perspectives, reflecting different paradigms, to address the multidimensional nature of problem situations. We cannot use all the various systems approaches at once, but they can be deployed appropriately in the course of interventions to promote overall improvement in the problem situations managers confront. Issues concerning efficiency, efficacy, effectiveness, mutual understanding, viability, antifragility, sustainability, empowerment, emancipation, and the interrelationships between them are identified and tackled as they assume importance in the problem situation as it evolves and the intervention proceeds. Critical systems thinking cannot do everything, and second-order analysis should make us modest about what it is possible to achieve using systems approaches. But, it can help.

No doubt critical systems thinking will seem difficult. But then, managerial work is becoming more complex and diverse. Most managers are likely to find themselves, on a regular basis, confronted by messes made up of interacting issues such as the need to increase productivity, become more market-centered, improve communications, adopt fairer recruitment and promotion strategies, and motivate a diverse workforce. They will also find themselves having to prioritize between the demands made on them because of lack of time and resources. They cannot tackle them all at once. This all seems like common sense. Critical systems thinkers are in tune with this common sense. They recognize that excellent organizational performance depends on managers paying attention to improving goal seeking and viability, exploring purposes, ensuring fairness, and promoting diversity. They have, at their disposal, a multimethodology, critical systems practice, which can make the best use of the different systems approaches that are available. It provides the means to tackle, in a holistic, creative, and balanced manner, the "messes" and "wicked problems," which managers confront. Critical systems thinking can assist them in responding to the demands of twenty-first century complexity. It can provide us with responsible leadership for a complex world.

It is one of the satisfactions of systems thinking that many of its insights transfer between levels. We find that the same imperatives for improvement, highlighted by critical systems thinking, are relevant at the world, societal, organizational, and even individual levels. Few of us will get the opportunity to practice our systems thinking on the world stage. But managers of organizations have a real opportunity to influence events for the better if they push their institutions to serve all their stakeholders. Most managers, as well as doing a good job now, want to make things better for future generations. Even as individuals we can hope, because of interconnectedness, that kind, well-directed actions will have a resonance beyond our immediate environments and, in some way, contribute to improvement in the social and natural worlds we live in together. This is a nice thought which I have not expressed very well. Hemingway did better in *For Whom the Bell Tolls*:

> Today is only one day in all the days that will ever be. But what will happen in all the other days that ever come can depend on what you do today.
>
> *(1940, Chapter 43, paragraph 5)*

Index

Critical Systems Thinking and the Management of Complexity, First Edition. Michael C. Jackson.
© 2019 John Wiley & Sons Ltd. Published 2019 by John Wiley & Sons Ltd.